THERMAL PLASMA TORCHES

THERMAL PLASMA TORCHES
Design, Characteristics, Applications

edited by

M.F. Zhukov and I.M. Zasypkin

Viva Books

New Delhi | Mumbai | Chennai | Kolkata | Bangalore | Hyderabad | Kochi

First Indian Edition 2008

VIVA BOOKS PRIVATE LIMITED

- 4737/23 Ansari Road, Daryaganj, New Delhi 110 002
  E-mail: vivadelhi@vivagroupindia.net, Tel. 42242200
- Plot No. 76, Service Industries, Shirvane, Sector 1, Nerul, Navi Mumbai 400 706
  E-mail: vivamumbai@vivagroupindia.net, Tel. 27721273, 27721274
- Jamals Fazal Chambers, 26 Greams Road, Chennai 600 006
  E-mail: vivachennai@vivagroupindia.net, Tel. 28290304, 28294241
- B-103, Jindal Towers, 21/1A/3 Darga Road, Kolkata 700 017
  E-mail: vivakolkata@vivagroupindia.net, Tel. 22836381, 22816713
- 7, Sovereign Park Apartments, 56-58, K. R. Road, Basavanagudi, Bangalore 560 004
  E-mail: vivabangalore@vivagroupindia.net, Tel. 26607409, 26607410
- 101-102 Mughal Marc Apartments, 3-4-637 to 641, Narayanguda, Hyderabad 500 029
  E-mail: vivahyderabad@vivagroupindia.net, Tel. 27564481, 27564482
- First Floor, Beevi Towers, SRM Road, Kaloor, Kochi 682 018, Kerala
  E-mail: vivakochi@vivagroupindia.net, Tel: 0484-2403055, 2403056

www.vivagroupindia.com

Published by arrangement with

Cambridge International Science Publishing
7, Meadow Walk, Great Abington
Cambridge CB1 6AZ, UK

ISBN-10: 81-309-0948-0
ISBN-13: 978-81-309-0948-6

Published by Vinod Vasishtha for Viva Books Private Limited, 4737/23, Ansari Road, Daryaganj, New Delhi-110 002. Printed and bound by Anand Sons, Delhi - 110 092.

Preface

This book deals with a large number of investigations, directly or indirectly associated with the calculation and design of low-temperature plasma generators (electric arc plasma torches) and plasma-processing reactors. Electric arc gas heaters are systems in which electric energy is converted into thermal energy by means of the generation of Joule heat in the discharge. Heating of the gas in these systems takes place mainly as a result of heat conductivity and convective heat exchange between the arc and the gas flow.

The interest in the investigations and application of the electric arc is caused by:

- high concentration of energy in the small volume of plasma;
- high rate of the chemical reactions, so that it is possible to produce high-productivity apparatus-reactors;
- the possibility of stationary heating of the gas to the mean mass temperature of the order of $15 \cdot 10^3$ K at a pressure of up to 20 MPa;
- high efficiency of the transformation of electrical energy into thermal energy with a relatively simple apparatus;
- reliability and stability of operation of equipment;
- the possibility of heating almost any gases: reduction, oxidation, inert gases and mixtures;
- simple automation of controlling the operating regime of the electrical arc;
- small size and small metal requirement of plasma technology.

The electric arc was produced for the first time in 1802 by Professor V.V. Petrov at the Medical-Surgical Academy in St Petersburg. Only after 100 years, at the beginning of the 20th century, systems appeared in industry using the arc for removing nitrogen oxides from air in the process of production of nitric acid. The plasma torches, constructed on the basis of the circuits proposed by Birkeland and Eide, Pauling and Siebert used alternating current. In the systems constructed by Sencher, a direct current electric arc 7 m long burnt in a vertical pipe blown with air.

In the 30s, the method of production of acetylene from natural gas using the electric arc was introduced in Germany. The direct current electric arc more than 1 m long, stabilised with a twisted gas flow, ran at a voltage of 7000 V and a current intensity of up to 1000 A. At present, the method is used in a number of countries.

Special attention to electric arc generators was evident at the end of the 50s because of the need to heat gases in aerodynamic pipes, in modelling of the flight of aircraft at supersonic speed and examination of the conditions of entry of space systems into the atmosphere of the earth and other planets.

In the 60s, the centre of gravity of technical applications of plasma torches was transferred at increasing rate to chemical, metallurgical and other conventional and new branches of industry. The low-temperature plasma, with the properties as the high concentration of energy in the small volume, high temperature and rates of the chemical reactions, etc, attracted attention mainly because of the possibility of constructing completely new high-productivity apparatus and technologies.

It can already be said at the present time that the low-temperature plasma is an important element of industrial technologies enabling processes with the extremely high rates to be achieved. This is not possible in the normal conditions.

In plasma technology and science, the interests of the fundamental and applied sciences are closely linked with production. The application of low-temperature plasma is a characteristic phenomenon of modern production and plasma torches represent a powerful tool in a number of industries.

Plasma technology creates suitable conditions for processes with closed cycles and this creates optimum conditions for solving the global problem, i.e. reducing the extent of contamination of the environment.

It is also important to note the application of plasma torches in plasma spraying which is a new rapidly developing branch of industry.

The semiempirical methods of calculating the electrical and thermal characteristics of linear plasma torches, developed at the Department of Plasma Dynamics of the Institute of Theoretical and Applied Mechanics of the Siberian division of the Russian Academy of Sciences, are based on the experimental determination of criterial relationships representing the basis of engineering methods of calculating plasma torches and selecting the parameters of the plasma source.

The further expansion of application in industry of plasma technologies is associated with improving all characteristics of plasma torches and electric arc reactors; with increasing the operating life of electrodes (the most heavily thermally stressed the sections of the plasma torches) by 1–2 orders of magnitude, i.e. by increasing the duration of continuous service of plasma torches to many hundreds or even thousands of hours; increasing the thermal efficiency; application of working gases of different chemical composition, taking into account the specific features of the technological process and ensuring the maximum extraction of the target product.

In this book, special attention is given to electric arc plasma torches-reactors, designed for processing solid materials. The point is that they are subject to specific requirements, the main of which are: high productivity, low consumption of the working gas and high consumption of the material of the solid phase. In addition to this, in the reactors of this type, it is necessary to combine organically the possibility of simultaneous occurrence in a large volume of the chamber of the chemical and electrophysical processes. These requires efficient filling of the reaction volume by the electric arc moving in the space at a relatively high rate under the effect of the external magnetic field of special topology.

Regardless of the externally simple design of the plasma torch, the latter is characterised by complicated physical processes of electromagnetic, thermal and aerodynamic nature; physical processes in the near-electrode regions of the arc discharge, on the surface of the electrode and inside the crystal lattice of the metal from which they are produced. In order to understand these processes, it was necessary to carry out systematic experimental investigations of a large number of phenomena in the electric discharge chamber which determine the electrical, thermal and erosion characteristics of the plasma torch [1].

The large variety and complexity of the processes in the electrical arc, in interaction of the arc spot with the walls of the channel, with the intrinsic and external magnetic fields, delay the theoretical investigations of the behaviour of the arc in the plasma torches using direct and alternating currents with different circuits. This also explains special attention given to experimental studies.

The experiments have made it possible to obtain a relatively large amount of information on the most important physical processes in the discharge chamber, the energy characteristics of the arc in different gases, heat exchange between the arc, the hot gas and the wall, and also on the methods of protecting the wall against high-intensity heat flows. In the book, there are data on the processes developing in the body of the electrode and increasing the erosion rate. Special attention is given to circuits of plasma-processing reactors designed for plasma chemical synthesis of gas media, and also for processing powder materials. Information is provided on the similarity criteria characteristic of the processes in the electric discharge chamber, which were used as a basis for the generalisation of the electrical and thermal characteristics of the plasma torches.

At present, plasma torches of the linear, coaxial, combined, multi-arc and other types using both alternating and direct currents have been developed. The variety of the systems is determined by technological applications. The power range varies from hundreds of watts to many thousand kilowatts.

The authors hope that the book will be useful to both technologists using plasma torches in different technical applications and researchers concerned with the examination of the physical processes in plasma torches and striving for further improvement of their electrical, thermal and erosion characteristics.

Contents

Chapter 1

Brief description of thermal plasma and electric heating of gas

The term 'plasma torch' (or plasmatron) according to the currently valid terminology, is apparatus, designed for the production of low-temperature plasma, i.e., the gas heated to a temperature of $(3–50)\cdot 10^3$ K. At present, heating of the gas by the electric arc is the most widely used method of producing low-temperature plasma.

1.1. FORMATION OF THE ELECTRIC ARC AND THE PROPERTIES OF ARC PLASMA

Arc discharge may form either in the separation of the initially contacting electrical contacts or in a spark breakdown of the gap between the contacts, or by transition from the glow discharge with increasing current intensity. The typical dependence of the cathode drop of the potential on current intensity in transition from glow to arc discharge is shown in Fig. 1.1. This transition is characterised by a large decrease of the cathode drop of the potential with increasing current intensity with a simultaneous decrease of the overall voltage drop.

If the cathode voltage drop in a glow discharge is approximately of the order of 100 V or more, in an arc discharge it is only 10–15 V. The reason for this difference is in different processes of transfer of electricity in the near-cathode regions and different methods of transferring the energy of the electrical field to the gas. Emission of electrons from the cathode in a glow discharge takes place as a result of bombardment of the cathode with the ions accelerated in a strong near-cathode field, and also as a result of the photoeffect from the radiation of the gas in the discharge. After receiving the required portion of kinetic energy in impact of an ion or a photon, the electron is capable of overcoming the force barrier and leave

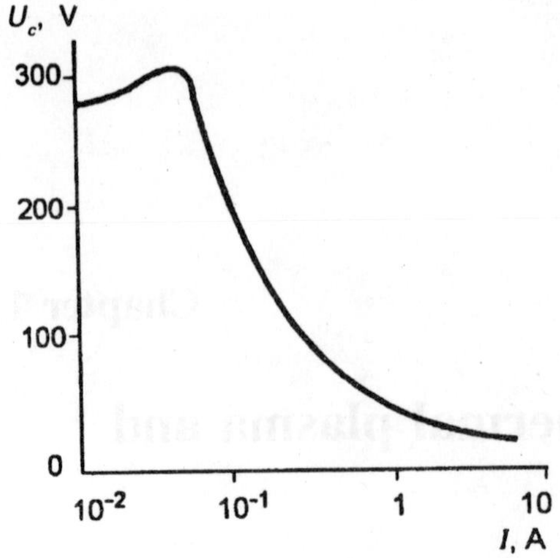

Fig. 1.1. The dependence of the cathode voltage drop on current intensity in transition from glow to arc discharge.

the metal. Subsequently, it is accelerated in the near-cathode electrical field to the energy sufficient for impact ionisation of the atoms, maintaining at the same time the process of emission of the electrons from the cathode.

If the intensity of current discharges increases, the increase of the number of electron impacts increases the temperature of the gas in the vicinity of the cathode and, from some moment, thermal ionisation starts to play the main role in ionisation of the gas. In thermal ionisation, the temperature of the electrons is close to the temperature of the ions and neutral particles. Consequently, it is necessary to ensure a large potential drop in the vicinity of the cathode on which the electrons acquired high energy. The exit of the electron from the cathode takes place now mainly by the mechanisms of auto-electronic (at a low cathode temperature) or thermoelectronic emission (at a high cathode temperature).

The form of the electrical discharge, formed at high current densities and characterised by a small cathode potential drop, is referred to as the electric arc. There are high-pressure and low-pressure arcs.

In the column of a high-pressure arc, the temperature of the electrons and heavy particles (ions and neutrals) is similar at every given point of the discharge, i.e. the conditions in the plasma of the arc are close to the conditions of local thermodynamic equilibrium (the plasma is quasi-isothermal). However, the absolute local thermodynamic equilibrium in arc plasma is not reached because the energy of the electrical field is transferred mainly to the electrons and, subsequently, through collisions to heavy particles. The conditions in which the arc plasma may be regarded as quasi-isothermal will now be estimated.

It is assumed that the energy, acquired by the electrons in the electrical field of the arc, is completely transferred to the heavy particles through elastic collisions:

$$\sigma E^2 = \frac{3}{2}k\left(T_e - T_g\right)\delta \nu_{eg}\, n_e, \tag{1.1}$$

where σ is the conductivity of the plasma, equal to $e^2 \cdot \lambda_e n_e / m_e v_e$, e is the electron charge, λ_e is the free path of the electrons in the gas, determined from the concentration of all components of the plasma and the collision cross-section Q_{ek}; $\lambda_e = \left(\sum_k n_k Q_{ek}\right)^{-1}$; v_e is the thermal velocity of the electron; T_e and T_g are the temperatures of the electrons and heavy particles; $\delta = 2\, m_e/m_g$ is the fraction of the energy transferred by the electron in an elastic impact (m_e and m_g are the masses of the electrons and heavy particles); $\nu_{eg} = v_e/\delta_e$ is the frequency of collisions of the electrons with the heavy particles; n_e is the electron concentration. The equation (1.1) can be transformed to the following form:

$$\frac{T_e - T_g}{T_e} = \frac{3\pi}{32}\left(\frac{\lambda_e e E^2}{\frac{3}{2}kT_e}\right)^2 \frac{m_g}{m_e}. \tag{1.2}$$

The complex in the brackets has the meaning of the ratio of the energy, acquired by the electron in the electrical field, to the kinetic energy of thermal motion. It may be seen that the high strength of the field E and the reduced pressure prevent the establishment of an equilibrium. For example, in near-electrode regions of the high-pressure arc, the high strength of the field causes that the condition of quasi-isothermal nature is not fulfilled. In the plasma of air and metal vapours, the equilibrium in the positive column of the arc is established at a pressure of $p \geq 1$ atm. In an inert gas, because of the dominance of photoprocesses (radiation is not absorbed in plasma), the quasi-isothermal condition is established only at high current intensities. For example, in argon plasma, the equilibrium is established at currents higher than 10 A and in helium plasma at currents higher than 200 A. The introduction of metal vapours in a sufficient amount (>1%) into the arc plasma results in a more rapid establishment of equilibrium.

In an arc discharge, the total number of the ions is equal to the total number of the electrons (for single ionisation) because these particles are produced from neutral particles in the same amount. Generally speaking, there may be processes resulting in the situation in which the number of the charged particles of some sign is greater than the number of other particles in individual small volumes of the discharge. These processes may include, for example, diffusion of the electrons, separation of the charges in a strong electrical field, etc. However, the forces formed in separation of the charges are so high that in almost all cases the arc discharge is quasi-neutral, i.e. local concentrations of the ions and electrons are close to each other. The conditions in which quasi-neutrality forms in the arc discharge plasma will be estimated. We use the Poisson equation:

$$\operatorname{div} E = \frac{e}{\varepsilon_e}(n_i - n_e), \tag{1.3}$$

where n_i and n_e is the concentration of the ions and electrons; ε_e is the dielectric constant. Since the strength of the electrical field in the central part of the arc column is almost constant along the radius, the value div$E$ should be estimated from the variation of the strength of the electrical field along the axis of the channel. Defining the knowingly large variation of the strength of the order of 10^3 V/cm, we obtain $n_i - n_e \approx 10^8$ cm^{-3} which is evidently far from the observed concentration values ($\sim 10^{14}$ and higher). It should be mentioned that in the peripheral zones of the arc discharge, where the temperature is low but there is a steep radial gradient of the strength of the electrical field, the separation of the charges may be quite significant. In particular, this is evident in the case in which the gas layer, heated by the arc, makes contact with the walls of the channel with a potential different from the arc potential. In visual examination of the arc, for example, in a cooled cylindrical channel between two electrodes positioned on the axis of the channel, there is a large number of special features [1].

The column of the arc is constricted and homogeneous in the axial direction. In the vicinity of the electrodes, the degree of arc constriction is greater and the diameter of the arc in the vicinity of the cathode is usually smaller than in the vicinity of the anode. The physical processes, leading to constriction of the arc in the vicinity of the electrode surfaces, are associated with the conditions in which the electrons are found, and with the characteristics of the arc (the nature of these processes has not as yet been completely explained). The near-electrode jets formed as a result of the axial gradient of magnetic pressure in the arc, play a significant role. The constriction of the arc column at the current intensity of up to 10^4 A is of the thermal nature and is associated with the removal of thermal energy from the central part of the column to the periphery. In the arc burning in the cooled channel without a gas flow, heat removal takes place mainly as a result of molecular heat conductivity. In a freely burning arc, the heat is removed by free convection. In longitudinal blowing of the gas (direction of movement of the gas is parallel to the axis of the arc), as in the case of transverse blowing (the direction of movement of the gas is normal to the axis of the arc), heat is removed by forcec laminar or turbulent convection. At a high current intensity, the intrinsic magnetic field of the arc is strong and results in additional (magnetic) constriction of the column, i.e. the pinch effect.

When discussing the diameter of the arc filament, it is necessary to take into account the fact that the measurements of this parameter give ambiguous results. This is associated with the continuous change of the parameters of the arc in the cross-section. For example, the current-conducting diameter of the filament can be determined as a diameter of the region through which the amount of electricity, differing from the total intensity of the discharge current by some small value (which, after all is conventional), passes. On the other hand, the effective current-conducting diameter can be determined from the ratio of the total conductivity of the arc to its maximum conductivity (on the axis of the arc). It is also important to distinguish (Fig. 1.2) between the luminous diameter of the arc, for example, on the basis of the point of the maximum decrease of darkening of the sheet on which the arc is photographed, and the current-conducting diameter determined by the point at which the local conductivity of the plasma is halved in comparison with the maximum value.

The measurements of the current-conducting radius of the arc make it possible to find the mean density of the current in the arc. Depending on the type of gas and the arcing conditions, the mean density of current in the arc column is in the range $10–10^3$ A/cm^2, whereas the mean density of the current on the cathode is $10^3–10^8$ A/cm^2, and on the anode it is $10^4–10^5$ A/cm^2. However, it may be possible to generate artificially the optimum conditions for the arc column in which the current density is considerably higher than the conventional current density (igniting, for example, an arc with a high current intensity in a capillary) but in the majority of cases in practice, the density of current in the arc column is in the given range.

The distribution of temperature in the cross-section of the arc column has the form qualitatively shown in Fig. 1.3. In the axial zone, temperature T is very high ($1\cdot10^4\div2\cdot10^4$ K). Temperature rapidly decreases in the direction to the periphery of the column approximately along

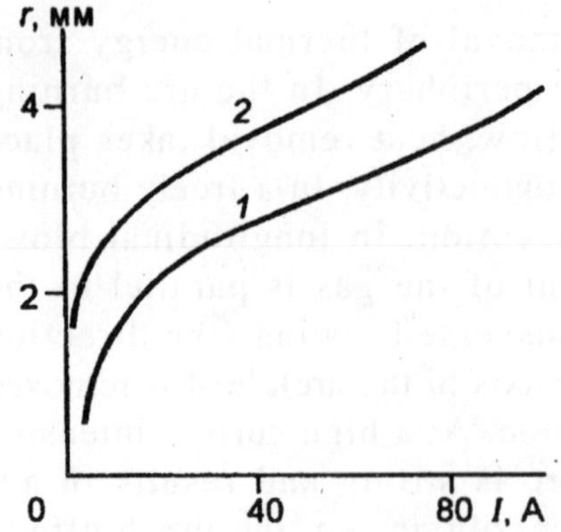

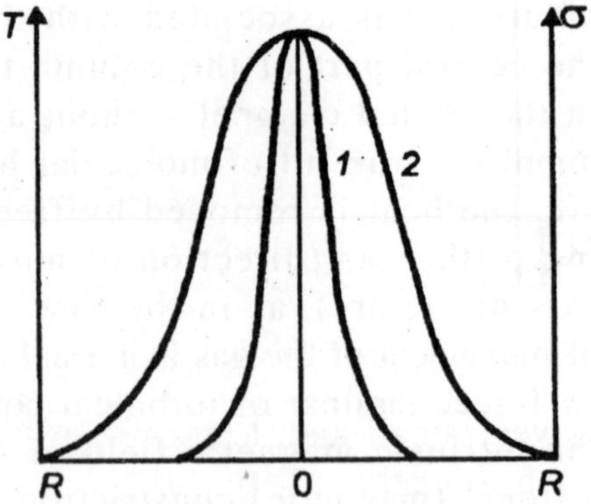

Fig. 1.2. Dependence of the current-conducting radius (1) and the radius of glow (2) of the arc, burning in argon, on current.
Fig. 1.3. (right) Schematic distribution of conductivity (1) and temperature T (2) in the cross-section of the arc column, related to the maximum values of σ_m and T_m.

the curves of a second order, and in the region of the walls of the channel it is characterised by the logarithmic distribution. The conductivity of plasma σ which determines at the given value of the current density the strength of the electrical field, decreases even more appreciably in the cross-section of the arc column than temperature. This is associated with the exponential dependence of conductivity on temperature. The distribution of the density of the radiant flux has the form approximately corresponding to the distribution of conductivity because of the exponential dependence of the radiant flux on temperature with the exponent considerably higher than unity.

The distribution of the potential along the length of the column of the arc discharge, not subjected to secondary effects (Fig. 1.4), is basically uniform, i.e. the strength of the electrical field is approximately constant. However, as shown later, the effect of the external conditions (the flow of the gas, the magnetic field, the walls of the channel) may result in a large change along the length of the column. In the near-electrode regions with the length δ_c and δ_a, the potential rapidly changes as a result of the processes of transfer of electricity between the column of the arc and the electrode. These regions are characterised by the disruption of thermal equilibrium and quasi-neutrality of arc plasma. The transfer of electricity in the near-cathode region takes place by means of the electrons emitted from the surface of the cathode, and the ions, arriving from the arc column. In the near-anode region, electricity is transferred mainly by the electrons, leaving the arc column for the anode.

The length of the near-electrode zones of distribution of the charges

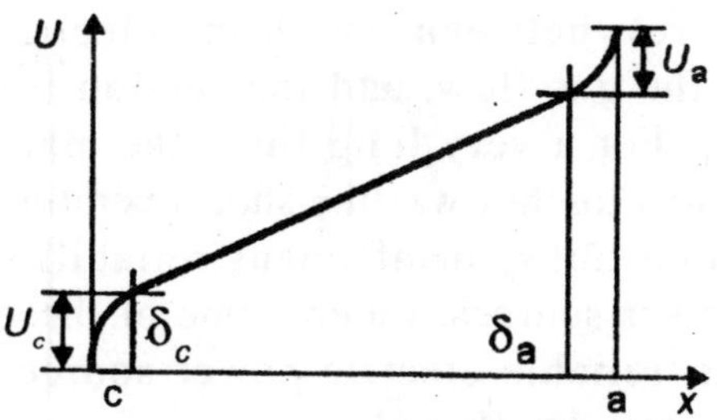

Fig. 1.4. Schematic distribution of potential along the arc (U_c, U_a are the cathode and anode voltage drops).

is very small and, according to the estimates, has the order of several path lengths of the particles (at the atmospheric pressure of ~10^{-4} m). This shows clearly that the strength of the electrical field in the near-electrode regions should be very high. For example, in the immediate vicinity of the cathode surface, the strength of the electrical field is estimated by the value of the order of 10^6–10^8 V/cm. Consequently, auto-electron emission (or emission by the field) of electrons from the cold cathode may be possible.

The strength of the electrical field in the arc column depends strongly on the diameter of the discharge channel, current intensity, the type and regime of the gas flow and a number of other conditions. For example, at the atmospheric pressure, the channel diameter of 1 cm and a current intensity of 100 A, the characteristic values of the strength of the electrical field for different gases are as follows (V/cm): argon 5–8, nitrogen 10–15, helium 15–20, hydrogen 30–50. The strength of the electrical field depends to some extent on the atomic number of matter: with increasing number, the strength decreases. Thus, if the discharge channel is characterised by the transition from the laminar turbulent flow, the strength may increase several times; maintaining, by means of the magnetic field, the arc column across the gas flow, it is possible to obtain the strength of the electrical field of the order of 50–100 V/cm.

The most important electrical characteristics of the arc is the volt–ampere characteristic (VAC). The form of this characteristic determines the selection of the parameters of the power source for the arc and the electrical efficiency of arc equipment.

1.2. ELECTRIC ARC GAS HEATERS – PLASMA TORCHES

As already mentioned in the introduction, the electrical arc has been studied for more than 20 years, and the first electric arc gas heaters appeared at the start of the 20th century. These devices included the main elements, characteristic of the currently available plasma systems: the electrodes (two or more), between which the electric arc burns, the chamber, restricting the gas flow, and the section for introduction of the working gas [2]. For a very long time, the main reason delaying the application of plasma torches was the short operating life, poor reproducibility of the conditions, insufficient reliability of equipment, including electrical power sources. Later, some of these problems were solved, in particular, reliable electric power sources for alternating and direct current were developed.

A large number of designs of both electric arc preheaters and plasma torches, using high-frequency current, microwave current, laser and other systems of heating the gas have been developed. We shall describe only the electric arc DC and AC plasma torches, used widely in various applications in science and technology.

Regardless of the existence of a large number of design solutions of plasma torches, resulting from different areas of application and electric power sources used for these applications, these systems are based on a limited number of principal schemes differing from each other mainly in the methods of stabilisation of the discharge. If we disregard coaxial plasma torches and some AC plasma torches, examined in [1, 3], we obtain the most widely used type of plasma torches–linear plasma torches.

In the linear plasma torches, the electrodes (rod, tubular, cylindrical, etc) are situated on the same line, directed along the gas flow. The simplest circuit of such a plasma torch is shown in Fig. 1.5. The electric discharge chamber of the plasma torch is formed by the internal (end) electrode 1, the cylindrical output electrode 2 and the insulator 3 placed between them and acting also as a device for introducing the working gas. The electrical arc 4 is ignited between the internal and output electrodes. The working gas is supplied into the channel with the flow rate G through the supply device in the electric insulator 3 through radial or tangential orifices with the circumferential component of velocity. Under the effect of the axial component of the speed of the gas flow, the closing (radial) section of the arc moves along the channel. This is accompanied by an increase of arc length and arc voltage. This increase of the arc length and of voltage is restricted by the shunting process, i.e. by the electrical breakdown between the arc and the wall of the electrode (this phenomenon is examined in greater detail in chapter 2). Consequently, the mean arc length, referred to as the self-setting arc length, is established. This length also depends on current density, the diameter of the discharge chamber, the type and pressure of gas, channel geometry, and other factors.

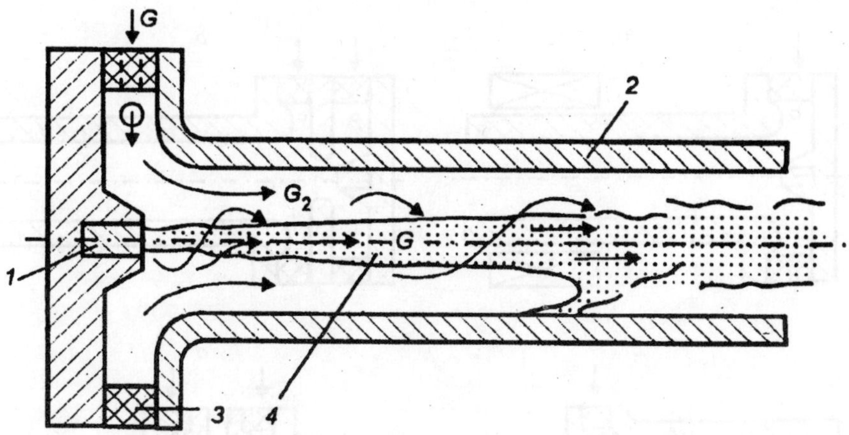

Fig. 1.5. The scheme of the linear single-chamber plasma torch with the self-setting arc length.

Part of the working gas, blown between the electrodes, penetrate into the arc column (G_1 in Fig. 1.5) and requires, as a result of generation of Joule heat, the temperature equal to the temperature of the arc in these conditions. The remaining gas G_2 flows in the channel between the arc and the wall or, more accurately, between the thin thermal boundary layer, formed along the electrically conducting part of the arc, and the wall of the electric discharge chamber. This gas is heated only slightly because there is no convective heat exchange between the arc and the main gas flow. The thermal boundary layer 'blocks' heat exchange. The interaction of the arc with the flow starts in the shunting zone, i.e. in the zone in which the thermal and boundary wall layers come together (for more details, see chapter 2), in which intensive mixing of the cold and high-temperature flows takes place. A plasma flow with a high-temperature core and the temperature profile rapidly decreasing in direction to the periphery, forms at the exit from the plasma torch.

Because of their simple design, the plasma torches with the self-setting arc length are used widely. Several circuits of the plasma torches of this type are presented in Fig. 1.6.

The VAC of the arc, as already mentioned, is the most important energy characteristics of the plasma torch. For the arc with the self-setting length, the characteristic is drooping (curve 1, Fig. 1.7), because the increase of current results in a decrease of the arc length and, consequently, arc voltage also decreases. The drooping VAC create certain difficulties in the matching of the arc with the electric power source. For example, in the case of non-regulated power sources with

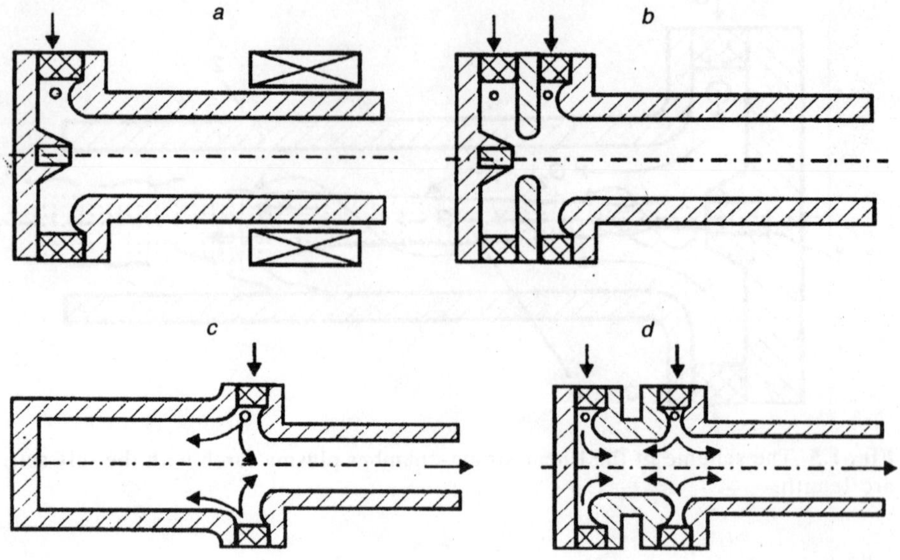

Fig. 1.6. Some schemes of plasma torches with the self-setting arc length. *a*) the single-chamber torch with an internal flat end electrode; *b*) two-chamber torch with an internal flat end electrode; *c*) single-chamber torch with a cup-shaped internal electrode; *d*) two-chamber torch with a cylindrical tubular internal electrode.

a hard characteristic, to ensure stable arcing, a ballast rheostat should be included in the circuit. However, this reduces the electrical efficiency of the plasma system. Another shortcoming of this plasma torch circuit is the high level of pulsations of arc voltage, especially at low currents, determined by large-scale shunting.

These shortcomings may be eliminated by fixing (using some procedure) the mean arc length in a specific range of variation of current density. In the cylindrical channel, this may be achieved by, for example, sudden expansion of the channel from the diameter d_2 to $d_3 > d_2$ at the end of the cylindrical electrode, i.e. by producing a ledge. The VAC of the arc in this case is lower than that of the arc with the self-setting length and is U-shaped (curve 2, Fig. 1.7). Naturally, if a specific current level is exceeded, the U–I characteristic changes (merges) into the characteristic of the arc with the self-setting length in the channel with the diameter $d = d_2$. The radial section of the arc is elongated into the channel with diameter d_2.

Of many methods of fixing the mean arc length in the cylindrical channel, Fig. 1.8 shows only two. Diagram *a* corresponds to the fixation of the arc length by a direct ledge, *b* with a direct ledge and subsequent reduction of the width of the cross-section of the

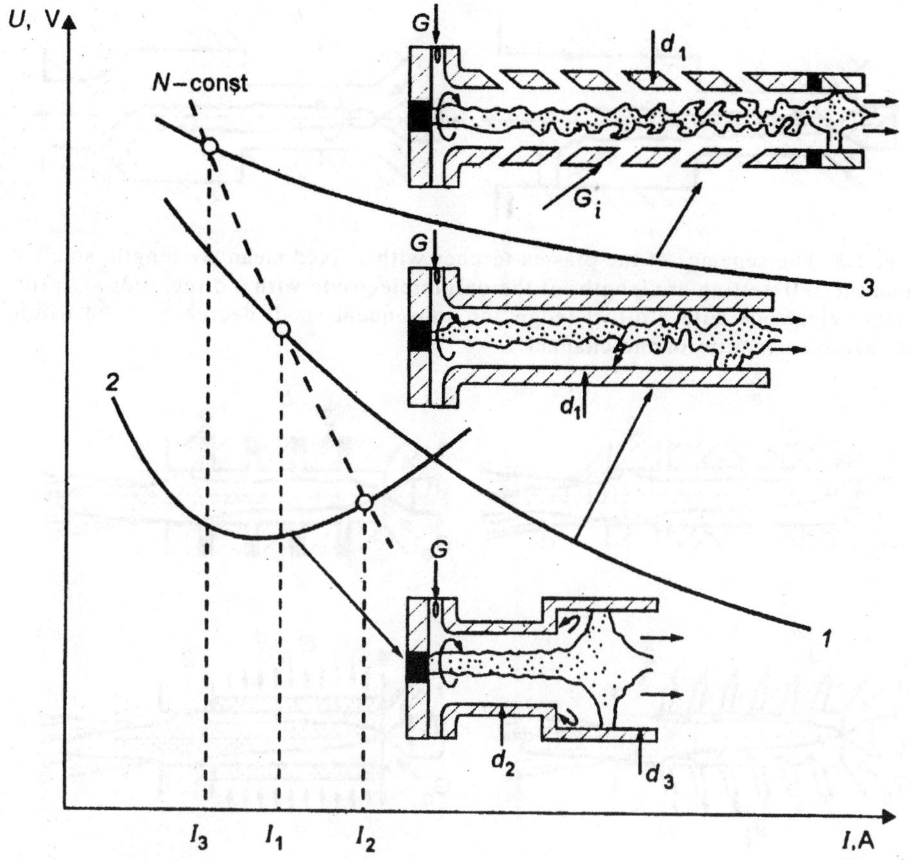

Fig. 1.7. Volt–ampere characteristics of the arc of the three principal plasma torch schemes.

channel. In all cases, the characteristics (VAC, thermal) of the arc with the self-setting length represent the upper limit for the characteristics of the arc in the plasma torches of this system.

The plasma torches with the fixation of the arc length by the direct ledge are also used at present because they are simple, reliable in service and do not have many of the shortcomings typical of the plasma torches with the self-setting arc length.

In the plasma torch with the arc length smaller than the self-setting length, it is not possible to produce plasma jets with a temperature higher than in the channel with the self-setting arc length. The high-enthalpy plasma flows can be produced only if the arc voltage greatly exceeds the arc voltage with the self-setting length at the same values of current and other parameters. This is achieved if an insulating electrode insert is placed between the electrodes. The insert prevents the shortening of the arc with increasing current (Fig. 1.9). This may

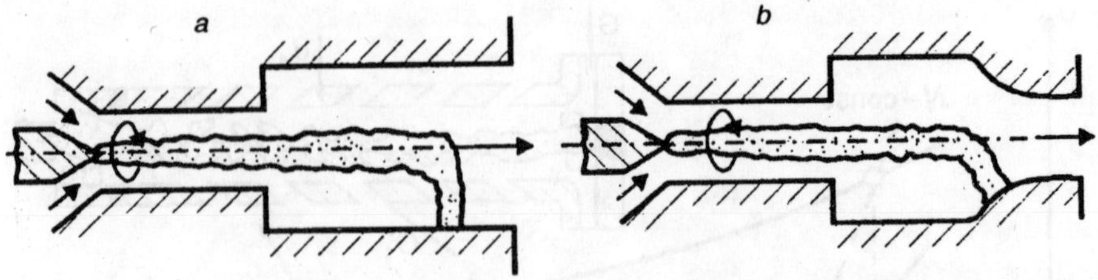

Fig. 1.8. The schemes of the plasma torches with a fixed mean arc length, smaller than the self-setting arc length. *a*) the output electrode with a direct ledge; *b*) the output electrode with a direct ledge and subsequent small decrease of the width of the cross-section of the channel.

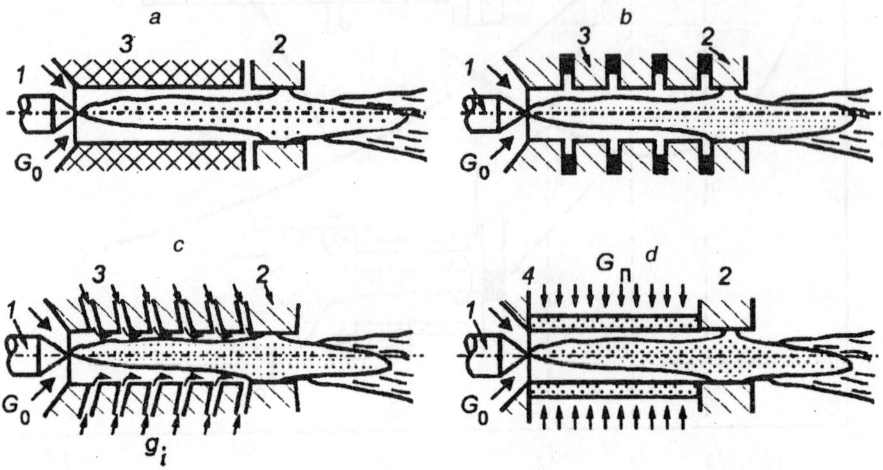

Fig. 1.9. The schemes of the plasma torches with the arc length greater than the self-setting length. *a*) the inter-electrode insert (IEI) produced from an electric insulation heat resistant materials; *b*) IEI produced from metallic cylindrical sections electrically insulated from each other and from the electrode; *c*) IEI identical to the scheme b with the gas supply distributed into the gaps between the sections; *d*) IEI produced from a porous material through which the gas is blown. 1) the internal electrode; 2) the output electrode; 3) IEI section; 4) porous insert.

be a cylindrical insert produced from an insulating material (*a*), a cylindrical insert produced from metallic discs-sections the isolated from each other and the electrodes (*b*), the same insert with the gas flow distributed in the inter-sectional gaps (*c*), the inter-electrode insert produced from porous materials with blowing part of the working gas through it (*d*), etc.

The VAC of the arc in the plasma torches of this type (curve 3, Fig. 1.7) is situated above the characteristic of the arc of the two previously discussed systems. It may be rising, hard, or slightly drooping in a wide current range.

If we compare the powers generated in the arc in the plasma torches of these systems, it may be seen that the equal powers are obtained at different current intensities ($I_2 > I_1 > I_3$, Fig. 1.7). Each system has its advantages in the given range of the parameters on the *U–I* plane.

Figure 1.7 shows that the given three circuits of the plasma torches include the entire plane of the *U–I*-characteristic, i.e., it is possible to generate any required VAC of the arc, selecting a plasma torch with one of these systems.

When describing the linear plasma torches, no mention has been made of the nature of working current. The characteristics, presented in Fig. 1.7, are typical of both DC plasma torches and single-phase basic plasma torches with industrial frequency [1, 2]. Even the three-phase AC plasma torches contain the main elements of these schemes [3].

The method of supplying the working gas into the channel was not discussed separately because these plasma torches may usually operate not only with the tangential but also axial supply of the working gas into the discharge chamber, especially if the problem of stabilisation of the arc spots on the axis of the internal electrode is solved.

As shown in the following chapters, the knowledge of the fundamental physical processes, taking place in the electric discharge chamber of the linear DC and AC plasma torches has made it possible to propose a simple classification of these torches. The special features of the interaction of the arc and of the gas flow blown onto the arc determine the mean length of the arc as the main parameters classification. Consequently, it has been possible to reduce all the linear plasma torches with greatly differing designs to three main groups [4]:

1. The plasma torches with the self-setting arc length;
2. The plasma torches with a fixed arc length, smaller than the self-setting length;
3. The plasma torches with a fixed arc length, longer than the self-setting length.

The other schemes, in particular, the scheme of the two-jet plasma torch, which is being used on an increasing scale in technological processes, are in fact variations of these three schemes.

Chapter 2

Electrophysical and aerodynamic processes in a plasma torch

2.1. SPECIAL FEATURES OF THE FLOW OF COLD GAS IN A LONG CYLINDRICAL CHANNEL

Prior to examining the special features of burning of an electrical arc in a long cylindrical channel of a plasma torch, attention will be given to the flow of a cold gas in the channel. Of greatest interest is the distribution of the degree of turbulence of the flow along the axis of the channel both in the case of smooth walls and with slits simulating intersectional gaps in the plasma torch with inter-electrode inserts (REV); it is also important to know the distribution of the mean axial velocity.

Turbulent motion is charactersied by the intensity of motion; the measure of relative intensity (degree of turbulence) is the intensity of turbulent pulsations:

$$\varepsilon=\sqrt{1/3(\overline{u}'^2+\overline{\upsilon}'^2+\overline{w}'^2)}/\overline{u}.$$

here $\bar{u}'^2$, $\bar{\upsilon}'^2$, $\bar{w}'^2$ – are the mean in respect of time squares of turbulent pulsations of the components of velocity in the direction of the coordinate axes; u is the velocity of gas at the given point averaged-out in respect of time.

Pulsation motion can be described quite accurately by mean of some average values. They include the degree of turbulence ε and the characteristic scale L (components of turbulence). If L is small in comparison with the dimensions of the body, then to describe the pulsation motion of the flow it is sufficient to know only the degree of turbulence ε. In the electric arc chamber in its initial section, the characteristic linear scale of the component of the turbulence (its

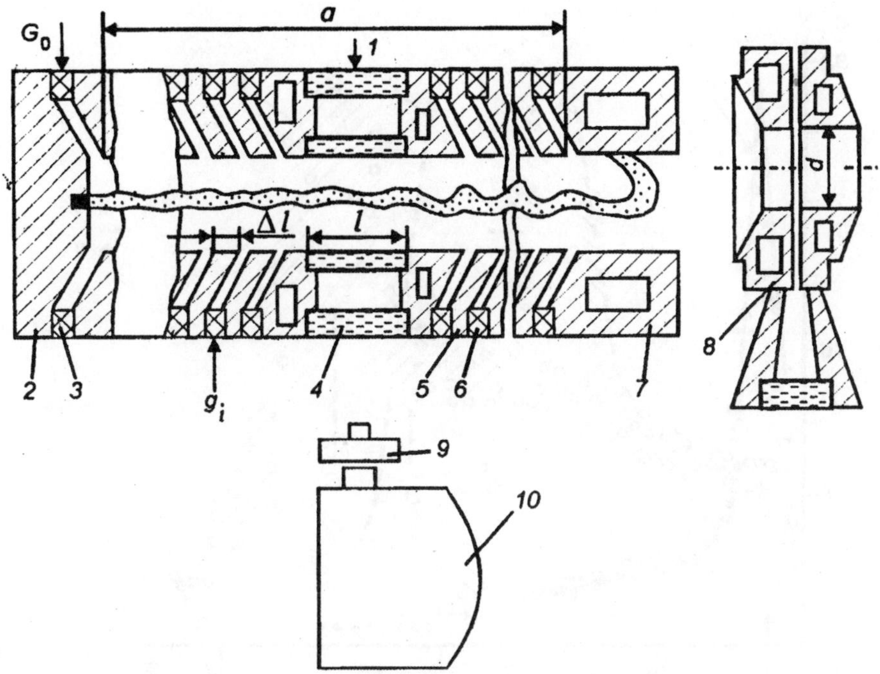

Fig. 2.1. The diagram of the plasma torch for examining the pulsation characteristics and taking photographs of the arc. *1*) Supply of gas for cooling quartz glass; *2,7*) the end output electrodes, respectively; *3*) near-electrode vortex chamber; *4*) optical section with a quartz insert; *5*) the section of the inter-electrode insert; *6*) inter-sectional twisting rings; *8*) optical section with a transverse slit; *9*) gate; *10*) super fast photorecording device (SFR–M).

diameter) is comparable with the characteristic size of the chamber–the diameter of the chamber channel, but considerably greater than the arc diameter. Evidently, in the turbulence section of the channel, L is comparable with the characteristic dimension of the arc. Because of difficulties in determining L, especially in burning of the electric arc, we shall confine ourselves at the moment to explaining the distribution of ε along the channel axes.

Investigations were carried out on a model of a linear plasma torch with vortex stabilisation of the arc (Fig.2.1): internal channel diameter $d=10\cdot10^{-3}$ m, the relative length of the IEI $\bar{a} = a/d = 32 \div 55$; in the experiments, air was supplied into the gaps between the sections. Simulating the plasma torch with the self-setting arc length, the sectional insert was replaced with a smooth pipe with the length $\bar{l} = l/d = 72$. If photographs were to be taken of the arc, quartz tubes or sections of the IEI with a transverse slit were installed in the characteristic sections of the channels.

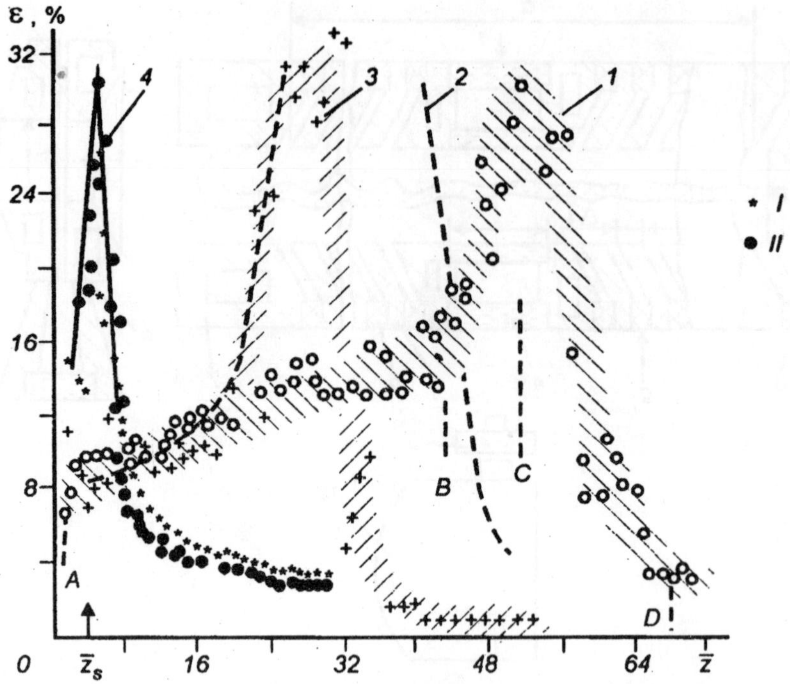

Fig. 2.2. The distribution of ε along the axis of the cylindrical channel of the plasma torch ($d = 10 \cdot 10^{-3}$ m; $G = 5 \cdot 10^{-3}$ kg/s). *1*) Channel with a smooth wall, $l = 77$; 2–4) sectional channel, $\bar{a} = 32$ (*2* – g_1=0, *3* – $g_1 = 0.5 \cdot 10^{-3}$ kg/s; *4* – $m_s = 1.0$ in the section $\bar{z}_s = 4.3$): I – $g_1 = 0$; II – $g_1 = 1 \cdot 10^{-3}$ kg/s.

Figure 2.2 shows the experimental data on the distribution of the degree of turbulence, ε, of the gas flow with the circumferential component of the velocity in the channel during the flow of the gas in both the channel with the smooth surface and in a sectional channel. The Reynolds number, calculated from the mean mass consumption of the gas and the channel diameter, greatly exceeds the critical value. This means that the flow of gas at entry into the channel is turbulent. Since a twisting device, a powerful turbuliser of gas flow, was installed at entry into the cylindrical chamber, the initial degree of turbulence of the flow was very high and reached 6–10% (in section *A* for the curves 1–3). Subsequently, along the flow, the degree of turbulence increased independently of the condition of the surface of the channel walls. This was in agreement with the results of investigations of other authors, for example [1]. For a channel with smooth walls (curve *1*, there is a distinctive increase of the intensity of turbulence which corresponds to the zone of closure of the turbulent boundary layer, formed on the channel wall at its outlet and developing downwards along the flow, and this is obsevered at the distance $\bar{z} \approx \frac{z}{d} = 40 \div 50$ (section B); the coordinate $\bar{z}$ is counted from the entry section in channel A. The section of the channel with length AB is refered to as the initial section ($\bar{z}_i$). The results are in good agreement with the calculated and experimental data which have been reviewed in detail in [2]. The value of ε, reaching its maximum level (section C), subsequently decreases downwards along the flow to the values which determine near-wall turbulence ($\varepsilon \approx 3$–5%, $\bar{z}_H \sim 65$, section D). Section $BD = \Delta \bar{z}_i$, in which the turbulent flow develops (starts in section B and ends in D), is the transition section. Behind the section D the gas flow is highly turbulent.

Thus, the channel contains three sections corresponding to three characteristic types of flow: initial, transition and highly turbulent. There are also studies, for example [3], in which data are given on the characteristics of the gas flow with the circumferential component of the velocity and moving in a long cylindrical pipe. The results of investigations are in qualitative agreement with those described in this chapter.

If the channel consists of sections (see Fig.2.2, curves *2* and *3*), the length of the initial section is reduced. This is associated with accelerated increase of the boundary layer on the surface of the channel, and in the first case (curve *2*) the gas is not supplied into the gaps between the sections, whereas in the second case is was supplied (curve *3*). The directions of the circumferencial component of the velocity of the gas, blown into the main section positioned at the end electrode, and into the interaction vortex chambers (is not specified otherwise) coincides. This gas supply will be referred to as accompanying. However when the direction of the circumferential components of the velocity are opposite, the gas suuplied is referred to as counter supply. When evaluating the effect of the consumption of the gas, g_i, supplied to the intersectional gaps, we use the dimensionless parameter $m_i = (\rho u)_i / (\rho u)_0$. Here the indices 0 and i relate to the parameters of the flow in the channel and the i-th intersectional slit, respectively.

As indicated by Fig. 2.2, changing the size of the intersectional blowing g_i, we can vary in a wide range the relative length of the initial section of the flow $\bar{z}_i$. However the quantity g_i is restricted by the total intersectional consumption of the working gas which does not exceed the flow rate through the plasma torch, determined by the technological process. Therefore, the quantity g_i, used in practice, ensures only ventillation of the gap in order to reduce the thermal losses to the surfaces of the sections and protection of electric insulators against overheating because of convective heat exchange. On the other hand, becuase of a decrease of the temperature of gas in the intersectional gap, the breakdown voltage between the sections increases; the difference of the potential between them, especially at the end of the electric arc chamber, may reach tens or even hundreds of volts. The problem of gas shielding on the surfaces of the sections, which are in contact with the high temperature gas, against high-intensity heat flows will be examined separately in chapter 6.

We shall discuss the effect of the local accompanying blowing of the gas g_s into a single intersectional slit, situated in the section $\bar{z}_s \leq \bar{z}_H$ on the distribution of the quantity ε along the channel axis (Fig.2.3). In the gas blowing section $\bar{z}_s$ (indicated by the arrow in the figure) there is a small surge of the values of ε on the background of the large number of experimental points corresponding to the case $g_i = 0$ (cross-hatched region). However, already at a distance of two-three guages downwards along the flow from the blowing cross-section the mean degree of turbulence coincides with the appropriate characheteristics of the 'non-disturbed' flow. Thus, the accompanying blowing of part of the working gas into a single intersectional gap, even at a relatively high consumption of the gas, has no significant effect on the form of the curve $\varepsilon = f(\bar{z})$. This is associated with a relatively weak effect of the gas flow blown in the same direction on the boundary layers. There is only local thicknening of the layer, i.e a decrease of the width of the 'flow through' section of the channel, and consequently, the increase of ε and the flow speed of the gas in the constriction section.

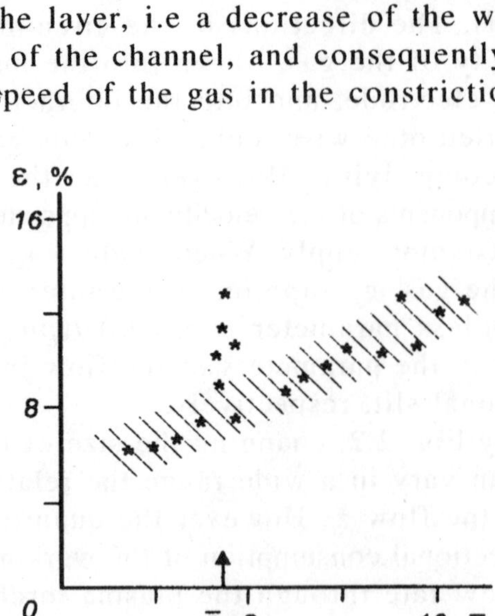

Fig. 2.3. The distribution of ε along the cylindrical section of channel in blowing the gas in the same direction ($d = 10 \cdot 10^{-3}$ m; $g_1 = 0$; $m_s = 0.3$; $\bar{z}_s = 6.9$).

The effect of the counter gas flow will be examined. Figure 2.4 shows the dependence $\varepsilon = f(\bar{z})$, for the counter flow of the gas in the section $\bar{z}_s$ of the initial part close to the entry to the channel; the value of the blowing parameter $m_s = 1.0$. Attention should be given to the rapid increase of the thickness of the boundary layer even at low values of m_s (see [4]). The transition section starts practically outside the blowing cross-section. The value $m_s = 1$ is close to optimum; a further increase of this value causes the reversed effect [4]. In the case of the counter flow of the gas and $m_s = 1$ the intersectional gas flow rate $g_i(m_i)$ blown along the flow behind the section $\bar{z}_s$, has almost no effect on the distribution $\varepsilon = f(\bar{z})$.

The level of turbulence of the gas flow ε in the section of the developed turbulent flow is determined in all likelihood by the surface roughness of the channel wall and by the presence or abscence of blowing of the gas through the intersectional slits, and this level is in the range 3–5%.

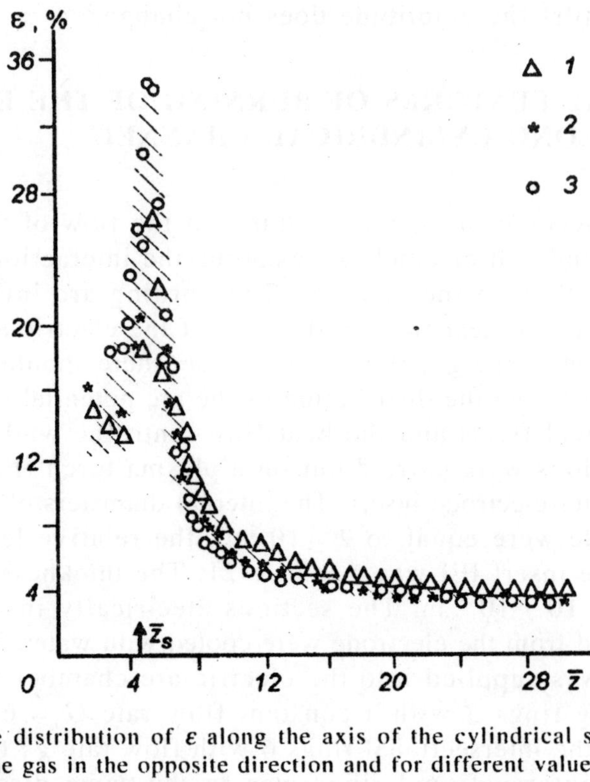

Fig. 2.4. The distribution of ε along the axis of the cylindrical sectional channel in blowing the gas in the opposite direction and for different values of g_i. ($d = 10 \cdot 10^{-3}$ m; $G_0 = 5 \cdot 10^{-3}$ kg/s; $\bar{a} = 32$; $\bar{z}_s = 4.3$; $m_s \approx 1.0$; g_i, kg/s; *1* – 0; 2 –0.5 · 10^{-3}; *3* – 1 · 10^{-3}.

It is interesting to compare the pulsation characteristics of the gas flow in different sections along the channel axis [4]. In the vicinity of the section with maximum ε, there are relatively low frequency pulsations of the flow (5÷10 kHz) with a high amplitude. In the interior of the flow, subjected to low frequency oscillations with a high amplitude, during movement along the flow, high-frequency oscillations with a considerably smaller (many times) amplitude are initiated. The section of developed turbulent flow is characterised mainly by high freqeuncy pulsations (of the order of 20 kHz) with a low amplitude.

In [5] it is noted that the artificially developed turbulence attenuates very rapidly and the value ε is approximately the same, regardless of the level of initial perturbation. The value of ε is in the range 4–5%, which is in agreement with the previously discussed data.

As indicated by a number of studies, the value of ε on the axis of the cylindrical channel is minimum and increases in the direction to the periphery. In the vicinity of the wall there is a low maximum which rapidly decreases with further approach to the surface of the wall and, at the same time, the frequency of pulsations of the flow decreases whilst the amplitude does not change.

2.2. SPECIAL FEATURES OF BURNING OF THE ELECTRIC ARC IN A LONG CYLINDRICAL CHANNEL

Taking into account the special features of the flow of the cold gas in a long cylindrical channel, we examine the interaction of the arc with the gas flow in the channel. The burning arc influences the distributrion of the heat flow in the wall of the electric arc chamber in the direction of the gas flow. In this case, there should be a closer relationship between the distribution of the arc potential (the strength of the electrical field) and the heat flows into the wall.

Investigations were carried out on a plasma torch (Fig.2.1) with a sectional inter-electrode insert. The internal diameters of the sections and the anode were equal to $2 \cdot 10^{-2}$ m, the relative length of the interelectrode insert IEI was $\bar{a} = 20 \div 21$. The thickness of a single section $\Delta l = 10 \cdot 10^{-3}$ m. The sections electrically insulated from each other and from the electrode were cooled with water. The working gas – air – was supplied into the electric arc chamber through the main twisting rings *3* with a constant flow rate $G_0 = 6 \cdot 10^{-3}$ kg/s and through the intersectional rings *6* with flow rate g_i; the gas flow through all intersectional rings was in the same direction. One of the intersectional rings, set in the intitial section of the IEI channel in the section $\bar{z}_s$, is used for supplying the gas with counter twisting and the flow rate g_s, regulated in a wide range. This supply of the gas, as indicated in the previous section, makes it possible to control the thickness of the boundary turbulent layer and, consequently, the relative length of the initial section $\bar{z}_n$ and the length of the turbulent zone $\bar{z}_t$ at the selected length of the insert *a*.

The appplication of the quartz pipe *4* with the length $l = 42 \cdot 10^{-3}$ m and the wall thickness of $(2.5 \div 3) \cdot 10^{-3}$ m, secured between two specially profiled and water-cooled copper sections, made it possible to carry out qualitative and, in a number of cases, quantitative investigations of pulsations of the arc. The internal diameter of the quartz pipe was the same as that of the section. To prevent overheating, the external surface of the pipe was cooled with a flow of cold air *1*, and the internal surface (on the side of the hot gas), was shielded with a gas screen formed by the cold working gas supplied into the intersectional slit in front of the optical section. With this film shielding it was possible to examine the arc in the section of the developed turbulent flow where the density of the heat flow is so high that the quartz glass would soften without specially organised protection.

Examination of the arc at different gas flow conditions during a single start up of the plasma torch (i.e. without the movement of the opotical section) was possible as a result of using the counter blowing with the parameter n_s varied from 0 to 1. The section of the electric arc chamber with the length of $27 \cdot 10^{-3}$ m was photographed. The time dependence of the glow of the arc was examined by recording an element of the arc through a transverse slit with the width of $2.5 \cdot 10^{-3}$ m. (A slit with the length of $1 \cdot 10^{-3}$m was set in the camera in the examined case). The slit with a funnel covered with quartz glass on the outside, was produced in the water-cooled cross-section *8* with the thickness $\Delta l = 24 \cdot 10^{-3}$ m. The arc was photographed with SFR-1M superhigh speed camera. The application of additional gates enabled a time delay from $1.7 \cdot 10^{-2}$ to $1 \cdot 10^{-3}$ s and the speed of rotation of the mirror was varied from $3.75 \cdot 10^{-3}$ to $6 \cdot 10^{4}$ rpm. In the 'time lens' regime, the maximum speed of rotation of the mirror and the double lens insert, the rate of recording was $2.5 \cdot 10^{5}$ frames/s, the speed of time development of the image was 750 m/s. To improve the resolution power of the entire system, the camera was placed at a distance of $(200 \div 250) \cdot 10^{3}$ m, from the object, and high sensitivity films RF-3, Izopanchrom T-24 and T-22 werre used; Zh-17 light filter was used in some cases.

High-speed photographic examination of the arc was accompanied by the determination of the strength of the electrical field of

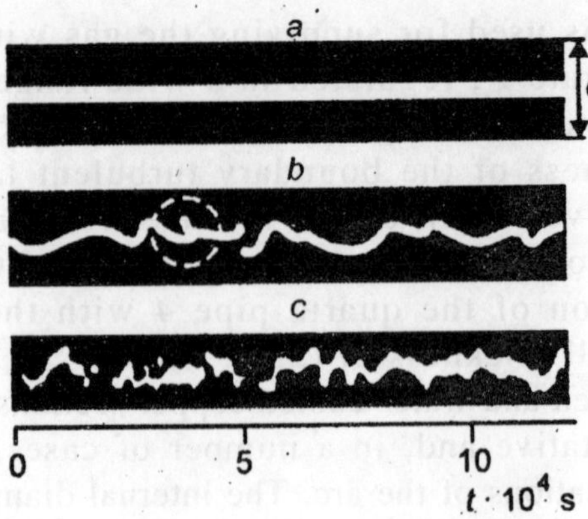

Fig. 2.5. Development of the glow of the element of the arc on the characteristic sections of the channel. *a*) initial section; $\bar{z}_s = 5.5$, $m_s = 0$, $I = 100$ A; *b*) transition section $\bar{z}_s = 535$, $m_s = 1.0$, $\bar{z}_s = 3$, $I = 100$ A; *c*) section of developed turbulent flow of the gas $\bar{z}_s = 15$, $\bar{z}_s = 3$, $m_s \approx 1.0$, $I = 180$ A.

the arc and the heat flows into the wall of the discharge channel (more details on these measurements are given in chapters 5 and 6).

According to the time scan of the glow of the arc element in different sections of the channel [6], the initial section (Fig.2.5, *b*) shows shows no transverse pulsations of the arc, and the scan of the arc is a straight band. Evidently, in the presence of twisting of the gas the electric arc is fixed quite efficiently in space (at the axis of the discharge chamber) by the Archimedes force. The transition section (Fig.2.5*b*) is characterised by radial oscillations of the arc element. In addition to this, one can also see the simultaneous existence of two branches of the arc (circled). Finally Fig. 2.5*c* shows information on the transverse oscillations of the arc and their frequency in the third characteristic section of the gas flow – turbulent.

Thus, even the qualitative examination of the problem of radial oscillations of the arc along the length of the electric discharge chamber confirms the conclusions made in section 2.1 in examining the distribution of the degree of turbulence ε along the channel axis, according to which in the case of burning of the arc in the channel there are three characteristic zones of the gas flow.

What is the variation of the average longitudinal component of the strength of the electrical field of the arc E along the channel? Its magnitude depends on the channel diameter, the gas flow rate and pressure, intensity of current and a number of other controlling parameters. As an example, Fig. 2.6 shows the distribution of E along a long electric discharge chamber. Three characteristic zones are clearly visibly along the curve. The first of them (*I*) corresponds

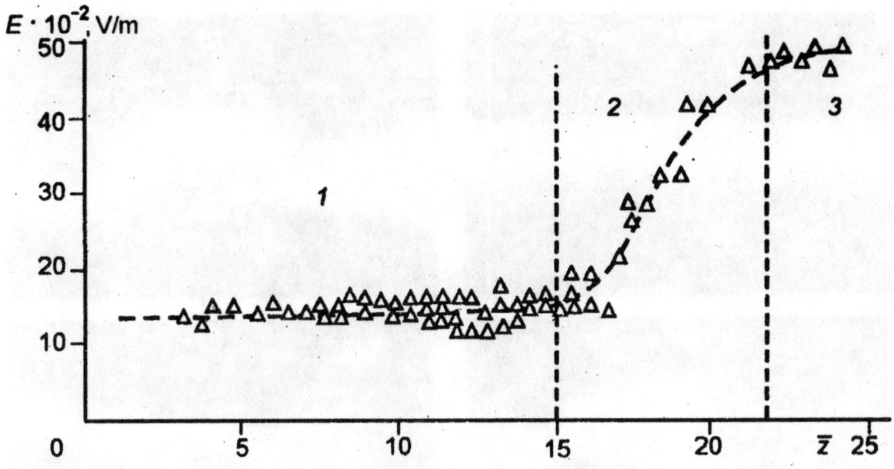

Fig. 2.6. Distribution of the strength of the electrical field of the arc along the electric discharge chamber.

to a constant strength of the field. The second zone (*2*) is characterised by an increase of the strength along the channel. It is followed by the third zone (*3*) with a constant value of E, if the gas flow rate, pressure and channel diameter remain constant along the direction of the flow.

Figure 2.7 shows photographs of the arc (made through a quartz tube) running in appropriate characteristic sections of the channel. In the initial section (*a*) the arc does not have any transverse pulsations. In the initial part of the transition section (*b*) they are already clearly visible. At the end of the transition section and in the developed turbulent section (*c*) the radial oscillations are clearly visible. In addition to this, the 'arc–arc' shunting is quite distinctive resulting in the splitting of the arc into a number of current-conducting channels changing with time (*c* and *d*). In detailed analysis of the shape of the arc, running in the channel, and also in the immersed jet (this will be dicussed later) it is clear that two processes develop simulataneously in the transition section of the flow:

a) A periodic process, determined by the appearance of the helical form of the arc and by magnetohydrodynamic instability of the arc column as an integral unit;

b) random pulsations, i.e. the oscillations of the arc in relation to the channel axis with a small amplitude of deviation caused by wall turbulence.

We examine a problem of the thermal layer of the arc because this is related directly to the given phenomenon. Figure 2.8 shows

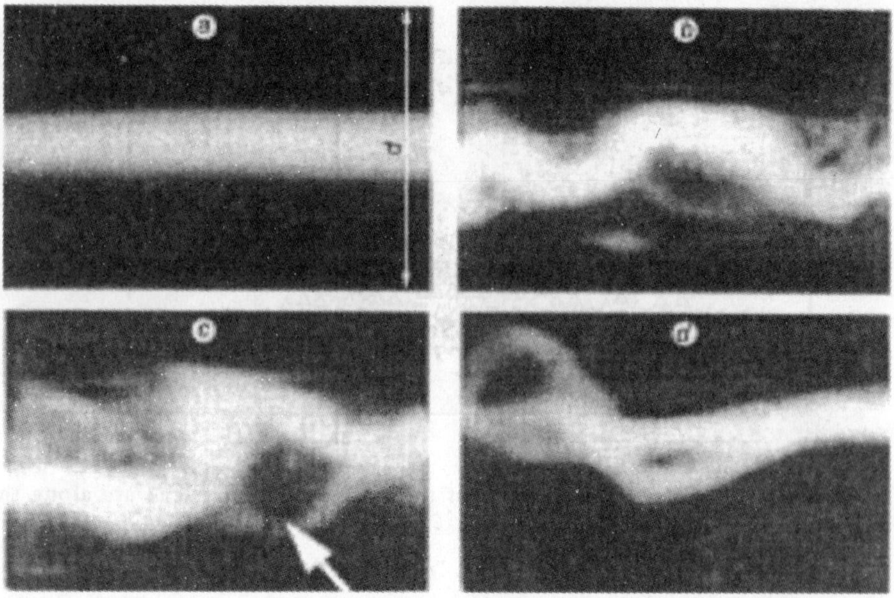

Fig. 2.7. Photographs of the arc in the individual sections of the channel ($d = 2 \cdot 10^{-2}$ m, $G = 15 \cdot 10^{-3}$ kg/s, $I = 100$ A, $\tau = 8 \cdot 10^{-6}$ s). *a*) the arc in the initial section of the gas flow; *b*) random oscillations of the arc in the transition section; *c*) 'arc–arc' shunting; *d*) splitting of the arc.

the scheme of interaction of the arc with the surrounding gas (*a*), Topler (*b*) and schlieren (*c*) photographs of the arc, running in an immersed jet. Since the arc column is characterised by high radiation intensity, its diameter in all likelihood is close to the recorded arc diameter of the arc. Therefore, the dimension r_0 may be regarded conventionally as the radius of the current-conducting channel (Fig.2.8*a*). This zone borders directly with the thermal layer of the arc. The external boundary of the layer is quite distinctive (Fig.2.8*b*) and is determined in experiments on the basis of the minimum of illumination on the Schieren photographs. The thermal layer of the arc is characterised by lower radiation intensity, and its radial size δ depends to a greater extent (than r_0) on the velocity of the external flow (in this case on the gas flow rate). The form of the external boundary of the thermal layer of the arc (Fig.2.9) also depends strongly on the flow speed (gas comsumption).

The estimate of the enthalpy of the gas in the region of the maximum density gradient [7] shows that the 'thickness' of the thermal layer in the radial direction where the gas temperature decreases from $T \approx 5000$ K to the temperature of the environment, is small and does not exceed 1.5 ÷ 2 mm. Calculations of the radius of the boundary

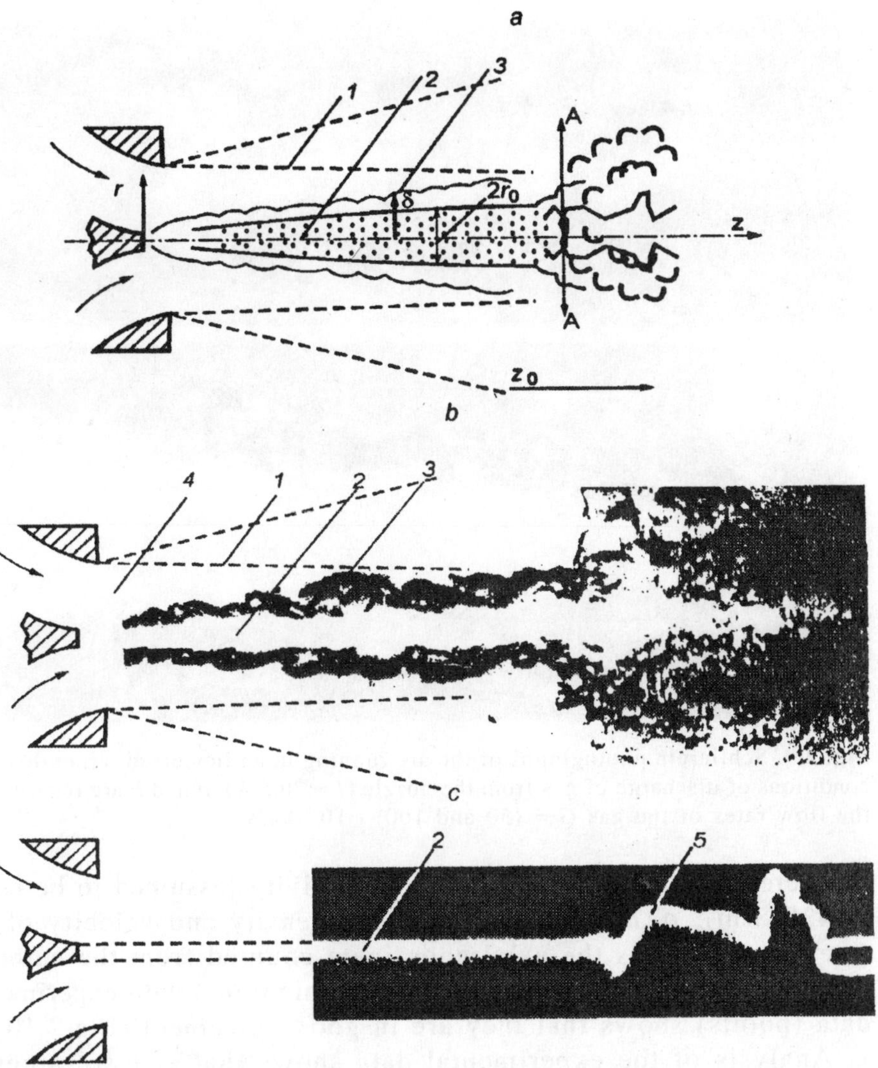

Fig. 2.8. Interaction of the arc with the surrounding gas (*a*); Tepler (*b*) and schlieren (*c*) photographs of the the arc running in an air jet discharged into the immersed space (the circumferential component of the gas velocity is equal to zero). *1*) the boundary of the jet core; *2*) electrical arc; *3*) the boundary of the thermal layer of the arc; *4*) external gas flow; *5*) turbulent section of the jet with the arc.

of the thermal layer, carried out by the numerical method in [8], give the following dependence:

$$\eta = 2.82 \cdot \xi^{0.315},$$

where $\eta = (\delta/l)2\pi(\lambda_0 h_0 \sigma_0 / C_{p_0})^{0.5}$; $\xi = (z/l^2)h_0\sigma_0\lambda_0^2\pi^2 \cdot (\rho_\infty u_\infty C_{p_0}^2)^{-1}$.

Here λ_0, h_0, C_{p_0} are the characteristic values of heat conductivity, enthalpy and heat capacity of the free flow; σ_0 is the

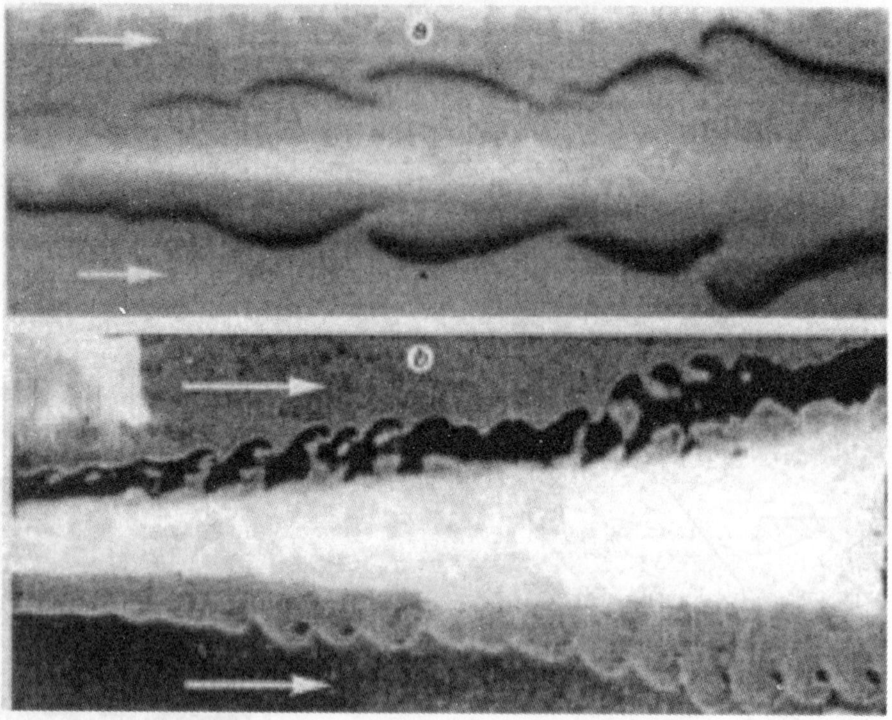

Fig. 2.9. Schlierein photographs of the arc running in an immersed jet, at different conditions of discharge of gas from the nozzle (I = 70.5 A). a and b are respectively the flow rates of the gas G = (50 and 100) · 10^{-3} kg/s.

characteristic value of electrical conductivity (assumed to be equal to 430 S/m); ρ_∞,u_∞ is respectively the density and velocity of the free gas flow; z is the axial coordinate counted from the cathode. A comparison of the calculated values (straightline) with experimental data (points) shows that they are in good agreement (Fig.2.10).

Analysis of the experimental data shows that at low values of the parameter ξ (this corresponds in the examined case to flow speeds, for example, for u = 124 m/s) the values of δ are greatly scattered as a result of the perturbation of the boundary of the thermal layer with increasing speed and due to difficulties in determining its true thickness. For example, at low speeds (Fig.2.9) the boundary has the form of a relatively smooth curve. With increase of the flow speed axisymmetric perturbations form and develop along the boudary and are identical to the wave on the surface of two media with different density which penetrate deeper and deeper into the thermal layer. At strong perturbations the boundary of the thermal layer of the arc was determined on the basis of the maximum and minimum δ, which is expressed by the appropriate signs in Fig. 2.10 [9].

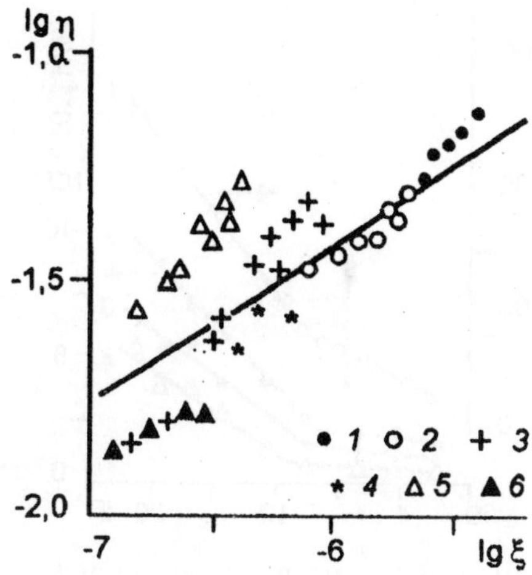

Fig. 2.10. Comparison of the calculated and experimental values of the boundary of the thermal layer.
1) U∞ = 12 m/s; *2*) 24.8 m/s; *3,4*) 62 m/s, *3* in respect of maximum and *4* in respect of minimum; *5,6*) 124 m/s, *5* in respect of maximum and *6* in respect of minimum.

The dependence was obtained in the section of the stable arcing. In the zone of contact of the boundary thermal layers the instability of the arc column starts to develop. In individual cases (Fig.2.8 *b*) the process of displacement of the thermal layer of the arc with the external region of the jet is of the explosive nature as a result of the formation, in the potential zone, of the jet of local axisymmetric deformation of the arc column moving in the direction of the flow at a speed of 15 ÷ 20 m/s. Increasing in the volume, the jet deforms the boundary of the thermal layer and approaching the area of contract of two boundary layers, it appearts to explode. Processing of the films of movement of the arc shows thast the speed of propagation of the bending peturbation in the turbulent zone in the axial direction approximately corresponds to the speed of the external flow.

Naturally, the interaction between arc and the gas flow is reflected in the distribution of the heat flow into the channel wall along the electric arc chamber. One should expect a close relationship between the distribution of the arc potential $V(\bar{z})$ (strength) and the heat flow $\bar{G}(\bar{z})$. Actually, comparison of the curves *1* and *3* in Fig.2.11 shows that the coordinates of the start of the rapid increase of the arc potential nad of the heat flow are approximately identical. As shown by further investigations, with increase or decrease of current intensity, the coordinate of increase of $\bar{Q}(\bar{z})$ is also displaced, as a result of the change in the thickness of the thermal layer.

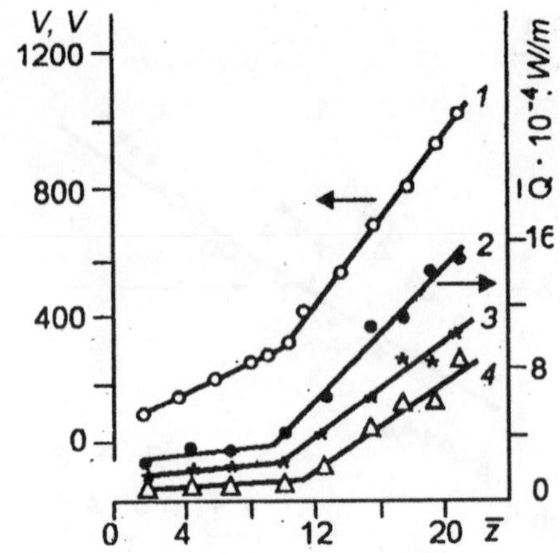

Fig. 2.11. The distribution of the arc potential (*1*) and heat flows (*2–4*) along the channel. Working gas – air. $d = 10 \cdot 10^{-3}$ m; $\bar{a} = 22.1$; $G = 15 \cdot 10^{-3}$ kg/s; $g_1 = 0.7 \cdot 10^{-3}$ kg/s, I, A: *1*) 120; *2*) 150; *3*) 120; *4*) 90.

The results of a large number of investigations of the distribution, along the channel axis, of the degree of turbulence of the cold flow, the strength of the electrical field of the arc and the heat flow into the channel wall, and also in examination by the optical methods of pulsations of the arc running in the channel or the inmersed arc provide a basis for constructing the scheme of the gas flow in a long cylindrical channel in the presence of the arc.

It should be mentioned that of highest interest for practice are the flow conditions in which the Reynolds numbers calculated on the basis of the input parameters of the cold gas are relatively high, and in the absence of the arc the flow in the channel is known to be turbulent.

The simplest scheme of the gas along the channel may be described as follows (Fig.2.12). In the initial section *AB* whose length is determined by the area of contact of the thermal layer of the arc *3* and the turbulent boundary layer *1*, developed on the channel wall, the arc burns in the laminar flow. Schlieren photography of the arc in the initial section visualizes the thermal boundary layer *3* and the electrically conducting arc zone *2*. In the section *BC* the thermal layer is disrupted. This process is efficiently recorded by a high-speed photorecording system if the arc burns in a quartz channel. Starting in section *B*, the arc column, (i.e., the region in which the main part of the electric current flows) starts to interact with the turbulent gas flow. The section *BCD* differs by the fact that it is characterised by gradual transition to the developed turbulent flow (transition zone). Finally, a steady turbulent flow is found in *DE*.

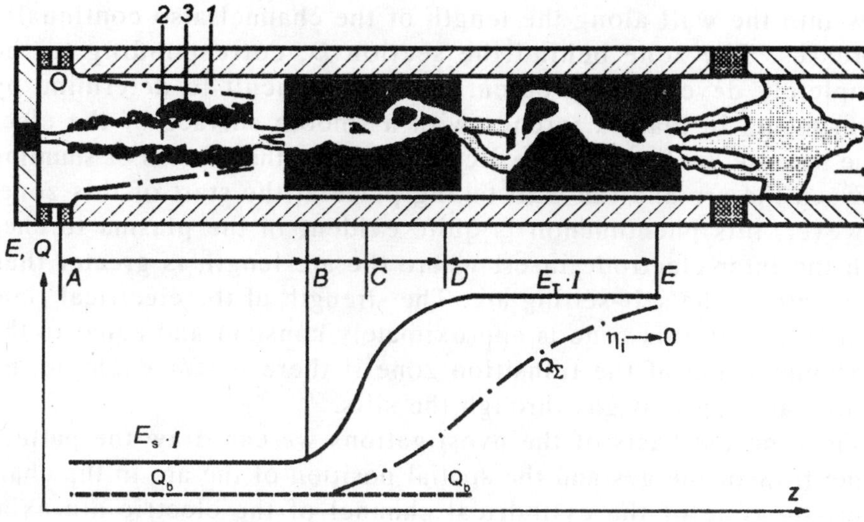

Fig. 2.12. Structure of interaction of the arc with the surrounding gas.

The characteristic special features of the initial zone (without the near-electrode region) are the constancy of the strength of the electrical field of the arc E_i along the channel which is experimentally confirmed by different independent methods; A weak relationship of E_i with the gas flow rate and the absence of transverse oscillations of the arc.

The special feature of the electric arc running in the transition zone *BCD* is the monotonic increase of the strength of the electrical field in the direction of the gas flow. Evidently, this was caused both by the intensification of the removal of thermal energy from the arc and by an increase of its actual length becuase of the marked distortion of the arc column over the measuring base (i.e. in the section between the centres of the adjacent measuring sections, with different electrical potentials). As already mentioned, in the initial section of the channel the arc is stable along the axis of the channel and has no transverse oscillations (Fig.2.7*a*). At the begining and end of the section of the transition gas flow (Fig.2.7*b*, *c*) there are not only radial oscillations of the arc but also splitting of the arc into two current-conducting channels caused by electrical breakdown (shunting) in the arc loop. In the developed turbulent section (Fig.2.7*d*), radial oscillations are even more distinctive, like the processes of development and disappearance of the current-conducting channels. The mean value of the strength of the arc at the end of the transition zone is several times higher than the strength in the initial section. Starting in section *B* (Fig.2.12), the intensity of the heat flow into the wall along the length of the channel also continually increases. The zone behind the section *D*, corresponding to the completely developed turbulent flow, is difficult to determine by experiments in plasma torches with a smooth surface of the electrode because the length of the arc is limited by the process of shunting between the arc and the wall taking place at the start of this zone. However, this phenomenon is quite evident in the plasma torches with the inter-electrode insert where the arc length is greater than the length of the self-setting arc. The strength of the electrical field of the arc E_t in this zone is approximately constant and equal to the maximum value of the transition zone if there is, for example, no additional supply of gas through the slits.

Thus, on the basis of the nvestigations we can draw the pattern of the flow of the gas and the spatial position of the arc in the characteristic zone of the cylindrical channel of the electric arc axial plasma torch. The flow zones themselves have a complicated structure and specific boundary conditions that require further detailed examination.

2.3. SPEED AND PULSATION CHARACTERISTICS OF ARC ELEMENTS

The photographs of the electrical arc running in the characteristic sections of a long electrical arc channel, presented in the previous section, provide qualitative information on the processes taking place. In the experiments concerned with the examination of the plasma flows, the optical methods of recording the movement of heterogeneities have been used widely.

Some characteristics of the plasma flow were determined by the analysis of the movement and pulsations of elements of the arc [4]. Photographs of the arc, produced using SFR-1M camera, were used for determining the mean and pulsation velocities of movement of the arc elements and also the frequency of their pulsations.

The axial radial speed of movement of the characteristic perturbed elements of the arc were computed from the displacement of the boundaries in the appropriate directions which was regarded as the given time period between the frames on succesive frames of CFR-films.

The pulsation and mean velocities of displacement of the arc elements were determined by the standard methods of processing the results of measurements, and the mean speed of movement of the peturbation boundary u_m was determined as the mean arithmetic value of the

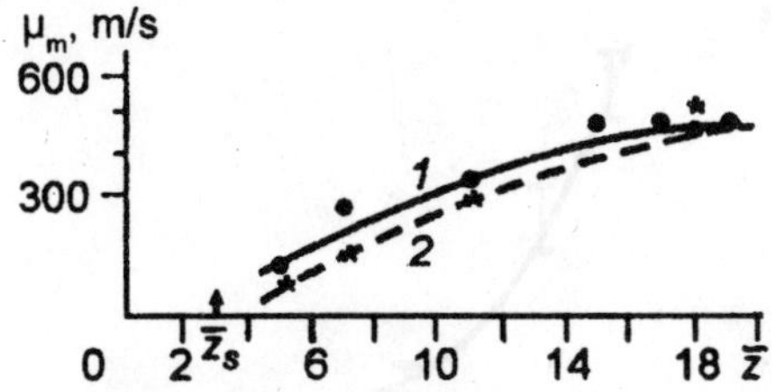

Fig. 2.13. Distribution of the mean speed of movement of the elements of the arc on the axis of the channel (*1*) and the mean mass speed of the gas flow (2) along the discharge chamber. $d = 20 \cdot 10^{-3}$ m; $\bar{a} = 20$; $G_0 = 6 \cdot 10^{-3}$ kg/s; $I = 100$ A; $\bar{z}_s = 3$; $g_s = 6.3 \cdot 10^{-3}$ kg/s ($m_s \approx 1.0$).

individual measurements of the speed in the selected direction.

Figure 2.13 shows the distribution (curve *1*) along the axis of the channel of the average speed of movement of the elements distributed along the axis. Curve *2* characterises the distribution of the mean mass speed of the high-temperature gas in a pipe taking into account heat generation by the electric arc and heat loses in the wall of the discharge chamber. Experiments show that slight divergence of the curves is found only in the transition section of the channel, but in the area of the developed turbulent flow the curves are almost identical. This indicates the propagation of the arc elements along the channel with approximately the mean mass speed of the flow. The difference in the speeds in the transition section is associated with a large difference in the maximum speed of the gas on the axis of the channel in comparison with the mean mass speed, because the process of displacement of the cold and high temperature gas flows is not yet complete.

The availablity of a large number of successive photographs and time sweeps of the arc made it possible to determine not only the mean but also pulsation components of the speed of movement of the arc elements in the examined sections of the channel assuming that the element of the arc moves with the pulsation speed.

The mean quadratic deviation is determined using the equation

$$\sigma_{ni} = \left[\sum_{i=1}^{n} (\Delta a_i)^2 / n(n-1) \right]^{0.5},$$

where n is the number of measurements; Δa_i is the absolute deviation of the i-th measurement from the mean value. For the aqxial pulsations $\Delta a_i = u'_i$, for radial pulsations $\Delta a_i = v'_i$. The measure of turbulence of the flow along the selectred direction is represented by $\varepsilon_z = \sigma_{nzi}/u_m$ and $\varepsilon_r = \sigma_{nri}/u_m$. The turbulence of the flow is

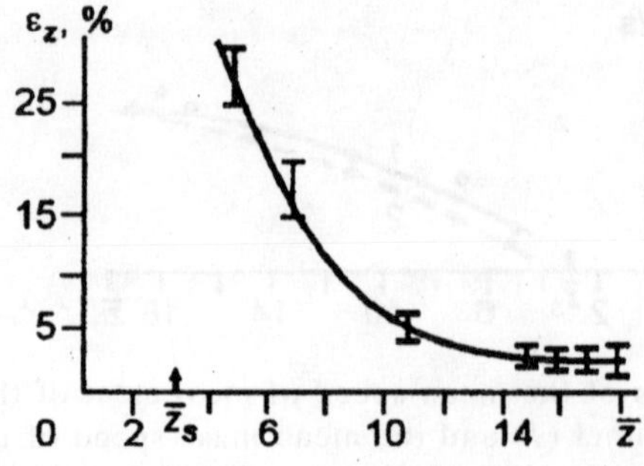

Fig. 2.14. Distribution of the axial component ε_z of the degree of turbulence of the gas flow with the arc along the channel. $d = 20 \cdot 10^{-3}$ m; $G = 18.5 \cdot 10^{-3}$ kg/s; $I = 100$ A; $\bar{z}_s = 3$; $G_0 = 6.1 \cdot 10^{-3}$ kg/s, $g_s = 6.3 \cdot 10^{-3}$ kg/s; $m_s = 1$; $\bar{a} = 20$.

determined on the whole from the equation $\varepsilon = [0.5(\varepsilon_z^2 + \varepsilon_r^2)]^{0.5}$. The characteristic frequencies of the oscillations of the arc were calculated from the time sweeps of their images.

The ditribution of ε_z is shown in Fig. 2.14. It is interesting to know that in the vicinity of the blowing zone of the gas ε_z reaches 25% and more, i.e. the same value as in cold blowing. Subsequently, the degree of turbulence rapidely decreases along the direction of the flow and behind the section $\bar{z} \approx 12$ reaches the level of 3–4 %. The distribution of the radial component of the degree of turbulence ε_r along the axis of the channel is the same.

Distribution of the total degree of turbulence ε along the axis of the channel is shown in Fig. 2.15. To facilitate analysis, the curves were displaced along the $\bar{z}$ axis by the value $\bar{z}_s$, where $\bar{z}_s$ is the ordinate

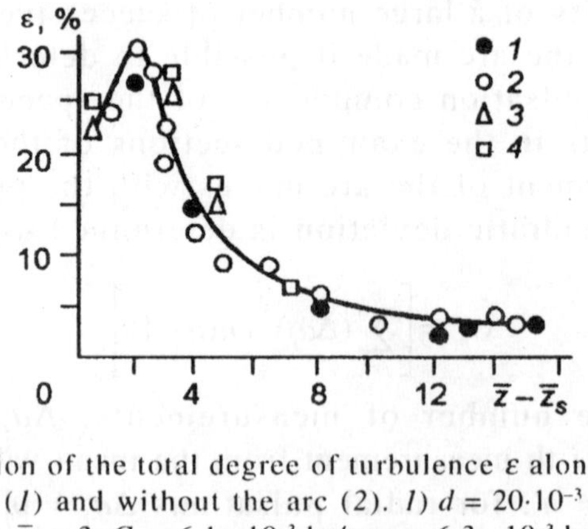

Fig. 2.15. Distribution of the total degree of turbulence ε along the channel in the presence of the arc (*1*) and without the arc (*2*). *1*) $d = 20 \cdot 10^{-3}$ m; $G = 18.5 \cdot 10^{-3}$ kg/s; $I = 100$ and 180 A; $\bar{z}_s = 3$; $G_0 = 6.1 \cdot 10^{-3}$ kg/s; $g_s = 6.3 \cdot 10^{-3}$ kg/s; $m_s = 1.0$; $g_3 = 2.3 \cdot 10^{-3}$ kg/s; $\bar{a} = 20$; 2) $\bar{z}_s = 4$; $G_0 = 5 \cdot 10^{-3}$ kg/s; $m_s = 1.0$; $\bar{a} = 32$; $d = 10 \cdot 10^{-3}$ m; $G = 10 \cdot 10^{-3}$ kg/s, $g_i = (0 \div 1) \cdot 10^{-3}$ kg/s; *3,4*) data from [10].

of the section of introduction of the turbulizing gas. The data were obtained by combining different measurement methods. The points *2* are the degree of turbulence recorded using a thermal anemometer in a twisted cold gas flow at the axis of the channel; points *1*– the results of processing of consecutive photographs of the arc column through quartz glass by high-speed photography [6]; point *3*, *4* are the results of processing and calculating the mean value and dispersion of the difference of the potentials between the two sections of the arc [10]. If some points were determined by the method of thermal anemometery used widely in gas dynamics and others by the contactless and relatively time consuming methods, the last set of the points is an example of the qualitatively new application of the classical probe measurements.

Regardless of the difference of the working parameters and conditions of measurements, all the three methods give the results that are in relatively good agreement with each other, especially in the section of the developed turbulent flow. The small difference between the value of ε in the transition section may be caused by the fact that the degree of turbulence, measured by the last method, is averaged out in respect of the cross section of the channel. Agreement is expected only if the value of ε is approximately constant in the entire range occupied by the arc in this section. Thus, the application of advance computing methods and new methods of processing the results of measurements by classical methods may provide additional information on the interaction between the electric arc and the gas flow.

Attention will now be given to the frequency characteristics of the arc, using the timesweep of the brightness of the arc [4]. As mentioned previously, in the initial section of the channel there are no large radial pulsations of the arc. However, the transition section is already characterised by radial deflection from the axis with a relatively low frequency of oscillations. In the zone of developed turbulent flow the frequency of oscillations increases and the amplitude slightly decreases. Examining the density of darkening of the photofilm, it may be seen that it is smaller for the arc running in the section of the developed turbulent flow and in this case the visible luminous diameter of the arc isalso smaller.

The frequency characteristics of the pulsations of the arc column in different characteristic sections of the channel differ. The transition section is characterised mainly by low-frequency arc oscillations (500÷1000 Hz). Oscillations with the frequency of 4÷5 kHz are superposed on them. In the section of the developed turbulent flow there are mainly pulsations with a frequency of 10÷50 kHz. These frequencies coincide with the characheteristic frequencies of pulsations of the cold gas flow. Thus, the electrical arc does not change greatly the frequency characteristis of the pulsations of the flow in the investigated range of the parameters.

Thus, it may be assumed that the pulsation characteristics of the gas flow with the electric arc are determined by the wall characteristic of the gas flow because in this and other (cold) cases, these characteristics (geometrical dimensions of the electric arc chamber, the Reynolds number of the flow, calculated from the viscosity of the gas at the wall temperature) are similar. In the experiments, the wall temperature was approximately 300 K for the flow without the arc and approxiamtely 400 K with the arc.

It may also be concluded that the pulsation characteristics of the electric arc are determined mainly by the pulsation characteristics of the gas flow. The inherent electromagnetic forces have no significant effect on the pulsation characteristics of the arc in the investigated current range (to 180 A). This conclusions confirms the assumption on the hydrodynamic nature of the interaction of the electric arc with the gas flow at a relatively low current intensity made in [7] and other studies when calculating turbulent electric arc.

2.4. TOMOGRAPHIC INVESTIGATIONS OF THE ELECTRIC ARC

2.4.1. Brief review

In physical investigations of plasma objects it is often necessary to examine the formations with a complicated structure. This factor greatly complicates the problems of diagnostics and requires development of specific methods and equipment. To examine objects of complicated form it is necessary to use tomographic diagnostic methods [11].

Depnding on the nature of the specific problem, the restoration of the structure of the object may be based on recording the beam of electrons, ions (including protons and α-particles) neutrons, photons and sound waves. The restoration of the internal structure of the object and its projections obtained as a result of illumination of the object from different directions or using its intrinsic radiation is the subject of computing tomography (CT).

Since plasma diagnostics using tomography is the subject of increasing attention of scientists, it is useful to examine breifly the studies in which the algorithims of computing tomography have been used in specific plasma investigations or numerical modelling, oriented to the fully specified experimental equipment, has been carried out.

The first publication, concerned with the application of emission tomography of optically fine plasma of arbitrary shape without a symmetric plane appeared in 1968 [12] (in earlier studies of these authors, attention was given to the configuration with a symmetry plane). This study describes a stationary freely running arc at an argon pressure close to atmospheric ($p = 1.1 \cdot 10^5$ Pa), in the transverse magnetic field of induction $B = 30 \cdot 10^{-4}$ T, current $I = 400$ A. Photographic recording was used. Examination was carried out from 15 directions uniformly distributed in a sector with the angle of 180°, and 73 measurements were taken in each direction. The images were reconstructed using the algorithm based on the expansion of the signals using special polynomials, orthogonal in relation to the direction of rotation. Temperature was determined by the methods of absolute intensity of continuum. In the study, the isoterms were constructed in two longitudinal and one transverse section of the arc.

The authors of [13, 14] described for the first time the results of detailed investigations of the temperature fields of a stationary argon arc onto which a gas was blown in the transverse direction, in relation to the flow rate and arc current intensity. Tomographic measurements were taken in equipment including a system of mirrors, an interference light filter and a camera. Reconstruction was carried outr using the data taken from eight directions non-uniformaly distributed within the limits of the angle of 90°, using the MacDonald algorithm in a variant with mirror symmetry. Temperature was determined using a special method on the baiss of the absolute intensity of continuum asuming local thermodynamic equilibrium (LTE). The authors established that an increase of the gas flow rate reduces the tranverse cross-section and increases the maximum plasma temperature.

The first description of a six-direction plasma tomograph and also of the method of measurements of the temperature fields of turbulent plasma, using the tomograph and tomographic images, were published in [15, 16].

The stationary arc with a transverse nitrogen flow and stabilised with a transverse magnetic field, was investigated in [17]. Measurements were taken using a spectrograph in seven directions with a uniform step in respect of the angular variable of 15° (mirror symmetry was typical of the distribution). Up to 80 measurements were taken in each direction. The measured temperature fields were used for constructing the field of velocity and for detecting vortex zones and the points of deceleration of the flow.

In [18] the tomgraphic method was used to measure the temperature of the plasma of an atmospheric pressure arc running in a mixture of 40% H_2+ 60% N_2 at a current intensity of 5 A and moving under the effect of a rotating magnetic field at a frequency of 15÷16 Hz. 24 projections were obtained from one direction at an angle of 360° assuming that the plasma remains stationary in rotation in the intrinsic reference system. The number of counts was N = 200. The gas temperature was determined from the absolute intensity of the line H_β (in the approximation of partial LTE), and the results of measurements of the strength of the electrical field of plasma E.

A toroidal arc, freely burning in argon at atmospheric pressure between two plane-parallel disks and maintained in equilibrium by intrinsic and external (vertical) magnetic fields, was described in [19]. Irradiation of argon was recorded in continuum (λ = 443 ± 5 nm) in the angle range from 0 to 90° with a step of 15°.

This gave the fields of temperature for different currents and radii of arc and the velocity fields were calculated.

In [20] the authors reported for the first time on the construction of a plasma tomograph with information inputted behind the output plane-parallel package of light guides and the photomatrix into the computer. The counting time was 0.2 μs. The non-stationary plasma of complicated configuration was described in [21, 22] (for more detailed results see below). The investigation into plasma tomography and the review of the algorithms of plasma tomography were published in [23].

These schemes of emission plasma diagnostics (2- and 6-view) were examined in [24]. This study also gives the results of processing the data obtained in measurement of radiation for a helical argon arc in a longitudinal magnetic field.

2.4.2. Experimental investigations of a non-stationary electric arch plasma

To calculate the electrical arc in a plasma torch it is necessary to obtain experimental data on the physical processes, taking place in the arc plasma in different discharge conditions, on the effect of the local and integral characteristics of the plasma of the parameters such as type and pressure of gas in the chamber, the flow rate of the gas, the externally applied magnetic field, etc.

Attention will be given to the behaviour of a plasma filament with current when the intrinsic magnetic field of current cannot be ignored, and the role of factors, stabilizing the position of the filament on the axis of the plasma channel (heat conductivity and viscosity), is small. It is assumed that at some moment the filament is randomly deformed, in the simplest case by bending or stretching. In bending as a result of different density of the force lines of the azimuthal magnetic field from the internal and external sides of the filament a magnetic gradient appears in this zone and random deformation increases under the effect of this pressure. The bending of the plasma filament may take place with different probability in the form of right-handed or left-handed screw deformation.

A plasma filament with the current characterised by finite conductivity is placed in a longitudinal magnetic field. The appearance of a helical perturbation in the plasma is accompanied by the formation of the Lorentz force, normal to the direction of current in the magnetic field. If the Lorentz force is directed to the centre of the chamber, the plasma filament is stabilised, if it is directed to the wall, deformation continues to develop.

Figure 2.16*a* shows how the longitudinal magnetic field stabilises the perturbation rotating in the anticlockwise direction (if examined in the direction of the field) developing a perturbation with the opposite direction of rotation (Fig.2.16*b*).

The results of examination of the effect of the gas flow rate and the longitudinal magnetic field on the form of temperature fields and the electrical characteristics of arc-plasma in a cylindrical channel are discussed below [21, 22].

An electric discharge in a sectional cylindrical channel 90 cm long was examined. Each section was 1.4 cm long, internal diameter $d = 3$ cm. The cathode was water-cooled tungsten rod with the tip angle of 60°, the anode was made of copper. The working gas (argon) was supplied from the side of the cathode, and its flow rate was varied in the range $G = 0.034 \div 12.7$ g/s, the gas pressure in the chamber was maintained on the level $p = 1 \cdot 10^5$ Pa, arc current $I = 100 \div 130$ A. In a number of experiments, the central part of the arc was placed in the longitudinal magnetic field generated by two solenoids with a total length of 30 cm. The magnetic induction B on the axis of the solenoids varied from 0 to 0.44 T. It was found that if the electrodes were placed in the region of the strong magnetic field, they failed very rapidly because of pulsations of the arc spot. Consequently, the composition of the plasma is disrupted. To prevent the effect of the electrode material on the plasma, the electrodes were moved 30 cm from the end of the magnetic coils.

The temperature fields of the plasma of complicated form were measured by the method of emission tomography. The transverse projections of the intensity of radiation of the plasma simultaneously from several directions were recorded using a plasma tomograph (Fig.2.17) in the form of a system consisting of an optical disk, which

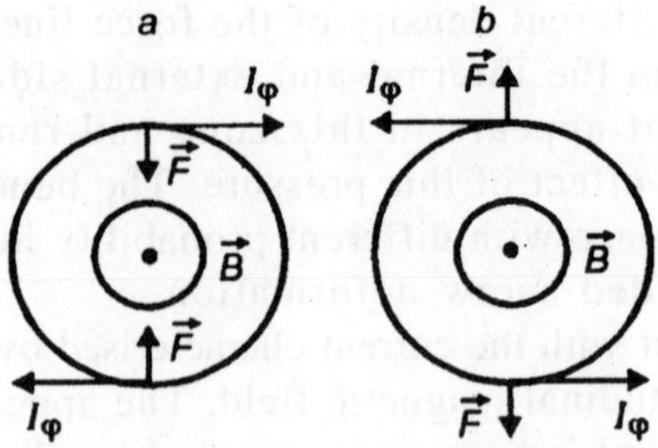

Fig. 2.16. Effect of the longitudinal magnetic field on the plasma filament with current (on the arc). *a*) magnetic field stabilises helical perturbation; *b*) supports the development of helical perturbation.

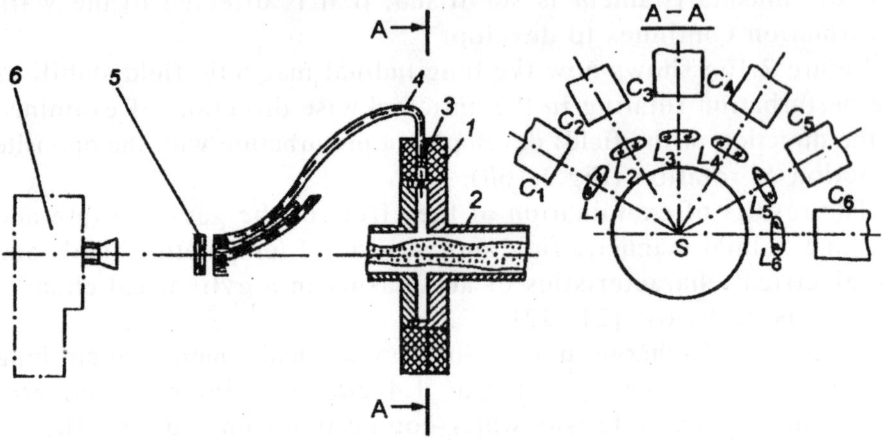

Fig. 2.17. Diagram of measurement with a plasma tomograph. *1*) measuring disk; *2*) electric arc chamber; *3*) lens; *4*) light guide; *5*) light filters; *6*) cine camera.

was part of the plasma channel and contained twelve windows, distributed at 30° steps around the circumference of the disk, lenses, light guides, a system of light filters, and a cine camera. The optical disk was placed between the coils of the solenoids. The cross section of the channel was focused, using short-focus lenses L_n, from n directions to the end surfaces of the corresponding light guides C_n, from the appropriate ends of the light guides, assembled in a block. The examined pattern was recorded using a photographic or cine camera. For optically transparent plasma it is sufficient to take measurements only from 6 initial light guides, because the projections, obtained from the opposite directions, are identical. The cine films in Fig. 2.18 show the variation of the pattern, recorded in the block of the light guides, in relation to the position and the form of the filament in the cross section of the chamber S. It also shows the patterns recorded in the block of the light guides for the appropriate positions of the plasma filament.

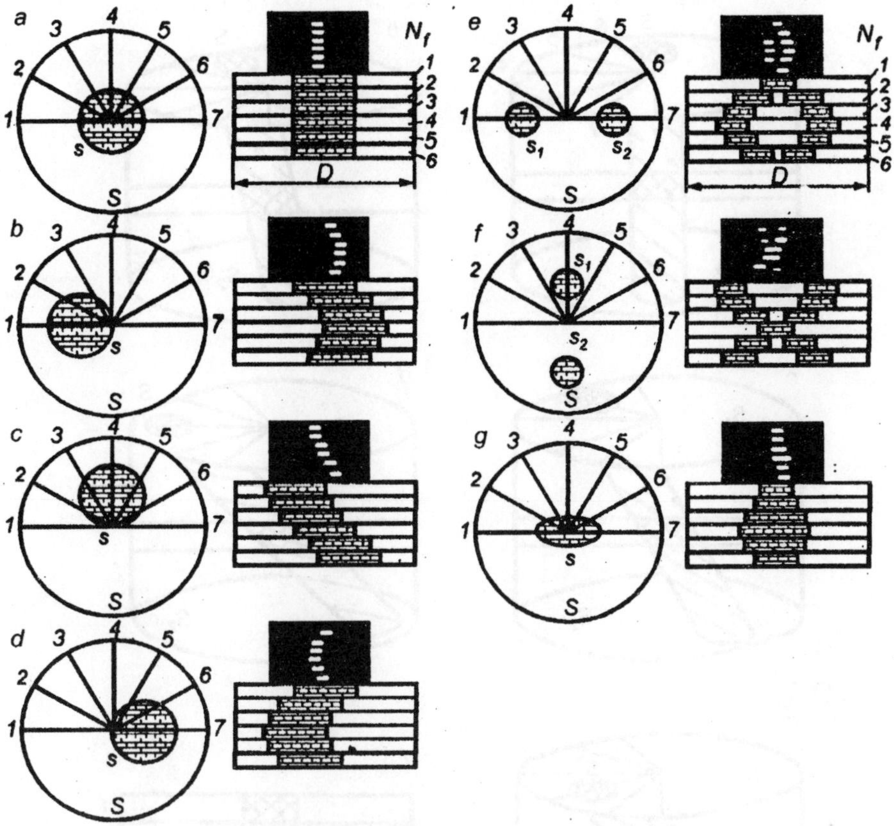

Fig. 2.18. Film frames on the block of light guides. *a*) The arc filament is situated on the axis of the chamber; *b–d*) the filament is displaced from the axis; *e–f*) split into two cords; *g*) expanded in another direction. *1–6* are the beams showing the direction of examination and the number of the light guides N_f.

If the filament *s* between the two sections S_i and S_2 is inclined in relation to the axis of the electric chamber and is situated in such a position that in the direction *1* (Fig 2.19) we see the pattern shown in Fig 2.19*a*, then in other directions we observe the patterns shown in Fig 2.19*e*. The angle of inclination of the filament depends on the examination direction. The 'pitch' of the helical arc can be determined from the maximum angle of inclination.

The patterns, obtained in the block of the light guides, were also used to determine the temperature fields of the plasma. In the later case, the transverse projections of radiation were measured in a narrow spectrum range with the half-width $\Delta\lambda = 5$ nm with the maximum at the wavelength of $\lambda = 465$ nm indicated by the system consisting of interference and glass light filters. It was assumed that the

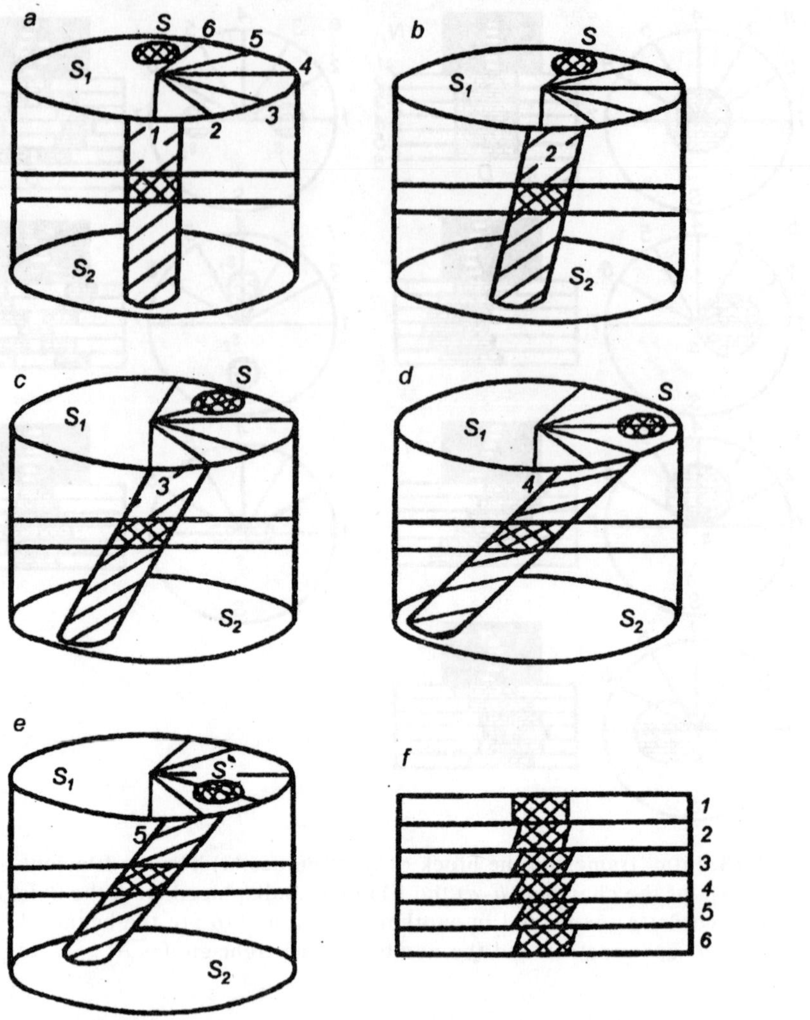

Fig. 2.19. Patterns observed on the block of the light guides when the electric arc filament in is inclined in relation to the axis of the arc chamber.

registered signal is determined only by the continuous spectrum of argon. The energy calibration of the measuring channel was carried out by the conventional method using SI-10-300 strip tungsten lamp.

The emission coefficients of plasma $\varepsilon(x, y)$ were calculated using the patterns obtained in the block of the light guides using the RICSS2 algorithm. The transition from the calculated coefficients to plasma temperature was realised using the relationship [25]

$$\varepsilon_\lambda(T) = A(n_e^2 / \lambda^2 \sqrt{T})\xi(\lambda, T). \tag{2.4.1}$$

Here n_e is the concentration of plasma electrons; λ is the radiation wavelength; T is plasma temperature; A is a constant which depends on the selection of the system of units. The multiplier $\xi(\lambda, T)$ takes into account the fact that the argon atoms are not similar to the hydrogen atoms. The values of the multiplier presented in [26] were used in the calculation they were obtained with the accuracy of 25 % and coincide with the data obtained in other studies. The relationship (2.4.1) holds for the plasma in LTE (local thermal equilibrium). It was shown in a number of studies that in argon plasma at the atmospheric pressure LTE is detected at a temperature of $T \geq 8200$ K. For these temperatures, plasma composition was calculated using Saha's equations, the equations of state and macroscopic neutrality. The results were in good agreement with the data in [27] and were used in (2.4.1) when constructing the dependence $\varepsilon(T)$ for $\lambda = 465$ nm of the continuous argon spectrum. The emission coefficient of argon continuum in the region of measurements is independent, within the error range, of the wavelength and, consequently, may be regarded as constant and equal to the coefficient of radiation at the wave length $\lambda = 465$ nm.

In [28] in analysis of a large number of studies it was found that the emission characteristics of conductivity of the plasma in the investigated range are determined mainly by the properties of the central core of the arc. It is also reported that these characteristics are not influenced by the presence of non-equilibrium in the wall regions.

To compare the temperature fields, the effective strength of the electrical field of the plasma $\langle E\rangle$, calculated from the following equation:

$$\langle E\rangle = I / \int_s \sigma(x, y)dxdy. \tag{2.4.2}$$

was used here. I is arc current, $\sigma(x, y)$ is the conductivity of plasma (in calculations, the values of $\sigma(T)$ were taken from [27]); S is the integration range given by the measured temperature field. The method used in this work (in particular, photographic recording) was characterised, in the temperature range $(8 \div 10)\cdot 10^3$ K for the argon plasma, by the error of determination of the radiation coefficient of 20–30 %. The error in evaluation of the temperature did not exceed 5–6% which equals ±500 K.

However, the error of calculation of $\langle E\rangle$ using the results of spectral measurement is far more complicated because, firstly, it is linked in a non-linear manner with the error of determination of the temperature field and, secondly, for cases in which the temperature field is divided into several maxima, the error depends on the accuracy of detachment of the current-free region of the plasma from the region through which the current flows.

Unfortunately, the authors of [21, 22] did not compare the value of $\langle E \rangle$ with the experimentally measured values of the strength of the electrical field of the arc E in argon for the given conditions and, consequently, the values of $\langle E \rangle$ can be used only for analysis of the results presented in the following section.

Figure 2.20 shows the dependence of arc voltage on the gas flow rate through the channel recorded at an arc current of I = 100 A in the absence of the external longitudinal magnetic field. On the curve, the authors of [21, 22] defined three characteristic sections. In the first of them–in the flow rate range $G = 0.034 \div 0.255$ g/s, Re = $70 \div 600$ (the Reynolds number was determined on the basis of the diameter of the channel and the gas parameters at entry into the channel [29]) with an increase of the flow rate in the general voltage in the arc rapidly decreases, in the second section $G = 0.25 \div 4.4$ g/s, Re = $600 \div 10^4$ – voltage is almost constant, and in the third section $G = 4.4 \div 12.7$ g/s, Re = $10^4 \div 3{\cdot}10^4$ – increases.

The resultant dependence $U = f(G)$ may be explained by examining the behaviour of the temperature field in different cross-sections of the electrical arc.

As mentioned previously, the temperature field was restored on the basis of the measured tomographic projections. Averaging was carried out for an arc length of 0.25 cm. Exposure time was 50 μs. Figures 2.21 and 2.22 give the temperature fields recorded in different cross sections of the arc at an argon flow rate of 0.034 g/s. In the vicinity of the cathode surface the diameter of the plasma filament is small (Fig.2.21*a*) and, in addition to this, the filament

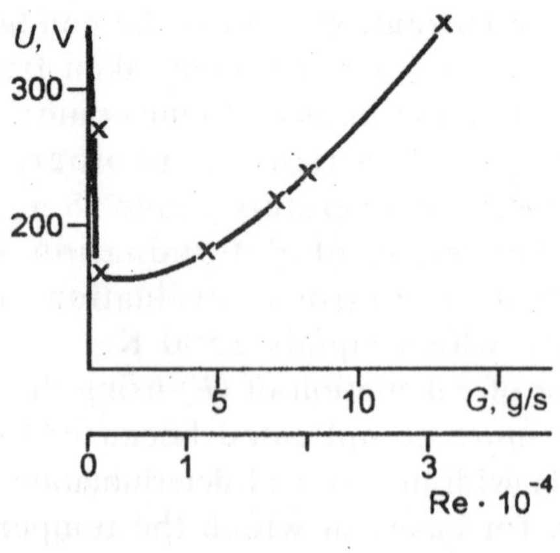

Fig. 2.20. Dependence of the arc voltage U on the flow rate of argon G. Arc length 80 cm.

is displaced from the axis of the chamber, the temperature field has no axial symmetry but is stationary in time, and the maximum temperature T_{max} = 12 480 K, $\langle E \rangle$ = 2.87 V/cm. With an increase of the distance from the cathode, the cross sectional area of the filament increases. Starting at the cross section z = 5 cm (Fig.2.21*b*) the temperature fields and $\langle E \rangle$ change with time (here and in the rest of the section, the isotherms are counted from the external arc inside the filament.

With further increase of the distance from the cathode (Fig.2.22*a*, z = 10 cm), the displacement of maximum temperature from the axis of the chamber increases and the temperature field is greatly deformed. At some moments, the isolines are stretched in one of the directions indicating the development in the plasma of a perturbation with the mode m = 2. If the temperature fields (Fig.2.22*a*) are examined successively in time, it may be seen that the plasma filament

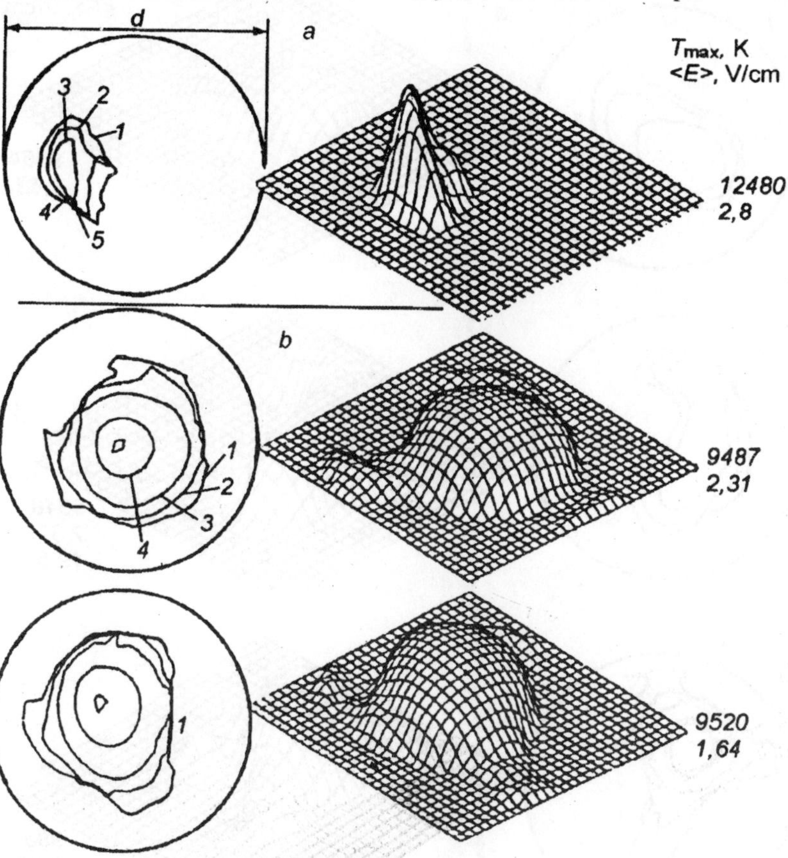

Fig. 2.21. Temperature fields of plasma in the two sections of the arc. *a*) z = 0.1 cm; *b*) z = 5 cm; the values of the isotherms for *a*: *1*) 11000 K; 2) 11500; *3*) 12000; *4*) 12500; *5*) 13000; *b*: *1*) 8500; 2) 8800; *3*) 9100; *4*) 9400; gas – argon, G = 0.034 *g/s*.

in the cross section of the electric arc chamber moves in a random manner.

In the sections situated further away from the cathode, the arc filament was split in two or more channels. Figure 2.22 *b* (z = 20 cm) shows clearly the development of the process of splitting in time. It is also important to note the decrease of T_{max} in some cases in splitting of the arc (see frame 7 in Fig.2.22, *b*).

The appearance of the non-stationary temperature fields was accompanied by the change with time of the strength of the electrical field of the plasma and by the increase of the mean arithme-
t i c
$\langle E \rangle_m$. The results shown in Figs.21 and 2.22, are presented in Fig.

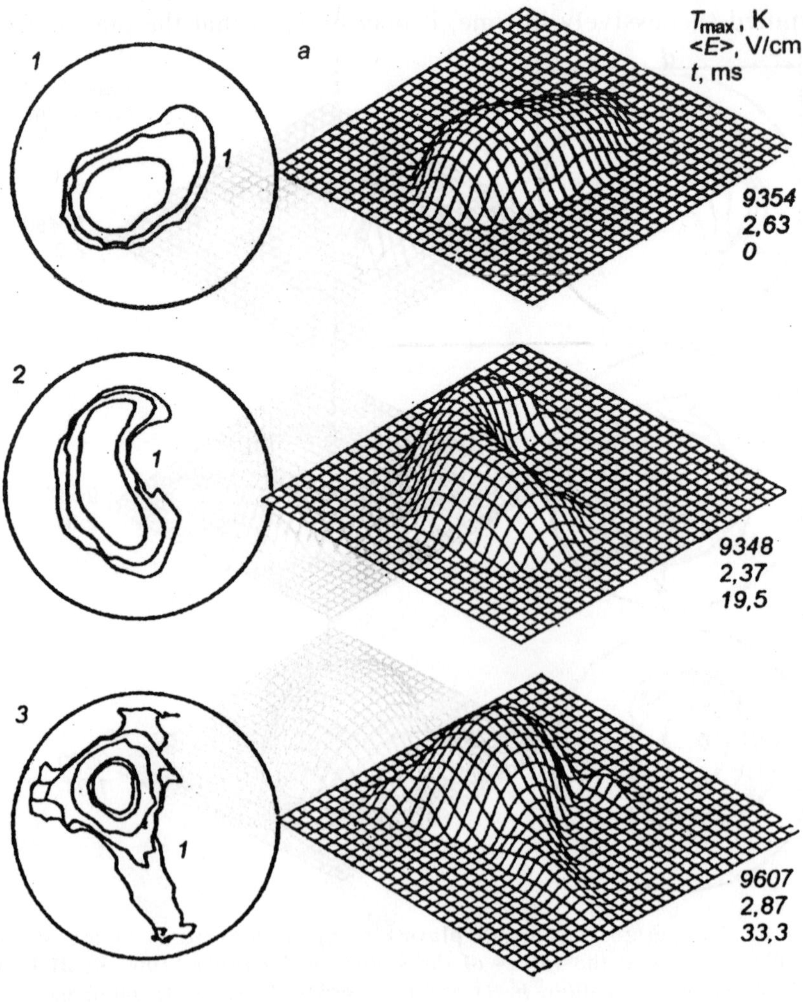

Start of **Fig. 2.22.**

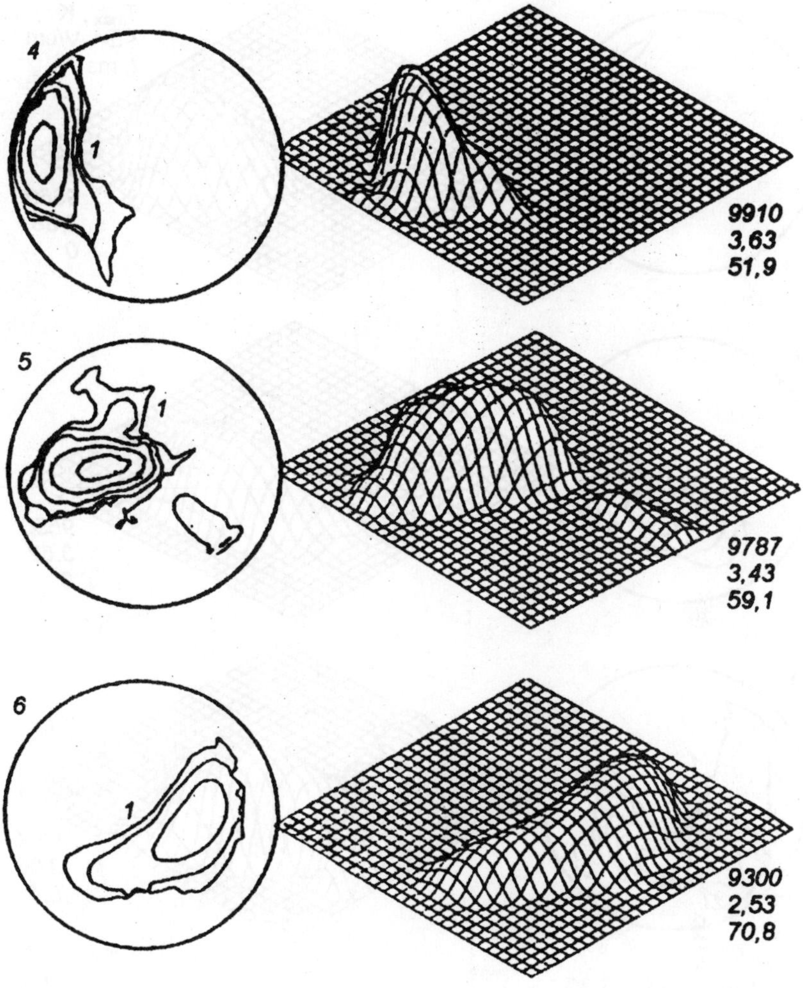

Continuation of **Fig. 2.22.**

2.23 in the form of the graph (curve *1*). The open symbols indicate the values of $\langle E \rangle$, obtained at different moments of time, the full symbols $\langle E \rangle_m$. With increase of the distance of the investigated section from the cathode, the range of the variations of $\langle E \rangle$ and the value $\langle E \rangle_m$ increases, but already at $z \geq 10$ cm for $G = 0.034$ g/s $\langle E \rangle_m$ and the scatter of the variations $\langle E \rangle$ remain approximately constant along the channel.

An increase of the gas flow rate is accompanied by a narrowing of the temperature field at the cathode and by an increase of T_{max} and $\langle E \rangle$ in this region. However, the general nature of deformation of the cross section of the arc filament is identical with that described previously. It is important to mention only the displace-

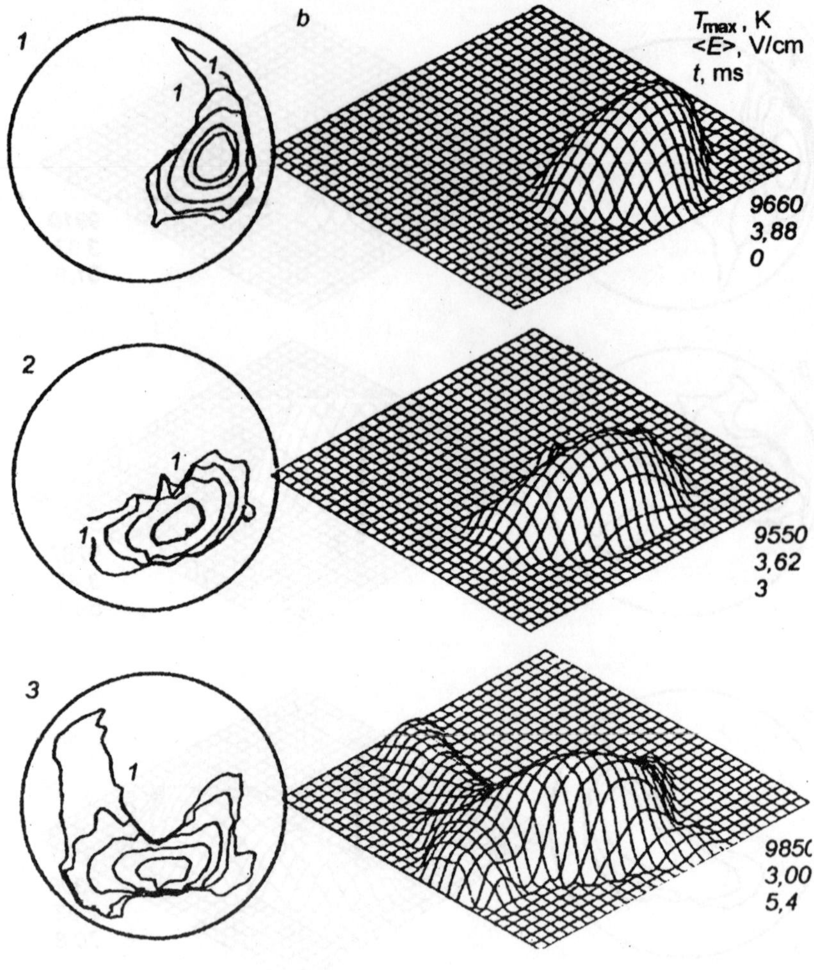

Continuation of **Fig. 2.22.**

ment of the area of appearance of non-stationary temperature fields along the flow. Thus, the length of the non-perturbed section of the arc filament increases. In fact, if at $G = 0.034$ g/s strong radial pulsations of the temperature field are found at $z = 10$ cm, then at $G = 0.175$ g/s they appear only at $z = 30$ cm. The same cross section shows large changes of $\langle E \rangle$ with time (curve *2* in Fig.2.23).

Since the length of the non-perturbed section of the arc increases with increasing gas flow rate, at some flow rates the arc filament should remain stationary over the entire examined length. At the parameters discussed previously, the filament remains stationary and occupies the central-symmetric position in the channel to $z \sim 55$ cm at $G = 0.25$ g/s (curve *3*, Fig.2.23) (at higher values of z investi-

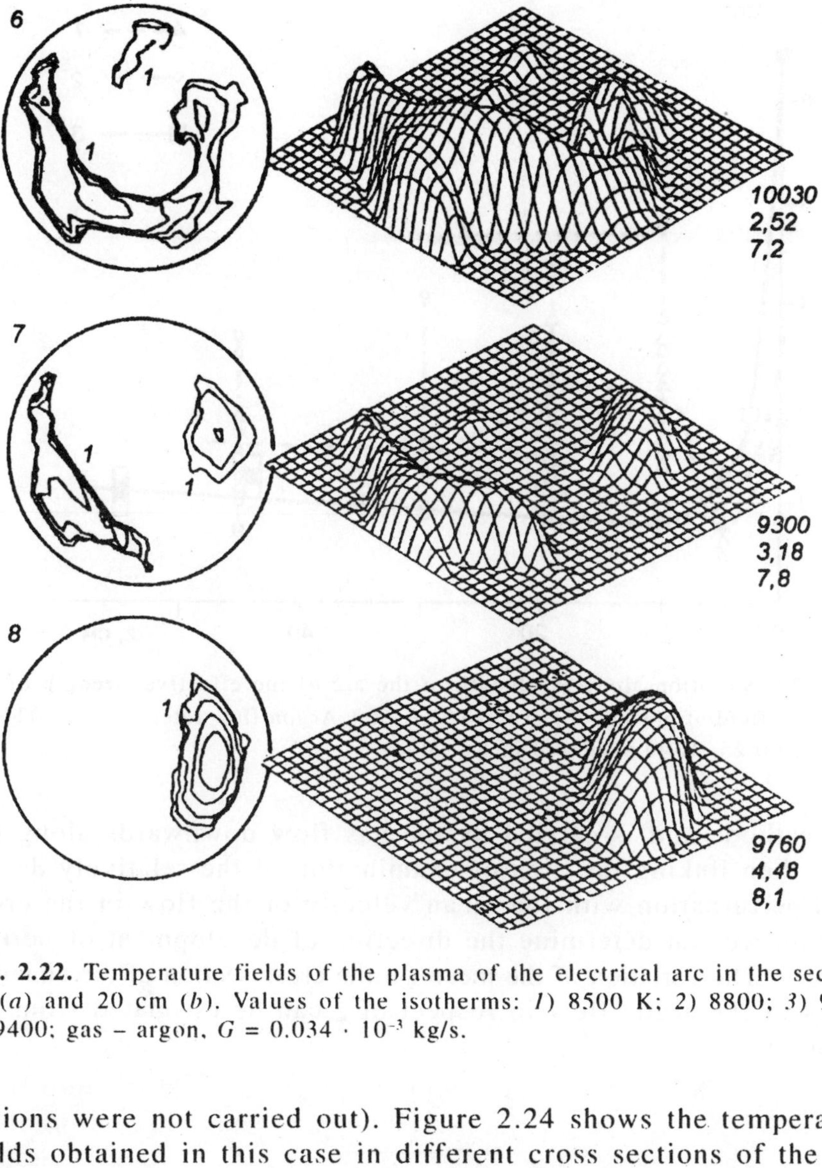

Fig. 2.22. Temperature fields of the plasma of the electrical arc in the sections 10 (*a*) and 20 cm (*b*). Values of the isotherms: *1*) 8500 K; *2*) 8800; *3*) 9100; *4*) 9400; gas – argon, $G = 0.034 \cdot 10^{-3}$ kg/s.

gations were not carried out). Figure 2.24 shows the temperature fields obtained in this case in different cross sections of the arc. It may be seen that the temperature field is axisymmetric over a large part of the arc column, and the length of the initial thermal section (according to the terminology used in [30]) is 10–20 cm. This is 2–3 times higher than the value obtained by approximation [30].

The displacement of the coordinate of the area of appearance of non stationary non symmetric temperature fields with increase of the gas flow rate leads to the conclusion according to which the movement of plasma along the flow is accompanied by the development of perturbation. If it is assumed that a pertubation forms

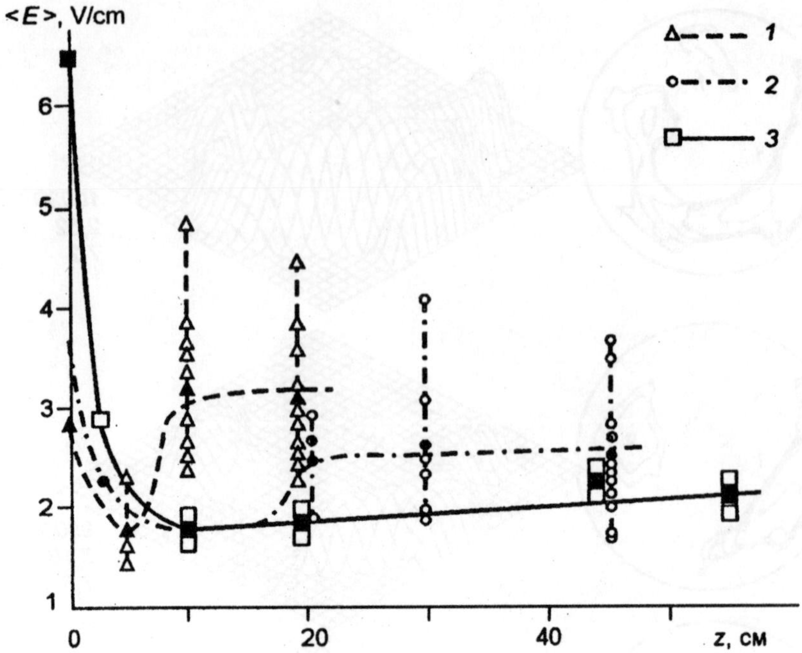

Fig. 2.23. Variation along the length of the arc of the effective strength of the electrical field of plasma $\langle E \rangle$ at Re = 70 ÷ 600. Argon flow rate, g/s: *1*) 0.034; *1*) 0.175; *3*) 0.25. Solid symbol/0 mean values.

at a cathode and is carried by the gas flow downwards along the flow, then linking the area of examination of the relatively developed perturbation with the mean velocity of the flow in the cross section, we can determine the direction of development of perturbation τ. The variation of the mean (in the cross section of the channel) speed of the argon flow in respect of z can be estimated from the equation:

$$\upsilon_m = \frac{4\int_0^d \upsilon r dr}{d^2},$$

using for this purpose the radial profile of the speed [30] and the variation in respect of z of the speed of the flow along the axis of the channel [28] measured in identical conditions. Table 2.1 gives the values of υ_m and the duration of development of perturbations:

$$\tau = \int_0^l \frac{dz}{\upsilon_m(z)}$$

with the modes m = 1 and 2 for different G. It may be seen that

τ, obtained at low flow rates (G = 0.034 ÷ 0.175), coincide. The large difference of τ at G = 0.25 g/s may be explained by the fact that this flow rate was characterised by very small displacement of the filament from the axis of the chamber, and evidently, the areas of detection of the perturbations were determined inaccurately.

The data shown in Fig.2.23 indicate that in the stationary section of the arc filament $\langle E \rangle$ is lower than in the non-stationary section. Since an increase of the gas flow rate increases the size of the stationary section, the arc voltage should decrease because of the decrease of the 'technical' strength of the electrical field. This was indeed observed in the flow rate range G = 0.25 ÷ 4.41 g/s (see Fig.2.20). In the flow rate range G = 0.25 ÷ 4.4 g/s, the arc filament is evidently stationary to z = 55 ÷ 70 cm and, therefore, arc voltage does not change in this region. At flow rates higher than 4.4 g/s the flow in the channel becomes turbulent, the arc filament deviates from the axis of the chamber and may be split into a number of channels, and the arc voltage increases with increase of the gas flow rate.

The effect of the longitudinal magnetic field on the integral and

Table 2.1. Duration of development of perturbation $\tau_{m=1}$, $\tau_{m=2}$ at different argon flow rates

G, g/s	l_{in}, cm	z, cm	υ_m, cm/s	$l_{m=1}$, cm	$\tau_{m=1}$, s	$l_{m=2}$, cm	$\tau_{m=2}$, s
0.034	1.7	0	2.7	5	0.18	10	0.40
		0.41	14.2				
		0.85	19.3				
		1.75	22.4				
0.175	8.7	0	14.1	20	0.13	30	0.43
		2.1	74.9				
		4.4	100.2				
		8.7	115.5				
0.25	12.5	0	20.2	55	0.38	–	–
		3	107				
		6.3	143.4				
		12.5	165				

Comment: d = 3 cm; p = 0.1 MPa; I = 100 A; l_{in} is the length of the initial hydrodynamic section of the arc; υ_m is the mean argon flow rate in the cross section; $l_{m=1}$, $l_{m=2}$ is the distance from the cathode to the section in which a relatively developed perturbation with modes m = 1 and m = 2, respectively, is found local parameters of the plasma was also investigated. Experiments were carried out at an argon flow rate of 0.25 g/s, and the other parameters were the same as mentioned previously. The magnetic fields were generated by two solenoids with a total length of 30 cm and superposed on the central part of the arc. As already mentioned, at these parameters the arc filament is stationary, at least to z = 55cm.

Figure 2.25 shows the dependence of arc voltage on the induction of the longitudinal and magnetic field. It is important to note the large increase of U at $B = 0 \div 0.03$ T. The behaviour of the temperature field in this case will be examined. It is well known that at $B = 0$ in the section z = 45 cm, the temperature field is axisymmetric,

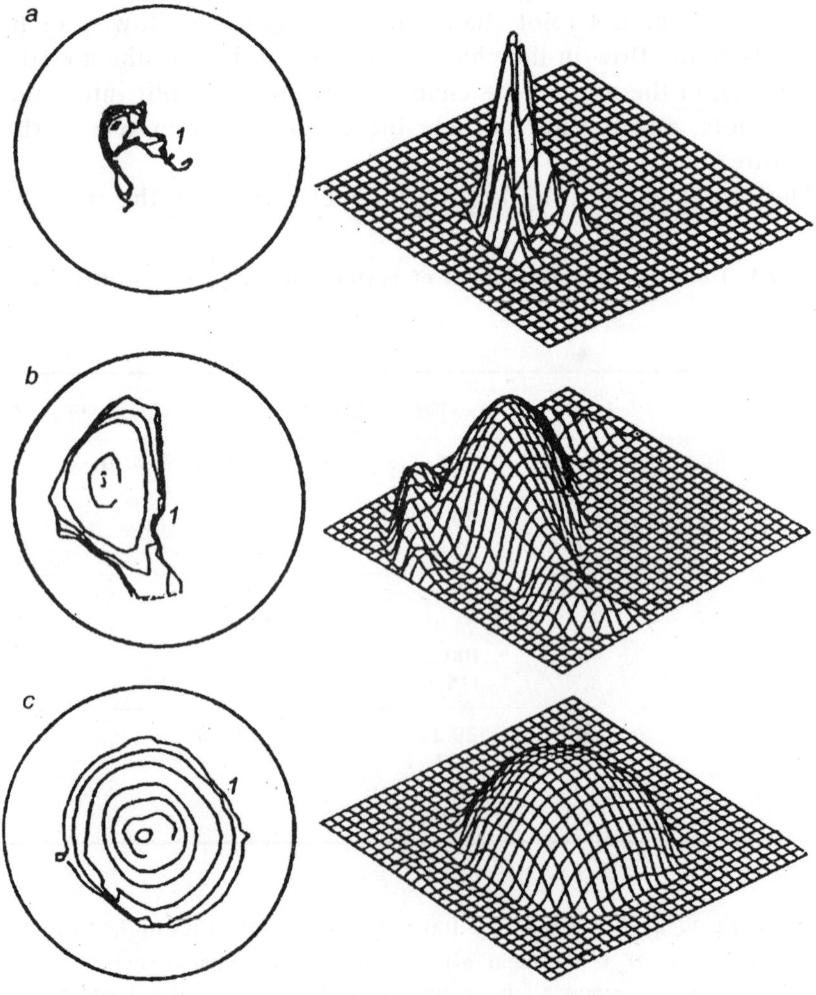

Start of **Fig. 2.24.**

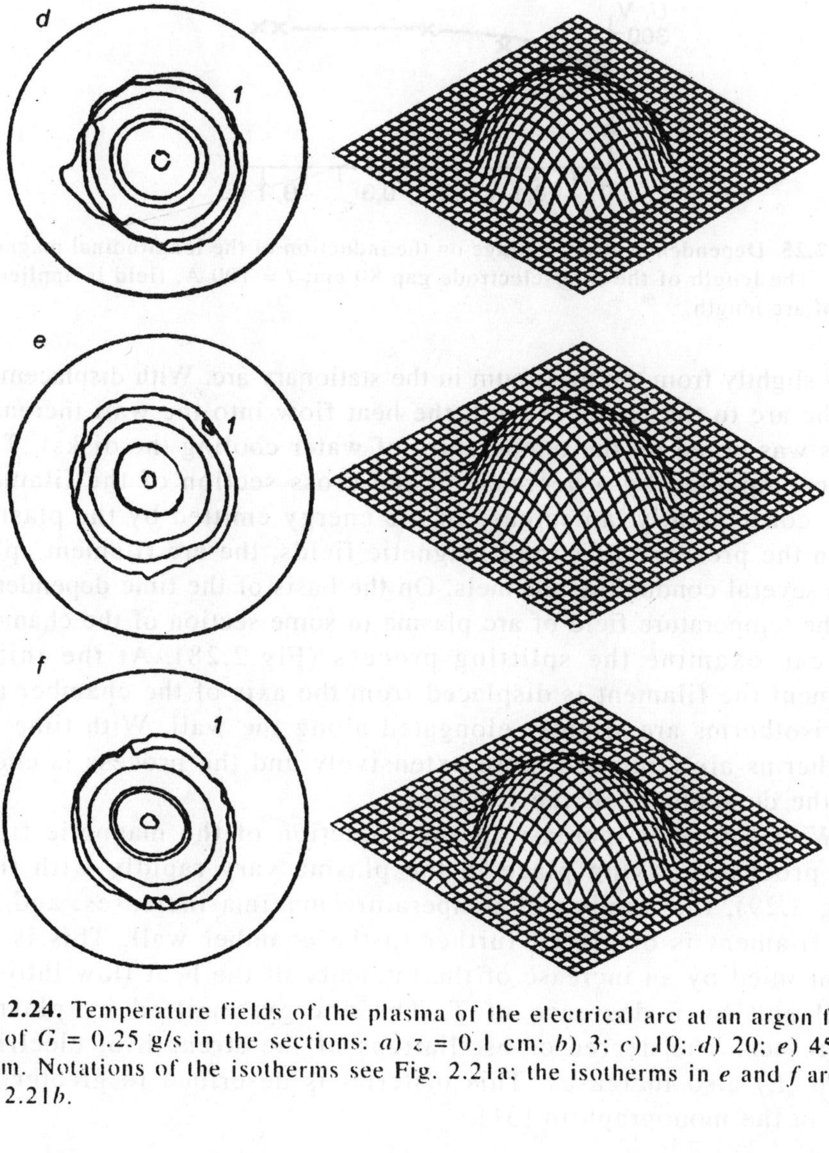

Fig. 2.24. Temperature fields of the plasma of the electrical arc at an argon flow rate of G = 0.25 g/s in the sections: *a*) z = 0.1 cm; *b*) 3; *c*) 10; *d*) 20; *e*) 45; *f*) 55 cm. Notations of the isotherms see Fig. 2.21a; the isotherms in *e* and *f* are in Fig. 2.21*b*.

but already in the field B = 0.01 T (Fig.2.26) the axial symmetry of the filament is disrupted. The temperature field resembles a lentil, elongated along the chamber wall. With time, this 'lentil' rotates around the axis of the chamber (Fig.2.27). The speed of rotation of the maximum temperature varies from 75 to 170 rev/s. The values of T_{max} and $\langle E \rangle$ were determined for the temperature distribution obtained in a single rotation of the filament around the axis of the arc chamber. Although T_{max} changes even during the single rotation, its changes are in the error range of the measurements and the temperature differs

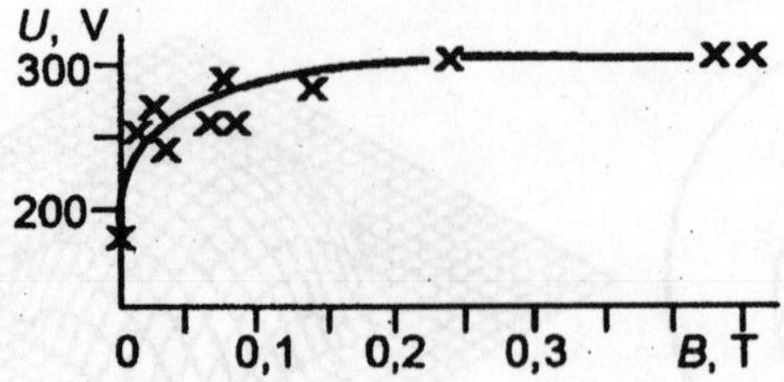

Fig. 2.25. Dependence of arc voltage on the induction of the longitudinal magnetic field. The length of the inter-electrode gap 80 cm; I = 100 A, field is applied to 1/3 of arc length.

only slightly from the maximum in the stationary arc. With displacement of the arc to the chamber wall, the heat flow into the wall increases (this was noted on the temperature of water cooling the disks). This is accompanied by a decrease of the cross section of the filament and, consequently, a decrease of the energy emitted by the plasma.

In the presence of strong magnetic fields, the arc filament splits into several conducting channels. On the basis of the time dependence of the temperature field of arc plasma in some section of the channels we can examine the splitting process (Fig.2.28). At the initial moment the filament is displaced from the axis of the chamber and the isotherms are slightly elongated along the wall. With time the isotherms are stretched more extensively and the process is ended by the detachment of the filament.

With a further increase of the induction of the magnetic field, the processes taking place in the plasma vary rapidly with time (Fig 2.29), the number of temperature maxima increases, and the arc filament is displaced further to the chamber wall. This is accompanied by an increase of the intensity of the heat flow into the wall and by a decrease of T. The energy, emitted by plasma, decreases. The degree of oscillations of the strength of electrical field $\langle E \rangle$ also increases. This material is described in greater detail in the monograph in [31].

2.5. SHUNTING

2.5.1. Qualitative pattern

The most characteristic electrophysical processes in the discharge chamber of the linear electric arc plasma torch is shunting, i.e. the electrical breakdown between the arc column and the wall of the chamber or in the arc loop. There is large-scale and small-scale shunting. The former includes (Fig. 2.30) the shunting (2) between the main arc column (*1*) and the chamber wall. This determines the arc length

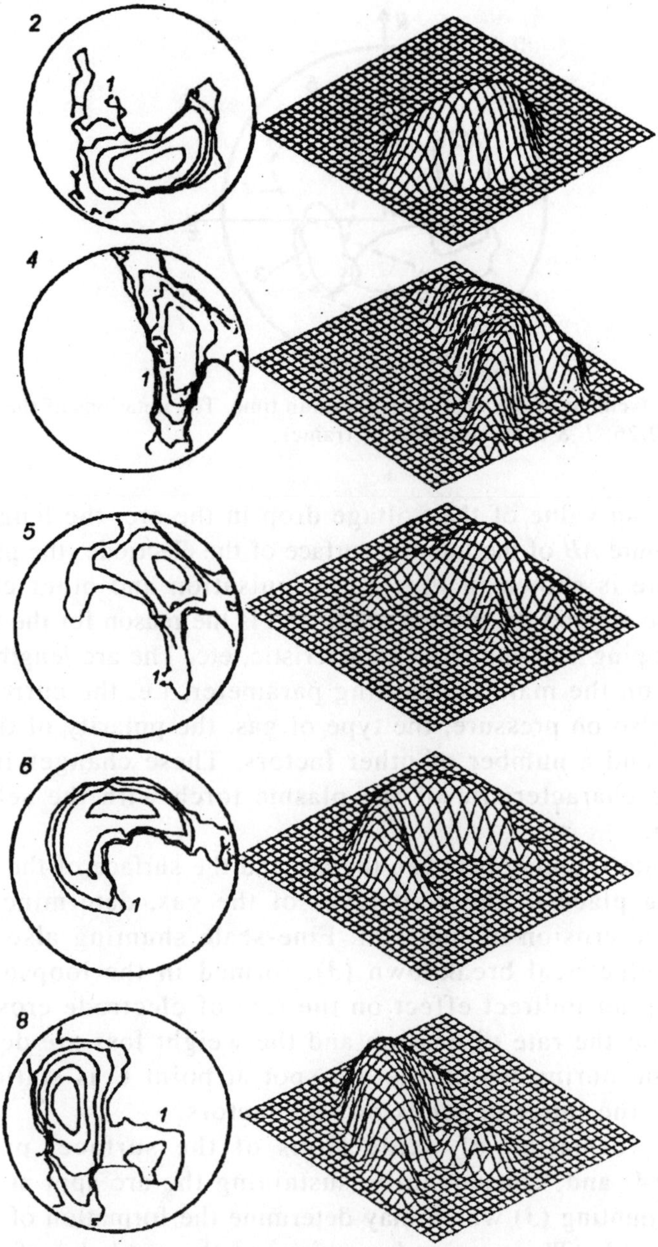

Fig. 2.26. Time dependence of the temperature field of the plasma of the electrical arc in the section z = 45 cm. I = 100 A; p = 1.10^5 Pa; B = 0.001 T; the notation of the isotherms is shown in Fig.2.21, *b*, gas – argon, G = 0.25 g/s.

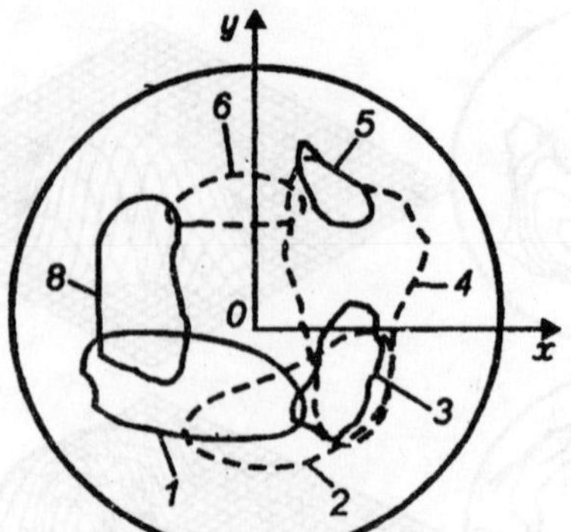

Fig. 2.27. Movement of the 9200 K isotherm in time. The notations of the parameters are in Fig. 2.26, *1–8* is the number of frames.

and the mean value of the voltage drop in the arc, the length of the zone of failure *AB* of the internal surface of the electrode (the photograph of the zone is shown in Fig. 2.31), pulsation and other characteristics of the arc and the plasma torch, and is the reason for the formation of the drooping volt–ampere characteristic, etc. The arc length depends primarily on the main controlling parameter, i.e. the current intensity, and also on pressure, the type of gas, the polarity of the output electrode and a number of other factors. These changes in the arc length are characteristic of the plasma torch with the self-setting arc length.

Fine-scale shunting between the arc and the surface of the electrode (*4*), taking place in the wall layer of the gas, determines mainly the specific erosion of material. Fine-scale shunting also includes 'arc–arc' electrical breakdown (*3*), formed in the loop of the arc and having an indirect effect on the rate of electrode erosion. The point is that the rate of erosion and the weight loss are determined by the time during which the arc spot at point *C* is stationary. In particular, the latter depends on two factors:

1. The formation of oxide films of the surface, preventing shunting (*4*) and, consequently, sustaining the arc spot at point *C*.

2. On shunting (*3*) which may determine the formation of electrical breakdown (*4*). The type and contours of the eroded surface of the copper output electrode–anode are shown in Fig. 2.31.

We examine the qualitative pattern of the large-scale shunting of the arc in the output electrode of a single chamber plasma torch (Fig. 2.32). In analysis, it is assumed that the voltage of the power source is considerably higher than the arc voltage.

It is assumed that at some moment of time t_1 the arc occupies

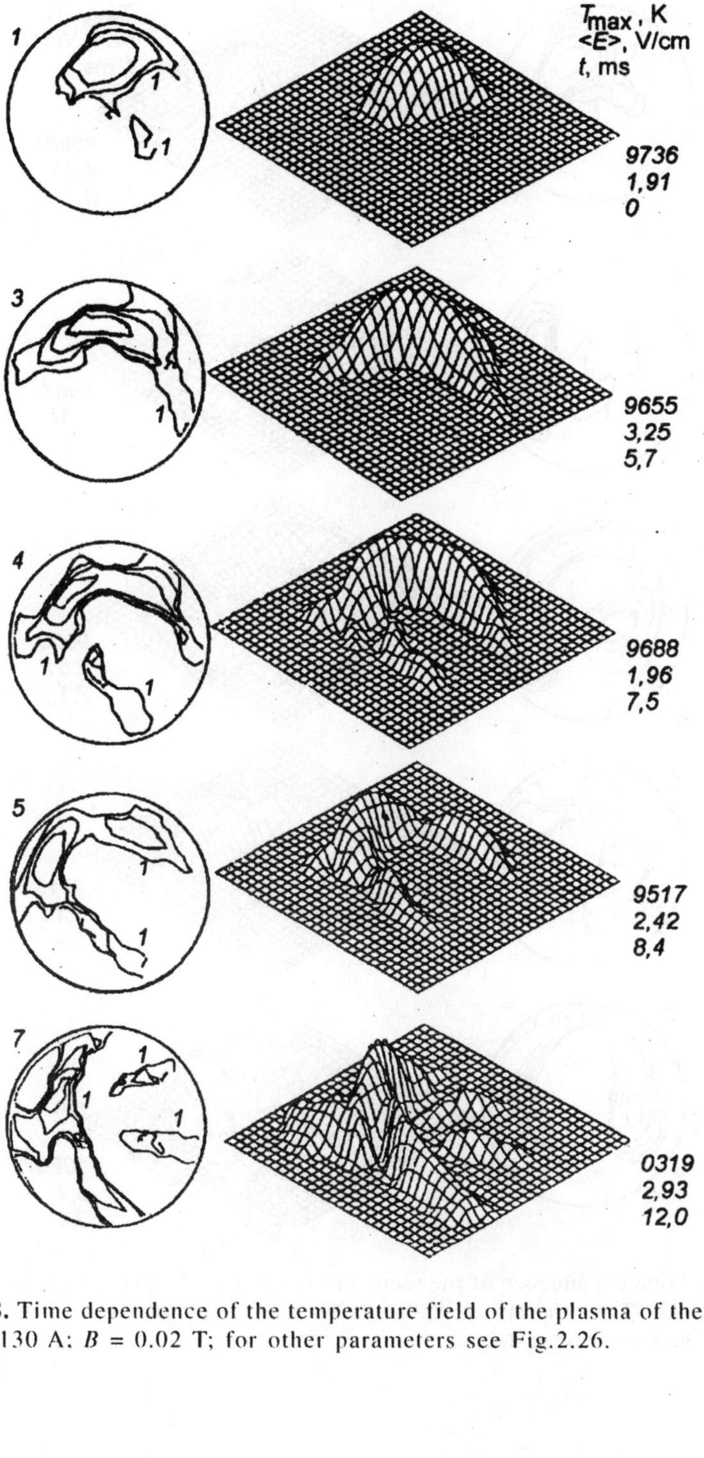

Fig. 2.28. Time dependence of the temperature field of the plasma of the electrical arc. I = 130 A; B = 0.02 T; for other parameters see Fig.2.26.

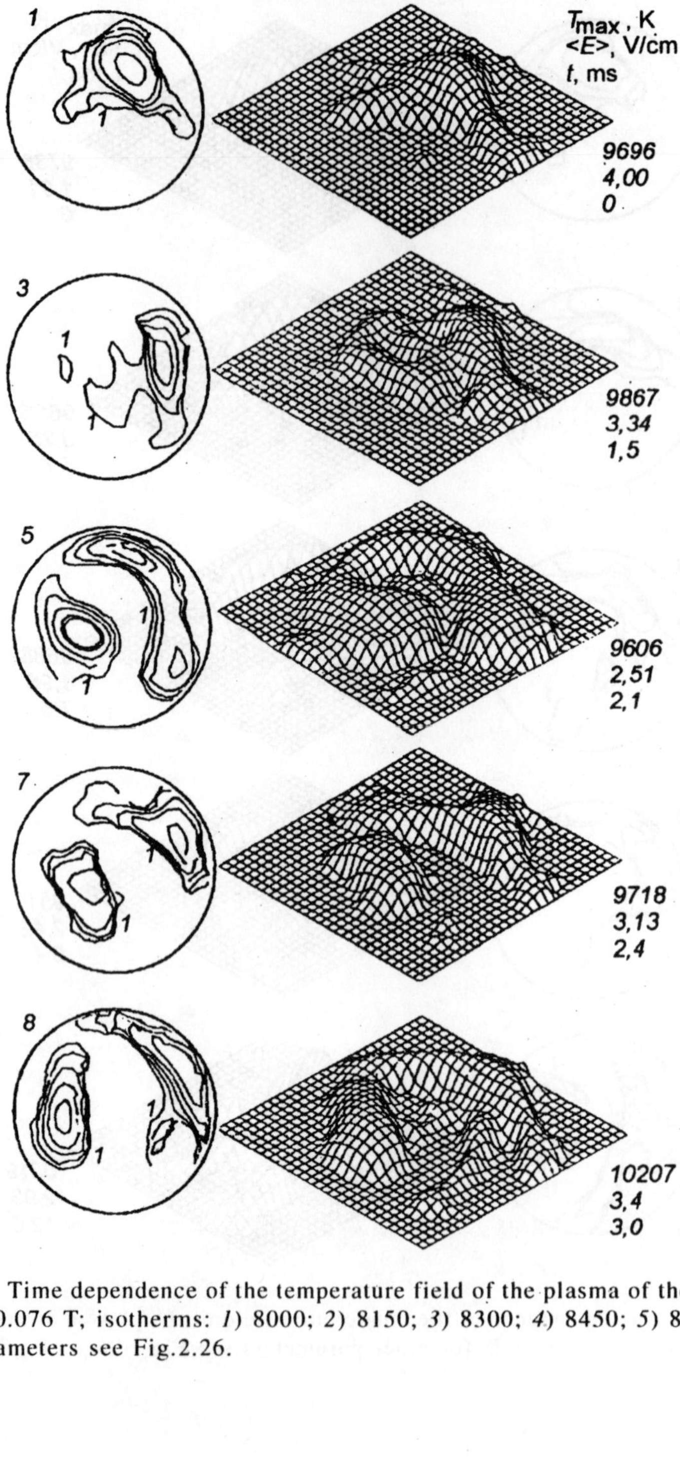

Fig. 2.29. Time dependence of the temperature field of the plasma of the electrical arc. B = 0.076 T; isotherms: *1*) 8000; *2*) 8150; *3*) 8300; *4*) 8450; *5*) 8600 K, for other parameters see Fig.2.26.

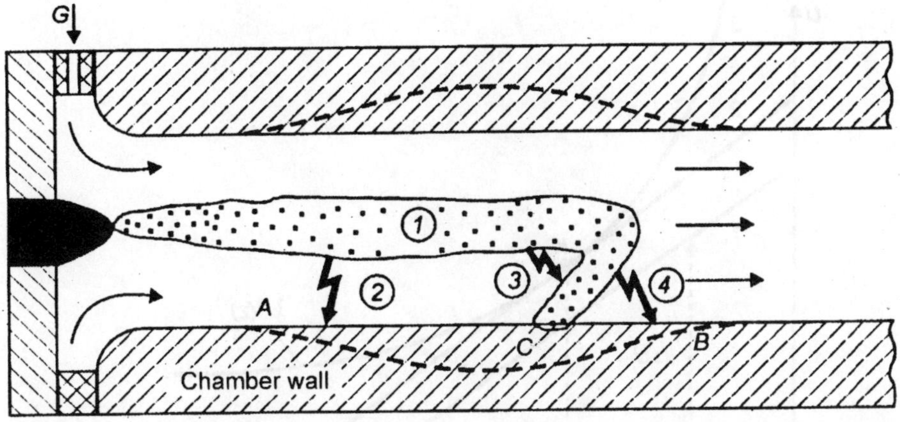

Fig. 2.30. Principal diagram of shunting of the electrical arc in the channel of the plasma torch.

Fig. 2.31. The contour of the eroded surface of a cooper output electrode - anode.

the position *ABC*. Under the effect of aerodynamic and electrodynamic forces, the section of the arc *AB* travels in the flow direction and, consequently, the arc length and voltage increase because they are linked together by the relationship:

$$U = \Delta U_e + \int_0^{l(t)} E(l)dl.$$

Here ΔU_e is the sum of the near-electrode potential drops; $E(l)$ is the strength of the electrical field; $l(t)$ is the arc length at the given moment of time. This makes it possible to explain the shunting processes in the case in which the strength of the electrical field of the arc, situated on the axis, is $E(l)$ = const. To simplify considerations, it is assumed that the potential of the arc electrode is equal to zero, and the origin of the coordinate z is selected at the end of the electrode (point *C*). Consequently, the distribution of the arc potential along the axis z for the moment of time t_1 corresponds qualitatively to curve

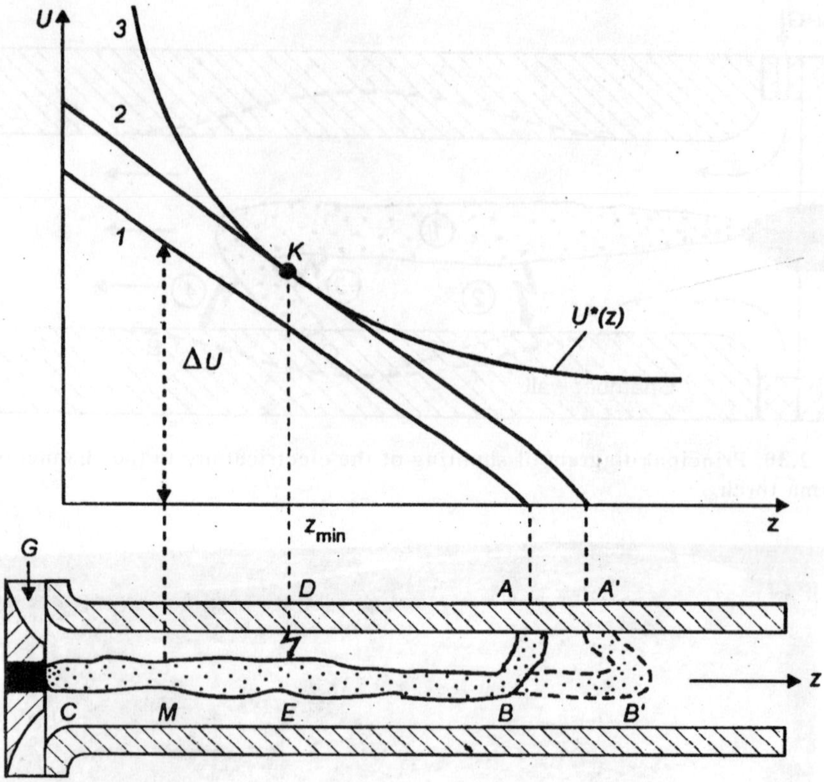

Fig. 2.32. Qualitative pattern of the formation of an electrical breakdown (shunting) between the arc and the wall of the electrode of the plasma torch.

1. Voltage U^*, required for a breakdown, changes along the axis z in accordance with the curve *3*. Because of the increase of the mean mass temperature, the voltage decreases in the direction of the flow. The arbitrarily selected point of the arc column M with the coordinate z and the surface of the electrode are linked by the potential difference:

$$\Delta U(z) = U(t) - Ez.$$

Under the effect of the applied potential difference, a breakdown may take place between the arc and the wall in some cross-section of the channel. For this purpose, it is necessary that $\Delta U(z) \geq U^*$.

It is clear that the essential condition for the moment of time t_1 is not fulfilled in any cross-section of the channel. At some subsequent moment of time t_2, the arc may occupy the position $A'B'C'$ in which the curve *2* of the distribution of the arc potential along the axis and the curve *3* have the common contact point. In this case, the quantity $\Delta U(z)$ in the section of the channel *DE* is equal to breakdown voltage. A breakdown takes place between the arc column and the electrode and this breakdown can develop in a short period of time in the transverse arc channel. With the appearance of the new channel, the channel *A'B'E'* starts to disappear because of the redistribution of the current in accordance with the electrical resistance of the branches. The newly formed radial section of the arc is 'washed away' by the flow and the shunting process is repeated.

The existence of the shunting mechanism has also been verified and confirmed by different methods; one of these methods is the oscillographic registration of arc voltage. Figure 2.33*a* shows the oscillogram including two periods of large-scale shunting (Fig. 2.30, *2*) with the pulsation amplitude ΔU_1. The oscillogram also shows clearly the pulsation of voltage of a smaller amplitude ΔU_2, determined by fine-scale shunting (Fig. 2.30, *3* and *4*). This distinctive nature of the shunting process is observed in cases in which the gas flow in the chamber is close to laminar. In the operating conditions of the plasma torch used in practice, the gas flow is more complicated.

The amplitude and frequency of pulsations of arc voltage U in large-scale shunting depend on the variation of the current intensity and the constant gas flow rate (Fig. 2.33, *b, c*). With increasing gas flow rate the amplitude of pulsations decreases and frequency increases. If the current intensity is maintained constant, but the gas flow rate increases, ΔU_1 increases and frequency decreases.

Small-scale shunting may be efficiently visualised if the experiments are conducted in a flat long the discharge channel with transparent side walls.

Figure 2.34 shows the frames of high-speed filming of the process [7]. At the initial moment of time (frames 1–6), the arc spot is stationary and the closing section subjected to the effect of the gas flow and the intrinsic magnetic field has the complicated form of the continuously deformed spatial loop. This period of time is characterised by small-scale shunting in the arc–arc loop (frames 4–6), disappearance of the individual sections and by the formation of new ones. The loop pulsates and is deformed until the arc–wall small-scale shunting takes place. Frame 7 shows clearly the transfer of shunting to the upper surface of the electrode; at the moment of time, corresponding to the frame 8, not only the development of the new electric arc channel but also the disappearance of the exist-

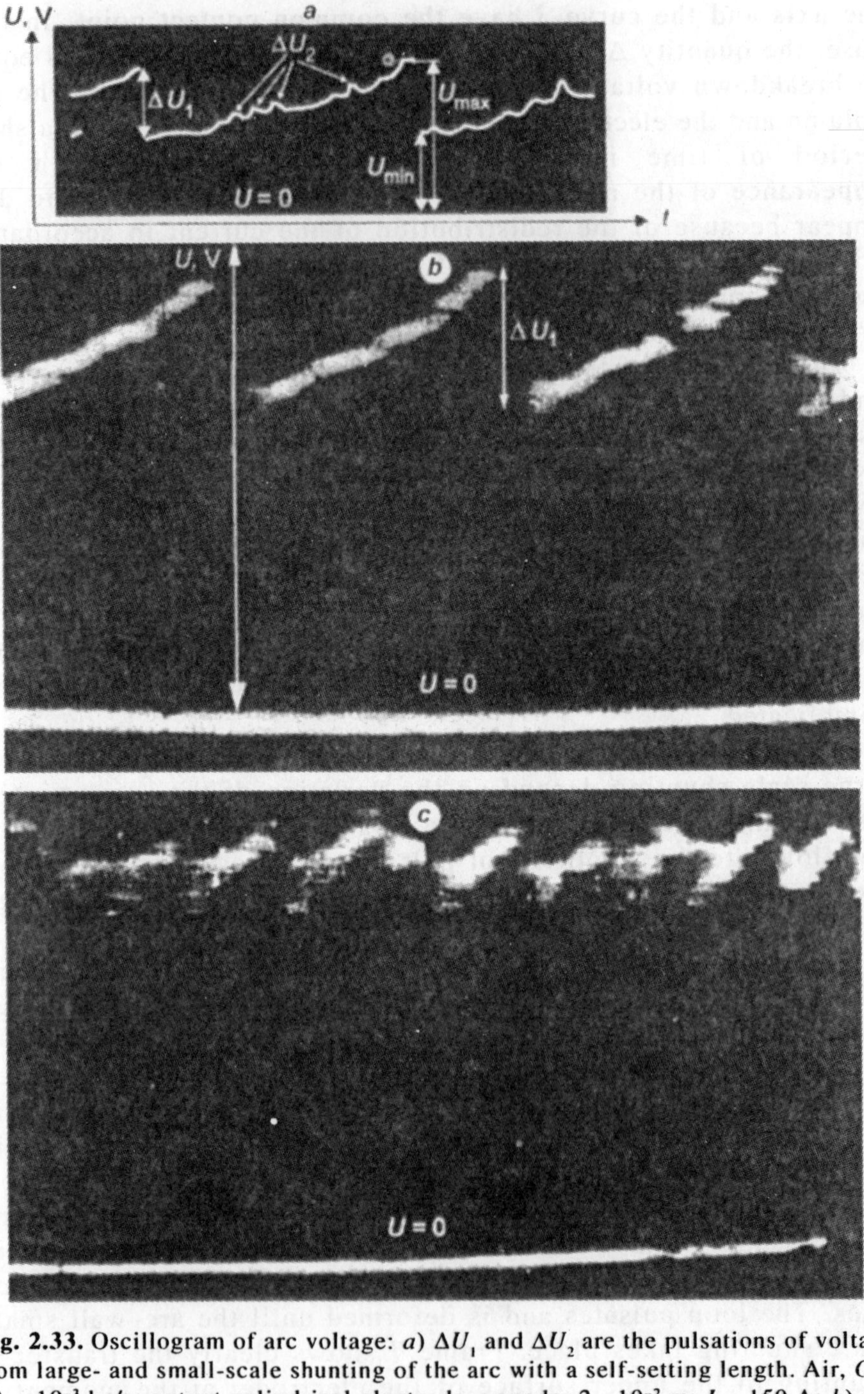

Fig. 2.33. Oscillogram of arc voltage: *a*) ΔU_1 and ΔU_2 are the pulsations of voltage from large- and small-scale shunting of the arc with a self-setting length. Air, $G = 10 \cdot 10^{-3}$ kg/s; output cylindrical electrode – anode; $2 \cdot 10^{-2}$ m; $I = 150$ A; *b*) $I = 50$ A; $G = 14 \cdot 10^{-3}$ kg/s; *c*) $I = 150$ A; $G = 14 \cdot 10^{-3}$ kg/s.

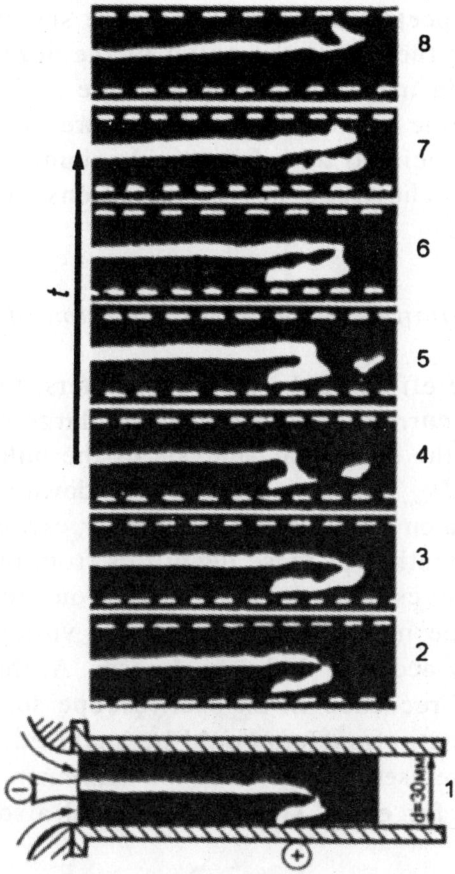

Fig. 2.34. Development of the electric arc discharge in a flat channel. The arrow indicates the directional movement of the film.

ing loop has been completed. These processes are then repeated. The process of moving of the closing section of the arc down along the flow by means of wall shunting is restricted only by large-scale shunting.

As shown previously, small-scale shunting causes additional pulsations of arc voltage. The amplitude and frequency of these pulsations differ by approximately an order of magnitude from those of the pulsations caused by large-scale shunting.

Arc shunting results not only in pulsations of arc length but also in a change of the speed of the gas flow and the temperature of the flow and, consequently, the luminosity of the plasma, recorded at the outlet of the plasma torch nozzle, may change (Fig. 2.35). Since the luminosity of plasma changes appreciably with the variation of the composition and temperature of the plasma, it maybe assumed that the pulsations of the luminosity can be recorded using optical and spectral devices. The figure shows the recording of the luminosity of the gas at the outlet of the nozzle (from the side) of a phase AC plasma torch during a single period of arcing when the output electrode is the cathode. There are distinctive pulsations of arc length, associated with large-scale shunting, and also small-scale pulsations. The frequency of pulsations in the latter case is considerably higher.

2.5.2. Some qualitative results of examination of the shunting process

The nature of the effect of the flow parameters, the geometry of the channel and arc current on the process of large-scale shunting can be explained by the statistical analysis of the pulsation components of arcing voltage: U_{max} and U_{min}, and also breakdown voltage U^*. Analysis is based on the data on the dispersion, asymmetry, excess and the correlation coefficient of deviation of these quantities from their mean value.

Quantitative investigations were carried out on a single-chamber plasma torch. The mean current I and arc voltage $\bar{U}$ were measured using highly accurate dial-type devices. At the same time, these parameters were recorded in an oscilloscope so that the pulsation components of the parameters could be analysed. The oscillograms (Fig. 2.32, *a*) were used to determine the maximum U_{max} and minimum U_{min} arc voltage for every shunting act, and also the mean values were calculated:

$$\bar{U}_{max} = \frac{1}{n}\sum^{n} U_{max}, \bar{U}_{min} = \frac{1}{n}\sum^{n} U_{min}$$

together with the mean breakdown voltage:

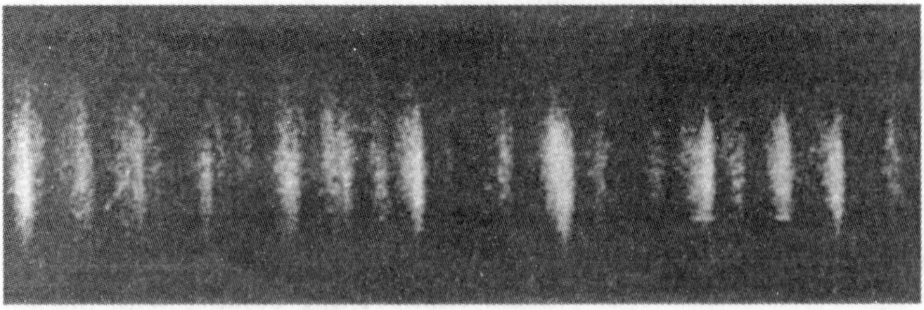

Fig. 2.35. Recording of the glow of the gas at the outlet of the plasma torch nozzle. both large-scale and small-scale pulsations of the arcs are clearly visible.

$$\bar{U}^* = \frac{1}{n}\sum^{n}(U_{max} - U_{min}),$$

where n is the number of measurements.

In addition to this, the RMS deviations were determined:

$$\bar{\sigma}_{max} = \sqrt{\frac{\sum^{n}(U_{max} - \bar{U}_{max})^2}{n}}$$

(similarly for $\bar{\sigma}_{min}$ and σ^*).

Analysis shows that the distribution of the required quantities may differ from the normal distribution. It was therefore necessary to calculate asymmetry and excess. The asymmetry of the distribution function

$$\alpha_{max} = \frac{\sum^{n}(U_{max} - \bar{U}_{max})^3}{n\bar{\sigma}^3_{max}}$$

(Similarly for α_{max} and α^*). The excess

$$i_{max} = \frac{1}{n\sigma^4_{max}}\sum^{n}(U_{max} - \bar{U}_{max})^4 - 3$$

(similarly for i_{max} and i^*_{max}). Analysis of the experimental material was carried out mainly only for the pulsations determined by large-scale shunting, and assuming that the shunting process is ergodic, i.e. independent of time. The latter has been confirmed at the agreement between the required mean value and the dispersion with increase of the number of samples.

$$K_{1,2} = \frac{\sum^{n}(U_1 - \bar{U}_1)(U_2 - \bar{U}_2)}{\bar{\sigma}_1\bar{\sigma}_2 n}$$

The calculated moments ($\bar{\upsilon}, \bar{\sigma}, \alpha_i$) for each of the characteristic values of voltage and the functions of the density of distribution of probability:

$$f_{max} = \frac{1}{\bar{\sigma}_{max}\sqrt{2\pi}} e - \frac{(U_{max} - \bar{U}_{max})^2}{2\bar{\sigma}^2_{max}} = \frac{n_i}{nh_\sigma}$$

were used for the verification of the resultant distributions for the similarity to the functions of normal distribution using the Pearson criterion (also for f_{min}, f^*). Here h_σ is the deviation step, n_i is the number of shunting acts used for the calculation of the mean and the dispersion; n is the number of measurements.

The distribution curves f_{max} proved to be quite similar to the normal distribution function in almost all examined ranges of operation of the plasma torch (Fig. 2.36*a*). A different pattern was detected for f^* (Fig. 2.36*b*) and f_{min}. In the case of low and relatively high consumptions ($G = 0.8$ and 12 g/s, respectively), the distribution curves are similar to the normal law, although asymmetry is observed. The dispersion is small. In the regime corresponding to the intermediate gas flow rate (G = 6 g/s), the dispersion rapidly increases and the distribution curve shows two peaks.

The formation of double peaks on the distribution curves f_{min} and f^* and the absence of double peaks in the case of f_{max} indicates two greatly different conditions of arc shunting. The transition from one regime to another is accompanied by a change in the level of the shunting voltage U^* and minimum arc voltage U_{min}. The relationship between these quantities in transition from one regime to another is also indicated by the correlation factor which is close to unity in this case. Outside these flow rate ranges, the coefficient is considerably lower than unity. These considerations are clearly illustrated by the oscillograms of voltages and photographic recording of arc length pulsations Δl obtained for a plasma torch with a longitudinal slit (Fig. 2.37). At low gas flow rates (G = 0.8 g/s), the shunting process is characterised by high stability, a high amplitude of pulsations of voltage (*1*) and arc length (*2*). With

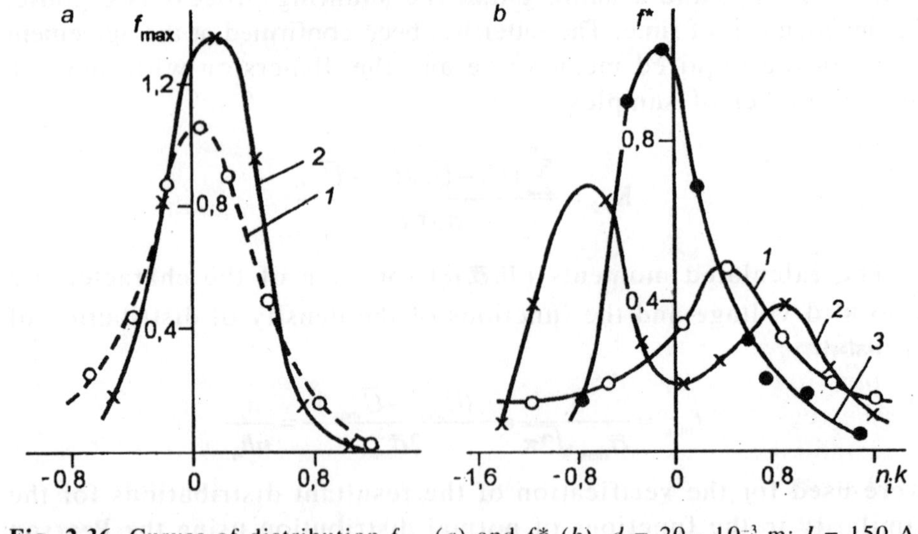

Fig. 2.36. Curves of distribution f_{max} (*a*) and f^* (*b*). $d = 20 \cdot 10^{-3}$ m: I = 150 A: K = 0.08 V^{-1}. *a*) *G*, kg/s: *1*) $12 \cdot 10^{-3}$; *2*) $6 \cdot 10^{-3}$. *b*) *G*, kg/s: *1*) $12 \cdot 10^{-3}$; *2*) $6 \cdot 10^{-3}$; *3*) $0.8 \cdot 10^{-3}$.

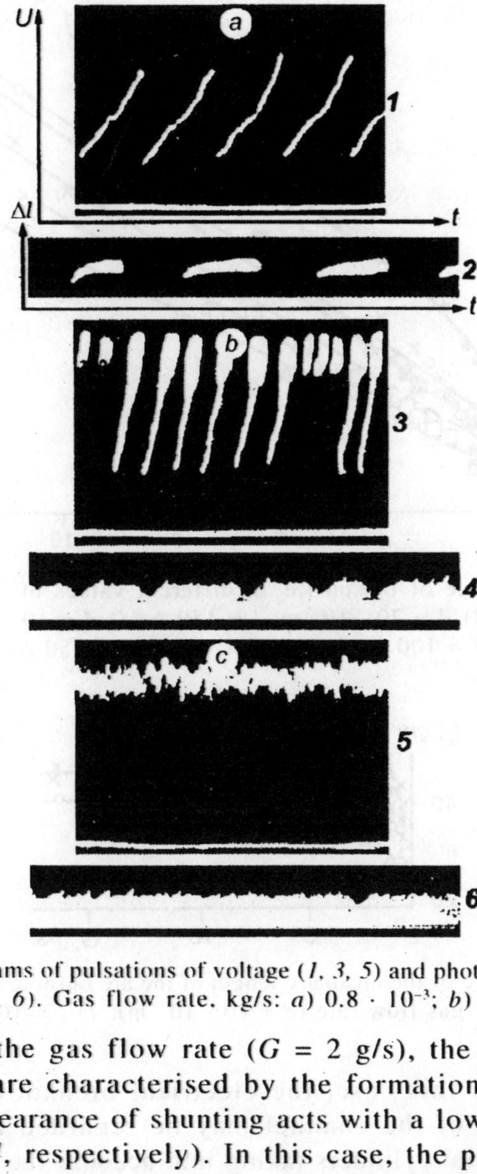

Fig. 2.37. Oscillograms of pulsations of voltage (*1, 3, 5*) and photographs of pulsations of arc length (*2, 4, 6*). Gas flow rate, kg/s: *a*) $0.8 \cdot 10^{-3}$; *b*) 2.10^{-3}; *c*) $12 \cdot 10^{-3}$.

an increase of the gas flow rate (G = 2 g/s), the areas of the existing process are characterised by the formation of a new process, i.e. the appearance of shunting acts with a lower amplitude U* and Δl (*3* and *4*, respectively). In this case, the probability of appearance of these quantities is of the order of 0.2. With increasing flow rate, the probability increases and, finally, at G = 12 g/s the regime (*5, 6*) with a low amplitude of the values U^* and Δl is established. These effects also explain the double peak form of the distribution of f_{min} and f^*.

The transition from one shunting regime to another is determined by the change in the nature of the gas flow and may be explained as follows [32–34]. At a low gas flow rate, the flow in the entire channel is laminar and the position of the arc in the vicinity of the

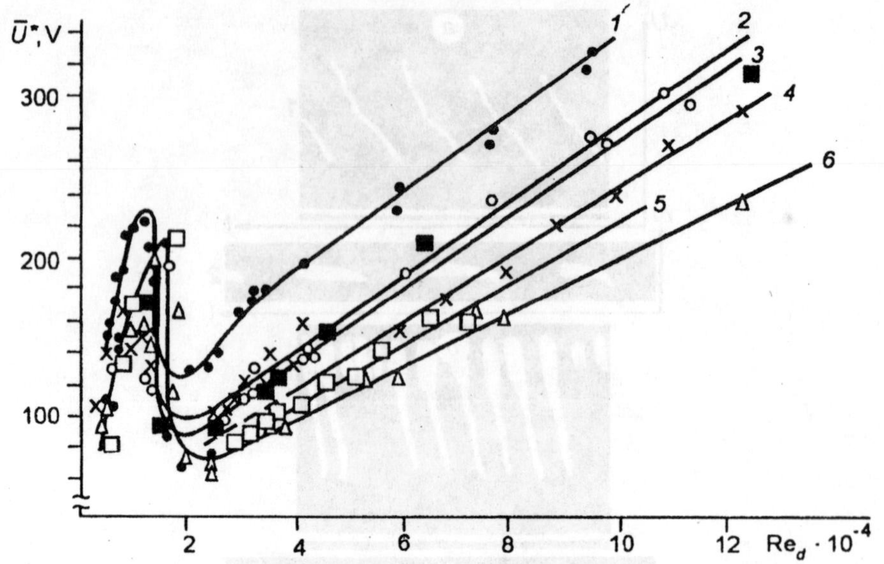

Fig. 2.38. Dependence of $\bar{U}^*$ on Re_d at different values of d and I. *1*) $d = 20 \cdot 10^{-3}$ m; $I = 100$ A; 2) $d = 20 \cdot 10^{-3}$ m; $I = 150$ A; *3*) $d = 10 \cdot 10^{-3}$ m; $I = 100$ A; *4*) $d = 15 \cdot 10^{-3}$ m; $I = 100$ A; *5*) $d = 10 \cdot 10^{-3}$ m; $I = 150$ A; *6*) $d = 15 \cdot 10^{-3}$ m; $I = 150$ A.

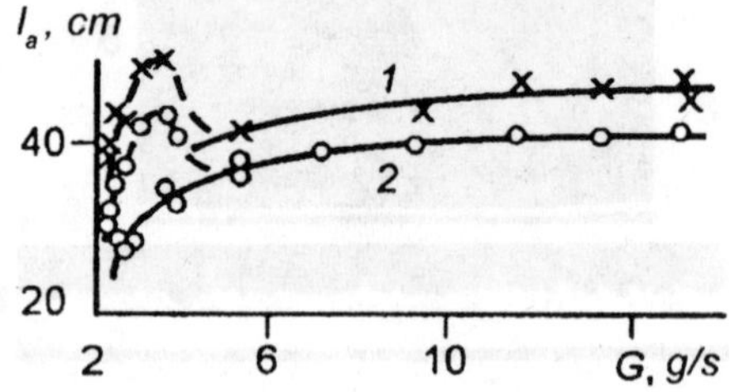

Fig. 2.39. Dependence of the mean arc length of the arc burning in a single-chamber plasma torch, on the gas flow rate ($d = 20 \cdot 10^{-3}$ m). *1*) $I = 100$ A; *2*) $I = 150$ A.

axis is stable. In this case, the electrical breakdown between the arc and the wall of the channel may be regarded as a breakdown between two coaxial cylinders taking into account the special features, introduced by the arc [35, 36]. With increase of the gas flow rate of the transition to the turbulent flow regime takes place. The interaction of the arc with the turbulent flow results in transverse oscillations of the arc. Consequently, the distance from the arc to the wall is shortened, the temperature fields in the cross-section of the channel are equalised and, in the final analysis, breakdown voltage decreases. Figure 2.38 shows the dependence of the mean breakdown voltage U^* on the number $Re_d = \frac{u \cdot d}{\nu}$, calculated for the cold gas. The curves reflect the sharp boundary between the two flow regimes. The Reynolds number of the transition may be assumed to be constant with a sufficient degree of accuracy and equal to $Re_d = 1.4 \cdot 10^4$. The result is regarded as the direct confirm ion of the hypothesis on the gas-dynamic nature of the variation of the nature of shunting.

In the experiments, the mean arc length l_g was also determined on the traces of erosion left by the arc spot on the electrode surface. Its dependence on the gas flow rate is illustrated by the graphs shown in Fig. 2.39. At the Reynolds numbers close to the transition numbers ($G \sim 5$ g/s), there is a large change of the form of the dependence.

It is interesting to estimate the thickness of the breakdown distance δ^* for the turbulent and laminar regimes. We accept a relatively accurate hypothesis according to which the entire breakdown voltage is applied to a thin wall layer of the cold gas, whose thickness in a general case may also depend on the gas flow conditions. In this formulation, in the examination of the conditions and the relationships of the breakdown, it is necessary to take into account the surface curvature assuming that the breakdown takes place between two flat electrodes, one of which is metallic. Consequently, in the determination of the breakdown distance δ^* it is possible to use the empirical dependence for the breakdown voltage in small gaps [36]:

$$\bar{U}^* = 3{,}33.10^4 \left(\frac{\rho}{\rho_0} \delta^* \right)^{0.9},$$

where ρ_0 is the density of air in the normal conditions.

Introducing the approximation $\frac{\rho}{\rho_0} = \left(\frac{h}{h_0} \right)^{-1}$, which is valid up to a temperature of approximately 4000 K, and assuming that the controlling value is the mean enthalpy h in the breakdown cross-section (disregarding heat losses through the electrodes because they are relatively small in this section):

$$h = \frac{I\bar{U}_{min}}{G},$$

the equation for $\bar{U}^*$ maybe presented in the following form:

$$\bar{U}^* = 3.33 \cdot 10^4 \left(\frac{G h_0}{I\bar{U}_{min}} \delta^* \right)^{0.9}.$$

Processing of the measured values of $\bar{U}^*$ in relation to the enthalpy

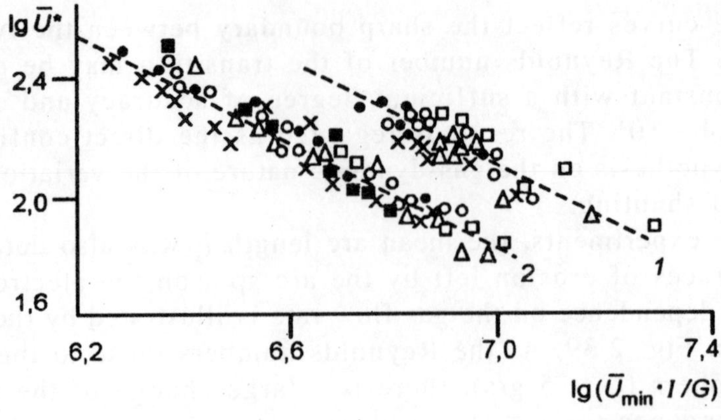

Fig. 2.40. Dependence of $\bar{U}$* on the enthalpy of the flow in the breakdown cross section. *1,2*) Laminar and turbulent regime, respectively.

of the flow in the breakdown cross-section shows (Fig. 2.40) that these values are described quite efficiently by the last dependence if it is assumed for the turbulent regime that $\delta^*_t = 3.3 \cdot 10^{-2}$ cm, and for the laminar regime $\delta^*_l = 6.9 \cdot 10^{-2}$ cm. The order of the resultant values of δ* is in agreement with the results published in [37] and confirm the assumption according to which the development of a breakdown between the arc and the channel wall is determined by the processes in the thin near-electrode layer of the cold gas whose size depends on the flow conditions in the channel of the plasma torch.

The results can be used to propose the following model of development of a breakdown in the gap between the arc and the wall in arcing in a laminar gas flow. The gap is conventionally divided into three characteristic regions. The first is the equilibrium region, including the arc column and its vicinity in which the concentration of the charged particle is equilibrium and the condition of quasi-neutrality is fulfilled. The second region is the diffusion region bordering with the equilibrium region in which the concentration of charged particles is non-equilibrium and is determined by ambipolar diffusion in the field of concentration, temperature and in the electrical field; the quasi-neutrality is established in this region. The third region is the region in the vicinity of the electrode to which the entire difference of the potentials between the arc and the wall in the given cross-section is applied. This region is characterised by the non-equilibrium concentration of the electrons extracted by the electrical field from the diffusion region or from the electrode (depending on the polarity of connection of the output electrode).

The initial stage of the development of the shunting process is based on the breakdown of the near-electrode region in which the energy of the electrons is sufficient for ionisation. The breakdown process with subsequent formation of the shunting arc channel in the zone is evidently of the avalanche-like nature, because the dependence of electrical conductivity on the temperature of the medium at relatively low temperatures is exponential. In the diffusion region, the controlling process of the development of breakdown is the thermal process, and the nature of this process is close to thermal breakdown.

In arcing in a turbulent flow characterised by large transverse pulsations of the arc, it is necessary to examine a more complicated model taking into account the effect of the pulsations of the arc from the viewpoint of the variation of the temperature distribution in the cross-section of the channel and the probability of formation of the breakdown with closing of the arc on the electrode.

The close relationship between the pulsations of arc voltage and turbulent transverse pulsations of the flow is indicated by the experimentally determined dependence of the shunting frequency f on the quantity G/d^3, inverse in relation to the characteristic timescale $t = \frac{d}{u}$ in the turbulent movement of the gas in the pipe (Fig. 2.41). Similar results may be explained if it is assumed that the time between consecutive breakdowns in arcing in the turbulence flow is associated with the duration of passage of the section of the arc from the axis of the channel of the the wall of the electrode under the effect of transverse pulsations of the velocity by the relationship $t \sim 0.5d/\upsilon'$, where d is the electrode diameter, υ' is the radial pulsation component of the velocity of the gas flow. Since in the turbulent flow $\upsilon' \sim \bar{u}$, then $t \sim \frac{d}{\bar{u}} \sim \frac{d^3}{G}$. Thus, the shunting frequency is $f = \Phi\left(\frac{G}{d^3}\right)$.

Examination of the relationships and special features of the arcing in a single-chamber plasma torch shows that the electrical characteristics of the arc, the stability of arcing and the shunting process greatly depend on the conditions and special features of the gas flow. Detailed examination of this process is not possible without detailed knowledge of the mechanism of interaction of the arc with the gas flow.

2.5.3. Electric discharge between solids

In section 2.5.2, attention was given to one of the fundamental electrophysical processes, used as a basis for understanding and further expansion of our knowledge of the complicated processes taking place

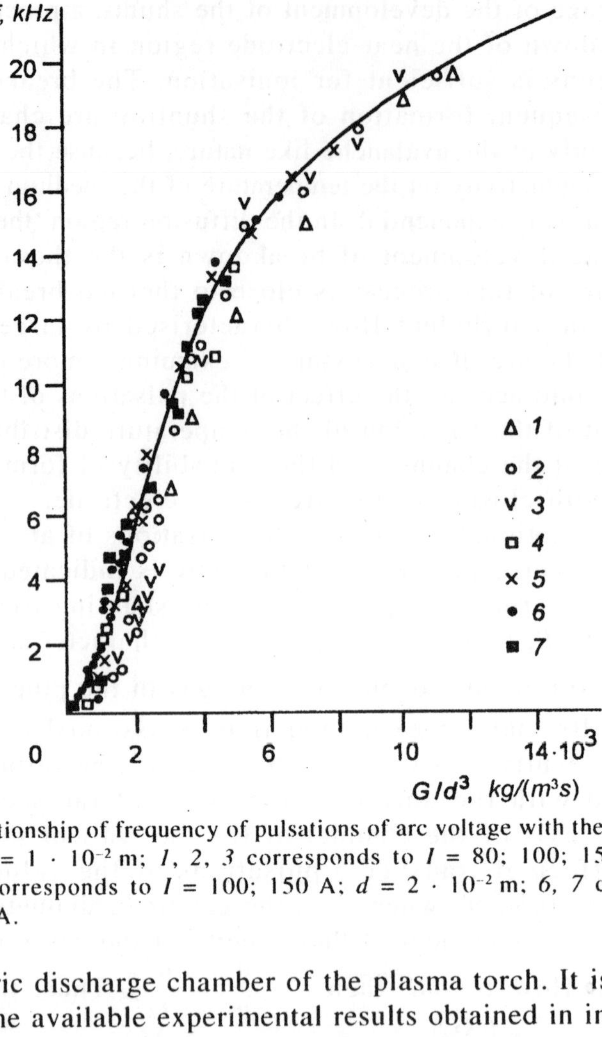

Fig. 2.41. Relationship of frequency of pulsations of arc voltage with the characteristics time scale. $d = 1 \cdot 10^{-2}$ m; *1, 2, 3* corresponds to I = 80; 100; 150 A; $d = 1.5 \cdot 10^{-2}$ m; *4, 5* corresponds to I = 100; 150 A; $d = 2 \cdot 10^{-2}$ m; *6, 7* corresponds to I = 100; 150 A.

in the electric discharge chamber of the plasma torch. It is also useful to present the available experimental results obtained in investigations carried out to determine the dependence of breakdown voltage on a number of factors which possibly, unable to the experts to understand or obtained further information on the electrical breakdown in different operating conditions of the plasma torch in order to find new methods of reducing the rate of erosion of the electrodes.

1. Gas breakdown voltage between two metallic electrodes

The most characteristic electrophysical process in the arcing chamber, i.e. shunting, is an electrical breakdown between the wall of the chamber and the arc which determines the arc length, electrode erosion and influences other processes. Usually, the following parameters are used to describe the breakdown phenomenon between the metallic electrodes: breakdown voltage U^*, the pressure in the inter-electrode gap p, and the distance between the electrodes Δz, described by the Pashen law. Figure 2.42 shows the dependences of breakdown voltage $U^* = f(p, \Delta z)$ in different gases for wide (*a*) and small (*b*) ranges of the values of the product $(p \cdot \Delta z)$.

The electrodes were made of platinum [38]. One of the curves in Fig. 2.42*a* provides information on the effect of a small addition of argon to neon on breakdown in the gas, i.e. a large decrease of breakdown voltage is recorded. The special feature must be taken into account in further investigations of breakdown voltage because the understanding of the physical processes, leading to this type of

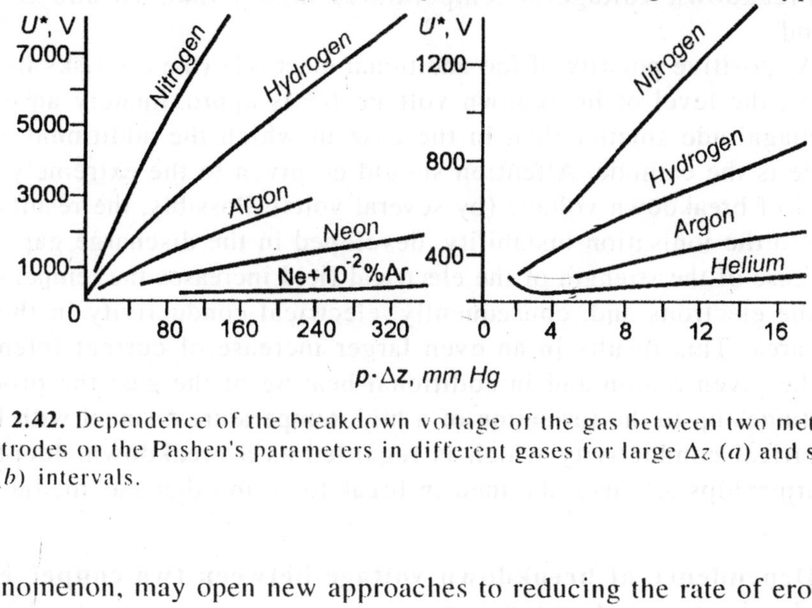

Fig. 2.42. Dependence of the breakdown voltage of the gas between two metallic electrodes on the Pashen's parameters in different gases for large Δz (*a*) and small Δz (*b*) intervals.

phenomenon, may open new approaches to reducing the rate of erosion of electrodes.

2. Electrical breakdown between the arc and a cold electrode [39].

The large-scale shunting which determines the mean length of the self-setting arc in a linear plasma torch is linked with the breakdown voltage U^* between the arc and the surface of the electrode whose value depends on the polarity of the output electrode (section), the diameter of the discharge chamber d, the mean mass temperature of the gas T_{mean} in the cross-section of the electrical breakdown or the additional electrode and the type of working gas.

Two temperature dependences of the breakdown voltage in argon between the arc and the additional electrode (a.e.), used either as a cathode or anode, are shown in Fig. 2.43. The voltage of different polarity was supplied to the additional electrode from an additional pulsed power source connected to the main anode and ensuring a linear increase of voltage at a rate of 10^5 V/s. The mean mass temperature of the arc was measured by changing the current intensity and gas flow rate.

At negative polarity of the additional electrode (solid circles in Fig. 2.43) and temperatures of up to 10 000 K, the level of the breakdown voltage corresponds to the cathode potential drop in glow discharge (a large scatter of the values of the experimental data is explained by possible hydrodynamic pulsations of the arc column). A decrease of breakdown voltage at temperatures higher than 10 000 K was found.

As positive polarity of the additional electrode (the asterisks in Fig. 2.43), the level of breakdown voltage U^* is approximately an order of magnitude smaller than in the case in which the additional electrode is the cathode. Attention should be given to the extremely low level of breakdown voltage (by several volts). Possibly, the results are due to the ionisation instability, developed in the discharge gap. The increase of the strength of the electrical field increases the temperature of the electrons and, consequently, electrical conductivity in the local area. This results in an even larger increase of current intensity in the given region and in additional heating of the gas; the process continues up to the formation of a high-temperature channel with high electrical conductivity which is regarded as a breakdown. Identical relationships are also obtained in breakdown in other gas media.

3. Dependence of breakdown voltage between two copper electrodes in air on temperature

The experimental equipment is described in Fig. 2.44 [39]: the diameter of copper water-cooled cylindrical electrodes was 10 mm, the distance between the electrodes Δz; the heated gas from the plasma torch was supplied here. The circumferential component of the velocity of the twisted gas flow at the outlet from the nozzle of the plasma torch is equal to zero because of the opposite direction of the tangential velocities of the gas supplied into the vortex chamber with the flow rate G_2 and G_3. The results show that the increase of the mean mass temperature of the gas results in a large decrease of the breakdown voltage between the electrodes (Fig. 2.45).

These results are important, in particular, for better understanding of the breakdown mechanism, with the selection of the gap between the sections in the plasma torch with the electrode insert. Calculations show (Fig. 2.46, solid line) that breakdown voltage U^* for

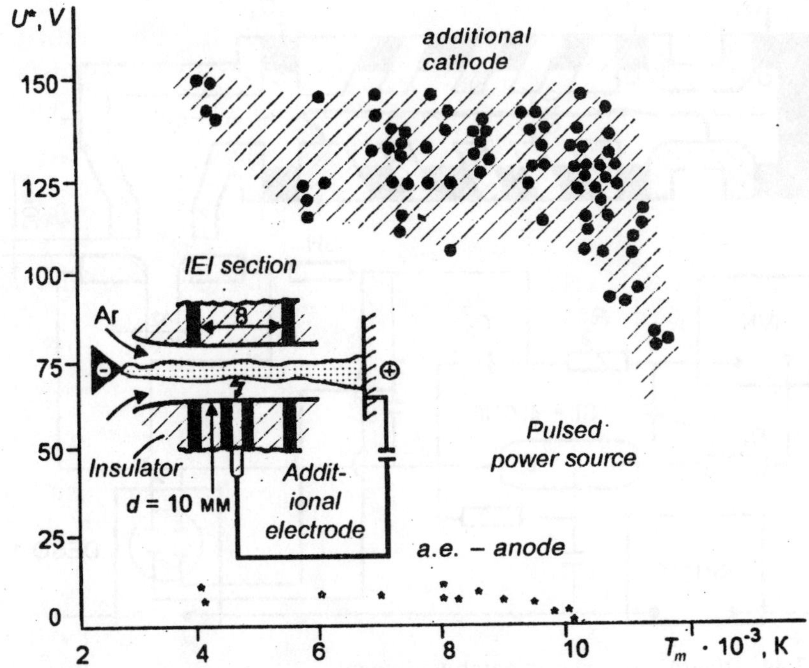

Fig. 2.43. Breakdown voltage between the arc and the cold electrode in relation to the mean mass temperature of the gas in the breakdown section.

heated air between two Rogovski-type tungsten electrodes is determined quite satisfactorily from the generalised Pashen law up to the temperature T_{mean} ~ 2200 K (p = 10^5 Pa, T_0 = 300 K, Δz = (0.5 ÷ 5) · 10^{-3} m). High temperatures are characterised by large deviations from this law (Fig. 2.46, points *1–3*): the experimental data correspond to different values of the length of the discharge gap Δz at the pressure in the discharge gap of p = 10^5 Pa.

The breakdown voltage for the air, heated to T_{mean} = 3300 K, for the gap of Δz = 5 · 10^{-3} m, was more than four times lower than that predicted by the Pashen law. The experimental data, presented in Fig. 2.46, may be used for estimating the breakdown potential of air in the temperature range 2000 ÷ 3500 K.

2.6. PULSATIONS OF THE 'RADIAL' SECTION OF THE ARC IN THE OUTPUT ELECTRODE OF AN AXIAL PLASMA TORCH

The pulsations of the arc and gas flow were investigated using photographic methods. The output electrode *1* of the investigated

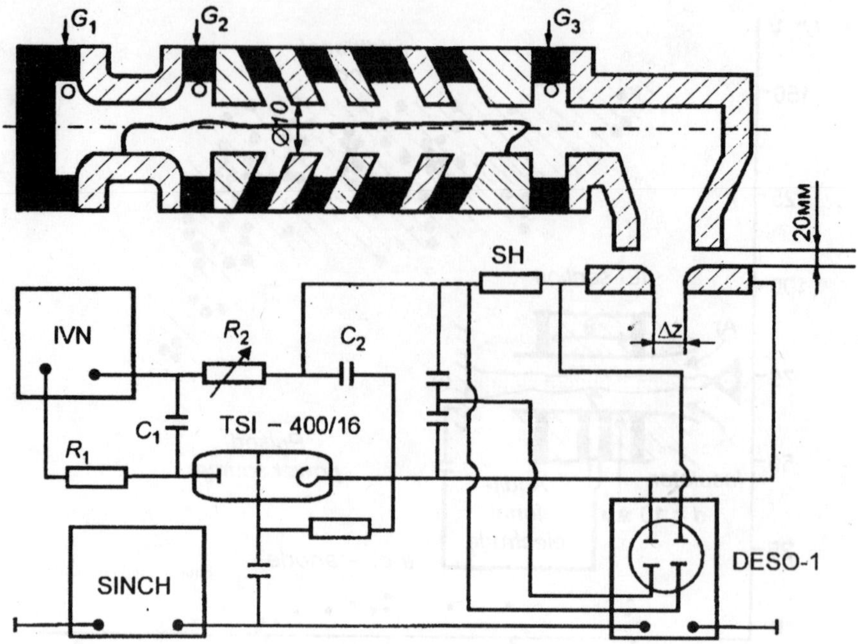

Fig. 2.44. Diagram of experimental equipment.

single-chamber axial plasma torch with gas-vortex stabilisation of the arc (Fig. 2.47) contained a narrow longitudinal slit *AB* for examination of the electrical discharge. The minimum section of the width of the slit in the internal surface of the electrode was 0.3 mm, the length of the slit approximately corresponded to the length of the electrode which was cooled with water. On the external surface of the electrode, the slit was covered with a silicate glass sheet.

The image of the slit *AB* was projected using the objective *2* on the moving film *3*. Scanning was carried out using a drum-type photographic device at a speed of 50 rpm/s. Exposure time was 1/50 s. The experiments confirm the presence of pulsations of the 'end' of the arc in the plasma torch with vortex stabilisation. It has been shown that the nature of pulsations in the plasma torch of the selected scheme depends on the polarity of the electrodes, the intensity of current and on a number of other factors, and is qualitatively identical in the systems of powering the arc with both direct and alternating current of industrial frequency [7].

In the previously described experiments, the plasma torch operated with single-phase AC. Consequently, during a single rotation of the drum, it was possible to record oscillations of the ends of the arc of both polarities with other conditions being constant (Fig. 2.48,

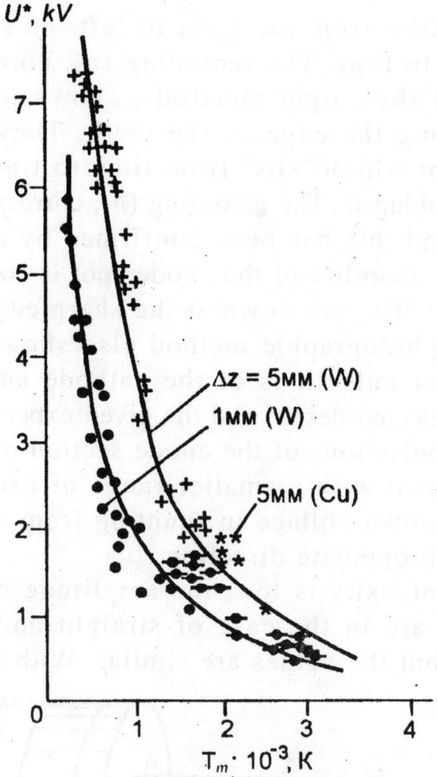

Fig. 2.45. Dependence of the breakdown voltage between two electrodes in air on each temperature.

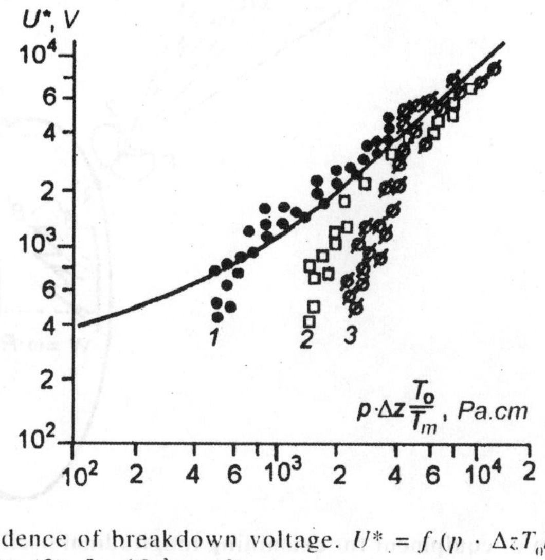

Fig. 2.46. Dependence of breakdown voltage. $U^* = f(p \cdot \Delta z T_0/T_m)$. $\Delta z = 1 \cdot 10^{-3}$ m (*1*); $3 \cdot 10^{-3}$ m (*2*); $5 \cdot 10^{-3}$ m (*3*).

movement of the film from the right to left, movement of the gas flow from bottom to top). The recording (*a*), corresponding to the inverse polarity of the output electrode, shows solid lines extending sometimes along the edge of the teeth. They reflect the path of the cathode spot which 'sits' from time to time on the edge of the slit and moves along it. The recording (*b*), corresponding to straight polarity, shows (and this has been confirmed by more detailed experiments), that the mobility of the anode spot is considerably higher. The anode spot never slows down at the sharp edge of the slit. Examination by the photographic method also showed the difference in the frequency of pulsations of the cathode and anode sections of the arc for the selected design and the given experimental conditions. The frequency of pulsations of the anode section of the arc is higher and this is associated with a smaller range of oscillations because of the lower breakdown voltage in shunting from the arc to the wall in comparison with opposite direction.

If the current intensity is low, the amplitude of the oscillations of the end of the arc in the case of straight and reverse polarity is relatively high but the values are similar. With increasing current

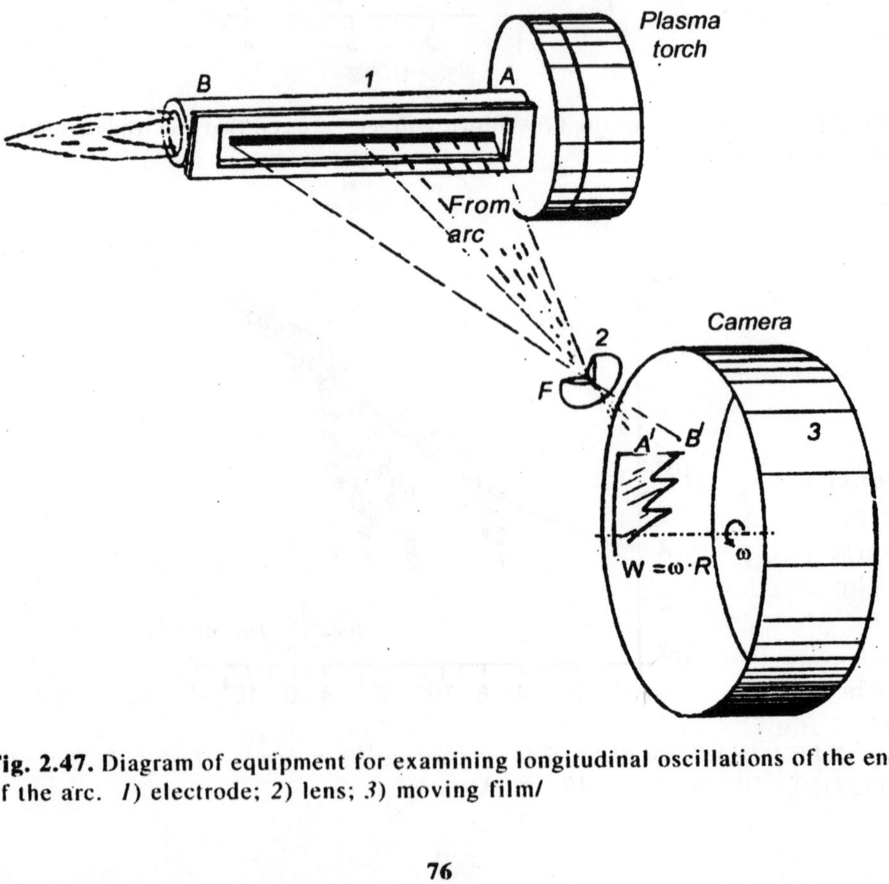

Fig. 2.47. Diagram of equipment for examining longitudinal oscillations of the end of the arc. *1*) electrode; *2*) lens; *3*) moving film/

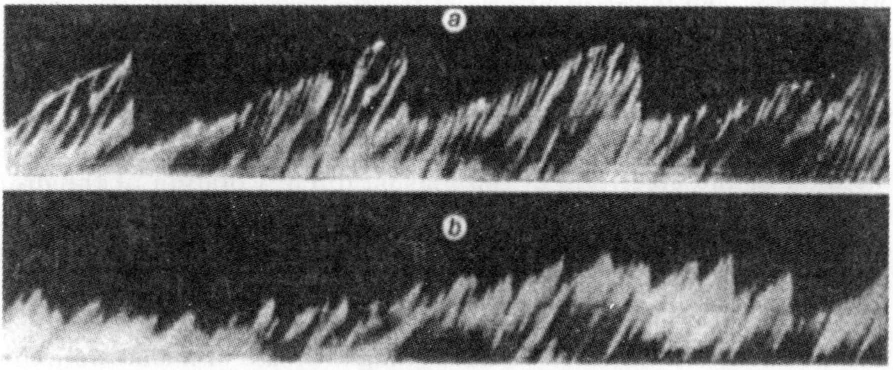

Fig. 2.48. Recording of the oscillations of the 'end' of the arc at reverse (*a*) and straight (*b*) polarity of connecting the output electrode.

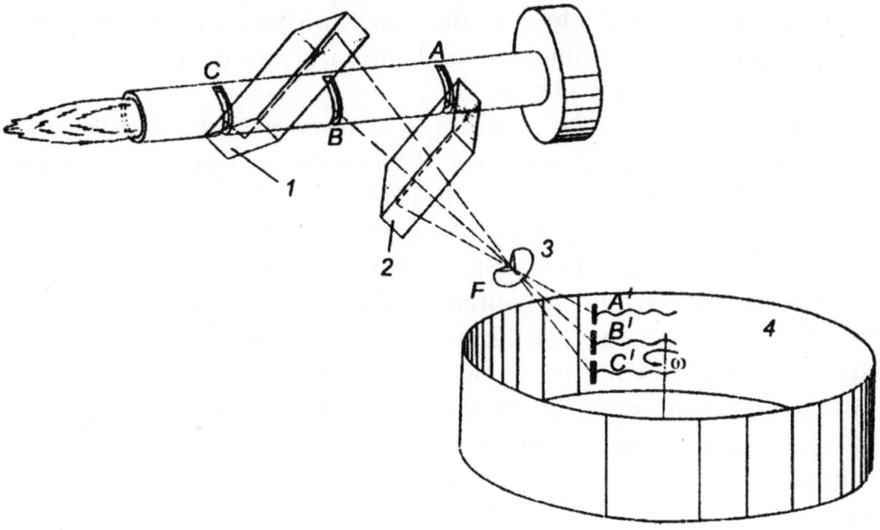

Fig. 2.49. Diagram of equipment for examining oscillations of the individual sections of the arc.

intensity the amplitude of the oscillations of the end of the arc at straight polarity greatly decreases, whereas in the case of reverse polarity this process is less pronounced.

The transverse oscillations of the arc are also easily recorded by the photographic method, if transverse slits are made in the output electrode. The diagram of such equipment is shown in Fig. 2.49. The body of the cylindrical electrode contains three narrow transverse slits *A*, *B* and *C*, closed on the outside with transparent sheets made of mica in order to avoid escape of the gas. In the investigated experiments, the distance between the slits was selected equal to

$30 \cdot 10^{-3}$ m, and the vertical size of the slit corresponds to the internal diameter of the electrode. Using the optical system, consisting of the prisms *1*, *2*, and the lens *3*, the image of the slits is projected onto the drum with the film *4* in such a manner that they were situated along a single generating line of the drum (A', B', C'), normal to the axis of the electrode of the plasma torch. Rotation of the drum is accompanied by synchronous sweep of the luminosity of the sections of the arc situated behind the three slits. According to the results obtained in the experiments with the longitudinal slit, the reflected glow of the internal (back in relation to the slit) surface of the electrode, which is a 'harmful' background, is not high because of the low reflection coefficient. At the centre, the brightness of the glow of the arc and of the electrode surface, if the arc spot is situated on the electrode surface at the given moment of time, is considerably greater than the brightness of the glow of the gas. Consequently, it was possible to carry out synchronous visualisation of the positions of the arc column and the electrode spot.

Figure 2.50 shows the recordings of transverse oscillations of the arc column in three sections for the reverse and straight polarities of the output electrode during a single period of passage of alternating current at a frequency of 50 Hz. The following conclusions may be made on the basis of the results. The transverse oscillations of the arc have the amplitude and frequency which change along the arc length. In the initial section of the arc chamber, the amplitude of the oscillations is not high and equals 0.5 ÷ 1 mm. This indicates that the position of the arc column in the space is stable. Along the flow, the column becomes less stable, the amplitude of transverse oscillations of the column increases, as indicated by the upper left recording. The left recordings show special features of the behaviour of the arc in three sections *A*, *B*, *C* for the case in which the output electrode is used as the cathode. The arc is shunted in the cross-section, close to the third slit *C*, as indicated by the

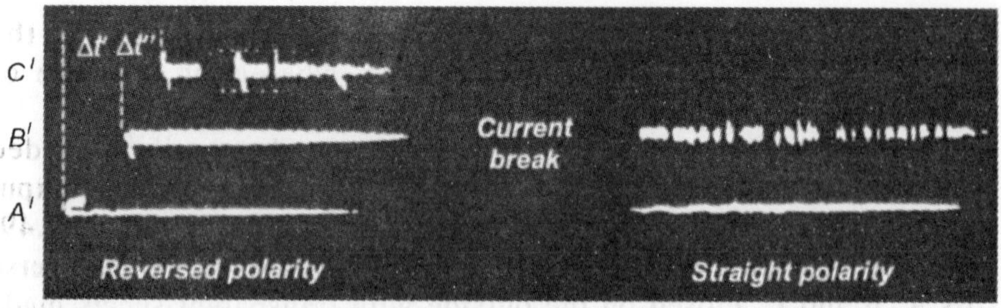

Fig. 2.50. Recording of transverse oscillations of the individual sections of the arc in three cross sections of a phase plasma torch.

upper discontinuous paths. The bright transverse surges are the glow of the near-electrode sections of the arc, passing away from the slit.

The right recordings show special features of arcing when the output electrode is used as the anode. They indicate that shunting already takes place in the region of the second slit, i.e. the arc with the given polarity of connection of the output electrode is considerably shorter (by more than $30 \cdot 10^{-3}$ m). The dark place between the two recordings, relating to the reverse and straight polarity, corresponds to a break in current.

The variation of the luminous diameter of the arc in relation to current intensity is also evident. When the polarity is changed, the current intensity passes through zero; on the recording, this is reflected by the reduction of the luminous diameter of the arc column. The delayed time of appearance of the arc in individual slits in relation to the previous value ($\Delta t'$, $\Delta t''$) makes it possible to find the mean velocity of movement of the near-axial closing section of the arc along the flow.

Using the transverse slit, it is possible to detect the deflection of the loop of the output end of the arc. In these experimental conditions, its value is equal to one gauge size of the electrode. Identical conclusions are also obtained from the analysis of deformation of the loop in a flat channel (Fig. 2.34).

Attention will now be given to the movement of the radial section of the arc in axial gas-vortex plasma torches which is determined not only by the longitudinal components of the velocity of the flow but also by the circumferential component. The effect of this component of the velocity of the movement of the closing section of the arc was investigated in experimental equipment including the two-chamber DC plasma torch and a high-speed cine camera (Fig. 2.51). The design of the plasma torch enabled examination of the movement of the radial sections of the arc in both the internal *4* and the output *5* electrode through quartz glass in the back cover of the plasma torch. By selecting the appropriate ratio of the flow rate of the gas through the vortex chamber, the large-scale shunting in the internal electrode was eliminated. In this case, the arc spot travelled along the narrow band approximately 3–4 mm wide in the zone of zero wall axial velocity. This explain the possibility of carrying out high-quality filming of radial sections of the arc.

In order to avoid the superposition of the images of the near-electrode sections of the arc in the internal and output electrodes and to determine unambiguously which of the arcs belongs to the

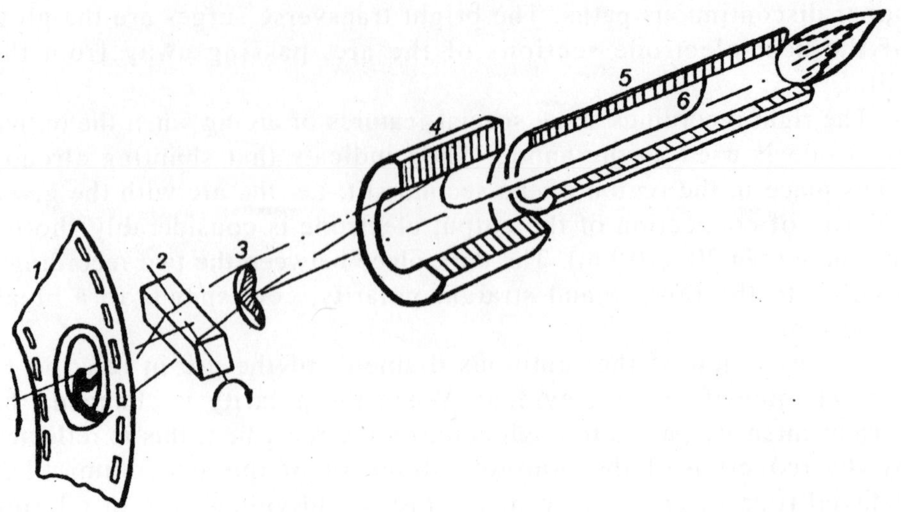

Fig. 2.51. Diagram of equipment for taking photographs of radial (closing) sections of the arc. *1*) Cine film; *2*) prism; *3*) lens; *4,5*) internal and output electrodes; *6*) electric arc.

internal electrode and which to the external one, the diameter of the end electrode was selected slightly larger than the diameter of the output electrode.

The arc was photographed with a high-speed camera (with the upper limit of 5000 frame/second) and using a high-speed photographic recording device operating in the 'time lens' regime. In the first case, one rotation of the radial section of the arc was displayed on 4–5 frames or more. This was sufficient for explaining the mean velocity and the nature of displacement of the near-electrode section of the arc. For better characterisation of the core of the arc and removing background from the glowing gas, it was necessary to use different combinations of light filters.

A characteristic photograph of the radial section of the arc, obtained using the 'time lens' (recording speed 124 000 frame/s), is shown in Fig. 2.52*a*. The form of the section resembles a helix with the convex section facing the side of movement of the gas. A decrease of the diameter of the arc column in the vicinity of the electrode surface is evident. The central part of the photograph corresponds to the projection of the entire positive arc column situated along the axis of the electrode. During rotational movement of the arc, the internal electrode may show microshunting between the wall and the

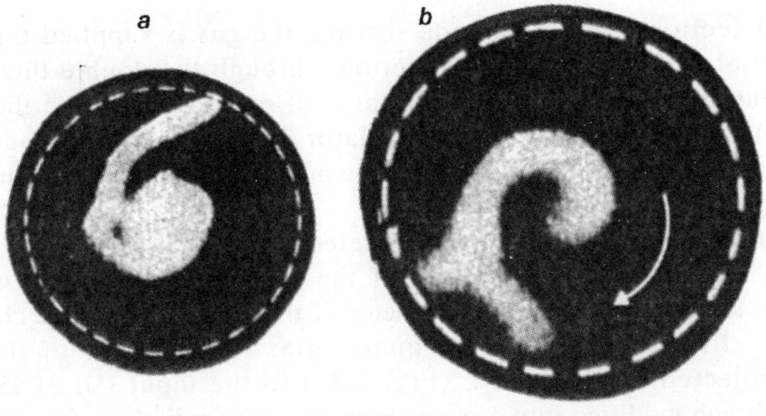

Fig. 2.52. Photograph of the radial section of the arc in the absence of microshunting between the arc and the electrode wall (*a*) and in shunting on the electrode wall (*b*).

adjacent section of the arc (Fig. 2.52*b*). The probability of microshunting increases with a decrease of pressure and increase of current.

2.7. SELF-OSCILLATIONS OF THE PARAMETERS OF THE ELECTRIC ARC

The parameters of the electric arc, running in the cylindrical channel and subjected to the effect of the gas flow, are usually non-stationary. The variation with time of the strength of the electrical field, temperature and gas pressure is caused by dynamic processes, such as shunting, and also by the formation of oscillations of the discharge parameters of the acoustic and magnetohydrodynamic nature. They generate a wide spectrum of pulsations of the brightness of the recorded radiation of the arc and the jet, and have been studied in a number of investigations [40–42]. The interest in this phenomenon is caused by the need to take into account pulsations when determining the arc temperature. In addition to this, investigations of the pulsations of radiation lead to understanding of the reasons causing these pulsations and, consequently, is an additional source of information on the complicated processes, taking place in electric arc systems.

The results presented below [43] were obtained in examination of a plasma torch with a sectional inter-electrode insert (IEI) and with blowing of the gas with the rate g_i into the gaps between the sections (Fig. 2.53). The devices used for twisting the gas flow have the same diameter, $D_0 = 5 \cdot 10^{-2}$ m, for the cathode section and the IEI sections. In the cathode section, the gas is supplied through tangential orifices, and in the sections through the double thread of the right-angle section where the angle of exit in relation to the axis z is 20°. The thickness of the insulator *2* between the cathode and the first section, and also between the adjacent sections was constant and equal to $2 \cdot 10^{-3}$ m.

To transfer radiation *J* from the electric discharge channel a slit $2 \cdot 10^{-3}$ m wide was cut in one of the sections. The height of the slit was similar to the internal diameter of the electric discharge channel *d*. The slit was covered with quartz glass. The image of the arc was projected by the lens *L* (Fig. 2.54) to the input slit of ISP-30 spectrograph modified into a monochromator. The radiation of continuum at a length of 393 nm was recorded using FEU-29 photoelectric multiplier. The transverse slit *K*, moved by an electric drive, was placed in the plane of the inlet slit of the ISP-30 spectrograph. The signal from the photoelectric multiplier was transferred through a current multiplier to an N-115 oscilloscope or through UZ-29 multiplier to the analyser of the spectrum of frequency of the sound range SK-4-26. To examine the behaviour of the arc in the space of the discharge channel, investigations were carried out using a SKS-1M high-speed cine camera in continuous scanning regime.

The experiments were conducted using a plasma torch with a relative

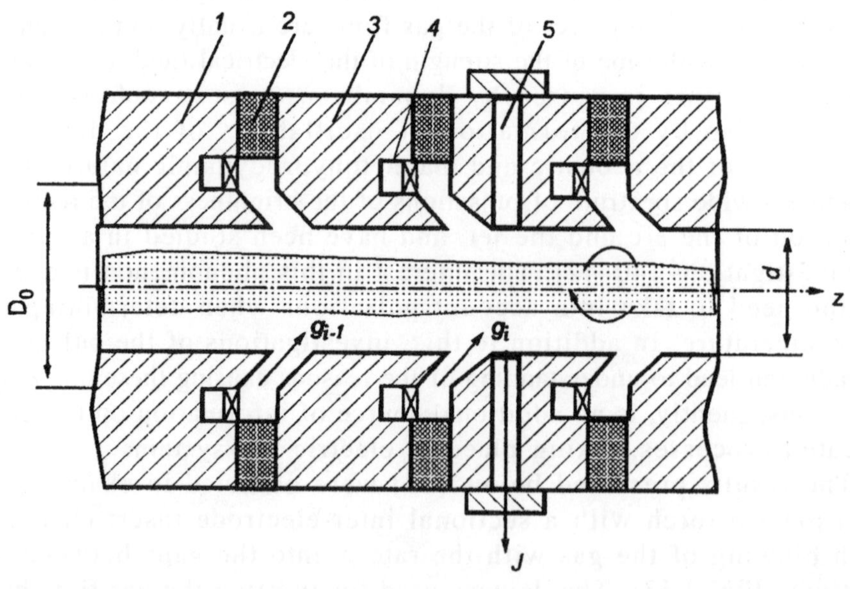

Fig. 2.53. Sectional channel. *1,3*) Sections of the inter-electrode insert; *2*) insulator; *4*) twisting device; *5*) section with a slit.

length of the electric discharge channel of $z/d < 10$, i.e. smaller than the length of the initial section. Electric power was supplied by a generator with the intrinsic frequency of pulsations of idle voltage of ν_0 = 1350 Hz and the amplitude not higher than 1%. The range of variation of arc current was I = 100 ÷ 600 A, the total flow rate of the gas in the sections was G_i = 0.5 ÷ 3.5 g/s, and air was used as the working gas.

The recorded pulsations of brightness are clearly divided into two groups. The first group includes high-frequency pulsations

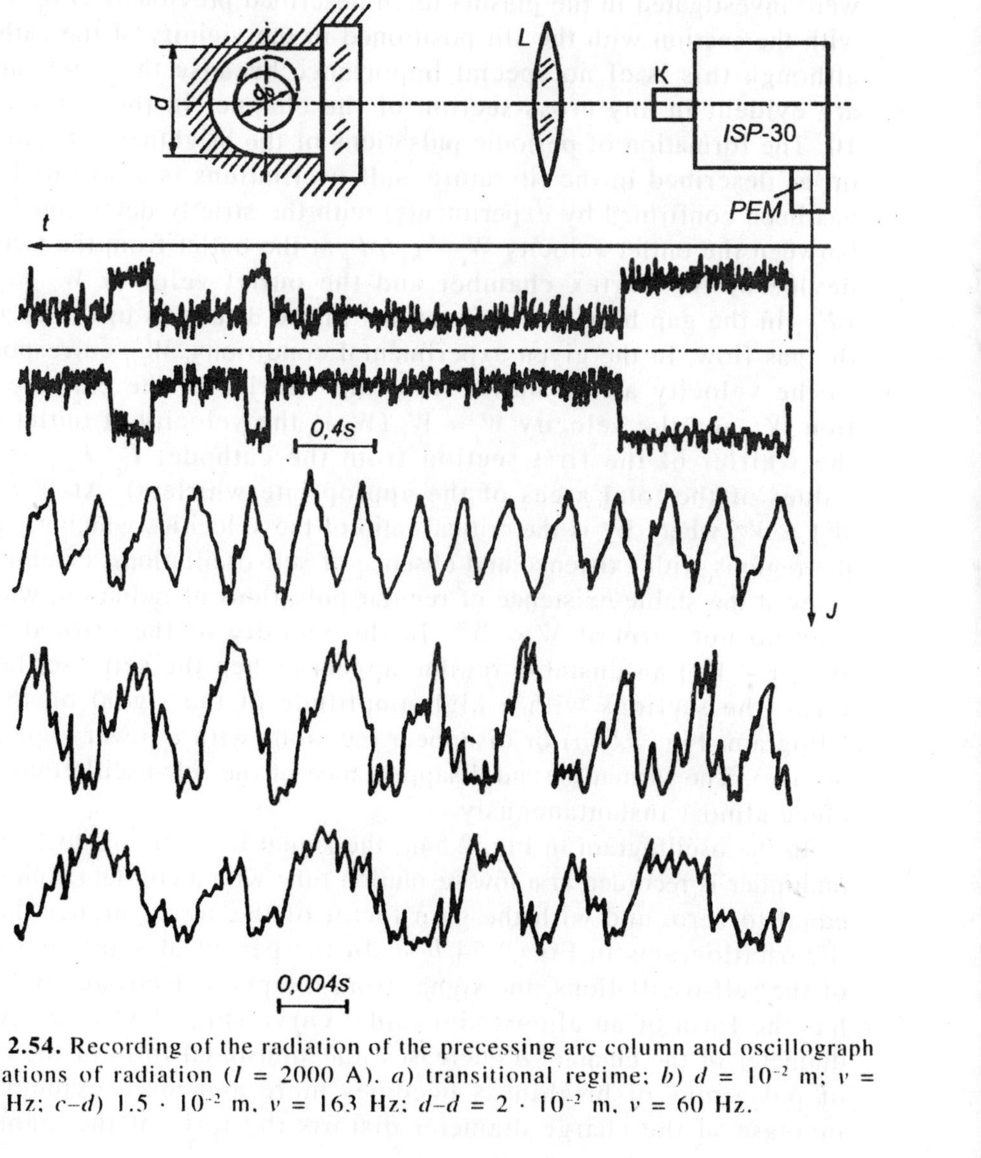

Fig. 2.54. Recording of the radiation of the precessing arc column and oscillograph pulsations of radiation (I = 2000 A). *a*) transitional regime; *b*) $d = 10^{-2}$ m; ν = 350 Hz; *c–d*) $1.5 \cdot 10^{-2}$ m, ν = 163 Hz; *d–d* = $2 \cdot 10^{-2}$ m, ν = 60 Hz.

(Fig. 2.54*a*). The physical nature of these pulsations is clear – shunting processes in the output electrode. This is confirmed by the analysis of the spectra of the frequency of these pulsations produced in SK-4-26 analyser; the maximum amplitude of pulsations of the complex spectrum of frequencies in the range 1–2 kHz. These data are in good agreement with the results of previous investigations [40, 41].

The second group (Fig. 2.54*b*) includes the regime with almost periodic pulsations and the frequency of 50 ÷ 1500 Hz. These pulsations were investigated in the plasma torch described previously (Fig. 2.54), with the section with the slit positioned in the vicinity of the cathode, although this is of no special importance because these pulsations are evident in any cross-section of the channel in the range $z/d <$ 10. The formation of periodic pulsations of the brightness of radiation or, as described in the literature, self-oscillations is associated (this has been confirmed by experiments) with the strictly determined ratio between the outlet velocity $W_i = g_i/\rho F_i$ at the outlet from the twisting device of the vortex chamber and the outlet velocity $W_{i-1} = g_{i-1}/\rho F_{i-1}$ in the gap between the sections in the direction upwards along the gas flow. In the given experimental conditions, W_{i-1} corresponded to the velocity at the outlet from the whirler of the cathode section W_c, and the velocity $W_i = W_s$ (W_s is the velocity at outlet from the whirler of the first section from the cathode; F_i, F_{i-1} are the values of the total areas of the appropriate whirlers). At $\bar{W} = (W_c/W_s) > \bar{W}^*$, where $\bar{W}^*$ is the critical ratio of the velocities which separates the regions with existence and absence of self-oscillations, examination showed the stable existence of regular pulsations of radiation, whereas they do not form at $\bar{W} < \bar{W}^*$. In the vicinity of the critical value $\bar{W}^*$ ($\bar{W} \approx \bar{W}^*$) an unstable regime appears when the self-oscillations form (the sections with a high amplitude of the signal on the oscillogram, Fig. 2.54*a*) or disappear (sections with a lower signal amplitude). The formation and disappearance of the self-oscillations takes place almost instantaneously.

In the oscillogram in Fig. 2.54*a*, the signal from the photoelectronic multiplier is recorded at a low resolution time with a constant component equal to zero, and with the gain factor of the signal higher than on the oscillograms in Fig. 2.54 *b–d*. In the period of stable existence of the self-oscillations, the signal from the photoelectronic multiplier has the form of an almost sinusoidal curve (Fig. 2.54 *b–d*). As the diameter of the channel *d* decreases, the sinusoidal form of the curves of pulsations of brightness becomes more and more evident. The increase of the charge diameter distorts the form of the signal. At

$d = 2 \cdot 10^{-2}$ m the frequency spectrum of the signal becomes more complicated because of the appearance of another harmonics as a result of the formation, in the arc column, of almost periodic increases or decreases of density (in respect of luminosity) in the direction downloads along the flow. This is also confirmed by the films recorded in SKS-1M.

With a decrease of the channel diameter the frequency of self-oscillations ν increases since $\nu \approx I/d^2$. However, according to the experimental results, this is not directly linked with the variation of the axial velocity. For example, an increase of the flow rate of the gas g_c through the whirler of the cathode section has almost no effect on the value of ν. In the range of stable existence of the self-oscillations, the frequency depends in a linear manner on the velocity W_s, i.e. the Strouhal number Sh = $\nu d/W_s$ for the fixed channel diameter is a constant value; this is characteristic of the self-oscillation processes of the vortex devices. However, when the channel diameter is changed, the Strouhal number also changes.

Figure 2.55 shows, in the coordinates W_c–W_s, experimental points

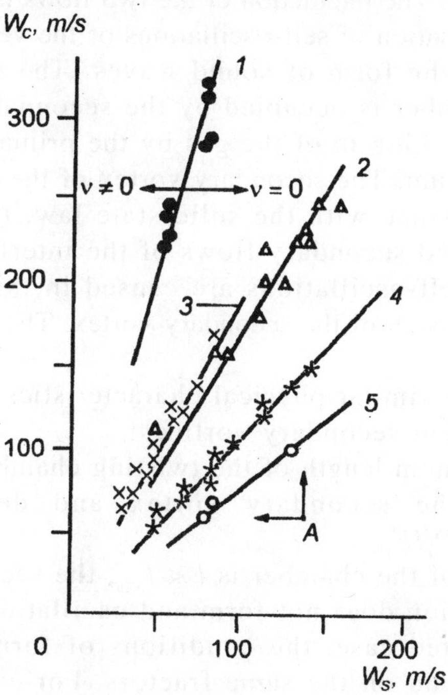

Fig. 2.55. Boundary conditions of formation of self-oscillations (I = 200 A, $G_i = 0.48 \cdot 10^{-3} \div 2.2 \cdot 10^{-3}$ kg/s, $z = 3 \cdot 10^{-2}$ m). *1)* $F_c = 3.5 \cdot 10^{-6}$ m², $F_s = 4.5 \cdot 10^{-6}$ m², $\bar{F}$ = 0.778; *2)* $7 \cdot 10^{-6}$, $4.5 \cdot 10^{-6}$; 1.556; *3)* $15.2 \cdot 10^{-6}$; $8.8 \cdot 10^{-6}$; 1.73; *4)* $15.2 \cdot 10^{-6}$; $4.5 \cdot 10^{-6}$; 3.378; *5)* $15.2 \cdot 10^{-6}$; $4.5 \cdot 10^{-6}$; 3.378; for *1,2,3,4* – $d = 1 \cdot 10^{-2}$ m; for *5* – $1.5 \cdot 10^{-2}$ m.

corresponding to the transition of the arc from the stable state to the self-oscillatory regime. The right half-plane in relation to each of the straight lines *1–5* is the region of absence of self-oscillations (the name used for its state) with $\nu = 0$, the left half-plane is the region of existence of these self-oscillations ($\nu \neq 0$). The transition from the stable state (for example, from point *A* on the W_c–W_s plane) to the unstable state is possible by two mechanisms: firstly, at the fixed value of the velocity W_s by the increase of W_c; secondly, at the fixed value of the velocity of the gas in the whirler of the cathode section W_c – by the decrease of W_s. At $\bar{W} < \bar{W}^*$, the high-temperature ($T > 10^4$ K) arc column with the diameter d_0 starts to rotate around the channel axis. Similar precession movement was detected in the examination of combustion in vortex chambers, in operation of vortex sound generators and, as reported in [44], is characteristic of twisted flows. Usually, these conditions are characterised by the formation of a peripheral twisted flow of the type of potential vortex, and of the internal flow slightly rotating in accordance with the solid-state rule. The interaction of the two flows in specific conditions results in the formation of self-oscillations of the vector of gas velocity and pressure in the form of sound waves. The central part of the long vortex chamber is occupied by the secondary flow, formed as a result of the sucking in of the gas by the primary vortex from the surrounding medium. The secondary vortex of the laminated structure rotates in accordance with the solid-state law; the axial velocities of the primary and secondary flows of the interface have the same direction. The self-oscillations are caused in this case by the formation of a precession of the secondary vortex. The essential conditions for this part:

1) identical or similar physical characteristics of the gas (liquid) of the primary and secondary vortices;

2) some minimum length of the twisting chamber l_{min} ensuring the formation of the secondary vortex and determined by the parameter $A = D_0^2/F_i$.

If the length of the chamber is $l < l_{min}$, the secondary vortex with quasi-solid rotation does not form and oscillations are not excited. In the investigated case, the conditions of formation of the self-oscillations depend on the same fractors. For example, the formation of self-oscillations requires some minimum velocity W_c, which determines the speed of rotation of the internal core of the flow (the arc column in the present case).

As indicated by Fig. 2.55, the straight lines *1–4*, constructed for the same diameter *d*, separate in respect of the parameter $\bar{F}$ =

F_c/F_s. This behaviour of the transition boundary may be explained by the effect of friction in the whirlers on the true velocity of rotation of the internal and external flows W_i. The true velocity of rotation of the internal flow W'_c depends on the velocity W_c, and the true velocity of the peripheral flow W'_s depends on velocity W'_s. The values of these velocity is in the vicinity of the boundary which separates of the internal and external flows, are proportional to the delivery velocities at exit from the whirler and are linked with them through the parameter A, i.e. $W'_c = W'_c f(D_0^2/F_c)$. The same is also valid for the velocity W'_s of the external flow. It may be seen that the geometry of the whirler has no effect on the functional relationship between the true and delivery velocities – in both whirlers it has the same form (for W'_c and W'_s). Since the parameter D_0 for both whirlers is the same, the true values of the velocity in the vicinity of the interaction boundary in both flows are proportional to the efficient sections of the whirlers. This also results in the previously mentioned detachment of the dependences *1–4* in respect of the parameter $\bar{F} = F_c/F_s$. In the immediate vicinity of the straight line *4*, obtained for the plasma torch with the channel diameter of $1 \cdot 10^{-2}$ m, there are the points of transition of the regimes for the channel diameter $1.5 \cdot 10^{-2}$ m (straight line *5*) at $\bar{F}$ = idem. At $\bar{F} = 1$, self-oscillations can form at the value of the tangential velocity of the internal flow close to the triple value of the velocity of the external flow: $W_c = 2.8W_s$. The process of transition from the re-

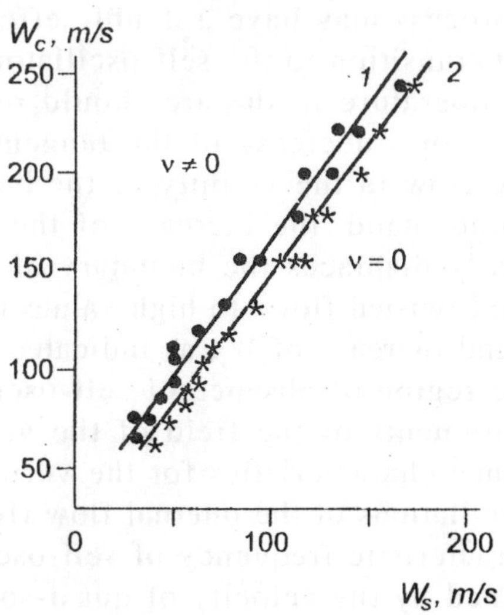

Fig. 2.56. Hysteresis of the formation of self-oscillations of the arc column ($d = 1 \cdot 10^{-2}$ m, I = 200 A). *1*) formation of oscillations, 2) disappearance of oscillations.

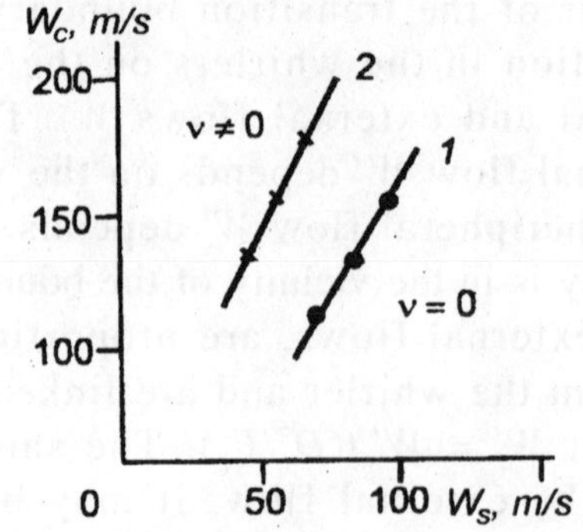

Fig. 2.57. Boundary conditions of formation of self-oscillations for different currents ($d = 1 \cdot 10^{-2}$ m, $\bar{F} = 1.73$). *1)* $I = 150$ A; *2)* 400 A.

gime $v = 0$ to the regime $v \neq 0$ and back is characterised by a hysteresis phenomenon which is clearly recorded in the experiments (Fig. 2.56).

Arc current also influences the position of the boundary of transition from the stable to self-oscillatory regime (Fig. 2.57). The increase of arc current reduces the width of the region of existence of self-oscillations. The variation of current leads mainly to a change of the radial temperature distribution. For example, for currents of $I > 200$ A, the temperature profile is greatly graded in the near-axial region (close to the rectangular distribution). This is associated with the increase of the diameter of the arc column and temperature at the axis. This process may have a double effect on the position of the boundary of transition to the self-oscillatory regime. In fact, the increase of temperature in the arc should result, as indicated by estimates in [7], in a decrease of the tangential component of the velocity of the flow in the vicinity of the boundary of the arc column. On the other hand, the increase of the radial size of the arc column evidently displaces the boundary of interaction of the external and internal twisted flows to high values of W_s. Both factors – decrease of W_c and increase of W_s, as indicated by Fig. 2.57, lead to widening of the region of absence of self-oscillations ($v = 0$).

Detailed measurements of the field of the velocities and pressures, and turbulence characteristics for the vortex devices and the presence of self-oscillations of the internal flow (for example, in [7]) show that the characteristic frequency of self-oscillations is unambiguously determined by the velocity of quasi-solid rotation of the internal flow. Consequently, knowing of the frequency of self-oscillations and the diameter of the internal flow, it is possible to estimate the tangential velocity at the boundary of the flows. Assuming that the same pattern of the flow also occurs in the investigated cases, the tangential velocity at the boundary of the arc column will be investigated. For the regimes, described in Fig. 2.54 *b–d* (in this case, the radius of the arc column is close to $r_0 = d/4$), this velocity is low and does not exceed $W_{r_a=r_0} < 10$ m/s. Since the axial velocity in the arc column is $\bar{U} > 100$ m/s, the ratio $W_{r_a=r_0} \ll 1$, i.e. the flow of the gas in the arc column may be regarded as almost completely axial and the effect of twisting on the distribution of static pressure in the arc column may be ignored. These results are of interest in the simulation of the electrical arc in the described conditions.

In addition to these results characterising the hydrodynamics of the twisted gas flow, stabilising the electric arc, they are also important from the viewpoint of the possibility of controlling the parameter of the jet leaving the plasma torch and ensuring stable arcing. In the experiments, it has been noted that the conditions with regular pulsations are accompanied by changes in the sound of the jet leaving the plasma torch, and also by the presence of short-time breakdowns of the arc to the section. The linear plasma torches often use separate input of different gases (for example, shielding of the cathode with an inert gas), i.e. there are 2 or more vortex chambers. It is clear that the previously mentioned effects, associated with the possibility of the formation of self-oscillations in this case, are important from the viewpoint of practice and must be taken into account in the development of plasma systems.

2.8. AERODYNAMICS OF THE INTERNAL ELECTRODE

The single-chamber plasma torch with a dead-end cup-shaped internal electrode (Fig. 2.58 *a, b*) and a two-chamber plasma torch (Fig. 2.58*c*) are used widely in industry. This is associated with the fact that the working medium in these plasma torches may be represented by many gases in comparison with the single-chamber plasma torch with an internal end electrode; in addition, the service life of the former is quite long. Further improvement of these plasma torches depends primarily on deeper understanding of the aerodynamics of the gas flow in the cavity of the internal electrode which has a strong effect on the electrical, erosion, pulsation and other characteristics of the plasma torch.

The presence of a closed cup-shaped electrode, and also of the two vortex chambers, results in the complicated pattern of the gas flow in the cavity of the end electrode (Fig. 2.58 *d, e*) and, consequently, in a more complicated dependence of the arc voltage and of the spatial position of the arc on the flow rate of the gas and the geometrical characteristics of the plasma torch [7].

The aerodynamics of the internal electrode was investigated on models produced from polished organic glass. The dimensions of the vortex chamber, the area of the inlet orifices in the chamber and the diameter of the end and output electrodes in these experiments were varied in a relatively wide range. The length of the end electrode did not exceed 20 length gages, and that of the output electrode was equal to or greater than the relative length of the end electrode.

The gas flow was visualised by different methods: by the introduction of smoke into the flow, sand particles or liquid jets, by the supply of oil through special orifices on the internal surface of the cylinder, by oil coloured with graphite. In some cases, oil and sand particles were introduced into the cavity of the cup prior to the start of the experiments. In visualisation using a liquid, the best results were obtained using a system of drainage orifices situated along the generating line of the electrode. The results of visualisation of the gas flow using smoke and also examination of the movement of sand particles, oil film or individual droplets, and the data obtained in the analysis of the values regarding the distribution of the static pressure were used to determine the aerodynamics of the flow in the cavity of the internal electrode. In order to decode the conditions characterised by high instability, the process was filmed.

The aim of formulation of these experiments with cold blowing only in this case could be regarded as achievable only if it would be possible to identify it with hot tests (with arcing). For this purpose, a series of experiments was carried out with arcs running in the plasma torches whose electric discharge chambers were in the form of exact copies of transparent models. In addition to the measurement of the electrical parameters of the arc, special attention was given to the position of the radial section of the arc in the internal electrode. Experiments were carried out on electrodes with a longitudinal slit which made it possible to produce photographic recording of the movement of the 'end' of the electrode along the axis of the channel, as already mentioned in section 2.6. The traces, left by the arc spot on the carefully cleaned surface of the electrode, made it possible to determine the zone of displacement of the spot.

Figure 2.58*a* shows schematically the spatial pattern of the flow of the gas in the cavity of the internal electrode in the absence of additional supply of gas at the end (single-chamber variant). The decrease of static pressure along the radius in the cross-section of the vortex chamber determines the inflow of part of the gas sup-

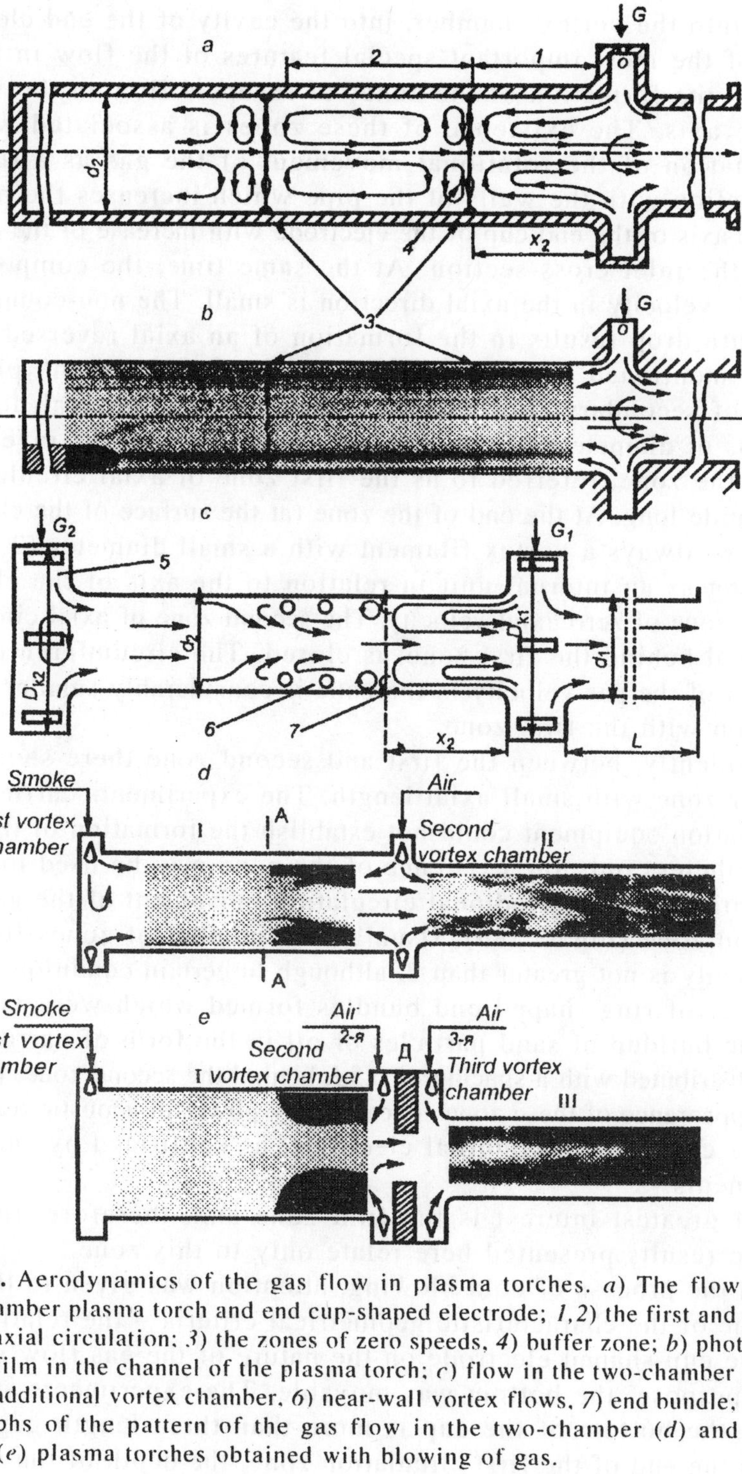

Fig. 2.58. Aerodynamics of the gas flow in plasma torches. *a*) The flow in the single-chamber plasma torch and end cup-shaped electrode; *1,2*) the first and second zones of axial circulation; *3*) the zones of zero speeds, *4*) buffer zone; *b*) photograph of an oil film in the channel of the plasma torch; *c*) flow in the two-chamber plasma torch: *5*) additional vortex chamber, *6*) near-wall vortex flows, *7*) end bundle; typical photographs of the pattern of the gas flow in the two-chamber (*d*) and three-chamber (*e*) plasma torches obtained with blowing of gas.

plied into the vortex chamber, into the cavity of the end electrode. One of the most important special features of the flow in the end pipe is the formation of the zones of reversed flow in the vicinity of the axis. The existence of these zones is associated with the attenuation of the rotational movement of the gas as a result of its friction with the walls of the pipe which increases the pressure on the axis of the end cup of the electrode with increase of the distance from the inlet cross-section. At the same time, the component of the gas velocity in the axial direction is small. The non-compensated pressure drop results in the formation of an axial reversed flow.

As mentioned in a number of investigations, in Rank pipes the zone of secondary flow may have the length of several diameters to tens of diameters of the pipe. As shown later, in the investigated case, the zone, referred to as the first zone of axial circulation, is also quite long. At the end of the zone (at the surface of the electrode) there is always a vortex filament with a small diameter (2–3 mm), rotating as an internal unit in relation to the axis of the electrode in the zone of zero axial velocity. The second zone of axial circulation, situated behind the first zone, is closed. The circumferential component of the gas velocity in the zone is considerably smaller in comparison with the first zone.

Evidently, between the first and second zone there should be a buffer zone with small axial length. The experiments carried out in simulation equipment could not establish the formation of this zone, nevertheless only the existence of the zone may be used to explain the direction of meriodional circulation movement of the gas in the second zone (Fig. 2.58*a*). Usually, the number of zones formed in the cavity is not greater than 2, although in certain conditions a larger number of ring-shaped end bundles formed which were visualised by the buildup of sand particles or oil in the form of narrow bands and distributed with a spacing of $0.5d_2$ behind the second zone. However, the appearance of these zones is caused by powerful acoustic oscillations in the cavity of the internal electrode, as confirmed by special experiments.

Of greatest interest is the first zone and, therefore, the qualitative results presented here relate only to this zone.

In the process of cold blowing, attention was given to the effect of one of the characteristic geometrical criteria – the relative depth of the cup-shaped electrode on the nature of the gas flow in it. For this purpose, the bottom was movable. The experiments show that until the bottom of the cup is more than three length gages away from the end of the first circulation zone, the depth of the electrode has almost no effect on the nature of the gas flow in it. On approaching the critical zone the first circulation zone rapidly fills the entire space. In the reversed process, i.e. increase of the depth of the electrode, the flow pattern is also restored immediately but there is a small hysteresis.

The aerodynamics of the gas flow in a dead-end electrode influences the special position of the arc in the electrode. In the case of low current, the radial section of the arc together with the arc spot are arrested in front of the first buffer zone, if large-scale shunting does not take place prior to this. The spot moves along a narrow band and this is clearly indicated by erosion of the electrode material. With increase of the current the ponderomotive forces, caused by the interaction of the intrinsic magnetic field of the axial part of the arc with the radial section, may be higher than the aerodynamic forces, maintaining the arc spot around the buffer zone. In this case, the radial section of the arc forms a loop with the convex part in the direction of the second circulation zone. At this configuration of the closing section, shunting may take place from the loop of the arc to the channel wall. The radial section of the arc, which penetrated in this manner into the second zone, moves in the direction to the bottom of the electrode and, in the final analysis, short circuits with the bottom of the electrode if the movement of the closing section is not restricted by large-scale shunting.

The pattern of the gas flow in the two-chamber plasma torch (Fig. 2.58*c*) in the absence of the gas flow through the additional vortex chamber ($G_2 = 0$) is obviously identical with that described previously. However, the supply of even a small amount of gas ($0 < \bar{G} = G_2/G_1 < 0.05$) at $G = G_1 + G_2 =$ const changes the pattern of the flow, in particular in the second zone. Firstly, a flow, directed in the direction of the output electrode, appears in the entire section from the end of the additional vortex chamber to the first zone. The vortex filament of the second zone disappears. A further increase of $\bar{G}$ results in the formation of specific near-wall vortex flows (position *6* in Fig. 2.58*c*). At $\bar{G} > 0.1$ all special features in the second zone disappear. Only the filament 7, with a complicated configuration, situated at the wall in the area of contact of the two flows remains.

Of greatest interest is the regime corresponding to the ratio $\bar{G} > 0.1$, because in this case it is possible to carry out extensive regulation of the length of the first circulation zone $\bar{x}_2 = x_2/d_2$ by changing the relative flow rate $\bar{G}$. The range $\bar{G} = 0.2 \div 0.3$ is characterised by the pulsations of the vortex filament in relation to the stable position. It should be mentioned that they can also form in the second range of variation of $\bar{G}$. This depends on the inlet velocities of the flows travelling into the main and additional vortex chambers, and on the ratio of the diameters D_{c1}/d_1 and D_{c2}/d_2. The investigated pulsations are characterised by the variation of their voltage in relation to the stable position with a subsequent decrease of the amplitude of oscillations with increase of $\bar{G}$. When $\bar{G} \to 1$, the first zone becomes shorter and the end filament moves to the outlet edge of the internal electrode.

The visualisation of the flow in the discharge channels of the two- and three-chamber plasma torches (Fig. 2.58 *d, e*) confirms the complicated flow pattern, including the presence of stable circulation zones and vortex formations in the near-wall regions (Fig. 2.58*d*).

Some quantitative results of 'cold blowing' will now be investigated. As already shown, the formation of the first circulation zone in the cavity of the internal electrode is determined by the presence of the radial pressure gradient formed as a result of the vortex movement of the flow organised in the chamber by means of the tangential supply of the gas to the chamber. The pressure gradient for the zone of the potential vortex is a function of the gas flow rate and the geometrical characteristics of the vortex chamber:

$$\frac{dp}{dr} = \rho\frac{w^2}{2};\ w = \frac{\Gamma}{2\pi r};\ \Gamma = 2\pi r_c u_{in};\ u_{in} = \frac{G}{\rho_{in} F_{in}},$$

Here w the tangential component of velocity; u_{in} is the velocity of the gas in the inlet orifices (slits) of the twisting ring; r_c is the radius of the vortex chamber; r is the actual radius; F_{in} is the area of all input orifices of the twisting ring.

The increase of the rate of supply of the gas into the vortex chamber increases the pressure gradient leading in turn to the intensification of the ejection of the gas from the cavity of the end electrode and, consequently, to the expansion of the first zone of axial circulation of the flow.

The nature of movement of the boundary layer is determined by the friction coefficient and the process of interaction of the boundary and near-axial gas flows moving in the opposite directions. The variation of the gas flow rate $\bar{G}$ has a complicated effect on the process in the electrode cavity. On the one hand, the increase of the gas flow rate should increase the velocity in the vortex chamber and, consequently, the pressure gradient in the potential zone of the vortex. On the other hand, the opposite process, associated with the losses of pressure in the output electrode takes place:

$$\Delta p = \frac{L}{d_1} \lambda \rho \frac{u^2}{2},$$

which is proportional to the square of velocity. Here u is the velocity of the gas in the output electrode; λ is the friction coefficient which depends on the number Re. The total effect of these factors may be determined only by experiments. Thus, even the simple analysis of the factors, influencing the flow in the cavity of the end electrode, shows that the wavelength of the return flow $\bar{x}_2$ (Fig. 2.58 *a, c*) is a function of many parameters:

$$\bar{x}_2 = f(G, F_{in}, D_c; d_1, d_2, ...).$$

The effect of some of them will be investigated.

Figure 2.59 shows the dependence of $\bar{x}_2$ on F_{in} and G = 4 g/s, the constant ratios $\bar{d} = d_2/d_1$, D_{c_2}/d_2 and subsonic velocities of supply of the gas into the vortex chamber of a single-chamber plasma torch. In accordance with the previous considerations, the increase of F_{in} reduces the length of the first zone. If $\bar{d} \ll 1$, the first zone almost does not form and, in addition to this, the gas flow in the electric arc chamber is highly unstable resulting in strong pulsations of the electrical and gas-dynamic parameters.

The indirect effect on $\bar{x}_2$ of the variation of static pressure at the end of the output electrode in throttling of the channel is shown in Fig. 2.60. The values along the abscissa are the total pressure (not the static pressure) which can be easily measured in the experiments in the preliminary chamber in front of the twisting ring. The graphs, shown in Figs. 2.59 and 2.60, clearly indicate the reduction of the depth of penetration of the first circulation zone with a decrease of the rate of supply of the gas into the vortex cham-

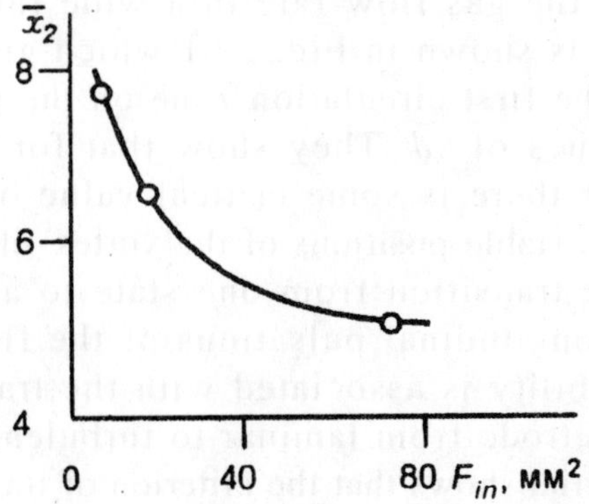

Fig. 2.59. Dependence of function $\bar{x}_2$ on the area of the input orifices of the vortex chamber.

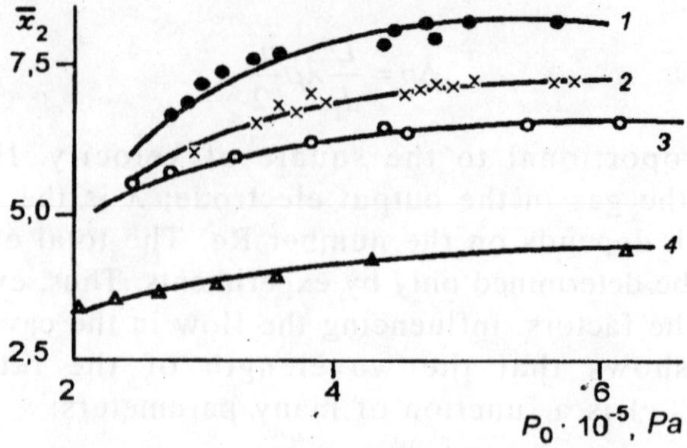

Fig. 2.60. Effect of the pressure on $\bar{x}_2$ at different values of $\bar{d}$. 1 – 4) corresponds to $\bar{d}$ = 1; 1.24; 1.33; 2.

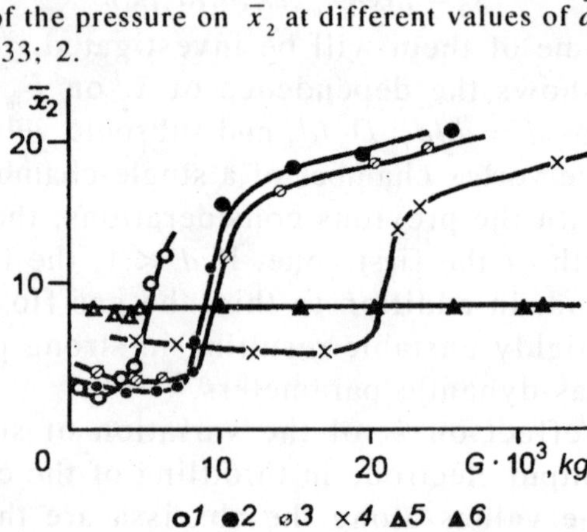

Fig. 2.61. Effect of the gas flow rate on $\bar{d}$ at different values of $\bar{d}$.
$\bar{d}$: *1* – 2) (10 mm/5 mm); 2 – 2) (20 mm/10 mm); *3*) 1.5 (1.5 mm/10 mm); *4*) 1.33 (20 mm/15 mm); *5*) 1(10 mm/10 mm); *6*) 1(20 mm/20 mm);

ber by some method.

The effect of the gas flow rate in a wide range of variation on the parameter $\bar{x}_2$ is shown in Fig. 2.61 which gives the dependence of the depth of the first circulation zone on the flow rate of the gas for different values of $\bar{d}$. They show that for every value of the relative diameter there is some critical value of the gas flow rate G_{cr} separating two stable positions of the vortex filament with different levels of $\bar{x}_2$. The transition from one state to another is accompanied by strong longitudinal pulsations of the flow. It has been assumed that instability is associated with the transition of the flow in the output electrode from laminar to turbulent. Processing of the experimental material shows that the criterion of transition from unstable position to another may be represented by the product $\mathrm{Re}_{d_1} \cdot \bar{d}$, where

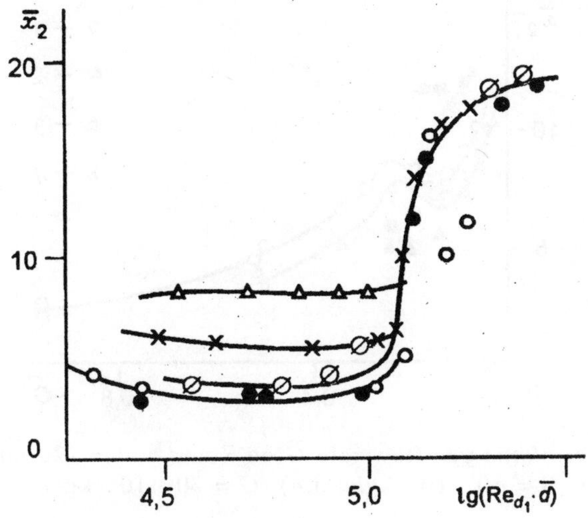

Fig. 2.62. Generalisation of measurements of $\bar{x}_2$ in respect of the complex $\mathrm{Re}_{d1}\cdot\bar{d}$. for designation see Fig.2.61.

the number $\mathrm{Re}_{d_1}\bar{d}=\frac{\rho u d_1}{\mu}$ is determined from the axial component of the velocity of the flow in the output electrode. As indicated by Fig. 2.62, at $\mathrm{Re}_{d_1}=1.2\cdot10^5$ the instability is maximum and the transition from one level of the first circulation zone $\bar{x}_2$ to the other takes place. It may also be mentioned that for each of the stable zone, the parameter $\bar{x}_2$ depends only slightly on the gas flow rate and is only a function of $\bar{d}$.

Attention will now be given to some of the quantitative results obtained in 'cold' blowing of a two-chamber plasma torch. The length of the first circulation zone in the plasma torch can be varied by changing the ratio of the gas flow rate through the vortex chamber. Consequently, this made it possible to regulate the position of the radial section of the arc and of the arc spot in the internal electrode situated in the zero velocity zone (the vortex filament of the first zone).

Figure 2.63 shows the dependence of $\bar{x}_2$ on $\bar{G}$ for two values of the total gas flow rate, differing by a factor of 3. For the selected range of variation of the flow rate it was found that its absolute value has only a slight effect on the length of the first circulation zone. The increase of the ratio $\bar{G}$ decreases the value of $\bar{x}_2$. The investigated curve contains a characteristic section, reflecting the strong instability of the vortex filament ($\bar{G}\approx0.1\div0.3$). In Fig. 2.63, it is indicated by the experimental points outside the curves. They can be used to estimate the amplitude and direction of ejection (one-

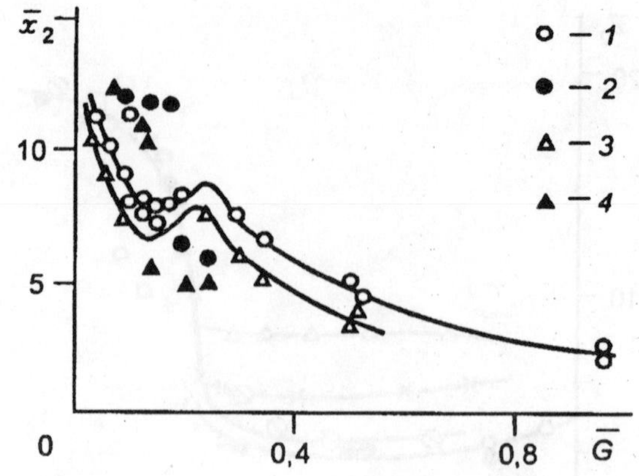

Fig. 2.63. Effect of the gas flow rate $\bar{G}$ on $\bar{x}_2$. ($\bar{d}$ = 1 = (20 mm/ 20 mm). F_{in}= 36.8 mm²). *1,2*) $G = 10 \cdot 10^{-3}$ kg/s; *3,4*) $G = 30 \cdot 10^{-3}$ kg/s;

sided) in relation to some stable (minimum or maximum) position. The ratio $\bar{G}$ = 0.2 ÷ 0.3 is critical, and if this value is exceeded, the position of the filament is stable up to complete disappearance of the vortex zone.

Hot blowing, carried out on actual structures, shows that the movement of the arc spot in the internal electrode of the single- or two-chamber plasma torches corresponds to the observed pattern of the flow in the case of cold blowing. In the presence of low current, the position of the arc spot is determined by the vortex filament. The longitudinal width of the eroded strip of the metal does not exceed 2–3 mm. In this case, the distribution of the heat flow from the arc spot on the surface of the electrode is relatively uniform and the temperature field in the electrode may be approximately calculated. In the case of relatively high current, the radial section of the arc in the single-chamber plasma torch may jump into the second zone under the effect of the difference in the ponderomotive and aerodynamic forces. The mean velocity of the axial flow in the zone, determined on the basis of the photographs, is order of 5 m/s, of the same order of magnitude as the circumferential component of velocity. Therefore, the heat flow through the at spot may be regarded as localised resulting in rapid failure of the electrode. The conditions in which the radial section of the arc is either closed with the end of the electrode or shunting takes place, as mentioned previously, may also occur.

Since the transition of the arc through the buffer zone is determined, generally speaking, not only by the effect of the electrodynamic forces but also by possible pulsations of the flow, then in this meaning the

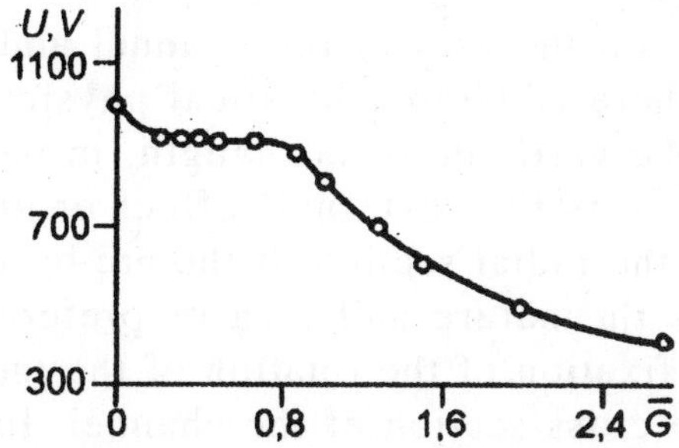

Fig. 2.64. Effect of the redistribution of the gas flow rate on arc voltage in the two-chamber plasma torch ($\bar{d}$ = 1 (20 mm/ 20 mm); I = 100 A; $G = 16 \cdot 10^{-3}$ kg/s).

arcing conditions in the single-chamber plasma torch may be regarded as less stable in comparison with those in the two-chamber plasma torch. In the latter, the results of hot tests confirmed the critical value of the ratio of the flow rates $\bar{G} = 0.2 \div 0.3$. The arc spot pulsates in the range $(2 \div 3)d_2$. The variation of the nature of gas flow in the cavity of the end electrode influences arc voltage (Fig. 2.64). However, it should be mentioned that, regardless of the large variation of the length of the first circulation zone with the increase of $\bar{G}$ from 0.2 to 0.8, the arc voltage may be regarded as almost constant. This is associated with the rearrangement of the flow both from the cavity of the end output electrode. In particular, this explains the completely satisfactory general form of the volt–ampere characteristic of the arcs running in the two-chamber plasma torches at different values of $\bar{G}$. At $\bar{G} > 1$, when the first circulation zone disappears and the arc spot moves only on the inlet edge of the end electrode, the voltage starts to drop rapidly as a result of many reasons, with one of the reasons being the earlier shunting of the arc in the output electrode.

These results show that the process of movement of the arc in the internal electrode is determined by the aerodynamics of the flow which is sufficiently close for both cold and hot conditions. In regulation of the position of the arc spot in the axial direction by varying the ratio of the flow rates $\bar{G} = G_2/G_1$ at almost constant arc voltage, there are considerable potential possibilities of increasing greatly the lifetime of the end electrode.

2.9. AERODYNAMICS OF THE CYLINDRICAL OUTPUT ELECTRODE WITH SUDDEN EXPANSION

In plasma generators of different schemes with gas-vortex stabilisation of the arc column on the axis of the channel and 'fixation' of the mean arc length there is another identical physical process leading to restriction of the variation of arc length, in particular shunting. The only difference is the 'external' effect on the arc, for example, in sustaining the radial section of the arc by the magnetic field which determines the nature and area of preferential shunting or a relatively strict fixation of the rotation of the radial section of the arc in some given cross-section of the channel. In many designs of plasma generators, the mean arc length is fixed using cylindrical output electrodes with sudden expansion of the efficient cross-section of the channel. When explaining the physical reasons leading to the constant conditions in the space of the shunting area of the arc in a wide range of the variation of the controlling parameters, such as arc current and gas flow rate, it is necessary to take into account primarily the gas-dynamic special features of the gas flow. In the presence of a ledge in the channel the flow always separates behind the cross-section of sudden expansion with the formation of the zone of the recirculation flow. Any detachment zone represents a powerful source of turbulence increasing the intensity of turbulent pulsations and supporting equalisation of the field of temperature, concentration, velocity and other parameters.

What are the mechanisms of interaction between the main flow and the detachment zone, the nature of gas flow inside the zone and its interaction with the main flow, the distribution of the heat transfer coefficients along the wall of the pipe behind the ledge, which determines the heated losses?

The following brief review is based on the experiments relating to turbulent flow and heat exchange behind the ledge both in a flat channel and in a circular pipe.

As shown by a large number of investigations of flat and axisymmetric non-twisted flows, the area immediately behind the ledge is characterised by the formation of a closed zone of recirculation flow, with the diagram shown in Fig. 2.65*a*. There are three characteristic regions in the zone: two stationary vortex regions I and II, and the nonstationary region III. Supplying the gas to the cavern from the main flow takes place mainly in the external boundary of region III; the outflow from the cavern (approximately the same mass) takes place through the boundary of region I, which is in contact with the external flow. Turbulence forms mainly along the same boundary. Turbulent pulsations are transferred by the averaged-out flow along the current lines, gradually attenuating and diffusing to different sides from them. Thus, in the direction downwards along the flow from

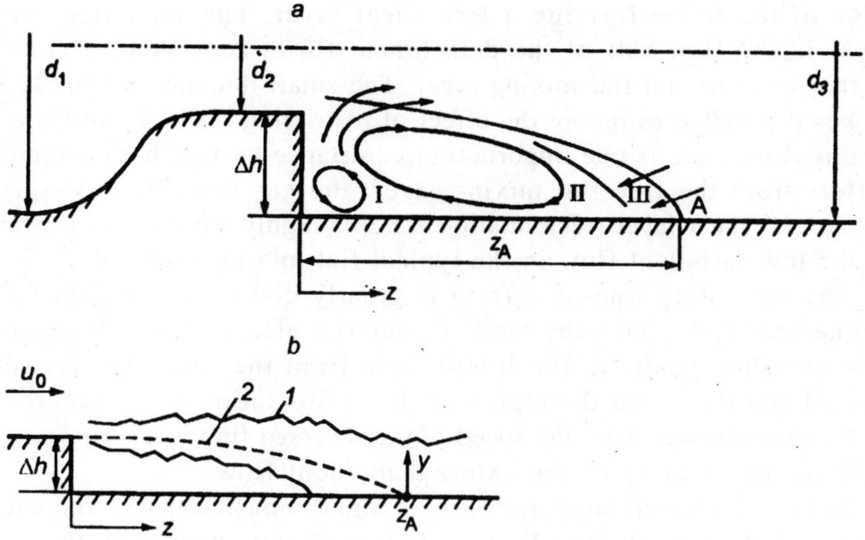

Fig. 2.65. Aerodynamics of the gas flow in a pipe behind the ledge. *1*) the boundary of the shear layer; *2*) separating current line.

the ledge the transverse transfer of the amount of motion and heat takes place from the ledge between the adjacent jets. The point of closure of the cavern *A* (its coordinate is z_A) is unstable.

The flow behind the ledge is one of the simplest detachment flows, determined by the marked variation of the geometry of the solid. However, regardless of apparent simplicity, long-term history of their examination and extensive use in engineering practice, the calculations of connecting shear layers in a wide range of variation of the parameters of the flow has not as yet been completely explained. The reason for this situation is the general state of the theory of turbulent detachment flows, and also the fact that despite the vast number of experimental investigations, there are a very small number of systematic data on the effect of parameters characterising the connected flows.

Below, we present some of the results indicating the complicated nature of this type of flow which may be used only as an orientation point in the search for the optimum design and control of the processes of transferring turbulent detachment flows. Special attention is given to flat flows.

Main special features of the flow behind the ledge

The structure of the flow field behind the ledge is relatively complicated (Fig. 2.65*b*). The arriving boundary layer separates from the sharp edge of the ledge forming a free shear layer. The separated shear layer in the first half of the detachment flows zone is very similar to the conventional flat mixing layer. The small thickness of the layer makes it possible to ignore the effect of restricting walls. Nevertheless, in this case, there is one important circumstance by which the situation differs from the free flat mixing layer: the gas from the low-speed side of the shear layer (recirculation zone) is highly turbulent in contrast to the low-turbulent flow in the typical flat mixing layer.

The separating line of current is greatly distorted in front of the connected flows onto the wall. Under the effect of a strong positive pressure gradient, the liquid flow from the shear layer is deflected and travels to the region of the recirculation flow. According to the experimental data, the speed of the reversed flow is approximately 20% of the velocity of the external incident flow.

In the attachment zone, the flow is highly nonstationary. The shear layer develops in the conditions of strong interaction with the wall under the stabilising effect of the curvature of the current lines and the positive pressure gradient.

Behind the attachment zone, a new sublayer of the boundary layer starts to grow in the connected shear layer. Measurements taken by different authors show that the external part of the connected shear layer retains the characteristics of the free layer at the distance of the order of 50 heights of the ledge Δh down along the flow behind the attachment point, i.e. large-scale vortices, formed in the separated free shear layer, are retained.

It is important to know the extreme difficulties in the measurement of the flow characteristics behind the ledge. This is caused by high turbulence of the flow, and also by frequent changes in the direction of movement of the liquid, especially in one of the most important regions of the flow, the attachment region.

The length of the region of the recirculation flow is one of the most important parameters of the investigated flows. According to the data obtained by different authors, the values of the length change from 4.9 to 8.2 heights of the ledge. Analysing these investigations, the authors of [45] defined:

a. The effect of the state of the boundary layer (turbulent or laminar) in detachment). The data obtained by these authors indicate the strong effect of the state on the length of the region of recirculation flow (Fig. 2.66). It is justified to assume that the flow does not depend on the Reynolds number, when the boundary layer becomes completely turbulent;

b. The experimental results for the effect of the thickness of

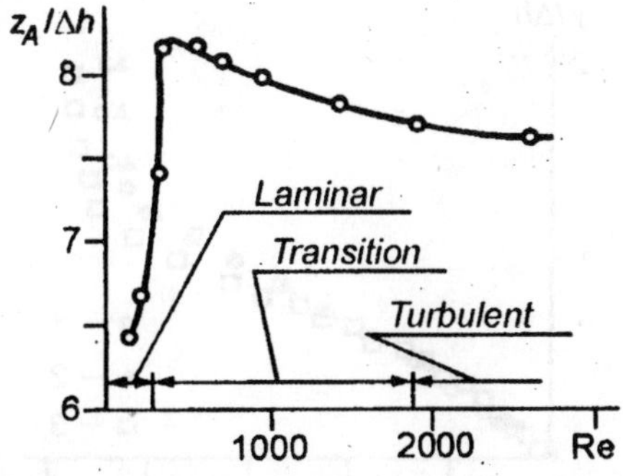

Fig. 2.66. Dependence of the relative length of the region of perpendicular flow on the state of the separating boundary layer. The number Re was calculated in respect of the thickness of the pulse loss.

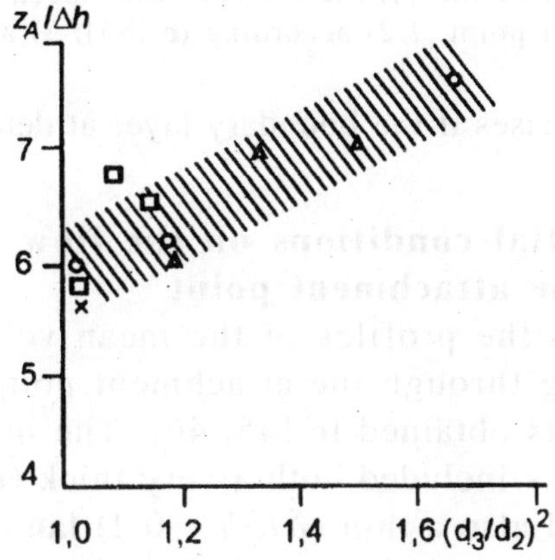

Fig. 2.67. Dependence of $z_A/\Delta h$ on the ratio of the areas at expansion of the channel.

the separated boundary layer δ_s to make it possible to draw unambiguous conclusions;

c. The currently available small number of data published by different authors relate to the case of a completely turbulent detached boundary layer (Fig. 2.67). They can be used to draw a conclusion on the linear dependence of quantity $z_A/\Delta h$ on the ratio of the areas in the expansion of the channel and increase of the length of the region of the recirculation flow with increase of this ratio;

d. Systematic investigations of the effect of the profile pressure gradient in the attachment zone have not been carried out;

e. Investigations of the ratio of the width of the channel to the height of the ledge (clogging up of the channel) show that at the values of the ratio greater than 10, the effect of the ratio may be ignored. In the case of less extensive clogging, the length of the region of recirculation flow increases, if the boundary layer at detachment

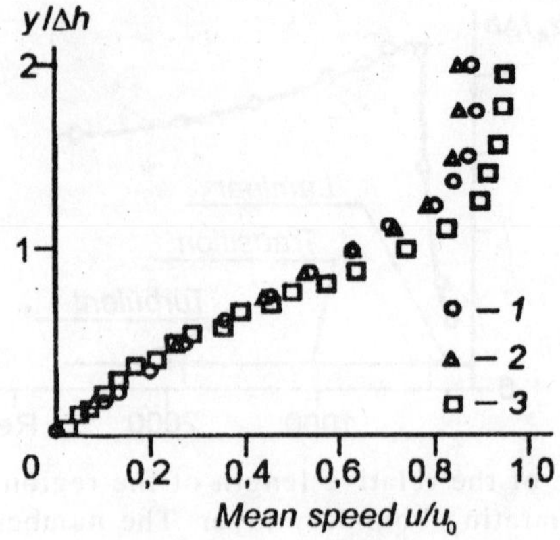

Fig. 2.68. Comparison of the profiles of the flow speed in the section passing through the attachment point. *1,2*) according to [45]; *3*) according to [46].

is laminar, and decreases if the boundary layer at detachment is turbulent.

The effect of initial conditions on the flow characteristics in the vicinity of the attachment point

Figure 2.68 shows the profiles of the mean velocity of the gas in the section passing through the attachment point, according to the experimental results obtained in [45, 46]. The initial conditions and in these experiments included both a very thick ($\delta_s/\Delta h$ = 2) turbulent boundary layer and also a thin ($\delta_s/\Delta h \approx 0.2$) laminar boundary layer. The length of the region of recirculation flow varied from 5 to 7.9 of the height of the ledge. Examination shows a good agreement of the data obtained in different experiments. Also, regardless of the large difference in the initial conditions, there is agreement in the profiles of the turbulent tangential stress. Thus, the mentioned experimental investigations show that the effect of the initial conditions weakens in the attachment zone of the shear layer.

The authors of [47] and [48] published the results of investigations of the effect of the geometry of the ledge on the structure of the turbulent detachment flow. Figure 2.69 shows the current lines in the flow around a step with different angles of inclination α [47] at Re = 47 000. Here, the Reynolds number was calculated from the maximum value of the velocity in the inlet channel and is height h at $(h + \Delta h)/\Delta h)$ = 1.48. The relative coordinate of the attachment point $z_A/\Delta h$ as a function of the angle of inclination is shown in Fig. 2.70, which indicates that $z_A/\Delta h$ decreases very slowly with a decrease of the angle of inclination α in the range from 90 to 25°. For the angles $\alpha \leq 15°$ and the Reynolds numbers Re > 33000 the

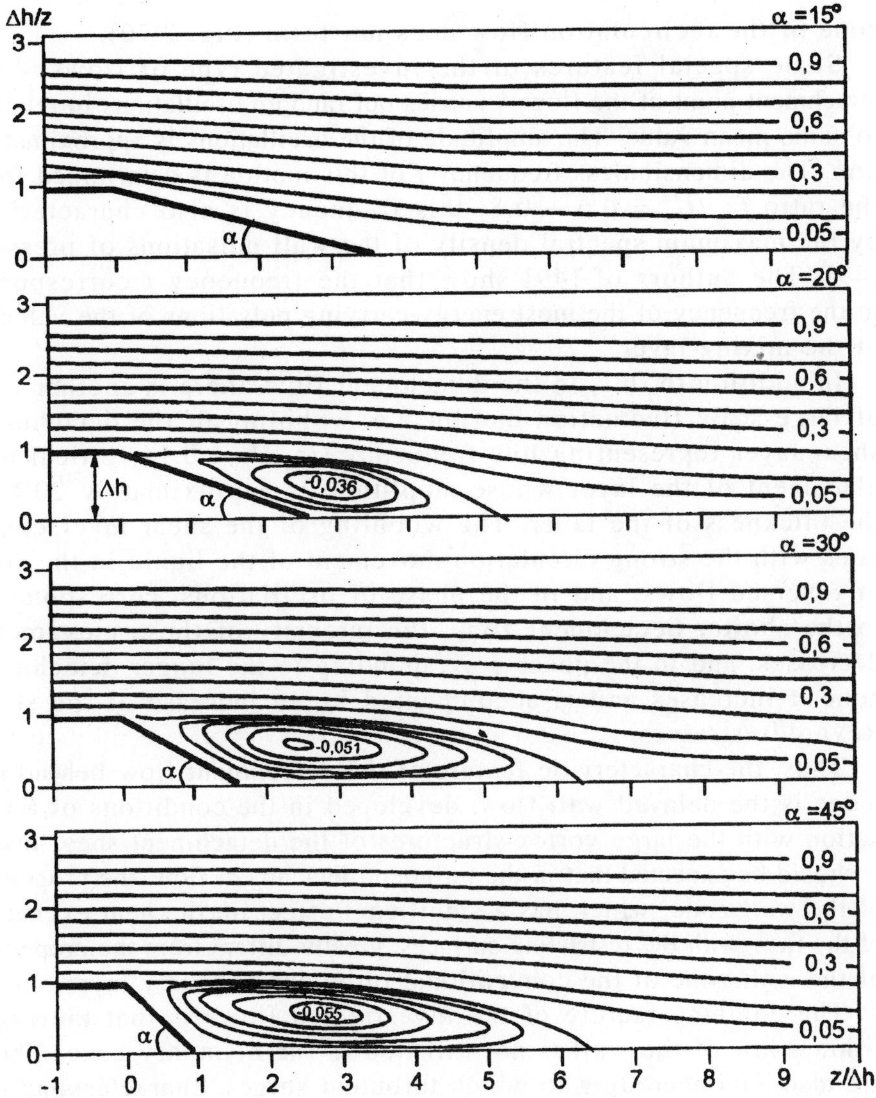

Fig. 2.69. Current lines in flow around a step with different angles of inclination.

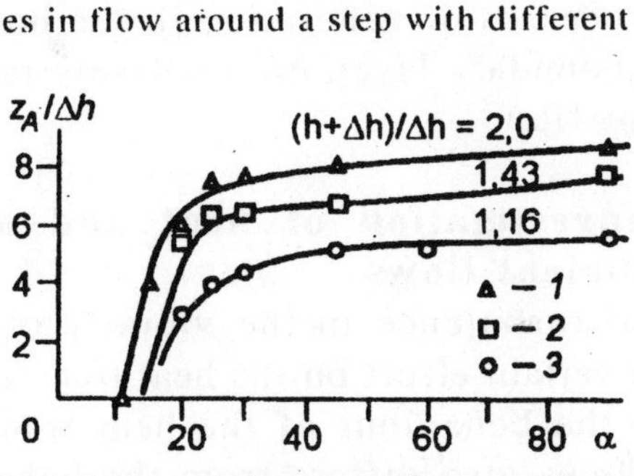

Fig. 2.70. Coordinates of the attachment point as a function of angle α. *1, 2*) according to the data of authors of [47]; *3*) according to [48].

zone of the recirculation flow does not form (Fig. 2.70).

Some special features of the investigated type of flows – the attachment point of the flow *A* carries out random oscillations in relation to some mean value. The amplitude of the oscillations is approximately $2\Delta h$. The dimensionless frequency f of this motion is determined from the ratio $fz_A / U_0 = 0.6 \div 0.8$; this frequency is also characterised by the maximum spectral density of the wall pulsations of pressure [49]. The authors of [49] show that the frequency f corresponds to the frequency of the most energy-carrying pulsations of the velocity of the mixing layer.

In addition to this, in the detachment flow, there is another type of large-scale fluctuation motion, i.e. wobbling of the detachment shear layer representing low-frequency ($fz_A/U_0 < 0.1$) vertical displacement of the layer whose amplitude is approximately 20% of the thickness of the layer. The wobbling of the shear layer correlates with the strong circulation movement of the liquid in the zone of reversed flows, and in the phase of oscillations, corresponding to the shorter detachment zone, the intensity of these movements decreases, and in the phase corresponding to the longer detachment zone it increases and is accompanied by an increase of the shear Reynolds stresses.

Thus, the characteristic feature of the detachment flow behind the ledge is the delayed wall flow, developed in the conditions of interaction with the large vortex structures of the detachment shear layer. It should be expected that in the turbulent detachment flow, the properties of the wall zone, which has a significant effect on the heat exchange of the flow with the restricting surfaces, greatly differs from the properties of the wall zone of the conventional turbulent boundary layer.

The common feature of the investigated flows is that the logarithmic law of the wall is not fulfilled in the boundary zone. There are also data according to which turbulent surges, characterising the activity of the processes of generation in the boundary zone of the normal turbulent boundary layer, are relatively rare in the wall zone of the detachment flow.

Experimental investigation of heat and mass transfer in detachment turbulent flows

Since the state of turbulence in the vicinity of the heat-transferring surface has a certain effect on the heat transfer process, it should be expected that the behaviour of the heat transfer coefficient in the detachment flows also differs from the behaviour of the identical quantity in the conventional turbulent boundary layer. It should be mentioned that in the turbulent boundary layer on a flat plate and in a circular pipe the laws of heat transfer for fluids with moderate Reynolds numbers have the following form:

$$Nu_z = 0.029 Re_z^{0.8} Pr^{0.4}, \quad Nu_d = 0.021 Re_d^{0.8} Pr^{0.43},$$

where $Nu_z = \alpha_w z / \lambda$, $Nu_d = \alpha_d / \lambda$; $Re = u_0 z / \nu$, $Re_d = u_m d / \nu$; α and ν is the heat conductivity and kinematic viscosity of the liquid; w is the coefficient of all heat transfer; d is the diameter of the pipe; z is the distance from the leading edge of the plate; u_0, u_m is the velocity of the incident flow and the mean consumption speed, respectively.

On the basis of a large number of experimental investigations, the following relationship was found for the heat transfer coefficient in the detachment turbulent flows [50]:

$$Nu = C\,Re^{2/3},$$

where constant C depends on the thermophysical properties of the heat carrier (Prandtl number Pr), the geometrical configuration of the flow, the state of the incident flow and the selection of the characteristic scale of the length and velocity in the criteria Nu and Re.

This law of the degree 2/3 was proposed in several studies [51, 52] in the processing of eperimental data on the heat and mass transfer in the detachment zone of the cylinder in the airflow at different values of the overloading coefficient. Subsequently, a large number of systematic investigations of heat transfer in detachment zones of solids with poor flow-around were carried out. In a generalised study [53] the authors analyzed the data of 44 literature sources containing the results of measurements of heat and mass transfer in detachment flows. The results were used to propose a correlation for the maximum heat transfer coefficient in the attachment point:

$$Nu = 0.19 Re^{2/3} Pr^{1/3},$$

where $Nu = \alpha_{w.max} z_A / \lambda$; $Re = u_0 z_A / \nu$; z_A is the distance from the detachment point to the attachment point (the length of the recirculation zone). This dependence is also valid for the flows with a fixed detachment point.

Thus, the dependence of the heat transfer coefficient on the Reynolds number in the detachment turbulent flows has a different form in comparison with the conventional turbulent boundary layer.

The data will also be presented obtained in the experimental examination of the length of the recirculation and heat exchange zone [54] in the discharge of high-temperature gas into a suddenly

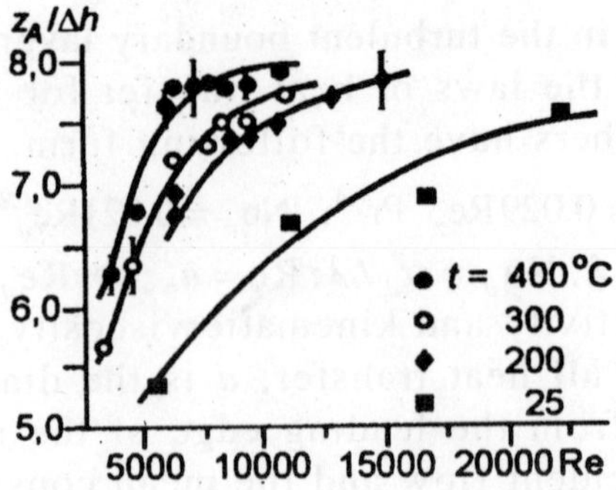

Fig.2.71. Relative length of the zone of recirculation as a function of the Reynolds number.

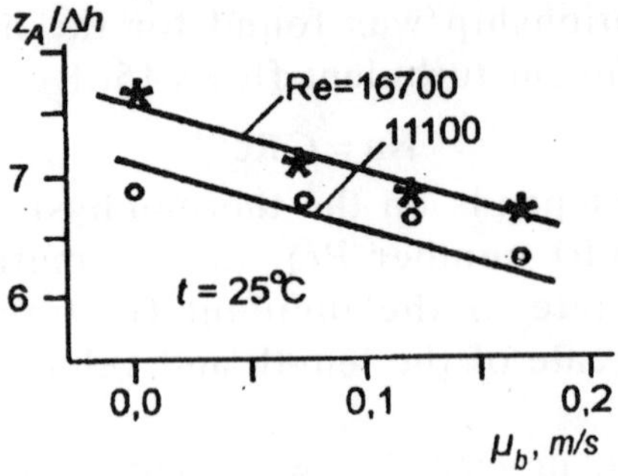

Fig. 2.72. Effect of the speed of blowing the gas through the porous wall on the length of the recirculation zone.

expanding channel with homogeneous blowing in of cold air through a porous insert in the wall of the channel immediately after expansion, i.e. behind the edge of the step. The length of the porous insert was $20\Delta h$. The results obtained in [54] show that the main factors, determining the size of the recirculation and heat transfer zone, are the Reynolds number of the incident flow and the inlet temperature (the Reynolds number was calculated from the velocity of the incident flow of the height of the flat inlet channel). Figure 2.71 shows the effect of the Reynolds number on the relative length of the recirculation zone $z_A/\Delta h$ for the incident flows with different temperature. For the high-temperature flows, the size of the zone rapidly increases with increasing Reynolds number to the value Re ~7000. Subsequently, the change of the recirculation zone becomes insignificant. At a constant mass flow rate of the blown air, the incident flow with higher temperature generates a longer recirculation zone. The volume consumption of the gas blown through the porous insert is 250 l/min. The length of the recirculation flow in the case of a cold flow is considerably smaller than for a hot flow, for the

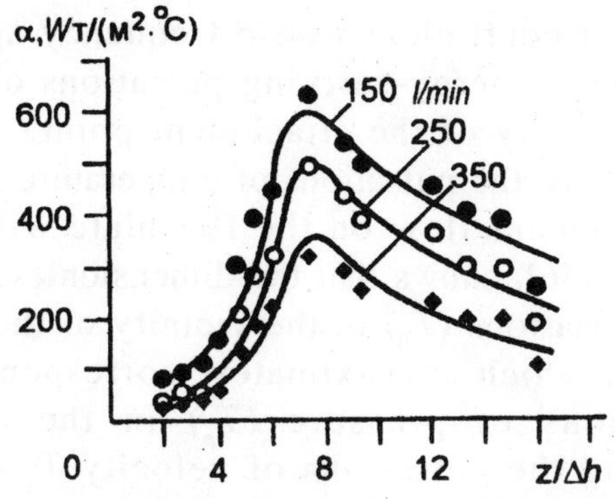

Fig. 2.73. Distribution of the local coefficient of heat transfer behind the edge of a step for different intensifies of blowing the cold gas through the porous walls.

same values of the Reynolds number.

The effect of the speed of blowing gas through the porous insert on the size of the recirculation zone is shown in Fig. 2.72. Here, the volume flow rate of the gas is equal to 0; 150; 250 and 350 l/min. When the flow rate of the gas is increased, the decrease of $z_A/\Delta h$ for the cold incident flow is large (in the figure, this corresponds to t = 25°C). The identical situation is recorded in the case of high-temperature flows but the rate of decrease of the size of the recirculation zone is greater at relatively small Reynolds numbers, and at Re>8600, this effect is again insignificant. Finally, Fig. 2.73 shows the effect of intensity of blowing in the gas through the pores wall on the heat transfer coefficient behind the edge of the step.

In analysis of the process of heat and mass transfer in turbulent detachment flows, it is important to take into account the non-stationary nature of the process determined by the nonstationary nature of the turbulent flow in the recirculation zone. Special features of the behaviour of the instantaneous coefficient of heat transfer in the vicinity of the attachment point of the detachment flow behind the ledge were investigated in [55] in the conditions of a constant heat flow on the wall. A special sensor was used to determine the instantaneous position of the attachment point which, as found later, oscillates around the mean value $\bar{z} = z_A$ with the mean quadratic deviation of the coordinate $\sigma_x = 1.0\Delta h$. The time sweep of the instantaneous coefficient of heat transfer, presented in the above studies, confirms the presence of large-scale quasi-periodic pulsations of Nusselt number with the characteristic period T, comparable with the characteristic period of wobbling of the detachment zone.

On the background of large-scale pulsations, there are fine peaks of the heat transfer coefficient whose frequency approximately corresponds to the most energy-carrying pulsations of the velocity and pressure in the vicinity of the attachment point.

Spectral analysis of the pulsations of temperature in the recirculation zone of the detachment flow on the flat plate with a blunted front end, carried out in [56], shows that the dimensionless integral timescale of temperature pulsations (T_θ) in the vicinity of the attachment point is $u_0 T_\theta / z_A \approx 0.24$, which approximately corresponds to the integral scale of pulsations of pressure (T_p) in the attachment point $u_0 T_p / z_A \approx 0.22$ and the pulsations of velocity T_u in the detachment mixing layer $u_0 T_u / z_A \approx 0.18$ [57]. These data indicate that the heat and mass transfer in the detachment flow is fully determined by the condition of the turbulent pulsations of the velocity field in the vicinity of the wall.

The review of the properties of the turbulent detachment flows demonstrates the complicated nature of the problem of modelling hydrodynamics and heat and mass exchange in these flows. The main difficulty is the principal difference of the characteristics of wall turbulence in the detachment flows in relation to the characteristics of wall turbulence in the conventional turbulent boundary layer (TBL), in particular, the absence of similarity in respect of the dynamic velocity and the non-fulfillment of the logarithmic law of the wall. The latter circumstance greatly complicates numerical modelling, because the assumption on the validity of the logarithmic profile of the mean velocity in the vicinity of the wall

$$u_+ = k^{-1} \ln y_+ + B,$$

where k = 0.4, B = 5.0, would enable us to avoid calculating the flow up to the wall viscous sublayer whose thickness is at least several orders of magnitude smaller than the characteristic scale of the external flow and is comparable with the local Kolmogorov length scale.

It should be mentioned that the wall zone in the detachment flows plays a more passive role in the hydrodynamic pattern of the flow in comparison with the wall zone of the conventional TBL, because it is subjected to the effect of large turbulence structures of the detachment shear layer, whereas in the conventional TBL, the wall region is the region of generation of its turbulence. Exact consideration of the special features of the turbulent flow in the walls of in calculation of the large-scale structure of the flow is not always compulsory, especially in cases in which it is necessary to calculate only the integral characteristics of the mean velocity field, such as the distance to the attachment point, etc.

The starting point for the processing and generalisation of the experimental data, and also for finding methods of controlling the transfer processes in turbulent detachment flows may be the physical model developed in [58, 59], and the number of main asymptotic laws for the averaged-out and statistical flow characteristics, determining the basis of the model. This theory is based on the representation of the flow in the walls zone as the flow subjected to intensive instantaneous accelerations, induced by large vortex flow structures.

Attention will now be given to the very interesting and extensive experimental material on the local heat transfer along a pipe behind a ledge at high gas temperatures (argon) presented in [60]. The investigations show that, in this case also, the general structure of the gas flow differs only slightly from the flow at moderate temperatures. We present several main parameters of the investigated channel (Fig. 2.74) and the flow: d_2 = 19 mm, d_3 = 49.5 mm, $l = \bar{l}_3/d_3$ is the length of the pipe in length gages equal to 9.7 and 3.1, the Mach number M = 0.11 at entry into the expanding channel. Investigations were carried out into the flows both with and without twisting. The mean input enthalpy varied in the range from 5560 to 18 400 kJ/kg, static pressure from $0.11 \cdot 10^5$ to $0.3 \cdot 10^5$ Pa, and the Re number, determined from the diameter of the pipe and the viscosity of the gas at entry into the pipe, from 210 to 450. STOPPED

The main results of the experiments (Fig. 2.74) show that the specific heat flow q into the wall of the channel initially increases, reaching the maximum value at some distance from the ledge (in the vicinity of z_A) and subsequently decreases downwards along the flow. The large inflow of heat at the end of the recirculation zone is largely determined by the small thickness of the boundary layer in the vicinity of the attachment point, and also by a high temperature (enthalpy) gradient in the direction, normal to the surface. In the section of increase of the specific heat flow there is an increase of pressure. This is in agreement with the generally accepted fact of the increase of static pressure in the equalisation of the field of velocities in a cylindrical mixing chamber. Comparing the distribution curves of q and p_3/p_2 (Fig. 2.74 *a, b*), it may be concluded that mixing is almost completed in the zone of closure of the cavern on the wall. With a decrease of the Reynolds number the attachment point is slightly displaced downwards along the flow. Attention should be given to the important result: the distribution of the ratio of total enthalpy in the attachment section to the total enthalpy at inlet into the pipe h_{0_3}/h_{0_2} indicates not only the high total heat

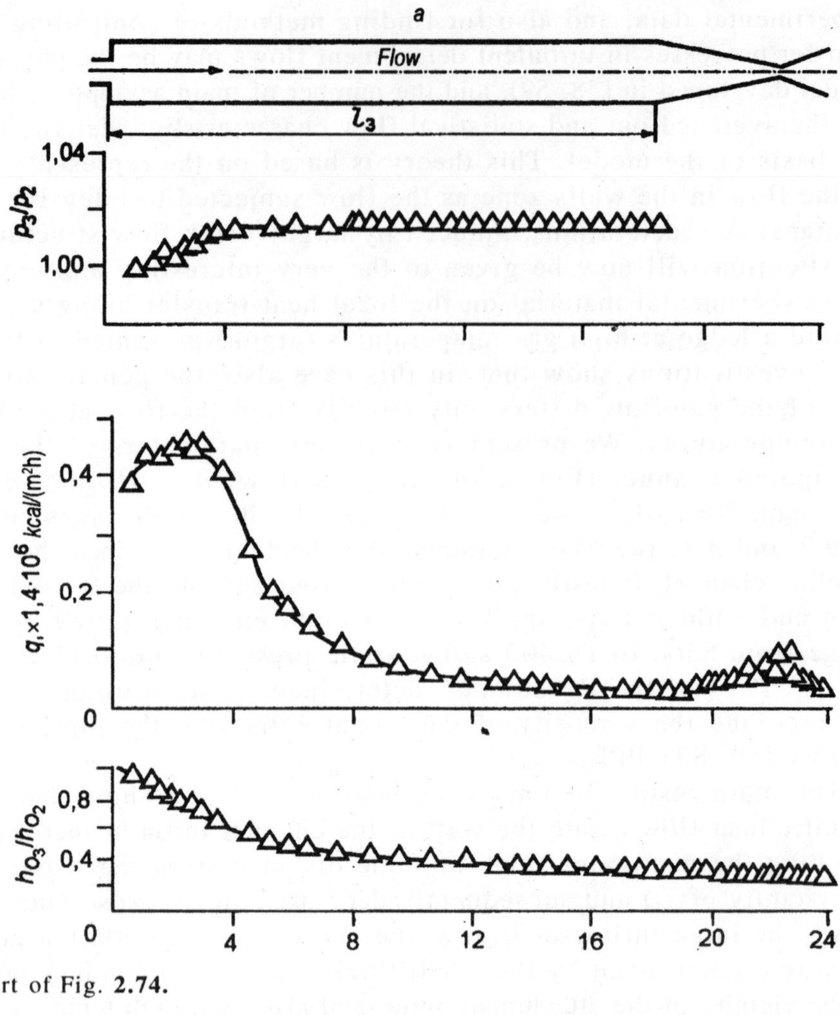

Start of Fig. 2.74.

losses behind the ledge but also high heat losses through the wall in the stalling zone. At the pipe length of 9.7 length gages, the energy losses in the pipe are approximately 80% of the energy at the output and, according to the authors of [60], convective heat exchange is the dominant process in the heat transfer process. The fraction of the cavern represents 20–30% of the losses of the initial thermal energy. Both results must be taken into account in the calculation and design of plasma torches with a stepped electrode.

In the case of the tangential (Fig. 2.74*b*) and radial (Fig. 2.74*a*) supply of the gas, the general form of the curves is the same. However, in the first case, the maximum of the heat flow is expressed more clearly (although on the basis of the value it differs only slightly from the maximum heat transfer in the supply of gas without twisting),

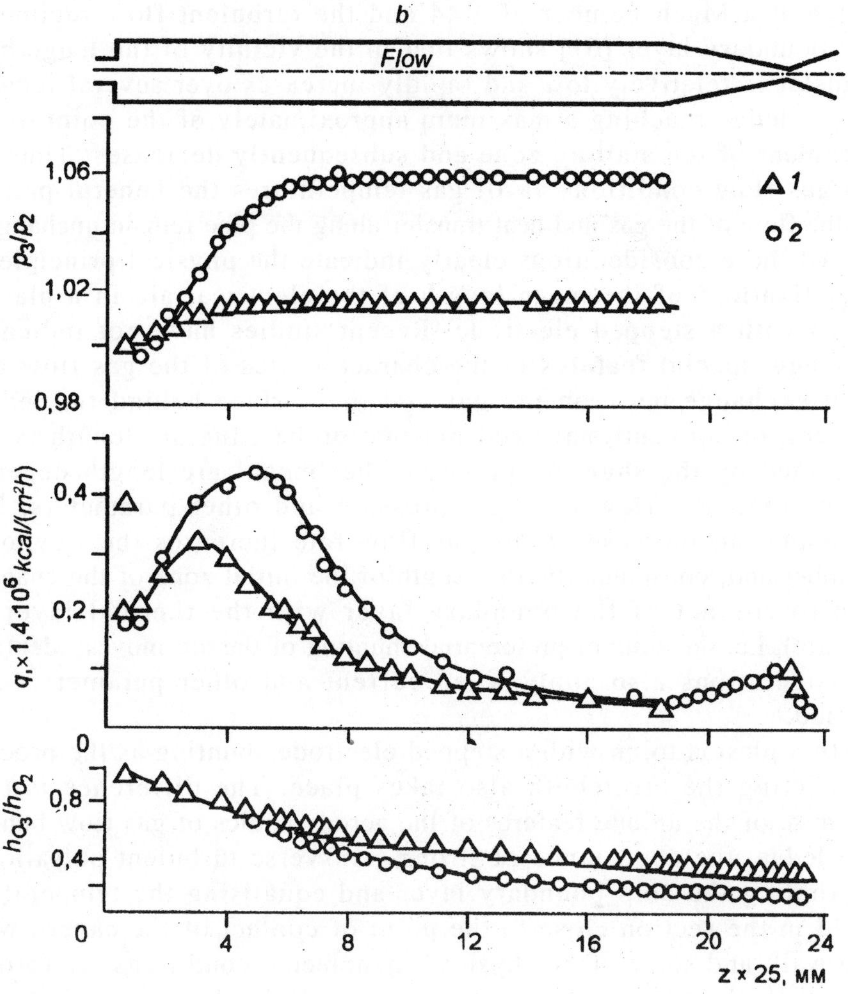

Fig. 2.74. Variation of different parameters of the flow along the length of the pipe and the nozzle (l_3/d_3 = 9.7). *a*) the radial supply of gas (φ = 0); h_{0_2} = 5480 kJ/kg, p_2 = 0.28 · 10^5 Pa, Re_{d_2} = 450; *b*) tangential supply of gas ($\varphi \neq 0$): *1*) h_{0_2} = 6080 kJ/kg, p_2 = 0.29 · 10^5 Pa, Re_{d_2} = 420; *2*) h_{0_2} = 1000 kJ/kg, p_2 = 0.17 · 10^5 Pa, Re_{d_2} = 330.

and the rate of restoration of pressure at the attachment point is higher.

For a short pipe ($\bar{l}_3$ = 3.1, $\omega \neq 0$, the tangential supply of the gas), the results are identical even if attachment takes place in the vicinity of the and other cylindrical section of the pipe.

These data, especially the data on the distribution of heat flows, are also valid when the gas velocity is supersonic (flat problem). The heat transfer coefficient α measured for this case behind the ledge at a Mach number of 2.44 and the turbulent flow regime in the boundary layer [61] shows that in the vicinity of the ledge heat transfer is relatively low and rapidly increases over several lengths of the ledge reaching a maximum approximately of the point of attachment of the stalling zone and subsequently decreases. Thus, in all gas flow conditions at all gas temperatures the general pattern of the flow of the gas and heat transfer along the pipe remain unchanged.

All these considerations clearly indicate the physical principle of the 'fixation' of the mean length of the electrical arc in a plasma torch with a stepped electrode. Recent studies have not indicated any new special features in the characteristics of the gas flow and heat exchange between the gas and the surface behind the ledge. In fact, in conventional axial plasma torches, the arc length is determined by the shunting process. The 'mean' arc length depends on current, gas flow rate, gas pressure and other parameters. For example, an increase of the gas flow rate increases the Reynolds number and, consequently, the length of the initial zone of the channel (up to contact of the boundary layer with the thermal layer of the arc), i.e. the zone of preferential shunting of the arc moves; identical considerations also apply when current and other parameters are varied.

In a plasma torch with a stepped electrode shunting as the process restricting the arc length also takes place. The difference is that because of the unique features of the aerodynamics of gas flow behind the ledge, the appearance of strong transverse turbulent pulsations, disrupting the cold boundary layer and equalising the temperature field in the section close to the point of contact of the cavern with the wall, and some other physical phenomena, conditions are formed in which the zone of the end of the cavern and the zone immediately behind it are the areas of preferential large scale arc–wall shunting in a wide range of the variation of the controlling parameters.

The results of a large number of experiments with the electrical arc have completely confirmed this. This may be illustrated by the profile pattern of the eroded surface of a copper outlet electrode–anode (Fig. 2.31), and recorded along the generating line of the cylinder. The anode operated for approximately 300 hours in air at a mean value of direct current of 650 A and at atmospheric pressure. It is clearly evident that as a result of random fluctuations of the flow and the arc, caused by different reasons, the shunting zone has a fully defined length with some statistical law of 'visits' of the arc spot to the surface of the electrode (in the experiment its maximum value was $(4 \div 5)\Delta h$). However, as clearly indicated by the profile diagram, the maximum erosion of material is recorded in the zone of the highest heat flows, i.e. in the section in which the equalisation of the field of velocity and temperature (end of the cavern) has been already completed. This is associated with the maximum frequency of large-scale arc–wall shunting in the given section.

Thus, the material presented in this section indicates that the 'fixation' of the mean arc length in the plasma torch with a stepped electrode is determined by the temperature field and hydrodynamic parameters of the flow.

The result of measurements of the distribution of the density of the heat flow along the wall of the channel behind the ledge, the data on the velocity field and intensity of turbulent pulsations in different sections of the channel, reports on the aerodynamics of the flow of the gas in the cavern and its linear dimensions and, finally, the profile diagram of the surface of the eroded output electrode which operated at high current intensities for a long period of time – all these are factors indicate the existence of a strong relationship between the nature of the gas flow in the channel and shunting, describes the physical principle of the process taking place and helps to produce plasma torches with a constant mean arc length in a wide range of the variation of the controlling parameters, such as at current, gas flow rate, gas pressure, etc.

The task requiring solution in the investigations of the plasma torches with a stepped electrode is the search for methods of reducing the heat losses in the electrode wall behind the ledge whilst maintaining the favourable conditions of arc shunting on the surface of the electrode at the end of the cavern. This makes it possible to increase greatly the thermal efficiency of the plasma torch.

Chapter 3

Mathematical methods of investigating arc discharges

The large number of applications of arc discharges are associated with the application of the arc column in which the main part of thermal and radiant energy is generated. At a high pressure (atmospheric and higher), the physical processes taking place in the arc column also determine the behaviour of the arc as a whole. Significant advances have been made in understanding the physical processes taking place in the electric arc plasma. This has resulted in the development of methods of mathematical modelling of the arc column [1–4].

To construct a theoretical model, it is necessary to solve in particular the problems of the equilibrium of plasma, the nature of transfer of radiation in plasma, the controlling mechanisms of transfer of energy and pulse, etc. Taking this into account, a mathematical model has been constructed based in a general case on a complicated nonlinear system of equations of radiation magnetic gas dynamics with the appropriate boundary-value conditions. To close the system, it is necessary to calculate or select the transfer coefficients and the thermodynamic quantities of plasma as a function of temperature and pressure.

At present, theoretical investigations of arc discharges take place in two main directions differing in the degree of detailisation of the processes, the examination methods and the accuracy of the results.

The first direction is based on the analytical methods which make it possible, with rational simplification of the problem, to obtain explicit and adequate relationships between the main parameters of the discharge. Although this approach cannot be applied efficiently for describing individual details of the processes in the electric discharges, the results are important both for determining the physical pattern of the plasma flows and for direct application in evaluating the parameters of the plasma torch.

Obtaining information in the area of electric arc plasma, the development of numerical method and greater possibilities of computing techniques have resulted in the development of a second direction– numerical or computer modelling. It can be used for a more detailed description of the processes in arc discharges, to take into account gas-dynamic field, electromagnetic forces, resulting in the pinch effect and acceleration of plasma, the transfer of radiation in plasma, nonequilibrium, and other effects.

Examination of processes taking place in the vicinity of the electrode took place independently to a certain degree, and the prospects for calculating arc discharges 'from electrode to electrode' appeared only in recent years. This problem can be solved using the experimental-theoretical approach based on the combined application of analytical and numerical models and, if necessary, the experimental data for formulation of the initial and boundary conditions. This makes it possible to close the problem and obtain the most complete and, in many cases, sufficient realistic description of the electric arc discharge, including the zone in the vicinity of the electrode.

3.1. MAIN EQUATIONS OF ELECTRIC ARC PLASMA

The high-pressure electric arc plasma is characterised by a complicated complex of mutually related gas-dynamic, thermal and electromagnetic processes. In a general case, it is described by a system of equations of radiation magnetic gas dynamics (MGD), including the laws of conservation of mass, pulse and energy, and the equations of electrodynamics and radiation transfer. It is assumed that the following hypotheses are fulfilled:

– continuity of the medium according to which any infinitely small volume of the medium is occupied by the matter;

– the ideal nature of the electric arc plasma.

Prior to writing the equations for electric arc plasma, we shall examine briefly other principal assumptions used in the description of the arc column, at the pressure of the order of atmospheric pressure.

Local thermodynamic equilibrium of the plasma

It is assumed that the plasma is in the state of complete thermodynamic equilibrium if: the velocity distribution of the particles is described by the Maxwell function, the population of the energy levels of the atoms and ions is described by the Boltzmann function, the spectral intensity of radiation is calculated using the Planck equation, and the composition of plasma using Saha's equation [1–3]. However, the complete thermodynamic equilibrium may be found only in the space of homogeneous, stationary, optically dense plasma when the collisional and radiation processes are equilibrium.

Real electric arc plasma is far away from the thermodynamically equilibrium plasma because of the presence in the former of the temperature and concentration gradients, separation of the temperature of different components of the plasma, and the processes of radiation transfer. It is described widely using the assumption on the local thermodynamic equilibrium (LTE) of the plasma. It is assumed that although the entire volume of plasma is not in the thermodynamic equilibrium, its individual microscopically small particles are in the equilibrium state. Consequently, it is possible to introduce, in the framework of the continuity of the medium, the concept of local equilibrium in small parts of the plasma system, characterising them by the local values of temperature, pressure, density and other thermodynamic parameters. It is also assumed that the temperatures of all particles are equal to the same value which is also the temperature of the plasma.

To fulfil the assumptions on the LTE, the frequency of collisions of the plasma components must be sufficiently high [5] to ensure that the Maxwell distribution is restored in the transition of the elementary volume from one region of the plasma to another. In this case:

– the electrons managed to transfer a large part of energy, received from the electrical field, to heavy particles;

– the ionisation processes are almost completely equalised by recombination;

– the large part of the excited atoms transfers its energy during collisions.

For the plasma in which the particle distribution greatly differs from Maxwell's distribution or the temperatures of the components of the plasma differ from each other, it is necessary to use the concept of partial local thermodynamic equilibrium (PLTE). This plasma is described using multi-temperature models, in particular the two-temperature model, when the temperatures of the electrons and heavy particles differ.

The volume nature of plasma radiation

The transfer of radiation in plasma is complicated and at a sufficiently high pressure of the working gas, the intensity of the discharge current and the geometrical dimensions of the plasma torch there may be extensive re-absorption of radiation [6]. To describe this transfer, it is necessary to use of the equations of transfer of radiant energy at the known dependences of the absorption coefficient on the frequency of radiation, temperature and plasma pressure. This greatly complicates the solution of the plasma equations because of inter-linking of the processes of transfer radiation and temperature and velocity fields [7].

Therefore, to describe the radiant losses of energy by the arc discharge at pressures of the order of atmospheric pressure, it is often necessary to use the assumption on the volume nature of plasma radiation. This makes it possible to simplify the mathematical formulation of the problem and its analysis, but the region of applicability of the model is restricted to a specific range of the plasma parameters.

3.1.1. The system of MGD equations

Taking into account the above considerations, the system of the MGD equations for describing the laminar flow of the equilibrium optically fine electric arc plasma maybe presented in the following form [1–4]:

–the continuity equation (conservation of mass):

$$\frac{\partial\rho}{\partial t}+\operatorname{div}(\rho\vec{V})=0; \tag{3.1}$$

–the equation of motion (conservation of pulse)

$$\begin{aligned}\rho\frac{\partial\vec{V}}{\partial t}+\rho(\vec{V}\operatorname{grad})\,\vec{V}&=(\rho-\rho_\infty)\vec{g}+\vec{E}\operatorname{div}\vec{D}+\vec{j}\times\vec{B}-\\&-\operatorname{grad}\left(p+\frac{2}{3}\mu\operatorname{div}\vec{V}\right)+2\operatorname{div}(\mu\dot{\vec{S}});\end{aligned} \tag{3.2}$$

–the energy equation

$$\begin{aligned}\rho\frac{\partial}{\partial t}\left(h+\frac{V^2}{2}\right)-\frac{\partial p}{\partial t}+\rho\vec{V}\operatorname{grad}\left(h+\frac{V^2}{2}\right)&=\vec{j}\cdot\vec{E}-\varphi+\\+\operatorname{div}\left(2\mu\vec{V}\dot{S}-\frac{2}{3}\mu\vec{V}\operatorname{div}\vec{V}+\frac{\lambda}{c_p}\operatorname{grad}h\right).\end{aligned} \tag{3.3}$$

The distribution of the external and intrinsic electromagnetic fields, generated by currents in the plasma, is described by Maxwell equations:

$$\text{rot}\vec{E} + \frac{\partial \vec{B}}{\partial t} = 0,\ \text{rot}\vec{H} = \vec{j} + \frac{\partial \vec{B}}{\partial t},\ \text{div}\vec{B} = 0,\ \text{div}\vec{D} = 0. \tag{3.4}$$

The system (3.1)–(3.4) is supplemented by the generalised Ohm law linking the density of current $\vec{j}$, the strength of the electrical field $\vec{E}$ and magnetic induction $\vec{B}$:

$$\vec{E} + \vec{V} \times \vec{B} = \frac{\vec{j}}{\sigma} + \frac{1}{en_e}(\vec{j} \times \vec{B} - \text{grad}\ p_e). \tag{3.5}$$

The equations are closed by the equation of state $p = R_0\rho T/M$, where M is molecular weight; R_0 is the gas constant. The coefficients of transfer and thermodynamic parameters, included in the equations, are the known functions of temperature T and pressure p.

When writing equations (3.1)–(3.5), the following notations were used: V – velocity, t – time, p – gas pressure, ρ – mass density, σ – electrical conductivity, λ – heat conductivity, μ – viscosity, φ – the volume density of radiation, h – specific enthalpy, c_p – specific heat capacity at constant pressure, g – free fall acceleration, e, n_e, p_e – the charge, concentration and partial pressure of the electrons, $\dot{S}$ – the tensor the strain rates with the components $S_{ik} = (\partial V_i/\partial x_k + \partial V_k/\partial x_i)/2$, where V_i, V_k are the components of the velocity vector.

Magnetic induction B and the strength of the magnetic field H, electric induction D and the strength of the electrical field E are linked by the relationships:

$$\vec{B} = \mu_0\vec{H};\ \vec{D} = \varepsilon_0\vec{E}.$$

These equations permit certain simplifications valid for the majority of the plasma processes in electric arc systems [1–4].

In the equation of motion (3.2), it is possible to select the Coulomb force, and also because $\rho \ll \rho_\infty$ the Archimedes force. Thus, for the plasma velocity, characteristic of arc generators, $V \approx 100$ m/s, the Archimedes number is Ar ~10^{-2}. However, it should be mentioned that in the case of low-current arcs, running in a free atmosphere, the Archimedes force must be taken into account because in this case the Archimedes force determines the pulse transfer.

In the energy equation (3.3) for arc plasma at a Mach number M < 0.3 the components, taking into account kinetic energy and its dissipation because of viscosity, are small [8].

The Ohm law (3.5) can be greatly simplified. Estimates show [1] that in this equation there are low values of the density of the current of the induced electrical field, the density of Hall current and the density of current, determined by the gradient of electronic pressure, in comparison with the density of current in the arc. Therefore, for the electric discharges, the Ohm law is usually used in the simplest form:

$$\vec{j} = \sigma \vec{E}. \tag{3.6}$$

Taking into account the above considerations, the system (3.1)–(3.5) for the stationary axisymmetric flow of plasma in the absence of the external magnetic field in the cylindrical coordinate system (r, z) is described by the equations:

$$\begin{gathered}
\frac{1}{r}\frac{\partial}{\partial r}(\rho v r) + \frac{\partial}{\partial z}(\rho u) = 0; \\
\rho v \frac{\partial (wr)}{\partial r} + \rho u \frac{\partial (wr)}{\partial z} = \frac{1}{r}\frac{\partial}{\partial r}\left[r\mu \left(\frac{\partial (wr)}{\partial r} - 2w \right) \right] + \frac{\partial}{\partial z}\left[\mu \frac{\partial (wr)}{\partial z} \right]; \\
\rho v \frac{\partial v}{\partial r} + \rho u \frac{\partial v}{\partial z} - \rho \frac{w^2}{r} = -\frac{\partial p}{\partial r} - j_2 B_\varphi + \frac{2}{r}\left(\mu r \frac{\partial v}{\partial r} \right) - \\
-\frac{2\mu v}{r^2} + \frac{\partial}{\partial z}\left[\mu \left(\frac{\partial u}{\partial r} + \frac{\partial v}{\partial z} \right) \right] - \frac{\partial}{\partial r}\left[\frac{2}{3}\mu \left(\frac{1}{r}\frac{\partial v r}{\partial r} + \frac{\partial u}{\partial z} \right) \right]; \\
\rho v \frac{\partial u}{\partial r} + \rho u \frac{\partial u}{\partial z} = -\frac{\partial p}{\partial z} - j_r B_\varphi + \frac{1}{r}\frac{\partial}{\partial r}\left[\mu r \left(\frac{\partial u}{\partial r} + \frac{\partial v}{\partial z} \right) \right] - \\
-\frac{\partial}{\partial z}\left[\frac{2}{3}\mu \left(\frac{1}{r}\frac{\partial v r}{\partial r} + \frac{\partial u}{\partial z} \right) + 2\frac{\partial}{\partial z}\left(u \frac{\partial u}{\partial z} \right) \right]; \\
\rho v \frac{\partial h}{\partial r} + \rho u \frac{\partial h}{\partial z} = j_r E_z + j_z E_r - \varphi + \frac{1}{r}\frac{\partial}{\partial r}\left(r\lambda \frac{\partial T}{\partial r} \right) + \frac{\partial}{\partial z}\left(\lambda \frac{\partial T}{\partial z} \right); \\
\frac{\partial E_r}{\partial z} - \frac{\partial E_z}{\partial r} = 0; \\
\frac{1}{r}\frac{\partial H_\varphi}{\partial r} = j_z; \quad -\frac{\partial H_\varphi}{\partial z} = j_r; \\
j_z = \sigma E_z; \quad j_r = \sigma E_r;
\end{gathered} \tag{3.7}$$

The boundary conditions have the form:

–the symmetry conditions:

$$\begin{gathered}
r = 0,\ z > 0: \quad v = 0, \frac{\partial u}{\partial r} = 0, \quad w = 0, \\
\frac{\partial T}{\partial r} = 0, \quad \frac{\partial E_z}{\partial r} = 0, \quad H_\varphi = 0;
\end{gathered} \tag{3.8}$$

–for the freely burning arc (the conditions of smooth contact with the surrounding medium):

$$r \to \infty, z > 0: \quad u \to 0, v \to 0, w \to 0, T \to T_\infty, p \to p_\infty; \tag{3.9}$$

–for the arc in the channel (the conditions in the input and output cross sections of the calculation area):

$$\begin{aligned} r = R, z > 0: &\quad u = 0, T = T_R, v = 0, p = p_R, w = 0; \\ z = 0: &\quad u = 0^0(r), v = v^0(r), p = p^0(r); \\ &\quad T = T^0(r), E_r = E_r^0(r), w = w^0(r); \\ z = L: &\quad u = u_1(r), h = h_1(r), w = w_1(r). \end{aligned}$$

The above equations are presented for the laminar flow of the electric arc plasma in the LTE state. At the same time, in many plasma systems, the plasma flow is turbulent and this may have a strong effect on all thermophysical, gas-dynamic and electrical parameters of the discharge. In addition to this, in the case of relatively low arc currents, the state of the plasma in the vicinity of the cold walls of the channel and electrodes may greatly differ from the equilibrium state. These problems, which require separate examination, will be discussed later.

3.1.2. Approximation of the MGD boundary layer

Further simplification of the MGD equations of the electric arc is associated with specific arcing conditions. In the case of relatively long arcs, running in a longitudinal flow or in a free atmosphere, the variation of the main parameters in the radial direction takes place at a considerably higher rate than in the axial direction. Consequently, we can transfer to the equations of the electric arc boundary layer which are derived and substantiated in [1–4]:

– the continuity equation

$$\frac{\partial}{\partial r}(\rho v r) + \frac{\partial}{\partial z}(\rho u r) = 0; \tag{3.10}$$

– the equation of motion

$$\rho u \frac{\partial u}{\partial z} + \rho v \frac{\partial u}{\partial r} = -\frac{\partial p}{\partial z} + \frac{1}{r}\frac{\partial}{\partial r}\left(r\mu \frac{\partial u}{\partial r} \right); \tag{3.11}$$

– the energy equation

$$\rho u c_p \frac{\partial T}{\partial z} + \rho v c_p \frac{\partial T}{\partial r} = \sigma E^2 - \varphi + \frac{1}{r}\frac{\partial}{\partial r}\left(r\lambda \frac{\partial T}{\partial r} \right). \tag{3.12}$$

the strength of the electrical field, the radial distribution of pressure and strength of the magnetic field in the arc column are determined by the relationships:

$$E = I/2\pi \int_0^\delta \sigma r dr; \tag{3.13}$$

$$p(r) = P_R + \mu_0 E \int_r^\delta \sigma H dr + \frac{\mu_0 H^2}{2}; \tag{3.14}$$

$$H(r) = \frac{E}{r} \int_0^r \sigma r dr. \tag{3.15}$$

The boundary conditions for the freely running arc without blowing any gas have the form:

$$\begin{aligned} &r = 0: v = 0, \frac{\partial u}{\partial r} = 0, \frac{\partial T}{\partial r} = 0; \\ &r = \delta: u = 0, T = T_\infty; \\ &z = 0: u = u^0(r), T = T^0(r). \end{aligned} \tag{3.16}$$

The side boundary $\delta = \delta(z)$ of the freely running arc on which the conditions of smooth contact with the surrounding medium are specified, is represented by the highest of the two coordinates δ_T, δ_u in the conditions:

$$\left.\frac{\partial T}{\partial r}\right|_{r=\delta_T} = 0, \quad \left.\frac{\partial u}{\partial r}\right|_{r=\delta_u} = 0.$$

3.1.3. Integral relationships

The equations of electric arc plasma in the differential form are relatively complicated for qualitative and quantitative analysis. Therefore, integral equations are used in many cases. These equations may be derived both on the basis of general laws of mechanics applicable to some volume of the plasma and the appropriate differential equations. The integral equations of continuity, motion and energy may be determined in the following form [1]:

$$\begin{aligned} &K = K_0 + \frac{\mu_0 I^2}{4\pi}\left(\ln\frac{\delta}{\delta^0} - \int_0^\delta \frac{I^2(r)}{I^2}\frac{dr}{r}\Big|_{z=0}^{z} \right), \\ &dG/dz = -2\pi\rho_\delta \delta v_\delta, \\ &dQ/dz = IE - F, \end{aligned} \tag{3.17}$$

where the enthalpy flow Q, the radiant energy flux F and the flow rate of the gas G in the arc are determined by the relationships:

$$Q = 2\pi \int_0^{\delta} \rho u(h - h_{\delta}) r dr,\ F = 2\pi, \int_0^{\delta} \varphi r dr,$$

$$G = 2\pi \int_0^{\delta} \rho u r dr. \tag{3.18}$$

The equations (3.17) and (3.18) are used for constructing various integral models of the arc column [1–4].

3.2. ANALYTICAL MODELS OF ARC DISCHARGE

The possibility of the analytical description of the electric arc plasma is determined mainly by the geometry of the arc column which, in turn, depends on the external conditions. In practice, the discharge is controlled on the basis of the external gas-dynamic and magnetic fields in which the form of the arc is a relatively complicated and may be spatially three-dimensional. At the same time, the conditions in which axisymmetric discharges form are encountered in many cases. In a number of cases, the arc column may be characterised by cylindrical symmetry.

3.2.1. The distribution of temperature in cylindrical arcs

Equations for the cylindrical arc

The stationary electrical arc, running in a cylindrical channel, is the simplest plasma object from the viewpoint of theoretical description. This is associated with the fact that in the case of a sufficiently long length of the channel, the latter is characterised by the formation of an axially homogeneus cylindrically symmetric arc column whose properties are not influenced by the electrodes. The radial distribution of temperature in such an arc is described by the equation of energy balance known as the Elenbaas–Heller equation:

$$-\frac{1}{r}\frac{d}{dr}\left(r\frac{dS}{dr}\right) = \sigma(S)E^2 - \varphi(S),\ S = \int_0^T \lambda(T)dT. \tag{3.19}$$

Here S is the potential of the heat flow, which is an unambiguous function of temperature; E is the strength of the electrical field which has only the axial component $E = E_z$ which is independent of the radial coordinate. The equation describes the steady process in which the Joule heat, generated in the arc, less the losses through radiation, is transferred to the walls of the channel by heat conductivity.

For analysis of equation (3.19), it is recommended to use the boundary condition written for the axis of the arc:

$$r=0: \quad S=S_0, dS/dr=0, \tag{3.20}$$

and the radius of the channel R or the strength E is determined by the additional condition $S(r = R) = S_R$, where S_R corresponds to the temperature of the channel walls T_R.

The total arc current is calculated from the Ohm law:

$$I = 2\pi E \int_0^R \sigma r dr. \tag{3.21}$$

The analytical solution of the equations (3.19)–(3.21) can be determined only by using simplifying assumptions associated primarily with the different approximation of the nonlinear plasma transfer coefficient.

The channel model

The channel model of the arc column has been used most extensively and developed efficiently. This model provides the simplest relationships between its parameters (see the review in [1]). The model is constructed on the basis of experimental investigations showing that in the case of relatively efficient cooling of the channel walls, the arc is constricted and occupies a relatively small region around the axis. Consequently, it is fully justified to assume that the main part of current also flows through this high-temperature plasma channel. The simplest model is the division of the arc column into the internal conducting channel with the radius r_* in which σ = const, and the external non-conducting channel, where $\sigma = 0$:

$$\sigma = \bar{\sigma}, 0 \le r \le r_*;$$

$$\sigma = 0, r_* \le 0 \le R.$$

Assuming that the entire radiation leaves the electrically conducting channel, equation (3.19) has the following solution

$$S(r) = S_0 - (S_0 - S_*) r^2 / r_*^2, 0 \le r \le r_*;$$

$$S(r) = S_* - (S_0 - S_*) \ln(r^2 / r_*^2), r_* \le r \le R,$$

describing the parabolic distribution $S(r)$ in the conducting region and the logarithmic decrease at the periphery.

The channel model gives the following relationships for determining the arc parameters:

$$r_*^2 = R^2 \exp(-\frac{S_*}{S_0 - S_*});$$

$$I = \pi r_*^2 \overline{\sigma} E;$$

$$4(S_0 - S_*) = r_*^2 (\overline{\sigma} E^2 - \overline{\varphi}).$$

The open form of the system of equations of the channel model required discussion and search for the additional relationship [9–13], starting with the principle of the Steenbeck minimum [14]. Analysis, carried out in [15] on the basis of the variational principle shows that, regardless of the method of determination, the additional relationships are reduced to the equations determining different approximations of the real dependences $\sigma(S)$ and $\varphi(S)$ by step functions. The comparison of the results of different channel models and the example of the arcs with atmospheric pressure in argon shows [1] that the model [12] gives the most suitable estimates of the arc column parameters.

The non-linear models

The nonlinear form of the plasma properties may be taken into account using exponential approximations of the dependences $\sigma(S)$ and $\varphi(S)$ [16–18]. For example, neglecting radiation in (3.19), and representing the dependence $\sigma(S)$ in the form [16]

$$\sigma(S) = (S/a)^{1/k},$$

gives the following relationships for the arc parameters:

$$E = \mu_1 a^{1/2k} S_0^{(k-1)/2k} / R, \quad I = 2\pi \lambda_1 R S_0^{(k+1)/2k} / \mu_1 a^{1/2k}.$$

Here μ_1 is the first root of the solution $s(x)$ of the dimensionless equation (3.19), $s = S/S_0$, $x = \mu_1 r/R$, and

$$\lambda_1 = \mu_1^2 \int_0^1 S^{1/k} x dx.$$

However, the models discussed previously do not provide the entire range of the solutions of (3.19) which are realised at the actual properties of the plasma and may differ qualitatively.

In the arcing conditions with the dominant heat conductivity, the profiles $T(r)$ are approximately parabolic and may contain 'excrescence', caused by the non-monotonic nature of heat conductivity $\lambda(T)$. A suitable example of this type of arc are arcs in nitrogen with atmospheric pressure at $T < 12\,000$ K [19].

The increase of plasma temperature in the energy balance increases the contribution of radiation and increases the strength of its effect on the form of $T(r)$. The primary information on the nature of the radial distribution of temperature follows from the differential equation of energy at the axis:

$$-2\frac{d^2S}{dr^2}\Big|_{r=0} = \sigma_0 E^2 - \varphi_0,$$

where $\sigma_0 = \sigma(S_0)$, $\varphi_0 = \varphi(S_0)$. It may be seen that the sign of the equation $\sigma_0 E^2 - \varphi_0$ determines the sign of the second derivative d^2S/dr^2, i.e. the sign of curvature of the profile in the immediate vicinity of the arc axis. If the intensity of Joule heating in the near-axial region is greater than the losses of energy of the radiation, temperature decreases with increase of the distance from the axis, and vice versa.

The assumption on the maximum temperature at the axis of the arc indicates that the strength of the electrical field should satisfy the condition:

$$E^2 > \varphi_0 / \sigma_0. \tag{3.22}$$

The form of the profile $T(r)$ of the optically thin arcs depends strongly on the nature of variation of the complex φ/σ with temperature [19]. If φ/σ increases with increasing T, the axial temperature of the arc for any given value of the strength E is restricted by the relationship (3.22). In this case, the distributions $T(r)$ are close to isothermal distributions with a wide centre, with approximately constant temperature. One of the examples of this type of arc is the air high-current high-pressure plasma. For gases with a decreasing temperature dependence of φ/σ, the value of T_0 for the given value of E is higher than in the case of the temperature at which $E^2 = \varphi/\sigma$. In this case, the profile $T(r)$ may consist of a narrow central core with a sharp decrease of temperature, i.e., the 'constricted' type of arc. A suitable example is a low current, low-pressure arc in the vapours of rare-earth elements.

Analysis of the variation of the components φ and σE^2 in the arc shows [7] that depending on the given values of E and T_0, equation (3.19) may have both a solution to $T(r)$ monotonically decreasing from the axis to the periphery and differing in the degree of filling of the profiles, and also diverging or oscillating solutions which do not satisfy the condition $r = R$: $T = T_R \ll T_0$ (Fig. 3.1). The latter appear because of the existence in the plasma of areas in which locally $\varphi > \sigma E^2$ and, generally speaking, they have no relationship with the description of the real arc.

Qualitative analysis

Qualitative analysis of equation (3.19) with the arbitrary dependences $\sigma(S)$ and $\varphi(S)$ makes it possible to reply to the question of the existence in the arc of both non-monotonic profiles $S(r)$ and $S(r = R) = 0$ [20], and also of the solutions with $S = 0$ at $r \to \infty$, describing a freely running cylindrical arc. In this case, for analysis we can use the functions

$$V(S,E) = \int_0^S (\sigma E^2 - \varphi)dS.$$

The characteristic dependences $V(S)$ for different values of the strength of the field E are shown in Fig. 3.2.

For the known dependence $V(S, E)$ and the given values of S_0 and E it is possible, in many cases, to determine (without calculations) the qualitative form of the distribution $S(r)$ for the cylindrical arc column. In the case of higher values $E > E_*$ (for example, for the argon at atmospheric pressure $E_* \approx 11.3$ V/cm), when $V(S, E)$ is only an increasing function of S (Fig. 3.2, curve 1), for any value on axis S_0 there are the monotonically decreasing solutions $S(r)$. Without taking into account radiation in equation (3.19) this holds for any $E > 0$. For $V(S, E)$ of type *3*, characteristic of the relatively low values of E, the solutions are realised at $S_0 < S_F$. If $S_0 = S_F$, then $S(r) = \text{const}$ (isothermal profile), and at $S_0 > S_F$, the potential of the heat flow S increases with increasing r (diverging solutions, not satisfying the condition $S(r=R)=0$). Analysis of the more complicated dependences $V(S, E)$, constructed on the phase plane $(S, dS/dr)$, makes it possible

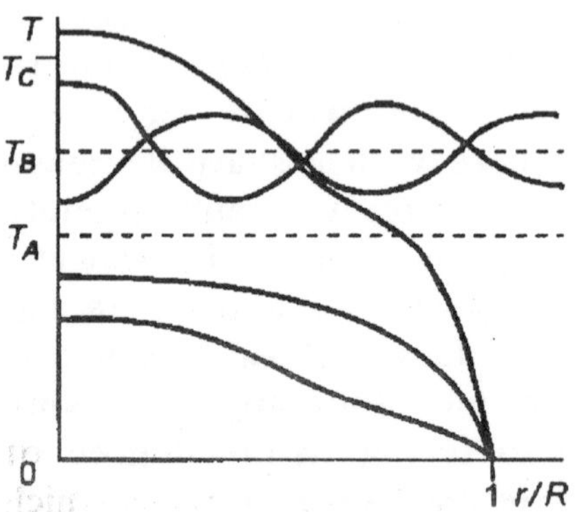

Fig. 3.1. The qualitative form of the temperature profiles of the cylindrical arc, corresponding to different regions of the dependence $\sigma E^2(T)$.

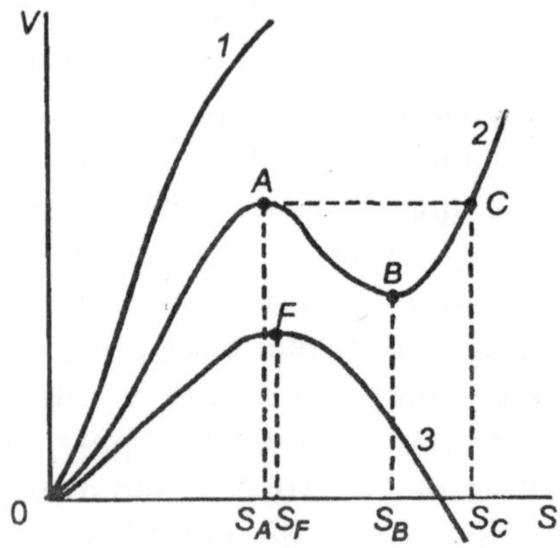

Fig. 3.2. Characteristic dependences of the $V(S)$ function for different values of E.

to conclude that in the case of the arbitrary dependences $\sigma(S)$ and $\varphi(S)$, the Elenbaas–Heller equation (3.19) does not have any non-monotonic solutions $S(r)$, satisfying the condition $S(r=R)=0$.

In addition to this, quantitative investigations have shown that at specific properties of the plasma (in particular, with increased volume radiation at low temperatures, where $dV/dS < 0$), equation (3.90) may have solutions satisfying the boundary condition for the freely running arc $S(r \to \infty) = 0$ (if this condition is fulfilled, the condition $dS/dr = 0$ is also valid). This case is clearly indicated by the example of exponential dependences

$$\sigma(S) = k_\sigma S^m,\ \varphi(S) = k_\varphi S^n, m > n.$$

In this case, equation (3.90) has analytical solutions, differing in the relationships between the exponents m and n:

a) $n = (1 + m)/2$, $m > n > 1$;

$$S(r) = S_0/(1 + ar^2)^{2/m-1}, \tag{3.23}$$

where

$$a = \frac{(1-m)^2(1+m)k_\varphi^2}{16k_\sigma E^2}, S_0 = \left[\frac{(1+m)k_\varphi}{2k_\sigma E^2}\right]^{2/(m-1)}.$$

Here the condition $S = d\,S/dr = 0$ is fulfilled at $r \to \infty$;

b) $n = 2m - 1$, $1/2 > m < 1$, $0 < n < 1$:

$$S(r) = S_0(1 - r^2/R^2)^{1/(1-m)}, \tag{3.24}$$

where

$$R = \frac{2\sqrt{k_\varphi}}{(1-m)\sqrt{mr}\,k_\sigma E^2}, S_0 = \left(\frac{k_\varphi}{mk_\sigma E^2}\right)^{1/(1-m)}.$$

In this case, the condition $S = dS/dr = 0$ is fulfilled at finite $r = R$. The qualitative form of the solutions of (3.23) and (3.24) is shown in Fig. 3.3 (the curves 1 and 2, respectively). These solutions describe the arc column, for which the entire amount of Joule heat is transferred by volume radiation. Since these arcs may burn also in the absence of walls, they may be referred to as arcs stabilised by radiation.

Ambiguity and stability of the solutions

In the case of sufficiently high arc currents, plasma radiation is so significant that in a certain part of the arc channel the generation of Joule heat and the losses of energy by radiation are

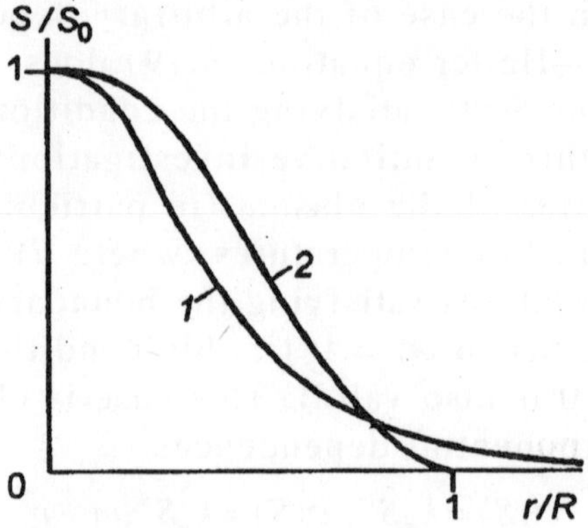

Fig. 3.3. The qualitative form of the solution of (3.23) and (3.24) at which the condition $dS/dr \to 0$ is fulfilled at $S \to 0$.

locally equalised. For this region of the are column, we have equation $\sigma E_*^2 = \varphi$, from which we obtain $E_* = \sqrt{\varphi/\sigma}$. The existence of the maximum of the temperature dependence of the function $\sqrt{\varphi/\sigma}$ and the associated complicated evolution of the form of the profile $T(r)$ with the variation of the axial temperature may result in the formation of special features of the volt–ampere characteristics of the arc [19].

Using the arc in hydrogen as an example, we examine a case in which the definition of current ambiguously determines the arcing conditions. For hydrogen, the function $\sqrt{\varphi/\sigma}$ initially increases with increasing temperature and subsequently decreases (Fig. 3.4*a*). This dependence may result in the formation of two stable arcing conditions: the first one – low temperature conditions on the increasing part of the VAC, stabilised by radiation, and the second one on the decreasing part, stabilised by heat conductivity.

Figure 3.4*b* shows the VAC of the hysteresis form obtained by solving the equations (3.19)–(3.21) [21]. It may be seen that in a specific current range there are three possible solutions with different values of E and T_0 (Fig. 3.4*c*). The effect of formation of loops on the E–I characteristics is stronger with increasing channel radius accompanied by a decrease of the significance of the relative role of heat conductivity, and with increasing gas pressure, increasing the losses through radiation.

The hysteresis form of the VAC leads to the formation of unstable arcing conditions. Examination shows that the profiles $T(r)$ *1* and *3* are stable, and the constricted profile *2* is unstable [22].

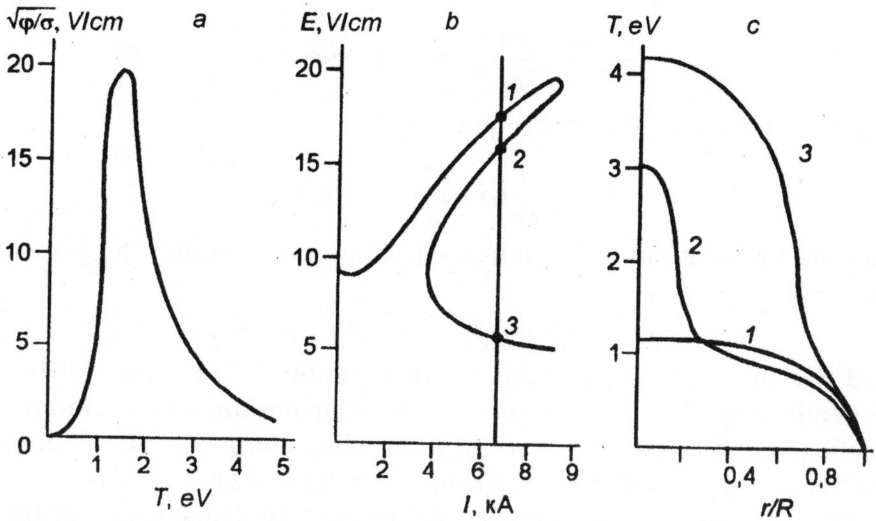

Fig. 3.4. The dependence $\sqrt{\varphi/\sigma}$ (T), $E(I)$, $T(r/R)$ for the hydrogen arc at atmospheric pressure (R = 2 cm).

Limiting characteristics

Assuming that the entire energy from the arc column is transferred by radiation and the temperature profile $T(r)$ is homogeneous, from the relationship $\sigma_0 E^2 = \varphi_0$ it is possible to determine the values of the strength E and current I in such an arc [23]:

$$E = \sqrt{\varphi_0/\sigma_0},\ I = \pi R^2 \sqrt{\sigma_0 \varphi_0}.$$

These formulae determine the limiting radiation E–I characteristics, restricting the range of the solutions of the system of equations (3.19) and (3.21).

The nonmonotonic temperature profiles

Since the actual values of thermal and electrical conductivity and volume radiation of plasma depend on gas pressure, the variation of gas pressure along the radius influences the radial variation of the profile $T(r)$ in the arc. Even small pressure gradients may result in a qualitative change of the profile.

In the case of pressures, slightly differing from pressure p_∞, it may be assumed that the heat conductivity and electrical conductivity of the plasma are independent of pressure, and the volume density of radiation changes as follows:

$$\varphi(S, p) = \varphi(S)(p/p_\infty)^2. \qquad (3.25)$$

The qualitative and numerical analyses of the energy equation

$$-\frac{1}{r}\frac{d}{dr}\left(r\frac{dS}{dr}\right)=\sigma(S)E^2-\varphi(S,p)$$

together with the Maxwell equation

$$J_z=\frac{1}{r}\frac{d}{dr}(rH_\varphi),\ r=0:\ H_\varphi=0,$$

and the momentum conservation equation, determining the pressure gradient

$$dp/dr=2\ \mu_o j_z H_\varphi,\quad r=R:\quad p=p_\infty,$$

and the Ohm law (3.21) taking into account (3.25), show [20] that depending on the arc parameters, we can obtain a large range of qualitatively differing solutions $S(r)$, including non-monotonic solutions, satisfying the boundary condition $S(r = R) = 0$ (Fig. 3.5).

For the non-monotonic profiles $S(r)$ of the type *2* and *3*, the condition $\sigma E^2<\varphi$ is satisfied in the near-axial region. With increase of the distance from the axis, the pressure in plasma decreases as a result of the pinch effect and, because of the dependence $\varphi(p)$, this results in the condition $\sigma E^2<\varphi$. The non-monotonic distributions $T(r)$ in the arc with the dependences of all transfer coefficients of the plasma onpressure taken into account, are presented in [24].

The dynamics of the arc stabilises by radiation

Analysis of the non-stationary energy equation

$$\frac{1}{\chi}\frac{\partial S}{\partial t}=\frac{1}{r}\frac{\partial}{\partial r}\left(r\frac{\partial}{\partial r}\right)+\frac{\sigma I^2(t)}{\left(2\pi\int_0^\infty \sigma r dr\right)^2}-\varphi \qquad (3.26)$$

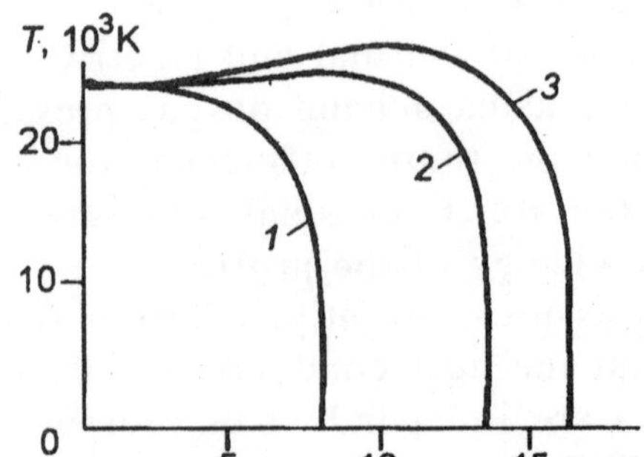

Fig. 3.5. Radial profiles of temperature $T(r)$ of the argon arc at p_∞ = 1 atm. *1*) I = 2.35 kA, p_0 = 1.03 atm; 2) I = 7.8 kA, p_0 = 1.1 atm; *3*) I = 11.3 kA, p_0 = 1.15 atm.

with the boundary conditions

$$r=0:\ \partial S(r,t)/\partial r=0;\ r\to\infty:\ S(r,t)\to 0$$

makes it possible to find solutions describing the dynamics of the arc stabilised by radiation [25]. In approximation of the plasma properties

$$\chi(S)=k_\chi S^n,\ \sigma(S)=k_\sigma S^{1-n},\ \varphi(S)=k_\varphi S^{1-2n},\ 0<n\le 1/2$$

equation (3.26) is reduced to the system of ordinary differential equations

$$\begin{aligned} &da/d\tau=a^{2n-1}x^{-4/n}i^2(\tau)-a^{1+n},\\ &\frac{2}{n(1-n)}\frac{dx_R}{d\tau}=a^n x_R-a^{-n}x_R^{-1}, \end{aligned} \tag{3.27}$$

which described, for the given current $i(\tau)$, the evolution of the dimensionless profile

$$y(x,\tau)=\begin{cases} a(\tau)[x_R^2(\tau)-x^2]^{1/n}, & 0\le x\le x_R(\tau),\\ 0, & x_R(\tau)\le x<\infty. \end{cases}$$

In the given relationships, the following notation is used:

$$y=S/S_*,\ x=r/R_*,\ x_R=R/R_*,\ \tau=t/t_*;$$

$$S_*^{2+n}=\frac{n^2k_\varphi I_m^2}{16\pi^2(1-n)k_\sigma},\ R_*=\frac{2}{n}\sqrt{\frac{1-n}{k_\varphi}}S_*^n,\ t_*=\frac{(1-n)S_*^n}{k_\varphi k_\chi}.$$

In a DC circuit (i = const), the system (3.27) describes the transfer of the arcing conditions to the stationary state with the parameters:

$$x_s=i^{2n/(2+n)},\ a_s=i^{-2/(2+n)},\ e_s=i^{-n/(2+n)},$$

where $e = E/E_*$, $E_* = 4\pi S_*/(nI_*)$. The static VAC of such an arc decreases. The stability of the stationary state depends on the nonlinearity parameter n and the type of electrical circuit in which the arc burns. In the simplest case, for a circuit consisting of the arc and a power source with the VAC of the type ie^α = const, $\alpha\ge 0$, there are three areas (Fig. 3.6) in which the behaviour of the arc differs and is determined by the type of the state of equilibrium of the system (3.27). The regions *1* and *2* correspond to the stable focus and the section in which the arc reaches the stationary regime regardless of the initial condition. In region *3* (the equilibrium point – saddle) the arc cannot show stable burning. For the stepped form of the VAC, the parameters of such an arc may show self-oscillations [25].

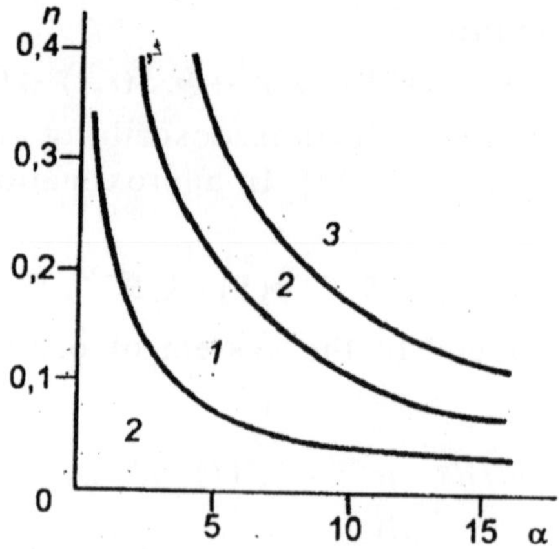

Fig. 3.6. Regions of different states of equilibrium of the dynamic system (3.27) at the VAC of the power source of the type $ie^{\alpha} = 1$. *1*) the region of the stable focus; *2*) the stable section; *3*) the saddle region.

3.2.2. The dynamics of the long arc in external fields

In most cases, the geometry and characteristics of the arc are controlled by the external gas-dynamic and magnetic field. If the direction of the effect of these fields does not coincide with the longitudinal axis of the discharge, the arc starts to move in the transverse direction. In certain conditions, the arc may assume a new steady position which differs from the initial non-perturbed position. The shape of the arc in the new condition depends on its parameters, the geometry of the external field, the method of arc stabilisation and others reasons, and maybe spatially three-dimensional. To determine the form of the arc in a general case, it is necessary to solve the system of three-dimensional magneto-hydrodynamic equations and this is associated with certain difficulties. At the same time, the considerations of the physical pattern of movement of the discharge, developed in [26, 27], make it possible to simplify greatly the solution of the problem.

According to these investigations, the movement of the arc is the displacement in its temperature field which takes place as a result of the composition of two speeds: gas-dynamic speed of movement of the plasma in the arc and the speed of displacement of the temperature field in relation to the gas (thermal wave). The latter type of movement is caused by sliding of the arc and is determined by the asymmetry of energy generation and the heat flow in the non-distorted arc. Analysis of these processes is based on the energy equation of the arc, and its form may be determined as the position of the characteristic isotherm in space. The solution of the problem for a steady arc, which is identical with the flat curve, was presented in [27]. Taking this approach into account, we present the model of the spatial–time dynamics of the arc in external fields [28].

For this purpose, we examine the energy equation:

$$\frac{1}{\chi}\left[\frac{\partial S}{\partial t}+(\vec{V}\cdot\vec{\nabla})S\right]=\Delta S+\sigma E^2-\varphi \tag{3.28}$$

and the Maxwell equation

$$\operatorname{rot}\vec{E}=-\partial\vec{B}/\partial t. \tag{3.29}$$

The form of the arc is given by the curve, representing the geometrical area of the points of the maximum plasma temperature. We select the orthogonal coordinate system linked with the arc and determined by three orthogonal unit vectors: normal $\vec{\nu}(l, t)$, tangential $\vec{\tau}(l, t)$ and binormal $\vec{\beta}(l, t)$ where l is the actual arc length (Fig. 3.7). Replacing the coordinates n, b directed along the vectors $\vec{\nu}$, $\vec{\beta}$, by the coordinates ρ, ω:

$$n=\rho\cos\theta,\ b=\rho\sin\theta,\ \theta=\omega-\int_0^s\kappa(s)ds,$$

the metrics may be written in the following form:

$$dl^2=d\rho^2+\rho^2d\omega^2+(1-k\rho\cos\theta)^2ds^2,$$

where $k(l, t)$ and k (l, t) is the curvature and twisting of the line.

At moderate speeds of movement of the arc, the moving tem-

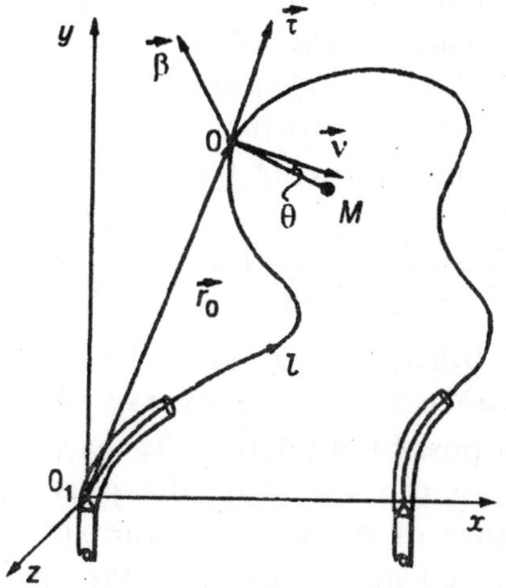

Fig. 3.7. The coordinate system for the model of the long arc.

perature field of the arc changes only slightly in the vicinity of maximum temperature. Consequently, on the basis of the approximate solution of equation (3.28), the following equation may be written for the region $T \approx T_{max}$:

$$S(\rho) = S_0 - a\rho^2 + O(\rho^4 / R^4), \tag{3.30}$$

where R is the characteristic transverse dimensions of the arc, and from equation (3.29) we obtain

$$E_1 = E_0 /(1 - k\rho\cos\theta) \approx E_0(1 + k\rho\cos\theta) + O(\rho^2 / R^2),$$
$$E_p = E_\omega = 0, \tag{3.31}$$

where E_0 is the strength of the electrical field on the line $T \approx T_{max}$.

Substituting (3.30) and (3.31) into (3.28), we obtain a system of equations for the components of the speed:

$$u_v(l,t) = -k(l,t)\chi_0\left(1 + \frac{4}{1 - \varphi_0 / \sigma_0 E_0^2}\right),$$
$$u_\beta(l,t) = 0. \tag{3.32}$$

These equations linked together of the relative speed of the gas u, which in the examined coordinate system is the speed of sliding of the maximum isotherm, of the instantaneous local curvature k of the arc and is parameters φ_0, σ_0, E_0 on the line of maximum temperature.

The application of the relationships of differential geometry for the local curvature of the curve and the directing cosines makes it possible to transfer to the differential form of writing equation (3.32), describing the spatial evolution of the form of the arc at the known speed of the plasma in the zone of its maximum temperature [28, 29]. In particular, for the plane geometry of the problem with the form of the curve defining the form $y = y\ (x,\ \tau)$, the dynamics equation has the following form:

$$\frac{\partial y}{\partial \tau} = \frac{\partial^2 y / \partial x^2}{1 + (\partial y / \partial x)^2} + u_y - u_x \frac{\partial y}{\partial x'}$$

where $\tau = At/d^2$, $u = Ud/A$, $A = \chi_0\ [1 + 4\ (1 - \varphi_0/\sigma_0 E^2)]$.

Analysis of the pattern of the transverse flow of the cold gas around the arc makes it possible to derive the approximate relationship $u_0 \approx u_\infty(\rho_\infty/\rho_0)^{1/2}$, linking the speed of the plasma u_0 in the zone of maximum temperature of the arc with the speed of the flow u_∞. For the arc in the external magnetic field, the speed of the plasma may be estimated from the condition of compensation of the ampere and viscous forces:

$$\vec{j}\times\vec{B}=\mu_0\Delta\vec{U},$$

where μ_0 with the viscosity of the plasma at $T = T_{max}$.

These equations can be used for of the analytical and numerical analysis of the dynamics of the form of the arc in gas-dynamic and magnetic fields of different geometry (Fig. 3.8) [28, 29].

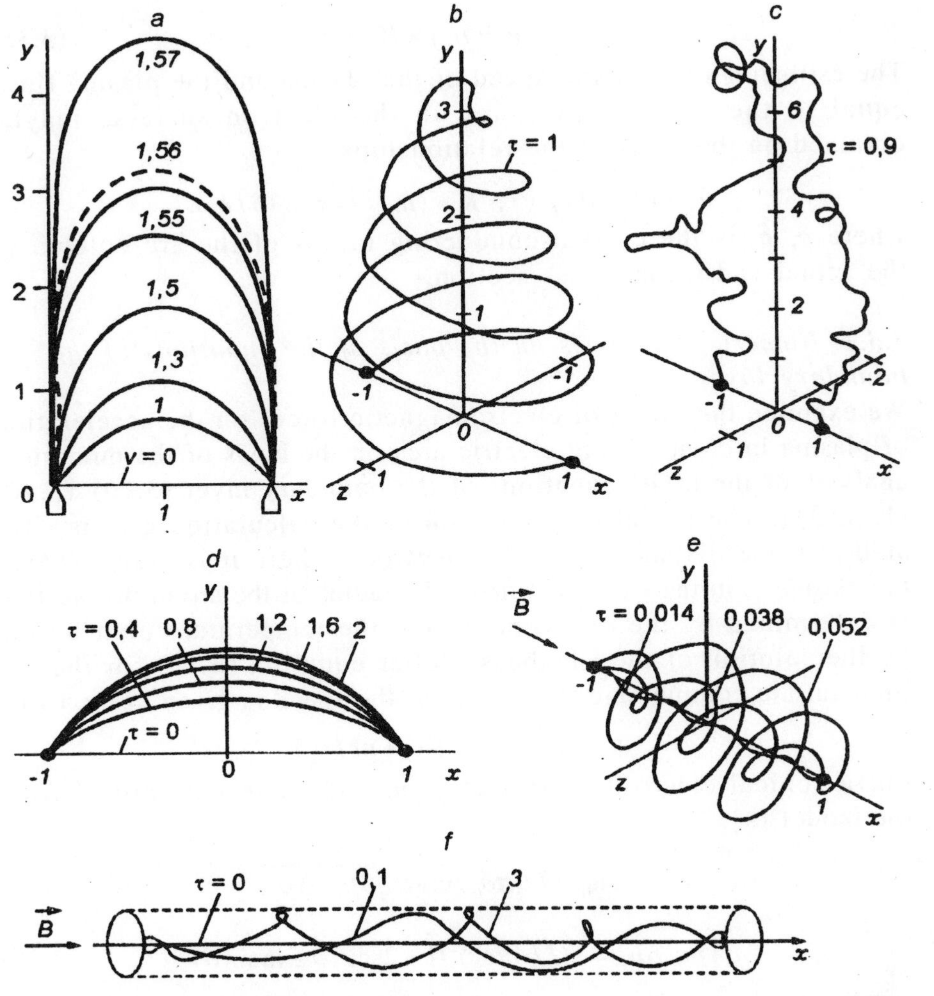

Fig. 3.8. The dynamics of long low-current arcs in the transverse (*a*), twisted (*b*), and pulsed (*c*) gas flows and in the transverse (*d*) and longitudinals (*e*, *f*) external magnetic fields.

3.3. EFFECT OF ELECTROMAGNETIC FORCES ON THE FORMATION OF PLASMA FLOWS IN ARCS

The experimental results [30] show that the plasma flows with the speed of several hundreds of metres, the second flow in the direction from the electoral the surface. The main reason for the formation of the flows in high current arcs are the electromagnetic (ampere) forces:

$$\vec{F} = \mu_0 \vec{j} \times \vec{H}. \quad (3.33)$$

The estimates of the axial speed of the plasma and the plasma flow, equal to the reactive pressure on the electron surface, maybe obtained on the basis of the relationships:

$$u_0 = (2\mu_0 jI / \rho)^{1/2}, \quad K = (\mu_0 I^2 / 4\pi)\ln(\delta / \delta_0),$$

where δ, δ_0 is the current-conducting radius of the arc column in the actual and initial cross-section.

3.3.1. Numerical analysis on the basis of the equations of the boundary layer

We examine the effect of electromagnetic forces on the acceleration of plasma in open-current electric arcs on the basis of the numerical analysis of the MGD equations of the boundary layer (3.10)–(3.16) [1, 4, 31]. The initial cross-section of the calculation grid is situated at some distance from the electrode where it is assumed that the plasma is in the LTE condition. The radius of the arc in this section is determined by the experiments and the temperature distribution by the solution of the Elenbaas–Heller equation (3.19) for the cylindrical arc column, and the profile of the speed is given in the form:

$$u = u_0(1 - r/\delta_0)^n (1 + nr/\delta_0).$$

For a conical electrode, parameters u_0 and n are determined from the model [4]:

$$u_0 = I/(\pi\delta_0)\,\mathrm{ctg}\theta\sqrt{5\mu_0/6\rho_0},$$

$$\int_0^{r_k} \rho u^2 r dr = -(I/4\pi\delta_0)^2(1 + 2\sec^2\theta - \ln\sin\theta). \quad (3.34)$$

For a flat electrode (n = 1), the axial value of the speed is calculated from the equation [4]

$$u_0^2 = \frac{\mu_0 I^2}{8\pi^2}\left[\ln\frac{\delta_0}{\delta_e} + \frac{1}{2} - \int_0^{\delta_0}\left(\frac{2\pi E}{r}\int_0^r \sigma r dr\right)^2 \frac{dr}{r}\right] / \int_0^{\delta_0} \rho\left(1 - \frac{r^2}{\delta^2}\right)^2 r dr,$$

in which on the basis of the experimental data the following values are selected for the electrode surface:

$$\delta_e = 0{,}065\,\sqrt{I},\ j_e = \text{const},\ \ u(r) = 0.$$

The calculated axial variations of the characteristics of the high-current arc in argon at atmospheric pressure at a current of I = 200 A, δ_0 = 1.5 mm, p_δ = 0.1 MPa, are shown in Fig. 3.9. It may be seen that they are in satisfactory agreement with the experimental data [32–35]. The axial speed for a conical electrode is almost twice the speed for the flat electrode and rapidly decreases with increase of the distance from the electrode. At the tip of the conical electrode, the degree of filling of the profile of the speed (3.34), corresponds to the value $n \cong 10$, i.e., the electromagnetic forces form a narrower near-axial high-intensity plasma flow in comparison with the flat (n = 1) or spherical electrodes.

To determine the main mechanisms of acceleration of the plasma by the electric arc, we examine the following numerical values:

- variant 1: the basic, the gas – argon, I = 200 A, δ_0 = 1.5 mm;
- variant 2: viscosity of argon reduced by a factor of 10;
- variant 3: viscosity of argon increased by a factor of 10;
- variant 4, electromagnetic forces are switched off ($\mu_0 \vec{j} \times \vec{H} = 0$).

Figure 3.10 shows that the main mechanism of acceleration of the plasma in the electric arc are the electromagnetic forces. In particular, these forces are considerable in the first calculation layers from the initial cross-section ($z/\delta_0 < 1$), characterised by high plasma temperatures of $T \approx 25\,000$ K, and consequently, low values of the density and viscosity of the gas. This region is characterised by the rapid expansion of the arc column and a decrease of the density of electric current.

The effect of the electromagnetic forces is reflected in:

– firstly, through the interaction of the radial component of the density of electrical current with the intrinsic magnetic field

$$F_{1z} = \mu_0 j_r H_\varphi = -\mu_0 H_\varphi \partial H_\varphi / \partial z =$$
$$= -\,\partial(\mu_0 H_\varphi^2 / 2)/\partial z = -\partial p_m / \partial z,$$

resulting in the nonuniformity of magnetic pressure $p_m = \mu_0 H_\varphi^2/2$; along the longitudinal axis;

– secondly, because of the nonuniform electromagnetic compression of the current-conducting arc column (pinch effect) as a result of the force

$$F_{2z} = -\partial p / \partial z = \partial(\mu_0 \int_r^\delta j_z H_\varphi dr)/\partial z.$$

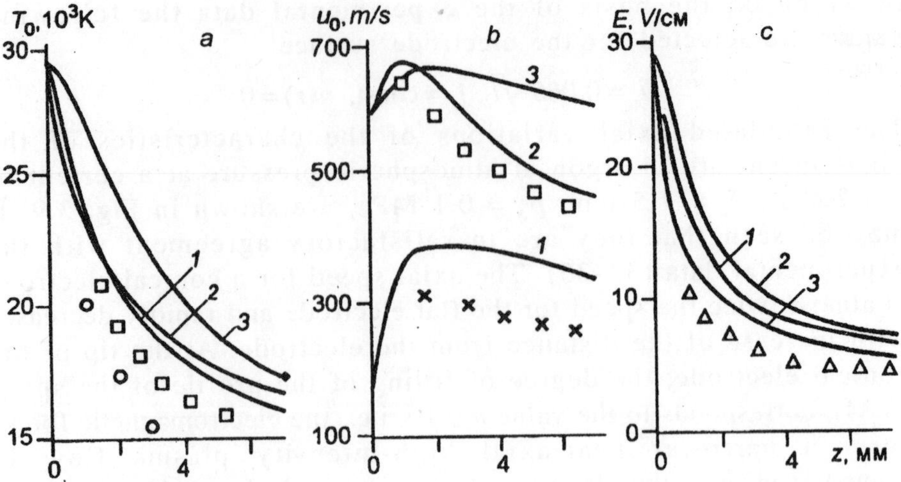

Fig. 3.9. The axial distributions of the characteristics of the flow of the argon plasma at I = 200 A, δ_0 = 1.5 mm. *1*) the cylindrical electrode, n = 1; *2*) the conical electrode with the angle at the tip of θ = 60°, n = 10; *3*) the cylindrical electrode, n = 10. The experiments carried out using the data: × [32], □[33], O[34], Δ[35].

The total moment of the plasma flow

$$K = 2\pi\int_0^{\delta}\rho u^2 r dr = K_0 + \mu_0\frac{I^2}{4\pi} + \left[\ln\frac{\delta}{\delta_0} + \int_0^{\delta}\frac{I^2(r)}{I^2}\frac{dr}{r}\Big|_0^z\right]$$

is determined only by the positive component of electromagnetic force $F_{1z} > 0$. The component of the force F_{2z} redistributes the electromagnetic pulse, constant in the cross-section of the arc: accelerates plasma in the axial region, where $F_{2z} < 0$ and decelerates, by the same value, in its peripheral part, where $F_{2z} > 0$. The electromagnetic forces pull the surrounding gas into the arc column, heat the gas and pump it in the axial direction (Fig. 3.10). This process, together with electromagnetic compression (pinch effect) results in the constriction of the arc column as a result of the dynamic pressure, compression of the column by the flow of the gas and cooling of the peripheral regions of the arc by the incoming cold gas (thermal pinch effect). On the whole, the amount of gas in the arc is determined by the effect of electromagnetic and viscous forces (Fig. 3.9).

At the distance $z/\delta_0 \approx 1{\div}2$ from the initial calculation section, the electromagnetic forces are comparable with the viscous forces and the axial velocity reaches the maximum value. At $z/\delta_0 > 2$, the viscous forces become controlling in the formation of the gas-dynamic flow pattern. They redistribute the total electromagnetic pulse in the cross-

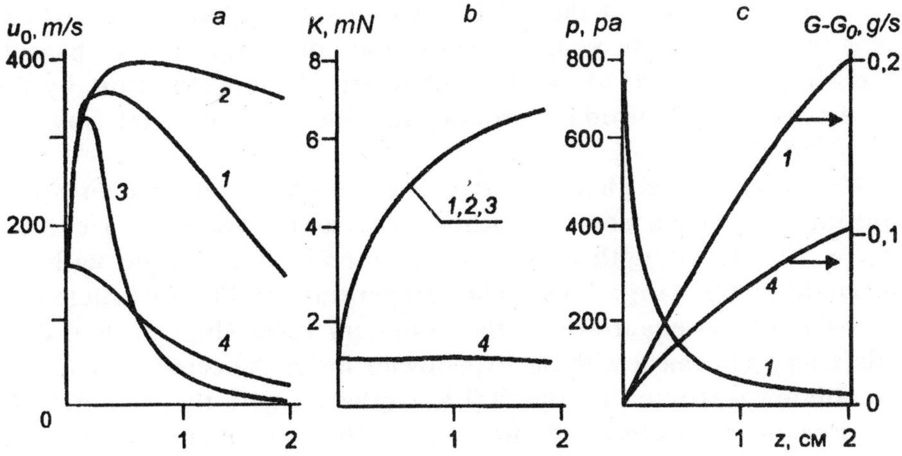

Fig. 3.10. Axial variation of the characteristics of the argon arc at I = 200 A, δ_0 = 1.5 mm. *1*) the main variant; *2, 3*) the viscosity of argon reduced and increased by a factor of 10, respectively; *4*) calculations carried out not taking into account electromagnetic forces.

section of the arc; slow down the plasma flow in the axial region and cause the gas at the periphery of the arc to move by the same pulse. This can be clearly seen by comparison of the calculation variants *1–3* in Fig. 3.5. At $z/\delta_0 < 1$, the values of the axial speed in all cases are almost identical indicating that the role of viscous forces is insignificant. At $z/\delta_0 > 1$ in plasma with a high viscosity, the axial speed decreases relatively rapidly, whereas in the case of low viscosity, the speed decreases slowly.

When ignoring the effect of electromagnetic forces (Fig. 3.10, variant *4*), the total pulse of the current is practically constant and equal to the initial pulse, the axial speed decreases with increasing coordinate z directly from the initial section, the transverse size of the arc increases and in the section $z/\delta_0 \approx 7$ it is twice the value in variant *1*. The consumption of gas in the arc is determined by the initial pulse and the effect of viscous forces.

3.3.2. Numerical analysis on the basis of a system of MGD equations

The results of numerical modelling show [1–4] that the description of the characteristics of the elongated electric arc using the MGD equations of the boundary layer is in satisfactory agreement with the experimental data, if directed vortex-free flows form. In simulation of short electrical arcs with the complicated electrode geometry, taking into account the vortex and reverse flows, it is necessary to use the complete system of the MGD equations. Calculations carried out using the system of the characteristics of arc discharges in the channels of plasma torches [1] show the possibility of the generation, by the arc discharge, of toroidal vortices, cathode and anode jet flows of the plasma.

We examine the characteristics of a high-current arc in argon running between a conical cathode and a flat anode at a current of I = 200 A, a length of L = 1 cm, a conical electrode with the tip angle of 60° (Fig. 3.11). The temperature field, constructed on the basis of the equations of the boundary layer (Fig. 3.11*a*), is in satisfactory agreement with the experiments inside the current-conducting arc column restricted by the 500 K isotherm, with the exception of the area in the vicinity of the anode. The agreement for the gas-dynamic characteristics of the flow is less satisfactory. This is associated with the fact that in the description of the arc in the approximation of the boundary layer, it is assumed that the second electrode (anode) is situated at an infinite distance. Therefore, no account is made of the interaction of the plasma flow with the anode jet, the deceleration of the flow on the surface of the anode and the spreading of the flow in the radial direction.

The spreading of the heated gas on the surface of the anode results in the situation in which the dimensions of the current-conducting region of the channel and, consequently, current density, the strength of the electrical field, and the pressure above the surface are lower in comparison with the anode. The electromagnetic forces change of the direction of their effect to the opposite direction (from the surface of the anode) and started to inhibit the plasma flow. This creates suitable conditions for the formation of the anode plasma jet transferring heat from the anode surface and reducing the heat flow from the arc column to the surface. The area of interaction of the anode and cathode jets is characterised by the increase of the transverse dimensions of the arc and the formation of the typical bell-shaped form of the arc.

To determine the nature of the MGD flows in the high-current electric arc, in the experiments carried out in [36] the arc column was passed through a copper cooled diaphragm with a diameter of 2.5 mm, 4 mm thick. We examine the characteristics of this arc at I = 200 A (Fig. 3.12) [1]. It may be seen that the region of the diaphragm is characterised by a high temperature of $T \approx 30\,000$ K and high-intensity plasma flows leave the orifices in the symmetric fashion at a speed of approximately 300 m/s which, as in the experiment, collide with the electrode jets, forming the

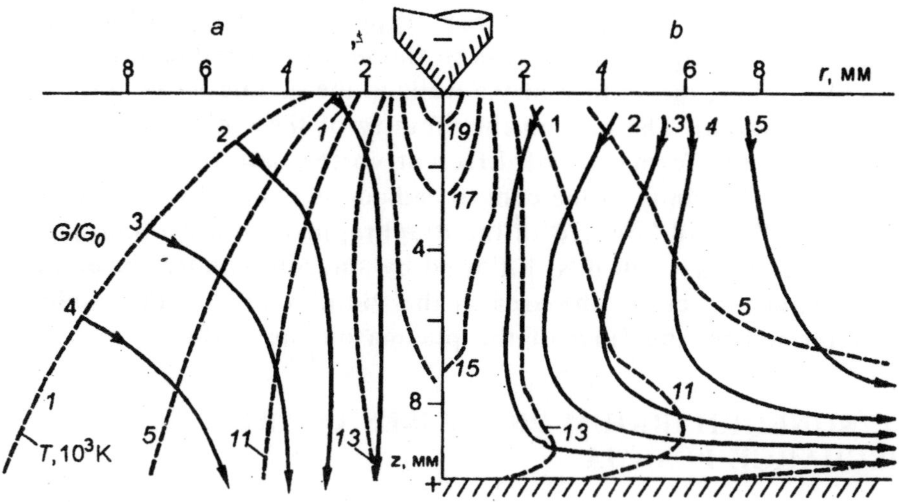

Fig. 3.11. Temperature fields and the current lines of the gas of the argon arc at I = 200 A. *a*) the elongated arc; *b*) the short arc.

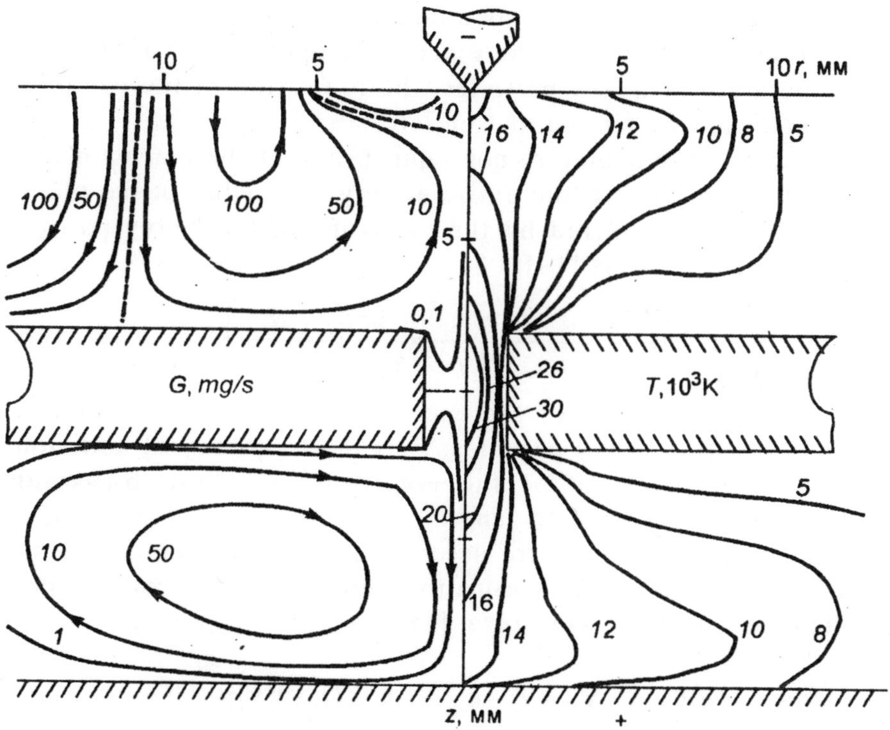

Fig. 3.12. The temperature fields and the current lines in the argon arc running in a slit with a diaphragm at I = 200 A.

configuration of the 'plasma tray' type. The supply of the gas inside the diaphragm at the walls in the area with reduced pressure takes place along the surface of the walls to the middle of the thickness of the diaphragm. The area in the vicinity of the electrodes is characterised by the formation of different vortex patterns of the flow: the area in the vicinity of the conical cathode is typical by the collision of the jets from the orifice in the diaphragm with a cathode jet of lower intensity ($u \approx 250$ m/s) followed by spreading under some angle in the radial direction; the area in the vicinity of the flat anode is characterised by the flow of the plasma on the surface.

3.4. NONEQUILIBRIUM PROCESSES IN ARC DISCHARGE PLASMA

At reduced pressure and arc current (for example, in argon at $I < 50$ A, in helium at $I < 200$ A) the state of the plasma in the vicinity of the electrode and channel walls deviates from the equilibrium state [37, 38]: the electron temperature is higher than the temperature of heavy particles, the ionisation of the gas is not locally balanced by recombination, etc.

The degree of temperature non-equilibrium of the plasma may be evaluated assuming that the entire energy dissipated in the plasma (Joule heat) is transferred by the electrons to the heavy particles as a result of collisions [14]:

$$1-\frac{T}{T_e}=\frac{\sigma E^2}{\delta_e n_e v_e 3kT_e/2}=\frac{3\pi}{16\delta_e}\left(\frac{l_e eE}{3kT_e/2}\right),$$

where δ_e is the fraction of the energy transferred by the electron in collision with a heavy particles; v_e is the frequency of collisions; n_e is the concentration of the electrons; k is the Boltzmann constant; l_e is the free path length.

The system of equations for describing the arc column taking into account the deviation of the plasma from the temperature and ionisation equilibria has the following form [1]:

$$\begin{gathered}\nabla\left(\frac{5}{2}kn_eT_e\vec{V}_e\right)=\nabla\lambda_e\nabla T_e-\varphi_e-U_I\dot{n}_e-Q_e+\vec{V}\nabla p_e+\vec{j}\vec{E},\\ \nabla\left(\frac{5}{2}k(n_i-n_a)T\vec{V}\right)=\nabla\lambda\nabla T+Q_e+\vec{V}\nabla p,\ \nabla(n_e\vec{V}_e)=\dot{n}_e,\\ \nabla(\rho\vec{V})=0,\ \rho(\vec{V}\nabla\vec{V})=-\nabla p+\mu_0\vec{j}\times\vec{H}+\nabla\tau_{ik},\end{gathered}\tag{3.35}$$

$$\nabla \times \vec{H} = \vec{j}, \nabla \times \vec{E} = 0, \vec{j} = \sigma \vec{E}, \ \vec{V}_e = \vec{V} + \vec{V}_d + \vec{V}_a + \vec{V}_t,$$

$$n_e = K_I n_e n_a - K_r n_e^2 n_i, \ \rho = m_a (n_i + n_a),$$

$$\delta_e = 2m_e / m_a, \ Q_e = \delta_e v_e n_e \frac{3}{2} k (T_e - T),$$

where $\vec{V}_a = -D_a \nabla(\ln n_e)$, $\vec{V}_t = -D_a \nabla(\ln T_e)$, $\vec{V}_d = \sigma E / en_e$ are the velocities of ambipolar diffusion, thermal diffusion and electron drift; U_I is the ionisation potential; K_I, K_r are the constants of impact ionisation and three-particle recombination.

We examine the characteristics of the plasma flow of atmospheric pressure argon obtained on the basis of the equations (3.35) for the experimental conditions [38]: current $I = 25 \div 300$ A, channel diameter $d = 0.5 \div 3$ cm, gas flow rate $G = 0 \div 3$ g/s (Fig. 3.13, Table 3.1). For comparison, we present the results of similar calculations in the framework of the equilibrium plasma model (3.7). As indicated by the calculation results, the distributions $T_e(r)$, E, $u(r)$, dp/dz, obtained using the LTE model, are in better agreement with the experimental data in comparison with the values calculated using the equilibrium plasma model. The calculated profile of the equilibrium temperature is always higher at the axis of the arc and lower at the periphery in relation to the electron temperature, since at $I/d < 15$ A/mm, the equilibrium model of the plasma gives a narrower current-conducting arc channel in comparison with the two-temperature model of experiments, this also determines higher values of the strength of the electrical field (Table 3.1).

In a measurements and in the PLTE plasma model, the temperature of the electrons at the periphery of the arc is always higher than the temperature of the heavy particles, the difference at the channel

Table 3.1. The characteristics of the argon arc at atmospheric pressure

Experiment conditions	I, A	d = 5mm			d = 3cm				
		T_{e0}, 10^3 K	E, V/cm	u, m/s	T_{e0}, 10^3 K	T_{eR}, 10^3 K	E, V/cm	u/G, m/g	$dp/(dz \cdot G)$ Pa · s/(g · cm)
LTE model	75	13.7	10.4	15	9.7	0.3	2.1	23.5	0.69
PLTE model		12.8	9.2	7	8.8	5.3	1.8	24.3	0.72
Experiment [38]		12	10	7.5	9.2	5	1.7	25	0.65
LTE model	225	21.5	15.2	55	10.5	0.3	2.3	36.6	1.35
PLTE model		16.2	16.7	15	10.3	6	2.1	37.3	1.39
Experiment [38]		15	21	18.5	10	6.3	2	36	1.28

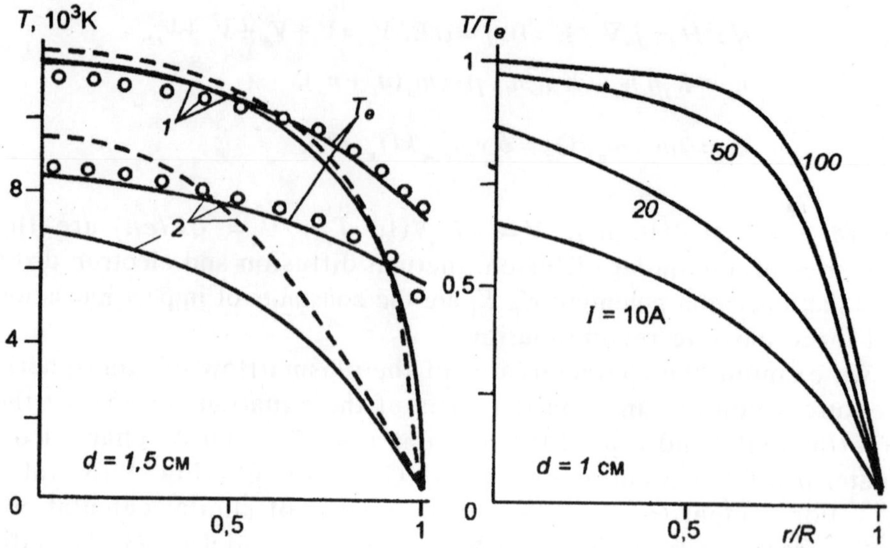

Fig. 3.13. Radial distribution of the temperature of the electrodes and heavy particles and the degree of temperature non-equilibrium of the plasma in the cross-section of the channel. *I*, A: *1*) 150; *2*) 75, broken lines – equilibrium temperature, circles – experimental data [38].

walls reaches 5000 K and increases with increasing *I/d*. At *I/d* < 15 A/mm, calculations of the arc using the LTE model of the plasma result, on the other hand, in low values of the strength of the electrical field, because the radii of the current-conducting channels in the models and in the experiment are comparable, and the distribution of electrical conductivity in the cross-section of the arc is completely determined by the temperature field. The agreement becomes less satisfactory with increasing *I/d* and *d* and is caused by the increase of the role of the re-absorption of radiation in the energy balance which results in a decrease of the temperature non-equilibrium of the arc plasma on the axis of the arc and in the increase in this parameter at the arc periphery.

The profile of equilibrium temperature is close to the distribution of temperature of the heavy particles in the vicinity of the walls. This results in satisfactory agreement between the calculated in experimental gas-dynamic characteristics of the plasma flow at *I/d* < 15 A/mm. With a decrease of *I/d* < 10 A/mm, the region of difference between the temperatures T_e and T in the two-temperature model of the plasma extends from the channel walls to the axis (Fig. 3.13). This is caused by a decrease of temperature and concentration of the electrons and by a decrease of the frequency of collisions of the electrons with the heavy particles. At $I/d < 2.5$ A/mm, the calculated values of T_e and T become lower than in the measurements (Table 3.1). Evidently, this is associated with the definition of the sections of collisions by the functions of electron temperature and by the fact that the kinetic processes in the plasma are not taken into account efficiently.

Thus, the two-temperature plasma model results in satisfactory agreement with the experimental data at $2.5 \leq I/d \leq 50$ A/mm, equilibrium at $10 \leq I/d \leq 50$ A/mm. At $I/d \leq 10$ A/mm, the results obtained on the basis of the LTE and PLTE models of the plasma are in almost complete agreement with each other in the current-conducting channel of the arc, with the exception of the peripheral region in which the electron temperature is always higher than the temperature of heavy particles. The deviation of plasma from ionisation equilibrium for the given initial parameters has no significant effect on the thermal and electromagnetic characteristics of plasma. In comparison with the results obtained on the basis of Saha's equation, the distribution of the concentration of the particles in the cross-section of the channel changes appreciably, for example, the value of n_e decreases several times at the axis of the arc and is several orders of magnitude higher on the periphery. This is in agreement with the measurements taken in [38].

The description of the developing arc flow on the basis of the equations of the boundary layer taking into account the deviation of plasma from the temperature and ionisation equilibria was published in [1, 2, 39], where the intrinsic electromagnetic forces are also taken into account. In [2], using the equations (3.35), the authors carried out the numerical analysis of the flow in the initial section of the channel of the plasma torch with the axial gas flow. It has been reported that in order to compare the calculated and experimental results, it is necessary to ensure adequate boundary conditions because the effect of these conditions is evident along the entire length of the initial section.

The authors of [31] calculated the arc on the surface of a flat electrode. When defining the boundary conditions, it was assumed that the temperature profile of the heavy particles in the vicinity of the electrode is identical with the distribution of temperature on the end surface of the electrode:

$$T = (T_k - T_R)\exp(-r^2/R_k^2) + T_R,$$

where T_k is the melting point of the electrode; R_k is the radius of the arc on the electrode, determined in the experiments. It was thus possible to calculate the characteristics of the arc from the electrode without using the model of near-electrode processes and melting of the electrode. The distribution of temperature and electron concentration is determined by solving one-dimensional equations (3.35), and the speed of plasma is equal to zero. As indicated by the calculation results (Fig. 3.14), when the arc current is increased from 50 to 200 A the electron temperature changes only slightly (from 10,000 to 12,000 K), in the non-isothermal region is shifted to the electrode in the direction opposite to the direction of current; $z = 50R_k/I$. At $I/2r_* > 10$ A/mm, the plasma is almost equilibrium in the current-conducting channel of the arc, with the exception of the peripheral region.

When the transverse dimensions of the arc are comparable with the longitudinal dimensions, the effect of the electrodes on the plasma characteristics is quite strong. The authors of [40] calculated a short arc from the outlet of an electrode attachment to a flat anode on the basis of a system of enriched the equations taking into account the temperature nonequilibrium of plasma. The authors of [1, 31] also took into account the processes associated with ambipolar and thermal diffusion, and calculated the arc in a narrow slit.

The authors of [31] calculating the arc in atmospheric pressure argon from the outlet of a cathode attachment to a flat copper anode (T_k = 1600 K), with the initial data obtained in the experiments in

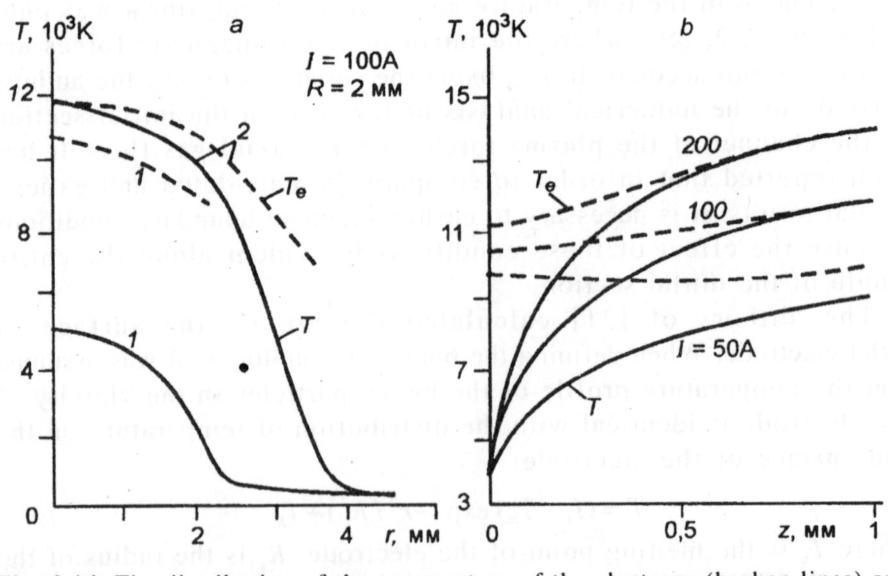

Fig. 3.14. The distribution of the temperature of the electrons (broken lines) and heavy particles (solid lines) at z = 0 (*1*) and 10 mm (*2*) (*a*) and along the axis of the arc (*b*).

[36]. Figure 3.15 shows that the flow pattern and the pattern of heating the gas in short electrical arcs differ qualitatively from the processes in the long arcs: in the calculated region, toroidal vortices form with a specific direction of rotation which depends on the dimensions of the arc on the anode. At R_a = 8 mm, the plasma flow leaving the nozzle expands to r_{*max} = 15 mm, is accelerated by electromagnetic forces, flows on the anode surface and spreads in the radial direction.

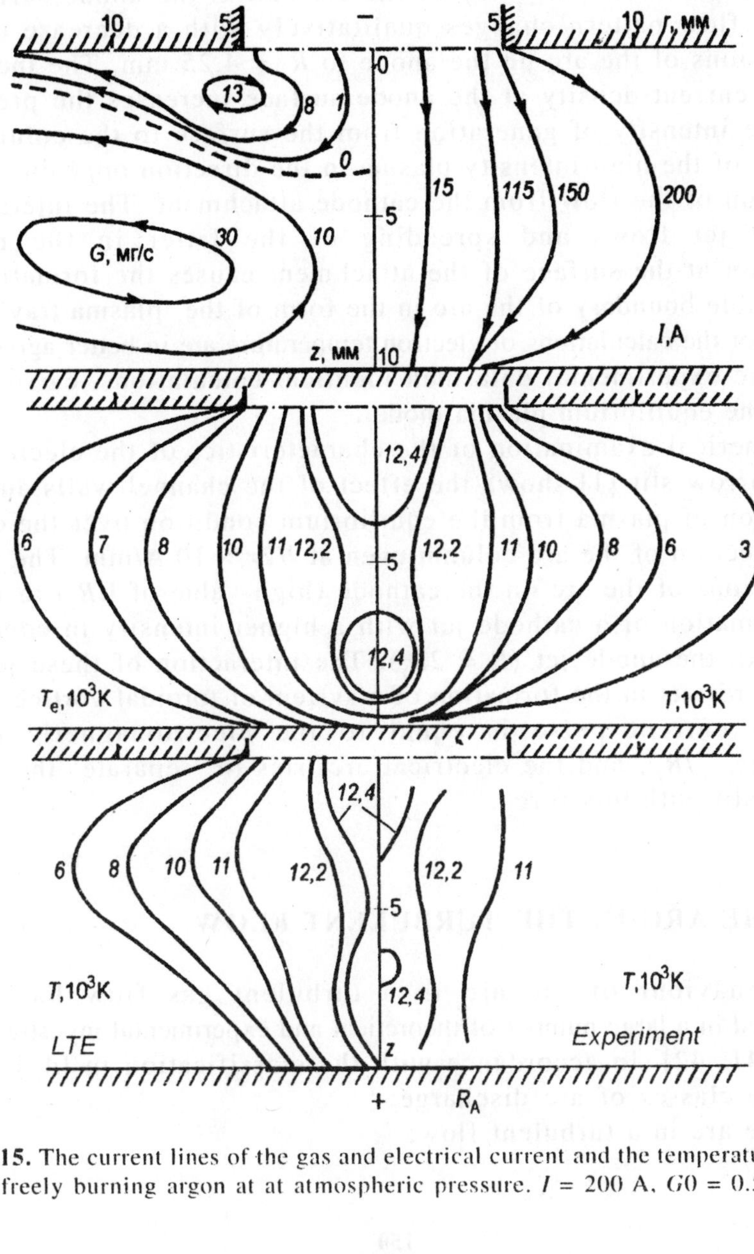

Fig. 3.15. The current lines of the gas and electrical current and the temperature fields in the freely burning argon at at atmospheric pressure. I = 200 A, $G0$ = 0.5 g/min.

The decrease of the dimensions of the arc in the direction to the surface of the anode increases the pressure in the current channel as a result of intrinsic electromagnetic compression (pinch effect). This results in the deceleration of the plasma flow.to the surface. The interaction of the plasma flow from the attachment with the 'weak' anode jet results in the transfer of energy to the region of collision of the jets with the formation of the typical the bell shape of the visible area of the stem of the arc. A large temperature difference is detected at the periphery of the arc and at the anode surface.

The flow pattern changes qualitatively with a decrease of the dimensions of the arc on the anode to $R_a = 4.25$ mm. The increase of the current density at the anode surface increases the pressure and the intensity of generation from the surface to the column of the arc of the high-intensity plasma in the direction opposite to the direction of the flow from the cathode attachment. The interaction of the jet flows and spreading of the latter in the radial direction at the surface of the attachment causes the formation of the visible boundary of the arc in the form of the 'plasma tray'. The results of the calculations of electron temperature are in better agreement with the experimental data then identical calculations, carried out using the equilibrium plasma model.

Numerical examination of the characteristics of the electric arc in a narrow slit [1] shows the effect of the channel walls and the deviation of plasma from the equilibrium condition over the entire cross-section of the arc column even at $I/2r_* > 10$ A/mm. The small dimensions of the arc on the cathode (high value of I/R_k) result in the formation of a cathode jet with a higher intensity in comparison with the anode jet ($R = 2R_k$). The interaction of these jets in the slit results in the formation of a system of toroidal vortices. The plasma current pulse is equal to the electromagnetic pulse $\mu I^2 \ln |r_{*max}/R_k|$, and the electrical arc tries to 'separate' the walls of the slit with this force.

3.5. THE ARC IN THE TURBULENT FLOW

The behaviour of the arc in a turbulent gas flow has been examined in a large number of theoretical and experimental investigations [1, 2, 41, 42]. In accordance with the classification in [42] there are two classes of arc discharge:

– the arc in a turbulent flow;

– a turbulent arc.

The first concept includes the case of a laminar arc with a turbulent flow-around, and cases in which fine-scale turbulence penetrates into the current-conducting channel of a stabilised arc. This class of discharge permits modelling on the basis of the equations of radiation magnetic gas dynamics.

The term 'turbulent arc' is used for discharges, interacting with large-scale turbulence. In this case, it is necessary to apply a probability, statistical description.

In most cases, calculations of the arcs in turbulent flows are carried out using semi-empirical turbulence theories, supplemented by hypotheses and experimental data on the behaviour of some physical quantities. The models, based on the application of the concept of 'mixing path' relate to the first order models. In a number of cases, it is necessary to use multi-parameter models, for example, the models of transfer of turbulence scale, turbulent kinetic energy, etc [43].

3.5.1. Turbulence model

We examine a steady motion of the gas in which an electric arc burns [44]. It is assumed that this movement is described by equations (3.7). In the presence of turbulence, the actual instantaneous values of velocity, temperature, density and pressure of gas, electrical conductivity, the strength of the electrical field and other characteristics of the flow show continuous random deviations from some stationary mean values. Using the Reynolds approach, the pulsating quantities may be represented in the form:

$$\varphi = \overline{\varphi} + \varphi'(t), \tag{3.36}$$

where φ is the instantaneous value of some quantity; $\overline{\varphi}$ is its value averaged out with the respect to time; φ' (t) is the pulsation of a quantity. It is assumed that the pulsations are small in comparison with the average values, and the latter depend only slightly on the averaging method.

We examine an incompressible gas, neglecting the pulsations of pressure and electromagnetic quantities and assuming that only speed and temperature are pulsating. Substituting (3.36) into (3.7) and carrying out averaging with subsequent application of the Bussinesq law according to which the dependence of turbulent tangential stresses on the mean strain rate coincides in the form with the Newton equation for laminar tangential stresses, it may be seen that the system (3.7) retains its form for the averaged-out turbulent flow, if the viscosity and heat conductivity are presented by the sum of laminar and turbulent viscosity and heat conductivity, respectively:

$$\eta = \eta_l + \eta_t, \quad \lambda = \lambda_l + \lambda_t. \tag{3.37}$$

To average out η_t and λ_t, it is necessary to use empirical data and the appropriate turbulence theories which are of semi-empirical nature. We present the equations of these theories, used in the calculations of electric arc plasma flows [44].

The Prandtl model of the mixing path is used in the vicinity of the channel walls to obtain the profile of turbulent viscosity. In accordance with this theory, the turbulent viscosity and heat conductivity are equal to:

$$\eta_t = \rho l_u^2 \left|\frac{\partial u}{\partial r}\right|; \; \lambda_t = \rho c_p l_u l_t \left|\frac{\partial u}{\partial r}\right|; \; l_u = \mathrm{K}(R - r). \tag{3.38}$$

Here i_u, l_t are the length of the mixing path for the pulse and heat content; K = 0.41 is the Karman constant. The turbulent Prandtl number, characterising the relationship between the turbulent viscosity and heat conductivity:

$$\mathrm{Pr} = \frac{\eta_t c_p}{\lambda_t} = \frac{l_u}{l_t}, \tag{3.39}$$

is usually close to 1 and, therefore, it is assumed that $l_u = l_t$.

Using the Prandtl–Kolmogorov relationship:

$$\eta_t = C_\eta \rho k^2 / \varepsilon \tag{3.40}$$

we determine the wall profile of the rate of dissipation of turbulent energy. Here $k = \sum_{i=1}^{3} (u_i')^2 / 2$ is the kinetic energy of turbulent pulsation; $\varepsilon = \frac{\eta_t}{\rho} \sum_{i,k} \left(\frac{\partial u_i'}{\partial x_k}\right)^2$ is the rate of dissociation of turbulent energy; C_η is an empirical constant.

The k–ε model is used to calculate the gas flow and temperature distribution in the entire region, with the exception of a narrow wall layer [43]. In this case, the turbulent viscosity is given by the Prandtl–Kolmogorov equation (3.40). The turbulent heat conductivity coefficient is determined from the equation (3.39). The equations for determining the fields $k(r, z)$ and $\varepsilon(r, z)$ are derived from the system of equations for the pulsation components which is derived together with the system for the average values when using the Reynolds procedure, and have the following form:

$$\frac{\partial}{\partial z}(\rho ukr)+\frac{\partial}{\partial r}(\rho \upsilon kr)-\frac{\partial}{\partial z}\left[\Gamma_k r\frac{\partial k}{\partial z}\right]-\frac{\partial}{\partial r}\left[\Gamma_k r\frac{\partial k}{\partial r}\right]-rS_k=0,$$
$$\frac{\partial}{\partial z}(\rho u\varepsilon r)+\frac{\partial}{\partial r}(\rho \upsilon\varepsilon r)-\frac{\partial}{\partial z}\left[\Gamma_k r\frac{\partial \varepsilon}{\partial z}\right]-\frac{\partial}{\partial r}\left[\Gamma_k r\frac{\partial \varepsilon}{\partial \varepsilon}\right]-rS_\varepsilon=0, \qquad (3.41)$$

Here

$$\Gamma_k=\eta_l+\frac{1}{\sigma_k}-\eta_t,\ \Gamma_\varepsilon=\eta_l+\frac{1}{\sigma_\varepsilon}-\eta_t,$$

$$S_k=\eta_t G-\varepsilon,\ S_\varepsilon=\frac{\varepsilon}{k}(C_{\varepsilon 1}\eta_t G-C_{\varepsilon 2}\varepsilon),$$

$$G=2\left[\left(\frac{\partial u}{\partial z}\right)^2+\left(\frac{\partial \upsilon}{\partial r}\right)^2+\left(\frac{\upsilon}{r}\right)^2\right]+\left[\frac{\partial u}{\partial r}+\frac{\partial \upsilon}{\partial z}\right]^2.$$

The empirical constants of the model: $\sigma_k = 1$; $\sigma_\varepsilon = 1.3$; $C_{\varepsilon l} = 1.44$; $C_{\varepsilon 2} = 1.92$.

The application of the multi-parameter model results in better understanding of the dynamics and spatial evolution of turbulence in the electric arc flow. The solution of equations (3.41) makes it possible to detect the role of the processes of generation and dissipation, diffusion and convective transfer of turbulent energy in the thermal field of the arc in each specific case. Consequently, it is possible to describe more complicated turbulent flows and obtain better agreement with the experiments in comparison with simple first order semi-empirical models.

It is important to note difficulties in the optimum selection of the constants of the models because of the absence of essential experimental information on specific plasma flows in the arc and difficulties in defining the boundary condition for ε, caused by the non-isotropic nature of turbulence in the vicinity of the walls.

We examine the characteristics of an electric arc in a cylindrical channel, with a turbulent gas flow blown onto it, obtained using the k–ε model in [44]. The calculation conditions: the radius of the channel R = 3 mm, arc current I = 100 A, gas flow rate G = 3 g/s, plasma forming gas – argon, the thermodynamic and transfer properties of the gas were taken from [1].

The boundary conditions are given in the following form:

$$z=0:\ T=T^0(r),\ u=u^0(r),\ \frac{av}{az}=0,\ k=k^0(r),\ \varepsilon=\varepsilon^0(r);$$

$$z=L:\ \frac{\partial^2 T}{\partial z^2}=0,\ \frac{\partial^2 u}{\partial z^2}=0,\ \frac{\partial^2 k}{\partial z^2}=0,\ \frac{\partial^2 \varepsilon}{\partial z^2}=0,\ p=p^0;$$

$$r=0:\ \frac{\partial T}{\partial r}=0,\ \frac{\partial u}{\partial r}=0,\ \frac{\partial k}{\partial r}=0,\ \frac{\partial \varepsilon}{\partial r}=0;$$

$$r=R:\ T=T_w,\ u=0,\ v=0,\ k=0.$$

The boundary value ε was selected in the vicinity of the wall from the distribution η_t using equation (3.40).

3.5.2. Analysis of the results

The results of the calculations are presented in Fig. 3.16–3.18. Figure 3.16*a* shows clearly the acceleration of the gas and plasma in the axial direction, caused by a pressure gradient, the increase of the value $\partial u/\partial r$ in the vicinity of the wall with increasing coordinate z and, finally, the formation of a boundary layer turbulent flow with $\frac{\partial u}{\partial r} \gg \frac{\partial u}{\partial z}$. Figure 3.16*b* shows that the calculation region is characterised by the dominant effect of the flow of the gas into the arc and the displacement of the gas to the outside is observed only in the vicinity of the input cross-section at the periphery of the discharge. The maximum of the radial speed is displaced to the outer boundary of the arc with increasing value of z. The displacement and flow of the gas into the arc in this case may be explained by one reason: the temperature dependence of gas density. At the periphery of the arc in the vicinity of the initial cross-section, the displacement of the plasma is associated with the heating of the external gas, but with increase of z the heating of the outer flow and cooling of the internal regions of the arc (Fig. 3.17) result in the formation of a reverse flow into the central regions of the discharge. We also note the fulfilment of the condition $v \ll u$, typical of the boundary layer, at $z/R > 4$.

The characteristic feature of the temperature field in Fig. 3.17 is the large variation of temperature in the radial direction in the vicinity of the wall, and a small variation in the near-axial zone. This distribution is typical of turbulent low-thermal flows. It is also important to note the large variation of temperature in the direction of the axis z in the vicinity of the initial cross-section, indicating the problems with the application of the approximation of the boundary layer in this area.

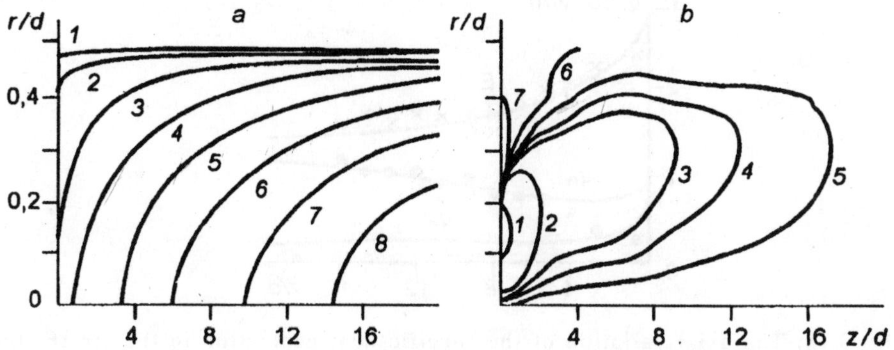

Fig. 3.16. Isolines of the axial (*a*) and radial (*b*) velocity of the arc in the turbulent flow. *a*) *u*, m/s: 200 (*1*), 1600 (8), the step between the isolines 200 m/s; *b*) υ, m/s: – 25(*1*), – 10 (*2*), – 5 (*3*), – 3.3 (*4*), – 1.7 (*5*), 0 (*6*), 5 (*7*).

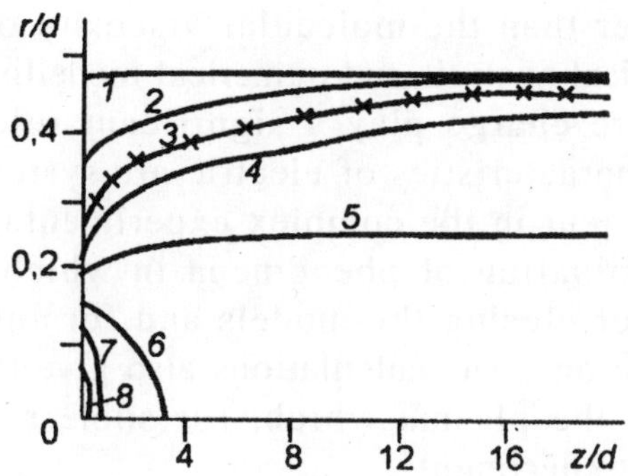

Fig. 3.17. Temperature field. T, K: 2000 (1), 14,000 (7), the step between the isolines 2000 K, the crosses indicate the boundary of the arc.

Figure 3.18 shows satisfactory agreement of the calculated results with the experimental data for the strength of the electrical field on the heat flow to the wall q_w. The difference of the values of q_w in the vicinity of the initial section is evidently associated with the definition of the initial conditions, which are not completely adequate to the actual conditions. It is important to note a tendency for the convergence of the curves for energy generation *IE* and q_w, indicating the process of establishment of the developed flow regime. However, in the investigated case the length of the channel is insufficient for the establishment of this regime in the final analysis.

Analysis of the calculation results shows that the kinetic energy of turbulence is generated most intensively in the region $0.6 < r/R < 0.9$, and is transferred along z by the convective gas flow and by diffusion of the wall. The maximum dissipation of turbulence energy is situated in the vicinity of the walls. Turbulent viscosity is two orders

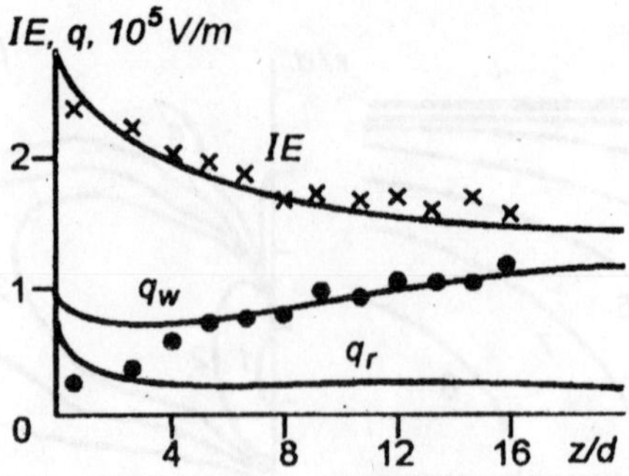

Fig. 3.18. The axial variation of the specific heat generation in the arc IE, the heat flow into the wall qw and the radiant flux qr. The crosses and circles-the experimental data from [45].

of magnitude higher than the molecular viscosity of cold argon.

Thus, the theoretical analysis and numerical modelling of the physical processes in the arc charge play a significant role as a means of investigating the characteristics of electric arc systems. This method has been most efficient in the complex experimental–theoretical approach to the examination of phenomena in which the experimental data are used for closing the models and formulating the initial and boundary conditions. The calculations also give the characteristics and parameters of the plasma which, for some reason, cannot be determined in the experiments.

An important task is the expansion of the application of the theoretical and theoretical–experimental methods of investigating electric arc systems in engineering practice. For this purpose, it is necessary to develop physical and mathematical models, combining the simplicity and lower labour content of the calculations with sufficient reliability and accuracy of the results.

Chapter 4

Modelling of processes in electric arc plasma torches

4.1. CONCEPT OF MODELLING OF PROCESSES

Since the theory of the electric arc is not yet capable of providing a method for the accurate calculation of electric arc plasma torches because it is difficult to take into account all the processes taking place in them, in the development and design of plasma torches it is often necessary to use experimental data obtained in the examination of electric arcs. However, the simple extrapolation of these data to the non-investigated ranges of the parameters in the development of more powerful plasma torches is associated with considerable errors and, in principle, is not efficient. The realisation of experimental investigations every time in new conditions is very time-consuming and expensive, especially in the area of high powers. Therefore, it is necessary to answer the question: how to use, for solving new problems, the available results obtained on less powerful systems? The answer to this question is provided by the theory of similarity and dimensions and its section is referred to as modelling.

Modelling is the development of methods which can be used to replace the natural phenomenon which is of interest by the examination of a similar phenomenon on models on a smaller scale with subsequent application of the results in different conditions. The method was developed a long time ago and was initially used in hydrodynamics and thermal engineering [1–5], and in the last couple of decades it has been used in plasma dynamics [6–11, etc].

The main idea of modelling is to use the results of experiments with models to predict effects, their numerical values and the relationships taking place in the natural conditions. Thus, examination of the natural phenomenon is replaced by examination of a physically similar phenomenon which is more convenient, simpler, faster and cheaper to realise. However, it should always be remembered that the examination of general quantitative relationships of the phenomenon may be useful only on the condition that it is based on a relatively large volume of information. The quantitative examination is preceded by the long process of formation of physical considerations.

The essential and sufficient condition of similarity of two processes is the equality of numerical values of some set of the determined parametric complexes, are referred to as similarity criteria. Methods have been developed for finding these criteria. However, it is insufficient to determine only the similarity of two phenomena. The main task and purpose of modelling in plasma dynamics is the generalisation of the results, obtained in the examination of the models, and presentation in the form of mathematical expressions, with the arguments represented by the previously mentioned similarity criteria. In reference to electric arc plasma torches, the problem or generalisation of the experimental data is based on determination, for each type of plasma torch, of generalised equations of volt–ampere and thermal characteristics of the arcs, in relation to the controlling parameters. These two equations represent the basis of design operations and, at present, they cannot be derived by other methods, for example, analytical methods.

4.2. METHODS FOR DETERMINING SIMILARITY CRITERIA

There are two main approaches to determining the similarity criteria: systematic and parametric. The systematic method uses a system of equations describing the investigated phenomenon and, subsequently, the π-theorem of similarity theory is applied.

The parametric method of determination of the similarity criteria is based on the physical knowledge of the given phenomenon and the maximum possible consideration of all controlling parameters with subsequent application of the π-theorem of the dimensional theory to the set of the values. If the systematic method is relatively well-known and, to some extent, standard, the parametric method requires considerable knowledge, intuition, etc.

Analysis of the similarity conditions, based on a specific system of the fundamental equations of the process, even if it is not possible to solve them successively, provides information which is more substantial than that provided by the elementary analysis of dimensions. However, there is always a question of the efficiency of the selected system of equations and boundary-value conditions. Therefore, these two methods are used together with the development of the theory of the electric arc.

According to similarity theory, the physical phenomena are similar only if they satisfy the same (identical) closed systems of equations and boundary-value conditions.

A system of equations, defining a specific physical phenomenon, after representation in the dimensionless (relative) form becomes the representative of not only one specific phenomenon but of an entire class of similar phenomenon.

The analysis of dimensions in the pure form can be used in the case of information restricted to the list of the fundamental parameters and physical properties, influencing the course of the given process. However, in this case, there is not sufficient substantiation for selecting the number of similarity criteria and, even more so, for the analysis of possible relationships between them.

From the dimensionless system of equations it is necessary to specify complex and simplex similarity criteria essential for generalisation. The complex similarity criteria are composed from the similar quantities with different dimensions. The simplex dimensionless similarity criteria are composed from homogeneous quantities, i.e. the quantities of the same dimensions. For example, the similarity simplexes include the known number π equal to the ratio of the length of the circumference to diameter, and the dimensionless coordinate $\bar{z} = z/d$, where d is the channel diameter, z is the axial coordinate.

According to the main π-theorem of similarity theory, the solution of a system of equations, determining the class of similar phenomena, may be represented in the form of arbitrary dependences on the complexes and similarity simplexes. The most suitable form of finding such a solution is the representation of the solution in the form of the product of the powers of these similarity complexes and simplexes.

The parametric method

The application of the method will be illustrated on an example of the flow of a viscous incompressible fluid in a pipe. It is assumed that the flow is determined by the following main parameters: the pipe diameter d, the length of the pipe l, pressure gradient Δp, viscosity of the fluid ν, density ρ and speed of movement of the fluid υ. This is followed by assuming that there is an equation which links the previously mentioned parameters:

$$f(p, \nu, \upsilon, \rho, d, l) = 0. \tag{4.1}$$

According to the π-theorem, if a phenomenon is determined by n-dimensional quantities, where k is the number of primary quantities and $(n-k)$ is the number of secondary quantities determined using k primary dimensions, this phenomenon may be represented in the form of the relationship of $(n-k)$ dimensionless criteria, composed of the initial quantities in different powers.

We shall use the theorem for finding similarity criteria in the case of a viscous flow of an incompressible fluid in a pipe.

In equation (4.9) there are six dimensional quantities ($n = 6$), and three primary dimensions were used in the formulation of these quantities: metre, kilogram, second ($k = 3$). According to π-theorem, this equation may be represented in the form of a relationship between $(n-k) = 6-3 = 3$ dimensionless quantities.

In a general form, the dimensionless similarity criterion is expressed as the product of n-dimensional quantities in different powers:

$$K_i = p^{n_1} \cdot \nu^{n_2} \cdot \upsilon^{n_3} \cdot \rho^{n_4} \cdot d^{n_5} \cdot l^{n_6}. \tag{4.2}$$

The dimensional quantities in this expression will be replaced by appropriate primary dimensions: L (m), M (kg) and T (s), using the generally accepted rule according to which the inclusion of a physical quantities in square brackets denotes its dimension:

$$[p] = M \cdot L^{-1}T^{-2}, \quad [\nu] = L^2T^{-1}, \quad [\upsilon] = L \cdot T^{-1},$$
$$[\rho] = ML^{-3}, \quad [d] = L, \quad [l] = L.$$

Since the criterion is a dimensionless quantity, its dimensions should be $[K_i] = 1$. Thus, from equation (4.2) we obtain

$$[K_i] = (M \cdot L^{-1} \cdot T^{-2})^{n_1} (L^2T^{-1})^{n_2} \times$$
$$\times (LT^{-1})^{n_3} (ML^{-3})^{n_4} (L)^{n_5} (L)^{n_6} = 1.$$

Opening the brackets and combining the homogeneous terms, gives:

$$M^{n_1+n_4} L^{-n_1+2n_2+n_3-3n_4+n_5+n_6} T^{-2n_1-n_2-n_3} = 1.$$

This shows that the condition of the dimensionless form of the required similarity criteria (complexes or simplexes) is the equality to zero of the sum of the powers at each of the primary measurement units:

$$n_1 + n_4 = 0,$$
$$-n_1 + 2n_2 + n_3 - 3n_4 + n_5 + n_6 = 0,$$
$$-2n_1 - n_2 - n_3 = 0.$$

Consequently,

$$
\begin{aligned}
n_1 &= -n_4, \\
n_3 &= 2n_4 - n_2, \\
n_6 &= -n_2 - n_5.
\end{aligned} \tag{4.3}
$$

Substituting the values of n_1, n_3 and n_6 from (4.3) into equation (4.2) of the i-th similarity criterion, we obtain

$$K_i = p^{-n_4} \nu^{n_2} \upsilon^{2n_4 - n_2} \rho^{n_4} d^{n_5} l^{-n_2 - n_5}. \tag{4.4}$$

In accordance with the rules of linear algebra [12] when the number of unknown quantities in the system of equations is larger than the number of equations, the selection of quantities n_2, n_4 and n_5 in equation (4.4) is arbitrary. To simplify calculations, each quantity will be given successively the value equal to unity, and the others will be equated to zero.

Thus, if $n_2 = 1$, $n_4 = n_5 = 0$, then

$$K_1 = \nu^1 \upsilon^{-1} \cdot l^{-1} = \frac{\nu}{\upsilon l} \equiv \frac{1}{\text{Re}}.$$

If $n_4 = 1$, and $n_2 = n_5 = 0$, then

$$K_2 = p^{-1} \upsilon^2 \rho^1 = \rho \upsilon^2 / p \equiv \text{Eu}.$$

If $n_5 = 1$, and $n_2 = n_4 = 0$, then

$$K_3 = d^1 \cdot l^{-1} = d / l.$$

Thus, we obtain the well-known complex similarity criteria, the Reynolds and Euler numbers, and also the simplex similarity criterion $K_3 = d/l$.

According to the similarity theory, the general solution of the equation of movement of the fluid may be presented in this case in the form of a functional dependence on the similarity criteria:

$$f(\text{Eu}, \text{Re}, d / l) = 0.$$

This equation is referred to as a criterial equation. Usually, the equation is solved in relation to an undetermined criterion. For example, if the determined quantity is the pressure gradient, the undetermined criterion is the Euler number and, consequently, the criterial dependence has the form:

$$\text{Eu} = \Phi(\text{Re}, d / l). \tag{4.5}$$

The systematic method

The systematic method is based on the homogeneity principle typical of all the physical equations. According to this principle, the terms of these equations have always the same dimensions. There are several variants of the systematic method. We examine one of them, the method of making equations dimensionless.

As an example, using the method, we determine a set of similarity criteria, characterising the movement of an incompressible viscous fluid in a pipe. This equation is the Navier–Stokes equation:

$$(\bar{v}, \text{grad})\bar{v} = g - \frac{1}{\rho}\,\text{grad}\,p + \nu\nabla^2\bar{v}, \tag{4.6}$$

where the symbols, used for shortening the form, are used in accordance with the vector theory [13]:

$$(\bar{v}, \text{grad})\bar{v} = \bar{i}\left(v_x\frac{\partial v_x}{\partial x} + v_y\frac{\partial v_x}{\partial y} + v_z\frac{\partial v_x}{\partial z}\right) +$$

$$+\bar{j}\left(v_x\frac{\partial v_y}{\partial x} + v_y\frac{\partial v_y}{\partial y} + v_z\frac{\partial v_y}{\partial z}\right) + \bar{k}\left(v_x\frac{\partial v_z}{\partial x} + v_y\frac{\partial v_z}{\partial y} + v_z\frac{\partial v_z}{\partial z}\right);$$

$$\text{grad}\,p = \bar{i}\frac{\partial p}{\partial x} + \bar{j}\frac{\partial p}{\partial y} + \bar{k}\frac{\partial p}{\partial z};$$

$$\nabla^2 v \equiv \Delta v = \bar{i}\left(\frac{\partial^2 v_x}{\partial x^2} + \frac{\partial^2 v_x}{\partial y^2} + \frac{\partial^2 v_x}{\partial z^2}\right) +$$

$$+\bar{j}\left(\frac{\partial^2 v_y}{\partial x^2} + \frac{\partial^2 v_y}{\partial y^2} + \frac{\partial^2 v_y}{\partial z^2}\right) +$$

$$+\bar{k}\left(\frac{\partial^2 v_z}{\partial x^2} + \frac{\partial^2 v_z}{\partial y^2} + \frac{\partial^2 v_z}{\partial z^2}\right),$$

$\bar{i}, \bar{j}, \bar{k}$ are the unit vectors of the Cartesian coordinate system. We introduce dimensionless parameters for speed v, density ρ, gravitational acceleration g, viscosity ν, coordinates x, y, z and pressure p:

$$\frac{v}{v_0} = v^*, \frac{\rho}{\rho_0} = \rho^*, \frac{g}{g_0} = g^*, \frac{\nu}{\nu_0} = \nu^*, \frac{x_i}{L} = x_i^*, \frac{p}{p_0} = p^*.$$

Here u_0, r_0, g_0, v_0, L, p_0 at the characteristic values of the flow parameters. We introduce these values into equation (4.6) and present it in the new form:

$$\frac{v_0^2}{L}(v^*, \text{grad})v^* = g_0 g^* - \frac{p_0}{\rho_0 L}\cdot\frac{1}{\rho^*}\,\text{grad}\,p^* + \frac{\nu_0 v_0}{L^2}\nabla^2 v^*.$$

Separating all terms of equation (4.6) by $\frac{v_0^2}{L}$, we obtain

$$(v^*, \text{grad})v^* = \frac{g_0 L}{v_0^2} g^* - \frac{p_0}{\rho_0 v_0^2} \frac{1}{\rho^*} \text{grad}\, p^* + \frac{\nu_0}{v_0 L} \nabla^2 v^*,$$

in which each term contains dimensionless complexes-generally known a similarity criteria:

$$\frac{g_0 L}{v_0^2} = \text{Fr (Froode number)}, \frac{p_0}{\rho_0 v_0^2} = \text{Eu (Euler number)},$$

$$\frac{v_0 L}{\nu_0} = \text{Re (Reynolds number)}.$$

Thus, the systematic method makes it possible determine dimensionless relationships including the values of the parameters of the process and the physical characteristics of the medium. These dimensionless complexes are then used as similarity criteria in accordance with the dimensionality theory.

The systematic method, based on a specific system of fundamental equations of the process, even if these equations cannot be solved of successively, provides a considerably larger amount of information than elementary analysis in the parametric approach. In particular, this is clearly evident in the examination of electric arc processes accompanied by different physical phenomena. Nevertheless, when searching for a similarity criteria in plasma dynamics, it is convenient to use both methods: the selection of the one of the methods is determined by the knowledge of physical processes and by the possibility of describing these processes by corresponding equations.

4.3. SIMILARITY CRITERIA OF ELECTRIC ARC PROCESSES

The system of equations, describing approximately the processes taking place in the discharge chamber of the electric arc plasma torch, includes:

– the equation of motion $\rho\frac{\partial v}{\partial t} + \rho(v, \text{grad})v = \rho g + \text{grad}\, p$

$+[\overline{jB}] + \rho\nu\nabla^2 v,$

– the equation of continuity of the flow $G = \int \rho v dS,$

– the equation of continuity of the current $I = \int j dS,$

– the equation of energy per unit length of the arc column

$$\frac{\pi D^2}{4}\rho \upsilon \,\text{grad}\,(h+\frac{\upsilon^2}{2})+\pi D\alpha(T-T_w)=$$

$$=\pi d_a \sigma_i \varepsilon_i T_a^4+\frac{4I^2}{\pi\sigma d_a}+\pi d_a \lambda \,\text{grad}\,T,$$

– the equation of rotation of the magnetic field $\bar{B}=\mu_M \bar{j}$,

– the Ohm law $j = sE$,

– the equation of the potential difference $U = \int Edx$,

– the shunting condition [7, 14] $a_1 \frac{Ex}{pD} > U_i$,

where

$$[\bar{j}\ \bar{B}]=\bar{i}(j_y B_z - j_z B_y)+\bar{j}(j_z B_x - j_x B_z)+\bar{k}(j_x B_y - j_y B_x);$$

$$\text{rot}\,\bar{B}=i\left(\frac{\partial B_z}{\partial y}-\frac{\partial B_y}{dz}\right)+\bar{j}\left(\frac{\partial B_x}{\partial z}-\frac{\partial B_z}{\partial x}\right)+\bar{k}\left(\frac{\partial B_y}{\partial x}-\frac{\partial B_x}{\partial y}\right);$$

μ_M is the magnetic permittivity of matter, H/m; U_i is the ionisation potential of the atoms; B is magnetic induction, T; $a_1 = 8kT/\pi d^2$, $k = 1.38 \cdot 10^{-23}$ J/deg is the Boltzmann constant; d is the atom diameter.

In order to close the system, the latter should include: the dependence of density p, enthalpy h, electrical conductivity σ, heat conductivity λ, and the radiation coefficients ε on temperature T, pressure p, the type of gas, and also description of the boundary conditions which depend on the design of the plasma torch.

To determine the similarity criteria, all the equations of the system are represented in the dimensionless form by means of replacing dimensional parameters by dimensionless ones:

$$\upsilon^*=\frac{\upsilon}{\upsilon_0},\ T^*=\frac{T}{T_0},\ \rho^*=\frac{\rho}{\rho_0},\ p^*=\frac{p}{p_0},\ j^*=\frac{j}{j_0},$$

$$h^*=\frac{h}{h_0},\ \lambda^*=\frac{\lambda}{\lambda_0},\ E^*=\frac{E}{E_0},\ \sigma^*=\frac{\sigma}{\sigma_0},\ \nu^*=\frac{\nu}{\nu_0},\ \mu_M^*=\frac{\mu_M}{\mu_{M,0}},$$

$$\chi_i^*=\frac{\chi_i}{\chi_0},\ B^*=\frac{B}{B_0},\ U^*=\frac{U}{U_0},\ t^*=\frac{t}{\tau_0},\ I^*=\frac{I}{I_0},$$

$$D^*=\frac{D}{L},\ d^*=\frac{d}{L}.$$

After this operation, the system has the following form:

$$\frac{\rho_0 v_0}{\tau_0}\rho^*\frac{\partial v^*}{\partial t^*}+\frac{\rho_0 v_0^2}{L}\rho^*(v^*,\ \mathrm{grad})v^*=\rho_0 g_0\rho^* g^*+\frac{p_0}{L}\ \mathrm{grad}\,P^*+$$

$$+j_0 B_0[\bar{j}^*\bar{B}^*]+\frac{\rho_0 \nu_0 v_0}{L^2}\rho^*\nu^*\nabla^2 v^*, \qquad (a)$$

$$G_0 G^*=\rho_0 v_0 S_0\int\rho^* v^* dS^*, \qquad (b)$$

$$I_0 I^*=j_0 L^2\int j^* dS^*, \qquad (c)$$

$$\rho_0 L v_0 h_0\frac{\pi}{4}D^{*2}\ \rho^* v^*\ \mathrm{grad}\,h^*+\rho_0 L v_0^3\frac{\pi}{4}D^{*2}\ \rho^* v^*\,\mathrm{grad}\frac{v^{*2}}{2}+$$

$$1\ L\alpha_0 T_0\pi D^*\alpha^*(T^*-T_w^*)=L\sigma_i\varepsilon_i T_0^4\pi d_a^*\sigma_i^*\varepsilon_i^* T_a^{*4}+$$

$$1\ \frac{I_0^2}{\sigma_0 L^2}\frac{4}{\pi}\frac{I^{*2}}{\sigma^* d_a^{*2}}+\lambda_0 T_0\pi d_a^*\lambda^*\,\mathrm{grad}\ T^*, \qquad (d)$$

$$\frac{B_0}{L}\mathrm{rot}\,\bar{B}^*=\mu_{M,0}\,j_0\mu_M^*\bar{j}^*, \qquad (e)$$

$$j_0 j^*=\sigma_0 E_0\sigma^* E^*, \qquad (f)$$

$$U_0 U^*=E_0 L\int E^* dl^*, \qquad (g)$$

$$a_1\frac{E_0}{p_0}\frac{E^*\chi^*}{p^* d^*}>U_0 U_i^*, \qquad (h)$$

All the terms in equation (*a*) are divided by the scale coefficient at the inertia term $\frac{\rho_0 v_0^2}{L}$, in equation (*b*) by G_0, in equation (*c*) by I_0, in equation (*d*) by $\rho_0 L v_0 h_0$, in equation (*e*) by B_0/L, in equation (*f*) by j_0, in equation (*g*) by U_0, and in equation (*h*) by U_0.

After this operation, the equation become dimensionless:

$$\frac{L}{\tau_0 v_0}\rho^*\frac{\partial v^*}{\partial t^*}+\rho^*(v^*,\ \mathrm{grad})v^*=\frac{g_0 L}{v_0^2}\rho^* g^*+\frac{p_0}{\rho_0 v_0^2}\mathrm{grad}\,p^*+$$

$$1\ \frac{j_0 B_0 L}{\rho_0 v_0^2}[j^* B^*]+\frac{\nu_0}{v_0 L}\rho^*\nu^*\nabla^2 v^*, \qquad (a')$$

$$G^*=\frac{\rho_0 v_0 L^2}{G_0}\int\rho^* v^* dS^*, \qquad (b')$$

$$I^* = \frac{j_0 L^2}{I_0} \int j^* dS^*, \tag{c'}$$

$$\frac{\pi}{4} D^{*2} \rho^* \upsilon^* \mathrm{grad} h^* + \frac{\upsilon_0^2}{h_0} \frac{\pi}{4} D^{*2} \rho^* \upsilon^* \mathrm{grad} \frac{\upsilon^{*2}}{2} +$$

$$+ \frac{\alpha_0 T_0 \pi}{\rho_0 \upsilon_0 h_0} \alpha^* (T^* - T_w^*) = \frac{\sigma_n \varepsilon_n T_0^4}{\rho_0 \upsilon_0 h_0} 4 d_a^* \sigma_i^* \varepsilon_i^* T_a^{*4} +$$

$$+ \frac{I_0^2}{\sigma_0 L^3 \rho_0 \upsilon_0 h_0} \frac{4}{\pi} \frac{I^{*2}}{\sigma^* d_a^{*2}} + \frac{\lambda_0 T_0}{\rho_0 L \upsilon_0 h_0} \pi d_a^* \lambda^* \,\mathrm{grad}\, T^*, \tag{d'}$$

$$\mathrm{rot} \overline{B}^* = \frac{\mu_{\mathrm{M},0} j_0 L}{B_0} \mu_{\mathrm{M}}^* \overline{j}^*, \tag{e'}$$

$$j^* = \frac{\sigma_0 E_0}{j_0} \sigma^* E^*, \tag{f'}$$

$$U^* = \frac{E_0 L}{U_0} \int E^* dl^*, \tag{g'}$$

$$\frac{a_1 E_0}{p_0 U_0} \cdot \frac{E^* x^*}{p^* d^*} > U_i^*. \tag{h'}$$

The dimensionless coefficients in front of every term in the dimensionless equations are similarity criteria. We present them in the appropriate sequence and enumerate:

$$K_1 = \frac{L}{\tau_0 \upsilon_0},\ K_2 = \frac{g_0 L}{\upsilon_0^2},\ K_3 = \frac{p_0}{\rho_0 \upsilon_0^2},\ K_4 = \frac{j_0 B_0 L}{\rho_0 \upsilon_0^2},$$

$$K_5 = \frac{\nu_0}{\upsilon_0 L},\ K_6 = \frac{\rho_0 \upsilon_0 L^2}{G_0},\ K_7 = \frac{j_0 L^2}{I_0},\ K_8 = \frac{\upsilon_0^2}{h_0},$$

$$K_9 = \frac{\alpha_0 T_0}{\rho_0 \upsilon_0 h_0},\ K_{10} = \frac{\sigma_i \varepsilon_i T_0^4}{\rho_0 \upsilon_0 h_0},\ K_{11} = \frac{I_0^2}{\sigma_0 h_0 L^3 \rho_0 \upsilon_0},\ K_{12} = \frac{\lambda_0 T_0}{\rho_0 L \upsilon_0 h_0},$$

$$K_{13} = \frac{\mu_{\mathrm{M},0} j_0 L}{B_0},\ K_{14} = \frac{\sigma_0 E_0}{j_0},\ K_{15} = \frac{E_0 L}{U_0},\ K_{16} = \frac{a_1 E_0}{p_0 U_0}.$$

The system of criteria, describing electric arc plasma, is not restricted only to the above equations. Many of them are more suitable for practical application or have no physical meaning. Similarity theory shows that any combination of the criteria is also a similarity criterion. Consequently, the resultant system of criteria can be transformed

to the form more suitable for application. Some of the criteria will be transformed using different combinations:

$$K_{17} = K_{11} \cdot K_6 = I_0^2 / \sigma_0 h_0 G_0 L,\ K_{18} = K_9 / K_{12} = \alpha_0 L / \lambda_0,$$
$$K_{19} = K_{10} / K_{11} = \sigma_0 \sigma_i \varepsilon_i T_0^4 L^3 / I_0^2,$$
$$K_{20} = K_{10} \cdot K_6 = \sigma_u \varepsilon_i T_0^4 L^2 / h_0 G_0,$$
$$K_{21} = K_{15} / K_{16} = p_0 L / a_1,\ K_{22} = K_9 \cdot K_6 = \alpha_0 T_0 L^2 / h^0 G_0,$$
$$K_{23} = K_{14} \cdot K_7 = \alpha_0 E_0 L^2 / I_0,$$
$$K_{24} = K_4 / K_3 K_7 = B_0 I_0 / p_0 L,\ K_{25} = 1 / K_5 K_6 = G_0 / \rho_0 v_0 L,$$
$$K_{26} = K_4 K_{13} / K_3 K_7^2 = \mu_{M,0} I_0^2 / p_0 L^2,$$
$$K_{27} = K_{14} K_7 / K_{15} = \sigma_0 U_0 L / I_0,$$
$$K_{28} = K_4 K_6^2 / K_7 = B_0 I_0 L^3 \rho_0 / G_0^2.$$

The entire set of the resultant similarity criteria is divided into two groups: determining and determined. The group of the determining criteria includes criteria whose structure contains the regime parameters of the arc process, such as: L, I, B, G, p. The group includes the following criteria:

$$K_{17} = I_0^2 / \sigma_0 h_0 G_0 L,\ K_{21} = p_0 L / a_1,\ K_{25} = G_0 / \rho_0 v_0 L,$$
$$K_{26} = \mu_{m,0} I_0^2 / p_0 L^2,\ K_{24} = B_0 I_0 / p_0 L, \text{ or } K_{28} = B_0 I_0 L^3 \rho_0 / G_0^2.$$

The group of the determined criteria includes the numbers whose structure contains some of the required (determined) quantities, for example, U, E, α, etc. This group is formed by the criteria:

$$K_{27} = \sigma_0 U_0 L / I_0,\ K_{23} = \sigma_0 E_0 L^2 / I_0,$$
$$K_{22} = \alpha_0 T_0 L^2 / h_0 G_0,\ K_{20} = \sigma_u \varepsilon_i E T_0^4 L^2 / h_0 G_0,$$

where α is the heat transfer coefficient; σ_i=5.7·10^{-8} W/(m^2·deg^4) is the Stefan–Boltzmann constant; ε_t is the emissivity of total normal radiation.

4.4. PHYSICAL MEANING OF SIMILARITY CRITERIA

Since the similarity criteria were derived from homogeneous physical equations after making the latter dimensionless, it is natural that each criterion is responsible to a certain degree for a specific physical process or phenomenon. We shall try to describe here the content of the specific criteria which may play a significant role in the electric arc plasma torches and can be used in generalisation of the experimental results.

The criterion $K_{22} = \alpha_0 T_0 L^2 / h_0 G_0$ characterises the level of heat losses from the walls of the discharge chamber as a result of convection in comparison with the thermal power of the plasma jet. The criterion $K_{24} = B_0 I_0 / p_0 L$ compares the magnetic pressure in the arc, determined by the intrinsic magnetic field, with the gas-dynamic pressure. Since the magnetic pressure is manifested in the form of the pinch effect, its value is high where the diameter of the arc column is small and current density is high, i.e. in the areas of constriction of the arc. Generally, the cross-section of the are column rapidly decreases (constriction) in the vicinity of the electrodes and in small-diameter diaphragms. Therefore, this criterion should be taken into account only for short arcs or for arcs in a narrow channel where the diaphragm areas occupy a large part of the length. If there is no forced restriction of the discharge diameter and the arc length is considerably greater than the length of the zone in the vicinity of the electrode, the effect of the intrinsic magnetic field on the processes in the arc and the discharge properties may be ignored, i.e., the criterion K_{26} may be excluded from the system of the determining criteria.

The criterion $K_{20} = \sigma_t \varepsilon_t T_0^4 L^2 / h_0 G_0$ shows the fraction represented by the radiation energy of the arc in relation to the thermal energy of the plasma flow. This criterion is especially important at high currents when the temperature in the arc column is high.

The criterion $K_{19} = \sigma_0 \sigma_t \varepsilon_t T_0^4 L^3 / L_0^2$ is the fraction of the energy irradiated by the arc in relation to the Joule heat generation in the arc.

The criterion $K_{17} = I_0^2 / \sigma_0 L h_0 G_0$ is an energy criterion. This criterion determines the intensity of energy exchange between the column of the electrical arc and the heated medium. It shows the extent by which the power of heat generation of the arc N_a is greater than the thermal power of the jet N_t, i.e. characterises the efficiency of the plasma torch as a thermal system and may be interpreted as follows:

$$K_{17} \sim \frac{N_a}{N_t} = \frac{1}{\eta},$$

where $\eta = N_t / N_a$ is the thermal efficiency of the plasma torch.

$K_{18} = \alpha_0 L / \lambda_0$ is the Nu number (Nusselt). This number characterises the relationship between the intensity of heat transfer and the temperature field in the boundary layer and shows the number of times by which the convective heat transfer is greater than conductive heat transfer.

The $K_{25} = G_0/\rho_0 v_0 L$ is the Re (Reynolds) number which determines the relationship between the inertia forces of the flow and the viscosity forces. At the critical value Re_{cr} the laminar flow regime changes to turbulent. In flow in a pipe, the Reynolds number may be interpreted as the ratio of the kinetic power of the jet to the power of the friction forces.

In some cases, the energy criterion $K_{17} = I_0^2/\sigma_0 L h_0 G_0$ in generalisation of the volt–ampere characteristics of the arc is replaced by another criterion which is a combination of K_{17} and K_{25}:

$$K_{21} = \sqrt{K_{17} \cdot K_{25}} = \frac{I_0}{L}\sqrt{\frac{1}{\sigma_0 h_0 \rho_0 v_0}}.$$

The criterion $K_{21} = p_0 L/a_1 \sim 1/\text{Kn}$ is a number reciprocal to the Knudsen criterion $\text{Kn} = \lambda_e/L$. This number characterises the development of the electrophysical process of large-scale shunting in the discharge chamber of the plasma torch. The process is based on a breakdown between the arc and the chamber wall. Physically, K_{21} is the electrical strength of this gap showing the number of free path lengths of the electron which fit in it, since $\lambda_e \sim 1/p$. The criterion $K_{28} = B_0 I_0 L^3 \rho_0/G_0^2$ gives the relationship between the effect of the electromagnetic and aerodynamic forces on the arc. It should be taken into account in the sections of the electric arc discharge where the electromagnetic and inertia forces are comparable. This relates primarily, for example, to coaxial plasma torches with the rotation of the discharge under the effect of electromagnetic forces.

Criterion $K_8 = v_0^2/h_0$ is the relationship between the kinetic and thermal energy of the flow. It is proportional to M^2, where $M = v/a$ is the Mach number which characterises the extent of transformation of the heat content of the kinetic energy of the flow. Usually, in plasma dynamics, this criterion is not important because the kinetic energy of the flow in the zone of the arc discharge can be ignored in comparison with thermal energy. For plasma torches with vortex stabilisation this ratio is approximately 10% or lower, consequently, the number M has only a slight effect on the properties of the discharge and may be excluded from the number of determining parameters. However, in a number of systems, for example, in railgun accelerators, in the presence of a strong magnetic field (of the order of 1 T and higher) the speed of movement of the arc discharge may reach or even exceed the speed of sound and, therefore, the number M must be included in the system of the determining criteria.

The criterion $K_{27} = \sigma_0 U_0 L/I_0$ characterises the strength of the atom discharge and belongs to the number of determining criteria, exactly like $K_{23} = \sigma_0 E_0 L^2/I_0$ is the criterion of the strength of the electrical field of the arc.

The main experimental method of determination of the heat losses in the plasma torches is the calorimetric method. In this method, the physical nature of heat losses is not important: radiation, convective or conductive. For thermal estimates it is convenient to combine the criteria responsible for the heat losses: K_{10}, K_9 and K_{12} into a single criterion. The complex resulting from the summation of these criteria, multiplied by the number K_6, the presence of total heat losses Q_{loss} in the plasma torch related to the heat content of the plasma jet Q_j, i.e.

$$(K_9 + K_{10} + K_{12})K_6 = Q_{loss} / Q_j.$$

This relationship may be referred to as the coefficient of heat losses $\tilde{\eta} \equiv Q_{loss}/Q_j$. The thermal efficiency of the plasma torch as equipment converting the electrical energy to the concentrated form of thermal energy, is characterised, as mentioned previously, by the thermal coefficient of efficiency which is the main output parameter of the plasma torch. Thermal efficiency η and the coefficient of heat losses $\tilde{\eta}$ are linked by the relationship:

$$\tilde{\eta} = \frac{1-\eta}{\eta} \text{ or } \eta = \frac{1}{1+\tilde{\eta}}.$$

4.5. METHOD FOR GENERALISING EXPERIMENTAL RESULTS

The final aim of criterial processing of the experimental data is the determination, in the form of specific equations, of the generalised VAC of the arc for the plasma torches of the investigated class and generalised thermal characteristics. These equations, together with the data for the erosion of the cathode and anode in relation to the external conditions, and also with the data on the cooling system of the element of the plasma torch and a number of other data are used as a basis for designing and developing more efficient and advanced electric arc gas heaters.

In processing of experimental materials obtained in the same working medium (for example, in air) a general method has been developed for simplified expression of the similarity criteria in the form of dimensional complexes from the changing part of the criterion. In the methods, for the given gas, the values of the coefficients, reflecting the physical properties, are assumed to be constant and are transferred from the dimensionless similarity criteria. The dimensional complex, remaining after this processing, consists of the regime parameters of the process: *G, I, d, p*, etc. This measure is essential because at present time there are no universal generalised equations valid for all gases.

According to the above considerations, it is essential to disregard the physical properties of the selected criteria and representing them in a more suitable form. The characteristic size *L* of the plasma torches is usually represented by the diameter of the discharge chamber *d*.

The group of the determining criteria:

$$I^2/Gd,\ G/d,\ pd,\ BI/pd,\ I^2/pd^2 \text{ and so on}$$

The group of the determined criteria:

$$U,\ Ed,\ Ud/I,\ Ed^2/I,\ \tilde{\eta} \text{ and so on}$$

The well-known Pashen law of the electrical breakdown, which is a result of the application of the method of similarity theory and dimensional theory to electric discharge in a stationary gas, plays a significant role in the generalisation of the experimental data on the electric arc in the gas flow.

In the case of an electrical breakdown (shunting) between the electrode and the arc in the plasma torch with the gas flow, the breakdown voltage U_s should depend not only on the dimensional criterion *pd*, but also on the dimensional part of the Reynolds number and the energy criterion,i.e.

$$U_s = f(pd, I^2/Gd, G/d).$$

Thus, we obtain a set of determining and determined criteria playing the role of arguments and functions, respectively:

$$Ud/I = f_U(I^2/Gd,\ G/d,\ pd...),$$

$$\tilde{\eta} = f_{\tilde{\eta}}(I^2/Gd,\ G/d,\ pd...).$$

Usually, the approximating function is obtained in the form of the product of the powers of the similarity criteria:

$$Ud/I = A_U(I^2/Gd)^{n_1}(G/d)^{n_2}(pd)^{n_3}, \tag{4.7}$$

$$\tilde{\eta} = A_{\tilde{\eta}}(I^2/Gd)^{n_1'}(G/d)^{n_2'}(pd)^{n_3'}. \tag{4.8}$$

The method for determination of the exponents n_i at criteria K_i is examined on the example of determination of the exponents n_1, n_2 and n_3 in equation (4.7). Taking the logarithm of this equation

$$\lg\frac{Ud}{I}=\lg A_{v}+n_1\lg\frac{I^2}{Gd}+n_2\lg\frac{G}{d}+n_3\lg(pd). \qquad (4.9)$$

Subsequently, we vary in succession each of the criteria in the right-hand side of the equation with all other parameters constant. If the exponents at constant, equation (4.9) is degenerated into a straight line equation with the angular coefficient of the type:

$$\lg\frac{Ud}{I}=C_i+n_i\lg K_i,$$

from which we determine, by the graphical method, *i*-th exponent n_i and criterion K_i, etc.

After determining all exponents n_i from several experimental points, wecalculate the constant multiplier A_U and accept its mean value.

Since the combination of the criteria also represents a criterion, we compile the following combination of the criteria:

$$(I^2/Gd\cdot G/d)^{0.5}=I/d=K^{*}.$$

Using the criterion K^*, the generalised equation for the VAC of the arc

$$Ud/I=A_U(I^2/Gd)^{n_1}(G/d)^{n_2}(pd)^{n_3}$$

may be presented in a form more suitable for application resolved in relation to the arc voltage. For this purpose, the equation is additionally multiplied on the left and right by the number $K^* = I/d$ and after simple transformations we obtain:

$$U=A_U(I^2/Gd)^{n_1+0.5}(G/d)^{n_2+0.5}(pd)^{n_3}.$$

The criterial equations, obtained on the basis of generalisation of the experimental data, can can be used efficiently only in the limits in which the similarity criteria included in the given generalised equation were verified and are reliable. Extrapolation outside the limits of action of the similarity criteria is burdened with inaccuracies and even qualitative errors.

As shown previously, the number of the similarity criteria is large. The most important criteria should be selected from the group of these criteria in the generalisation of the experimental data. The selection principle is simple: if the addition of a new criterion provides a correction within the limits of accuracy of the experiments, it is not rational to introduce it. Consequently, as the accuracy of the experiments increases the efficiency of selection of the most important criteria, influencing the arc characteristics, also increases.

The accuracy of approximation of the unknown dependences is

also of considerable importance. If one equation is insufficient because of accuracy to describe the entire examined range of the parameters, the experimental curve is divided into a number of characteristic sections and a generalised formula is found for each section.

Examples of the application of criterial dependences for the generalisation of the characteristics of the electric arc in plasma torches of different systems are presented in the following chapter.

Chapter 5

Energy characteristics of the arc in different gases

The main energy characteristic of the arc burning in an electric arc gas heater, i.e. plasma torch (plasmatron), is the volt–ampere characteristic which determines, with other conditions being equal, the relationship between the arc voltage and current intensity in the arc. The classification of linear plasma torches, presented in chapter 1, is based on differences in the formation of the volt–ampere characteristics of the arc for the main circuit of the plasma torches. Knowledge of the volt–ampere characteristic of the arc and the possibility of calculating the characteristic for each specific case enable development of electric arc generators of the thermal plasma with the given vacuum parameters. Fundamentals of the calculation of the volt–ampere characteristics of the arc in plasma torches of different systems were described in the first attempts for the generalisation of the experimental results of investigations of the plasma torches in the criterional form [1–3]. The possibilities of using the similarity criteria for describing the processes in electric arc plasma torches have been described in chapter 4. Below, we examine specific examples of the application of criteria of complexes for the generalisation of the energy characteristics of the arc in different conditions.

5.1. GENERALISED VOLT–AMPERE CHARACTERISTICS OF THE ARC IN DIFFERENT GASES

As already mentioned in chapter 4, in analysis of the results of investigations of the plasma torches working with media of the same chemical composition, it is very efficient to replace the dimensionless similarity criteria by dimensional complexes that are most important in the investigated conditions. This is supported by the fact that the thermophysical and transfer properties of the gas in the dimensionless complexes and are selected, to simplify calculations, at some constant temperature characteristic of the arcing processes in the plasma torch [4]. Some of the results presented below are generalised in thi form.

The plasma torches with a self-setting ac length and the arc length fixed by a ledge have been studied most excessively in industry. This is due to simple design, reliability and the fact that other gas heaters are not suitable for this application. The result of a large number of investigations of the plasma torches of these systems [4–9] show they can be calculated using the generalised electrical and thermal characteristic. At the given geometry of the plasma torch and the same working gas these characteristics depend on a small number of the determining parameters.

The equation of the volt–ampere characteristic (VAC) of the DC air arc single-chamber plasma torches with a self-setting arc length and straight polarity of connection of the electrodes (output electron – anode) has the form:

$$U^{+} = 1290(I^{2}/Gd)^{-0.15}(G/d)^{0.30}(pd)^{0.25}. \tag{5.1}$$

The agreement between the experiments and the calculations is shown in Fig. 5.1. The maximum deviation of the experimental values from the calculated curve does not exceed 6–8% in the range of variation of the complexes:

$$I^{2}/Gd = 1\cdot 10^{7} \div 4\cdot 10^{10}\, A^{2}\cdot \mathrm{s/kg\cdot m});$$

$$G/d = 0.1 \div 2.0\ \mathrm{kg/(m\cdot s)};\quad pd = (5 \div 35)\cdot 10^{2}\,\mathrm{N/m}.$$

In the same plasma torch at a reverse polarity of connection of the electrodes (output electron-cathode) the equation of the volt–ampered characteristic of the arc is written in the form:

$$U^{-} = 1970\,(I^{2}/Gd)^{-0.17}(G/d)^{0.15}(pd)^{0.25}. \tag{5.2}$$

Comparison of the equations (5.1) and (5.2) shows that the difference of the U–I characteristics of the arc in the single-chamber plasma torch is greater than the quantitative difference associated with the difference in the processes of shunting of the arc in the output electrode at straight and reversed polarity. Different exponents at (G/d) (or number Re_d) reflect, in all likelihood, the stronger effect of shunting of the arc in the output electrode of the process of electrical breakdown at reversed polarity, i.e. breakdown from the cold cathode. Therefore, shunting takes place quite frequently in the loop of the arc, and the arc spot on the electrode remains

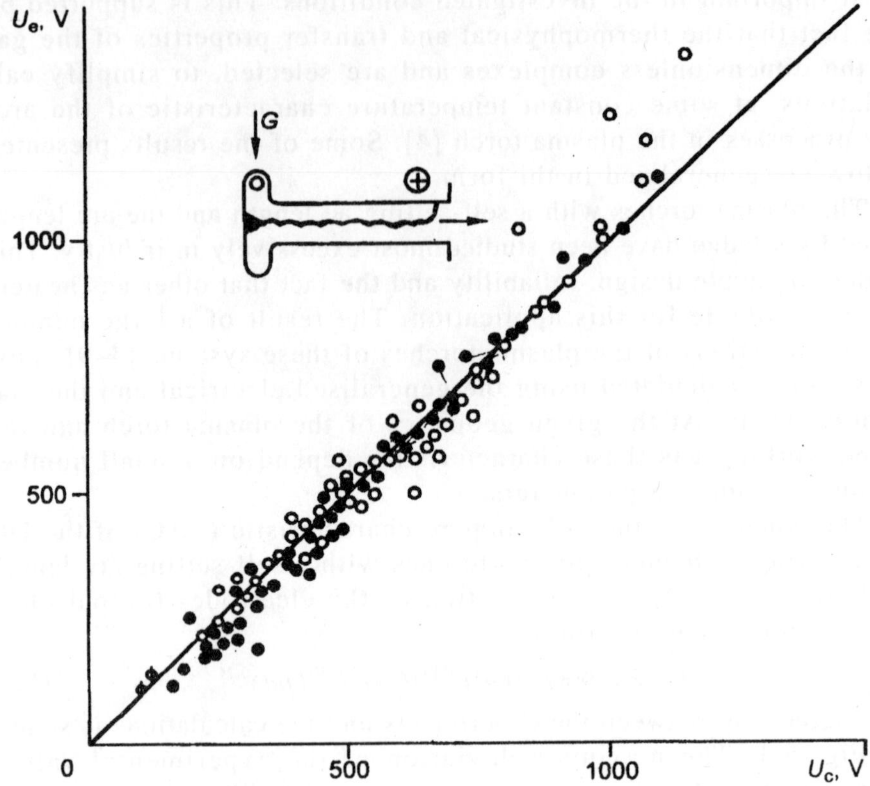

Fig. 5.1. Experimental data and the generalised volt–ampere characteristic of the single-chamber plasma torch with a self-setting arc length. U_e - experimental value of voltage; U_c – calculated from equation (5.1) [4].

stationary.

In the two-chamber plasma torch, the VAC of the air arc at straight polarity of connection of the electrodes is calculated from the equation [5]

$$U^+ = 1360(I^2 / Gd)^{-0.20}(G/d)^{0.25}(pd)^{-0.35} = 1360\varphi. \qquad (5.3)$$

Equation (5.3) of the maximum deviation of the experimental points from the calculation curve smaller than 12% holds in a very wide range of variation of the complexes:

$$I^2 / Gd = 1 \cdot 10^6 \div 4 \cdot 10^9 \ \text{A}^2\text{s}/(\text{kg} \cdot \text{m});$$

$$G/d = 5 \cdot 10^{-2} \div 26 \ \text{kg}/(\text{s} \cdot \text{m});$$

$$pd = 1 \cdot 10^3 \div 8 \cdot 10^5 \ \text{N}/\text{m}$$

and determining parameters:

$$I = 50 \div 5000 \text{ A}; G = 1 \cdot 10^{-3} \div 3.5 \text{ kg/s};$$
$$d = (5 \div 76) \cdot 10^{-3} \text{ m}; p = (1 \div 100) \cdot 10^5 \text{ Pa}.$$

It was also noted that at a current intensity higher than 300–400 A, the VAC of the arc for both polarities of connection of the electrodes merged almost completely into a single characteristic. Consequently, it is justified to use equation (5.2) as a single equation for calculating the arc at both polarities. Comparison of the experiments in the calculations using equation (5.3) is shown in Fig. 5.2 (for straight polarity)

The VAC of the arc, burning in a plasma torch with two-sided discharge in a wide range of variation of the determining parameters is satisfactory described by the equation:

$$U = 3060\,(I^2/Gd)^{-0.17}(G/d)^{0.12}(pd)^{0.25}. \tag{5.4}$$

This equation has exponents at different complexes similar to equation (5.2). The coefficient at the dimensional complexes in (5.4) is close to the sum of the coefficients of equations (5.2) and (5.1). Evidently, this is associated with the fact that the plasma torch with two-side discharge may be treated as a single-chamber plasma torch

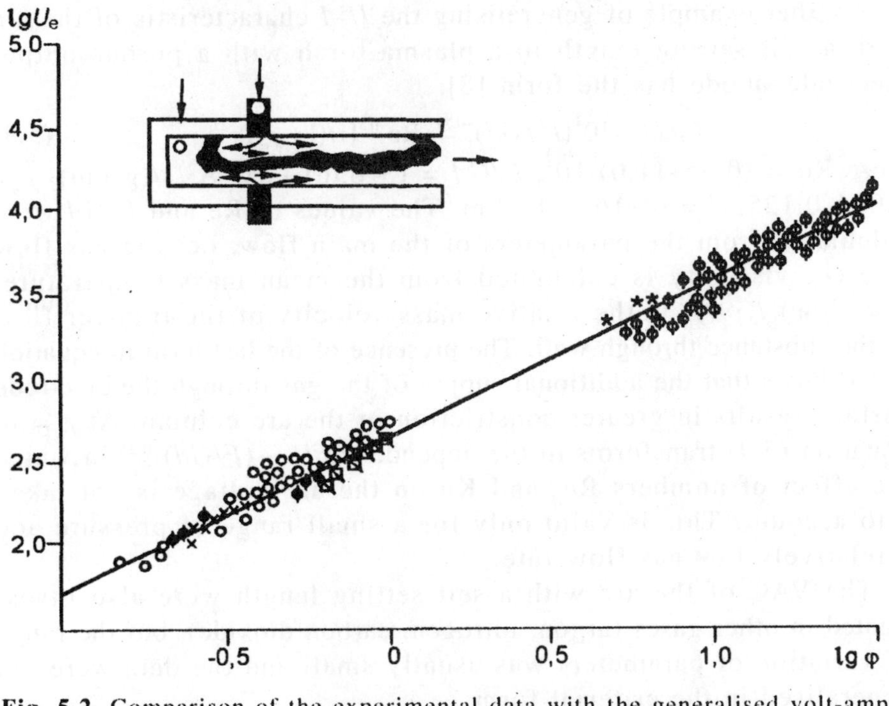

Fig. 5.2. Comparison of the experimental data with the generalised volt-ampere characteristic of the arc in the two-chamber plasma torch (equation (5.3) [5].

in which the arcs with the self-setting length are combined.

For the AC air arc with high-frequency, the VAC of the single-chamber plasma torch is presented in the form:

$$\tilde{U} = 3930(I^2/Gd)^{-0.18}(G/d)^{0.28}(pd)^{0.20} \tag{5.5}$$

in the range of the variation of the parameters

$$I^2/Gd = 10^7 \div 4\cdot 10^{10}\ \mathrm{A^2\cdot s/(kg\cdot m)};$$
$$G/d = 0{,}1 \div 20\ \mathrm{kg/(s\cdot m)};\ pd = 500 \div 3500\ \mathrm{N/m}.$$

For the two-chamber plasma torch we have

$$U = 2150\,(I^2/Gd)^{-0.15}(G/d)^{0.16}(pd)^{0.20}. \tag{5.6}$$

For a hydrogen arc [3] the equation of the VAC is written in the form:

$$U^+ = 9700\,(I^2/Gd)^{-0.20}(G/d)^{0.50}(pd)^{0.36}. \tag{5.7}$$

In methane [7], the VAC of the arc in a single-chamber plasma torch with a cup-shaped internal electrode has the form:

$$U^+ = 1.525\cdot 10^5 (I^2/Gd)^{-0.35}(G/d)^{0.35}(pd)^{0.185}(\bar{d})^{0.475}. \tag{5.8}$$

The range of variation of the determining parameters: I = 40÷1000 A, d = (1.2÷8.6)·10^{-2} m, G = 0.009÷0.525; P = (1÷1.8)·10^5 Pa, $\bar{d} = d_k/d$, $d_k \geq d$.

Another example of generalising the U–I characteristic of the arc with a self-setting length in a plasma torch with a porous output electrode–anode has the form [8]:

$$Ud/I = 10^4 (I^2/Gd)^{-0.75}\cdot \mathrm{Re}^{-0.5}[1/(1+\tilde{j}_w)]^{-2.6}. \tag{5.9}$$

Here Re = (0.35÷11.0)·10^3, I^2/Gd = (3÷656) · 10^2 A^2 s/(g cm), $\tilde{j}_w$= 0.014÷0.125; d = (4÷16) · 10^{-3} m. The values of Re and I^2/Gd were calculated from the parameters of the main flow, i.e. the gas flow rate G_0, viscosity is calculated from the mean mass temperature; $\tilde{j}_w = (\rho w)_w/(\rho w)_0$ is the relative mass velocity of the transfer flow of the substance through wall. The presence of the last term in equation (5.9) shows that the additional supply of the gas through the electrode surface results in greater constriction of the arc column. At $\tilde{j}_w = 0$, equation (5.9) transforms to the dependence $U \sim (I^2/Gd)^{-0.25}$ in which the effect of numbers Re_d and Kn on the arc voltage is not taken into account. This is valid only for a small range of pressure and a relatively low gas flow rate.

The VAC of the arc with a self-setting length were also investigated in other gases (argon, nitrogen, carbon dioxide), but the range of variation of parameters was usually small and the data were not generalised in the criterial form.

The VACs of the arc in the plasma torches with self-setting arc length were discussed previously. These equations were derived in the general form taking into account the main criteria. At the same time, in many applications, especially in the case of a narrow range of the variation of the parameters, it is necessary to modify the equations because any combination of the dimensional criteria is also a criterion. The varied part of the criterial complexes may be presented in the form I/d; I/G; $I/(Gd{\cdot}p)$, etc. For example, after replacing the complex I^2/Gd by $I/(Gd{\cdot}p)$, equation (5.1) has the form:

$$U^+ = 1290(I/Gd\,p)^{-0.30}(G/d)^{0.15}(pd)^{-0.05}. \tag{5.10}$$

At relatively small changes of G/d and pd only one complex of this equation $I/(Gd{\cdot}p)$ can be used for generalisation of the experimental data [4].

A suitable example of this approach to the generalisation of the VAC of the arc in different gases is the study [9] where the authors published the characteristics of the arc in a plasma torch with a cup-shaped internal and a cylindrical output electrode (Fig. 5.3). Here D is the diameter of the internal cup-shaped electrode, and $D \geq d$. This difference in the parameters is small by the results in addition of scattering of the generalised quantities. We shall discuss the case of the equal electrode diameters ($D = d$): Generalisation was carried out in the form:

$$Ud\sigma_0/I = A(I^2/Gd\sigma_0 h_0)^{-b}(\sqrt{\rho_0/p_0}\,pd^2/G)^c. \tag{5.11}$$

In this equation, the second co-multiplier is the combination of the Knudsen and Reynolds criteria in which the changing parts are $(pd)^{-1}$ and G/d, respectively. For different gases we obtain the following values of the coefficient A and exponents b and c, and also the maximum deviation of the experimental data from the calculation equation (Table 5.1).

This work is interesting because of the attempt to reduce the experimental data for different gases to a single equation. Evidently, this is not possible without taking the properties of the working gas into account. The authors of [9] proposed as the first approximation the power approximation of the electrical conductivity of the gases $\sigma = \sigma_0(h/h_0)^n$ (this parameter is responsible for the properties of the arc) and determined the values of the exponent n for different gases (they are given in the last column of Table 5.1.). Consequently, equation (5.11) acquired the additional multiplier n^{-k}. Finally, after processing the data available for different gases, the authors of [9] obtained the equation

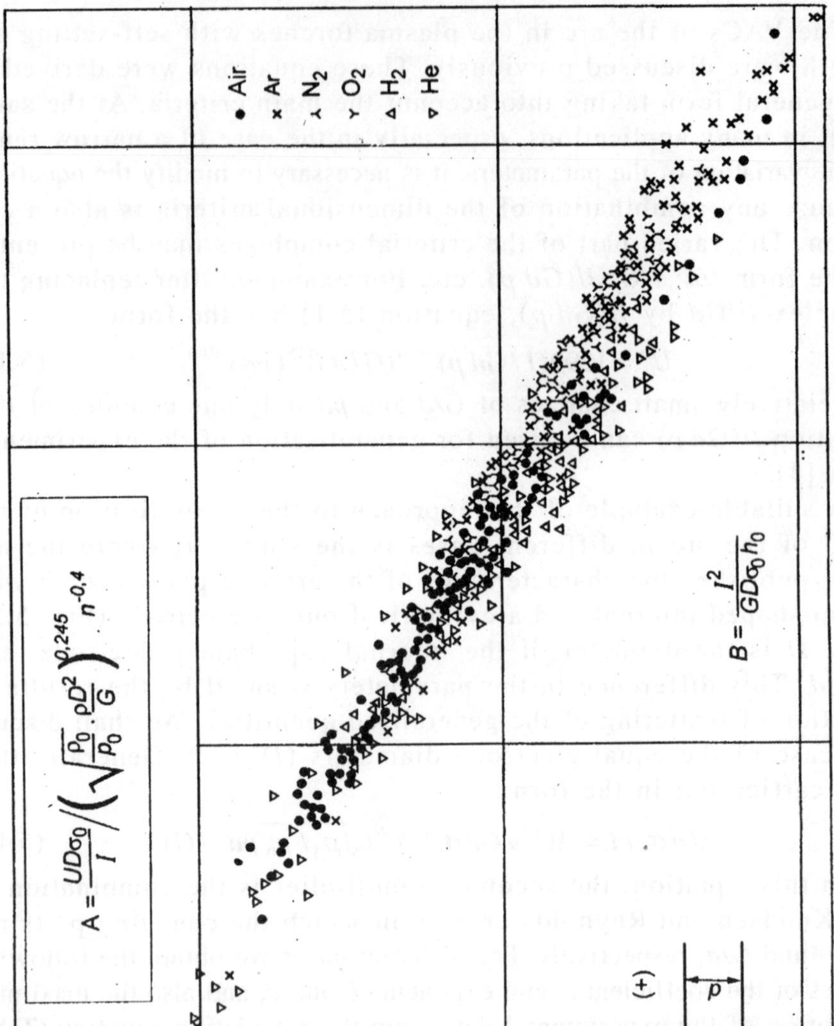

Fig. 5.3. Dependence of the complex $[(UD\sigma_0/I]:\left[\left(\sqrt{\rho_0/p_0}\right)\cdot\left(pD^2/G\right)\right]^{0.245}\cdot n^{-0.4}$, on the energy criteria for a linear plasma torch with vortex stabilisation [9]. $D = 0.01\div0.04$ m; $d = 0.008\div0.04$ m.

Table 5.1. Values of the structural coefficient and exponents of equation (5.1·1) for different gases.

Gas	*b*	*c*	*A*	Maximum deviation, %	*n*
N_2	0.7032	0.1625	0.5757	±17	1.21
H_2	0.6910	0.0936	0.6359	±16	1.38
Air	0.6016	0.2254	0.5105	±16	1.19
O_2	0.6172	0.1558	1.153	±5	0.81
He	0.6748	0.4025	0.1051	±41	0.514
Ar	0.6122	0.2360	0.5645	±14	0.48

$$U D\sigma_0 / I = 0.4293 (I^2 / G D\sigma_0 h_0)^{-0.6127} (\sqrt{\rho_0 / p_0}\, pD^2 / G)^{0.245} n^{-0.4}. \quad (5.12)$$

If we reduce the previously examined equations to the dependences on the unique criterial one-dimensional complexes, it should be noted that the difference is on the whole small, especially if we are concerned with the similar physical properties of the gases. A larger difference is obtained in the coefficients at criterial complexes determined by the selection of the values of the gas transfer coefficient, i.e. the values of gas enthalpy h_0, electrical conductivity σ_0 and viscosity μ_0, included in the dimensional criterial complexes. In [4] the determining values were the values of h_0 and μ_0 corresponding to the temperature of the gas at entry into the discharge chamber, and the electrical conductivity σ_0 is taken at a characteristic temperature corresponding to 1% of gas ionisation. In this selection, it is possible to take into account the hydrodynamic and electromagnetic effects in the arc and the surrounding gas flow. In [9], the determining parameter is the temperature in the region of inflection of the linear approximation of the time dependence $\sigma = \sigma_0 (h/h_0)^n$, which can be calculated with sufficient accuracy. However, in this case, the Reynolds criterion is not used in the explicit form for generalisation and it is necessary to use other relationships, without taking the hydrodynamic parameters of the gas flow into account. Since there is a strong relationship between the parameters of the gas flow and the arc characteristics, the approach [4] appears to be more suitable.

The results and equations were obtained quite a long time ago, in the period when considerable effort was made to obtain experimental data and find methods of generalising and engineering calculations of the characteristics of electric arc. These data are essential for both calculating and designing highly efficient thermal plasma generators for technological applications, and also for constructing the analytical model of the arc in a gas flow. Later, the methods of experimental investigations were improved and new approaches developed to processing and analysis of the results. In this respect, special attention should be given to studies [10, 11] concerned with the examination of the plasma torches with an end flat (or in the form of a truncated cone) electrode and self-setting length of the arc running in argon and nitrogen [10], and also with a cup-shaped internal electrode for the arc in air [11]. At the same time, measurements were taken of the integral characteristics of the arc such as current intensity, voltage, heat flow into the wall and pulsation characteristics: oscillations of voltage, current intensity, luminosity of plasma and acoustic oscillations of the jet. It is interesting to mention the results of generalisation of the VAC characteristics; the method of generalisation was taken from [1–4]. The determining criteria were:

$$S_U = Ud\sigma_0 / I; \quad S_i = I^2 / Gd\sigma_0 h_0; \quad \mathrm{Re} = G / \mu_0 d.$$

The values of the transfer coefficient were taken at 1% ionisation of working gases (Table 5.2).

The following dependence was obtained as a result of processing the VAC characteristics of the arc in nitrogen at different values of the gas flow rate and the channel diameter in the case of a cathode in the form of a truncated cone in [10]:

$$U d\sigma_0 / I = 4.95\,(I^2 / Gd\sigma_0 h_0)^{-0.654} (G / \mu_0 d)^{-0{,}327}. \tag{5.13}$$

Here, the range of variation of the determining parameters is small (for example, I = 200÷400 A), and the gas pressure in the channel differs ony slightly from atmoshperic pressure.

The equation obtained for a flat end cathode is slightly different:

$$U d\sigma_0 / I = 2.5 (I^2 / Gd\sigma_0 h_0)^{-0.643} (G / \mu_0 d)^{-0.137}. \tag{5.14}$$

The difference in the powers at the Reynolds number was attributed by the authors to the effect of twisting of the gas flow which differs for different types of cathode. In argon, the follow-

Table 5.2. Characteristic values of the physical parameters of the gases used in the calculation equations.

Parameter	N_2	Ar	Air
T_0, K	8600	9400	8600
σ_0, A^2s^3 / (kg · m^3)	1200	2350	1280
μ_0, kg/(m · s)	0.00022	0.000261	0.000222
k_0, J/(m · K)	2	0.487	1.61
h_0, J/kg	$45.8 \cdot 10^6$	$5.2 \cdot 10^6$	$42 \cdot 10^6$

ing dependence was obtained for a flat cathode:

$$Ud\sigma_0/I = 4.95(I^2/Gd\sigma_0 h_0)^{-0.565}(G/\mu_0 d)^{-0.183} \tag{5.15}$$

Further, the authors attempted to derive an equation suitable for argon and nitrogen, introducing a new dimensionless parameter, the Prandlt number Pr = $\mu_0 \cdot h_0/(k_0 \cdot T_0)$. In this variant, the Pr number was determined from the constant quantities characteristic of the gases and presented in Table 5.2, i.e., it is constant for every type of gas. The equation for calculating the VAC has the form:

$$Ud\sigma_0/I = 2.04(I^2/Gd\sigma_0 h_0)^{-0.57}(G/\mu_0 d)^{-0.12}(\mu_0 h_0/k_0 T_0)^{-0.386}. \tag{5.16}$$

With the accuracy of ±2.4%, this equation generalises the experimental data for argon and nitrogen.

It may easily be shown that the difference in the *U–I* characteristics of the arc for different gases, determined by the last term in equation (5.16), is constant and equal to ~2. Naturally, the equation can be used only in a narrow range of the variation of the parameters, investigated by the authors.

If we use the equations from [10] for the same relationships as in [4]: $U = f(I^2/Gd;\ G/d)$, the difference from the results obtained in [4] will be manifested only in the exponents and critical complexes, and the difference is quite small.

The investigation were also carried out on the plasma torches with a cup-shaped internal electrode in air [11]. The complex (D_c/D_a), i.e. the ratio of the cathode and anode diameters, was introduced for generalisation. At straight polarity of connection of the output electrode the following equation was obtained:

$$Ud_a\sigma_0/I = K(I^2/Gd_a\sigma h)^{-0.616}(G/\mu_0 d_a)^{-0.284}(d_c/d_a)^{-0.586}. \tag{5.17}$$

A similar equation was also found for reverse polarity. Coefficient *K* at straight polarity is equal to [12], in the case in reverse polarity it is 1175.

The authors compared the results with the equations for identical investigations carried out in [4] and showed that a difference is found only in the constant coefficients (this is natural because different determining temperatures were selected for the dimensionless parameters) and only in the second decimal point in the exponents at criterial complexes. Thus, new investigations, carried in equipment on a qualitatively high level, only improve the accuracy and confirm the previously published results and correctness of application of the selected criterial complexes for generalisation of the internal characteristics of the arc.

It was shown in chapter 2 that in the plasma torches on the vor-

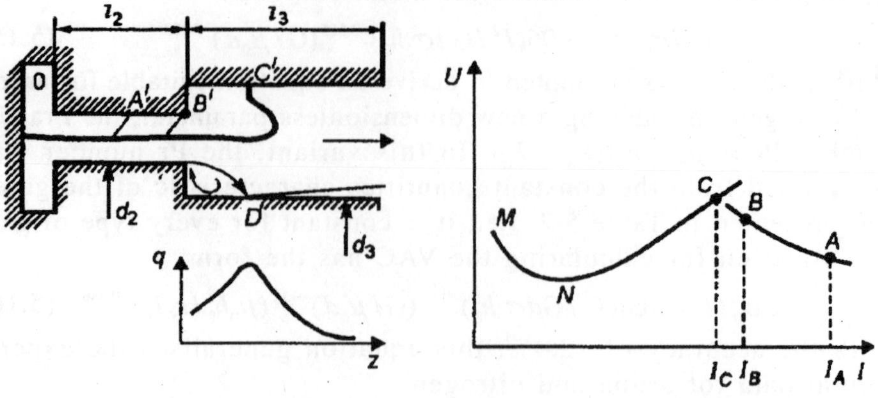

Fig. 5.4. Formation of the VAC of the arc in a plasma torch with a stepped electrode.

tex system with smooth electrodes, the mean arc length is determined by the electrical breakdown (shunting) between the arc column and the wall of the discharge chamber. Shunting is one of the reasons for the formation of a drooping VAC of the arc restricting the temperature of the heated gas, and requiring inclusion of a ballast resistance in the electrical circuit, etc. Naturally, it is desirable to create such conditions in a discharge chamber at which the VAC characteristic would be rising and controllable and this would ensure stable arching and electrical efficiency close to unity and removed the restrictions on the power input and gas temperature. In axial plasma torches, one of the methods of producing the rising VAC characteristics is the fixation of the mean arc length by some method in the relevant range of the working parameters, because the *E–I* characteristics in all gases are U-shaped. The shunting of the arc makes it possible to draw the conclusion on the possibility of constructing different types of vortex plasma torches with the fixation of the mean arc length. In one of them (for example, in plasma torches with an interelectrode insert) the arc length is greater than the self-setting arc length, and in plasma torches of the second type the arc length is smaller than the self-setting length.

The most widely used plasma torch of the second type is a plasma torch with a sudden expansion of the output electrode, a ledge [4]. The typical circuit of such a plasma torch, aerodynamics of the gas flow in the torch and conditions of the formation of the *U–I* characteristic of the arc are presented in Fig. 5.4.

In plasma torches with a smooth output electrode, the arc length is determined by the shunting process. This is also typical of the plasma torches with a step electrode, but there is a difference between them caused by the gas dynamics of the flow and heat exchange between the gas and the wall of the electrode behind the ledge (described in detail in chapter 2).

Fixation of the mean arc length is associated with the presence of a detachment zone, subsequent attachment of the flow to the surface of the electrode behind the ledge (zone D'), and destruction of the boundary layer in the convection region.

The qualitative investigations of the flow of the gas in a flat channel with a ledge, and also experimental examinatioof the flow and heat exchange in these channels have confirmed the existence of the previously mentioned zones of detachment and attachment of the flow to the surface of the channel behind th ledge, the recirculation zone between the ledge and the discharged jet, and also the presence of a maximum heat flow of q (z) in the area of contact of the jet, leaving the narrow section channel, with the surface of the channel behind the ledge (Fig. 5.4). These factors generate, behind the zone D', highly favourable conditions for the arc–electrode breakdown in a wide range of the variation of current intensity and localisation of the shunting zone of the arc in the channel with a diameter d_3. The experiments with the arc show that the end of the ledge and part of the surface of the electrode behind the ledge, approximately corresponding to the dead zone, have no traces of the effect of the arc spot. According to experimental results, the distance from the ledge to the start of the shunting zone is ~$5\Delta h$, where Δh is the height of the ledge.

Thus, the examine natured of the flow of the gas and its heat exchange with the wall of the channel behind the ledge determine the fixation of the mean arc length. Consequently, the VAC of the arc in a wide range of the values of current intensity up to I_C (Fig. 5.4 and 5.5) contains both the decreasing *MN* and rising *NC* sections, determined by the *E–I* characteristic of the arc (Fig. 5.4). With a further increase of current intensity ($I > I_C$), when the arc spot of the arc is suddenly 'bonded' initially with the edge of the ledge (point B') and, subsequently, with increasing I, changes to the channel with the diameter d_2 and starts to move in the direction against the flow, the VAC characteristic always contains a drooping section (curve *BA*).

Figure 5.5 shows the typical *U–I* characteristics of the arc for different flow rates of air. At the top, the rising sections of the VAC of the arc at the given values of G, d_2, p, are restricted by the VAC characteristic of the arc with the self-setting length in the channel $d = d_2$, as indicated by the shape of the curve *1* (the broken curve is here calculated from equation (5.3) for the given param-

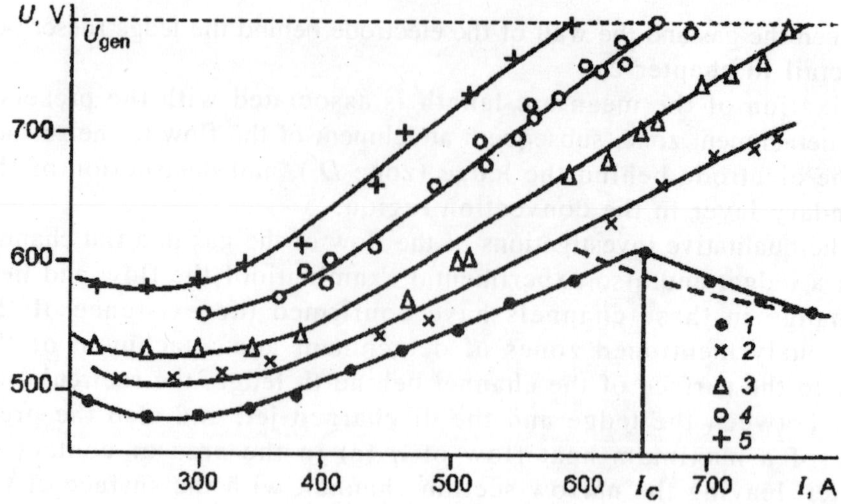

Fig. 5.5. Volt–ampere characteristics in the plasma torch with the arc length fixed by a ledge. Air, d_2=2.7·10^{-2} m; l_2 = 26·10^{-2} m; G = 40·10^{-3} kg/s (*1*), 60 (*2*), 80 (*3*), 100 (*4*), 120 (*5*).

eters). The boundary of the rising section on the left (bottom) is the minimum on the *U–I* characteristic. At the pressure in the chamber close to atmospheric pressure, the intensity of current, corresponding to this minimum, is determined from the relationship I/d_2 = const ≈ 10^4 A/m.

These experiments confirm the possibility of stable arcing in the ballast-free regime in operation from a power source with the rigid (U_{gen} = const) characteristic. For example, the curve *4* was determined by the variation of the voltage of the power source, and electrical efficiency of the source η_e is close to unity. The practical importance of the result is that it opens new possibilities for the simultaneous stable operation of several plasma torches from a single electric power source (tests were carried out on two power sources). This is very important for the creation of multiarc plasma systems (reactors) with the power of several tens of megawatt.

In the generalisation of the experimental data we shall restrict ourselves to examining the rising section of the VAC characteristic because the drooping section may be calculated from one of the previously mentioned equations, for example (5.1), at $d = d_3$ for the section *MN* (Fig. 5.4), at $d = d_2$ for the section *BA*.

The processing of the experimental material in the criterial form has made it possible to recommend the generalised equation for calculation of the rising section of the *U–I* characteristic of the arc with the accuracy of up to ±10%:

$$U = 4.55(1+4.6\cdot10^{-5} I/d_2)(G/d_2)^{0.22}(l_2/d_2)^{0.95}(pd_2)^{0.23}. \quad (5.18)$$

Equation (5.18) is verified in the following range of the variation of the determining complexes:

$$I/d_2 = 8\cdot10^3 \ldots 4\cdot10^4\,\text{A/m};\; G/d_2 = 0.8\ldots6.5\,\text{kg/(m}\cdot\text{s)};$$
$$l_2/d_2 = 5.6\ldots14.5;\; pd_2 = 2\cdot10^3\ldots40\cdot10^3\,\text{N/m}.$$

In contrast to the generalised VACs, for plasma torches with the self-setting arc length equation (5.18) includes the parametric criterion l_2/d_2; (I^2/Gd) was substituted by the complex (I/d_2) because this complex determines the position of the minimum of the voltage corresponding to the origin of the rising section of the U–I characteristic. In the experiments, the parametric criterion d_3/d_2 was varied in the range 1.8÷1.9, i.e. it was almost constant. In the case of the variation in a wider range, this criterion should be included in equation (5.18).

The mixture of air and natural gas (CH_4) in a plasma torch with a ledge was used in the experiments to obtain the following dependence for the rising section of the U–I characteristic:

$$U = 1.51(I/d_2)^{0.28}(G_\Sigma/d_2)^{0.22}(l_2/d_2)^{0.5}\times \\ \times[1+(G_{CH_4}/G_{air})^{0.8}](pd_2)^{0.23}, \quad (5.19)$$

which was verified for the following values of the parameters: G_{CH_4}/G_{air} = 0÷0.3, l = (35÷60) · 10^{-3} m, d_2 = 8 · 10^{-3} m, p = 1 · 10^5 Pa, $G_\Sigma = G_{CH_4} + G_{air}$ = (3÷9) · 10^{-3} kg/s, I = 200÷500 A; the arc voltage was varied in the range U = 200÷450 V. According to equation (5.90), U is characterised by a different dependence on (I/d) and the effect of the structural parameter (l_2/d_2) in comparison with the equation (5.18) is weaker. In this case, it is also necessary to add a co-multiplier, taking into account the mixing of the natural gas with air. Since these equations work in a limited range of the variation of the parameters and describe the rising section of the U–I characteristic, the difference between them is only in the selection of approximation of the curves (linear or power) and has only a slight effect on the accuracy of generalisation of the experimental data (compare equations (5.18) and (5.19)).

Because of the promising nature of the application of steam for various plasma chemical technologies as a high-temperature reagent, it is interesting to develop and investigate steam plasma torches [12]. The electric discharge chamber of the plasma torch for heating of steam is made in the form of a cone narrowing in the direction of the gas flow (confusor) with transition to the cylinder cross-sec-

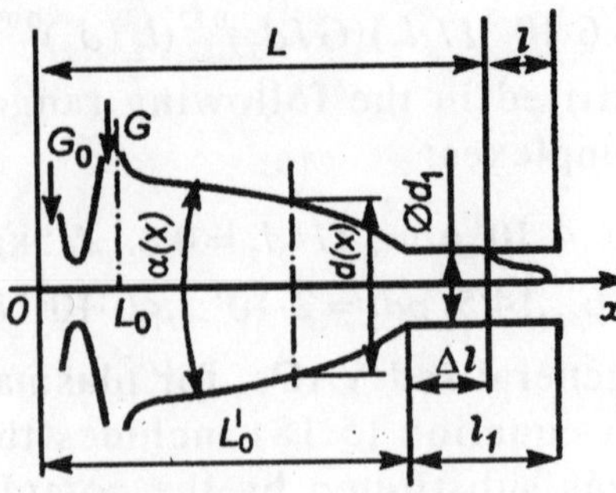

Fig. 5.6. Geometry of the discharge chamber of a water steam plasma torch indicating the main parameters.

tion. The anodic outlet section of the chamber may be smooth or contain a ledge. The generalised VAC of the arc, burning in steam, has the form which differs from the equations presented previously:

$$U = 70 + 17.6[1 + 0.5\exp(-G_0/0.025\cdot10^{-3})]\times \\ \times(I^2/GD)^{-0.13}(G/D)^{0.20}(pD)^{0.48}(L/D)^{1+\alpha/88.8}. \qquad (5.20)$$

Here $D = (1/(L-L_0))\int_{L_0}^{L} d(x)dx$ is the generalised diameter of the discharge chamber; $\alpha = (1/(L-L_0))\int_{L_0}^{L}\alpha(x)dx$ is the average total angle of narrowing of the flow part of the discharge chamber, the remaining parameters are shown in Fig. 5.6.

Equation (5.20) was verified at a pressure of $p \sim 1\cdot10^5$ Pa at outlet from a plasma torch in the following range of variation of the criteria and dimensionless parameters:

$$I^2/GD = (3.0\div367)\cdot10^8\ \mathrm{A^2s/(kg\cdot m)};$$
$$pD = (1.7\div4.9)\cdot10^3\ \mathrm{N/m};\ G/D = (0.017\div0.22)\ \mathrm{kg/m\cdot s};$$
$$\alpha = 0\div22^\circ, \bar{L} = L/D = 4.1\div13.5;\ D/d_1 = 1\div3.5.$$

Attention should be given to the presence, in equation (5.20), of a free member which, according to [13], is the sum of the near-electrode drops of the potential and voltage drop in the loop of the anodic section of the arc. This is a relatively rough approximation, especially at arc voltages of 200÷300 V and the variation of the intensity of arc current by several hundreds of amperes. The term in the square brackets takes into account the effect of blowing a shielding gas (argon) in front of the end cathode, which is relatively strong, i.e. arc voltage decreases by 1/3 in the presence of even a smaller weight amount of blown argon. The effect of blowing was verified in the relatively low-voltage plasma torches with the channel length not exceeding 5–7 length length gages. Possibly, with the increase of the arc length in the high-voltage steam plasma torches, the effect of the blowing of shielding gas decreases because of the detachment of argon from the arc column. The latter co-multiplier in this equation determines the effect of the relative length and form (narrowing) of the channel on arc voltage. The effect of other criterial complexes is approximately the same as that of the air arc, only the dimensional complex $(pD)^{0.48}$ has a higher exponent in comparison with previous equations (see equations (5.1)–(5.6)). In the cylindrical channel with a constant diameter, the form of equation (5.20) is greatly simplified:

$$U = 70 + 26.4(I^2/Gd)^{-0.13}(G/d)^{0.20}(pd)^{0.48}(L/d). \qquad (3.20a)$$

It should be mentioned that the equation (5.20a) contains, as a parameter, the relative arc length (L/d).

There is also a relatively large number of VAC of the arc in plasma torches with a small cylindrical output electrode or a ledge for argon, carbon dioxide and other gases [4, 7]. However, these characteristics are usually obtained in a narrow range of electric-gas dynamic parameters, for specific circuits of the plasma torches. Therefore, they are not usually generalised.

As an example, it is useful to mention the *U–I* characteristics of the arc burning in carbon dioxide in a plasma torch with a stepped electrode (Fig. 5.7). For all the gas flow rates there are both rising (shunting behind the ledge in the channel with $d = d_3$) and drooping

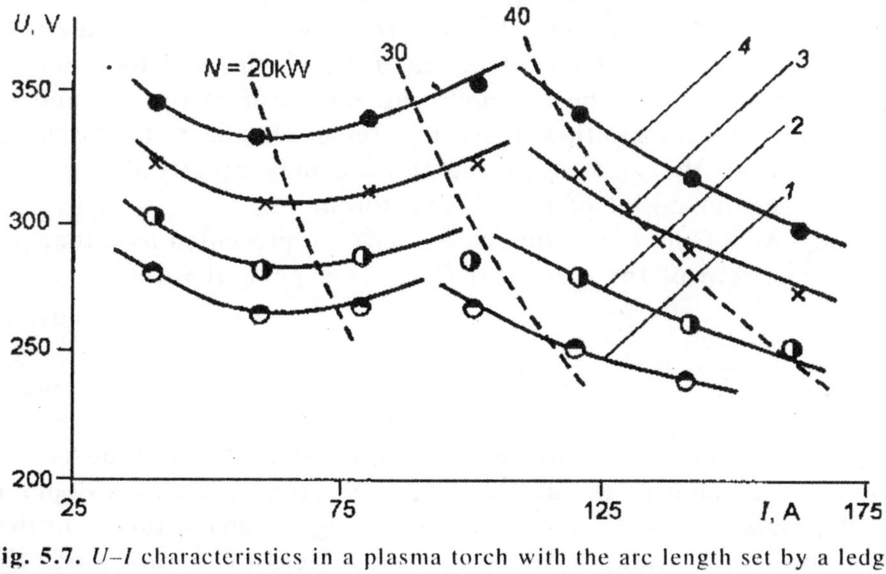

Fig. 5.7. *U–I* characteristics in a plasma torch with the arc length set by a ledge. CO_2, d_2 =0.5·10^{-2} m, l_2=4·10^{-2} m: G=1·10^{-3} kg/s (*1*), 2 (*2*), 3 (*3*), 4 (*4*).

(shunting in the channel with $d = d_2$) sections of the characteristic [14].

5.2. ENERGY CHARACTERISTICS OF THE ARC IN PLASMA TORCHES WITH INTER-ELECTRODE INSERTS

The equations presented previously for the generalisation in the criterial form of the VAC of the arc make it possible to calculate them in plasma torches of the first two systems using the classification proposed in chapter 1. The simple and useful form of generalisation results from the fact that almost in the entire arc length, with the exception of near-electrode sections whose contribution is small, the longitudinal component of the strength of the electrical field of the arc is maintained constant along the channel, and the pressure of the working gas in the channel differs only slightly from the pressure at outlet from the plasma torch.

A completely different pattern is found (chapter 2) in the plasma torches of the third system, i.e. with the arc length greater than for the arc with the self-setting length. These plasma torches include plasma torches with the inter-electrode inserts (IEI) of different design: with a sectioned IEI and with the distribution of the part of the working gas blown along the insert, with a gas-dynamic IEI (the gas-dynamic IEI is the inter-electrode inset with a large diameter ($D > d$) where the arc is stabilised at the axis of the channel by the vortex flow of the gas blown at the periphery of the insert), and with IEI made from porous materials, etc. In the presence of the inter-electrode insert, the strength of the electrical field of the arc does not remain constant along the channel, and the form of the arc and its interaction with the gas flow differ in different sections of the discharge chamber, and the pressure along the arc may greatly differ from the pressure at outlet of the plasma torch.

The VAC of the arc in this case may be represented as a function of the strength of the electrical field. In a general form:

$$U = \int_0^L E(z)dz, \tag{5.21}$$

where L is the length of the channel from the cathode to the anode attachment of the arc, and $E(z)$ is a function which depends on the main working parameters. Knowing the magnitude and distribution of the strength of the electrical field of the arc along the axis z in different arcing conditions makes it possible to optimise the selection of the circuit of the plasma generator from the viewpoint of increasing the heat content of the gas at the minimum losses of electric energy. At the same time, the data on the strength, obtained in the experiments, are used also for the verification of the analytical calculation models of the arc.

A relatively large number of methods and equipment have been developed for the measurement of the strength of the electrical field of the arc. The main of them have been developed in the 50s and 60s and are described in detail in [4].

One of the verified and efficient methods of determining the local strength of the electrical field is the probe method proposed by Langmuir in 1923 [15], for the examination of the characteristics of low-density plasma. The detailed electrical and optical investigations of operation of the probe in the high-pressure arc have made it possible to explain the mechanism of perturbation of the discharge by the probe and determine the optimum working conditions in which it is possible to obtain at least partial information on the investigated section of the arc, including the distribution of the arc potential along the length of the channel. As a result of selecting a suitable probe it is possible to minimise the disruption of the arc: in most cases, it is recommended to use a tungsten rod probe with a thickness of approximately 0.2 mm, moving along the discharge at a rate of 20–150 cm/s, depending on the experimental conditions. The error of measurements of the distribution of the potential by the probe method does not usually exceed 5%. One of the problems of direct measurements of the potential of the arc by these methods is the presence of the contact difference of the probe–plasma potential. Comparison of the individual methods of measurement shows that the contact potential difference is always constant along the arc and equal to approximately 2 V [15]. Consequently, it is possible to take measurements of the distribution of the potential (or of the potential difference) over a specific measuring base with a relatively high accuracy. The authors of [4] described the method of moving probes used for these purposes, i.e., a group of several probes introduced simultaneously into the discharge chamber [16]. Measurements were taken of the distribution of the potentials of the probes along the chamber and of the difference of the potentials of the adjacent probes. The results of the measurements obtained by the individual methods were in satisfactory agreement.

The authors of [17] proposed different methods of measurement of the distribution of the potential along stabilised (with no consumption) wall of the arc. The peripheral circular probes were represented by discs-sections, forming the inter-electrode insert, and were electrically isolated from each other and from the electrodes. In further stages, this method was developed and its procedure was justified in [18] for the arc on which gas was blown at a low rate. In the case of the argon arc it was shown that the floating potential, acquired by the section, corresponds to the potential of the section of the arc opposite one of the edges of the section. This displacement of the points of correspondence of the potentials is almost constant along the channel. The next stage of investigations was the application of the method of circular probes to arcs stabilised by the vortex gas flow in plasma torches with sectioned inter-electrode inserts [19]. In the study, special attention was given to examining the characteristics of a non-independent discharge formed between the arc and the section of the inter-electrode insert, and the floating potential, acquired by the section, and also to the effect of the dimensions of the section and charge leakage from the section on the magnitude of the floating potential. If the conductivity of the gas around the measuring section is sufficiently high, it may be used for examining non-stationary processes in the arc (the method of measurement of the arc potential was described in detail in [4]).

The same method was also developed further for the determination of the strength of the electrical field of the arc in a turbulent flow of different gases, including in the presence of the intersectional blowing of the gas [20]. Measurements were taken of both the distribution of the arc potential along the discharge channel and also of the potential difference of the adjacent sections. The resultant values of the strength of the electrical field were compared with the values measured by other methods in similar conditions. It has been established that the floating potential of the sections corresponds to the potential of the section of the arc enclosed inside the section. There have been many studies concerned with improvement of the method of measurement of the strength of the electrical field of the arc and improvement of the conditions of reliability of application of the method. The methods of diagnostics of thermal plasma have been described in detail in [21], and the methods of measurement of the strength of the electrical field in the previously mentioned monographs [4, 20].

5.2.1. Distribution of the strength of the electrical field of the arc in a long cylindrical channel

The majority of investigations of the integral and local characteristics of the electrical arc were carried out in axial plasma torches with a sectioned inter-electrode insert (IEI). The section is placed in the gap between the internal and outlet electrodes and consists of a set of disks-sections thermally and electrically insulated from each other and also from the electrodes. The diagram of a plasma torch with the main designations of its geometrical parameters is shown in Fig. 5.8. The figure also shows the diagram of measurement of the distribution of the potential and the strength of the electrical field of the arc along the IEI.

The working gas is supplied into the plasma torch in the vicinity of the outlet part of the internal electrode. If necessary, a small

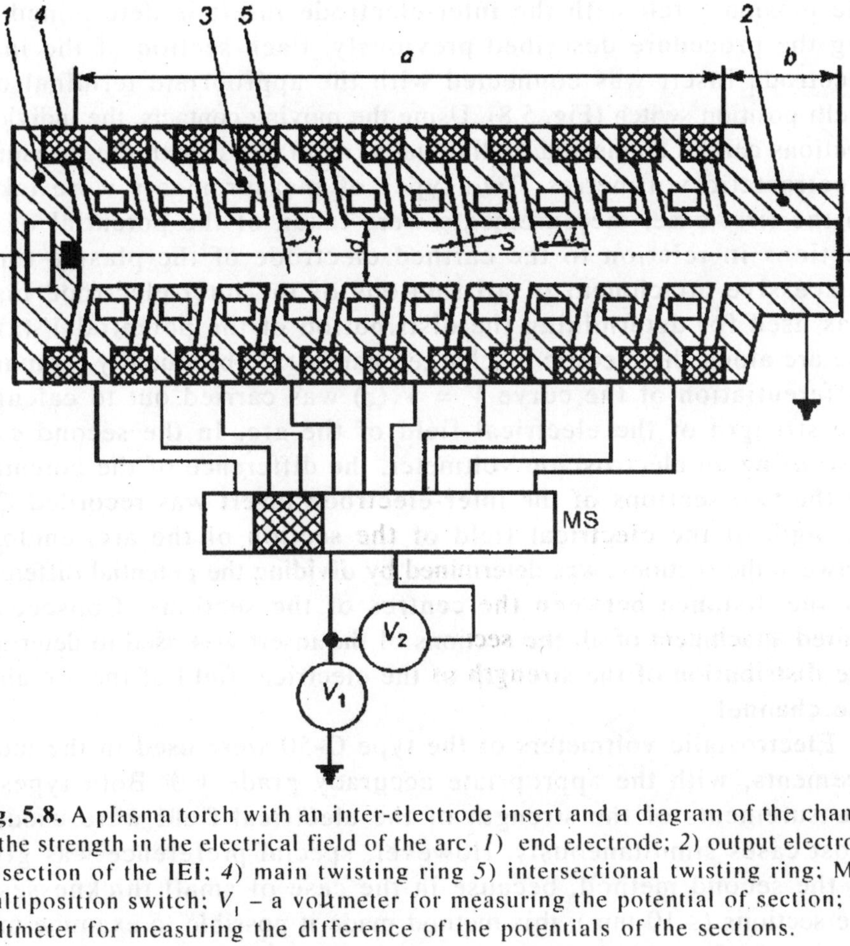

Fig. 5.8. A plasma torch with an inter-electrode insert and a diagram of the changes in the strength in the electrical field of the arc. *1*) end electrode; *2*) output electrode; *3*) section of the IEI; *4*) main twisting ring *5*) intersectional twisting ring; MS - multiposition switch; V_1 – a voltmeter for measuring the potential of section; V_2– voltmeter for measuring the difference of the potentials of the sections.

or a large part of the gas may be introduced into the discharge channel along the IEI through the inter-sectional gaps. In the majority of the experiments, the working gas was supplied into the channel with twisting, i.e., with the circumferential component of the flow rate *w*. The sections of the inter-electrode insert, with individual cooling with water, were used as the end probes in the measurement of the arc potential along the discharge chamber and also as calorimeters for the determination of the heat losses into the channel walls. The design of the IEI permits placing of the individual windows and the slits for optical investigations, positioning of the pressure sensors, different process, etc. Thus, the plasma torches with the inter-electrode insert may be used for a wide range of investigations of different characteristics of the electrical arc.

The strength of the electrical field of the arc in the channel of the plasma torch with the inter-electrode insert is determined using the procedure described previously. Each section of the inter-electrode insert was connected with the appropriate terminal of a multi-position switch (Fig. 5.8). Using the moving contacts, the individual sections can be connected, individually, or in pairs, with the measuring electrostatic voltmetres. Two types of measurements were taken. In the first case, measurements were taken of the potential of the sections in relation to the earthed electrode of the plasma torch. Successive attachment of all sections of the inter-electrode insert was used for determining the distribution of the potential $V(z)$ of the arc along the electric discharge chamber. Subsequently, graphical differentiation of the curve $V = V(z)$ was carried out to calculate the strength of the electrical field of the arc. In the second case, also using an electrostatic voltmeter, the difference of the potentials of the two sections of the inter-electrode insert was recorded. The strength of the electrical field of the section of the arc, enclosed between the sections, was determined by dividing the potential difference by the distance between the centres of the sections. Consecutive paired attachment of all the sections of the insert was used to determine the distribution of the strength of the electrical field of the arc along the channel.

Electrostatic voltmeters of the type C-50 were used in the measurements, with the appropriate accuracy grade 1.0. Both types of measurements of the strength of the electrical field were used, in most cases simultaneously. However, special preference was given to the second method, because in the case of small thicknesses of the sections ($\leq$ 10 mm), this method made it possible to examine more accurately the variation of the potential along the discharge chamber.

The main error in the measurements is caused by the determination of the gage length, in the present case the points of correspondence of the arc potentials and the section. The results of a large number of experiments showed that, in the majority of cases, especially in the section with the developed turbulent flow of the gas, the sections of the inter-electrode insert operate in the probing conditions and efficiently track the changes of the arc potential. The point of correspondence of the potentials of the arc and the sections is situated inside the section, mainly in its centre. In some cases, in particular in the case of reversed polarity of connection of the electrodes, the point of correspondence of the potentials is displaced downwards along the gas flow from the centre of the section, but this displacement is smooth in the entire channel without any sharp transitions and in the case of small thickness of the sections has almost no effect on the accuracy of determination of the measuring base. The total error of the measurements of the strength of the electrical field of the arc by these methods did not exceed 5–6%.

The diagram of the flow of the gas in a long cylindrical channel with the electric arc burning in it, was described in chapter 2. The diagram was proposed on the basis of a large number of experimental investigations of the electrical, thermal, optical, pulsation and a number of other characteristics of both the gas flow and also of the arc onto which the gas was blown [4, 20]. The electric arc, stabilised with a vortex gas flow, was investigated. In this case, at least in the initial section of the channel, the gas-dynamic forces prevail over electrodynamic forces and there is good agreement between the characteristics of the cold gas flow without the arc and the flow with the electric arc running in it.

Figure 2.12 in Chapter 2 shows the diagram of the flow of gas and the appropriate distribution of the strength of the electrical field of the arc (experimental data), and also the photographs of the arc, obtained using high-speed filming in different sections of the channel through quartz inserts between the sections of the interelectrode inserts. The measurements of the strength was carried out with the distributed blowing of the gas along the inter-electrode insert (a small amount of gas was blown). If the sections of the arc in the immediate vicinity of the electrodes are excluded from examination, the curves of the distribution of the strength of the electrical field of the arc along the channel $E(z)$ shows three distinctive sections corresponding to the sections shown in the diagram of the gas flow.

The information, presented in section 2 .2 in chapter 2, will be briefly repeated, stressing the correspondence between the strength of the electrical field and interaction of the arc with the gas flow. In the first initial section of the arc (from entry into the channel), Fig. 2.12, the arc is stabilised on the hydrodynamic axis of the gas flow. The strength of the electrical field E_s in the section is constant along the channel and relatively low. In the immediate vicinity of the electrode of there is the 'entry' section with the length of 1–2 length length gages, subjected to the effect of the cold flow of the gas entering this area. The strength of the electrical field in the section slightly increases in the direction to the end electrode. However, the contribution of the given section to the total arc voltage is small and in approximate calculations it is usually ignored.

The initial section on the $E(z)$ curve is followed by the section of monotonic increase of strength whose length in the investigated conditions in air did not usually exceed 4–6 length length gages. The transition sections followed by the section in which the strength of the electrical field is again approximately constant. This corresponds to the section of the developed turbulent gas flow. The photographs show clearly the formation and development of the pulsations of the arc in the transition section. The amplitude of pulsations almost reaches the diameter of the channel. It is followed by the formation of a flow in which the development of the regime of interaction of the arc with the gas flow, referred to as 'the electrical arc in the turbulent gas flow' [22, 23], is completed. Under the effect of the turbulence pulsations of the flow the arc randomly oscillates in space. These pulsations are maintained and developed further by the intrinsic electromagnetic forces of the arc. The arc column is split into several current-conducting channels and new branches of the arc appear and the old ones disappear. Naturally, in this case we can talk about only about some mean-static parameters of the arc. In particular, the strength of the electrical field, calculated as the ratio of the difference of the potentials of the probes-sections to the length of the measuring base, is not the true part of the averaged-out 'technical' strength.

In the section of the developed turbulent flow, the strength E_t may exceed E_s 2–3 times. Another contribution to the general voltage in the arc is provided by the section of the arc in the output electrode. Usually, the section is defined on the basis of the position of the zone of preferential attachment of the arc in this electrode because there is a distinctive arc loop, as in the case of the arc with the self-setting length. Thus, knowing the strength of the electrical field in the characteristic sections and the length of the sections, we can calculate the arc voltage taking the need for the contribution of near-electrode sections into account. The characteristics of the arc in the examined sections of the gas flow will be examined in greater detail.

5.2.2. Dependence of the strength of the electrical field of the arc on the determining parameters in the initial and transition sections of the channel

The results of measurements of the strength of the electrical field of the arc in the initial section of the channel E_i have been published in many investigations for different gases. Initially, we investigate E_i–I characteristics of the air arc, determined in a plasma torch with a fixed mean arc length using a ledge [24, 25] for the case of a wide range of the variation of the working parameters: d_2 = (2.0; 2.5; 3.0) · 10^{-2} m, G = 30÷90) · 10^{-3} kg/s; the intensity of arc current I reached 1500 A. The results of the measurements show that the strength E_i at I = const is almost constant along the channel, and the E_i–I characteristics of the arc is complicated (Fig. 5.9). On the whole, the U-shaped experimental characteristic contains, in the initial section $E = f(I)$, local maxima and minima, and currents higher than 800 A show, for the examined conditions, the gradation of the I_i–I characteristic (curves 1–3). It is interesting to compare this experimental curve with the empirical dependence, recorded in [26], for the arc running in a plasma torch with an inter-electrode insert:

$$E_i \cdot d = 3.26 \cdot 10^{-2} (G/d)^{0.15} (pd)^{0.13} \times \\ \times [355 - 10^{-2} I/d + 5.13 \cdot 10^{-7} (I/d)^2]. \tag{5.22}$$

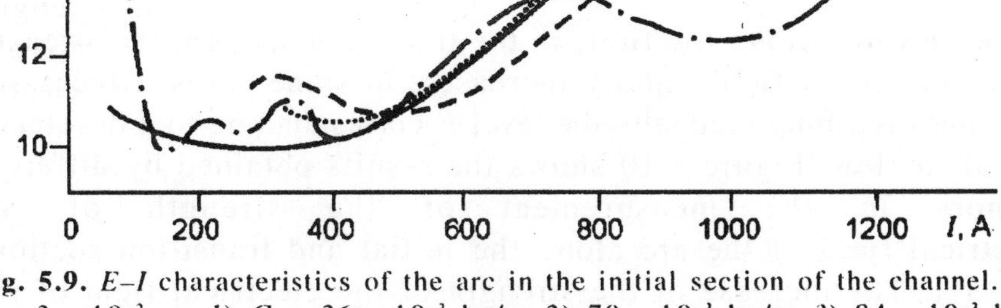

Fig. 5.9. E–I characteristics of the arc in the initial section of the channel. Air, d = 3 · 10^{-2} m. *1*) G = 36 · 10^{-3} kg/s; *2*) 70 · 10^{-3} kg/s; *3*) 84 · 10^{-3} kg/s; *4*) calculated from equation (5.22), G = 36·10^{-3} kg/s.

This formula was verified in the following range of variation of the parameters:

$$I=(50\div800)\,\text{A},\quad G=(1.5\div70)\cdot10^{-3}\,\text{kg/s},$$
$$p=(1\div4)\cdot10^{5}\,\text{Pa},\quad d=(0.5\div3.0)\cdot10^{-2}\,\text{m}.$$

Curve *4* in Fig. 5.9 is the result of calculation using equation (5.22) for the working parameters corresponding to the curve *1*. In this case, only the rising section of the E_i–I characteristic is generalised, of course, without taking the local extremum of curve *1* into account. It may be seen that up to a current intensity of $I \approx 700$A, which corresponds to $I/d \sim (2\div2.5)\cdot10^4$ A/m, the curve, calculated from equation (5.22) is similar to the given experimental dependence $E_i(I)$. Approximately the same agreement is also found in comparison with the data published in many other investigations (for more details [4, 20]) in which the experimental conditions differ in the schemes of the investigated plasma torches, the methods of gas supply into the discharge channel (one-, two- and three-dimensional plasma torches with a ledge and a smooth channel, plasma torches with an inter-electrode insert and different distribution of the supply of the gas along the insert, gas-dynamic inter-electrode insert, etc). Thus, on the basis of the available experimental data it may be concluded that equation (5.22) describes satisfactorily the strength of the electrical field of the air arc in the initial section of the channel for the given range of the parameters. Comparison with the data published in [25] with some other data also shows that at $I/d > 2\cdot10^4$ A/m the value of E_i changes only slightly with increase of current intensity and it may be evaluated with sufficient accuracy using the value of E_i at $I/d = 2\cdot10^4$ A/m. It is also interesting to note the weak dependence of the product $(E_i\cdot d)$ on the Reynolds and Knudsen numbers which is characteristic of arcing in a laminar gas flow when the removal of heat from the arc takes place as a result of the radiation and laminar heat exchange in the thin thermal layer of the arc (chapter 2).

Downwards along the flow, in the transition section, the strength of the electrical field rapidly increases; in some cases it increases 2–3 times reaching gradually the level E_t corresponding to a developed turbulent flow. Figure 5.10 shows the results obtained by different authors in the measurement of the strength of the electrical field of the arc along the initial and transition sections [21–31]. The increase of the strength of the electrical field of the arc in the smooth and sectioned channels with different widths of the slits takes place usually over a length of 4–6 length length gages

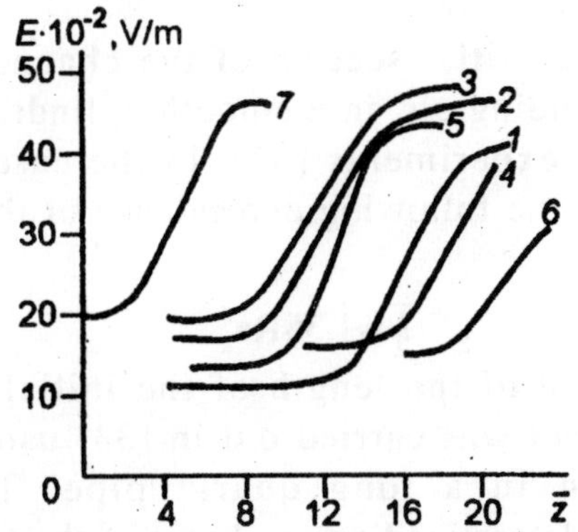

Fig. 5.10. Strength of the electrical field in the transition section of the channel. *1)* $d = 2 \cdot 10^{-2}$ m, $G = 30 \cdot 10^{-3}$ kg/s; $g_i = 0.5 \cdot 10^{-3}$ kg/s, $I = 120$ A [20]; *2)* $d = 2 \cdot 10^{-2}$ m, $G = 17.9 \cdot 10^{-3}$ kg/s, $I = 120$ A [31]; *3)* $d = 2 \cdot 10^{-2}$ m, $G = 8.5 \cdot 10^{-3}$ kg/s, $I = 120$ A [31]; *4)* $d = 2 \cdot 10^{-2}$ m, $G = 30 \cdot 10^{-3}$ kg/s, $I = 160$ A [28]; *5)* $d = 2 \cdot 10^{-2}$ m, $G = 26 \cdot 10^{-3}$ kg/s, $I = 100$ A [30]; *6)* $d = 2 \cdot 10^{-2}$ m, $G = 38 \cdot 10^{-3}$ kg/s, $I = 500 \div 700$ A [20]; *7)* $d = 1 \cdot 10^{-2}$ m, $G = 15 \cdot 10^{-3}$ kg/s, $I = 100$ A [27].

(at occurrence of 100÷200 A) with approximately a constant 'rate' of increase, approximately 5 V/cm over 1 cm of the length of the section. With increase of current intensity, the length of the transition section decreases, and the value of E_t, to which the strength of the electrical field depends, also decreases. Taking into account the available data, it is possible to determine some criteria of dependence for E_{tr} on the main regime parameters. However, the almost linear increase of the strength in the section, the weak dependence of the 'rate' of increase of the strength on the regime parameters and the short length of the transition section make it possible to approximate the strength of the electrical field E_{tr} by a linear dependence between E_i and E_t, accepting that the length of the section is equal to 4–6 length length gages.

To calculate the characteristics of the electrical arc in the long cylindrical channel, it is necessary to know the relative length of the given sections: initial ($\bar{l}_i = z_i/d$), transition ($\bar{l}_{tr} = \Delta z_{tr}/d$) and developed turbulent section ($\bar{l}_t\ \Delta z_t/d$). For the given length of the inter-electrode insert in the plasma torches with the insert, the ratio of the sections also determines the arc voltage. It is necessary to determine the length of the initial section in plasma torches of different systems because this length determines the voltage and the self-setting arc length in the smooth cylindrical output electrode, and also the required length of the channel up to the ledge in a plasma torch with a step output electrode, etc.

The length of the initial section of the channel $\bar{l}_i$ in the flow of diatomic gases, including air, in a smooth cylindrical pipe was found analytically and by experiments [32]. In the case of moderate temperatures of the gas, the following dependence of the Reynolds number Red was obtained:

$$\bar{l}_i = 1.35\,\mathrm{Re}_d^{0.25}. \tag{5.23}$$

The determination of the length of the initial section of the arc in the smooth channel was carried out in [33] using the photographs of the arc column in a long quartz pipe. The origin of the transition zone was determined on the basis of the formation of random oscillations of the arc column. In the study, the authors propose an empirical dependence of the relative length of the initial section on the Reynolds number of the gas flow at entry into the channel and on the energy criterion $\bar{I} = I/\left(d\cdot\sqrt{\mu h \sigma}\right)$:

$$\bar{l}_i = 1.435\,\mathrm{Re}_d^{0.27}/(1+1.3\cdot 10^{-3}\bar{I}^{1.1}). \tag{5.24}$$

Here $\mathrm{Re}_d = (\rho u)_0 d/\mu$; μ and h is the viscosity and enthalpy at the temperature of the gas at entry into the channel (T = 300 K); electrical conductivity σ in the case of air was calculated at T = 6400 K. The exponent at Re_d, equal to α = 0.27, was selected to generalise the experimental data with a minimum scatter.

In the channel of the plasma torch with a sectioned inter-electrode insert, the length of the initial section was determined on the basis of the start of the increase of the strength of the electrical field and heat losses into the wall of the channel, i.e. along the length of the section *AB* on the scheme in Fig. 2.12, chapter 2. Without the arc, the length of the section of the channel from entry into the channel to the area of closure of the wall the boundary layer was determined, in both the section and smooth channels, using a thermoanemometer on the basis of the start of the rapid increase of the degree of turbulence of the flow on the channel axis. The results of the measurements are presented in Fig. 5.11 which shows the dependence of the complex $(l_i/d)\,\mathrm{Re}_d^{-0.25}$ on parameter I for the section channel (curve *1*), and for comparison there are the calculated data from [33] for a smooth channel (curve *2*). The graph also shows the experimental points obtained using a thermoanemometer in the absence of the arc (I = 0) for the smooth and sectioned channels. Comparison shows that in the smooth channel, the length of the initial section is considerably greater in comparison with that in the sectioned channel in the same conditions. According to the experimental results, the length of the initial section in the sectioned channel de-

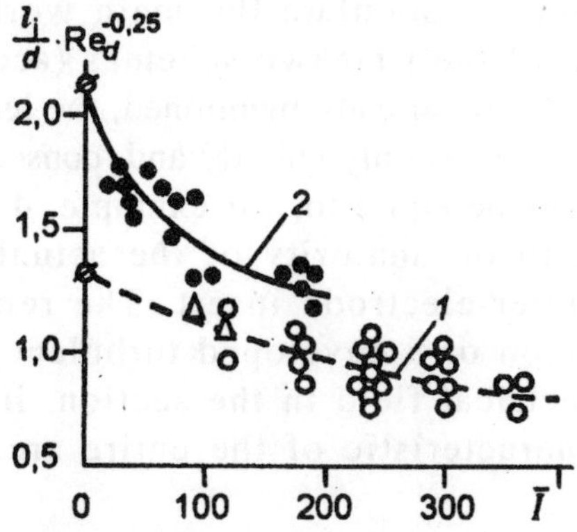

Fig. 5.11. Dependence of the complex $\bar{l}_i \mathrm{Re}_d^{-0.25}$ on I. *1*) IEI, O – $d = 1 \cdot 10^{-2}$ m, Δ – $2 \cdot 10^{-2}$ m; *2*) smooth channel • – data from [33], Ø – results of measurements with a thermoanemometer, $d = 1 \cdot 10^{-2}$ m, $I = 0$.

creases with increase of the width of the slits and depends only slightly on the presence of the accompanying inter-sectional gas supply. It may be concluded that, with other conditions being equal, the length of the initial section is determined by the rate of increase of the thickness of the boundary layer, i.e., by the surface roughness of the channel.

The experimental data for the sectioned channel are generalised by the dependence

$$\bar{l}_i = 1.35 \cdot \mathrm{Re}_d^{0.25} / (1 + 1.85 \cdot 10^{-3} \bar{I}), \tag{5.25}$$

with the accuracy to ±10%. The dependence was verified in the variation range $\mathrm{Re}_d = 10^4 \div 10^5$, $\bar{I} = 0 \div 400$. The numerator of the first part of the equation is the length of the initial section of the gas flow without the arc in the smooth pipe not taking the twisting of the gas in the channel into account (see (5.23)). This agreement may be accidental to a large degree and is explained by the weak effect of the twisting of the flow and the small width of the inter-sectional slits, because the effect of these factors is directly opposite. The numerator in the equation (5.25) determines the presence and effect of the thermal layer of the arc. Because of the constant and relatively small width of the slits in the experiments ($s = 1 \div 2$ mm), the effect of the slits is not presented in the explicit form. However, in the sectional channel with a large width of the slits, the initial section is shorter, i.e. generally speaking, the generalised dependence includes complex (s/d) in some form. It is also important to note the good agreement between the results of measurements, obtained using the thermoanemometer, of the length of the initial section of the cold flow with the measured values in the presence of the arc.

The data on the length of the initial section, calculated using equation

(5.25), may be used to calculate the main working parameters of the plasma torches of the first two schemes (according to the classification in chapter 1). As already mentioned, the length of the transition section is small and changes only slightly and, consequently, in evaluation it may be assumed to be equal to, for example, 4 length length gages which correspond to the majority of the actual system of plasma torches with the inter-electrode insert. The remaining part of the channel is the section of a developed turbulent flow. Knowing the strength of the electrical field in the section, it is possible to calculate the VAC characteristic of the entire arc.

5.2.3. Variation of arcing voltage by the gas-dynamic effect

The qualitative analysis of the behaviour of the arc in the turbulent gas flow, presented in chapter 2, shows that from the viewpoint of increasing the energy input into the arc it is convenient to ensure that the developed turbulent flow occupies a large part of the discharge channel. The gas flow may be turbulised by various methods, for example, by placing a ledge, introduction of different turbulizers into the channel, etc. Blowing the gas through the slits between the sections in the plasma torch with the inter-electrode insert enables the simplest turbulisation of the flow already in the initial section of the channels) [20].

Let us consider the variation of arc voltage in a plasma torch with an inter-electrode insert with constant relative length $\bar{a}$, with the boundary layer, developing in the initial section of the channel, affected by the working gas blown partially through only one of the intersectional slits with the coordinate $\bar{z}_s < \bar{z}_i$. The determining dimensionless gas-dynamic parameter is the blowing parameter $m_s = (\rho u)_s/(\rho u)_{0s}$. Here, the indexes 0*s* and *s* relate to the parameters of the flow in the section $\bar{z}_s$ in respectively the channel and the intersectional slits. The inter-electrode insert of the investigated plasma torch consisted of sections with a thickness of $(7 \div 21) \cdot 10^{-3}$ m, the inter-sectional gap $s = (1.5 \div 2) \cdot 10^{-3}$ m; the sections were distributed in groups in the order of decreasing thickness in the direction of the gas flow. The flow of the gas through the selected slit g_s was varied from 0 to $7.5 \cdot 10^{-3}$ kg/s, which corresponds to the variation of the parameter m_s from 0 to 2.3.

There are three possible variants of supplying the gas through the slit: two variants – along the tangent to the circumference, and one variant – in the radial direction. In turn, the supply of the gas along the tangent may coincide with the direction of the main flow

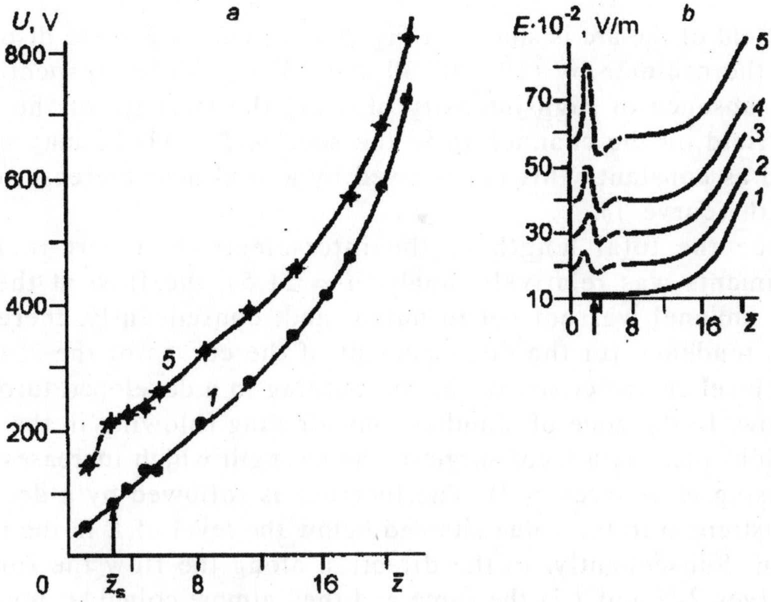

Fig. 5.12. Distribution of the potential (*a*) and the strength of the electrical field of the arc (*b*) along the axis of the channel with the gas blown in the same direction. $d = 2 \cdot 10^{-2}$ m; $\bar{a} = 21.5$; $\bar{z}_s = 3.2$, $I = 120$ A; $G = 30 \cdot 10^{-3}$ kg/s $g_i = 0.5 \cdot 10^{-3}$ kg/s, $G_0 + g_s = \text{const} = 15 \cdot 10^{-3}$ kg/s; 1–5 $- m_s = 0.08$; 0.18; 0.39; 0.62; 1.2, respectively.

travelling into the channel in the cathode zone (simultaneous supply) all in the opposite direction (opposite supply).

Initially, we examine the blowing of cold gas through a slit with twisting in the same direction. Figure 5.12*a* shows the distribution of the arc potential along the axis of the channel for different values of the blowing parameter. To improve the accuracy of examination of the curves, the scale is constructed for the curve *1* corresponding to the distribution of the potential along the arc without any intensive blowing ($m_s = 0.08$), and curve *5* is displaced along the ordinate by 100 V. At $m_s = 0.08$ the distribution of the arc potential along the initial section is linear; further, starting at $\bar{z} = 11$–12, the potential increases in a non-linear manner (the zone of contact of the boundary layer and of its mixing with the high-temperature gas). The increase of m_s in the blowing section is accompanied by a small increase of the potential (curve *5*). The length of the increase is small and at a distance of 3–4 length gages downwards along the flow from the blowing sections 1 and *5* are almost identical. The total arc voltage may be assumed to be constant in a wide range of variation of m_s. The corresponding distribution of the strength of the electrical field of the arc is shown in Fig. 5.12 *b*; curves *2–5* are displaced along the ordinate by (10; 20; 30 and 15) · 10^2 V/m, respectively. In the absence of high-intensity blowing the strength of the electrical field on the channel up to the section $\bar{z}$ = 11–12 may be regarded as constant. This is followed by a nonlinear increase of the strength (curve 1).

Since the total length of the inter-electrode insert in these experiments was relatively small ($\bar{a}$ = 21.5), the flow at the end of the channel was not yet turbulent and, consequently, there was only a tendency for the displacement of the curves of the strength to the level characteristic of the arc burning in a developed turbulent gas flow. In the zone of simultaneous blowing (blowing in the same direction) there is a local surge of the strength which increases with increasing m_s (curves *2–5*). The increase is followed by a decrease of the strength to the value situated below the level of E in the initial section. Subsequently, in the direction along the flow the form of the curves *2–5* and *1* is the same and they almost coincide. Identical results were obtained in examination of the arc in argon [34].

Analysis of the experimental material shows that the simultaneous blowing of the gas with the variation of m_s has only a small local effect on the strength of the electrical field in the vicinity of the blowing zone and this is possible only if the boundary layer interacts slightly with the blown gas and is displaced by the gas from the wall producing a unique local 'narrowing' of the channel increasing the value of E. The simultaneous blowing of the gas in other sections of the initial part of the channel has a similar effect on the strength of the electrical field.

What is the distribution of the potential if the gas is blown in the opposite direction? Examination of the variation of the degree of turbulence of the flow along the channel in this case indicates a decrease of the length of the initial section of the channel with increasing m_s.

The distributions $V(z)$ and $E(z)$ for different values of m_s are shown in Figs. 5.13 and 5.14. Already at relatively low values of m_s the start of increase of E is displaced in the direction of the blowing section (curve 2 in the graphs). At m_s = 1 the strength starts to increase in the blowing zone (curve *3*). Since the strength in the transition section depends only slightly on m_s, then with other conditions being equal, the length of the section with the developed turbulent flow increases with increasing m_s and this results in an increase of arc voltage.

In all likelihood, blowing in the opposite direction results in the intensification of mass exchange between the boundary layer and

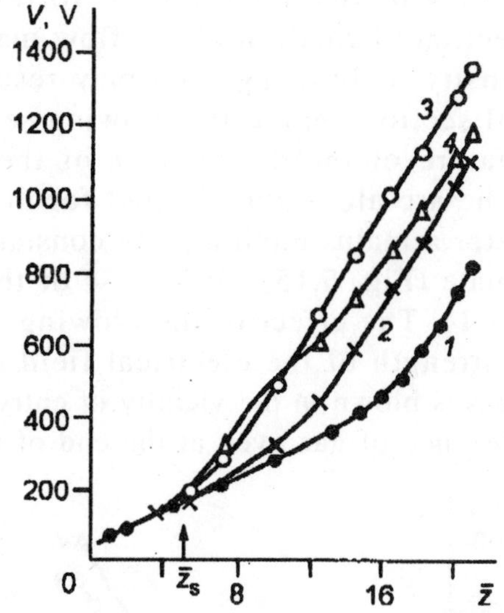

Fig. 5.13. Distribution of the arc potential along the axis of the channel with the gas blown in the opposite direction. 1–4 – m_s = 0.08; 0.37; 1.1; 2.1, receptively; z_s = 5. For the remaining symbols see Fig. 5.12.

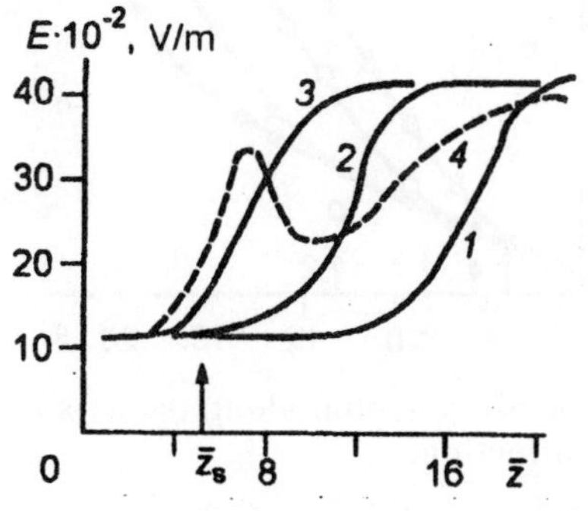

Fig. 5.14. Distribution of the strength of the electrical field of the arc along the channel with the gas blown in the opposite direction; for symbols see Fig.5.13.

the core of the flow. At m_s = 1 the coordinate $\bar{z}_i \approx \bar{z}_s$. It should be mentioned that at any value of m_s the value of E in the developed turbulent section remains on approximately the same level. At $m_s > 1$ the distribution $E(z)$ shows a local increase in the strength (curve *4* in Fig. 5.14), followed by a decrease and, subsequently, by a monotonic increase of the level of the strength in the developed turbulent flow. This distribution of the strength reduces the arc voltage (Fig. 5.13, curve *4*). Evidently, this is associated with overtwisting of the flow. Since the pulse of the blown gas directed along the tangent of the form prevails over the pulse of the main flow ($m_s > 1$), the stability of the arc may be disrupted by the vortex in the blowing section and the direction of rotation of the flow may change. In the case of strong intensity of blowing, this may result in the formation of a new initial section behind the blowing section.

The described nature of the distribution of the strength of the electrical field on the arc along the channel for different values of the blowing parameter remains qualitatively constant irrespective of the blowing coordinate (Fig. 5.15, curves *2–4* of the distribution of the potential at $m_s \approx 1$). The effect of the blowing parameter on the distribution of the strength of the electrical field in the channel is most marked if the gas is blown in the vicinity of entry into the channel. Nevertheless, the presence of gas even at the end of the initial section

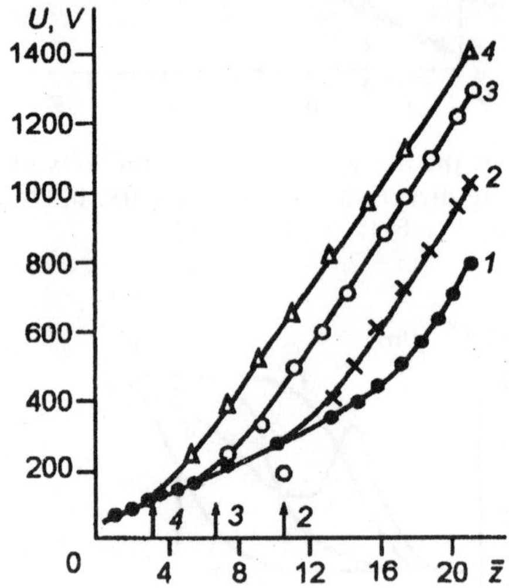

Fig. 5.15. Distribution of the arc potential along the axis of the channel with the gas blown in the opposite direction.

Curve No.	$\bar{z}_s$	$G_0 \cdot 10^3$, kg/s	$g_0 \cdot 10^3$, kg/s	m_s	U, V
1	–	15	0	0	1100
2	10.5	10	5	0.9	1330
3	6.8	10	5	1.0	1660
4	3.2	10	5	1.2	1760

in the case of a relatively low intensity and blowing in the same direction results in a large change of the electrical characteristics of the arc.

Analysis of the results presented in Fig. 5.16*a* (here U_0 is arc voltage at $m_s = 0.07$) enables the following conclusions to be drawn: *a*) the optimum voltage corresponds to approximately $m_s = 1$ which is in good agreement with the data on the distribution of the degree of turbulence of the gas flow and the strength of the electrical field of the arc along the axis of the electric arc chamber; *b*) as the value of $\overline{z}_s$ decreases, the effect of m_s becomes stronger; *c*) at $m_s>1$ arcing is unstable and in some cases the arc is extinguished, especially with increasing m_s. The graph, shown in the Fig. 5.16*b*, shows that the zone of counter blowing should not be placed in the vicinity of entry into the electric arc chamber ($\overline{z}_s < 2$) not at the end of the initial section of the channel ($\overline{z}_s \approx 12$). At low values of $\overline{z}_s$ the arc spot is destabilised on the cathode because of the disruption of twisting of the gas flow increasing the degree of erosion of the electrode material.

The third variant of the supply of gas – without twisting – was examined in [35]. Investigations were carried out on a plasma torch with a sectioned inter-electrode insert ($d = 15\cdot10^{-3}$ m). In order to ensure a stable position of the arc spot on the cathode, the gas with the flow rate of G_0 was introduced into the gap between the cathode and the first section with twisting, and in all subsequent slits it was introduced without twisting under the angle of ~30° in relation to the axis of the plasma torch. The distribution of the strength of the electrical field of the arc along the sectioned channel in different conditions of gas supply is shown in Fig. 5.17. Comparison of the curves *1* and *3* shows that the initial section with supply of the gas with accompanying twisting is considerably longer in comparison without twisting. According to the results of the effect on the strength of the arc, the supply of the gas under a small angle without twisting occupies an intermediate position between the supply of gas with

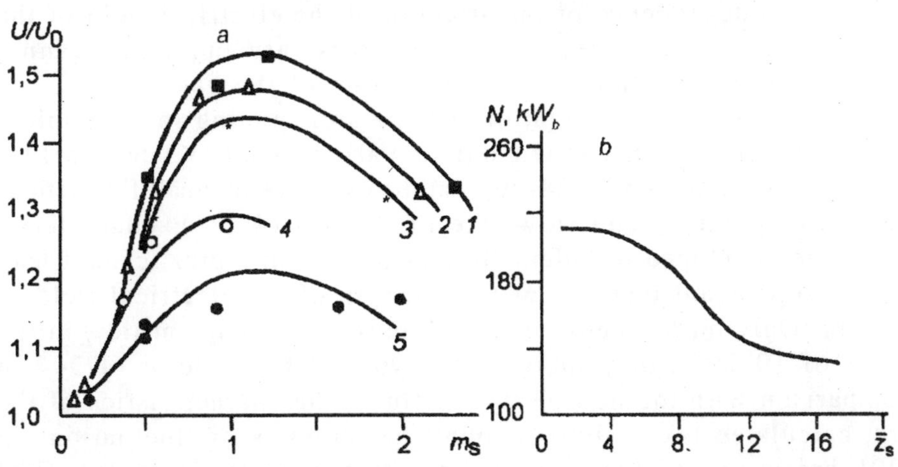

Fig. 5.16. Dependence of the relative voltage of the arc U/U_0 on m_s (*a*) and the arc power *N* on $\overline{z}_s$ (*b*) at $m_s = 1.0$. $d = 20\cdot10^{-3}$ m; $\overline{a} = 21.5$; $G = 30\cdot10^{-3}$ kg/s; $g_i = 0.5\cdot10^{-3}$ kg/s. $I = 120$ A; output electrode – cathode; 1–5) $\overline{z}_s = 3.2$; 5.0; 6.8; 8.7; 10.5 respectively.

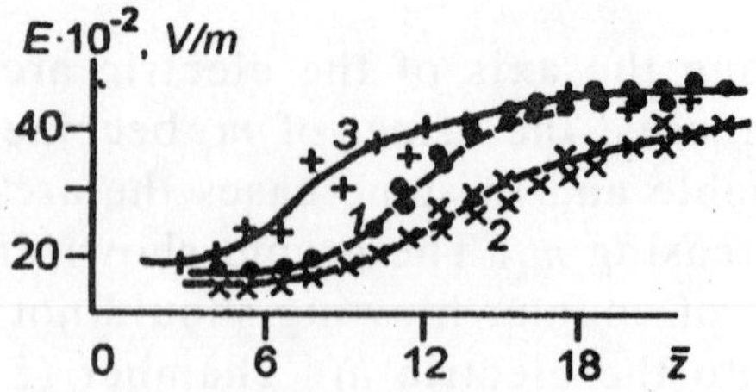

Fig. 5.17. The distribution of the strength of the electrical field on the arc along the axis of the channel with the gas supplied with and without twisting. $d = 15 \cdot 10^{-3}$ m; $G_0 = 1.5 \cdot 10^{-3}$ kg/s; $G_i = 17.9 \cdot 10^{-3}$ kg/s, $I = 120$ A; $G_n = \sum g_i$; *1,2*) twisting; *3*) without twisting ($G_n = 0.575\ G$ for the curves *1,3* and 0.27 for curve *2*).

twisting in the same and opposite directions.

The last method of the supply of the gas is effective, for example, in the introduction of dusted media when the effect of detachment of the solid particles is undesirable, and in a number of other cases.

5.2.4. Dependence of the strength of the electrical field of the arc on the determining parameters in the section of the developed turbulent flow of the gas

The power of a low-temperature plasma generator can be increased by a conventional method, i.e. increasing current, and also by increasing arc voltage, i.e. in the plasma torch with the inter-electrode insert the section of the electric arc channel with the developed turbulent flow will become more and more controlling. It is therefore necessary to find, on the basis of the experimental data, the generalised dependence of the strength of the electrical field of the arc on the main determining parameters: arc current, channel diameter, pressure, the type and flow rate of the gas.

The theoretical investigations of the arc, running in a turbulent gas flow, have been carried out in various studies such as [23, 36–39]. It was reported in [38–39] that the existence of small fluctuations of the temperature and flow rate of the gas (4–5%), characteristic of the developed turbulent flow of the gas in a pipe, cannot lead to any significant increase of the strength of the electrical field of the arc. Only more intensive fluctuations of these quantities, in the order of 10–20%, may increase the voltage by a factor of 3–4 in comparison with the non-perturbed flow. The characteristics of the arc, calculated taking into account the intensity of fluctuations in [39], are in satisfactory agreement with the experiments described in [40, 41]. Examination of the form of the arc, burning in the section of the channel with a developed turbulent flow [20], makes it possible to assume that the mechanism of the increase of the technical strength of the electrical field of the arc is, in all likelihood, not only the increase of the intensity of heat exchange between the arc and the gas, but also by the increase of the real arc length in the length gage section of the channel. At present, there is no complete theory of the electrical arc running in a developed turbulent gas flow, and the currently available models of the turbulent arc [23, 36, 37] do not have a sufficiently large experimental base and do not reflect fully the entire variety of the processes of interaction of the electrical arc with a turbulent gas flow. Therefore, for the development of the method of calculation of, in particular, the electrical characteristics of the arc in the plasma torch with the inter-electrode insert, it is necessary to generalise the experimental data on the strength of the electrical field of the arc in the section of the developed turbulent flow. One of the first attempts in this area was made by the authors of [31] but owing to the fact that the resulting equation did not include a controlling parameter such as gas pressure, the equation is partial and can be used only in the conditions (in respect of pressure) in which the experiments were carried out.

Examination of the dependence of the technical strength of the electrical field in the section of the developed turbulent flow on the determining parameters was carried out on a plasma torch with an inter-electrode insert (Fig. 5.8). The internal diameters of the investigated channels were $d = (10; 20; 30) \cdot 10^{-3}$ m. In the majority of the experiments, the diameters of the cylindrical output electrode-anode and the channel were identical. Anodes with $d_a = 14 \cdot 10^{-3}$ m were used only in the channel with $d = 10 \cdot 10^{-3}$ m. The relative length of the inter-electrode insert $\overline{a}$ was varied from 12 to 34. The thickness of the sections of the inter-electrode insert was $10 \cdot 10^{-3}$ m; at $d = 10 \cdot 10^{-3}$ m sections with a thickness of $16 \cdot 10^{-3}$ and $21 \cdot 10^{-3}$ m were also used. The gap between the sections was $(1 \div 2) \cdot 10^{-3}$ m. The sections of the inter-electrode insert were cooled with water. Part of the working gas with the flow rate G_0 was supplied through the vortex chamber into the electric discharge channel between the end electrode and the first section of the insert. The remaining gas was supplied to the vortex chambers between the sections. The flow rate of the gas g_i through a single twisting ring was varied in the range $(0 \div 1) \cdot 10^{-3}$ kg/s. In order to prevent breakdown between the last section of the insert and the anode, the gas flow rate was slightly increased: $g_a = (1 \div 3) \cdot 10^{-3}$ kg/s. In the majority of experiments, to increase the size of the section with the developed turbulent flow, a gas was supplied through the gap between the sections with the flow rate g_s at a distance of $\overline{z}_s$ = 1–5 length gages from entry into the electric arc chamber. The total gas flow rate through the plasma torch $G = G_0 + g_a + g_s + \sum g_i$ was varied from $6\cdot10^{-3}$ to $50\cdot10^{-3}$ kg/s. The experiments were carried out at arc currents of I = 40÷600 A.

In the generalisation of the integral characteristics of the arc with the self-setting length or the length fixed by a ledge, the determining parameters were represented by the pressure in the characteristic section (in the end) of the electric arc chamber, and the total gas flow rate. In generalisation of the strength of the electrical field of the arc it must be remembered that the pressure and flow rate of the gas, and also the channel diameter (if the electric arc chamber is not cylindrical) relate to the selected section of the channel. The latter must be especially stressed because in the plasma torches with the interelectrode insert the pressure and flow rate of the gas greatly change along the channel. This is clearly illustrated by the curves of distribution of the pressure shown in Fig. 5.18. The results of measurements showed that in the section of the developed turbulent flow of the gas (without taking the output electrode into account), the pressure decreases by almost a factor of 1.5.

The data on the electrical characteristics of the arc will now be discussed. Typical E_t–I characteristics of the arc for four values of the air flow rate are presented in Fig. 5.19. In the investigated current range, the characteristics decrease. The increase of the flow rate increases the strength of the electrical field. The same effect on the strength is exerted by the increase of gas pressure and a decrease of the channel diameter.

Selecting the dimensionless criteria in the generalisation of the experimental data, it was assumed that the effect of radiation and of the intrinsic magnetic field of the arc is small. Therefore, the determining parameters were represented by the arc current, the gas flow rate and pressure and also by the diameter of the electric arc chamber. The dimensionless criteria were:

$$S_E = 2(\sigma / \pi\mu h)^{0.5}(Ed); \quad S_I = 2(\pi\mu h\sigma)^{-0.5}(I/d);$$
$$\mathrm{Re}_d = 4G/(\pi d\mu); \quad \mathrm{Kn} = \mathrm{k}T/(Q\cdot p\cdot d).$$

Here μ, σ, h, T are the characteristic values of viscosity, electrical conductivity, enthalpy and temperature of the gas; k is the Boltzmann constant; Q is the effective scattering section of the electrons. The experimental material was generalised using the standard procedure [1, 2, 4] and the formula for the strength of the electrical field of

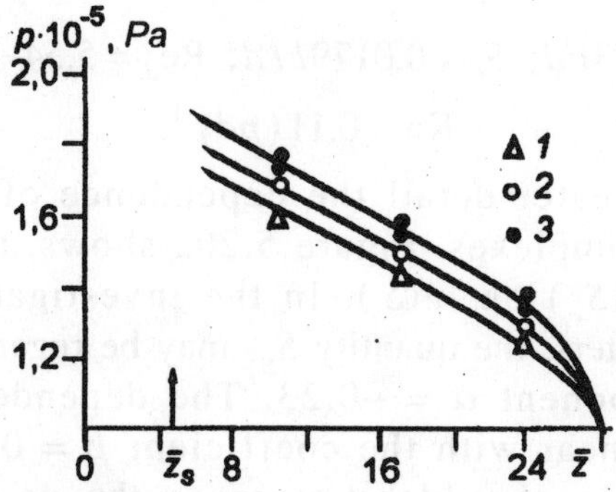

Fig. 5.18. Distribution of the gas pressure on the channel. $d = 20 \cdot 10^{-3}$ m; $\bar{a} = 25$; $\bar{b} = 3$; $\bar{z}_s = 4.5$; $m_s = 1.0$; $I = 100$ A; *1)* $G = 25 \cdot 10^{-3}$ kg/s, $g_i = 0$; 2) $G = 26.3 \cdot 10^{-4}$ kg/s, $g_i = 0.1 \cdot 10^{-3}$ kg/s; *3)* $G = (27.3 \div 27.8) \cdot 10^{-3}$ kg/s; $g_i = 0.4 \cdot 10^{-3}$ kg/s.

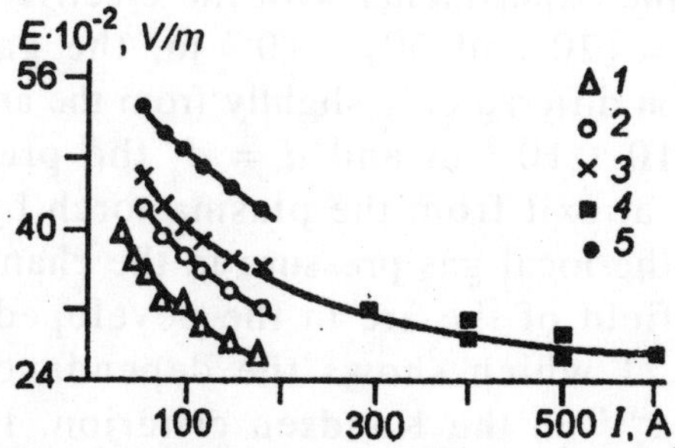

Fig. 5.19. E_t–I characteristics of the arc. $d = 20 \cdot 10^{-3}$ m; $\bar{a} = 20.25$; $\bar{z}_s = 2$; $m_s = 1.0$; $p = 1 \cdot 10^5$ Pa; *1)* $G = 14.8 \cdot 10^{-3}$ kg/s, $g_i = 0$; *2)* $G = 21.4 \cdot 10^{-3}$ kg/s, $g_i = 0.15 \cdot 10^{-3}$ kg/s; *3)* $G = 25.1 \cdot 10^{-3}$ kg/s, $g_i = 0.30 \cdot 10^{-3}$ kg/s; *4)* $G = 24.5 \cdot 10^{-3}$ kg/s, $g_i = 0.37 \cdot 10^{-3}$ kg/s, $\bar{a} = 14.3$, $G_0(N_2) = 6.0 \cdot 10^{-3}$ kg/s; *5)* $G = 36.9 \cdot 10^{-3}$ kg/s, $g_i = 0.54 \cdot 10^{-3}$ kg/s.

the arc was derived in the following form:

$$S_{E_T} = C \cdot S_I^{\alpha} \cdot \mathrm{Re}_d^{\beta} \cdot \mathrm{Kn}^{\gamma}. \tag{5.26}$$

In subsequent stages, the characteristic values of temperature, enthalpy, the viscosity and electrical conductivity of the gas were regarded as constant: $T = 400$ K; $h = 4 \cdot 10^5$ J/kg; $\mu = 2.3 \cdot 10^{-5}$ kg/(m · s). According to [4], the electrical conductivity of air at $T = 6400$ K was in this case $\sigma = 432$ S/m. The effective scattering section of the electrons in the arc, included in the Knudsen number, depends only slightly on temperature and may be assumed to be equal to $Q = 5 \cdot 10^{-20}$ m² in the case of air [42].

Taking these assumptions into account, if we examine only the changing parts of the criteria all complexes from the equation (5.26), we obtain

$$S_{E_T} = 7.73Ed;\ S_I = 0.0179I/d;\ \mathrm{Re}_d = 5.54 \cdot 10^4 G/d;$$
$$\mathrm{Kn} = 0.11\,(pd)^{-1}.$$

We examine in greater detail the dependence of S_{E_T} on the determining criterial complexes. Figure 5.20a shows, as an example, the dependence of $\lg(S_{e_t})$ on $\lg(S_i)$. In the investigated range of variation of the parameters, the quantity S_{E_T} may be regarded as proportional to S_i with the exponent $\alpha = -0.23$. The dependence of $\lg(S_{E_T})$ on $\lg(\mathrm{Re}_d)$ is also linear with the coefficient $\beta = 0.47$ (Fig. 5.20*b*).

Special attention should be given to the determination of the dependence of the strength of the electrical field of the arc on gas pressure. As already mentioned, in examination of E, the values of the determining parameters should be considered for the investigated cross-section. In the experiments with the electric arc chambers with the diameters $d = (20 \text{ and } 30) \cdot 10^{-3}$ m, the gas pressure in the measurement section differed only slightly from the atmospheric pressure, whereas at $d = 10 \cdot 10^{-3}$ m and $d = d_a$ the pressure was higher than the pressure at exit from the plasma torch by $(0{,}5 \div 0.7) \cdot 10^5$ Pa. The effect of the local gas pressure in the channel on the strength of the electrical field of the arc in the developed turbulent flow is shown in Fig. 5.21 which shows the dependence of the complex $A = S_{E_t} \cdot S_I^{0.23} \cdot \mathrm{Re}_d^{-0.47}$ on the Knudsen criterion. In the investigated range of the variation of the Knudsen number, the dependence of lg A on lg Kn should be regarded as linear with the coefficient $\gamma = -0.2$. The observed scatter of the experimental points is caused mainly by errors in the determination of pressure in the measurement zone.

For the approximate calculation of the strength of the electrical

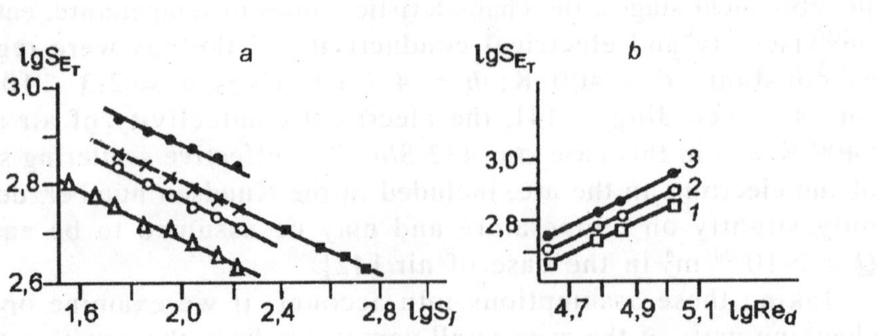

Fig. 5.20 Dependence of $\lg S_{E_T}$ on $\lg S_I$ (*a*) and on $\lg \mathrm{Re}_d$ (*b*). a) all parameters correspond to Fig.5.19; *b*–$d = 20 \cdot 10^{-3}$ m, $\bar{a} = 20.25$, $\bar{z}_x = 2$, $\bar{z} = 16 \div 20$; *1*) $S_I = 53.7$ (I = 60 A); *2*) $S_I = 89.5$ (I = 100 A); *3*) $S_I = 134$ (I = 150 A).

field of the arc running in air, the following equation has been proposed which generalises all the experimental data:

$$S_{E_T} = 1.34 \cdot S_I^{-0.23} \mathrm{Re}_d^{0.47} \mathrm{Kn}^{-0.2}. \tag{5.27}$$

In the measurement range of the criteria S_I = 35÷540, Re_d = (2.7÷11.0) · 10^4; Kn = 1.3÷11) ·10^{-5}, the relative deviation of the experimental points from the calculated curve does not exceed ±6%. The length of the base used for the measurement of the difference of the arc potentials, and also the variation of the flow rate and pressure of the gas in the base, are relatively small, so that it was possible to assume that the strength of the electrical field in the measurement base is constant. Figure 5.22 shows the dependence of S_{e_t} on the complex $\varphi = S_i^{-0.23}\ \mathrm{Re}_d^{0.47} \cdot \mathrm{Kn}^{-0.2}$.

Taking into account only the changing parts of the dimensionless criterial complexes, equation (5.27) has the form which is more suitable for technical calculations of the strength of the electrical field of the arc:

$$E_T \cdot d = 115(I/d)^{-0,23}(G/d)^{0,47}(pd)^{0.2}. \tag{5.28}$$

The satisfactory results obtained using equation (5.28) for the calculation of the distribution of the strength of the electrical field of the arc along the entire section of the developed turbulent flow are indicated by the curve shown in Fig. 5.23. The value of E_t was calculated from the local values of the flow rate and pressure of the gas [30]. At a large increase of the flow rate of the gas along the section of the developed turbulent flow, the relative deviation of the experimental points from the calculated curve does not exceed ±10% at a reliability of 0.95.

Thus, the E_t–I characteristic is drooping in the investigated range of the variation of the complex S_i. On the other side, E_i–I

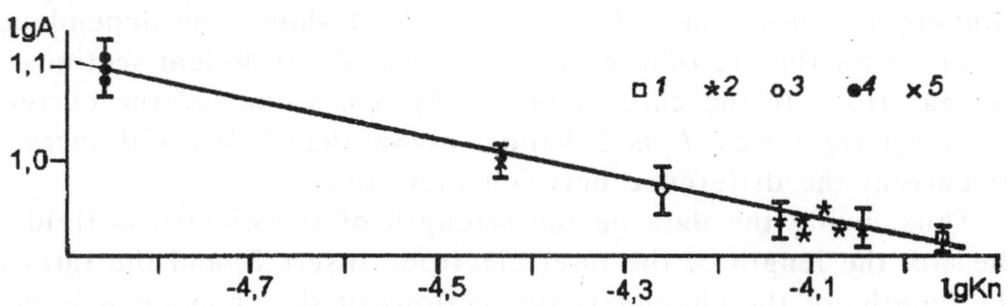

Fig. 5.21. Dependence of lg A and lg Kn. *1)* d = 10 · 10^{-3} m, d_a = 14 · 10^{-3} m, p = 1 · 10^5 Pa; 2) $d = d_a$ = 10 · 10^{-3} m, p = (1÷1.7) · 10^5 Pa; *3)* d = 20 · 10^{-3} m, p = 1 · 10^5 Pa; *4)* d = 7 · 10^{-3} m, p = 11.2 · 10^5 Pa [40]; *5)* d = 30 · 10^{-3} m; p = 1 · 10^5 Pa.

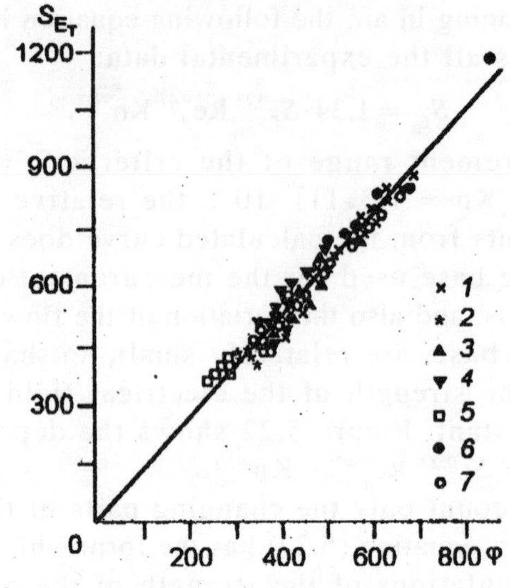

Fig. 5.22. Comparison of experimental data with the generalized $E_T - I$ characteristic of the arc. *1*) $d = 30 \cdot 10^{-3}$ m, $p = 1 \cdot 10^5$ Pa; 2) $d = 10 \cdot 10^{-3}$ m, $p = (1 \div 1.7) \cdot 10^5$ Pa; *3*) $d = 20 \cdot 10^{-3}$ m, $p = 1 \cdot 10^5$ Pa [35]; *4*) $d = 15 \cdot 10^{-3}$ m, $p = 1 \cdot 10^5$ Pa [35]; *5*) $d = 10 \cdot 10^{-3}$ m, $p = 1 \cdot 10^5$ Pa [35]; *6*) $d = 7 \cdot 10^{-3}$ m, $p = 11.2 \cdot 10^5$ Pa [40]; *7*) $d = 20 \cdot 10^{-3}$ m, $p = 1 \cdot 10^5$ Pa.

characteristic is *U*-shaped [3, 26] with the extended rising section. As shown in [25], the true dependence $E_i = f(I)$ is more complicated in comparison with the evaluation equation (5.22) which is valid in a relatively narrow range of the variation of the parameter I/d, namely: $4 \cdot 10^3 \leq I/d \leq 2 \cdot 10^4$ m. In [38, 43] it has been assumed that at high currents the strength $E_i \to E_t$. Without discussing the validity of the hypothesis and examining the physics of the phenomenon, leading to the convergence of the values of the strength of the electrical field in different sections of the channel, it will be shown that this convergence does take place. Figure 5.24 shows the dependence $E = f(I)$ for the arc running in the initial and turbulent sections of the gas flow. In the case of relatively low values of the current, as shown previously, E_t is 2–3 times higher than E_i but with increase of current the difference between them decreases.

Thus, using the data on the strength of the electrical field of the arc, the length of the inter-electrode insert $\overline{a}$ and the ratio of the lengths of the characteristic sections of the channel, it is possible to calculate the VAC of the arc. If $\overline{a}$ is slightly higher than $\overline{z}_i$, and there is no counter blowing, the VAC is *U*-shaped because the role of the turbulent section of the arc is not significant. If the value $\overline{a}$ is high or counter blowing 'does operate', the role of the

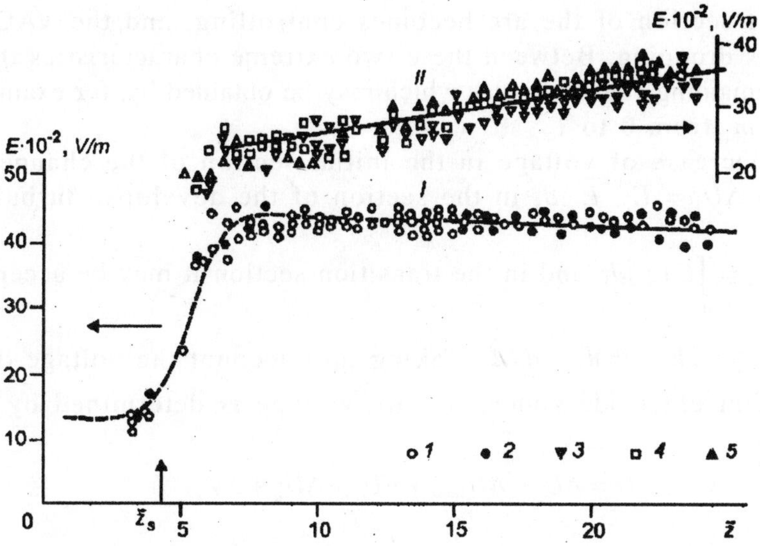

Fig. 5.23. Distribution of the strength of electrical field of the arc along the channel. $d = 20 \cdot 10^{-3}$ m, $\bar{a} = 25$; $\bar{z}_s = 4.5$; $I = 100$ A: $g_i = 0$, $G \approx 24.6 \cdot 10^{-3}$ kg/s; *1*) $m_s = 1.0$; *2*) $m_s = 1.1$; II: $g_i = 0.4 \cdot 10^{-3}$ kg/s, $G \approx 28.0 \cdot 10^{-3}$ kg/s; *3*) $m_s = 1.1$; *4*) $m_s = 1.43$; *5*) $m_s = 1.65$. Solid line – calculated from equation (5.28).

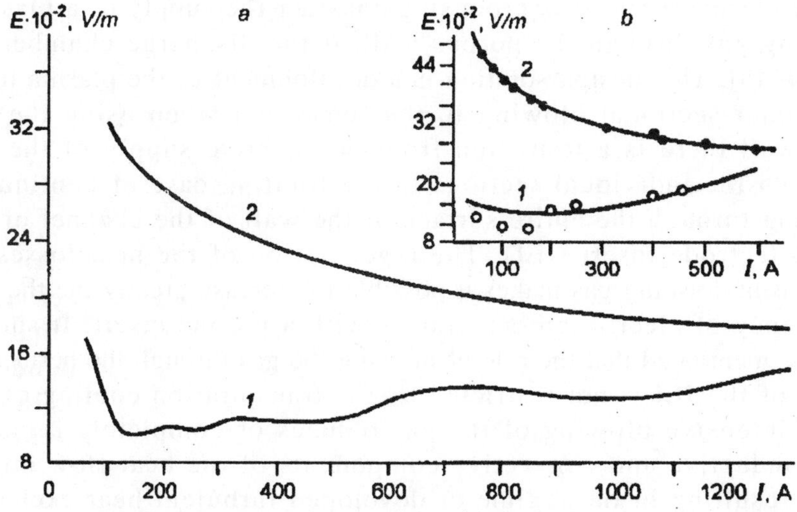

Fig. 5.24. Dependence of the strength of the electrical field of the arc on current. *a*) $d = 30 \cdot 10^{-3}$ m, $G = 36 \cdot 10^{3}$ kg/s, $p = 1 \cdot 10^{5}$ Pa; *1*) experimental data [25] for the initial section of the channel (recorded in an automatic recording device); *2*) Calculated from equation (5.28) for the turbulent section of the channel; *b*) $d = 20 \cdot 10^{-3}$ m, $G = 24.5 \cdot 10^{-3}$ kg/s, $p = 1 \cdot 10^{5}$ Pa; *1*) calculated from equation (5.22), circles - experimental points; *2*) experiments (turbulent section of the channel).

turbulent section of the arc becomes controlling, and the VAC of the arc is drooping. Between these two extreme characteristics there are all remaining characteristics which may be obtained by, for example, varying m_s from 0 to 1.

The decrease of voltage in the initial section of the channel is equal to $\Delta U_i = E_i \cdot \bar{l}_i \cdot d$, in the section of the developed turbulent flow $\Delta U_t = \int_0^{l_T} E_t(z_t)dz_t$ and in the transition section it may be accepted that $\Delta U_{trans} = (E_t - E_i)l_{trans} \cdot d/2$. Taking into account the voltage drop in the near-electrode zones, the arc voltage is determined by the equation:

$$U = \Delta U_i + \Delta U_{trans} + \Delta U_t + \Delta U_a + \Delta U_c.$$

5.3. THE ENERGY CHARACTERISTICS OF THE ARC IN A POROUS CHANNEL

For the effective hydrodynamic effect on the parameters of arc discharge in the plasma torch, it is promising consider the supply of a plasma-forming gas through the porous wall of the discharge chamber [4, 20, 44–48]. This design solution is a development of the plasma torch with inter-sectional blowing in the sense that when using the porous wall there is a transition from the discrete supply of the gas between the individual sections to the limiting case of continuous blowing through the entire surface of the wall of the channel of the inter-electrode insert (IEI). The regeneration of the heat losses by the plasma-forming gas makes it possible to increase greatly the thermal efficiency of electric arc generators with a porous insert. It should also be mentioned that the role of blowing the gas through the permeable walls of the IEI is not restricted by the transpiration cooling of the wall. Intensive blowing of the gas reduces or completely removes the conductive and convective components of the heat flow on the wall, resulting in the regime of developed turbulent heat exchange between the arc and the heated gas. This type of blowing has an active effect on the electrical parameters of the discharge and, primarily, on the strength of the electrical field on the arc.

The investigations carried out in [47] show that in the case of relatively low-intensity of blowing of the gas to the porous wall, the arc column is stabilised on the channel axis and the arc is split into several current-conducting channels. We examine the structure of the discharge for different types of blowing hydrogen and nitrogen. The authors of [47] noted efficient spatial stabilisation of the hydrogen arc in the case of low-intensity blowing through the porous wall. With increase of the specific blowing rate of the gas at a constant intensity of current, the discharge was constricted and the axial temperature increased from $13 \cdot 10^3$ to $16 \cdot 10^3$ K at constant values of the current intensity and gas flow rate, examination showed variations of the axial temperature with the amplitude $(1.5 \div 2.0) \cdot 10^3$ K with a period of 60–75 μs and with the amplitude of $(3 \div 4) \cdot 10^3$ K with a period of 300 μs. According to the estimates, the duration of formation of the equilibrium profile of temperature is 20–30 μs, and the time to establishment of the equilibrium profile of the concentration is 90–100 μs. The lifetime of the plasma in the equilibrium condition is ~165 μs. Comparison of the experimental values of n_e and T_e with the calculated (equilibrium) dependence $n_e\ (T_e)$ shows the deviation from the LTE both in respect of temperature and the concentration of electrons, with the deviation being outside the error range of the measurements. The deviation from the LTE increases with increasing gas flow rate.

In the case of blowing nitrogen into the discharge channel [49], the deviation from the thermal equilibrium is also recorded in a large part of the cross-section of the channel, and the deviation is represented by the fact that T_e and T_i are higher than T and reaches $1000 \div 2000$ K in the peripheral section of the channel.

In the case of high blowing rates of the gas, the situation is different [50]. High-speed filming shows that with the increase of the flow rate of nitrogen in the cross-section of the channel close to the exit cross-section, the arc column is divided into several current-conducting channels (Fig. 5.25). A continuous rearrangement of temperature profiles was recorded. This rearrangement takes place within

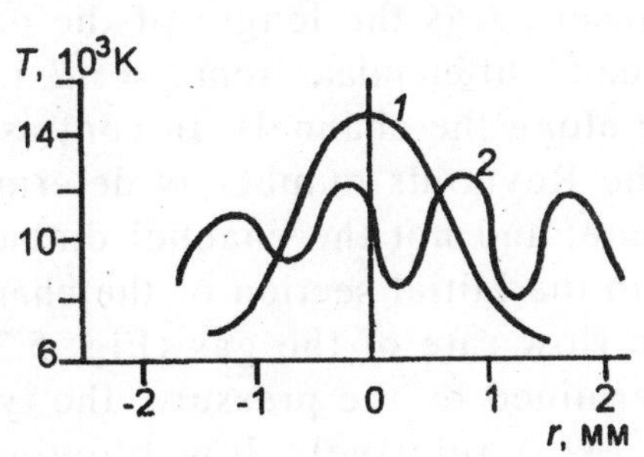

Fig. 5.25. Distribution of temperature in the cross section of the discharge channel with nitrogen blown through the porous insert. G = 0.18 kg/s, I = 280 A, $\bar{z}$ = z/d = 2.5 [50]: *1*) monoprofile; 2) multi-filament form.

the period of $10^{-4} \div 10^{-5}$ s. With increase of the gas flow rate the relative duration of existence of the discharge in the multi-filament form increases. In transition to the multicord form of the discharge, the characteristic temperature on the axis decreases from $(14 \div 16) \cdot 10^3$ K to $(10 \div 12) \cdot 10^3$ K. The distribution of temperature in the cross-section of the channel may be described as follows. In the central (current-conducting) part, there is a relatively homogeneous 'diffusion' zone with the electron temperature of $(6 \div 8) \cdot 10^3$ K characterised by the formation, displacement and disappearance of the 'constricted' current filaments. The identical 'diffusion' zone evidently forms in the traces of the filaments and undergoes radial oscillations together with the filaments. In addition to the radial displacement of the current filament, they also move in the helical manner. The identical situation is also found in the case of high-intensity blowing of hydrogen, H_2 and CO_2.

Measurements of the distribution of the potential and the strength of the electrical field of the arc in the permeable channel, carried out in the previously cited studies, shows that the increase of the flow rate increases the gas pressure in the arc channel, and also increases the drop of the potential along the length of the anode whose relative value reaches 30–40 % of the total drop. The strength of the electrical field E increases with increase of the axial gas flow rate ($G = \pi d \dot{m} z$, where $\dot{m}$ is the specific flow rate of the gas related to the internal surface of the channel).

The dependence of the strength of the electrical field of the arc on the Reynolds number of the gas flow is shown in Fig. 5.26 for different gases and blowing intensity through the permeable wall [51, 52]. The Reynolds number is determined as $\mathrm{Re} = 4\dot{m}\bar{l}/\mu$, where $\bar{l} = z/L$ is the relative coordinate along the porous insert, μ is the viscosity of the gas at inlet temperature, z is the actual coordinate along the porous insert, L is the length of the porous insert (coordinate $\bar{l}$ is introduced to differentiate from $\bar{z} = z/d$, i.e. the dimensionless relative coordinate along the channel). In contrast to the previously examined cases, the Reynolds number is determined in respect of the actual coordinate, and not the channel diameter.

The value of E in the initial section of the channel increases with the increase of the flow rate of the gas (Fig. 5.26, curves 1–4, 5–6, 7–8) and is determined by the pressure, the type of gas, and the channel diameter. At a relatively low blowing intensity of air (curve 1) to the critical value $\mathrm{Re}^* \approx 10^5$ the strength E is not high and may be regarded as proportional to $\sim\mathrm{Re}^{0.4}$. This value of the Reynolds number corresponds to the coordinate along the insert

$\bar{z} = z/d = 2.5 \div 3$. At Re > Re* the increase of the gas flow rate is accompanied by a rapid increase of E, starting from approximately the same cross-section of the channel. In the section $z/d = 4$, as confirmed by the authors of [51, 52], $E \sim Re^{0.8}$, and at the end of the porous insert ($z/d \sim 5$) $E \sim Re^{1.6}$. This anomalously high increase of E in the area of the output electrode is difficult to explain on the basis of the scheme of interaction of the arc with the turbulent gas flow (chapter 2). Possibly, the main role in the high values of the strength of the electrical field is played the fact that calculations are carried out to determine the technical strength, i.e. the results of measurements of the potential difference of the adjacent probes were divided by the distance between them. The presence of the arc loop in the output electrode and the large thickness of the cold layer of the gas between the arc and the wall had the unavoidable effect on the accuracy of measurements of the strength of the electrical field. It was therefore necessary to carry out identical measurements in the conditions in which the section of the arc in the output electrode did not influence the accuracy of measurements of the arc potential. These measurements were carried out in [53].

In order to explain the relationships in the distribution of the strength of the electrical field of the arc along the channel with the combined (permeable and non-permeable) walls, experiments were carried out using plasma torches with IEI [20]. The internal diameter of the electric arc chamber in the experiments was constant and equal to 2 cm. The specially developed block of the sections [54] which could be placed in any section of the channel, consisting of a set of po-

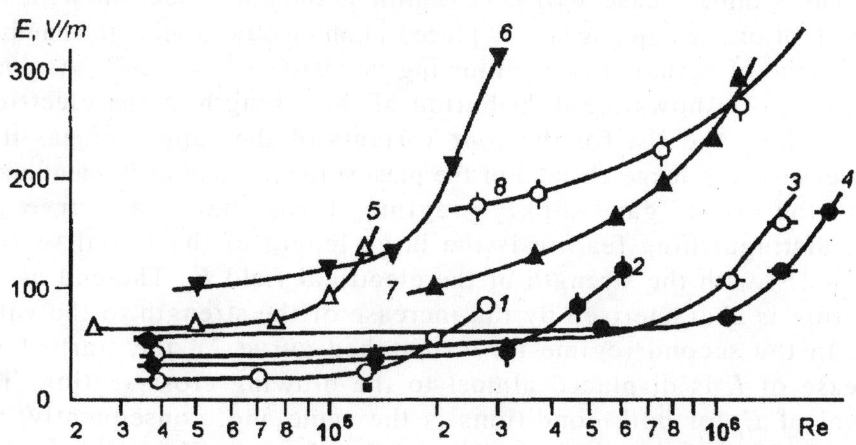

Fig. 5.26. Dependence of the strength of the electrical field of the arc on Re in blowing through the porous insert [51]. *1–4)* air; *5,6)* H_2; *7,8)* CO_2.

rous inserts (produced from foam cordierite) with the length of one length gage each, separated by non-permeable diaphragms. Two types of separating copper diaphragms were used: water-cooled diaphragms 2 cm thick and uncooled diaphragms, thickness approximately 3 mm. The number of the porous sections in the block was varied from 1 to 6. The total flow rate of the working gas (air) was varied in the range 25÷85 g/s so that it was possible to examine the effect of the intensity of blowing $\bar{g}_p = g_p/F$ in the range 0.5÷2 g/(s·cm²). Here g_p is the flow rate of the gas through the porous wall, F is the area of the internal surface of the wall. The majority of the experiments were carried out at $\bar{g}_p$ ~ 1.1 g/(s·cm²) and the arc current I = 120 A. The working gas was not supplied between the sections of the IEI in front of and behind the porous block. To realise the regime of developed turbulent flow of the gas, intensive counter blowing of the gas was supplied in front of the porous sections at a distance of 2–3 length gages from entry into the channel [3, 20]. The static pressure of the working gas in the plasma torch was determined in all experiments behind the porous block. The strength of the electrical field of the arc was measured in all sections of the channel: in front of the porous block, in the zone of the block, and behind the block downwards along the flow. The sections of the IEI and the diagrams of the porous block were used as the end probes in the measurement of the potential of the appropriate section of the arc. The measurement procedure was described previously, and the area of determination of the strength of the electrical field of the arc did not exceed ± 6 %.

The simplest case will be examined: only one section with the length of one length gageage is placed in an electric discharge chamber; the range of variation of the blowing intensity is $\bar{g}_p$ = 0.2÷2 g/(s·cm²). Figure 5.27 shows the distribution of the strength of the electrical field along the IEI for the four variants of the supply of gas into the electric discharge channel of the plasma torch. As already mentioned, for the first gas supply regime (cross-hatched curve 1) the distinguishing feature is the large length of the initial section ($\bar{l}_i$ ~ 15) with the strength of the electrical field E_i. The end of the section is characterised by the increase of the strength to the value E_t. In the second regime (cross-hatched curve 2), the start of increase of E is displaced almost to the blowing cross-section. The level of E_t for both conditions is the same and, consequently, the curves 1 and 2 merge at the end of the channel. The solid lines in Fig. 5.27 are the results of calculation of E_i and E_t, using equations (5.22) and (5.28), respectively. For the third and fourth regimes (points 3 and 4), the distribution curves of the strength extended to the level E_t behind the porous section. Regardless of the fact that the intensity of blowing through the porous insert is relatively high (g_p = 1.2 g/(s·cm²)), the values of the strength do not exceed E_t.

The form of the curve $E(z)$ remains qualitatively constant even if the gas is blown through three porous sections, situated at the distance of 1 length gage from each other in the initial section of the channel (Fig. 5.28, curve 1). The strength of the electrical field increases to the level E_t but does not exceed this level. When the sections are separated only by the thin non-cooled diaphragms, the blowing of gas may be regarded as almost continuous along the porous block (curve 2). In this case, the strength of the electrical field increases more rapidly because of the rapid increase of the flow rate of gas in the section in which the measurements are taken (compare curves 1 and 2). At the end of the porous block, the level of the strength is 15–20 % higher than the calculated values of E_t for the given conditions. In the direction along the flow the strength rapidly decreases to the level of E_t.

Thus, if the porous block is placed in the initial section of the channel, the blowing of gas through the block at a relatively high value of g_p accelerates the process of turbulisation of the flow (as

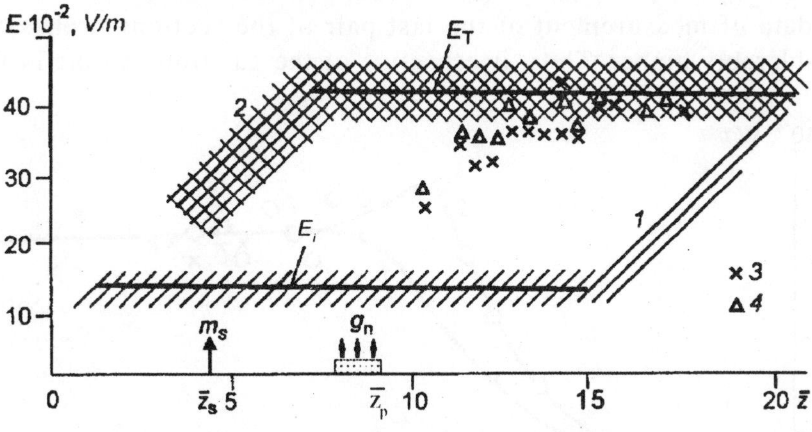

Fig. 5.27. Distribution of the strength of the electrical field of the arc along the discharge channel at d = 2 cm, G = 25 g/s, g_i = 0.1 g/s; I = 100÷120 A. *1*) sectioned channel with the distributed gas flow g_i = 0.1 g/s; *2*) blowing in the opposite direction with m_s ≈ 1.0 in the section $\bar{z}_s$ = 4.5; *3*) section channel with the gas blown through a porous insert with length $\bar{l}_p$ = 1 in the section $\bar{z}_p$ = 9, g_p = 15 g/s; $\bar{g}_p$ =1.2 g/(s · cm²); *4*) blowing in the opposite direction with m_s ≈ 1.0 in the section $\bar{z}_s$ = 4.5 and blowing through the porous insert $\bar{l}_i$ = 1; $\bar{z}_p$ = 9; g_p = 15 g/s; $\bar{g}_p$ 1.2 g/(s · cm²).

in the case of counter blowing into the IEI with non-permeable walls) and, consequently, the strength of the electrical field of the arc increases from the value of E_i in front of the first section to E_t and the end of the third section (Fig. 5.29, curve 1). However, if this block is situated in the zone of transition or developed turbulent gas flow, which in the given experiments was obtained by counter blowing of the gas in the section $z_s \sim 2$ at $m_s \sim 1$, the strength of the electrical field at the start of the zone of porous blowing is already close to the level corresponding to the value of E in the transition or developed turbulence section (Fig. 5.29, curve 2). At the end of thê porous insert, both curves almost completely merge with each other because the total flow rate of the gas and the pressure in both cases are approximately identical. The graph also gives the data obtained in [46] for similar values of the gas flow rate, current intensity and pressure (curve 3). For better understanding of the experimental data, the coordinate of the origin of the porous IEI (and, consequently, curve 3) is combined with the start of the block of the porous sections. It should be mentioned that the length of the channel with the porous walls in [46] equalled approximately 5 length gages, and the internal diameter was 2 cm. In the length of the first three length gages of the porous IEI, the curve 3 was situated between the curves 1 and 2 and is determined by the prior history of the development of the boundary layer; at the end of the IEI, the value of E (according to the data of measurement of the last pair of the sections-probes) was considerably higher. Thus, irrespective of the gas flow regime in front

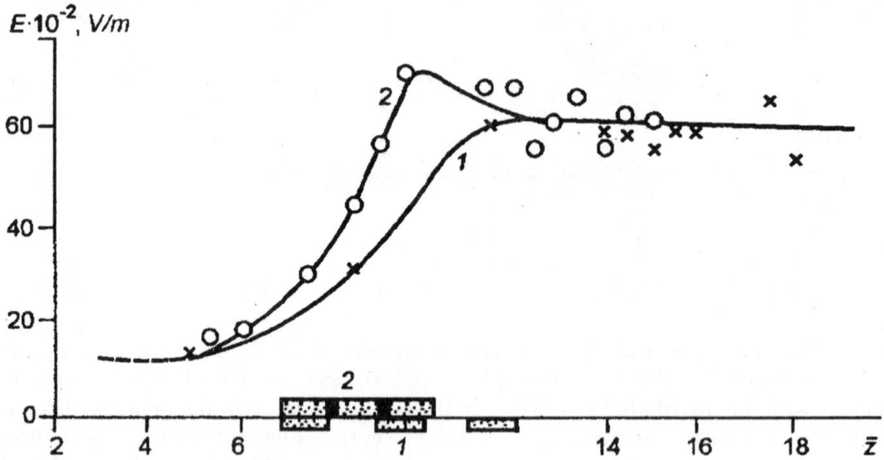

Fig. 5.28. Distribution of the strength of the electrical field along the channel in blowing of the gas through three seperate (*1*) and closely spaced (*2*) porous sections for I = 120 A, G_0 = 6 g/s, g_p = 50 g/s, $\bar{g}_p$ = 1.3 g/(s · cm²).

of the porous section of the channel in the first length gages, the turbulent flow regime forms or continues to develop and, consequently, the strength of the electrical field increases.

A further increase of the length of the porous block does not result in any qualitative change in the nature of distribution $E(z)$. For example, the graph in Fig. 5.30 gives the data on the strength of the electrical field of the arc along a block consisting of 1, 3, 4 or 6 porous sections whilst retaining the constant value of g_p. The results of all experiments are in relatively good agreement with each other, i.e. the strength in porous blowing does not depend on the length of the porous block but it depends on the flow rate and pressure of the gas in the given cross-section of the channel at a constant arc current intensity. At a distance of approximately 4 length gages from the start of the porous channel and the given value g_p, the rate of increase of the strength of the electrical field of the arc decreases and this is characteristic of the developed turbulent flow of the gas ($E_t \sim G^{\beta}$, and $\beta < 1$). According to the results of comparison, in this case the strength E is slightly higher than the strength of the electrical field of the arc in the developed turbulent flow of air in the plasma torch with the sectioned inter-electrode insert, calculated using equation (5.28) or analytically [55] (the calculated level of E_t at the end of the porous block corresponds to the horizontal section of the experimental curve behind the porous block). It may be seen that this value is 20% or more higher. In particular, this is associated with the lower (in comparison with the non-permeable channel) mean mass temperature of the gases

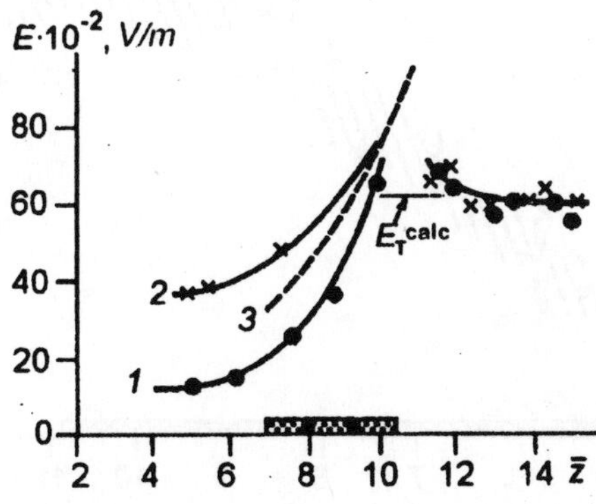

Fig. 5.29. Distribution of the strength of the electrical field along the channel at d = 2 cm, G_0 = 6 g/s, in a combined channel with three porous sections at I = 120 A, g_p = 1.1 g/(s · cm^2), $\bar{z}_p$ = 7, $\bar{l}_p$ = 3.5. *1*) p = 0.21 MPa, m_x = 0; 2) p = 0.24 MPa, m_x = 1, $\bar{z}_x$ = 2; *3*) for a continuous porous insert according to the data in [47] at I = 120 A, g_p = 1 g/(s·cm^2), $\bar{z}_p$ = 0, $\bar{l}_p$ = 5, p = 0.3 MPa (broken line).

surrounding the arc. As already mentioned, behind the short porous section ($\bar{l}_i < 3$) or in the case of a low intensity of blowing, the strength does not exceed E_t. The effect operates only at relatively high values of the length of the porous section ($\bar{l}_i > 3$) and the blowing intensity $g_p \geq 0.5$ g/(s · cm²).

The data in [46, 48], obtained for a porous channel with the length of approximately 5 length gages and at the values of g_p, gas pressure and at current intensity ($I \sim 200$ A) similar to those in the investigated case, are in good agreement with the results, with the exception of the point in the extreme position along the flow (see points 5 in Fig. 5.30*a*). The value of the strength at this point is considerably higher than the mean level, determined in [53]. This high value of E in the vicinity of the output electrode, as already mentioned several times, is associated with the fact that the true length of the section of the arc from the last section and to attachment to the output electrode is not available.

We examine the effect of the intensity of blowing the gas through the porous wall on the strength of the electrical field of the arc. Figure 5.31 shows the curves corresponding to different distributions of the blowing intensity in the sections, with the total gas flow rate through the entire block unchanged. Curve 2 was obtained for the same flow rate of the working gas from each of the six sections of the block, corresponding to $g_p = 1.0 \div 1.1$ g/(s·cm²). The distribution

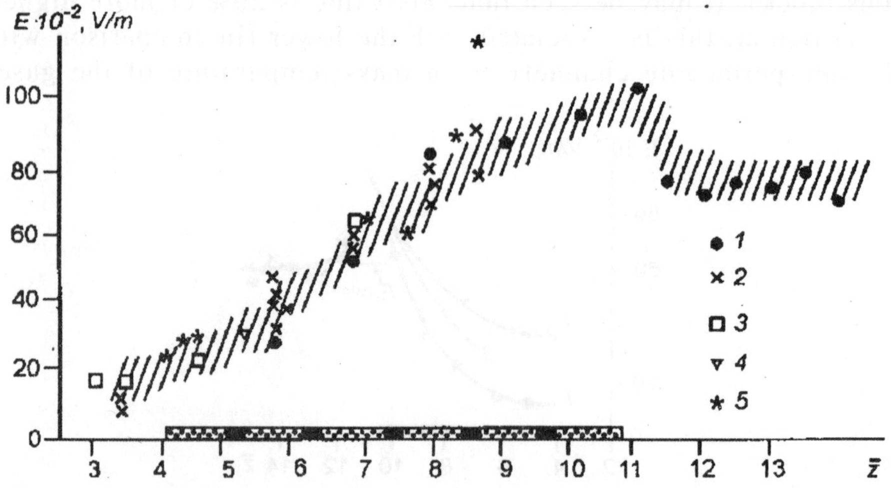

Fig. 5.30. Distribution of the strength of the electrical field of the arc along the channel in the blowing of the gas through the porous section at $G_0 = 6$ g/s, $g_p = (20 \div 80)$ g/s, $\bar{g}_p = 1.1$ g/(s · cm²), $I = 120$ A. *1*) six porous sections; 2) four; 3) three; *4*) one; *5*) data from [46] at $\bar{g}_p = 1$ g/(s · cm²), $\bar{l}_p = 5$, $I \sim 200$ A, $p = 0.3$ MPa.

of E at the increasing or decreasing (along the block) flow rate of the gas in the ratios of 1:2:3 or 3:2:1 respectively (the porous sections are combined in pairs) is illustrated by the curves 3 and 1. At the end of the porous block the value of E is almost the same for all the three distribution g_p in the sections of the porous block.

The effect of the length of the porous block and of the intensity of blowing at the constant total flow rate of the gas on the distribution of the strength of the electrical field along the channel is illustrated in Fig. 5.32. To facilitate comparison, the first sections of the blocks, consisting of three and six porous elements, are combined. Halving the length of the porous section of the channel at the same total flow rate (g_p is correspondingly doubled) has only a slight effect on the level of strength and the end of the permeable section where the developed turbulent flow already exists. There are changes only in the curvature of increase of E along the length of the block of porous sections because the length of the transition zone from E_i to E_t decreases.

For more detailed analysis of the processes taking place in the discharge chamber it is desirable to have information on the mean and pulsation characteristics of the turbulent gas flow and of the electrical arc in the gas flow. For this purpose, high-speed filming of the arc in different conditions was carried out. The time dependence of the illumination intensity of the element of the arc was regulated, as in [20], by SFR-1M photographic recording device in the regime of continuous sweep through a transverse slit 2.5 mm wide, closed with a quartz window. The slit, produced in a special section, was at a distance of ~15 mm from the end of the last porous insert. The frames 1–3 (Fig. 5.33*a*) were obtained in the recording through a slit situated behind the relatively short porous block (3.5 length gages), consisting of three sections. Frame 4 shows, for comparison, photo sweep of the element of the arc in the transition section of the flow in a plasma torch with a non-permeable sectioned IEI. There are many common features of the sweeps, in particular, the frequency of pulsation of the arc column is similar and the range of oscillations is comparable and almost equal to the channel diameter. It should be mentioned that the recorded luminous diameter of the arc in the frame 4 is slightly smaller in comparison with the frames 1–3, because of the installation of an additional diaphragm with a slit approximately 1 mm wide. Without the additional slit, the luminous diameter of the element of the arc in both cases would be approximately the same.

The photographs of the arc in the channel behind the porous section

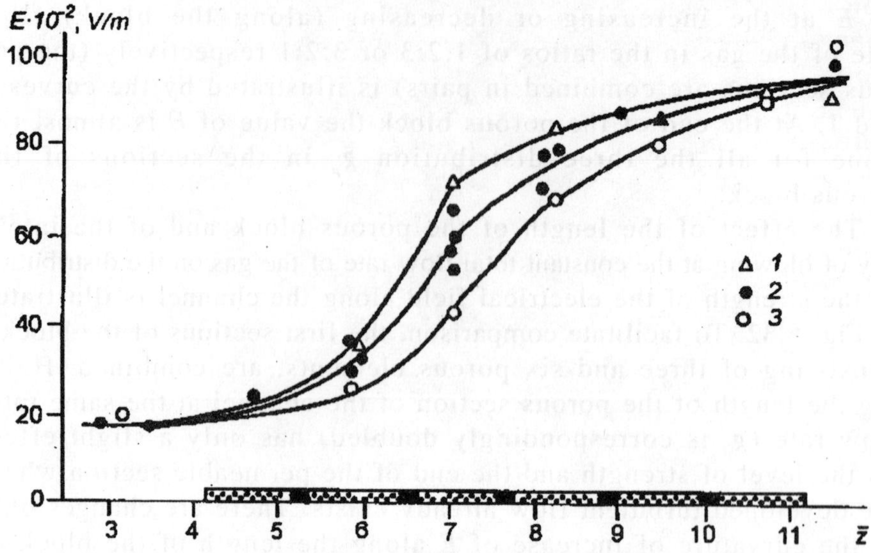

Fig. 5.31. Dependence $E(z)$ for different distribution of the gas flow through the porous sections. *1*) $\bar{g}_{pi}$ = 1.5; 1.5; 1.0; 1.0; 0.5; 0.5 g/(s · cm²); *2*) $\bar{g}_p$ = 1.0÷1.1; 3) $\bar{g}_{pi}$ = 0.5; 0.5; 1.0; 1.0; 1.5; 1.5.

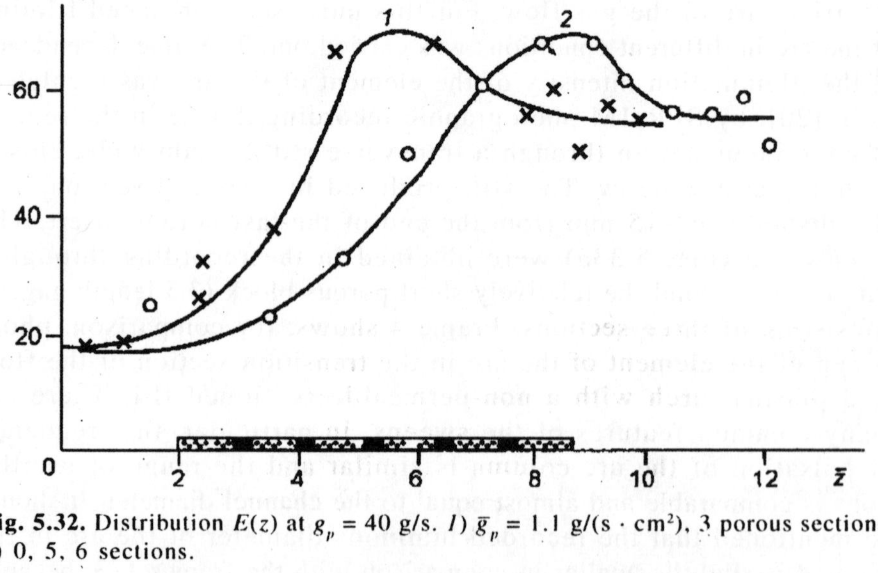

Fig. 5.32. Distribution $E(z)$ at g_p = 40 g/s. *1*) $\bar{g}_p$ = 1.1 g/(s · cm²), 3 porous sections; *2*) 0, 5, 6 sections.

(in particular, in the case of the developed turbulent gas flow) show that the arc is often split into two or more current-conducting channels (Fig. 5.33*b*). In this case, the length of the porous IEI is approximately 7 length gages (frames 1, 2). Frame 3 relates to the section of the developed turbulent flow in the non-permeable sectioned channel. There are many common features between the first and the last frame: the luminous diameter of the arc decreases, i.e. the extent of constriction of the current-conducting channel has increased [20, 46], the amplitude of oscillations of the arc column decreased, by the frequency remained approximately the same in both cases. The arc column was split into independent current-conducting channels, especially clearly visible in frame 2. However, the porous channel often shows high-frequency oscillations with a small amplitude (frames 1, 2) which is not observed in the non-permeable channel (frame 3).

Thus, the time scanning of the arc behind the short porous IEI (3.5 length gages) and at a relatively low values of g_p indicates the flow characteristic of the transition regime. In the case of longer porous IEI (7 length gages) for approximately the same values of g_p (or even slightly lower values), the scanning of the illumination intensity of the arc indicate the existence of a developed turbulent gas flow.

The results of processing of the photographs in the method described in [20] gave information of the mean frequency of oscillations of the arc column presented in Table 5.3. Here ε is the mean RMS error of the measurements.

For the porous inter-electrode insert with $\bar{l}_i = 3.5$, the controlling factors are the oscillations of the arc with the frequency of 20÷24 kHz which is close to the data for the porous channel [46, 50]. With the increase of the length of the inter-electrode insert ($\bar{l}_i = 7$) of the frequency of oscillations increases to 28÷30 kHz. In addition to this, there are also high-frequency (~100 kHz) oscillations of the arc, superposed on the main frequency. In the case of the IEI with non-permeable walls, the frequency of oscillations of the arc in the section of the developed turbulent flow is in the same range – the mean value is ~30 kHz. In the transition section, the frequency is slightly lower [20].

Using the data presented in Table 5.3, we can estimate the characteristic dimensions of turbulent vortices in the investigated case, i.e. the scale of turbulence. According to [56], the identical hydrodynamic situations in the non-stationary gas flow are described by the homochronicity criterion Ho = ut/L. Here u, t, L are the values of the speed, time and length, respectively. The product ut is some linear scale, which determines the turbulent flow in this case. The characteristic speed may be represented by the speed of sound a, the mean u and pulsation u' speed of the gas flow. Time t is given

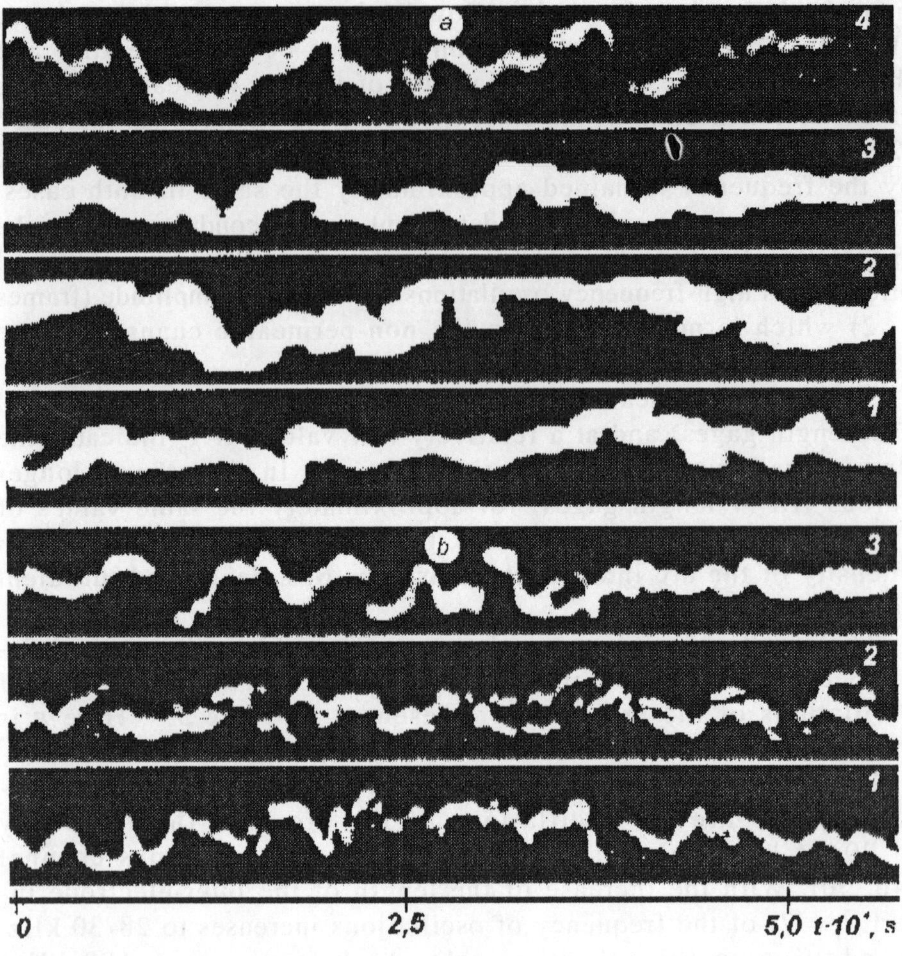

Fig. 5.33. Time sweep of the glow of an arc element in a channel behind the porous section with the length of 3.5 gages (*a*) and 7 gages (*b*) at d = 2 cm, I = 120 A. *a*) *1*) g_p = 34 g/s; $\bar{g}_p$ = 0.74 g/(s · cm^2); *2*) 46; 1.06; *3*) 80; 1.97; *4*) transition section of the channel in the plasma torch with a non–permeable section IEI at I = 100 A, G ~ 20 g/s; *b*) *1*) g_p = 49 g/s; $\bar{g}_p$ = 0.57 g/(s · cm^2); *2*) 64; 0.77; *3*) section of the developed turbulent flow of the gas in the plasma torch with a non-permeable sectioned IEI at I =120 A, G ~ 25 g/s.

by the previously mentioned frequencies of pulsations of the flow. In the flow of the gas in the pipe, the dimensions of the turbulent vortices change from maximum, associated with the size of the channel (diameter d and radius r) and minimum, determined by the viscosity properties of the flow).

For the estimates, we use three values of the frequency: f_1 = 20 kHz, f_2 = 30 kHz, f_3 = 100 kHz. Since in the investigated case the channel of the plasma torch receives air at room temperature (300 K) and at a high flow rate, the thickness of the layer of the cold gas in the wall region is relatively large. Therefore, the characteristic speed is the speed of sound a = 348 m/s at T = 300 K. Consequently, $L_1 = a/f_1$ = 17.4 mm, $L_2 = a/f_2$ = 11.6 mm. The scale L_1 is close to the diameter of the channel, L_2 to the radius of the channel. It was mentioned previously that the pulsations of the arc with the frequencies corresponding to L_1 are detected mainly at the length of the porous inter-electrode insert of 3.5 length gages. In the non-permeable channel, the frequencies correspond to the transition section of the flow. The range of the oscillations of the arc column is close to the diameter of the channel, which corresponds to the estimate. Behind the long porous inter-electrode insert (7 length gages) and in the section of the developed turbulent flow of the gas in the plasma torch with the inter-electrode insert with the non-permeable walls, the range of fluctuations of the arc column (as indicated by the photographs) is close to the radius of the channel. The characteristic scale L_2 is close to the radius of the channel. Thus, the frequency characteristics of the arc, burning in the plasma torch with the inter-electrode insert with permeable and non-permeable walls, confirmed the almost complete identity of the pattern of the gas flow and of its interaction with the arc.

We estimate the scale of the vortices, with the frequency of pulsations of ~ 100 kHz. The ratio $L_3 = a/f_3$ shows that L_3 = 3.5 mm. The resultant dimension is comparable with the visible luminous diam-

Table 5.3. Characteristic frequencies of pulsations of the arc in a porous channel

$\bar{l}_p$	$\bar{G}$, g/s	g_p, g/(s · cm²)	$f \pm \varepsilon$, kHz	Comment
3.5	34	0.74	21.9±1.8	
	45	1.01	23.4±2.8	
	46	1.06	20.6±2.8	
	63	1.55	22.5±1.9	
	80	1.97	23.9±3.8	
	20		29.2±3.9	Non-permeable channel, transition section
7	49	0.57	20±2	
	64	0.77	30±2	
	25		28.9±2.7	Non-permeable channel, Developed turbulent section
7	49	0.57	88.6±9.5	High-frequency oscillations
	64	0.77	109±10.6	

eter of the arc. It is possible that it is the minimum size of the turbulent vortices which still influence the arc. If the characteristic values are the values of the speed of discharge of the gas from the porous wall or of its pulsation component, the values of *ut* do not exceed fractions of a millimetre and, evidently, do not reflect the physical nature of the process taking place in the discharge chamber.

Thus, the measurements of the strength of the electrical field of the arc in the plasma torch with a porous inter-electrode insert of different length and also the high-speed filming of the arc show that when blowing the gas through the porous wall the processes taking place are the same as those in the flow of the gas in the sectioned inter-electrode insert with non-permeable walls. In the porous inter-electrode insert, a turbulent flow starts to form already at the gas blowing intensity of $g_p \geq 0.2$ g/(s·cm^2) at the start of the first porous section. The length of the transition section is usually 3–4 length gages. Subsequently, the flow changes to a developed turbulent flow. The slightly higher, in comparison with [20], level of the strength of the electrical field is explained by the lower (in comparison with the case of the arc in the channel with the non-permeable walls) mean mass temperature of the gas surrounding the arc, and also by the increase of the arc length as a result of high-frequency pulsations detected in examination of the arc behind the porous channel.

5.4. STRENGTH OF THE ELECTRICAL FIELD OF THE ARC IN HYDROGEN AND HYDROGEN-CONTAINING MEDIA

Analysing the data on the electrical characteristics of the hydrogen arc [47, 57–63], we obtain information on the mean strength of the electrical field of the arc. These data are conventionally divided into two groups. For example, in a number of studies [57–60] it is shown that in plasma torches with the diameter of the electric arc chamber of $d \sim 2 \cdot 10^{-2}$ m at the pressure close to atmospheric, the strength of the electrical field is $E = (15 \div 30) \cdot 10^2$ V/cm. At the same time, in plasma torches with the inter-electrode insert in the presence of the starting section with the diameter smaller than the diameter of the channel, the mean value of the strength of the electrical field in the same conditions is $(40 \div 50) \cdot 10^2$ V/m or greater [61]. In [47, 62] data were obtained in a plasma torch with an inter-electrode insert produced from porous ceramics at the gas flow rate G of up to 0.03 kg/s (these values are an order of magnitude higher than those mentioned in the previously cited studies). Here, the value of E is almost constant with current (I = 500÷600 A) and is proportional to $G^{0.8}$. No data have been published on the distribution of the strength of the electrical field of the arc along the channel, on the structure of the arc and its interaction with the gas flow in the above studies.

In order to investigate the distribution of the potential and strength of the electrical field of a hydrogen arc, the authors of [64] carried out experiments with a plasma torch with an inter-electrode insert with a diameter $d = 2\cdot10^{-2}$ and $3\cdot10^{-2}$ m, the relative length of the sectioned inter-electrode insert $\bar{a} = a/d$ of up to 18, and the output electrode $\bar{b} = b/d = 2÷3$. The diameter of the starting (first from the cathode) section $d_{s.s}$ was either equal to the diameter of the channel or smaller than the diameter. Measurements were taken at a total hydrogen flow rate of $G = (3÷7)\cdot10^{-3}$ kg/s, the pressure at the outlet from the plasma torch of $p = (1.0÷1.5)\cdot10^5$ Pa. The flow rate of hydrogen between the cathode and the starting section G_0 was $(1÷2)\cdot10^{-3}$ kg/s. Between the sections of the interelectrode insert with a diameter $d = 2\cdot10^{-2}$ m, the gas was supplied at a flow rate of $g_i = (0.3÷0.9)\cdot10^{-3}$ kg/s, and when $d = 3\cdot10^{-2}$ m, it was $g_i = (0.175÷0.35)\cdot10^{-3}$ kg/s. Arc current I changed from 300 to 700 A. The strength of the electrical field of the arc was determined on the basis of the previously described procedure is caused by the differentiation of the distribution of the potential of the sections along the inter-electrode insert and on the basis of the measurements of the difference in the potentials of the adjacent sections. The measurements were taken using electrostatic voltmeters. The distance between the centres of the sections (measuring base) was $2.4\cdot10^{-2}$ m at $d = 2\cdot10^{-2}$ m and $1\cdot10^{-2}$ m at $d = 3\cdot10^{-2}$ m. The instrument error of the measurements was ±6%.

The distribution of the strength of the electrical field of the hydrogen arc along the inter-electrode insert in the channel with the diameter $d = 2\cdot10^{-2}$ m with the relative length of up to 18 length gages [65] is presented in Fig. 5.34. The diameter of the starting section here is either equal to the diameter of the channel (curve *3*) or only slightly smaller than the diameter ($d_{s.s} = 1.2\cdot10^{-2}$ m, curves *1, 2*). In this case, the starting section, without causing any significant distortion, results only in a relatively small increase of the strength of the electrical field in the direction to the cathode. Qualitatively, the distribution of the strength of the electrical field of the arc in hydrogen along the channel corresponds to the distribution of the strength of the air arc shown in Fig. 2.12.

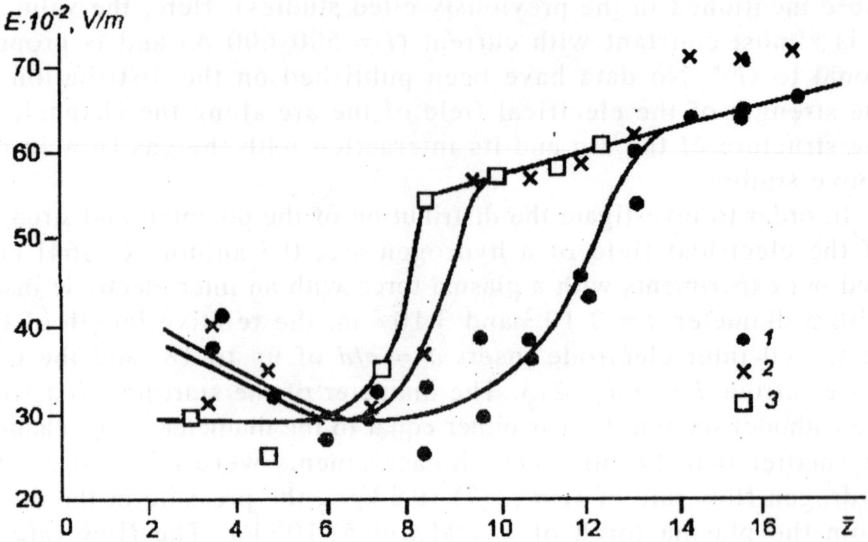

Fig. 5.34. Distribution of the strength of the electrical field of the arc along the channel with the diameter d = 0.02 m. *1,2*) $d_{s.s}$ = 0.012 m, l_c = 0.022 m, n (number of sections) = 15 ($\bar{a}$ = 18), G_0 = 1 · 10⁻³ g/s, g_p= 0.3 · 10⁻³g/s, Σg_i = 4.5 · 10⁻³ kg/s, I = 300 A and 400 A respectively; *3*) $d_{s.s}$ = d, n = 13, G_0 = 1 · 10⁻³ kg/s, Σg_i = 3.75 · 10⁻³ kg/s, I = 500 A.

The identical distribution of the strength of the electrical field of the arc along the interelectrode insert was obtained in the channel with the diameter $d = 3{\cdot}10^{-2}$ m in the absence of the starting section (Fig. 5.35a). At a length of the channel of approximately 7 length gages, there were only the initial and transition sections in the distribution E (z), and at a current intensity of I = 600 A the curve $E(z)$ moved closer to the level E_t (curve 3). The value E_i changes in the range from $17{\cdot}10^2$ V/m at a current intensity of I = 400 A to $27{\cdot}10^2$ V/m at I = 600 A (curves *1–3*). In the presence of a strong turbuliser of the flow such as the starting section with the ratio of the diameters $d/d_{s.s} \geq 2.5$, the entire length of the channel behind the starting section is characterised (Fig. 5.35*b*) by the strength typical of the developed turbulent gas flow. The level of the strength at the end of the transition section (curve 3 in Fig. 5.35*a*) and at the end of the developed turbulent flow (Fig. 5.35*b*) in the examined case is in good agreement.

We also present the distribution $E(z)$ in a channel up to 12 length gages and with the diameter $d = 2{\cdot}10^{-2}$ m at different hydrogen flow rates (Fig. 5.36). The behaviour of the curves is the same as in Fig. 5.35. At low hydrogen flow rates (Fig. 5.36*a*) the effect of the starting

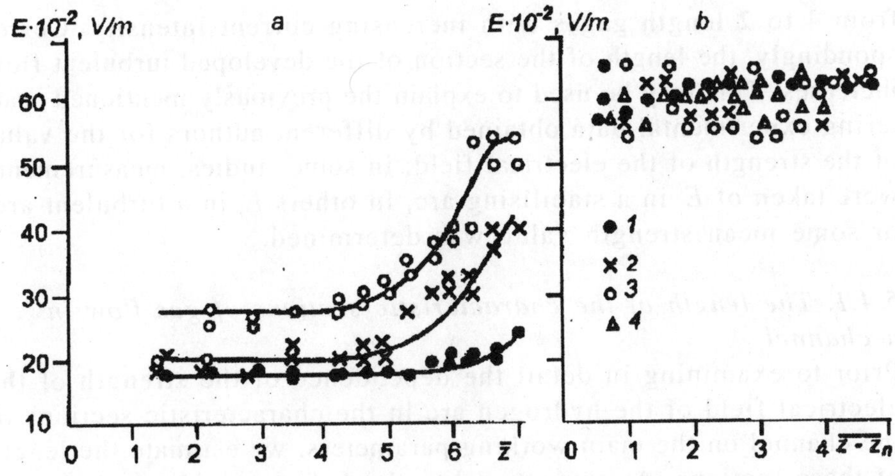

Fig. 5.35. Distribution of the strength of the electrical field of the arc along the channel with the diameter $d = 0.03$ m. *1)* $I = 400$ A; *2)* 500; *3)* 600; *4)* 700 A; *a)* $d_{s.s} = d$, $a = 0.23$ m. $G_0 = 1.25 \cdot 10^{-3}$ kg/s, $\Sigma g_i = 4.5 \cdot 10^{-3}$ kg/s; *b)* $d_{s.s} = 0.012$ m, $a = 0.12$ and 0.15 m, $G_0 = 1 \cdot 10^{-3}$ kg/s, $\Sigma g_i = 5 \cdot 10^{-3}$ kg/s;

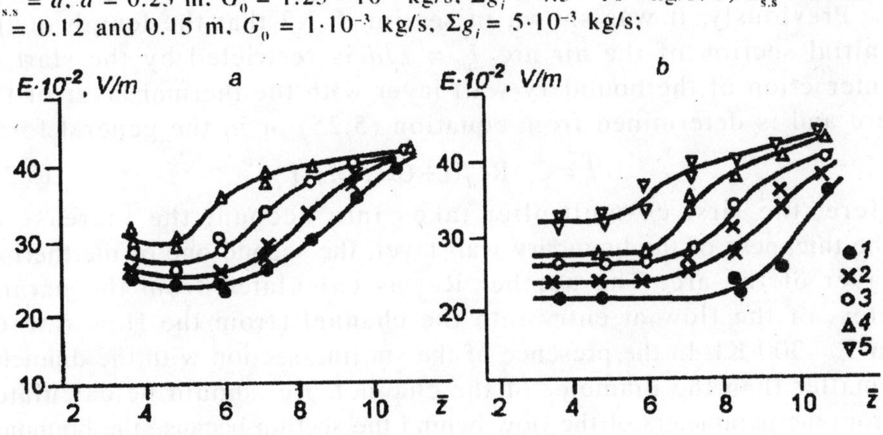

Fig. 5.36. Distribution of the strength of the electrical field of the arc along the channel. $d = 0.02$ m, $d_{s.s} = 0.012$ m, $n = 10$, $G_0 = 1. \cdot 10^{-3}$ kg/s, *1)* $I = 300$ A; *2)* 400; *3)* 500; *4)* 600; *5)* 700 A; *a)* $g_i = 0.3 \cdot 10^{-3}$ kg/s; $\Sigma g_i = 3 \cdot 10^{-3}$ kg/s, *b)* $g_i = 0.4 \cdot 10^{-3}$ kg/s, $\Sigma g_i = 4 \cdot 10^{-3}$ kg/s.

section on the strength at the start of the channel is stronger.

Thus, it may be concluded that in the investigated conditions in hydrogen, as in other gases, there are three characteristic sections in the distribution of the strength of the electrical field of the arc along the channel: initial, with the level $E_i \approx 30 \cdot 10^2$ V/m; transition, with increasing E_{trans}, and the section of developed turbulent flow, in which $E_t > 50 \cdot 10^2$ V/m. The length of the initial section is approximately 9 length gages at $I = 300$ A, and approximately 6 length gages at $I = 500$ A. The length of the transition section decreases from 4 to 2 length gages with increasing current intensity. Correspondingly, the length of the section of the developed turbulent flow increases. This may be used to explain the previously mentioned scattering experimental data obtained by different authors for the value of the strength of the electrical field: in some studies, measurements were taken of E_i in a stabilising arc, in others E_t in a turbulent arc, or some mean strength value was determined.

5.4.1. The length of the characteristic sections of gas flow in a channel

Prior to examining in detail the dependence of the strength of the electrical field of the hydrogen arc in the characteristic sections of the channel on the main working parameters, we estimate the length of these sections. At a constant length of the inter-electrode insert, it is sufficient to determine the length of the initial and transition sections.

Previously, it was shown in section 5.2.2 that the length of the initial section of the air arc $\overline{l}_i = z_i/d$ is restricted by the start of interaction of the boundary wall layer with the thermal layer of the arc and is determined from equation (5.25) or in the general form:

$$\overline{l}_{\text{i}} = C_1 \cdot \text{Re}_d^m [1 + C_2 (I/d)^n]^{-1}. \qquad (5.29)$$

Here, the first co-multiplier takes into account the increase of the thickness of the boundary wall layer, the second one of the thermal layer of the arc. The number Re_d as calculated from the parameters of the flow at entry into the channel (from the flow rate G_0 at $T = 300$ K). In the presence of the starting section with the diameter smaller than the diameter of the channel, Re_d should be calculated from the parameters of the flow behind the section because the boundary layer starts to develop behind the section. Consequently

$$\text{Re}_d = (\rho u)_1 \cdot d / \mu_{\text{w}} = 4G_1 / (\pi d \mu_{\text{w}}),$$

where $G_1 = G_0 + g_1$ is the flow rate of the gas behind the starting section; μ_{w} is the viscosity of the gas at the wall temperature. Here, we do not take into account the process of development of the thermal layer of the arc which starts from the interelectrode, but this assumption is fully acceptable for the estimates. Coefficient C_1 for air and other diatomic gases is 1.35 [20, 32]. The coefficient C_2 includes $\sqrt{\mu h \sigma}$, where μ, h, and σ as in section 5.2.2 are the characteristic values of viscosity, enthalpy and electrical conductivity of the gas. It is assumed that the criterial dependence (5.29) also holds for the hydrogen arc with gas-vortex stabilisation.

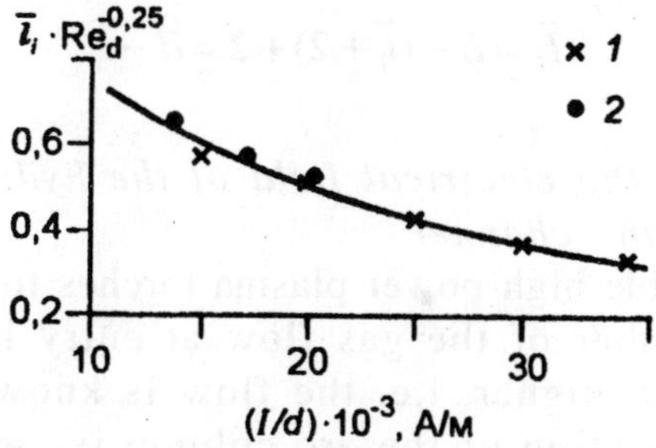

Fig. 5.37. Dependence of the length of the initial section of a hydrogen arc on (I/d). *1*) d = 0.02 m, $G_0 = (1 \div 1.5) \cdot 10^{-3}$ kg/s; 2) d = 0.03 m, $G_0 = 1 \cdot 10^{-3}$ kg/s.

The dependence of $\bar{l}_i \cdot Re_d^{-0.25}$ on I/d is presented in Fig. 5.37. The experimental values for different diameters of the channel in satisfactory agreement. The data for $d = 2 \cdot 10^{-2}$ m were obtained, as recommended previously, taking into account the length of the starting section $\bar{l}_s \approx 1.4$. The viscosity of hydrogen is determined on the basis of the conditions at entry into the channel, i.e. $\mu = 1.38 \cdot 10^{-5}$ kg/(m·s) at T = 300 K. As shown by the processing of the results of measurements, the value of the coefficient C_1 should be assumed to be equal to 1.35, as in the case of air. Without considering transformations, it should be mentioned that the mean value $C_2 \approx 8.3 \cdot 10^{-5}$, and the exponent $n = 1.0$. Thus, for the estimate of the length of the initial section of the hydrogen arc in the investigated range of the parameters, we propose the equation:

$$\bar{l}_i = 1.35 \cdot Re_d^{0.25}(1 + 8.3 \cdot 10^{-5} I/d)^{-1}. \tag{5.30}$$

The continuous curve in Fig. 5.37 was calculated using this equation.

We now return to Fig. 5.34–5.36. The transition section, characterised by the increase of the strength of the electrical field of the air, situated downwards along the flow behind the initial section. The length of the section decreases with increasing current intensity and is equal on average to approximately 2 length gages. Calculations can be carried out assuming that $\bar{l}_{trans} \approx 2$.

Thus, we have estimated the length of two sections. The length of the third section – the section of the developed turbulent flow of the gas in the channel – is equal to the length of the remaining section of the channel in which the arc burns, i.e. to some part of the interelectrode insert and the section to the zone of attachment of the arc in the output electrode. According to the experimental results, the length of the latter is usually 1–2 length gages. Finally, we obtain:

$$\bar{l}_T = \bar{a} - (\bar{l}_i + 2) + 2 = \bar{a} - \bar{l}_i.$$

5.4.2. Strength of the electrical field of the hydrogen arc in the initial section of the channel

In currently available high-power plasma torches for heating hydrogen, the Reynolds number of the gas flow at entry into the channel is usually $(3 \div 4) \cdot 10^4$ higher, i.e. the flow is known to be turbulent. The vortex stabilisation of the arc column is, according to the investigations, difficult in the initial section of the channel. Therefore, the measurement of the strength of the electrical field of the non-perturbed arc in the initial section of the channel of a relatively large diameter is associated with considerable difficulties. We examine table 5.4 which gives the experimental data on the value of E_i in channels with the diameter 0.02 and 0.03 m, and also in the starting section with $d_s = 1.2 \cdot 10^{-2}$ s (on the basis of the difference of the potentials of the cathode and the first section of the inter-electrode in the third). Measurements were taken at the gas pressure close to atmospheric pressure: $p = (1 \div 1.5) \cdot 10^5$ Pa. Processing of the data shows that E_i is approximately inversely proportional to the diameter of the channel. The product $E_i \cdot d$ depends only slightly on the parameter (G/d) (or on Re_d). Because of the small variation of the pressure of the gas it was not possible to examine the effect of the complex $(p \cdot d)$ on the quantity $E_i \cdot d$. The main parameter, affecting the strength, is the intensity of arc current, and investigations show the separation of the curves obtained for different diameters of the channel when constructing the dependence $E_i \cdot d = f(I/d)$. The dependence on current intensity was determined in the form $E_i \cdot d = f(I)$, and the experimental points were situated on the same curve. We determine the mean values of $E_i \cdot d$ for different values of current intensity (Table 5.5).

The dependence $E_i \cdot d = f(I)$ is shown in Fig. 5.38. In all likelihood, at a current intensity of approximately 200 A, the E_i–I characteristic of the arc has a minimum because according to the data obtained in certain studies, for example [57], at a current lower than 150 A the E–I characteristic of the arc drops and according to the curve in Fig. 5.38 for $I > 300$ A the strength of the electrical field increases.

If the data obtained in [62] are converted to other gas flow rates, corresponding to the investigated case, then the results fit satisfactorily the curve in Fig. 5.38. Using the methods described in [4] the dependence $E_i \cdot d = f(I)$ can be presented in the form of a series

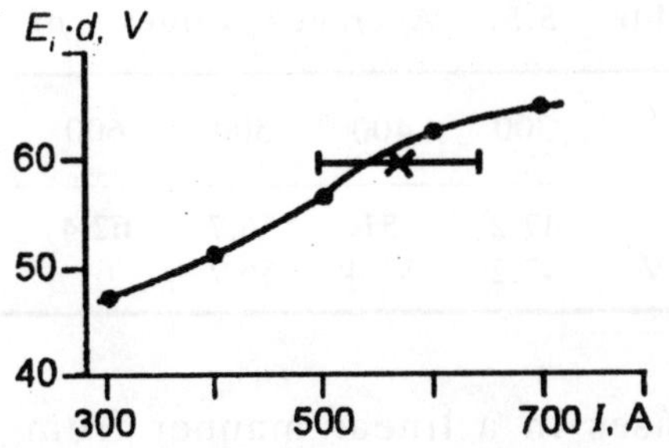

Fig. 5.38. Dependence of the $E_i \cdot d$ on I. d = 0.012; 0.02; 0.03 m; G = (1÷2) · 10^{-3} kg/s; p = (1÷1.5) · 10^5 Pa. Cross – the data from [62], I = 500÷650 A.

Table 5.4. Values of $E_i \cdot d$ for a hydrogen arc.

I, A	$d \cdot 10^2$, m	$G \cdot 10^3$, kg/s	$E_i \cdot d$, V	I, A	$d \cdot 10^2$, m	$G \cdot 10^3$, kg/s	$E_i \cdot d$, V
300	1.2	1	51	300	2	1–2	41.6
400	–	–	50	400	–	–	50
500	–	–	56	700	–	–	68.8
600	–	–	57	700	–	–	60.4
700	–	–	59	300	–	–	50
400	–	–	52.4	400	–	–	50
500	–	–	64	500	–	–	60
500	–	–	54.5	500	–	–	50
500	–	–	61.3	400	3	1–2	52.5
600	–	–	59	500	–	–	52.5
700	–	–	66.7	500	–	–	60
300	2	1–2	46	600	–	–	60
400	–	–	47.5	600	–	–	75
500	–	–	52	400	–	–	54.6
600	–	–	56	500	–	–	50
700	–	–	62.5	500	–	–	60
700	–	–	70.8				

if in respect of the negative degrees of current intensity I and we restrict ourselves to three terms to simplify calculations. Avoiding cumbersome transformations, the final result may be written in the following form:

$$E_i \cdot d = 94.7 - 2.6 \cdot 10^4 I^{-1} + 3.57 \cdot 10^6 I^{-2}. \qquad (5.31)$$

This equation approximates with sufficient accuracy the existing data in the current range I = 300÷700 A. The values of $E_i \cdot d$, calculated from equation (5.31) are presented in the last line of Table 5.5.

5.4.3. *Strength of the electrical field of the arc in a developed turbulent hydrogen flow*

In transition section it may be assumed with a sufficiently high reliability

Table 5.5. Average values of E_i

I, A	300	400	500	600	700
$E_i \cdot d$, V	47.2	51	56.7	62.4	64.7
$(E_i \cdot d)_{cal}$, V	47.2	51.4	56.7	61	64.7

that E_{trans} increases in a linear manner from the values E_i to the values of E_t characteristic of the given conditions, if E is represented by the mean value $E_{trans} = (E_i + E_t)/2$.

We examine the currently available experimental data for the strength E_t for the channels with different internal diameters $d = 0.02$ and 0.03 m. The distribution of E along the channel ($d = 0.02$ m), presented in Fig. 5.34, shows that E_t increases along the inter-electrode insert because of the increase of the flow rate of hydrogen, is independent of arc current, and in the examined case equals $(55 \div 65)\ 10^2$ V/m. In a channel with a diameter of 0.03 m, the data on the strength were obtained mainly in turbulisation of the flow behind the starting section (5.35*b*). Some data on the values of E_t at different flow rates and currents are presented in Fig. 5.39. The presented material does not indicate any dependence of E_t on the intensity of arc current. Actually, averaging (to remove the random error in measurements) the values of E_t in the sections with the length of 2–3 length gages shows that mean E_t, as at $d = 0.02$ m, is independent of current intensity (Fig. 5.40). Interesting information is obtained using the data on the measurements of the strength of the electrical field in high-current hydrogen using alternating current [66, 67]. At pressures close to atmospheric, in the current range $3 \div 4.5$ kA, the strength slightly increases with the increase of the gas flow rate and amounts to $(35 \div 50) \cdot 10^2$ V/m. It may be assumed that the strength of the electrical field of the hydrogen arc at the atmospheric pressure is approximately constant or slowly decreases with the increase of the intensity of arc current to several kiloamperes.

The dependence of the strength E_t on the flow rate of the gas G (Fig. 5.41) shows that E_t is only slightly linked with the flow rate and the scatter of the values is large. Processing the data (5.41) gives the following formulae: $E_t = 1.54 \cdot 10^4\ G^{0.17}$ or $E_t = 1.85 \cdot 10^4\ G^{0.2}$.

According to [66], at a gas flow rates of $G = (25 \div 118) \cdot 10^{-3}$ kg/s, the pressure $p = (1 \div 3.5) \cdot 10^5$ Pa and currents of $2800 \div 5400$ A, the strength is proportional to $G^{0.185}$. If we except this dependence of E on G, verified in a wide range of variation of the

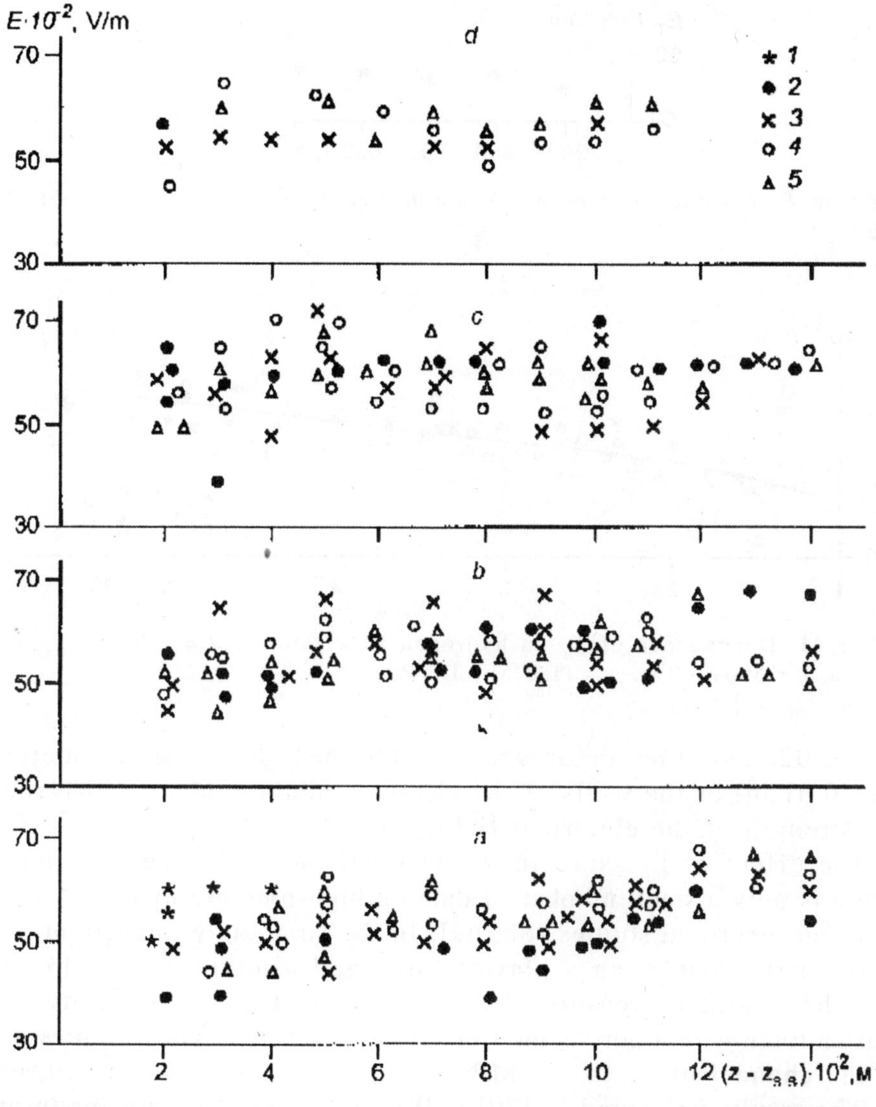

Fig. 5.39. Distribution of the strength of the electrical field of the arc in hydrogen along the IEI. d = 0.03 m, $d_{s.s}$ = 0.012 m, n = 12 and 15, l_s = 0.01 m. *1*) I = 300 A; *2*) 400; *3*) 500; *4*) 600; *5*) 700 A; *a*) $G_0 = 1 \cdot 10^{-3}$ kg/s; $\Sigma g_i = 3 \cdot 10^{-3}$ kg/s, *b*) $1 \cdot 10^{-3}$ kg/s, $4 \cdot 10^{-3}$ kg/s; *c*) $1 \cdot 10^{-3}$ kg/s, $5 \cdot 10^{-3}$ kg/s; *d*) $1.5 \cdot 10^{-3}$ kg/s, $4.5 \cdot 10^{-5}$ kg/s.

parameters, the following equation maybe presented:

$$E_T = 1.7 \cdot 10^4 \cdot G^{0.185}. \tag{5.32}$$

This equation does not include the intensity of arc current I. There is also no dependence on the channel diameter, i.e. the walls have no effect on the strength of the electrical field of the arc at

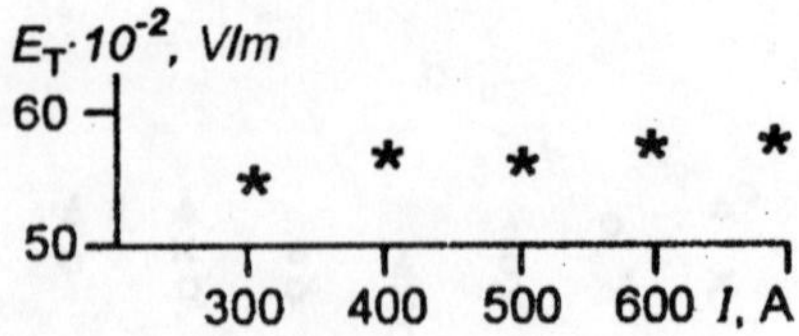

Fig. 5.40. E_T–I characteristic of the arc. d = 0.03 m; $d_{s.s}$ = 0.012 m, $G = (3 \div 3.5) \cdot 10^{-3}$ kg.s.

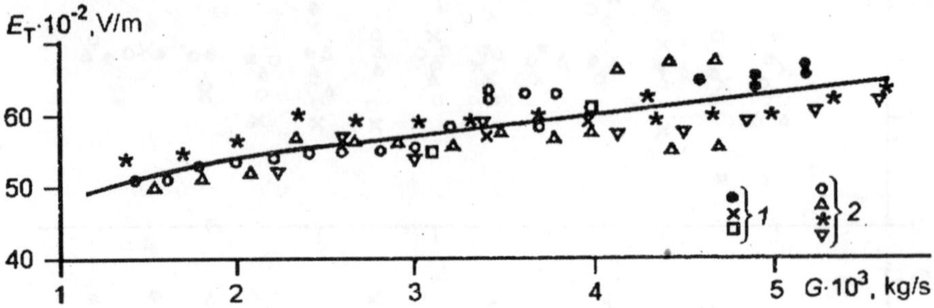

Fig. 5.41. Dependence of E_T on hydrogen flow rate. *1*) d = 0.02 m; *2*) d = 0.03 m; $I = 300 \div 700$ A; $p = (1 \div 1.5) \cdot 10^5$ Pa.

d = 0.02 and 0.03 m or greater [63, 66]. At small diameters (d < 0.01 m) of the walls of the channel should evidently influence the strength of the electrical field.

The effect of pressure in the channel on E_t will be examined. There is only a small number of data on high-pressure hydrogen arcs. One can mentione studies [68, 69]. In the first study, investigations were carried out using a plasma torch and a mixture of hydrogen with helium at a pressure of $4 \cdot 10^5$ Pa, in the second study – a plasma torch with a porous inter-electrode insert at a hydrogen pressure in the channel of up to 1.5 MPa.

The results presented in [70] will be analysed. Experiments were carried out in the already described plasma torch with a sectioned IEI with a diameter d = 0.03 m, the starting section $d_{s.s} = 1.2 \cdot 10^{-2}$ m, the length of the inter-electrode insert $a = 0.075 \div 0.13$ m (arc length $0.1 \div 0.15$ m), arc current intensity $I = 300 \div 700$ A, the gas flow rate $G = (3 \div 4) \cdot 10^{-3}$ kg/s, the pressure in the channel $(1 \div 6) \cdot 10^5$ Pa (Fig. 5.42*a*). With increase of pressure from $1 \div 10^5$ to $5 \div 10^5$ Pa, the strength of the electrical field of the hydrogen arc is approximately doubled. Here the values of the strength of the field [69] obtained in a porous channel at high gas flow rates are noted here (5).

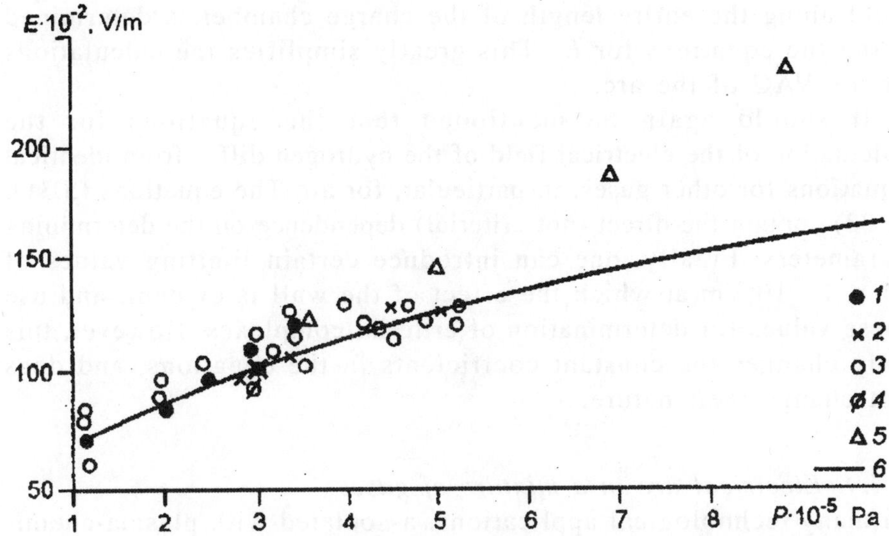

Fig. 5.42. Dependence of the strength of the electrical field of the arc on the pressure of hydrogen in the channel. *1–3*) $G = 3 \cdot 10^{-3}$ kg/s, I = 300÷600 A; *4*) $G = 4 \cdot 10^{-3}$ kg/s, I = 500 A; *5*) data from [69]; *6*) calculated from equation (5.33).

Using the values in Fig. 5.42, the equation (5.32) can be supplemented by the dependence of the strength of the electrical field on the gas pressure in the channel:

$$E_t = 190 \cdot G^{0.185} \cdot p^{0.4}. \tag{5.33}$$

The continuous curves in Fig. 5.42 shows the calculation of E_t using equation (5.33). The data published in [69] are also sufficiently described by this equation. The dependences of the strength of the electrical field on the gas pressure for high-current alternating current arcs, presented in [66] for pressures up to $4 \cdot 10^5$ Pa: $E \sim p^{0.536}$, and in [67]: $U \sim p^{0.416}$, are similar.

Thus, the following information is available on the strength of the electrical field of the arc in the turbulent hydrogen flow:

– in the investigated range of the parameters, the value of E_t is almost completely independent of arc current intensity;

– E_t depends only slightly on the gas flow rate and quite strongly on the pressure of hydrogen in the channel;

– E_t is independent of the channel diameter, i.e. the channel walls at $d > 1 \cdot 10^{-2}$ m have no influence on the electrical arc. In other gases, this is evident only at considerably larger channel diameters.

It is also important to take into account the fact that in the majority of commercial systems for heating hydrogen the arc burns in the conditions of turbulent gas flow, i.e. the strength of the electrical field along the entire length of the charge chamber is determined using the equations for E_t. This greatly simplifies the calculations of the VAC of the arc.

It should again be mentioned that the equations for the calculation of the electrical field of the hydrogen differ from identical equations for other gases, in particular, for air. The equations (5.31), (5.33) contain the direct (not criterial) dependence on the determining parameters. Finally, one can introduce certain limiting values of $d^* \leq 1 \cdot 10^{-2}$ m at which the effect of the wall is evident, and use these values for determination of criterisl complexes. However, this only changes the constant coefficients in the equations, and does not change their nature.

5.4.4. Electrical arc in a mixture of gases

In many technological applications, associated with plasma-chemical processes, hydrogen is regarded as the heat carrier and one of the reagents. It is often also necessary to heat a mixture of gases, for example, hydrogen with the addition of methane, air with the addition of methane, etc.

Usually, the literature contains scattered reports on the energy characteristics of the arc in the gas mixtures. The VAC of the arc are generalised, but usually only in a narrow range of the parameters. For example, in a mixture of hydrogen with natural gas with relatively small (up to 10–12 vol.%) additions of methane, the voltage and, correspondingly, the strength of the electrical field increase in proportion to the volume addition of methane. This is associated primarily with the chemical processes taking place in the mixture of gases at high temperatures, for example, with the formation of acetylene and its homologs in the mixture $H_2 + CH_4$. Equation (5.19) was derived previously for a mixture of air with natural gas to calculate the *U–I* characteristic which shows that in the initial section of the channel $E_{mix} \sim E_{air}$ $[1 + (G_{CH4}/G_{air})^{0.8}]$. In this case, the strength of the electrical field also increases with the increase of the amount of methane in the mixture. Identical results were obtained for the $CH_4 + O_2$ mixture [71]. The voltage in the arc (and, evidently, the strength of the electrical field) increases with a decrease of the oxygen content.

It is interesting to examine the data on the electrical arc in steam which may be regarded as a mixture of hydrogen and oxygen. For the arc running in steam, the VAC (see equations (5.20), (5.20*a*)) were obtained in a relatively wide range of variation of the parameters [12] in channels with confusor constriction and in a cylindrical channel. The processes of arcing in steam have been analysed [72, 73], and also the E–I-characteristics of air and steam arcs in the narrowing and expanding channels [73, 74], the E–I characteristics of the arc in steam in a cylindrical channel with a fixed arc length [72] assuming the constant strength of the arc along the channel.

Typical E–I-characteristics are shown in Fig. 5.43. Unfortunately, the accuracy of measurements of this type is not high and only in some quantitative estimates can be made. For example, Fig. 5.44 compares the E–I-characteristics, obtained in steam (curve *1*) and calculated for the air arc using equation (5.18) (curve *2*). Since the measurements of the strength of the electrical field in steam and in air in deriving equation (5.18) were taken using approximately the same methods (variation of the arc length with other parameters being constant), the form of the curves is in qualitative agreement, although the strength of the electrical field in steam is higher than in air. It is also important to mentioned the data on the strength of the electrical field of the arc in the vortex flow of steam with shielding the cathode with argon [73]. At a constant steam flow rate of approximately $1 \cdot 10^{-3}$ kg/s the addition of up to 25% of argon to steam reduces by 30–40% arc voltage and the mean strength, although with a further increase of the flow rate of argon the strength of the electrical field remains approximately constant (Fig. 5.45). The authors explain this decrease by the fact that in the near-cathode region the arc burns in the argon flow and this is followed by the mixing of the shielding and working gas and by gradual separation of argon in peripheral regions of the channel as a result of centrifugal forces. Since in this case we are concerned with short arcs (L ~ 0.1 m), the effect of the argon addition may be quite strong.

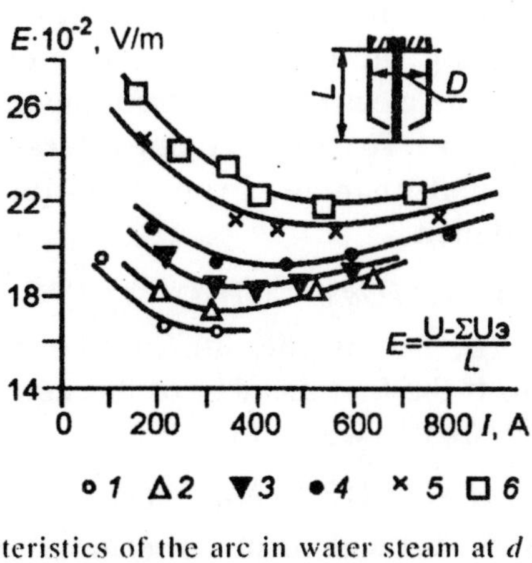

Fig. 5.43. E–I characteristics of the arc in water steam at d = 2 cm, L = 14.5 cm. *1*) G = 1.3 g/s; *2*) 2.1; *3*) 3.1; *4*) 4.3; *5*) 5.5; *6*) 6.8.

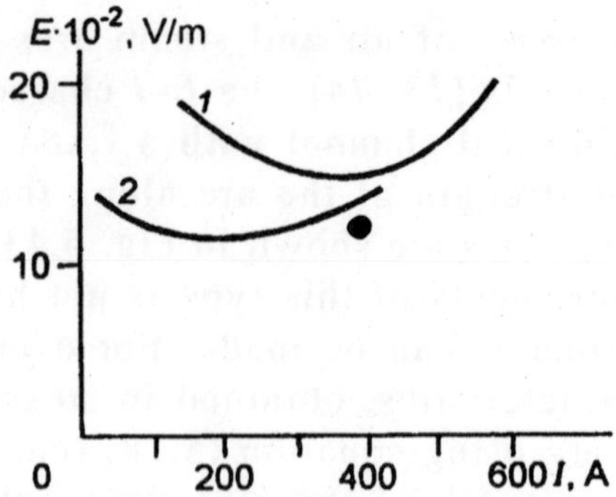

Fig. 5.44. *E–I*-characteristics of the arc at d = 1.9 cm, G = 5 g/s. *1*) steam (300°C); *2*) air (20°C), calculated from equation (5.18).

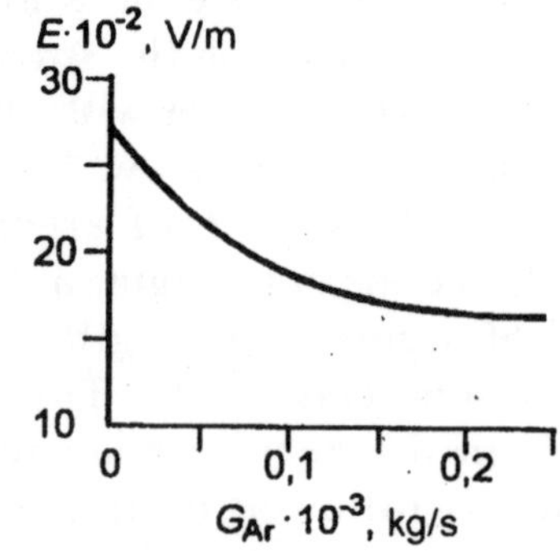

Fig. 5.45. Dependence of the mean strength of the electrical field of the arc in the steam on the argon flow rate used for shielding the cathode. d = 2 cm, G_{H_2O} = 1.3 g/s, I = 200 A.

Using the same assumption on the constancy of the strength of the electrical field along the channel, we can calculate the mean value of the strength of the electrical field from the previously described VAC of the arc in steam. To simplify considerations, we examine the case of a cylindrical channel. From equation (5.20*a*) we easily obtain the dependence

$$E \cdot d \sim (I^2 / Gd)^{-0.13} (G / d)^{0.20} (pd)^{0.48}, \tag{5.34}$$

Here $E = (U - \Sigma U_e)/L$, ΣU_e is the sum of the near-electrode decrease of the potential. If the first member of this equation is represented by $(I^2/Gd) = (I/d)^2 (G/d)^{-1}$, we obtain:

$$E \cdot d \div (I / d)^{-0.26} (G / d)^{0.33} (pd)^{0.48}. \tag{5.34a}$$

We compare equation (5.34*a*) with identical relationships for the turbulent flow of air and hydrogen:

$$E_t \cdot d \sim (I/d)^{-0.23}(G/d)^{0.47}(pd)^{0.2}, \qquad (5.28a)$$

$$E_t \cdot d^* \sim (G/d^*)^{0.185}(pd^*)^{0.4}. \qquad (5.33a)$$

Here for the hydrogen arc d^* is some limiting value of the diameter of the channel which was already discussed in section 5.4.3. The dependence (5.28*a*) for the air arc is, as already mentioned several times, the same as for the oxygen arc, the only difference is in the constant coefficient linked with the coefficient of transfer of the gases.

Comparison of the equations (5.34*a*), (5.28*a*) and (5.33*a*) shows clearly that we can separate the effect of the type of gases in steam on the strength of the electrical field of the steam arc. For example, parameter (I/d) has approximately the same exponents in the equations (5.34*a*) and (5.28*a*), i.e. the effect of the intensity of arc current is determined by the 'the oxygen component' of the working gas. Previously, it was also shown that the strength of the electrical field of the hydrogen arc is independent of the current intensity in the investigated range of the parameters. In turn, as the hydrogen and steam arcs show almost the same reaction to the variation of pressure in the channel: $E_{H_2O} \sim (pd)^{0.48}$; $E_{H_2} \sim (pd^*)^{0.4}$.

The effect of the flow rate of the gas (number Re_d) for hydrogen and oxygen arcs differs. In steam $E_{H_2} \sim (G/d^*)^{0.185}$; $E_{O_2} \sim (G/d)^{0.47}$, $E_{H_2O} \sim (G/d)^{0.33}$ i.e. the exponent at (G/d) is the mean between the dependences for the component gases. The strong effect of the turbulent flow of oxygen on the strength is compensated by the considerably weaker effect of the hydrogen flow.

Naturally, this is a very primitive analysis but it does make it possible to describe the steam plasma as a mixture of hydrogen and oxygen and determine the effect of the parameters on the energy characteristics of the arc. It is possible that the same approach can also be used for other mixtures if the characteristics of the gases in the mixture are known.

***

In this chapter, we analyzed the energy characteristics of the electric arc in different gases. We examined the VAC characteristics, *E–I* characteristics, carried out calculation engineering estimates of the dependence of the energy characteristics on the main working parameters of the plasma torch. Finally, not all the published results have been mentioned. In particular, insufficient attention has been given to gases with such as helium, argon, carbon dioxide and others. However, the data for these gases usually differ and accurate calculation equations have not as yet been derived.

Chapter 6

Heat exchange in the electric arc chamber of a linear plasma torch

One of the main problems in the electric arc heating of the gases the protection of the walls of the discharge chamber against the thermal effect of the high-temperature gas flow and the arc and also the maximum possible decrease of the rate of erosion of the electrons in the zone of the reference arc support. Thermal protection should be ensure, on the one side, normal functioning of elements of the plasma torch and, on the other side, the retention of the high thermal coefficient of efficiency. To solve this important problem, which determines the efficiency of heating the gas in the electric arc heater, it is necessary to examine in detail the thermal processes taking place in the column of the electric arc, and also heat exchange between the arc, the gas and the walls of the discharge chamber.

Taking into account the classification, presented in chapter 1, attention will be given initially to the integral the thermal characteristics of the plasma coating of the most widely used systems-with the self-setting arc length and with fixation of the arc length with a ledge. As already mentioned when describing the electrical characteristics of the plasma torch is of these types in chapter 5, in examination and generalisation of the characteristics can utilise relative simplicity of the processes of interaction of the arc with the gas flow and the electron the top in this case, to calculate the main parameters of the plasma Cote, it is sufficient to have the integral characteristics of the arc described by a small number of criteria or complexes [1].

6.1. INTEGRAL THERMAL CHARACTERISTICS OF PLASMA TORCHES WITH THE SELF-SETTING AND FIXED (USING A LEDGE) ARC LENGTH

The thermal characteristic of the plasma torches of these systems are presented in the form of the dependence of thermal efficiency on the main criterial complexes. In according to the definition, the thermal efficiency of the plasma torch is the ratio of the heat, carried by the gas from the plasma torch per unit time, to the arc power:

$$\eta = G \cdot \Delta h / (U \cdot I), \tag{6.1}$$

where Δh is the increase of the gas enthalpy in the plasma torch.

The value of η depends on the heat losses in the wall of the discharge chamber, i.e.

$$\eta = (U \cdot I - Q_p) / (U \cdot I). \tag{6.2}$$

Here Q_p are the total heat losses into the elements of the plasma torch. It was noted in chapter 4 that the measure of the thermal efficiency of the plasma torch is often represented by the value $\eta = (1 - \eta)/\eta$, which determines the ratio of the heat losses in the plasma torch to the heat content of the plasma jet, i.e. the integral coefficient of heat transfer.

The results of a large number of investigations of plasma torches of these systems show that in a general form their thermal characteristic may be written as the function of the main criteria complexes [1, 2]:

$$\tilde{\eta} = (1-\eta)/\eta = A(I^2/G \cdot d)^{\alpha}(G/d)^{\beta}(p/d)^{\gamma}(l/d)^{h}. \tag{6.3}$$

The constant multiplier A and the exponents at the complexes are determined in experiments for geometrically similar plasma torches of each system. For example, for the air plasma torches of the two-chamber type, the authors of [3] obtained the following dependence:

$$\tilde{\eta} = 1.08 \cdot 10^{-4}(I^2/G \cdot d)^{0.27}(G/d)^{-0.27}(pd)^{0.30}(l/d)^{0.5}, \tag{6.4}$$

This dependence was verified in a wide range of variation of current intensity (I = 50÷3600 A), gas flow rate (G = $1 \cdot 10^{-3}$÷2.2 kg/s), the diameter of the output electrode (d = $1 \cdot 10^{-2}$÷$7.6 \cdot 10^{-2}$ m). In this case, the values of l/d varied from 5 to 40; I^2/Gd = $5(10^6 \div 10^9)$ A^2 s/(kg m); G/d = 0.5÷56 kg/(sm); pd = $1 \cdot 10^3$÷$8 \cdot 10^5$ N/m.

The results of a large number of experiments show that this formula is valid (if the accuracy of ±10% is regarded as acceptable) for calculating the thermal characteristics of the plasma torches of the single-chamber type, the plasma torches with two-sided discharge, plasma torches with smooth and step output electrodes (in the latter case $\bar{l}$ refers to $\bar{l} = l_2/d_2 + l_3/d_3$), and DC and AC plasma torches. Thus, the thermal characteristic calculated from equation (46.4) refers to a large variety of the linear plasma torches with the self-setting and fixed (using a ledge) arc length. The graph, corresponding to this characteristic, the dependence of η on the dimensional complex $\Psi = (I^2/G \cdot d)^{0.27}(G/d)^{-0.27}(pd)^{0.30} \cdot (l/d)^{0.50}$ is shown in Fig. 6.1.

One of the most widely used working gases is, as already mentioned, hydrogen. This is associated with the application of this gas as an energy carrier and a reagent in many plasma-chemical processes. A sufficiently detailed review of the current state of the development of electric arc hydrogen heaters was presented in [11].

The thermal characteristics of the air plasma torch with the self-setting arc length in the range of variation of the complexes $G/I = 10^{-6}\div10^{-5}$ kg /(s ·A); $G/d = 0.04\div0.25$ kg/(s ·A); $pd = (1\div3)$ 10^3 N/m, obtained at $d = 1\cdot10^{-2}\div2\cdot10^{-2}$ m, $l = 0.1\div0.4$ m, are generalised by the dependence [12]:

$$\tilde{\eta} = 6.54\cdot10^{-8}(I^2/Gd)^{0.20}(G/d)^{-0.20}(pd)^{0.98}(l/d)^{1.38} \tag{6.5}$$

or

$$\tilde{\eta} = 6.54\cdot10^{-8}(I/G)^{0.4}(pd)^{0.98}(l/d)^{1.38}. \tag{6.5a}$$

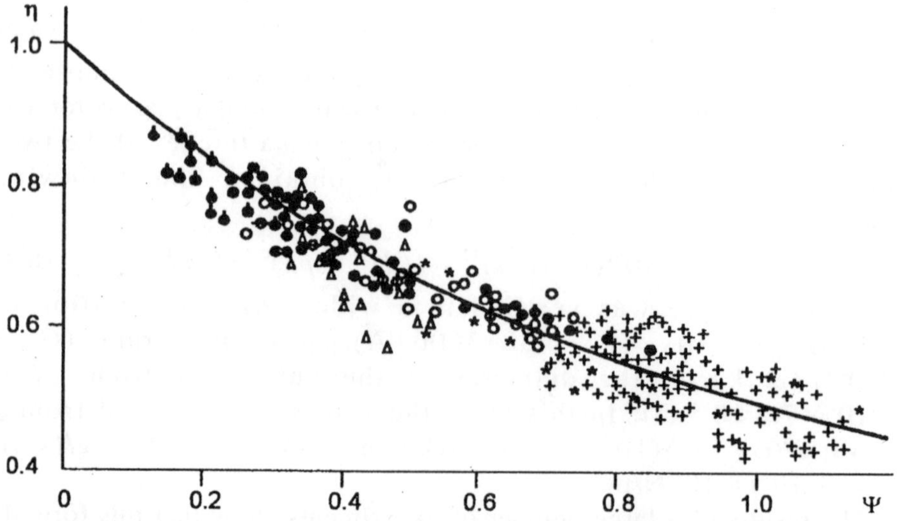

Fig. 6.1. Thermal characteristics of the two-chamber air AC and DC plasma torches experimental points – data from [3–10] solid curve – calculated from equation (6.4).

In this equation it is important to note the almost linear dependence of $\tilde{\eta}$ on the complex (pd) or on the gas pressure in the channel (at d = const), and a very strong dependence on the relative length of the channel. This is results in certain doubts regarding the correctness of generalisation of the experimental data, but no other generalisation has been carried out of the thermal characteristics of hydrogen arcs in the plasma torches of the first two systems, and the currently available results are quite close to the experimental data reported in [12].

In plasma torches with the mean arc length fixed with a ledge it is possible, according to [13], to modify the equation (6.5) for calculating the thermal characteristic to the form:

$$\tilde{\eta} = 6.54 \cdot 10^{-8} (I^2 / G d_2)^{0.20} (G / d_2)^{-0.20} \times \\ \times (p d_2)^{0.98} (l_2 / d_2 + l_3 / d_3)^{1.38}. \quad (6.6)$$

Using the equations for the VAC and thermal characteristics one can carry out engineering calculations of the linear plasma torches with the self-setting arc length and with the arc length fixed with a ledge for air (and with some error for nitrogen and oxygen) and also hydrogen [13]. For other gases, the experimental data in the criterial form have not been systematised. The studies [14, 15], mentioned in chapter 5, presented only the results of experiments which are in sufficient agreement with the previously presented data. The characteristics of the hydrogen arc in the study [16] were not generalised; in this study, the data were also close to those mentioned previously in this book.

The absence of the generalised characteristics for the plasma torches of these two systems is associated in all likelihood with the fact that in the majority of cases in realisation of any plasma process, attention is given to the total losses of energy, determined by the VAC of the plasma torch, and less attention is paid to the problems of optimisation of the energy losses.

6.2. HEAT LOSSES IN THE DISCHARGE CHAMBER OF THE PLASMA TORCH WITH THE INTER-ELECTRODE INSERT

The significance of the problems of thermal efficiency in the plasma torches of the third system, i.e., with the arc length greater than the self-setting length, is completely different. In most cases, these are high-power plasma torches, with the power of up to several megawatt and, consequently, the decrease of the heat losses in the walls of the discharge chamber by even several percent results in a larger gain in reducing the energy losses. The complicated electrophysical and thermal processes, taking place in the plasma torches of this type, were described in chapter 2. Detailed examination of the heat processes in the discharge chamber makes it possible to increase the thermal efficiency of the plasma torch and develop powerful, highly efficient electric arc gas heaters.

The plasma torches with the sectioned inter-electrode insert [17] are more suitable for examining heat exchange processes in the discharge chamber. The individual supply of water to the sections of the insert makes it possible to measure the intensity of the heat flows in the sections for different working conditions of the plasma torch, and the possibility of supplying part of the working gas (or different gases and mixtures) through the gaps between the sections makes it possible to organise in some way the gas screen of the walls of the discharge chamber. Changing the thickness of the sections, and also the length of the entire inter-electrode insert, it is possible to examine in considerable detail the variation of the heat losses into the walls of the discharge channel and determine the quantitative and qualitative characteristics of heat exchange. In some experiments, the individual sections or parts of the sections of the insert may be replaced by quartz windows, examination slits, special sections for taking gas samples, introducing probes, etc. This makes it possible to carry out the spectral and other investigations of the arc column, the thermal layer of the arc and heat exchange between them and the walls of the channel.

6.2.1. Heat losses in the plasma torch with gas vortex stabilisation of the arc

In the distribution of the heat losses along the long cylindrical electric arc chamber of the plasma torch with the inter-electrode insert, examination showed two characteristic sections: the first section from entry into the channel, corresponding to the initial section of gas flow with the heat flow into the wall approximately constant along the section, and the second section with rapidly increasing heat losses (Fig. 6.2*a*). Curves *1–4* corresponds to different flow rates of the gas through the gaps between the sections. Figure 6.2*b* shows the distribution of the local thermal efficiency η_i along the channel which is determined by the ratio of the heat losses per unit length of the section of the inter-electrode insert Q_i to energy generation in the appropriate sec-

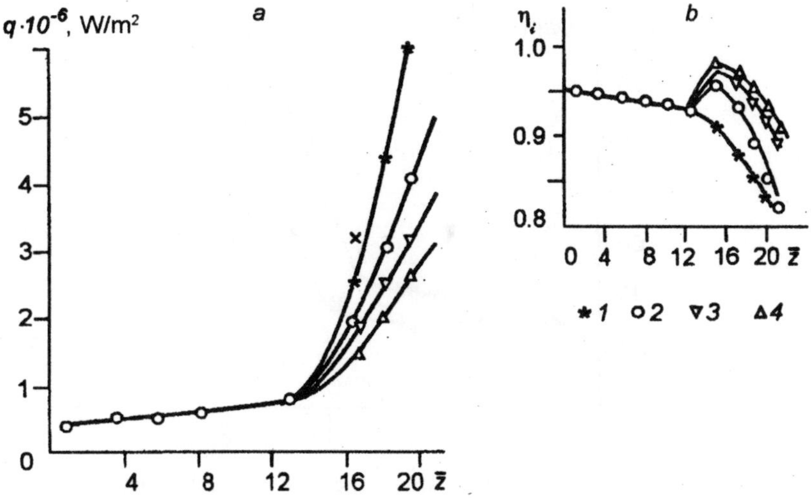

Fig. 6.2. Distribution of heat losses (a) and local thermal efficiency η_i (b) along the electric discharge chamber of a plasma torch with an IEI. $d = 1 \cdot 10^{-2}$ m, $\bar{a} = 23$, $G = 15 \cdot 10^{-3}$ kg/s, $I = 120$ A. g_i, kg/s: *1* – 0; *2* – $0.15 \cdot 10^{-3}$; *3* – $0.3 \cdot 10^{-3}$; *4* – $0.87 \cdot 10^{-3}$.

tion of the arc $E_i \cdot I$:

$$\eta_i = 1 - Q_i / (E_i \cdot I). \tag{6.7}$$

These data show that in the case of relatively low values of current, the heat losses in the initial section of the channel are not large and represent only several percent of the energy contribution to the arc. The losses slowly increase along the section and are almost completely independent of the gas flow rate between the sections.

At a distance of 13–14 gauges from the internal electrode, the heat flows into the channel walls started to increase rapidly, especially in the absence of blowing the gas between the sections (curve *1*, Fig. 6.2*a*). The presence of even lower intensity and blowing of the gas through the gaps between the sections reduces the heat losses in the sections of the interelectrode insert in this section (curves *2–4*). The local thermal efficiency (Fig. 6.2*b*) changes appropriately. The increase of η_i at the start of the second section in the presence of blowing the gas between the sections is determined by the increase of the energy input into the arc, associated with the increase of the strength of the electrical field (see chapter 5).

More detailed investigations of heat exchange in the initial section of the channel were carried out in [18]. It was shown that a small increase of the heat flow into the wall along the channel in the absence of blowing the gas between the sections (curve *1*, Fig. 6.3) takes place as a result of the increase of gas temperature in the boundary

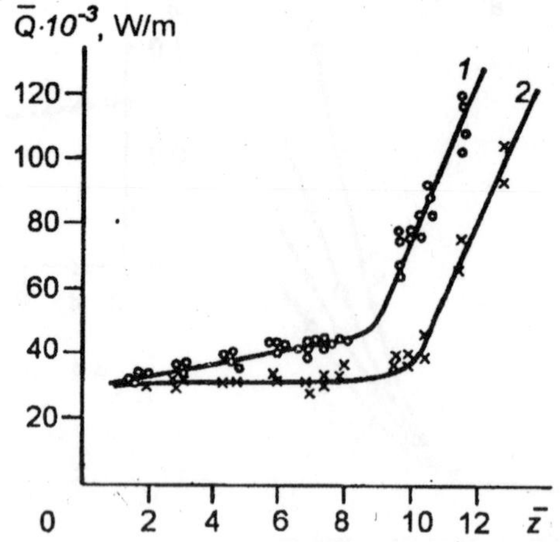

Fig. 6.3. Distribution of heat losses along the IEI. $d = 2\cdot10^{-2}$ m, $\bar{a} = 14$, $G_0 = 6\cdot10^{-3}$ kg/s, 150; g_i, kg/s: *1*– 0; *2* – $(0.3\div0.45)\cdot10^{-3}$ kg/s.

layer determined evidently by turbulent heat exchange with the main gas flow and, to some extent, by the absorption of arc radiation by the wall layer. When gas is blown between the sections of the inter-electrode insert (curves *2*), the gas temperature around the wall remains (starting at a specific flow rate g_i) approximately constant and, consequently, the heat flow in the initial section is also constant.

The increase of the arc current and gas pressure rapidly increase the level of the heat losses into the walls in the initial section of the channel, whereas the increase of the gas flow rate G has only a small effect on these parameters. Behind the initial section, the heat losses are determined both by the intensity of arc current and gas flow rate. The possibility of reducing the heat losses in the section is determined by the gas screen of the walls, produced by means of blowing the gas through the gaps between the sections.

If we examine the scheme of gas flow in the electric arc channel, described in chapter 2 then, taking into account the results shown in Figs. 6.2 and 6.3, we can make several assumptions regarding the heat transfer mechanism. In the initial section of the channel, the main contribution to the heat losses into the chamber wall comes from, in all likelihood, by arc radiation. The role of the remaining factors is small even in the absence of the blowing of the gas between the sections. In the transition section and in the zone of developed turbulent flows, convective losses, which increase in the direction along the flow, are added to the radiant heat losses. However, the sections are also characterised by a large increase of the energy input into the arc, i.e. the local efficiency of the plasma torch greatly increases (Fig. 6.2*b*). On the whole, the thermal efficiency of the

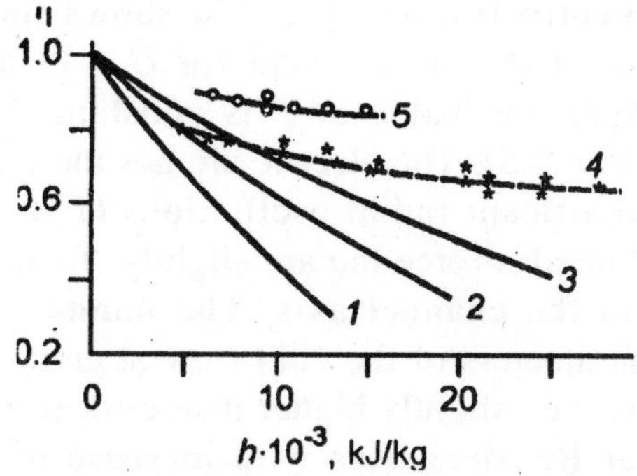

Fig. 6.4. Dependence of the thermal efficiency η of plasma torches of different schemes on the enthalpy of the gas at the outlet of the plasma torch h. 1– 3) two-chamber plasma torch with a self-setting arc length; calculated from equation (6.4) for l = 20; 10 and 5 respectively; 4) plasma torch with IEI, $d = 1 \cdot 10^{-2}$ m, $\bar{a} = 13 \div 17$, $G = (8 \div 15) \cdot 10^{-3}$ kg/s, $I = 50 \div 60$ A [19]; 5) plasma torch with IEI, $d = 1 \cdot 10^{-2}$ m, $\bar{a} = 22 \div 34$, $G = 15 \cdot 10^{-3}$ kg/s, $I = 60 \div 180$ A [20].

plasma torch with the inter-electrode insert is higher than in the plasma torches of the first two systems. They are compared in Fig. 6.4. Here the curves *1–3* show the calculation of the efficiency using equation (6.4) for plasma torches with cylindrical electrodes of different lengths, and the curves *4* and *5* are the experimental data for the plasma torches with the inter-electrode insert. The advantages of the plasma torches with the inter-electrode insert are especially evident with the increase of enthalpy (or temperature) of the gas at exit from the plasma torch [19, 20].

6.2.2. *The characteristics of the arc in the axial gas flow*

Previously, we discussed mainly the plasma torches with gas vortex stabilisation of the arc. The installation of an insulated insert between the electrodes makes it possible to ignite the electric arc in the plasma torch with axial gas supply and examine the interaction of the arc with the gas flow in the absence of radial pressure gradients stabilising the arc on the channel axis. Stable arcing in the discharge chamber of the plasma torch is possible only if the arc spot is fixed in some manner in the centre of the end flat or rod electrode.

The electrical and thermal characteristics of the arc in the axial flow of nitrogen, and also the time sweep of the image of the arc element through the transverse slit, situated approximately in the centre of the channel of the interelectrode insert, were obtained in [21].

The distribution of the strength of the electrical field of the arc along the inter-electrode insert with the length of approximately 7 gages for the nitrogen flow rates $G = (1; 2; 4) \cdot 10^{-3}$ kg/s is shown in Fig. 6.5

(curves *1, 2, 3*, respectively), and Fig. 6.6 shows the time sweep of the intensity of glow of the arc element for G = (1 and 2) · 10^{-3} kg/s. At G = 1 · 10^{-3} kg/s, the value of E is constant along the channel and is low (curve *1*, Fig. 6.5). The electric arc has the distinctive filament form without any significant radial oscillations (*1* in Fig. 6.6). Under the effect of the Archimedes force the arc slightly 'floats up', i.e. slightly moves upwards from the channel axis. The number Re_d, determined on the basis of the parameters of the cold flow at entry into the channel, is approximately 3500, i.e., slightly higher than critical. Downwards along the flow, the number Re_d decreases with increase of the temperature of the gas heated by the arc. In the section $\bar{z}_c$ = 4, in which the time sweep of the arc was determined, the number Re_d, calculated from the mean mass parameters of the flow, is approximately equal to 700, i.e. the arc burns in the laminar gas flow. The values of the strength of the electrical field and the time sweep of the glow intensity of the arc

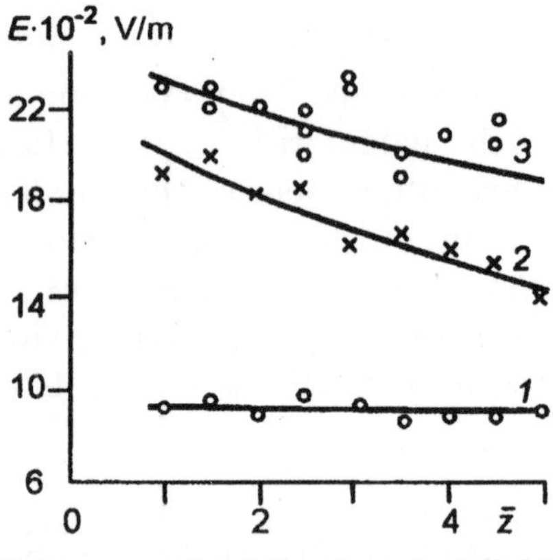

Fig. 6.5. Distribution of the strength of the electrical field of the arc in the axial flow of nitrogen along the channel. d = 20·10^{-3} m, d_a = 6 · 10^{-3} m, I = 140 A. G, kg/s: 1) 1·10^{-3}; 2) 2·10^{-3}; 3) 4·10^{-3}.

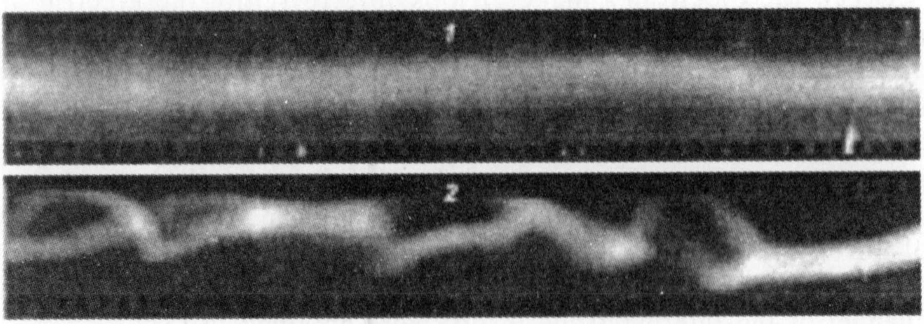

Fig. 6.6. Photosweep of the glow of the element of the arc through a transverse slit in the channel. d = 20·10^{-3} m, d_a = 6 · 10^{-3} m, $\bar{a}$ = 7, $\bar{z}_s \approx$ 4, S = 0.5·10^{-3} m, I = 140 A. G, kg/s: 1) 1·10^{-3}; 2) 2·10^{-3}.

confirmed this.

With increase of the gas flow rate Re_d of the flow increases. Since the non-twisted gas flow does not contain any radial pressure gradients, stabilising the arc along the channel axis, already at $G = 2 \cdot 10^{-3}$ kg/s, there are random transverse oscillations of the arc with the amplitude, comparable with the channel radius (*2* in Fig. 6.6*a*). Examination showed clearly the splitting of the arc column into several current-conducting channels. The number Re_d of the flow at entry into the channel in this case is approximately 7000, and in the zone of high-speed filming it is higher than 1000. Turbulence pulsations of the flow cause oscillations of the electric arc, and the stabilising effect of gas viscosity is insufficient to make the flow laminar. The strength of the electrical field at the start of the channel reaches $20 \cdot 10^2$ V/m and decreases in the direction along the flow to $14 \cdot 10^2$ V/m (curve *2*, Fig. 6.5). This is determined by the increase of the mean mass temperature of the gas, i.e. a decrease of the heat transfer from the arc. Increasing gas flow rate increases the frequency of pulsations of the arc column, and also increases E (curve *3*). These conditions correspond to burning of the arc in the turbulent gas flow. Similar photographs of the arc were obtained in [22].

Interesting information was obtained from the distribution of the energy input into the arc and heat losses along the discharge chamber (Fig. 6.7*a*). At $G = 1 \cdot 10^{-3}$ kg/s, the heat losses $\bar{Q}$, related to the unit length, in the first gages of the channel are determined by the radiation of the arc column and for nitrogen plasma equal up to 40% of the energy contribution $E \cdot I$ in the section. The convective heat losses are then added to the radiant losses. At a distance of approximately 6 lenght gages from entry into the channel, the heat losses are already close to the specific energy input to the arc, i.e. the local thermal efficiency of plasma torches downwards along the flow is close to zero. This shows that in the plasma torches with the inter-electrode insert, used for spraying powder materials, characterised by these conditions, the thermal efficiency of the nitrogen arc does not exceed 0.6. In this case, the length of the inter-electrode insert should not be greater than 5–6 length gages.

When the gas flow rate is increased to $2 \cdot 10^{-3}$ kg/s or greater (Fig. 6.7*b*), the length of the section in which the radiant heat losses are controlling, increases. The thermal losses start to increase only at the end of the channel. The specific energy contribution to the arc is high in the vicinity of the internal electrode and is almost halved at the end of the channel. Since the energy contribution in every

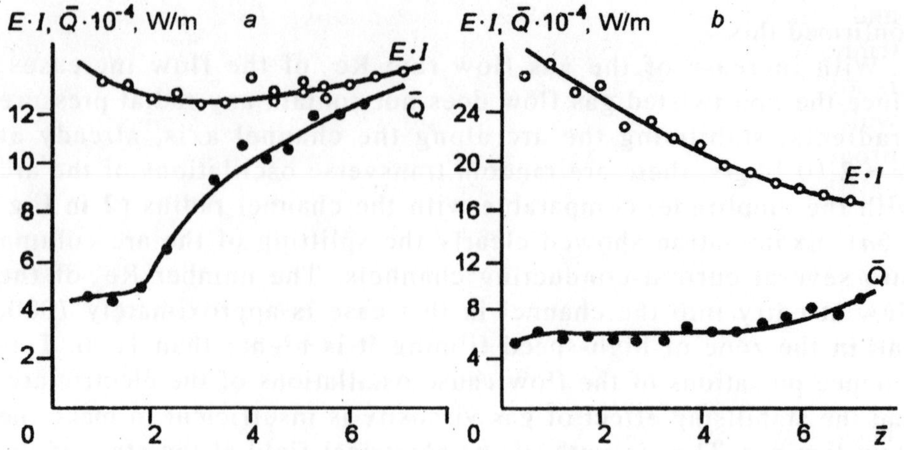

Fig. 6.7. Distribution of the heat losses ($\bar{Q}$), and energy input to the arc ($E \cdot I$) related to the unit length of the channel along the discharge chamber at axial supply of nitrogen. *a)* $G = 1 \cdot 10^{-3}$ kg/s; *b)* $G = 2 \cdot 10^{-3}$ kg/s. $d = 20 \cdot 10^{-3}$ m, $d_a = 6 \cdot 10^{-3}$ m, $\bar{a} = 7$, $I = 140$ A.

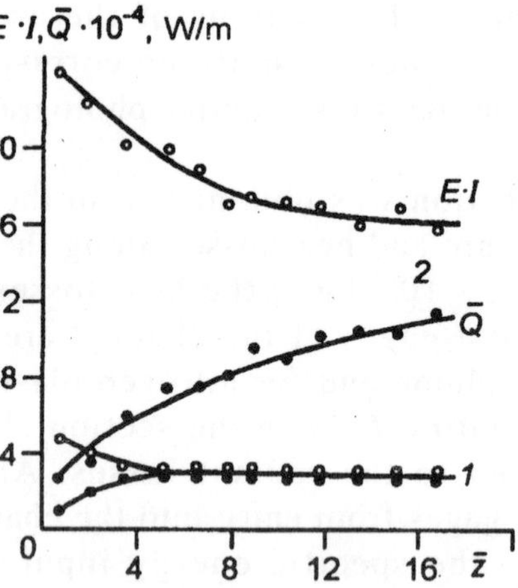

Fig. 6.8. Distribution of the energy input into the arc and heat losses, related to the unit length, along the discharge chamber for the arc in the axial argon flow. $d = 0.6 \cdot 10^{-2}$ m, $a = 0.12$ m; *1)* $I = 50$ A, $G = 0.1 \cdot 10^{-3}$ kg/s; 2) $I = 100$ A, $G = 3.0 \cdot 10^{-3}$ kg/s.

section of the arc is determined by the strength of the electrical field (curve *2* in Fig. 6.5 for the given case), the decrease of the energy contribution is explained by the same reasons as the decrease of the strength of the electrical field. Similar results were obtained in [23] for the arc in the axial argon flow (Fig. 6.8).

On the basis of these data it may be concluded that the axial channel of the plasma torch with the inter-electrode insert in both the twisted and axial flows of the working gas contains two sections in the distribution of heat losses. In the vicinity of entry into the channel, there is a section with a constant or slowly increasing heat flow into the wall. Further, the heat losses increase along the channel and approach the value of the energy contribution to the arc, i.e. the local thermal efficiency in the section tends to zero.

6.3. HEAT EXCHANGE OF THE ELECTRICAL ARC IN THE TURBULENT GAS FLOW WITH THE WALLS OF THE DISCHARGE CHAMBER

The examination of the heat exchange of the electrical arc with the walls of the channel and the increase of thermal efficiency of the plasma torches have been studied in a number of investigations. If in early investigations, which can be reviewed in [24], the role of radiation of the arc in the heat exchange with the wall is usually ignored and only the total losses are taken into account, in subsequent investigations (both experimental and analytical), the role of radiation heat losses is important [1, 17, 25, etc]. Attempts have been made to separate the heat transfer mechanisms and subtract from the total losses the losses caused by radiation, convective or conductive heat transfer processes. The role of these exchange mechanisms, their value in the energy balance, generated by the electrical arc, was examined briefly in chapter 3 (for greater detail see [26]). The experimental investigations of heat exchange in the discharge chamber of the plasma torches (mainly with the inter-electrode insert) have made it possible in a number of cases to determine the dependence on the main working parameters of the plasma torch and find analytical or engineering methods of calculating them.

According to the scheme of interaction of the electrical arc with the turbulent gas flow, described in chapter 2, there are two regions in which the heat exchange mechanism differs: it is the initial section of the channel and the section of the developed turbulent gas flow. We examine, from this position, the results of experimental investigations of the arc in the discharge chamber of the plasma torch with the inter-electrode insert.

6.3.1. Heat exchange in the initial section of the channel

We return to the distribution of the heat losses along the sectioned channel of the plasma torch with the inter-electrode insert, described in Figs. 6.2 and 6.3. At low values of current intensity (I = 100÷

200 A), the heat losses into the wall in the initial section of the channel are relatively small and, according to the data in [1, 17, 25, 27] equal, for the arc in air, 10–15% of the local energy contribution to the arc. In the absence of the inter-sectional blowing of the gas, the losses slowly increase along the initial section. When a gas is blown between the sections of the insert, and also according to the results of measurements of the heat flows in a porous channel or in a sectional channel, using a disc-calorimeter with the diameter large agenda channel diameter, in the measurements with special radiation detectors [1, 17, 25], the heat losses in the initial section are almost constant along the channel and equal to the heat losses in the initial sections of the inter-electrode insert (see, for example, Fig. 6.3). According to comparison [26], these heat losses are close to the calculated radiant flux from the arc (for the wavelength of $\lambda \geq 200$ nm). The increase of the heat losses along the initial section of the air arc in the absence of blowing between the sections is determined, as mentioned previously, by the thermal processes in the boundary wall layer of the gas. In particular, a significant role is played by absorption by cold air (or, more accurately, atmospheric oxygen) of ultraviolet radiation of the arc ($\lambda < 200$ nm), whose role is, as mentioned in [26], quite important. Heating of the boundary layer results in the appearance of the convective component of the heat flow which may equal 30–50% of the total heat flow into the wall in the given section (Fig. 6.3).

This component of the heat losses has been studied quite extensively in [27]. The calculation equations, proposed in [27] for estimating the convective component show that, in the majority of cases, in this section (up to $Re_z \sim 10^6$) there is heat exchange of the laminar gas flow with the chamber wall. It should be mentioned that the presence of the convective component results in certain differences in the measured (by different authors) losses into the channel wall in the initial section. According to the majority of authors, the heat flow into the wall is determined only by the radiation of the arc, i.e. $Q_w = Q_r$. However, in individual studies, for example in [25, 26] it has been attempted to separate Q_w and Q_r. For example, in [25] it has been reported that for the air arc $Q_w = (1.5\div1.7)Q_r$. In [17], the following relationship was proposed for calculating the heat flows into the wall at a pressure close to atmospheric:

$$\bar{Q}_w = 6.2 \cdot I^{1.6} (\mathrm{W/m}), \tag{6.8}$$

which averages the results of many investigations for air and nitrogen.

This dependence is similar to that derived in [27], where $Q_w \sim I^{1.5}$. If we take into account only the radiant component of the heat flow, the coefficient in the right-hand part of equation (6.8) decreases. For pressures up to 1 MPa and current intensity of up to 1 kA, to estimate the radiant heat flow into the wall in the initial section, we can use the equation generalising the results of [17, 18, 23, 28–36] and many other results:

$$Q_r/(p \cdot l) = 4.6 \cdot 10^{-5} I^{1.6}, \text{ W/(m} \cdot \text{Pa)}. \qquad (6.9)$$

The correspondence of the experimental data in a large number of investigations, collected in approximately 30 years of investigations, with the calculation curve is presented in Fig. 6.9. The graph gives the data on the heat flows into the wall in the initial section of the channel in the plasma torches of greatly differing systems, not only in air, but also in nitrogen, argon, oxygen and water vapours. Since in the experimental material the radiant and convective components of the heat flows were not separated in the majority of cases, the scatter of the points (Fig. 6.9) is determined primarily by differences in the convective component. At higher pressures, the dependence of Q_{rad} on p differs from linear dependence [17]. For example, in [37], for the pressures of $(50 \div 200) \cdot 10^5$ Pa it has been shown that $Q_r \sim p^{0.5}$.

The attempts to separate the convective component of the heat flow in the initial section have been made by many authors. The investigations have been reviewed in, for example, [17, 27]. The equations obtained for calculating convective heat exchange usually operate only in a narrow range of the parameters and make it possible to carry out the generalised estimates. Since the difference between the radiant and total heat flow in the channel wall in the this section is not very large, the proposed dependence (6.9) makes it possible to estimate the radiant heat losses in a wide range of the variation of the parameters and in plasma torches of any system and any geometrical dimensions. It may also be used for estimating the total heat losses in the initial section of the channel in these plasma torches.

6.3.2. Heat exchange in the section of the developed turbulent flow of gas

Downwards along the flow from the initial section of the gas flow, there is a zone of contact of the thermal layer of the arc with the wall boundary layer (see the scheme in Fig. 2.12). It is characterised by the rearrangement of the nature of gas flow. The process of mixing of the high-temperature gas from the thermal layer of the arc with

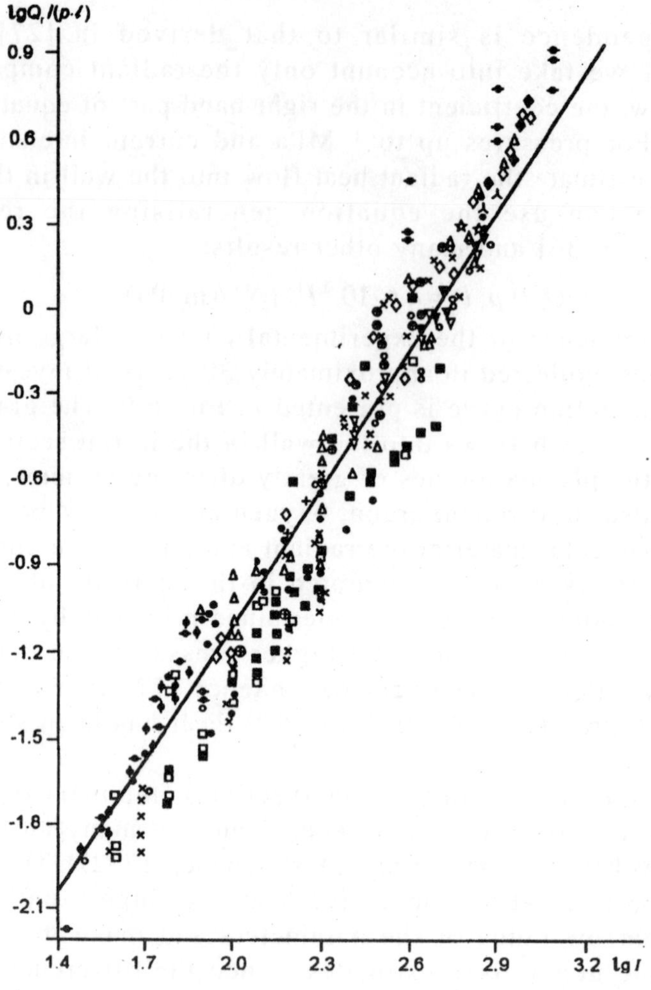

Fig. 6.9. Dependence of the radiant heat losses into the wall in the initial section of the channel on the working parameters of the arc (the data by the authors and also from studies [17–36] and many other studies) Solid curve – calculated from equation (6.9).

the cold wall gas starts here. The radial distribution of the velocity and profiles of gas pressure changes here. The vortex stabilisation of the arc column is disrupted because the regions of reduced pressure on the axis of the channel disappear.

Of the mixing of the cold and hot gas results in the situation in which the mean mass temperature of the gas in the wall layers increases and the convective heat flow starts to arrive at the channel wall, in addition to arc radiation. Along the length of several gages (transition region of the flow), the convective flow increases in a non-linear

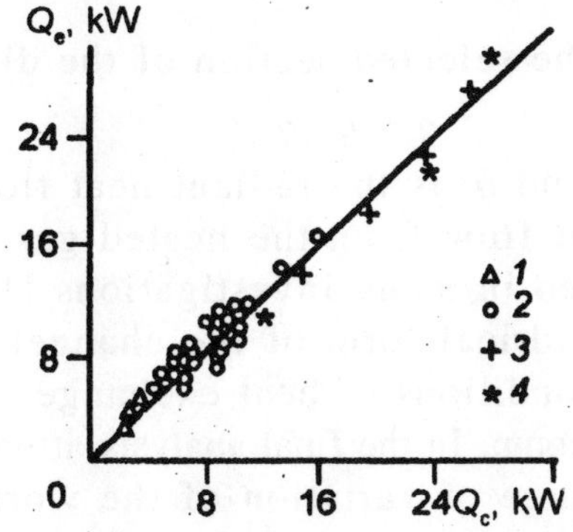

Fig. 6.10. Comparison of the experimental Q_e and calculated Q_c convective heat flows into the anode. *1*) N_2, $d = 2{-}10^{-2}$ m, $I = 60 \div 180$ A; *2*) air, $d = 2{-}10^{-2}$ m, $I = 40 \div 180$ A; *3*) air, $d = 2{-}10^{-2}$ m, $I = 300 \div 600$ A; *4*) H_2, $d = (2 \div 3) \cdot 10^{-2}$ m, $I = 300 \div 700$ A.

manner (Fig. 6.2*a*) and subsequently, this flow intensity becomes approximately proportional to the mean mass temperature of the gas which increases along the channel. The total heat flow, especially in the absence of inter-sectional gas blowing, rapidly increases, tending to the value of the specific energy contribution to the arc. This is shown clearly in Fig. 6.7 for the axial gas flow. As already mentioned previously (Fig. 6.2), even lower intensity inter-sectional blowing of the gas rapidly changes the heat losses into the wall in the developed turbulent section. Consequently, the results of investigations of the efficiency of gas screening of the wall of the discharge chamber are important for optimising the conditions of gas heating in this section.

Special attention will be given to the heat exchange between the arc, gas and the channel wall in the section of the developed turbulent gas flow. These investigations have been carried out in particular for the arc burning in air.

As shown in the previous section, the radiant heat flow from the arc into the wall the discharge chamber is approximately constant along the initial section. In the transition and developed turbulence sections it is, according to the measurements [17, 25, 26] on approximately the same level or may even decrease. This decrease is not large and, consequently, it may be assumed with a high degree of reliability in estimating heat exchange that the radiant flux into the wall remains approximately constant along the entire inter-electrode insert. It was mentioned in chapter 3 that the main role in the heat losses is played by the radiation and convective flux, and the role of conductive heat exchange is not large. Thus, knowing the value of the radiant losses, it may be assumed that the convective flux into the wall in the selected section of the discharge channel:

$$q_c = q - q_r, \tag{6.10}$$

where q is the total, and q_r is the radiant heat flow into the wall.

The convective heat flow from the heated gas into the channel wall has been evaluated in many investigations [1, 13, 17, 18, 27, 35, 36, etc]. The cylindrical form of the channel, the presence of high non-isothermal conditions of heat exchange, and other factors have been taken into account. In the final analysis, it has been established [17] that in a wide range of variation of the working parameters: the type of gas, temperature, pressure, the convective heat flow into the wall of the discharge chamber, may be calculated using the equations for the heat exchange of the gas flow with the wall of the cylindrical channel, obtained at moderate temperatures [38]:

$$q_c = \mathrm{St}(\rho u)_0 (h_0 - h_w), \tag{6.11}$$

$$\mathrm{St} = 0.023 \mathrm{Re}_d^{-0.20} \cdot \mathrm{Pr}^{-0.57}. \tag{6.12}$$

Since the equations (6.11) and (6.12) include the parameters of the gas flow, the main difficulty in the calculation is the selection of the determining gas temperature. As shown in [17, 27, 39], the starting point may be represented by the mean mass temperature of the heated gas. In this case, it is possible to ignore the effect of the temperature factor, and the numbers Re_d and Pr are determined from the following equation:

$$\mathrm{Re}_d = (\rho u)_0 \cdot d / \mu_0; \mathrm{Pr} = \mu_0 \cdot c_{p0} / \lambda_0.$$

Comparison of the results of calculations, using equations (6.11), (6.12), of the convective heat losses into the output electrode of the plasma

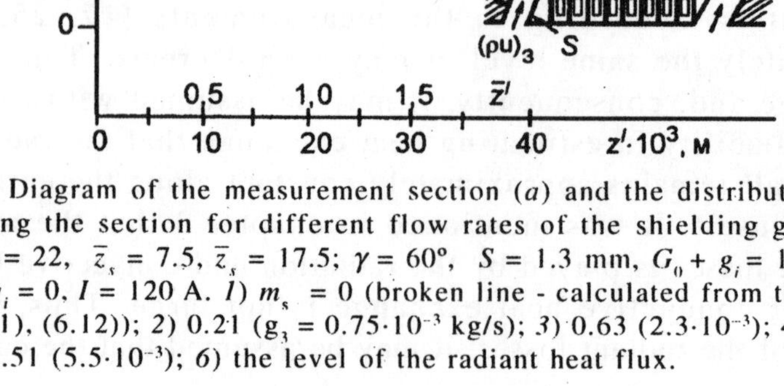

Fig. 6.11. Diagram of the measurement section (*a*) and the distribution of heat losses along the section for different flow rates of the shielding gas (*b*). d = 20 mm, $\bar{a}$ = 22, $\bar{z}_s$ = 7.5, $\bar{z}_s$ = 17.5; γ = 60°, S = 1.3 mm, $G_0 + g_i = 14 \cdot 10^{-3}$ kg/s, m_s = 1.0; g_i = 0, I = 120 A. *1*) m_s = 0 (broken line - calculated from the equation (6.9), (6.11), (6.12)); 2) 0.21 ($g_3 = 0.75 \cdot 10^{-3}$ kg/s); *3*) 0.63 ($2.3 \cdot 10^{-3}$); *4*) 1.0 ($3.6 \cdot 10^{-3}$); *5*) 1.51 ($5.5 \cdot 10^{-3}$); *6*) the level of the radiant heat flux.

torch with the inter-electrode insert and of the experimental data for individual gases (disregarding the flow through the anode spot of the arc) is presented in Fig. 6.10. For nitrogen and air, the experiments and calculated values are in good agreement in the entire investigated range of the parameters. In hydrogen, at temperatures higher than 3000 K, the experimental data are 20–25% higher than the calculated values. This difference is found in the dissociation temperature range characterised by the anomalous behaviour of the transfer coefficient in hydrogen [11].

Thus, in the section of the developed turbulent flow of the working gas in the plasma torch, the convective heat flow into the channel wall may be calculated (in any case for diatomic gases) using equations for heat exchange, obtained at moderate temperatures, if the determining parameter is represented at the mean mass temperature of the gas in the section in which the calculations are carried out. The radiant heat losses may be estimated using equation (6.9).

6.3.3. The efficiency of gas screen of the wall of the discharge chamber

The heat exchange between the high-temperature gas flow and the walls of the channel was investigated in [17, 35] using a measuring section consisting of a set of disks thermally insulated from each other (Fig. 6.11*a*). The thickness of the copper disc was 4 mm, the thickness of the heat- and electrical-insulating interlayer of fluoroplastic and mica was 0.3 mm, the number of discs in a set 12, the length of the entire section 54 mm, internal diameter 20 mm. The restricting (outer) discs were specially machined to have a profile for making contact with the adjacent sections of the inter-electrode insert. The gas was blown through a slit formed between the first disc and the next section. The main bulk of the experiments was carried out for the fixed position of the gage section and constant working parameters. The mean mass temperature of the working gas (air) in front of the gage section was approximately 3300 K. The individual supply of water to all sections of the inter-electrode insert and discs made it possible to carry out calorimetric measurements of the heat flows in them.

Figure 6.11*b* shows the distribution of heat losses along the gage section for different gas flow rates of the gas blown through the slit in front of the section. As in the investigations of the efficiency of gas screening at low temperatures [39], the parameter, characterising the intensity of the gas screen, was $m_{scr} = (\rho u)_{scr}/(\rho u)_0 = (d/4S)\cdot(g_{scr}/G_0)$. In the absence of blowing the shielding gas, the heat flow slowly increases along the gage section (points *1*). The broken curve here indicates the results of calculations of the heat flows into the gage section using equation $q = q_c + q_r$, where q_c is the convective heat flow, calculated from equation (6.11), q_r is the radiant flux into the wall calculated from equation (6.9). The level of the radiant heat flux (155 W/disc) is shown by the dot-and-dash curve *6* in the graph. It may be seen that there is good agreement between the experimental data and the calculated values in the absence of blowing the shielding gas. Blowing even a small amount of gas in front of the section greatly reduces the heat flows to the first discs (curve *2*). The increase of m_{scr} extends the effect of the screen to larger and larger numbers of discs (curves *3, 4*). At $m_{scr} = 1.51$, only the radiant heat flux (the section of the absolute screen) falls on the first discs of the gage section, and the effect of the gas screen is extended faraway behind the gage section (curve *5*).

The efficiency of shielding the walls with the gas screen is determined by the dimensionless ratios $\theta = (T_0 - T^*_w)/(T_0 - T_w)$ [39], T_0, T_w, T^*_w where T_0, T_w, T^*_w is the mean mass temperature of the gas, the wall temperature, and the adiabatic temperature of the wall in the absence of the screen, respectively. The ratio is based on the hypothesis, confirmed by experiments, according to which both in the absence and in presence of the gas screen the heat flow into the adiabatic wall is expressed by the same heat exchange law $q_c = \alpha\,(T^*_w - T_w)$.

In some studies, it has been assumed that $\theta = (T_0 - T^*_w)/(T_0 - T_s)$, where T_s is the temperature of the gas blown through the slit. In the case of a metallic, watercooled wall, at $T_0 \gg T_w$, the difference in the definitions is small.

In the case of low gas flow temperatures T_0, the quantity θ characterises the ratio of the convective gas flow, taken away by the screen $(q_c - q_{c.s})$ to the flow q_c in the absence of a restricting screen, i.e.

$$\theta' = (q_c - q_{c.s.})/q_c. \tag{6.13}$$

Here, there is a prime next of the value θ because we assume the constant heat transfer coefficient which is valid at low temperatures of the flow but requires clarification in the case of high temperatures and, correspondingly, large temperature differences T_0 and T^*_w. In this case, the relationship between θ and θ' is established by means of the temperature factor. In transition from temperature to gas enthalpy or to thermal flows into the wall, this incorrectness in the definition is removed.

Since it is quite difficult to determine the adiabatic wall temperature T^*_w in the conditions of combined cooling and high gas temperature, the measure of efficiency of the gas screen is represented here by the relationship (6.13) which makes it possible to use directly the results of measurements of the heat losses in the discs of the gage section.

The radiant heat flux into the wall of the gage section in the presence and absence of the gas screen may be assumed to be constant and, consequently, the equation (6.13) for the i-th disc of the gage section is reduced to the form:

$$\theta'_i = (Q_0 - Q_{scr})_i / (Q_0 - Q_r)_i \tag{6.14}$$

Here Q_0 and Q^*_{scr} are the total heat flows into the disc in the absence and presence of the shielding gas flow; Q_r is the radiant heat flux into the same disc. The relationship (6.14) is also used as the basis of processing of the experimental data.

The distribution of the efficiency of the gas screen, corresponding to the data in Fig. 6.11, is shown in Fig. 6.12. At $m_{scr} = 0.21$, the efficiency of the screen θ' does not exceed 0.6 even in the first disc (curve *1*), at $m_{scr} = 1.51$, convective heat flows into the first disks do not form (the absolute screen, $\theta' = 1$) and in the subsequent disks $\theta' > 0.5$ (curve *4*), i.e. a sufficiently effective screen is also extended to further sections of the inter-electrode insert along the flow.

The efficiency of the gas screen of the gage section depends not only on the gas flow rate supplied into the slit in front of the section, but also on the width of the slit (Fig. 6.13) because according to the definition $m_{scr} \sim 1/S$. Thus, θ' is determined by a number of dimensionless parameters of which most significant is the distance from the

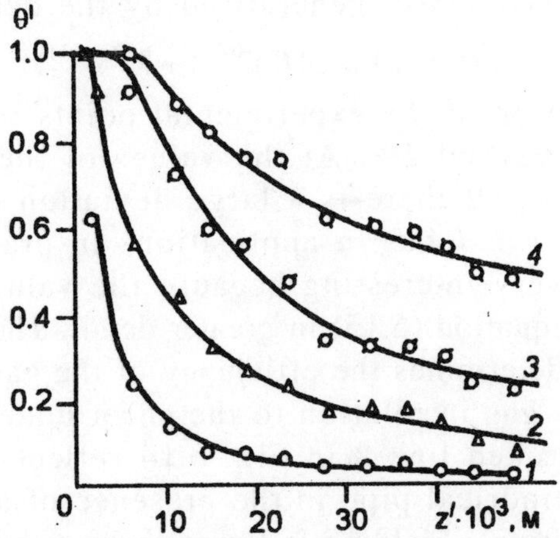

Fig. 6.12. Efficiency of the gas shielding along the gage section. d = 20 mm, $\bar{a}$ = 21.5, 7.5, $\bar{z}_{scr}$ = 17.5; S = 1.3 mm, γ = 60°; I = 120 A $G_0 + g_s$ = 14·10⁻³ kg/s; *1–4*) values of m_{scr}, equal to respectively 0.21; 0.63; 1.0; 1.51.

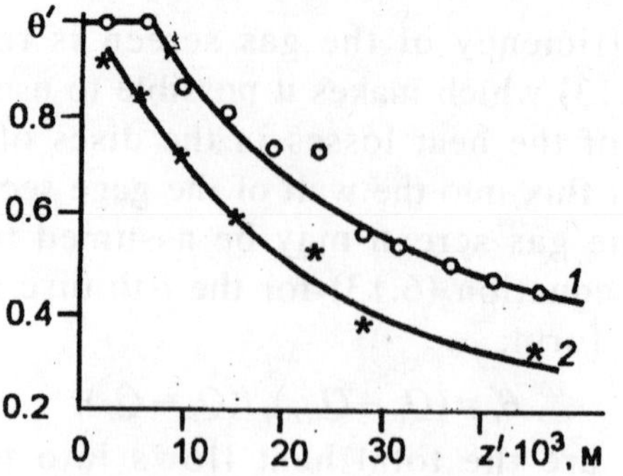

Fig. 6.13. Effect of the width of the slit S on θ'. d = 20 mm, $\bar{a}$ = 21.5, $\bar{z}_c$ = 7.5, $\bar{z}_{scr}$ = 17.5; G_0+g_s = 14·10⁻³ kg/s; g_{scr} = 4.8·10⁻³ kg/s; I = 120 A; γ = 60°; 1,2) the values of S are equal to 1.3 and 4.2 mm, respectively.

start of the gage section $\bar{z}' = (z' - z_1')/S$ (here z_1' is the length of the section of the absolute screen) and the blowing parameter ms.

In the case of moderate temperatures of the gas flow, the efficiency of the screen of the adiabatic and non-adiabatic walls is the function of a dimensionless criterial complex [39]:

$$K = (z' - z_1')\mathrm{Re}_s^{-0.25}/(m_s \cdot S),$$

where $\mathrm{Re}_s = (\rho u)_s \cdot S/\mu_0$.

The experimental data, obtained in examination of the gas screen of the gage section in a plasma torch with an inter-electrode insert at gas flow temperatures of 3300 K, are presented in Fig. 6.14 in the form of the dependence of θ' on K. Parameter m_{scr} in this case changes from 0.2 to 1.5, the width of the slit from 1.3 to 4.2 mm. These data are efficiently generalised by the dependence:

$$\theta' = (1 + 0.24K)^{-0.8}(1 + K^2)^{-0.14}. \qquad (6.15)$$

The RMS deviation of the experimental points from the calculated curve does not exceed 2%. At the values of the parameter of the gas screen $m_{scr} < 0.2$ there is a large deviation of the experimental data from curve *1* but in applications in practice these values of m_{scr} are not very interesting because the value of θ' is small.

We examine equation (6.15) in greater detail. The first co-multiplier in the equation determines the efficiency of the gas screen on a sheet when blowing a gas in relation to the sheet under the angle $\gamma = 0°$ [38, 39]. The broken line *2* in Fig. 6.14 reflects this relationship. Thus, in the cylindrical pipe in the presence of a twisted flow, the efficiency of thermal shielding is lower than in the case of the sheet, especially at high values of K. There are several reasons for this. One of them is associated with the cylindrical form of the channel,

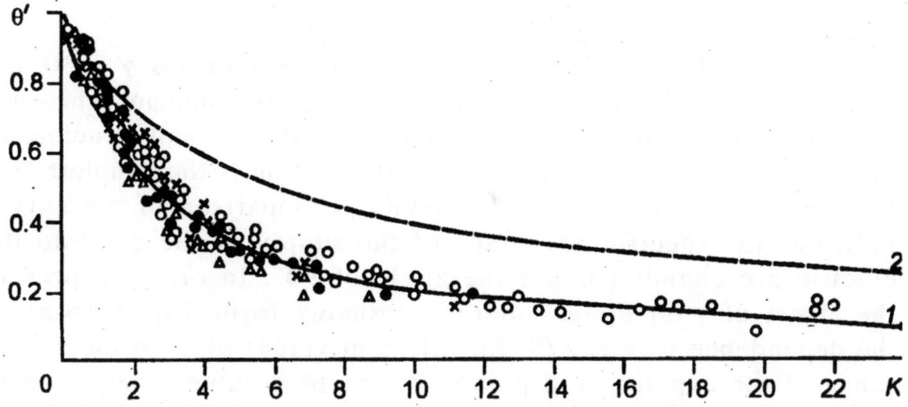

Fig. 6.14. Efficiency of boundary cooling. *1*) calculated from equation (6.15); *2*) calculated from the equations in [38, 39] for a flat sheet; O – S = 1.3 mm; X – 2.2; Δ – 3.2; • – 4.2.

as indicated in [40]. In addition to this, in the twisted flow, the efficiency of the screen should be generalised in respect of the current lines and, in this case, the experimental points would be closer to the broken curve. However, this is not possible because there are no data on the velocity v_φ in the investigated sections. Another reason is the large angle of blowing γ. It should be mentioned that in the case of gas screening of the sheet, the increase of the blowing angle of the shielding gas decreases the efficiency of the screen on the whole along the entire length of the sheet and for all values of m_{scr} [41]. It is naturally to assume the existence, in the electric arc chamber with the inter-electrode insert, of a relationship between the blowing angle of the gas γ and the thickness of displacement of the boundary layer along the flow behind the blowing section determined as the length of the section of the absolute screen, and also the intensity of mixing of the gas flows.

We examine the effects of the angle of blowing the shielding gas on the efficiency of the screen of the walls of the discharged chamber of the plasma torch with the inter-electrode insert. These data were obtained for the angle of blowing of γ = 60°. In the previously described experiments, the angle γ = 30, 45, 75 and 90°. The width of the slit, counted along the normal to the wall, was correspondingly (1.3; 1.2; 1.8 and 1.8) · 10^{-3} m. The total flow rate of the working gas (air) in the cross-section of shielding blowing was $G_0 + g_s = 22 \cdot 10^{-3}$ kg/s, the value of g_s varied from 0 to 6.5 · 10^{-3} kg/s, so that the blowing parameter m_{scr} could be vary from 0 to 1.1. The mean mass temperature of the gas in front of the measuring section was, as previously, 3000 K. The radiant heat flux to the single disc was approximately 165 W.

The dependences of θ' on z'/S for the two values $\gamma = 30$ and 90° are shown in Fig. 6.15. The angle $\gamma = 30°$ is the minimum achievable angle from the position of developing the still efficient structure of the section. At $\gamma = 90°$, the design of the section is the simplest,easy to manufacture and reliable in service. Comparison of the curves indicates the specific advantage of the supply of the gas into the electric arc chamber under the angle of 90°. In order to present the effect of γ on θ' in a more convincing form, Fig. 6.16 gives the dependence $\theta' = f(z'/S)$ for different values of γ. In the scatter range of the experimental points, it may be assumed that θ' is independent of γ if the angles are in the range $45° \leq \gamma \leq 90°$.

If we examine the distribution of the heat losses along the gage section when blowing the gas with similar parameters m_{scr}, but under different angles (Fig. 6.17), it may be seen that the most efficient method is the blowing of the gas under the angle of $\gamma = 75 \div 90°$. A significant contribution to the decrease of the heat losses is provided by the section with the absolute screen whose length increases with increase of the blowing angle (curves *2–4*).

We return again to the calculation of the heat losses along the gage section in the absence of boundary gas blowing. As in Fig. 6.11, the broken curve *1* in Fig. 6.17 is the calculated density of the heat flow into the discs of the measurement section. The radiant flux was taken from the experiments (calculation carried out using the equation (6.9) gives approximately the same value), and the convective flux was calculated using equations (6.11), (6.12).

Since the numbers Re_d and Pr include the parameters of the gas flow, the main difficulty in the calculation, as in the case of generalisation of the electrical characteristics of the arc, is the selection of the determining temperature of the gas. According to the results of specially formulated investigations [27, 39, 42], the starting point is represented by the mean mass temperature of the heated gas. In this case, the temperature factor is equal to unity. The numbers Re_d and Pr are determined as follows: $Re_d = (\rho u)_0\, d/\mu_0$; $Pr = \mu_0 c_{p0}/\lambda_0$. Here μ_0 is the viscosity at the mean mass temperature of deceleration of the gas. The dependence of the Prandtl number on temperature was calculated using handbook data [43]. As indicated by Figs. 6.11 and 6.17, the agreement between the results of the calculation of the experimental values is good.

The length of the section of the absolute screen, where $\theta' = 1$, increases with the increase of the angle and the intensity of blowing. At the blowing parameters close to unity and higher, the value of

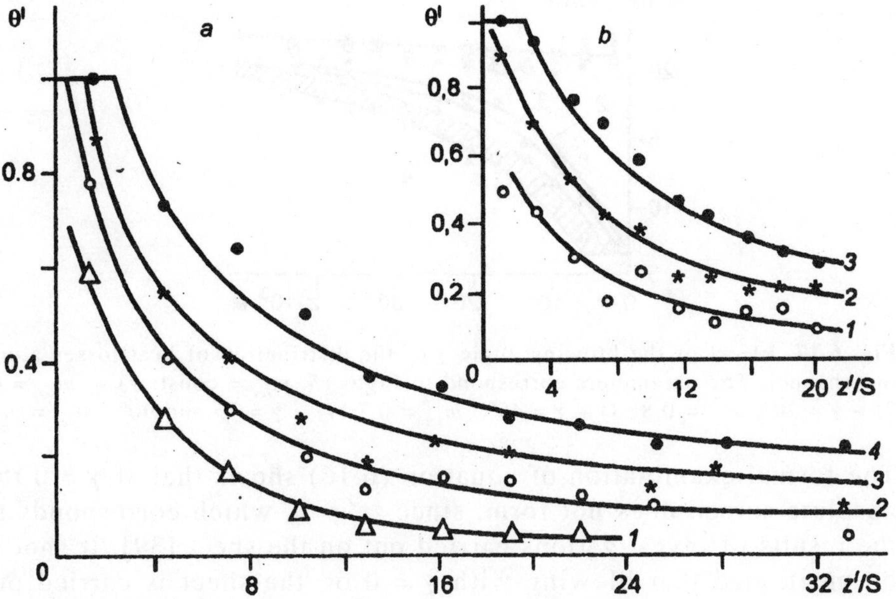

Fig. 6.15. Dependence of θ' on z'/S at $\gamma = 30°$ (*a*) and $90°$ (*b*). $d = 20\cdot10^{-3}$ m, $\bar{a} = 25$, $\bar{z}_s = 4.5$, $m_s = 10$, $\bar{z}_{scr} = 20$; $I = 120$ A; $G_0+g_s = 22.1\cdot10^{-3}$ kg/s; *a*) $g_i = 1.3\cdot10^{-3}$ m; *1–4*) $m_{scr} = 0.35$; 0.6; 0.8; 1.1 respectively; *b*) $S = 1.8\cdot10^{-3}$ m, 1–3) $m_s = 0.36$; 0.58; 0.8 respectively.

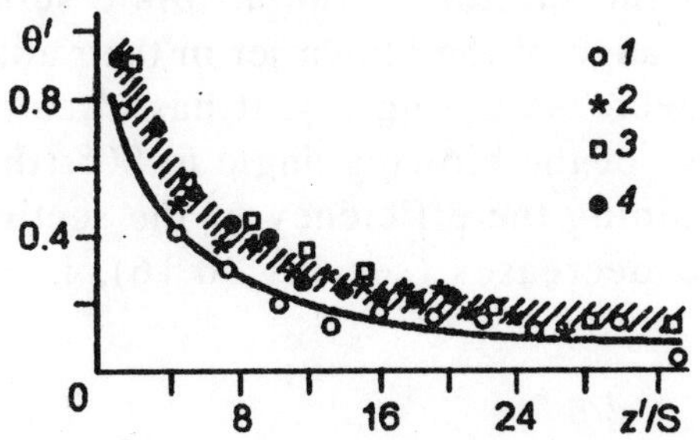

Fig. 6.16. Dependence of θ' on z'/S for different values of angle γ. Parameters correspond to Fig. 6.15.; $m_{scr} = 0.6$; *1–4*) $\gamma = 30$; 45; 60; 75 and 90°.

z_1' is comparable with the length of 1–2 discs of the measurement section, i.e. the contribution of the section to the total efficiency of the screen is very high. Using the data analysed in [39, 40] and some other investigations, it may be established that the length of the section of the absolute screen is proportional to $m_{scr}^{\alpha} \cdot \sin^{\beta}\gamma$. The processing [44] of the existing experimental material (Fig. 6.18) show the following dependence:

$$z_1'/S = 4.28 \cdot m_{scr}^2 \cdot \sin\gamma, \tag{6.16}$$

given for the variation range $\gamma = 30\div90°$, $S = 1\div5$ mm, $m_{scr} = 0.4\div1.5$.

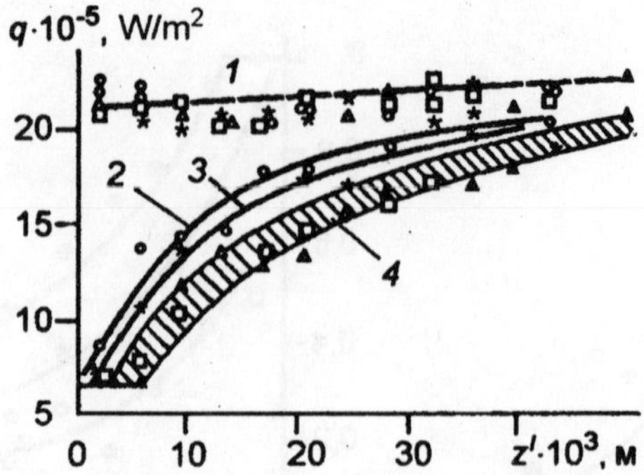

Fig. 6.17. Effect of the blowing angle γ of the distribution of heat losses along the channel. The parameters correspond to Fig.6.15. m_{scr} = const; *1*) – $m_{scr} = 0$; *2*) – $\gamma = 30°$, $m_{scr} = 0.8$; *3*) – $\gamma = 45°$, $m_{scr} = 0.7$; *4*) – $\gamma = 75$ and $90°$, $m_{scr} = 0.8$.

The formal examination of equation (6.16) shows that at $\gamma = 0$ the absolute screen does not form, since $z_1' = 0$, which corresponds to the results of investigations carried out on the sheet [39]. It should be mentioned that blowing with $\gamma = 0$ on the sheet is carried out through the slit parallel to the surface, i.e. there is a step whose height is equal to the width of the slit + the thickness of the upper wall. In this case, z_1' is determined by the propagation of the core of the blown jet. In the cylindrical cost and diameter channel. The length of the section of the absolute screen is determined here by the hitting range of the blown jet in the radial direction, which decreases with a decrease of angle γ. It has already been mentioned that with a decrease of the blowing angle to 30° (the minimum angle, obtained when retaining the efficiency of the section), the efficiency of the screen also decreases (see Fig. 6.16), i.e. there is a corre-

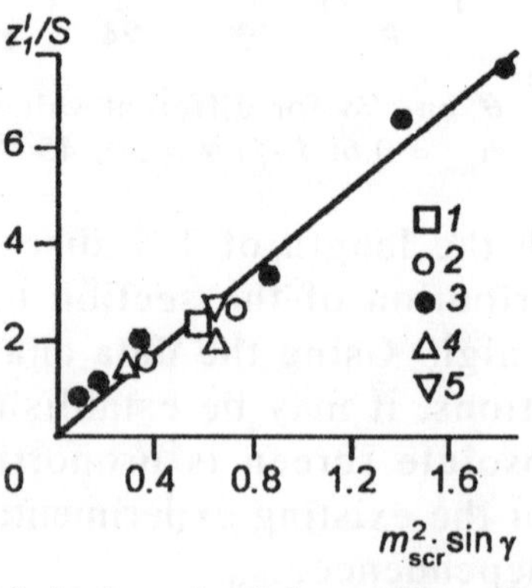

Fig. 6.18. Dependence of the length of the section of the absolute screen on the determining parameters. *1*) $\gamma = 30°$, $S = 1.3$ mm; 2) $\gamma = 45°$, $S = 1.2$ mm; *3*) $\gamma = 60°$, $S = 1.3 \div 4.2$ mm; *4*) $\gamma = 75°$, $S = 1.8$ mm; 5) $\gamma = 90°$, $S = 1.8$ mm.

spondence between the experimental data and the equation (6.16) in the investigated range of the parameters.

Taking into account the length of the section of the absolute screen, the authors of [44] constructed the dependences $\theta' = f(K)$ for every blowing angle of the gas (Fig. 6.19). As in the case of $\gamma = 60°$, the experimental values for each angle are generalised through parameter K by the single dependence for all values of m_{scr}. The dependence of θ' on the blowing angle for two values of K = 24 is shown in Fig. 6.20. It may be seen that the efficiency of the screen is maximum at $\gamma = 45 \div 60°$ and rapidly decreases at $\gamma = 30°$. At a blowing angle of 75÷90°the value of θ is also slightly lower than the maximum value, but as already mentioned in this case the length of the section of the absolute screen is maximum. Thus, in organising the gas screen in plasma torches with the inter-electrode insert it is preferred to blow the gas under the angle $\gamma \geq 45°$. From the viewpoint of designing the sections it is more convenient to blow the gas along the normal to the main flow, especially if the problem of protecting the insulating components against arc radiation is solved.

We also examined other methods of supplying the shielding gas, namely: radial (non-twisted) blowing of the gas through the inter-sectional slit, and the blowing of gas through the porous section of the discharge chamber [44]. In the former case, the shielding gas was supplied into the channel under the angle of $\gamma = 75°$ through a number of radial holes in the twisting ring. In the latter case, the gas was supplied using a section with an insert made of porous molybdenum, connected to the measurement section. The efficiency of the gas screen of the measurement section in the three methods of

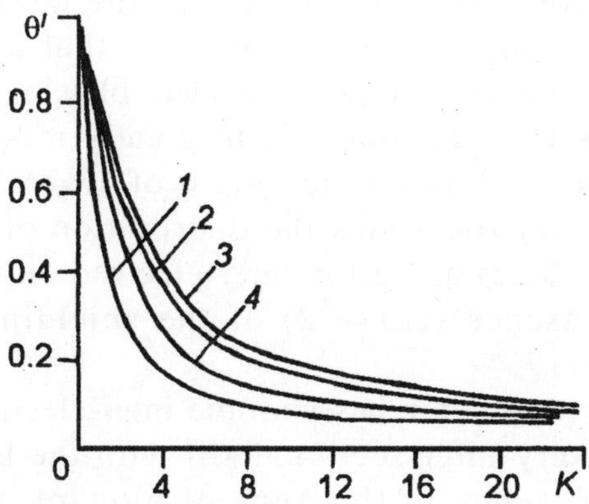

Fig. 6.19. Dependence of θ' on K for different blowing angles. *1)* $\gamma = 30°$; *2)* $\gamma = 45°$, *3)* $\gamma = 60°$, *4)* $\gamma = 75$ and 90°.

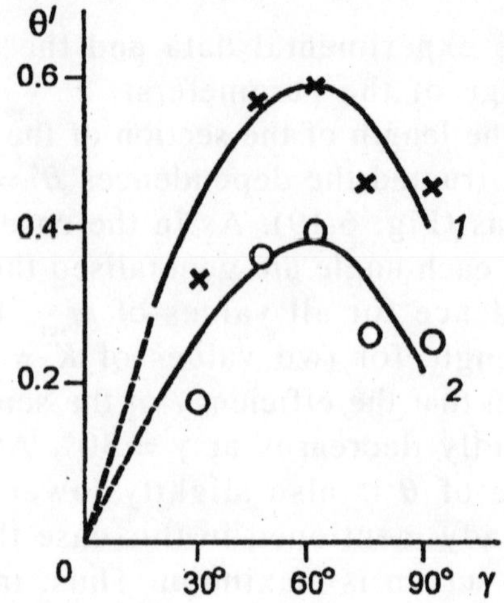

Fig. 6.20. Dependence of θ' on the blowing angle of the shielding gas γ for K = 2.0 (curve *1*) and 4.0 (curve *2*).

supplying the gas will be compared (Fig. 6.21). The data were obtained for approximately the same flow rate of the shielding gas. It may be seen that the efficiency of the gas screen in the case of the twisted main and blown flows (curve *1*) is higher than in the other two cases. The efficiency of the screen behind the porous ring (curve *3*), especially at the start of the measurement section, is higher than in the case of radial blowing through the slit (curve *2*). Evidently, the main role in reducing the efficiency of the screen in radial blowing and the supply of gas through the porous ring is played by better mixing of the hot main (twisted) and cold shielding (non-twisted) gas flows.

Previously, we discussed the single method of blowing the gas through the slit situated in subsection in the developed turbulence section of the gas flow. It has been shown that at the width of the inter-sectional slits $S = 1 \div 5$ mm, the boundary blowing with the parameter m_{scr} close to unity makes it possible to greatly reduce the convective heat flow into the wall along the length of several diameters of the channel. This clearly illustrates the distribution of the local thermal efficiency η_i (Fig. 6.22) along the inter-electrode insert in the absence (curve *1*) and presence (curve *2*) of the shielding gas blowing in the section $\bar{z}_{scr} = 17$.

Usually, in the plasma torches with the inter-electrode insert blowing is organised to every inter-sectional slit with the blowing parameter $m_i \sim 0.1$. The efficiency of this type of blowing for organising the gas screen is relatively low (compared Fig. 6.2*b* and 6.22). It is far more efficient, as shown by the experiments and indicated by Fig.

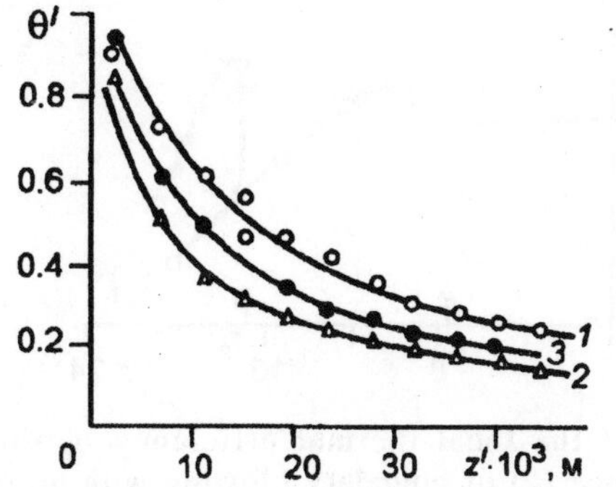

Fig. 6.21. The efficiency of the screen in different methods of supplying the shielding gas. d = 20 mm, $\bar{a}$ = 24, $\bar{z}_s$ = 4, $\bar{z}_{scr}$ = 18; $\bar{z}_p$ = 16.5; $G_0 + g_i = 22 \cdot 10^{-3}$ kg/s; $g_i = 0$; I = 150 A; 1) γ = 75°, $g_{scr} = 3.3 \cdot 10^{-3}$ kg/s; m_s = 0.44 tangential blowing; 2) γ = 75°; $g_{scr} = 3.3 \cdot 10^{-3}$ kg/s; m_{scr} = 0.44 radial blowing; 3) Δl_i = 28 mm; $g_i = 4.3 \cdot 10^{-3}$ kg/s - blowing through a porous band.

6.22, to organise the local blowing of the gas with the parameter $m_{scr} \sim 1$ at a distance of 3–4 length gages from each other. In fact, according to [44], for the same total flow rate of the gas, the thermal efficiency of the plasma torch with uniformly distributed blowing of the gas with q_i = 0.5 g/s was 0.76, and the efficiency of the same plasma torch in blowing of the shielding gas with $m_{scr} = 0.8 \div 1.2$ in four sections of the turbulent section increased to 0.83. Convective heat flows decreased by approximately 35%.

6.3.4. Distribution of current and heat exchange in the output electrode of the plasma torch with an inter-electrode insert

The output electrode as the element of the plasma torch with the inter-electrode insert with the highest thermal stresses provides a significant contribution to the general fraction of the heat losses. Sufficient investigations have been carried out into both the total heat losses, and the distribution of the specific heat flows along the channel of the output electrode for plasma torches with the self-setting arc length and the mean length of the arc fixed by the 'ledge'. The methods of increasing thermal efficiency [1, 24] have been developed, together with methods of increasing the duration of operation and the enthalpy of the heated gas. Especially important investigations have been carried out to examine the thermal characteristics of the plasma torches in aerospace investigations [37, 45, 46]. These studies have been reviewed in [46]. At present, the main area of application of powerful electric arc heaters of the gas is plasma chemistry and, consequently, the problems of increasing the thermal efficiency of

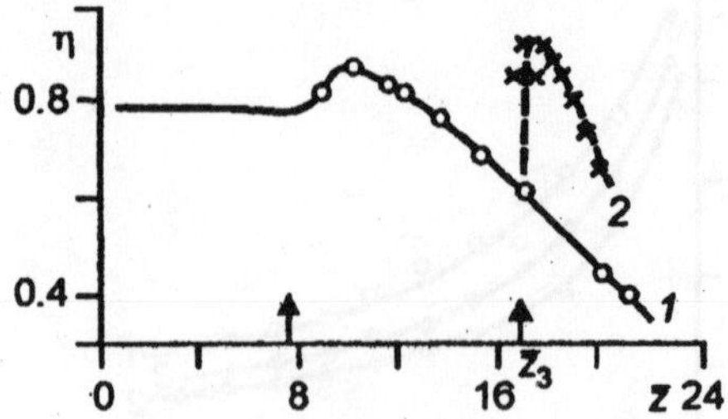

Fig. 6.22. Distribution of the local thermal efficiency η_i along the channel in the absence (*1*) and presence (2) of boundary blowing with m_{scr} = 1.0. d = 20 mm, $\bar{a}$ = 22, $\bar{z}_s$ = 7.5, $\bar{z}_{scr}$ = 17; S = 1.3 mm; $G_0 + g_s = 14\cdot10^{-3}$ kg/s; $g_{scr} = 3.6\cdot10^{-3}$ kg/s; I = 120 A.

the plasma torches, the operating time, and the reproducibility of the results relate to the group of the most important tasks [17, 47, 48]. Since the plasma torch with the inter-electrode insert is most promising both for obtaining high temperatures of the heating gas and for obtaining high unit power (10 MW or higher) [11, 17, 46], the investigations of heat exchange in the output electrode carried out to optimise the characteristics of the electrode are important [31].

The problem is solved by explaining the distribution of time-averaged relative current density along the anode for two characteristic gas flows in front of the electrode – transitional and developed turbulence. Both regimes are of considerable practical interest, especially the latter one, because it corresponds to the arcing conditions resulting in the maximum energy contribution to the arc. Knowing the conditions of the gas flow in front of the anode, we can determine in advance the zones of the working surface of the anode visited most frequently by the arc spot. Although the absolute value of the heat flow, supplied through the arc spot, is not high in comparison with the convective flow, it is usually concentrated in a small section of the anode surface. Consequently, the conventional density of the heat flow, related to the area of the section of the electrode visited by the arc spot, may be large. Therefore, local superheating of the walls is controlling in the evaluation of the efficiency of the electrode. On the other hand, the distribution of current density indicates the necessary minimum length of the electrode.

The experiments were carried out using a plasma torch with a cylindrical sectional anode with sub-sonic flow of the high-temperature gas in the channel (Fig. 6.23). The internal diameters of the section of the inter-electrode insert and of the anode are equal. The anode is assembled from copper water-cooled cylindrical discs with a thickness of $4\cdot10^{-3}$ m isolated from each other by asbestos interlayers; the number of disks in a set was 12. They are electrically closed with the positive

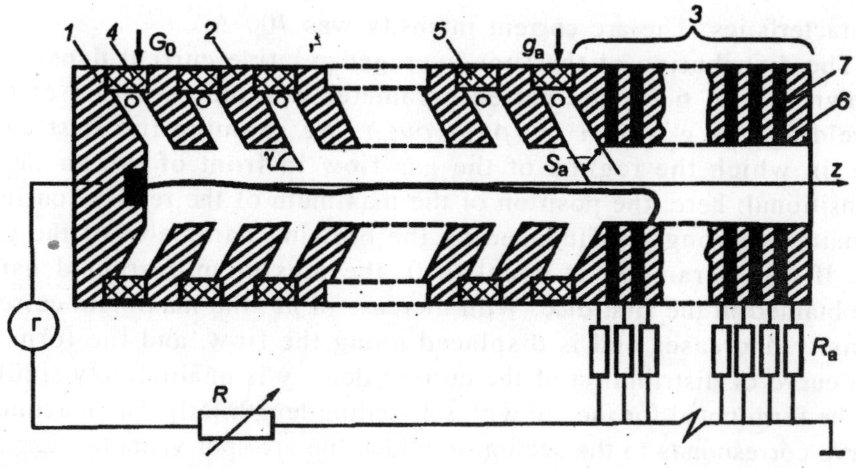

Fig. 6.23. Diagram of a plasma torch and electric power supply. *1*) cathode; *2*) IEI section, *3*) output electrode – anode *4*) first twisting ring; *5*) intersectional twisting ring *6*) disc, 7) insulator; *G* – power source, *R* – ballast resistance R_a – additional resistance.

pole of the electric power source through equal low-ohmic resistances, shunts (R_r = 0.014 Ohm) and a ballast rheostat. The distribution of current along the anode is determined by the measurement of the voltage drop in the shunts. In addition to this, in plasma torches of different geometrical dimensions, measurements were taken of the integral heat losses into the solid anodes with a relative length of 2, 3 and 6 length gages and the internal diameter d = (1.0; 2.0, and 3.0) · 10^{-2} m. The working gas was mainly air.

To investigate the effect of gas blowing in front of the anode on its electrical and thermal characteristics into a slit with the width of $S_a = 2 \cdot 10^{-3}$ m, formed by the surfaces of the last section and the anode, the gas was supplied in the tangential direction with a flow rate $g_a = (0 \div 7) \cdot 10^{-3}$ kg/s ($m_a = 0 \div 1.4$). The angle of blowing the gas into the electric arc chamber between the last section and the anode was $\gamma = 60°$. The individual supply of water to the sections of the inter-electrode insert and the discs of the anode made it possible to carry out calorimetric measurements of the heat losses in them and, consequently, determine the enthalpy of deceleration of the gas in front of the anode and the density of the heat flows along the anode. The difference in the temperatures of the cooling water was measured with a differential transistor thermal sensor [49].

The averaged-out relative current density and the density of heat losses into the discs of the anode were measured at currents up to 200 A. Nitrogen and hydrogen were also used in the determination of the integer of thermal characteristics. The arc current intensity was 700 A.

The distribution of the time-averaged relative current density is shown in Fig. 6.24: the varied parameter was the flow rate of the shielding gas g_a (or $m_a = (\rho u)_a/(\rho u)_0$). We examine the first case (*a*), in which the regime of the gas flow in front of the anode is transitional: here, the position of the maximum of the relative current density is strongly influenced by the blowing parameter of the gas m_a. If this parameter is equal to 0, the maximum current density is obtained in the first disc. With increase of m_a, the maximum current density decreases and is displaced along the flow, and the form of the curve of distribution of the current density is qualitatively similar to the form noted for the arc with self-setting length [50]. The maximum curve corresponds to the section in which the arc spot visits the surface of the anode most frequently. Attention will be given to the simultaneous increase of the mean (in respect of maximum $\bar{i}$) and maximum arc length with increase of g_a. This circumstance requires the use of

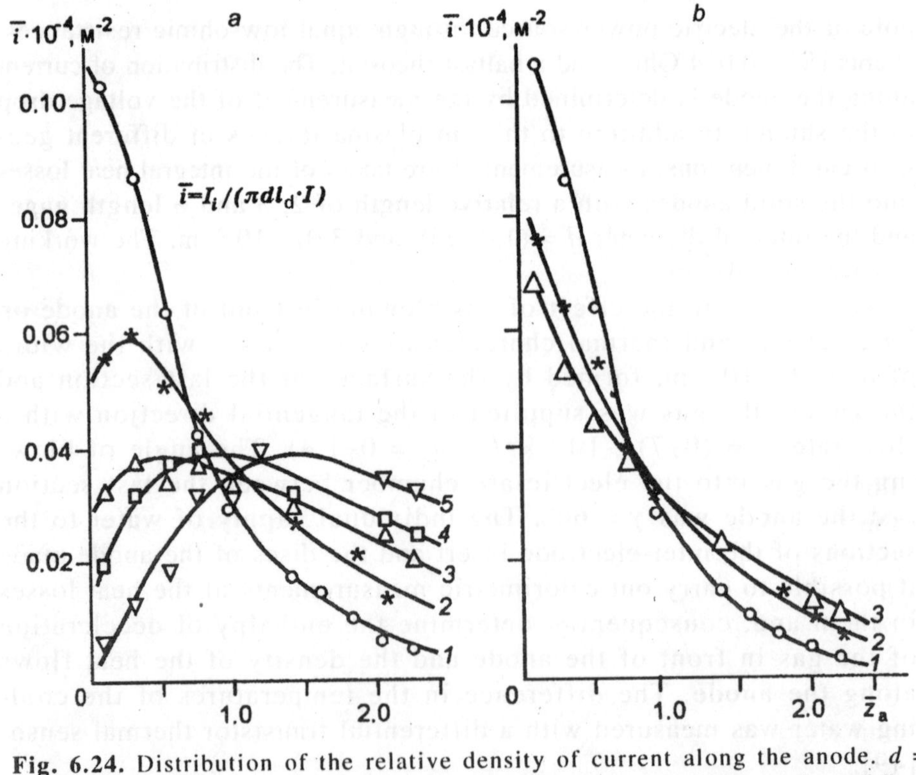

Fig. 6.24. Distribution of the relative density of current along the anode. d = $20\cdot10^{-3}$ m; $\bar{a}$ = 20.5, I = 90 A; G_0 = $10\cdot10^{-3}$ kg/s; g_i = 0. *a*) transition regime of flow in front of the anodes; *1–5*) values of m_a are equal to respectively 0; 0.32; 0.67; 0.98; 1, 2; *b*) developed turbulent flow of the gas in front of the anode; $\bar{z}_s$ = 7.5; m_s = 1.0; *1–3*) the values of m_a are equal to respectively 0; 0.37; 0.90.

long electrodes to ensure that the electrical arc does not move to the end of the anode which, in turn, would increase the heat losses and reduce the thermal efficiency of the system.

A completely different situation is found in the case of the developed turbulent gas flow in front of the anode (Fig. 6.24*b*). Here the blowing of the cold gas within the same limits of its variation as in the first regime has no longer any significant effect on the curve of distribution of the relative current density, especially of the regime corresponding to $m_a = 0$ is not considered. The highest current density is obtained in the first disc, and the main proportion of current is removed from the section of the electrode with the length of 1.5–2.0 gages. For example, for $m_a = 0.9$ ($g_a = 4.7 \cdot 10^{-3}$ kg/s) more than 90% of the arc current is taken from the surface of the anode with the length smaller than 2 gages, so that short output electrodes can be used and, consequently, the heat losses may be minimised. In order to prevent the movement of the arc outside the limits of the channel, it is desirable to install a solenoid with a magnetic circuit at the end of the electrode. It is also important to note the increase of the shunting frequency of the arc in the output electrodes by 1–2 orders of magnitude in comparison with the shunting frequency of the arc with the self-setting length, with other conditions being equal. This reduces the specific erosion of the electrode (reduces the time of arrest of the arc spot), and ensures the uniform distribution of erosion on the surface of the electrode thus increasing the service life of the anode.

In the investigated ranges of the parameters, the distribution of the relative current density along the anode does not depend on the value of total current in the investigated range of variation at a constant gas flow rate g_a (Fig. 6.25). This may be used as a basis for the evaluation approximation of the results to higher currents.

We know examine the distribution of the surface density of the heat flow along the channel, determined by the convective heat transfer mechanism, the heat flow through the anode spot and arc radiation. For the transitional regime of the gas flow it is shown in Fig. 6.26*a*. At $g_a = 0$ (curve *1*), the surface density of the heat flow, especially at the entry section of the anode, is considerably higher than the level of the losses determined by turbulent heat exchange q_t. This is associated with the displacement of the hot gas into the slit S_a, the inflow of heat through the arc spot, and by other factors. At $g_a \gg 0$, the density of the heat flow at entry into the electrode is lower in comparison with turbulent heat exchange; the effect of the gas screen is evident. However, this does not indicate any decrease of the total heat flow into the anode because

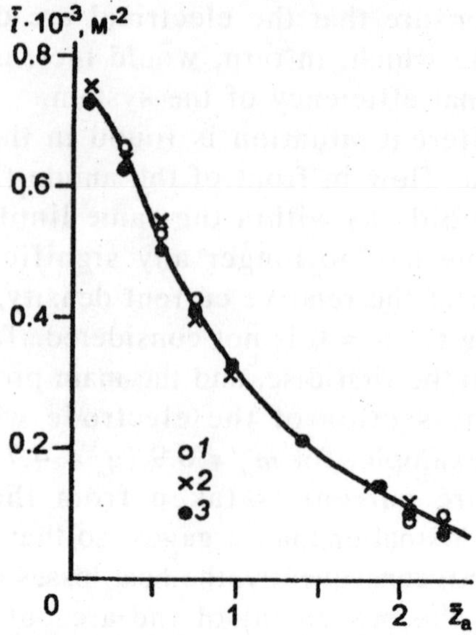

Fig. 6.25. Distribution of the relative flow density along the anode for developed turbulent gas flow $d = 20\cdot10^{-3}$ m; $\bar{a} = 20.5$; *1)* $\bar{z}_s = 7.5$; $G_0 = 10\cdot10^{-3}$ kg/s; $g_s = 3.6\cdot10^{-3}$ kg/s; $m_a = 0.37$; *1–3)* $I = 60$; 90; 120 A, respectively.

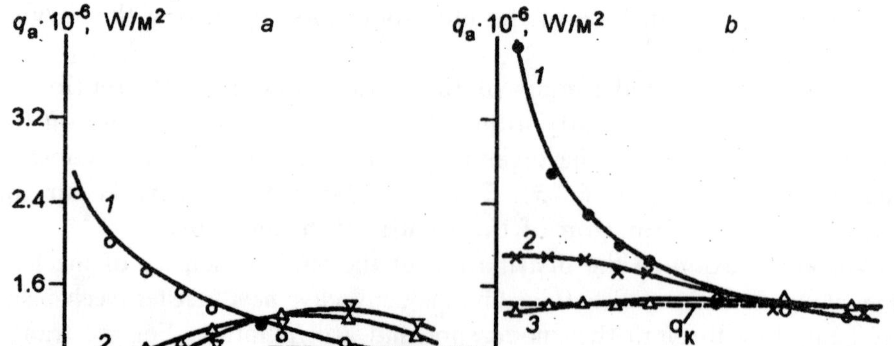

Fig. 6.26. Distribution of heat losses along the anode. $d = 20\cdot10^{-3}$ m; $\overline{a} = 20.5$, $I = 90$ A; $G_0=10\cdot10^{-3}$ kg/s; *a)* transition gas flow regime in front of the anode; *1 – 3)* the values of m_a are equal to respectively 0; 0.87; 1.2; *b)* developed turbulent flow of the gas in front of the anode $\bar{z}_s = 7.5$; $m_s = 1.0$; *1–3)* the values of m_a are equal to respectively 0; 0.36; 0.90. Broken line – calculated convective heat flow.

the required length of the electrode increases as a result of the displacement of the maximum attachment of the arc spot downwards along the flow (Fig. 6.24). Further, along axis z_a the value of q_a increases, reaches its maximum value and subsequently tends to the level of the heat flow, determined by turbulent heat exchange. The coordinates of the maximum heat losses and the current density approximately correspond to each other.

If the flow of the gas in front of entry into the anode is developed and turbulent, the surface density of the heat flow along the anode at all values of g_a is higher in comparison with turbulent heat exchange, or close to it (Fig. 6.26*b*). Only in the case of very strong blowing of the gas in front of the anode (curve *3*) the heat flow into the first disc is slightly lower than into the subsequent ones. Consequently, the gas screen (in the range of variation of m_a) has no significant effect on the decrease of the heat losses into the anode, especially if low flow rates are disregarded.

Thus, the blowing of the gas through the slit in front of the anode is essential only for increasing the electrical strength of the gap between the output electrode and the adjacent section and for preventing the displacement of the high-temperature gas into the slit. To ensure this, it is sufficient to obtain m_a~0.3. Stronger gas blowing is not rational.

As already mentioned, the heat flow into the anode is determined by turbulent heat exchange, arc radiation and the heat flow through the arc spot: $Q_a = Q_c + Q_r + Q_s$. The above distribution of current shows that in the case of the developed turbulent flow of the gas in the anode zone, the radiant heat flows from the arc must be taken into account only in the first length gage. According to estimates, in the investigated range of the parameters they do not exceed 1.5–2.0% of the total heat flow into the electrode, i.e., they are insignificant. The heat flow through the anode spot of the argon arc is determined from the equation [1, 51]:

$$Q_s \approx 6 \cdot I, \text{W}. \tag{6.17}$$

The validity of equation (6.17) for an air arc was specially verified in the current range (50÷200 A). The anode was made of copper. Taking into account the distribution of current along the anode (Fig. 6.24), the fraction of the heat losses through the anode spot, calculated using equation (6.17), may reach more than 20%. Consequently, they must be taken into account when evaluating the heat losses into the anode.

The results of calculation of the surface densities of heat losses into the anode are presented in Fig. 6.27. The surface density of the convective heat flow into the cylindrical output electrode was

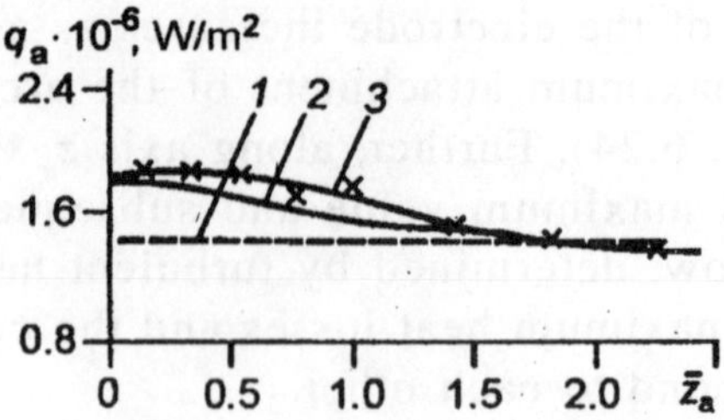

Fig. 6.27. Comparison of the results of calculating the heat losses along the anode with the experimental data. $d = 20 \cdot 10^{-3}$ m; $\bar{a} = 20.5$, $I = 90$ A; $G_0 = 10 \cdot 10^{-3}$ kg/s; $g_a = 2 \cdot 10^{-3}$ kg/s; $\bar{z}_s = 7.5$; $g_s = 3.6 \cdot 10^{-3}$ kg/s; *1* – calculated density of the convective heat flow; *2* – calculation taking the heat flow through the arc spot into account; *3* – experimental data.

calculated using equation (6.11) assuming that the supply of energy into the gas along the anode on the side of the arc is equal to zero, and the temperature and flow rate of the gas remained constant. Using the data on the distribution of current along the anode, we can determine the current I_i through the individual discs of the electrode and use the equation $q_s = 6I_i/(\pi d \Delta l)$ to determine the density of the heat flow from the anode spot of the arc which is added to the surface density of the convective heat flow. The distribution of the calculated surface density of the heat flow along the anode is illustrated by curve *2*. Comparison of the curves *2* and *3* shows that in the case of moderate blowing of the gas in front of the anode, the results of the calculations are in satisfactory qualitative and quantitative agreement with the experimentally measured heat losses into the wall of the electrode.

Since, as mentioned previously, the gas screen, organised by the supply of working gas into the slit in front of the anode, is not very efficient, the total heat losses in the output electrode may be reduced only by minimising the length of the electrode.

In order to determine the total heat flows into the cylindrical anodes of different relative length in the case of the developed turbulent flow in front of the anode, investigations were carried out using air, nitrogen and hydrogen arcs at currents of up to 700 A. The mean mass temperature of the gas in front of entry into the anode channel was $(3.0 \div 6.5) \cdot 10^3$ K. According to the experiments, the heat losses into the body of the output electrode with the length of up to 6 gages are directly proportional to the internal surface area of the electrode. In the case of large relative length of the anode, it is necessary to take into account the decrease of the mean mass temperature along the channel. Comparison of the experiments with the calculations using the equations for the convective heat exchange of the turbulent flow of the gas with the wall of the cylindrical channel was presented previously in Fig. 6.10.

Thus, for the developed turbulent gas flow in the output electrode-anode, we can note the following:

1. The main part of current is removed from the arc in the first gage;
2. The gas screen of the anode does not reduce the heat losses into the anode and is essential only for preventing the electrical breakdown in the anode–section gap, which is already obtained at $m_a \approx 0.30$;
3. In order to reduce the integral heat losses and increase the thermal efficiency of the plasma torch, the length of the anode should be restricted to two gages;
4. To prevent, in the latter case, the movement of the arc to the end of the short anode, it is desirable to install a solenoid with a magnetic circuit;
5. The surface density of the heat flow into the electrode (disregarding the flux through the anode spot) may be calculated with sufficient accuracy using the well-known equation for the heat exchange of the developed turbulent gas flow with the pipe wall.

6.3.5. Thermal efficiency of the plasma torch with the inter-electrode insert

In the above section, we presented the integral thermal characteristics of the plasma torches with self-setting and fixed (with a ledge) arc length and the empirical relationships, which make it possible to calculate their thermal efficiency. Data have also been presented on the heat losses in the plasma torch with the inter-electrode insert in different sections of the electric discharge chamber. It has been shown possible to calculate the heat losses. Less attention has been paid only to heat flows into the end internal electrode–cathode, because they are small and have almost no effect on thermal efficiency. The thermal processes in the cathode will be discussed in the following chapters.

For the plasma torches with the sectional inter-electrode insert, we can obtain the single analytical dependence of efficiency of the determining parameters, as indicated by the previously presented data. This is due to the fact that when determining the dependence $\eta = \eta\ (I^2/Gd, \ldots)$ in addition to the criteria already used in equation (6.3) it is necessary to add criteria taking into account boundary cooling, and the possibility of variation of the length of the section of the developed turbulent flow at $\bar{a}$ = const, etc. Therefore, the efficiency of the plasma torch with the inter-electrode insert must be calculated element by element. All the necessary data for this are already available.

However, to describe the advantages of the plasma torch with the inter-electrode insert in comparison with other linear plasma torches we return to Fig. 6.4 which gives the data on the efficiency of plasma torches with a smooth electrode (curves *1–3*) and two curves (*4* and *5*), characterising the dependence $\eta = \eta(h)$ for the plasma torches with the inter-electrode insert. In the case of low gas enthalpy, the coefficient of efficiency of the plasma torches of the two-chamber type and with the inter-electrode insert are relatively high and approximately equal to each other, i.e. there are no special advantages in the case of the plasma torch with the inter-electrode insert. The advantage of these torches becomes more evident with increase of the required enthalpy. At $h = 25 \cdot 10^3$ kJ/kg, the extent by which the η value of the plasma torch with the inter-electrode insert is higher than the same value of the two-chamber plasma torches more than 50% ($\bar{l} = 5$, curve *3*). At $\bar{l} > 5$, this advantage is even greater.

Attention will also be given to the increase of η of the plasma torches with the inter-electrode insert with the increase of the length of the insert $\bar{a}$, i.e. the length of the turbulence section of the channel Δz_t (curves *4* and *5* in Fig. *6.4*). In the specific case for the selected values of $\bar{a}$, g_i, $\bar{l}$, d, the efficiency coefficient increases by more than 20%. The reason for this is quite clear if we analyze the thermal efficiency of the plasma torch with the inter-electrode insert:

$$\eta = 1 - (\Sigma Qr_i + \Sigma Q_s + \Sigma Q_T)/[I(l_i E_i + l_s E_s + l_T E_T)]. \qquad (6.18)$$

For the simplest case, i.e. the constant heat losses in all sections of the channel equal to, for example, radiant losses $\bar{Q}_r$, and constant E_T – with the increase of the length of the turbulence section Δz_t the efficiency increases and tends to $\eta = 1 - [\bar{Q}_r/(I \cdot E_T)]$. The value $\bar{Q}_r/(I \cdot E_T) \ll 1$ and, consequently, in the case of small radiant heat losses the efficiency may be quite close to unity. In the real conditions, the distribution $\bar{Q}(z)$ is more complicated. The increase of η or, at least, its constancy with the increase of enthalpy is also ensured by the counter blowing of the gas, organised in the initial section of the channel, because this increases the length of the turbulence section of the channel Δz_t. For example, at $d = 20$ 10^{-2} m; $\bar{a} = 21.5$; $G = 30 \cdot 10^{-3}$ kg/s; $g_i = 0.5 \cdot 10^{-3}$ kg/s; I = 90 A, the enthalpy of the gas (air is the working gas) at outlet from the plasma torch is equal to approximately $3.1 \cdot 10^3$ kJ/kg, and η =

0.8. Intensive counter blowing of the gas (m_s = 1.0) at the distance $\bar{z}_s$= 7.5 increases the enthalpy by only 50% (to 4.6 · 10^3 kJ/kg) at almost constant efficiency (η = 0.79).

As mentioned previously, the thermal efficiency of the plasma torch with the inter-electrode insert is also greatly increased by the increase of the inter-sectional gas flow rate g_i in the turbulence section of the channel (Fig. 6.2) or by its redistribution, optimising the gas screen (Fig. 6.22).

Thus, the thermal efficiency of the plasma total with the inter-electrode insert is relatively high, and the designer is capable of varying η depending on the requirements of the technological process, making, if necessary, the design of the plasma torch more complicated for increasing efficiency.

6.4. ELECTRIC ARC GENERATOR OF LOW TEMPERATURE PLASMA WITH A GAS VORTEX INTER-ELECTRODE INSERT

Regardless of certain advances in the area of thermal protection of the walls of the sections of the inter-electrode insert, search is continuing for new methods of reducing the heat losses into the electric arc chamber, primarily the heat losses determined by convective heat exchange. This is associated with the fact that when using multi-slit gas greens, the efficiency of thermal protection as a result of high-intensity turbulent mixing at the interface between the cold and hot gases rapidly decreases in the direction downwards along the flow from the area of blowing the cooling gas, and the rate of decrease increases with the gas temperature [38]. In addition to this, the distribution of a large number of sections with the distributed supply of the cooling gas greatly complicates the design of the plasma torch.

One of the methods of reducing the convective heat losses is the suppression of turbulent pulsations at the interface between the hot and cold jet, for example, by enclosing the hot jet with the arc on its axis in a cold gas vortex, restricting the wall, with a positive density gradient along the radius. The solution was realised in a plasma torch with a gas-vortex inter-electrode insert [52, 53]. The vortex flow is restricted in a cylindrical pipe (*5*) (Fig. 6.28) whose diameter D is considerably greater than the internal diameter d_1 of the starting electrode (*2*). The peripheral orifices of slits with a rectangular section (*4*) are designed for the introduction of the cold gas with the flow rate G_3 oriented in such a manner that the jet enters the

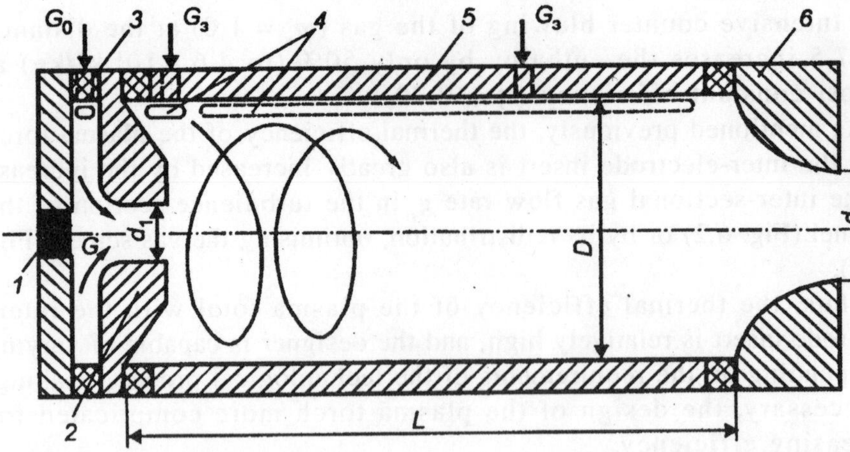

Fig. 6.28. Diagram of a plasma torch with a gas-vortex inter-electrode insert. *1*) end electrode; *2*) starting electrode; *3,4*) supply of gas *5*) casing of the gas vortex IEI; *6*) output electrode; G_0 and G_3 are respectively the flow rates of the gas through the additional and main vortex chamber.

chamber along the tangent to its internal surface. The twisted jet of the cold gas G_0 is introduced into the same chamber, and the direction of twisting the jet of the hot gas should be in the same direction or in the opposite direction in relation to the cold gas.

We examine qualitatively the aerodynamics of the gas flow in such a chamber. As an example, Fig. 6.29 gives the photographs of the central job, discharged into the vortex chamber at different values of d_0 and different forms of the contour of the output nozzle. In the simplest case (Fig. 6.29*a*) the contour of the nozzle is formed by a circle. In a more complicated variant (Fig. 6.29*b*) the nozzle is formed by two ledges, and $d_0'' < d_0' < D$ (d_0' and d_0'' – other diameters of successive sections of the output electrode with a variable cross-section). The cylindrical pipe of the vortex chamber is produced from silicate glass and is 0.35 m long. To visualise the jet, a small amount of tobacco fumes was added into the chamber. In this visualisation only the external boundaries of the jet can be easily seen.

The first gas blowing regime (*a*) is characterised by high stability in the space of the central jet which moves through the entire chamber with a distinctive boundary between the internal and external flow; the form of the jet is close to cylindrical. In certain conditions, even if the initial level of turbulence of the secondary flow is higher, the turbulence at the interface of two isothermal jets is greatly suppressed and this is qualitatively expressed in a small displacement of these jets. This small displacement of the gases is the result of the ef-

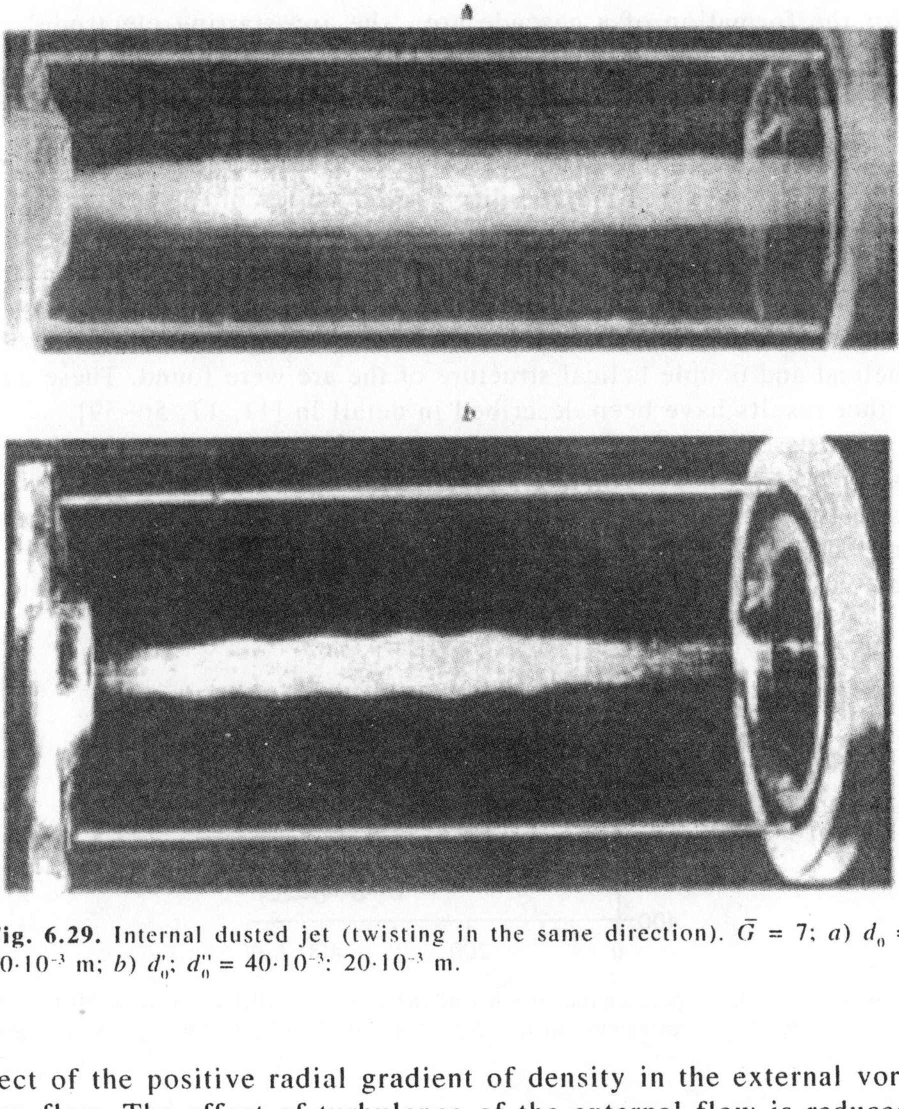

Fig. 6.29. Internal dusted jet (twisting in the same direction). $\bar{G}$ = 7; *a*) d_0 = $40{\cdot}10^{-3}$ m; *b*) d_0'; d_0'' = $40{\cdot}10^{-3}$: $20{\cdot}10^{-3}$ m.

fect of the positive radial gradient of density in the external vortex flow. The effect of turbulence of the external flow is reduced to increasing the extent of its penetration into the central jet [54, 55].

The second flow regime (*b*), i.e., the 'helical' regime, was also reported in [55]. A third regime is also possible (for example, at d_0 = $20{\cdot}10^{-3}$ m), characterised by the large expansion of the central jet in the radial direction directly at exit from the starting electrode with subsequent transition to a cylindrical jet with a smaller diameter with relatively distinctive boundaries. This form of the central jet already shows a certain 'risk' if the arc burns on the axis (or the high-temperature jet is there). On one hand, there are suitable conditions for the formation of a cascade arc: 'the arc–starting electrode' or even the 'arc–vortex chamber', on the other hand, the intensity of the heat flows into the electrode (*2*) in the vortex chamber (*5*) (Fig. 6.28) increases.

Detailed investigations were carried out to examine the structure of the flow of the gas in the discharge chamber of the plasma torch with the gas-vortex inter-electrode insert, investigations of the stability of gas vortices, and of the interaction of the wall vortex with the main gas flow. The structure of the electric arc in the vortex flow in a larger diameter chamber was investigated. The regimes with the stable helical and double helical structure of the arc were found. These and other results have been described in detail in [11, 17, 56–59].

We describe several characteristics of the arc, running in the plasma torch of the investigated system. Figure 6.30 shows the VAC of a direct current air arc for different values of the length of the electric arc chamber L. In the examined current range, the characteristics are drooping. The decrease of $\bar{G} = G_{scr}/G_0$ from 2 to 0.5 results

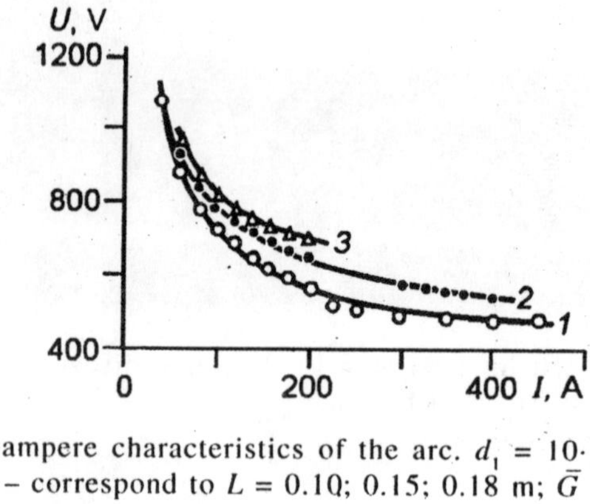

Fig. 6.30. Volt–ampere characteristics of the arc. $d_1 = 10{\cdot}10^{-3}$ m; $d_0 = 20{\cdot}10^{-3}$ m; $D = 0.15$ m; *1–3* – correspond to $L = 0.10$; 0.15; 0.18 m; $\bar{G} = 2.0$; $G_0 = 5{\cdot}10^{-3}$ kg/s.

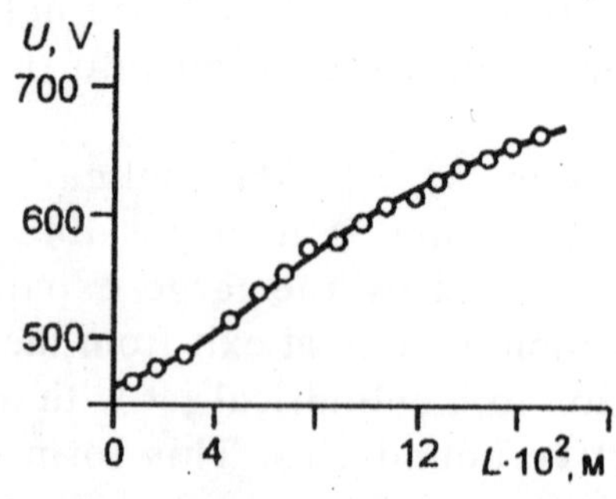

Fig. 6.31. Dependence of arc voltage on the length of the gas vortex IEI. $I = 300$ A; $\bar{G} = 3$; $G_{scr} = 15{\cdot}10^{-3}$ kg/s; $d_1 = 10{\cdot}10^{-3}$ m; $d_0 = 20{\cdot}10^{-3}$ m.

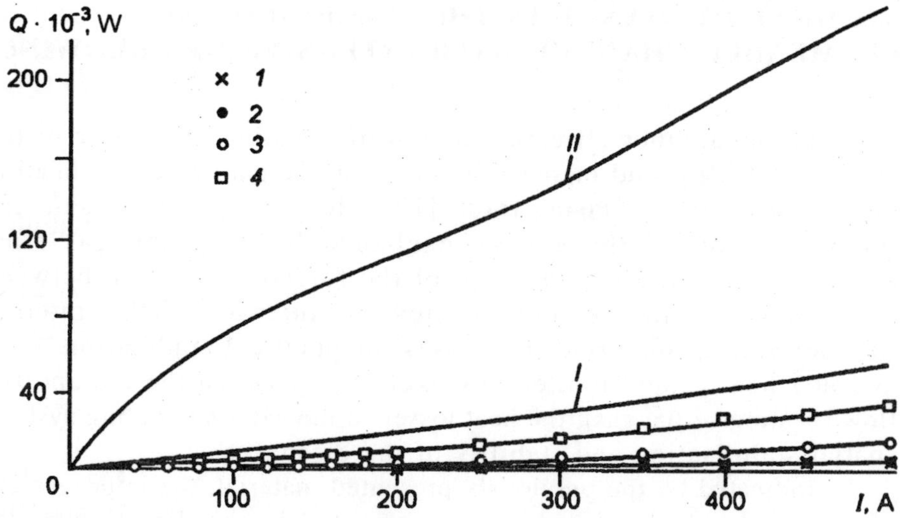

Fig. 6.32. Heat losses into the elements of the electric arc chamber $G_0 = 5 \cdot 10^{-3}$ kg/s; $\bar{G} = 2.0$ kg/s; $L = 0.1$ m; *1*) end electrode; *2*) body of the gas vortex IEI; *3*) starting electrode; *4*) output electrode; I) total losses; II) the power generated in the arc.

in a small change of the level of voltage. The results confirm indirectly the high stability of the central jet and its mixing with the external gas flow. The VACs of the AC arc are almost identical with the characteristics of the DC arc when the breaks in current are removed and continuous arcing is ensured using a high-frequency discharge in the current range (40÷200 A). The dependence pf arc voltage on the length of the gas vortex insert is indicated by the curve shown in Fig. 6.31; it is identical with the curve plotted in [60].

The intensity of the heat flow into the chamber correlates fully with the aerodynamic pattern of the gas flow, obtained on the basis of the data of 'cold' blowing. Calorimetric measurements showed that the heat losses through the wall of the vortex chamber are on the level of the radiant losses (Fig. 6.32). Consequently, the convective heat exchange between the central jet and the wall is almost non-existent, so that it is possible to develop a high efficiency plasma torch.

On the basis of the experimental results, this type of plasma torch may be regarded as highly promising for practical applications.

6.5. HEAT EXCHANGE IN THE COMBINED AND PERMEABLE CHANNEL WITH INTENSIVE GAS BLOWING

We examine another efficient method of boosting the power of the arc, gas enthalpy, and increasing the economic parameters of heating the gas in a linear plasma torch [17]. Here, we discuss again the intensification of the energy exchange between the gas flow and the electrical arc as a result of the forced interaction between the two. The method of organisation and intensity of this interaction determines important processes for practical application, such as Joule heat generation, energy exchange between the arc and the flow (heating of the gas), the heat losses to the external cooling system, spatial stabilisation and stability of the discharge.

As indicated by the previously presented material, the improvement of linear plasma torches has been oriented in the direction of intensification of the hydrodynamic and thermal interaction of the gas flow with the discharge. This resulted in the development of high-efficiency plasma torches with a sectional inter-electrode insert and the supply of the plasma forming gas into the gaps between the sections. The improvement of the power of the plasma torches and of the economic parameters, and also of the operating life of the electrodes may be achieved most efficiently by increasing the volt–ampere ratio of the discharge, the density of energy generation and by reducing of the heat losses which, in turn, requires solving the problems of protection of the wall of the electric discharge chamber against powerful heat flows, the realisation of the high values of the strength of the electrical field, and ensuring the electrical strength of the inter-electrode insert.

The satisfaction of these requirements in the plasma torch with the sectional water cooled channel without film shielding is associated with a number of restrictions. Firstly, when cooling the channel walls with a liquid, the power of the discharge is limited by the limiting heat flow supply through the wall to the cooling agent without overheating and fracturing the wall. Secondly, intensive external cooling in the presence of high-temperature gradients always results in high heat losses and a decrease of thermal efficiency. Even when using blowing between the sections, it is not possible to avoid completely the losses because of the transfer of heat by radiation and convective losses in the turbulent regime of arcing which is most advantageous from the viewpoint of the energy parameters, since the efficiency of film shielding rapidly decreases in the direction downwards along the flow. Thirdly, to prevent inter-sectional breakdown and the 'arc–section wall' breakdown, with increase of the strength of the electrical field it is necessary to reduce the length of the sections. This increases the number of the sections and greatly complicates the design and reduces the reliability of operation of the system.

To realise the effective hydrodynamic effect on the set of the parameters of the arc discharge, it is promising to use the plasma torch with the supply of plasma forming gas through the porous wall of the discharge channel [61]. As already mentioned in chapter 5, this plasma torch is a further development of the plasma torches with blowing between the sections in the sense that the application of the porous wall results in a transition from the discrete introduction of the gas between the individual sections to the limiting case of continuous blowing through the entire surface of the channel of the inter-electrode insert.

From the viewpoint of hydrodynamics, the porous wall plays in this case the role of a design element used for organising the supply of the plasma forming gas into the discharge zone, and the already naturally intensive radial flow of the gas isolates the discharge from the internal surface of the channel wall. From the viewpoint of the heat balance, the most porous wall plays the active positive role because the porous material is characterised by the regeneration of radiant heat losses and by the return of these losses to the main flow together with the blowing gas.

Whether reaching a specific intensity of blowing in the regime of displacement of the boundary wall, the thermal and dynamic interaction of the main gas flow with the wall of the channel is prevented and the heat losses resulting from the heat conductivity and convection decrease to zero, i.e. the thermal efficiency in the section of the inter-electrode insert on the condition of regeneration of radiant losses may be equal to almost 100%.

According to the experiments [62], the effect of radial blowing is manifested not only in the transpiration cooling of the channel wall but also in the active effect of blowing on the electrical parameters of the discharge and energy exchange in the gas flow. For example, strong blowing of the gas increases the strength of the electrical field of the arc because of a number of reasons, including the effect of constriction of the current-conducting channel in intensive heat exchange with the surrounding medium. The interaction between the transverse and longitudinal gas flows results in the realisation of the turbulent flow regime which, in turn, intensifies the energy exchange between the arc and the surrounding gas and also increases the strength of the electrical field.

Thus, the application of strong blowing of the gas through the porous wall of the channel makes it possible to solve simultaneously two tasks – intensification of energy generation in the arc as a result of the increase of the strength of the electrical field and stopping the heat losses into the walls of the discharge channel in the section of the inter-electrode insert as a result of regeneration of heat in the porous wall.

In certain conditions, high-intensity blowing results in the spatial stabilisation of the arc column, reduces the possibility of radial fluctuations of the column, and the flow of the blown cold gas increases the electrical strength of the arc–wall gap thus increasing the length and reducing the number of the sections of the inter-electrode insert for the channel made of electrically conducting materials. It should be mentioned that the use of porous channels made of non-conducting materials (ceramics) with transpiration cooling greatly increase arc voltage without sectioning the channel [63]. The application of porous materials with variable permeability along the length of the sections of the inter-electrode insert permits profiling the intensity of blowing along the length of the channel in accordance with a previously specified law. Porous blowing may be combined with the tangential twisting of the gas or the axial flow on the side of the end electrode.

The above special features indicate the extensive possibilities of the investigated system for aerodynamic control of the discharge parameters and realisation of the strong effect of the plasma-forming gas on the arc. It is promising for the development of high-efficiency small heaters with a high density of energy generation in the unit volume and high efficiency of conversion of electrical energy to the thermal energy of the gas flow.

There is a relatively large number of studies concerned with the examination of plasma torches with porous cooling of the walls. They include, for example, the study [64], mentioned in chapter 5, which describes not only the electrical but also thermal characteristics of a plasma torch with a self-setting arc length in a cylindrical permeable electrode. The following dependence was proposed for the thermal efficiency of the plasma torch:

$$(1-\eta)/\eta = 2.9 \cdot \mathrm{Re}^{-0.5} \cdot \mathrm{Kn}^{-0.15} \cdot \tilde{j}_w^{-0.25}, \tag{6.19}$$

This dependence was verified in the range of variation of the parameters Re = $(0.35 \div 11.0) \cdot 10^3$, Kn = $(0.14 \div 14.50) \cdot 10^{-4}$, $\tilde{j}_w = (\rho \upsilon)_w / (\rho \upsilon)_0$ = $0.014 \div 0.125$, $d = 0.4 \div 1.6$ cm. The deviation of the experimental points from the calculated curve does not exceed ±15% in this case.

On the basis of a large number of experimental data, the authors of [65] analysed porous cooling with special réference to electric gas heaters. Heat exchange in the electric arc chamber of the plasma torch has a number of special features the main of which are the presence of the electrical arc, interacting with the turbulent gas flow, and significant non-isothermal nature. Unfortunately, the majority of the experimental investigations of the plasma torches with porous cooling of the walls have been concerned mainly with the technical problems arising in electric arc heaters of this type. The processes of heat exchange in the porous wall, the efficiency of boundary cooling, and also the structure of the arc and its interaction with the gas flow in a permeable channel, have been studied less extensively. One can mention previously cited studies [17, 61–63] and some other studies. The problems of development of the electric arc gas heaters with a porous wall have been examined in sufficient detail in the monograph [66]. This monograph also gives the results of experimental investigations of these plasma torches and solution of problems of thermal shielding of the channel walls against the effect of the high-temperature gas flow. In this chapter, we present materials for the thermal shielding of the discharge chamber of the plasma torch not included in square back [66], and also the results of experimental examination of turbulent heat exchange in the permeable section of the electric arc chamber of the plasma torch with the inter-electrode insert.

Investigations were carried out for a channel diameter of $d = 20 \cdot 10^{-3}$ m, the length of the sectioned inter-electrode insert was almost 20 length gages. The porous section, installed at different sections of the electric arc chamber, had the form of a porous sleeve with the internal diameter $d = 20 \cdot 10^{-3}$ m, wall thickness $3 \cdot 10^{-3}$ m, and ranks of $28 \cdot 10^{-3}$ m, secured between the copper water cooled discs. The discs are specially profiled to make contact with the adjacent sections of the inter-electrode insert. To reduce the extent of leakage of heat through the ends of the porous sleeve, it is necessary to use three-layer interlayers consisting of two asbestos rings a thickness of $0.3 \cdot 10^{-3}$ m each, separated by a mica ring with a thickness of $0.1 \cdot 10^{-3}$ m. Molybdenum sleeves with a porosity of 60% were used in the experiments.

Air was used as the plasma-forming gas. The gas, cooling the porous insert, was air at low temperatures, at high temperatures it was nitrogen. Experiments are also carried out with blowing a foreign gas (helium in air). The flow rate of the cooling gas g_p was varied from $0.35 \cdot 10^{-3}$ to $7.20 \cdot 10^{-3}$ kg/s, which corresponded to the change of the specific flow rate of the cooling agent $\bar{g}_p$ from 0.2 to 4.2 kg/(m^2 s). No air was supplied between the sections of the inter-electrode insert.

In the examination of turbulent heat exchange the length of the initial section in the section $\bar{z}_s = 2.5$ was reduced by high-intensity counter blowing of the gas with the blowing parameter m_s close to unity. The mean mass temperature of the gas in the channel of the plasma torch was determined on the basis of the thermal balance and was equal to $(3.5\div5.0)\cdot10^3$ K in the investigated section. The Reynolds number of the main flow Re_d, calculated from the mean mass parameters, varied in the range $(3.9\div10.4)\cdot10^3$. The intensity of arc current also varied in the range $(100\div180)$ A.

The temperature of the outer surface of the porous sleeve was measured with chromel–alumel thermocouples with a diameter of $0.1\cdot10^{-3}$ m. The junction of the thermocouples was pressed to the surface using a special device. Preliminary experiments showed that the temperature of the outer surface of the wall is almost constant everywhere and, consequently, in subsequent investigations only the values obtained from one thermocouple, situated in the mean section of the porous sleeve, were considered. The recording device was a pyrometric voltmetre of the MPP-254 type, accuracy grade 1.0.

Regardless of the special measures, it has not been possible to eliminate completely the loss of heat through the ends of the porous sleeve. The heat flow into the side water-cooled discs of the porous section increased with a decrease of the flow rate of the cooling gas because of the increase of the temperature of the sleeve.

The heat flow into the porous sleeve was determined using the equation:

$$Q = Q_g + Q_{re} + Q_1. \tag{6.20}$$

Here here Q_g is the heat flow removed by the gas from the unit length of the porous wall; Q_1 is the loss of heat through the ends of the porous sleeve determined by calorimetric measurements of the heat flows into water cooled discs of the porous section, related to the unit length; Q_{re} is the gradient heat flow, emitted from the unit length of the external surface of the porous sleeve, equal to

$$Q_{re} = 2\pi r_2 \varepsilon \sigma T_2^4. \tag{6.21}$$

In equation (6.21) σ is the Stefan–Boltzmann constant, ε is the density of the porous material which according to the data published in [61] for molybdenum with a porosity of 60% is equal to approximately 0.6.

To determine the heat flow Q_g removed by the gas, it is necessary to know the temperature of the cooling agent at exit from the pores

T_{g1}. However, the direct measurement of this quantity during arcing is associated with considerable difficulties. T_{g1} may be calculated from the energy balance equation which, assuming the low-intensity heat flow in the axial direction, has the following form [61, 67, 68]:

$$(1/r)d[\lambda_w r(dT_w/dr)]/dr+[\tilde{g}_p c_p/(2\pi r)](dT_g/dr)=0, \qquad (6.22)$$

where $\tilde{g}_p$ is the flow rate of the cooling gas, related to the unit length of the wall. The heat conductivity of the wall material λ_w may be assumed to be almost completely constant in the given temperature gradient range [61]. Equation (6.22) includes two independent variables: the wall temperature T_w and the gas temperature T_g. The second equation linking T_w and T_g is the equation of local heat exchange between the wall and the cooling gas:

$$(1/r)\lambda_w d[r(dT_w/dr)]/dr=\alpha_v(T_w-T_g), \qquad (6.23)$$

where α_v is the volume coefficient of heat exchange.

In [67] it was assumed that the quantity α_v is so high that the gas temperature reaches the value T_w almost immediately after entry into the porous material. The validity of this assumption for the variation range of the parameters, examined in the present work, is confirmed by the analysis of the results of calculation of the temperature fields of the wall and the cooling agent, presented in [68]. At $T_w=T_g=T$, the solution of equation (6.22) is greatly simplified and for the boundary conditions $r = r_2$: $T = T_2$, $-2\pi\, r_2\, \lambda_w\, (dT/dr) = g_p c_p(T_2-T_\infty)+Q_{re}$ has the form [67]:

$$[T-T_\infty+Q_{re}/(\tilde{g}_p c_p)]/[T_2-T_\infty+Q_{re}/(\tilde{g}_p c_p)]=(r_2/r)^{\tilde{g}_p \cdot n} \qquad (6.24)$$

for $r_1 < r < r_2$. Here $n = c_p/(2\pi\lambda_w)$; T_∞ is the initial temperature of the cooling agent. Equation (6.24) makes it possible to determine the temperature of the internal surface of the porous wall T_1, if T_2, the flow rate of the cooling gas and the properties of the wall material are available. The values of λ_w according to the producer data are 26 W/(m·deg). Knowing T_1, it is possible to determine the total heat flow removed by the gas:

$$Q_r = \tilde{g}_p c_p(T_1-T_\infty). \qquad (6.25)$$

We now return to examination of the experimental material. Figure 6.33 shows the dependence of T_2 on the specific consumption of the cooling gas at $\bar{z}_p = 2$. The results of measurements show that at $\bar{z}_p = 8$ the temperature of the sleeve is slightly higher than in the previous case, for the same values of the arc current and the flow rate of the cooling agent. This is determined by the fact that when

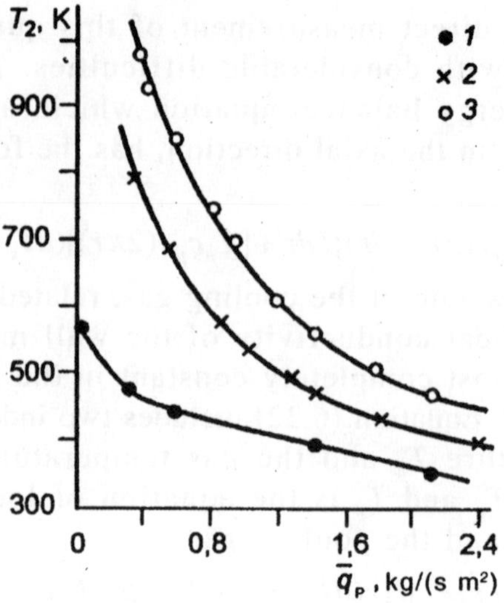

Fig. 6.33. Dependence of the temperature of the outer surface of a porous sleeve on the flow rate of the cooling agent. $\bar{a}$ = 14, $\bar{z}_s$ = 2; $G_0 = (5.2 \div 5.6) \cdot 10^{-3}$ kg/s; $g_s = (0.5 \div 4.3) \cdot 10^{-3}$ kg/s. I = 100 (*1*); 150 (*2*); 180 (*3*) A.

the porous section is located at a distance of two gages from the cathode, the gas supplied into the gap between the cathode and the first section of the inter-electrode insert, influences the cooling of the porous sleeve. The radiant heat flow from the arc to the wall remains approximately constant for different values of G_0. The increase of the gas flow rate into the gap between the cathode and the first section of the inter-electrode insert reduces T_2. This is explained by the fact that in addition to removing the heat by the gas, blown through the permeable surface, exchange of heat takes place between the hot porous section and the cold twisted gas flow in the channel of the plasma torch which becomes more intensive with increasing G_0. Because of this effect at relatively low intensities of the heat flow into the wall ($I \leq 100$ A) the porous sleeve cools down even if the supply of the cooling agent through the sleeve is interrupted (Fig. 6.33, curve *1*). At $\bar{z}_p = 8$, the variation of G_0 has no significant effect on the temperature of the external surface of the porous sleeve.

The heat balance in the porous sleeve taking into account the escape of heat through the end seals for different positions of the porous section at I = 150 A, is shown in Fig. 6.34. The escape of heat is related conventionally to the length of the porous insert for comparison with the components of the heat losses into the wall. The radiant heat flow, emitted by the external surface of the porous sleeve (curve *1*), is determined from equation (6.21). The radiant heat flow is approximately of the same intensity as the escape of heat through the end seals (curve *2*). These losses must be taken into account at T_2 > 800 K (for example, at T_2 = 1000 K Q_{re} = 2.7·10³ W/m). The heat flow, taken away by the cooling gas (curve *3*), was calculated using the equations (6.24) and (6.25). The lower values of the total heat losses into the wall at $\bar{z}_p = 2$ (compare the curves *4* in Fig. 6.34 *a* and *b*) are explained by the effect of the flow rate G_0 on the cooling of the porous sleeve. In the case of low gas flow rates in the gap between the cathode and the first section of the inter-electrode insert, when the effect of additional cooling is small, the total heat losses into the porous insert at $\bar{z}_p = 2$ and $\bar{z}_p = 8$ are the same and equal 23·10³ W/m for I = 150 A (Fig. 6.34*a*, curve *5*).

The small increase of the heat losses with a decrease of the flow rate of the cooling agent ($\bar{g}_p$ < 1 kg/(s·m²)) at $\bar{z}_p = 8$ (Fig. 6.34 *b*, curve 4) is caused evidently by the appearance of the convective component of the heat flow.

The heat losses into the porous insert at currents of 150 and 180 A, calculated using the heat balance equation, equal 2.3·10⁴ and 3.3·10⁴ W/m, respectively, and with the accuracy to 10% coincide with the measured values of the heat losses in the first section of the inter-electrode insert. Consequently, it may be concluded that it is possible to determine the heat flow in the permeable wall using the measured values of the temperature of the porous insert and the flow rate of the cooling agent of the basis of the model, proposing [67]. The method is especially suitable in the measurement of arc radiation because with appropriate gas blowing, the inten-

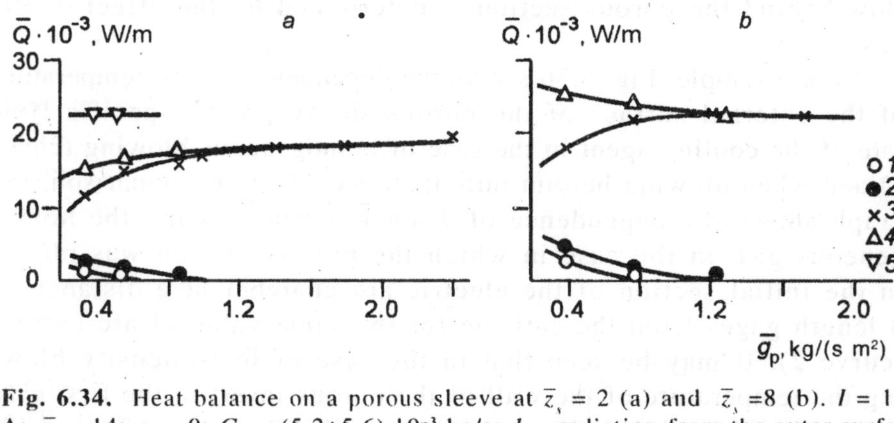

Fig. 6.34. Heat balance on a porous sleeve at $\bar{z}_s$ = 2 (a) and $\bar{z}_s$ =8 (b). I = 150 A; $\bar{a}$ = 14; g_i = 0; G_0 = (5.2÷5.6)·10⁻³ kg/s; *1* – radiation from the outer surface of the sleeve; *2*) escape of heat through seals; *3*) heat flow taken away by the cooling agent; *4*) total heat flow into the porous sleeve; *5*) total heat flow into the porous sleeve at G_0 = 1.6·10⁻³ kg/s.

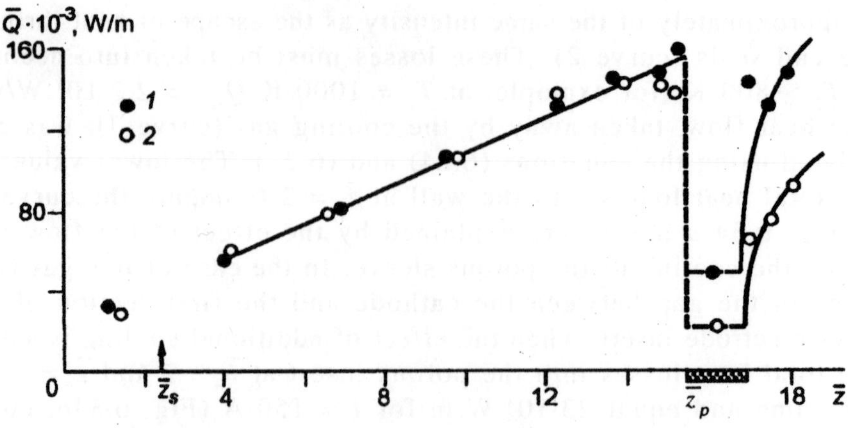

Fig. 6.35. Distribution of heat flows along the channel I = 150 A; $\bar{a}$ = 20; $\bar{z}_s$ = 25; $\bar{z}_p$ =15.5; Re_d = 3.9· 10^3; *1*) g_s = $1.7\cdot10^{-3}$ kg/s; *2*) g_s = $7.5\cdot10^{-3}$ kg/s.

sity of heat exchange, determined by the processes of heat conductivity and convection, tends to zero. Some points of the set of the points presented in Fig. 6.9 were obtained by this method.

We now discuss the results of examination of turbulent heat exchange. Figure 6.35 shows the distribution of heat losses along the channel in the presence of turbulised blowing whose coordinate is indicated by the arrow. As already mentioned, the heat flow in the initial section is determined by arc radiation. Immediately behind the blowing section, convective heat losses, whose value rapidly increases along the channel, are added to the radiant heat flow. Blowing of the gas to the porous sleeve results in a large decrease of the heat losses into the permeable wall. Heat exchange downwards along the flow behind the porous section is determined by the effect of the gas screen.

As an example, Fig. 6.36 shows the dependence of the temperature of the external surface of the porous sleeve on the specific flow rate of the cooling agent in the case of homogeneous blowing (curve *1*) and when blowing helium into air (curve *3*). For comparison, the graph shows the dependence of T_2 on $\bar{g}_p$ when blowing the homogeneous gas, in the case in which the porous section was placed in the initial section of the electric arc chamber at a distance of 8 length gages from the cathode for the same value of arc current (curve *2*). It may be seen that in the case of low-intensity blowing the temperature of the wall of the porous insert in the first case is considerably higher. With increase of the flow rate of the cooling gas, the curves *1* and *2* come close to each other since the convective component of the heat flow into the wall decreases in this

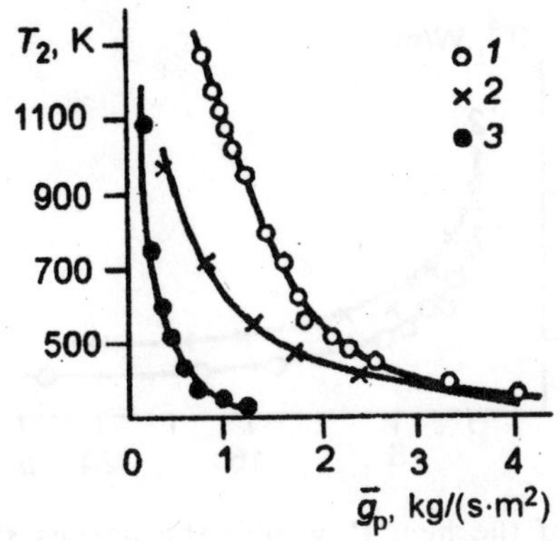

Fig. 6.36. Dependence of the temperature of the outer surface of the porous sleeve on the specific flow rate of the cooling agent. I = 150 A: g_s = (0.3÷7.0)·10^{-3} kg/$_s$ *1*); =15.5; $\bar{z}_p$ = 2.5; $Re_{d.}$ = 9.4· 10^3; *2*) $\bar{z}_p$ = 8; (initial section); *3*) $\bar{z}_p$ = 15.5; $\bar{z}_s$ = 2.5; Re_d = 9.4· 10^3 (helium into air).

case and the radiant fluxes are similar.

We introduce the relative variation of the Stanton number $\psi = St/St_0$, where St and St_0 are the Stanton numbers on the permeable and non-permeable surface, respectively. The comparison of dimensionless heat exchange coefficients should be carried out only for the same values of the Reynolds number. However, the selection of the characteristic value of this quantity is associated with certain problems caused by the selection of both the dimensions and the values of the physical properties of the gas, included in the Reynolds number. Within the framework of the theory, described in [39], it was shown that the characteristic Reynolds number at the standard value of St_0 is represented by $Re^{**} = \rho_0 u_0 \delta^{**}/\mu_w$, where δ^{**} is the thickness of the energy loss, μ_w is the viscosity at the wall temperature. With this definition of St_0 it is possible to take into account separately the effect of different perturbing factors on the variation of the relative law of heat exchange (non-isothermal nature, compressibility, transverse flow of matter, etc).

However, the processing of the experimental data for the stabilised gas flow in the pipe in accordance with Re** is slightly conventional [69]. Usually, in the processing of the experimental data in the examined flow regime it is necessary to use the mean parameters, and the characteristic Reynolds number is determined from the equation $Re_d = \rho u d/\mu_w$, where d is the channel diameter, μ_w is the viscosity of the mean temperature of the flow. It was shown in [70] that in the processing of the experimental data using the mean parameters, the effect of non-isothermal nature of heat exchange

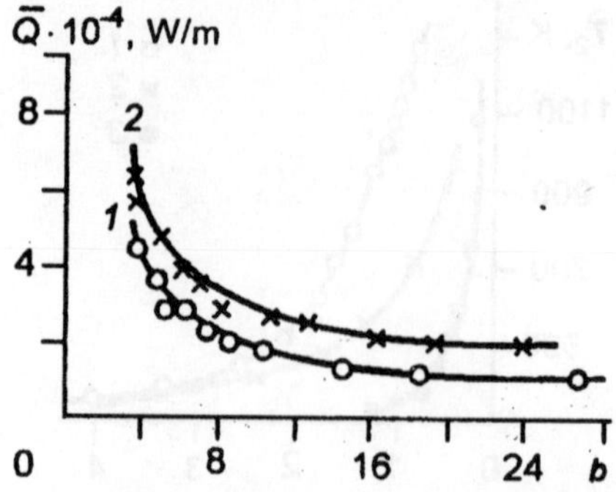

Fig. 6.37. Dependence of the heat flow into the porous sleeve on the blowing parameter. $\bar{a}$ = 20; $\bar{z}_s$ = 2.5; $\bar{z}_p$ = 15.5; *1)* I = 100 A; T_0 = 4100 K, Re_d = 4.6·10³; *2)* I = 150 A, T_0 = 5000 K, Re_d = 3.9·10³.

was not detected. With special reference to electric arc gas heaters, the calculation of convective heat flows was examined in detail in section 6.2.3, where the heat flow and the St number were calculated using equations (6.11) and (6.12), respectively, and the physical properties of the gas flow were determined on the basis of the mean mass parameters.

Figure 6 .37 shows the dependence of the heat flow into the porous sleeve on the blowing parameter: $b = \tilde{g}_p/(\rho u)_0 \cdot St_0$. It should be mentioned that in the case of high intensity blowing the heat flow into the permeable wall remains approximately constant, starting at some value of the blowing parameter, and is determined by arc radiation (supercritical blowing). With a decrease of b, the heat flow into the permeable wall starts to increase because the convective component of the heat flow appears. The minimum flow rate of the cooling agent is restricted by the heat resistance of the material of the porous wall.

Since the convective heat flow into the permeable wall at supercritical blowing parameters is equal to zero, this circumstance may be used in the determination of arc radiation in the turbulent gas flow. The values of the radiant heat flows for the arc in a developed turbulent flow, determined in this manner, are approximately a factor 1.5 lower than the radiant heat losses in the initial section (Fig. 6.35), i.e. the radiant heat flow from the turbulent arc decreases in comparison with the flow from the arc in the initial section of the electric arc chamber. This decrease was already mentioned previously in measurements by other methods.

Figure 6.38 shows the comparison of the experimental data with the calculated values using the equation derived in [39]:

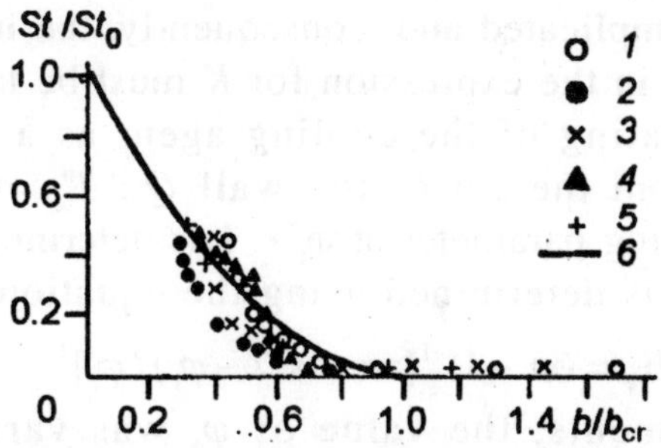

Fig. 6.38. Experimental data for heat exchange in the channel of a plasma torch with a permeable wall. $\bar{a} = 20$; $\bar{z}_s = 2.5$; $\bar{z}_p = 15.5$; $I = (100 \div 150)$ A; *1* – $T_0 = 4100$ K, $Re_d = 4.6 \cdot 10^3$; $\varphi = 0.08 \div 0.29$; *2* – $T_0 = 3300$ K, $Re_d = 10.6 \cdot 10^3$; $\varphi = 0.11 \div 0.35$; *3* – $T_0 = 5000$ K, $Re_d = 3.9 \cdot 10^3$; $\varphi = 0.08 \div 0.26$; *4* – $T_0 = 3950$ K, $Re_d = 9.4 \cdot 10^3$; $\varphi = 0.09 \div 0.34$; *5* – $T_0 = 5000$ K, $Re_d = 3.9 \cdot 10^3$; $\varphi = 0.51 \div 1.37$ (helium in air); calculated from equation (6.26).

$$\psi = [1 - b/b_{cr}(\varphi)]^2, \tag{6.26}$$

where $b_{cr}(\varphi)$ is the critical blowing parameter, which is a function of the non-isothermal nature. According to [39], at $\varphi < 1$

$$b_{cr}(\varphi) = (1-\varphi)^{-1}\left[\ln\frac{1+\sqrt{1-\varphi}}{1-\sqrt{1-\varphi}}\right]^2, \tag{6.27}$$

where $\varphi = T_1/T_0$ is the temperature factor.

The experimental value of the Stanton number was determined using the following equation:

$$\mathrm{St} = (Q - Q_r)/[\pi d(\rho u)_0(h_0 - h_1)]. \quad 6.28)$$

In equation (6.28), h_0 is the mean mass enthalpy of the gas; h_1 is the enthalpy of the gas at the temperature of the internal surface of the porous sleeve; Q_r is the radiant heat flow from the arc to the wall.

The experimental data for heat exchange with blowing of helium into air were processed using the procedure described in [69, 71], where it is shown that all the limiting equations, derived for the blowing of the homogeneous gas in the non-isothermal conditions, may also be applied to the blowing of a foreign gas, if the temperature factor is replaced by $\varphi_1 = \rho_0/\rho_w$. According to [71]:

$$\varphi_1 = \rho_0/\rho_w = \varphi[1 + (\bar{R} - 1)/(K + 1)], \tag{6.29}$$

where $\bar{R} = R_1/R_0$; $K = (c_{p1}/c_{p0})(T_1 - T_\infty)/(T_0 - T_1)$; R_1 to R_0, c_{p1} to c_{p0} are the gas constant and heat capacities of the blown and main gases, respectively. The above equations were obtained for the case of purely convective heat exchange. Parameter K in the supercritical blowing converts to zero. However, in this case, the irradiation-convective heat exchange is complicated and, consequently, the initial temperature of the blown gas T_∞ in the expression for K must be taken into account considering the heating of the cooling agent as a result of the radiant heat flow from the act of the wall Q_r: $T'_\infty = T_\infty + Q_r/(\tilde{g}_s c_p)$.

The critical blowing parameter at $\varphi_1 < 1$ is determined from equation (6.27), at $\varphi_1 > 1$ it is determined using the equation derived in [39]:

$$b_{cr} = (\varphi_1 - 1)^{-1}[\arccos(2 - \varphi_1)/\varphi_1]^2. \tag{6.30}$$

In these experiments, the value of φ_1 was varied in the range 0.51÷1.37.

As shown in Fig. 6.38, the experimental data are in satisfactory agreement with the results of calculations using equation (6.26), although the scatter of the experimental data is large.

The experiments show that the equation (6.26) may be used for estimating the turbulent heat exchange in the stabilisers gas flow in the channel of the plasma torch with the permeable wall in blowing both the homogeneous and foreign gases, if St_0 is determined from equation (6.12).

We now examine the results of examination of the gas screen of the walls of the discharge chamber of the plasma torch behind the porous section. The experimental procedure and the method of processing the experimental data were described in section 6.3. The distribution of the efficiency of the gas screen along the gage section at different gas flow rates through the porous section is shown in Fig. 6.39. The relative permeability of the wall $m_w = (\rho u_w)/(\rho u)_0 = \tilde{g}_s/(\rho u)_0$ varied from 0.022 to 0.056. It should be mentioned that the efficiency of film cooling behind the porous section decreases quite rapidly with a decrease of the flow rate of the shielding gas. This is also clearly

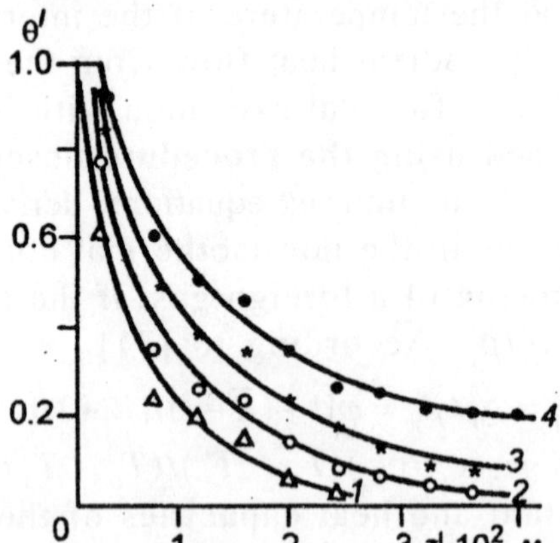

Fig. 6.39. Efficiency of the gas screen behind the porous section. I – 120 A; $\bar{a}$ = 24; $\bar{z}_s$ = 4.0; $\bar{z}_p$ = 16.5; Re_d = 1.4· 10⁴; d = 20·10⁻³ m; T_0 = 3300 K; 1–4) m_w = 0.022; 0.034; 0.044; 0.056 respectively.

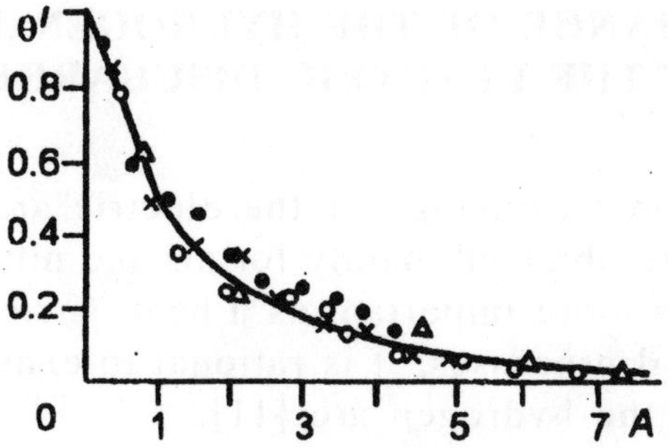

Fig. 6.40. Generalisation of experimental data on the efficiency of gas screen behind the porous section. For symbols see Fig. 6.39. Solid line – calculated from equation (6.31).

indicated in Fig. 6.35.

The experimental results obtained for boundary cooling can be generalised quite sufficiently if we use the dimensionless complex [39]:

$$A = \mathrm{Re}_{z'}[\mathrm{Re}_w(1+K_1)]^{-1.25}.$$

Here $\mathrm{Re}_{z'} = (\rho u)_0 z'/\mu_0$; $\mathrm{Re}_w = \bar{g}_s \Delta l_s/\mu_0$; $K_1 = (T_{w1} - T_\infty)/(T_0 - T_{w1})$. Figure 6.40 shows the experimental results reflecting the dependence of θ' on A. They are described quite efficiently by the equation:

$$\theta' = (1+0.25A)^{-0.8}(1+2A^2)^{-0.45}. \qquad (6.31)$$

The comparison of the efficiency of the gas screen in blowing the gas through the slit in front of the gage section and through the permeable section was made previously in Fig. 6.21. Regardless of the fact that the efficiency of the screen behind the porous ring is slightly lower than in blowing the gas through the slit, the blown gas completely reduces the heat flows to the porous insert. Taking this into account, the efficiency of film cooling of the walls in blowing the gas through the porous ring may be at least not lower than in blowing through the inter-sectional slits.

The experimental material presented in this paragraph can be used for estimating the film cooling of the walls of the discharge chamber of the plasma torch in the presence of a porous insert in part of the channel using the standard procedures [38, 39] suitable for relatively low temperatures of the gas flow. More details on thermal shielding may be obtained by examining the monograph [66].

6.6. HEAT EXCHANGE OF THE HYDROGEN ARC WITH THE WALLS OF THE ELECTRIC DISCHARGE CHAMBER

The data on the heat exchange of the electric arc with the walls of the channel were obtained mainly for air and nitrogen arcs. Since hydrogen becomes more important as a heat carrier and reagent in many technological processes, it is rational to examine the thermal characteristics of the hydrogen arc [11].

It is important to note the large number of calculations studies concerned with the analytical investigations of hydrogen plasma. They have been reviewed in [72–76]. However, because of the absence of reliable experimental results obtained in the investigations of the electric arc in hydrogen and, in particular, heat exchange in the hydrogen plasma, exact calculations are associated with difficulties. They are based on the measurement of transfer properties and optical characteristics of the plasma in consumption-free or capillary discharges in hydrogen which are far away from the actual conditions. The radiation of hydrogen plasma has been studied most extensively in the previously mentioned discharges, shock pipes, etc [74, 76, 77]. The experimental data on other types of heat transfer and, in particular, in turbulent hydrogen plasma, are not available.

Below, we present some data on the thermal characteristics of the electric arc in hydrogen. The thermal efficiency of the plasma torch is determined by the heat losses into all elements of the torch, i.e. $1-\eta = Q/(UI)$. Here η is thermal efficiency, UI is the power generated in the arc, Q are the heat losses in the plasma torch with the inter-electrode insert which can be determined from the equation:

$$Q = Q_{\rm cat} + Q_{\rm s.s.} + Q_{\rm IEI} + Q_{\rm a}.$$

In this equation $Q_{\rm cat}$, $Q_{\rm s.s.}$, $Q_{\rm iei}$, $Q_{\rm a}$ at the heat losses in the cathode, the starting section, the inter-electrode insert and the anode.

We examine the relationship between the heat losses into the elements of the plasma torch and its working parameters.

6.6.1. Heat flow into the end cathode

The heat flow into the internal end cathode of a plasma torch is determined mainly by the heat flows of the cathode arc spot [1]. The heat flow in the cathode from the hydrogen arc was investigated in [78]. The data obtained in [78] are shown in Fig. 6.41 (solid line). They are approximated by the dependence:

$$Q_{\rm cat} = 4.7I. \tag{6.32}$$

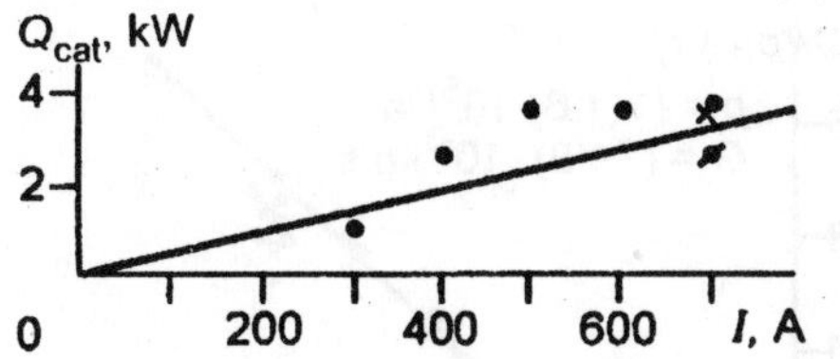

Fig. 6.41. Heat flow into the cathode straight line – calculated from equation (6.32 from [78]; experimental points – data [11].

The graph also shows the experimental data from [11]. Regardless of the large scatter, determined by the measurement area, they are close to the data measured in [78]. Thus, the heat losses into the internal end electrode-cathode increase linearly with an increase of the arc current intensity and are comparatively low (approximately 3 kW at I = 700 A).

6.6.2. The heat flow in the section of the inter-electrode insert and the starting electrode

Investigations were carried out on plasma torches with an inter-electrode insert with the internal diameter of the channel $d = 2 \cdot 10^{-2}$ and $3 \cdot 10^{-2}$ m, the starting section $d_{s.s} = (1.2\div1.4) \cdot 10^{-2}$ m.

The measurements of heat losses in the section of the inter-electrode insert show that the heat flow is approximately constant along the inter-electrode insert, independent of the channel diameter, and is determined by the arc current intensity and gas pressure, i.e. by the same parameters as the radiant heat flow in other gases [1, 17]. The data on the heat flows in several separate sections ($d = 3 \cdot 10^{-2}$ m, $l_c \sim 1 \cdot 10^{-2}$ m) are presented in Table 6.1.

The data were processesed by the procedure described in [13], i.e. by constructing the logarithmic dependence of the heat flow through the unit length of the channel $\bar{Q}$ (W/m), related to the pressure, in relation to the arc current intensity. The gas pressure in the measurement

Table 6.1

I, A	$G_0 \cdot 10^3$, kg/s	$G \cdot 10^3$, kg/s	$p \cdot 10^{25}$, Pa	Q_{10}, kW	Q_{11}, kW	Q_{14}, kW
300	1.25	7	1.08	0.41	0.39	0.41
400	1.25	7	1.16	0.95	0.97	0.98
500	1.25	7	1.17	1.62	1.62	1.62
600	1.25	7	1.20	2.26	2.24	2.30
700	1.25	7	1.26	3.21	3.05	3.01
700	1.25	8	1.56	3.20	3.55	3.07
700	1.25	8	1.46	3.03	3.23	3.90
700	1.25	8	1.63	3.13	3.32	2.84

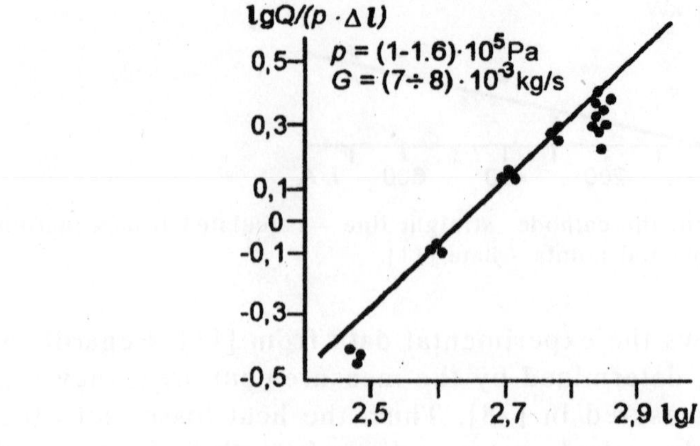

Fig. 6.42. Dependence of $\lg Q/(p\Delta l)$ on $\lg I$ (d = 3 × 10^{-2} m; $d_{s.s.}$=1.2 × 10^{-2} m).

zone was assumed to be equal to the pressure at exit from the plasma torch, which slightly increases the scatter of the data. The dependence lg ($\bar{Q}/p$) on lg$I$, constructed using the data presented in Table 6.1, is shown in Fig. 6.42. It may be seen that with the exception of the limiting currents where the scatter of the data is greater, the experimental data fit a curve, generalised by the equation:

$$\bar{Q} = 5.2 \cdot 10^{-6} I^2 p. \tag{6.33}$$

The heat losses into the starting section with the diameter 1.2·10^{-2} m and the length of 3.1·10^{-2} m are close to the values calculated using equation (6.33), but the data are greatly scattered, firstly as a result of the fact that in the experiments we determine the pressure at exit from the plasma torch and not in the zone of measurement of the heat flows. The effect of the flow rate and pressure of the gas on the heat losses in the walls of the channel of the hydrogen plasma torch will be investigated.

Figure 6.44 shows the results of measurements of the heat flows in the section of the inter-electrode insert at different gas flow rates. Points *1* were taken from the data in the previous graph, i.e., they correspond to G = (7÷8) · 10^{-3} kg/s and p = (1÷1.6) · 10^5 Pa. Points *2* were obtained at the same values but the gas flow rate was G = (6÷6.5) · 10^3 kg/s, points *3* at G = (5÷5.5) · 10^{-3} kg/s and, finally, the points *4* at G = (3÷4) · 10^{-3} kg/s and pressures up to 5 · 10^5 Pa. The data obtained for different gas flow rates differ, and halving the flow rate almost doubles the heat flows in the section. The curve I in Fig. 6.43 is calculated using equation (6.33), curve II was calculated from the same equation using the coefficient 7.4 · 10^{-6}, and the curve III at 9.5 · 10^{-6}. However, the tendency for the increase of the heat

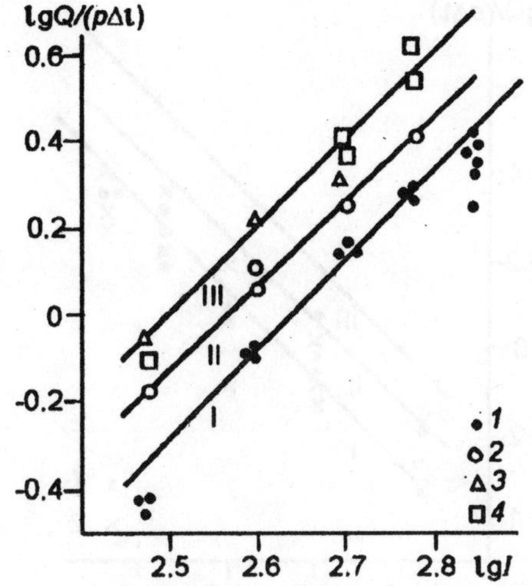

Fig. 6.43. Dependence of the heat flow into the section of the IEI on the arc current intensity for different hydrogen flow rates.

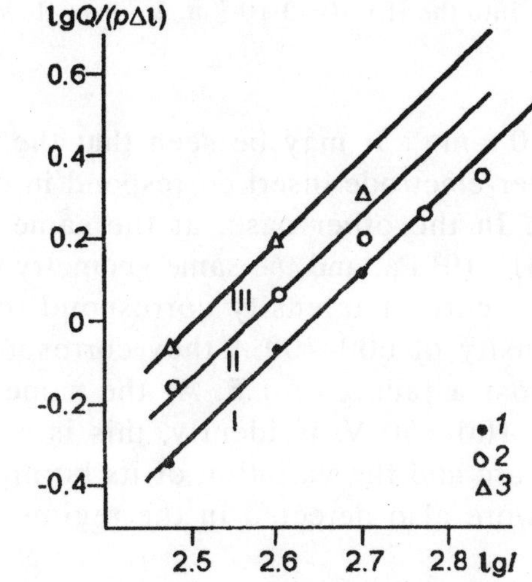

Fig. 6.44. Dependence of the heat flow into the section of the IEI on arc current intensity. Curves I – III correspond to the data in Fig 6.43. 1, 2 – $D = 7\cdot10^{-2}$ m, $G = (4.2\div4.25)\cdot10^{-3}$ kg/s; 3 – $d = 3\cdot10^{-2}$ m, $G = (5\div5.5)\cdot10^{-3}$ kg/s.

losses with the decrease of the gas flow rate is not recorded in all cases. For example, points *3* at low current intensity are situated on the curve III and with increase of current intensity they tend to curve II. This is shown in greater detail in Fig. 6.44, where the curves I–III are the same as in Fig. 6.43. The graph gives the results of measurements of the heat flows in a large-diameter inter-electrode insert ($D = 7 \cdot 10^{-2}$ m – points *1* and *2*) and with an interelectrode

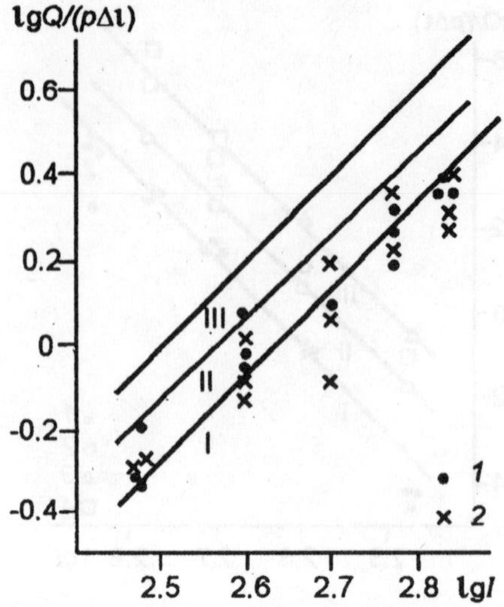

Fig. 6.45. Dependence of the heat flow into the section of the IEI and into the starting section of the arc current intensity. Curves I = III correspond to the data in Fig 6.43. 1 – heat flow into the IEI, $d = 3 \cdot 10^{-2}$ m, 2 – heat flow into the starting section $d_{s.s} = 1.4 \cdot 10^{-2}$ m.

insert with $d = 3 \cdot 10^{-2}$ m$^{-.3}$. It may be seen that the heat losses in the large-diameter inter-electrode insert correspond in one case in the accuracy to curve *1*. In the other case, at the same gas flow rate and pressure of $(1 \div 1.5) \cdot 10^5$ Pa, and the same geometry of the channel, the heat losses at low current intensity correspond to the curve II, and at a current intensity of 600÷700 A they correspond to curve I, i.e. decrease by almost a factor of 1.5. At the same time, the arc voltage decreases by 100÷150 V. Evidently, this is accompanied by rearrangement of the arc and the variation of its burning mechanism. Similar phenomena were also detected in the regime corresponding to points *3*.

The heat flow into the starting section with the diameter smaller than the diameter of the channel usually corresponds to the curve I (Fig. 6.45).

Thus, according to heat exchange in the electric discharge chamber, there are at least two conditions of burning of the hydrogen arc in the investigated range of the parameters ($d = (1 \div 10) \cdot 10^{-2}$ m, $G = (3 \div 8) \cdot 10^{-3}$ kg/s, $p = (1 \div 6) \cdot 10^5$ Pa, $I = 300 \div 700$ A). In one of these regimes the heat losses are constant along the channel and are determined from equation (6.33). In the other regime, the heat losses are also constant along the inter-electrode insert but increase with a decrease of the gas flow rate. An increase of the arc current intensity is accompanied by a tendency to transition from the second regime to the first one and, generally speaking, to a general decrease of the heat flow into the wall of the channel in comparison with that calculated from equation (6.33). For example, at a current intensity of 700 A, the experimental points are distributed mainly below the curve I (Fig. 6.42).

Thus, the heat losses in the section of the inter-electrode insert and the starting section from the arc in hydrogen at $d > 1\cdot10^{-2}$ m are approximately constant along the channel, are independent of the diameter of the channel, and determined by the arcing conditions. The minimum values of the heat losses may be evaluated using equation (6.33).

6.6.3. The heat flow into the output electrode – anode

The results of examination of the heat flows into the anodes from the arc, burning in different gases, were presented in the chapters 6.3.2 and 6.3.4. It was reported [31] that the heat losses in the cylindrical anode are determined by convective heat transfer, arc radiation and by the heat flow through the anode spot of the arc. Radiation plays a significant role in the section of the anode to the zone of attachment of the arc, i.e. over the length of 1–2 gages from entry into the electrode. In the zone of contact of the arc with the electrode, the heat flow through the anode spot is very important. Local heating in this area is very intensive and, moving only rapidly the spot on the electrode, it is possible to avoid melting of the surface of the electrode in the zone of attachment of the arc. The heat flow through the anode spot, according to [1], is:

$$Q_s = U_e I, \text{W}. \tag{6.34}$$

Here U_e is some effective value of the anode voltage drop. For the arc in a turbulent airflow or in a nitrogen flow $U_e \approx 6$ V. The same value of U_e may also be accepted for the hydrogen arc [31].

The convective heat flow into the cylindrical output electrode, as shown in section 6.3.2, may be calculated using equations (6.11) and (6.12). The experimental results are compared with the calculated data in Fig. 6.10. It may be seen that in hydrogen, the calculated results are in satisfactory agreement with the experimental data up to the mean mass temperatures of the gas of 3000 K. At higher temperatures, the experimental values of the heat flow are 20% or more higher than the calculated values. In all likelihood, at temperatures above 3000 K the heat conductivity plays an increasingly important role, and the maximum value of heat conductivity at 3800 K is an order of magnitude higher than at 3000 K. The density of the heat flows may reach 2 kW/cm^2 higher, i.e., it may approach the limiting values for the actual cooling systems ($q^* \sim 5$ kW/cm^2). The total value of the heat losses into the anode with the length of 3–4 gauges at a pressure of $(5 \div 6) \cdot 10^5$ Pa may reach 20% of more of the arc power. Evidently, at high thermal loads the optimisation of the cooling system is very important in the retention of the efficiency of the plasma torch.

The data on the heat flows into the walls of the channel of the hydrogen plasma torch with the inter-electrode insert, presented in the section, make it possible to evaluate the heat losses into all elements of the plasma torch, and together with the data on the energy characteristics (see chapter 5) they can be used to determine the thermal efficiency of the plasma torch and the efficiency of heating the gas in the torch.

6.7. GENERALISED THERMAL CHARACTERISTIC OF THE STEAM-VORTEX PLASMA TORCH

The steam plasma, used as a reagent and energy carrier, has a similar important role (in comparison with hydrogen) in the processes of processing carbon-containing initial materials, and also in the elimination of toxic chemical, medical and household waste. As mentioned previously, in particular in chapter 5, the development of steam plasma torches is associated with a number of difficulties. One of these problems is the organisation of the flow of steam in a channel without condensation on the wall [79]. On the other hand, using water or steam for cooling the walls of the working body opens considerable possibilities for increasing the thermal efficiency and efficiency of heating of steam plasma [80]. Investigations of the thermal characteristic of the plasma torchs for heating steam have been carried out in a large number of studies [79–81]. Measurements were taken mainly of the integral heat flows into the sections of the plasma torches, including flows into the confusor part of the discharge chamber with different constriction angles (the central angle from 0 to 22°), and the output electrodesteamanode in the presence and absence of the ledge in the anode (Fig. 6.46).

The processing of the results of investigations of the thermal characteristics of the steam-vortex plasma torches was carried out using the methods described previously in the form of the dependence of the coefficient of relative heat losses $\tilde{\eta} = (1 - \eta)/\eta$ on the main criterial complexes. The following equation was obtained:

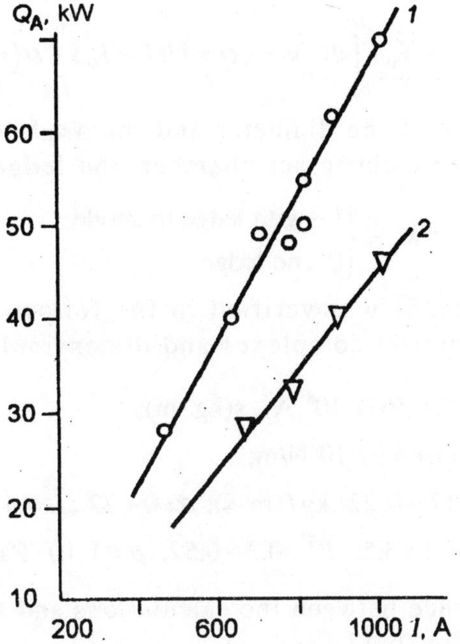

Fig. 6.46. Heat flows into the anode of the steam-vortex plasma torch in the presence (1) and absence (2) of a ledge in relation to the arc current intensity (steam flow rate $4\cdot10^{-3}$ kg/s).

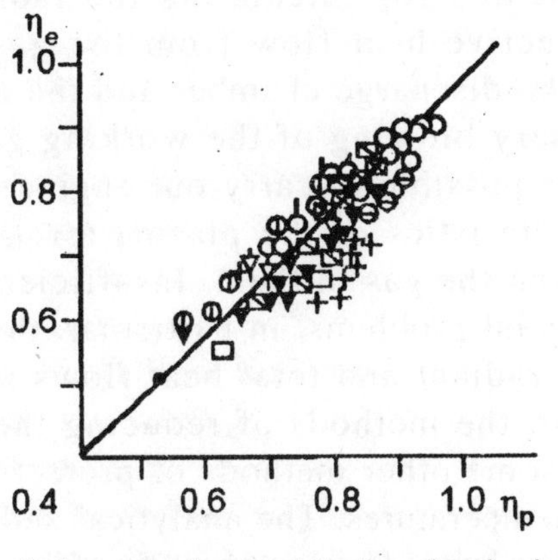

Fig. 6.47. Comparison of the experimental (η_e) and calculated (η_c) values of the thermal efficiency for the steam vortex plasma torches of different types.

$$\tilde{\eta} = 3.02\cdot10^{-6}(I^2/GD)^{0.32}(G/D)^{-0.57}(pD)^{0.40}\times$$
$$\times(1+1.2K_y)(1+\mathrm{tg}(\alpha/2))(\bar{T}/\bar{L})^{0.5}, \qquad (6.35)$$

where

$$D=1/(L-L_0)\int_{L_0}^{L} d(z)dz;\quad \alpha=1/(L-L_0)\int_{L_0}^{L}\alpha(z)dZ$$

At the mean values of the diameter and the angle of narrowing of the flow part of the electric act chamber, the ledge coefficient:

$$K_y=\begin{cases}1-\text{with ledge in anode}\\0-\text{no ledge}\end{cases}$$

The equation (6.35) was verified in the following range of the variation of the criterial complexes and dimensionless parameters:

$$(I^2/GD)=(3\div367)\cdot10^8\ \text{A}^2\cdot\text{s(kg}\cdot\text{m)};$$

$$pD=(1\div4.9)\cdot10^3\text{N/m};$$

$$G/D=(0.017\div0{,}22)\ \text{kg}/(\text{m}\cdot\text{s});\ \alpha=0\div22^{\circ};\bar{L}=4.1\div13.5;$$

$$D/d_1=1\div3.5;\ \bar{l}/\bar{L}=0.3\div0.52;\ p=1\cdot10^5\ \text{Pa}.$$

The correspondence between the calculations and the experimental data is shown in Fig. 6.47.

In this chapter, the results are presented of experimental investigations of heat exchange in the discharge chamber of electric arc heaters of gases of different systems. Empirical and semi-empirical relationships are presented for calculating the radiant heat flow from the arc, the convective heat flow from the gas heated by the arc into the walls of the discharge chamber and the efficiency of various methods of boundary blowing of the working gas into the channel. Consequently, it is possible to carry out engineering calculations of the thermal characteristics of the plasma torches and estimate the efficiency of heating the gas in them. Insufficient attention has been paid to several special problems, in particular, the analytical methods of calculating the radiant and total heat flows into the walls of the discharge chamber, the methods of reducing the radiant heat flows from the arc, and some other methods of protecting the walls against the effect of high temperatures. The analytical and calculation methods of examination have been developed quite efficiently in recent years and described in detail in the previous volumes of the low-temperature plasma series.

Chapter 7

Direct current linear plasma torches

In the introduction and in the first section of chapter 7, we discuss information on the schemes of linear plasma torches, which were partially described in chapter 1. However, it is regarded as essential to repeat this material because it is important for understanding the selection of the design and construction of the plasma torch.

The main types of design of the linear plasma torches and some of the characteristics are presented in the form of figures, schemes and graphs. This material is known to various degrees and meaning. Examination of the present in material will probably increase the knowledge on the state of the investigated problems and help the formation of approaches to the development of new schemes of the plasma torches and reactors, satisfying the current requirements on equipment for plasma-technological processes.

The electric arc gas heaters are low-temperature plasma generators, also referred to as plasmatrons, i.e. equipment in which the heat-generating element (electrical arc) is practically the only available means of stationary heating of the gas to high temperatures at the optimum transformation of electric energy to thermal energy by means of conductive, radiant and convective heat exchange. The advantages of electric arc plasma torches make it possible to use them efficiently in many branches of industry; some of them are as follows:

– the economic efficiency of the transformation of electrical energy to thermal energy by the currently available types of plasma torches, characterised by the high values of electrical and thermal efficiency;

– the reliability and stability of operation of electric arc equipment;

– the relatively long operating life of the electrodes expressed usually in hundreds of hours depending on the type of plasma torch and its application, the power of the electrical arc (current intensity) and the type of working gas;

– the wide range of the power of the constructed plasma torches–

from hundreds of watts to several megawatts;

– the possibility of heating almost any gas or mixture of gases, including reduction, oxidation, inert gases, used widely in different industrial technologies;

– simple automation of control of the operating regime of the electric arc;

– the small size and relatively small material requirement.

The plasma torches are interesting because of the possibility of efficient realisation of chemical, metallurgical and other processes, the construction of low-waste technologies, organisation of complex processing of initial materials, production of materials with completely new physical–mechanical and chemical properties, miniaturisation of industrial equipment. This is explained by the fact that at high temperatures, the rate of chemical reactions is many times higher than the rate of conventional technologies at the temperatures used at present, and this also relates to the travel speed of the products of chemical reactions in the reactor.

7.1. CLASSIFICATION OF LINEAR PLASMA TORCHES

The knowledge of the fundamental physical processes, taking place in the discharge chamber of the linear DC plasma torches, has made it possible to propose a simple classification scheme. The special features of the interaction of the arc with the gas blown onto the arc determine the arc length as the main parameter in this classification. Consequently, it has been possible to reduce the entire range of completely different designs of linear plasma torches to three large classes [1].

1. The plasma torches with the self-setting mean arc length L_a, which depends on the current intensity, the polarity of the output electrode, the type and consumption of working gas, the diameter of the chamber and the pressure in the chamber. The arc length is set by the mechanism of large-scale shunting. The plasma torches of this group with the solid output electrode have a drooping VAC of the arc (Fig. 7.1, curve 1).

2. The plasma torches with the fixed mean arc length L_a, i.e., the length is constant in a relatively wide range of the variation of current with the above-mentioned main parameters constant, and is always smaller than the self-setting length in the channel with the diameter d_2; it is determined by the aerodynamics of the flow behind the ledge. The VAC of the arc is U-shaped (curve 2).

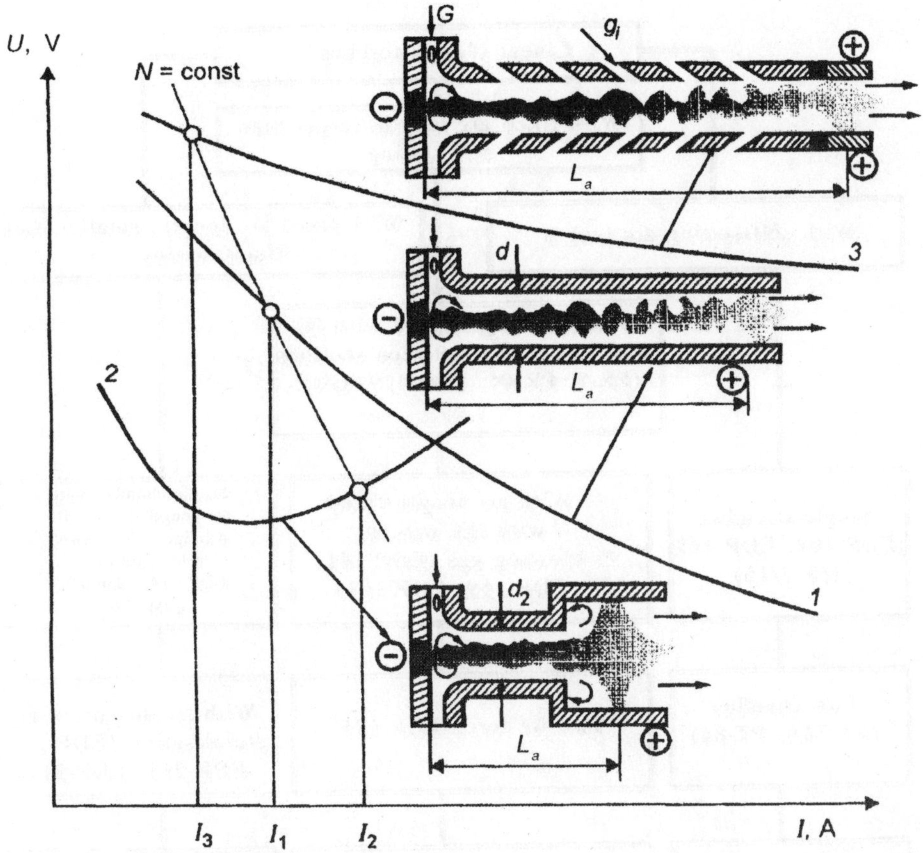

Fig.7.1. Classification of linear plasma torches. Volt–ampere characteristics of the arc of three types of linear plasma torch.

3. The plasma torches with the inter-electrode insert (IEI). In this case, the mean arc length L_{IEI} is also constant in a wide range of variation of current intensity, but $L_{IEI} > L$ because of the selection of the insert of the appropriate length. In the majority of the structures of the plasma torches, the inter-electrode insert has the form of a set of electrically insulating sections. The working gas, supplied into the chamber through the gaps between the sections, is designed for protecting the walls of the discharge chamber against the convective heat flows and preventing electrical breakdown between the sections. The VAC of the arc is slightly drooping (curve 3).

Figure 7.2 shows the names of the three groups of the plasma torches, developed at the Plasma Dynamics Section of The ITPM Institute, Russian Academy of Sciences in cooperation with the Scientific Research Institute of Chemical Engineering, Novosibirsk [2].

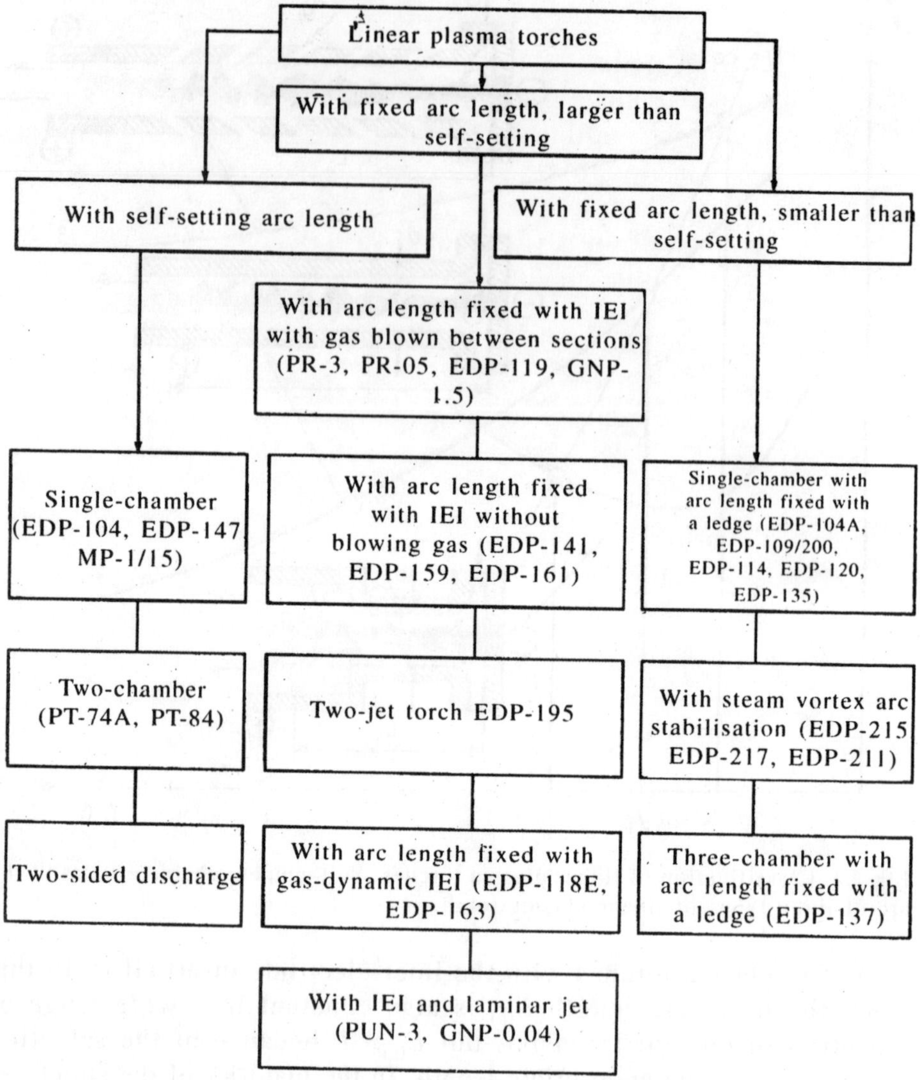

Fig.7.2. Table of terms of three classes of plasma torches.

7.2. PLASMA TORCHES WITH THE SELF-SETTING ARC LENGTH

7.2.1. Single-chamber plasma torches

They are of the simplest design and are reliable in service. There are several variants of single-chamber plasma torches:

– with a flat end electrode and a single vortex chamber. In this case, the material of the end electrode–cathode and the working gas must be compatible (for example, tungsten and inert gases, zirconium, hafnium and oxygen-containing gases) (Fig. 7.3*a*);

– with the auxiliary vortex chamber for the separation of the cathode material from the working gas into which the appropriate shielding gas, which does not react chemically with the material of the cathode, is supplied (Fig. 7.3*b*);

–with the cup-shaped closed end tubular copper electrode (Fig. 7.3*c*). The plane of rotation of the radial section of the arc A–A is determined by the special features of the flow of the gas in the tubular electrode, by the magnetic field of the solenoid, installed on the electrode, or by other influences.

In all variants, the output electrode is usually produced from copper.

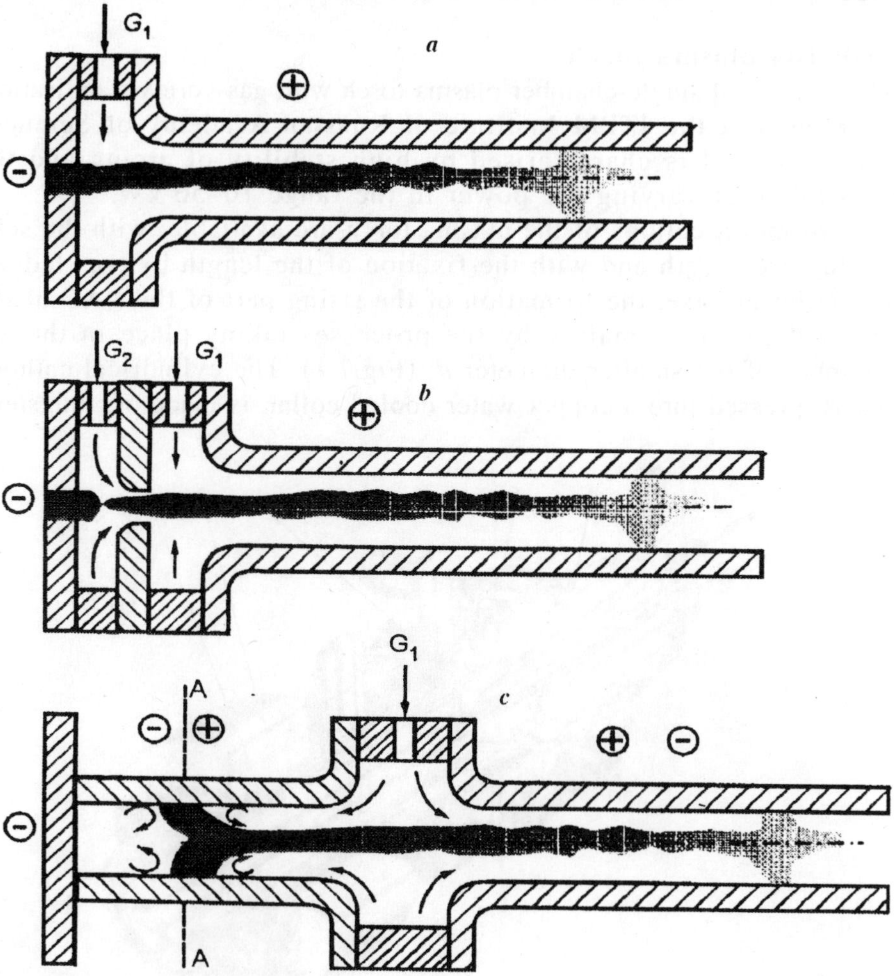

Fig. 7.3. Single-chamber plasma torches. a) with a flat end electrode; b) with a flat end electrode and an auxiliary vortex chamber; c) with a cup-shaped closed end tubular electrode.

However, for some processes, it is necessary to produce the electrode from a different material: cast iron, non-magnetic steel, a pseudo-alloy based on refractory metals, for example, tungsten with copper.

The internal diameter of the tubular electrode used for these applications is in the majority of cases constant along the length, although variants with conical electrodes have already been developed.

The VAC of the arc of all the examined plasma torches is drooping; when using a power source with a 'hard' VAC, a regulated ballast resistance is introduced into the electrical circuit of the arc, ensuring stable arcing.

EDP-104 plasma torch

This is a small single-chamber plasma torch with gas-vortex stabilisation, developed at the ITPM Institute of Russian Academy of Sciences (Fig. 7.4) and is characterised by high stability of arcing and the possibility of varying the power in the range 10–50 kW.

Two modifications of the plasma torch are available: with the self-setting arc length and with the fixation of the length by the 'ledge'. In the latter case, the formation of the rising part of the VAC of the arc is determined mainly by the processes taking place in the arc chamber of the smaller diameter d_2 (Fig.7.1). The cylindrical cathode insert, pressed into a copper water cooled collar, is made of: tungsten–

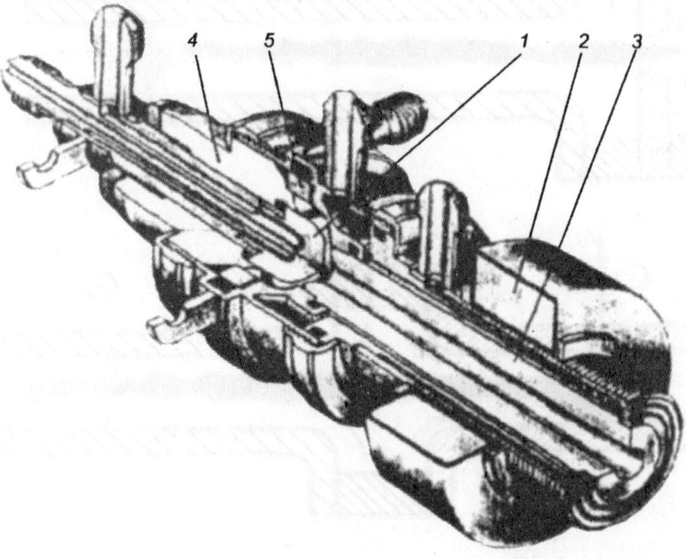

Fig. 7.4. EDP-104 plasma torch. 1) internal electrode; 2) output electrode; 3) permanent magnet (solenoid); 4) insulator; 5) working gas supply section.

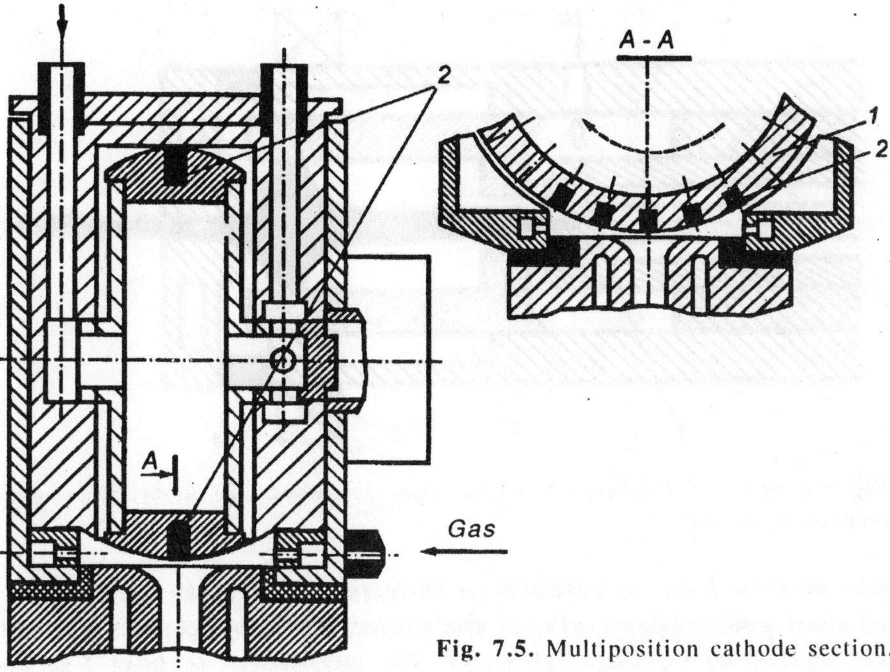

Fig. 7.5. Multiposition cathode section.

in argon, helium, nitrogen, hydrogen, and zirconium, hafnium for air, carbon dioxide, steam.

A multiposition electrode cathode section (Fig. 7.5) may also be used. The section greatly increases the duration of continuous operation of the plasma torch and may be used in any of the plasma torches of the single-chamber design [2]. It has the form of a copper watercooled drum *1* with the inserts made of emitting material *2* inserted so that they are flush with the edge of the drum. The number and distance between the inserts with the uniform distribution around the circumference are selected in accordance with the required duration of continuous operation, setting the step of the drive of the rotating mechanism, which activates the next insert.

MP-1/5 plasma microtorch

This microtorch generates a small diameter high-temperature jet (with the diameter not exceeding 1 mm).

A number of technological processes use plasma torches with the power of up to 1–3 kW. For example, for cutting cloth and thin-sheet materials, a plasma torch with a power of up to 1 kW was developed. It is a single-chamber plasma torch with vortex stabilisation of the arc (Fig. 7.6). Copper electrodes are used: the copper cath-

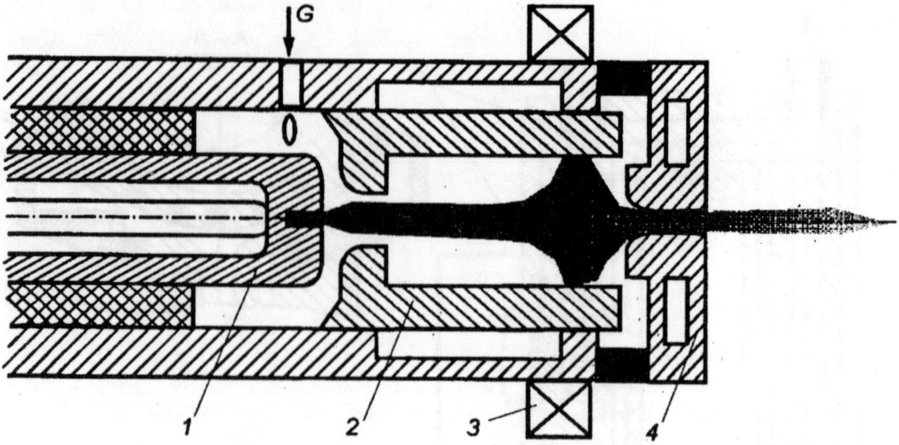

Fig. 7.6. MP-1/15 1 kW microplasma torch. 1) cathode; 2) anode; 3) permanent magnet; 4) nozzle.

ode section *1* has a hafnium or tungsten insert, when using air or an inert gas, respectively, as the plasma-forming gas. Components are cooled by commercial water. The permanent magnet *3* is used for moving the attachment part of the arc on the internal diameter of the anode *2*. The plasma jet is produced using the water cooled nozzle *4* (electrically insulated from the anode), with the diameter of the output orifice of up to 1 mm. The modified plasma microtorch uses the anode made of Cr18Ni10Ti stainless steel with indirect cooling (i.e., through the water cooled collar). An original electric power source (rectifier), connected directly into the mains with a voltage of 220 V, was developed for this plasma torch with a low arc voltage, Fig. 7.7.

The operating life of the electrodes in operation with inert gases is up to 50 h, in air it is not less than 8 h. The thermal efficiency of the plasma torch reaches 0.7.

The plasma microtorches also include the single-chamber plasma torch with a smooth electrode with the power of up to 3.5 kW [3]. The diagram of the discharge chamber is identical with the EDP-104 plasma torch. Air is used as the plasma forming gas. Figure 7.8 shows its VAC. The plasma torch may operate both at the atmospheric pressure (curve 1) and also at a higher pressure (curve 2). The efficiency of the plasma torch reaches 0.7. The plasma torch is used for plasma ignition of liquid and gas fuels in gas turbine engines [3].

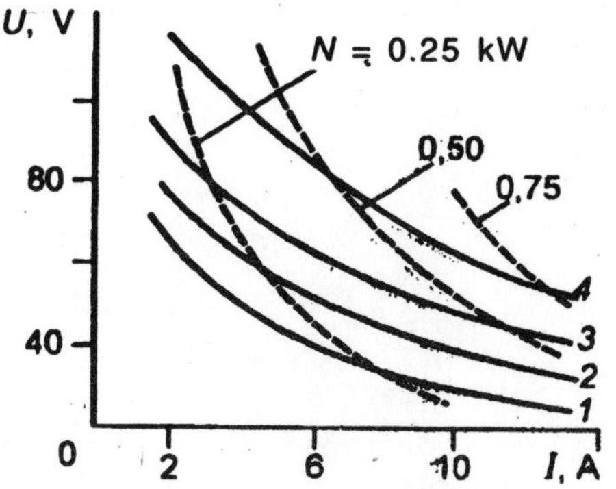

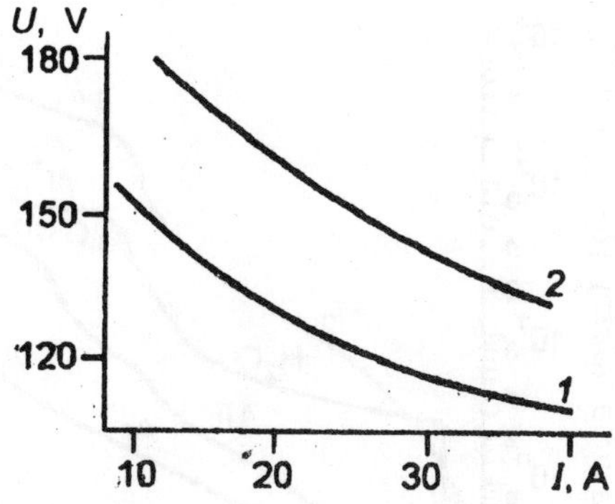

Fig. 7.7. Volt–ampere characteristics of the arc of the MO-1/15 plasma torch. Plasma-forming gas: 1.2) argon; 3,4) air. Gas flow rate, kg/s: 1) $1\cdot10^{-4}$; 2) $1.4\cdot10^{-4}$; 3) $1\cdot10^{-5}$; 4) $4.2\cdot10^{-5}$.

Fig. 7.8. (right) Volt–ampere characteristics of the arc of a plasma torch ($N \leq 3.5$ kW). Flow rate of plasma forming air $8\cdot10^{-4}$ kg/s: pressure in the discharge chamber, Pa: 1) 10^5; 2) $5\cdot10^5$.

The steam plasma torch

A large number of working schemes of the electric arc plasma torches using different gases have been developed throughout the world. They produce air, argon, hydrogen and other types of plasma. However, until recently, there were no plasma torches generating steam plasma, consisting of two components: hydrogen and oxygen. As regards the properties, this plasma differs quite distinctively from other gas media. The specific heat content of steam plasma is almost an order of magnitude higher than that of, for example, the air plasma at the same temperature. The oxidation–reduction properties of the steam plasma with its ecological efficiency and high specific heat content (Fig. 7.9) may be utilised in the gasification of coal [4], the elimination of processing of toxic substances and waste, and also in plasma cutting of metals and spraying of heat-resisting coating. In recent years, the steam plasma has been used in atomic power industry in production of nuclear fuel.

The steam-vortex linear plasma torches have been developed most efficiently in the group of water plasma generators.

In gas-vortex plasma torches, as already mentioned, the working medium is represented by air, nitrogen, hydrogen, argon and other gases whose properties are close to the ideal gas. To reduce the difference between the properties of water steam and the ideal gas, and use the steam in vortex plasma torches, it is necessary to:

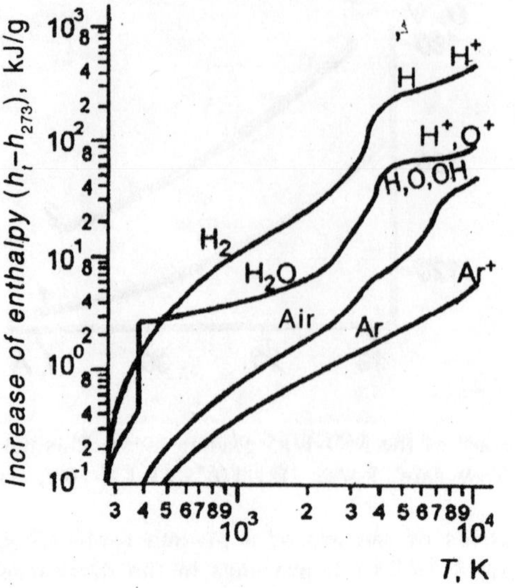

Fig. 7.9. Increase of the enthalpy of different gases with increase of their temperature.

–preheat the steam to 250–350°C;

–eliminate the reasons for condensation of steam on cold surfaces of the discharge chamber and associated effects inside the arc chamber.

To ensure stable burning of the electrical arc, the design of the steam–vortex plasma torch should satisfy the following three conditions [5]: the wall of the gas-discharge chamber should be hot, and the chamber should be of the confusor type; a damper should be placed in front of the discharge chamber in the water–steam system.

The ITPM Institute of the Siberian Division of the Russian Academy of Sciences developed a series of simple single-chamber steam–vortex plasma torches of different power (Table 7.1).

Prior to operation with water steam, the walls of the discharge chamber of the plasma torch should be heated to the temperature higher than the temperature of saturated steam. Heating is carried out using a plasma torch for 2–3 min in air followed by smooth transition to steam. After completing the operation, air should be added to the discharge chamber of the plasma torch for 3–5 min for complete removal of moisture. The arc in the plasma torch is ignited using an oscillator. The VAC of the arc of the steam–vortex plasma torches of different powers is shown in Fig. 7.10. The steam is supplied to the steam–vortex plasma during operation from a central steam system with steam preheated to 250–350 °C or, in the absence of such a system, using a special steam generator whose diagram is shown in Fig. 7.11. The operating principle of the steam generator is based on the utilisation of Joule heat generation in the walls of the steam-generating pipe through which water from the water cooling system of the steam gen-

Table 7.1. Main technical characteristics of steam vortex plasma torches [4]

Parameters	Plasma torch			
	EDP-166, EDP-215	EDP-217	EDP-211	EDP-201
Power, kW	10–70	60–150	200–500	400–1000
Steam plasma yield, g/s	0.5–3.0	1.0–5.0	5.0–30	10–60
Maximum arc current, A	250	500	800	800
Efficiency, %	50–70	60–70	60–75	70–80
Flow rate of shielding gas (nitrogen), g/s	–	0.5–0.7	0.5–1.0	0.5–1.0
Cathode life, h	30	100	100	100
Anode life, h	300	300	300	300
Weight, kg	1.25	12.5	42	82
Dimensions, m	0.192×0.11×0.26	0.34×0.26×0.2	0.64×0.27×0.25	0.84×0.34×0.30

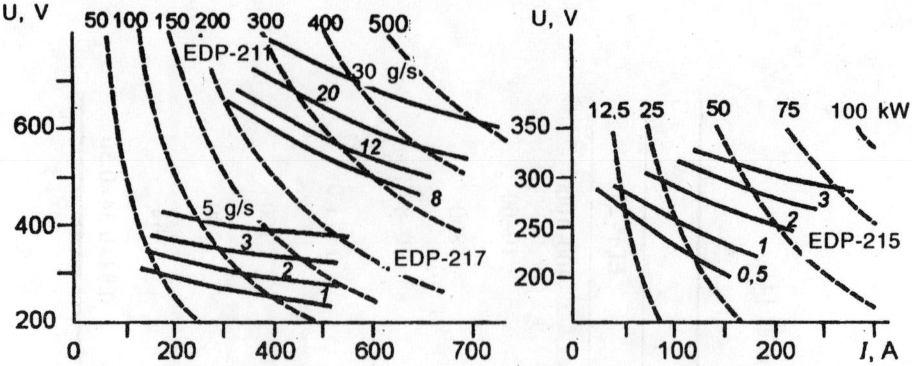

Fig. 7.10. Volt–ampere characteristics of the arc of steam vortex plasma torches EDP-217, EDP-211 and EDP-215 for different gas flow rates.

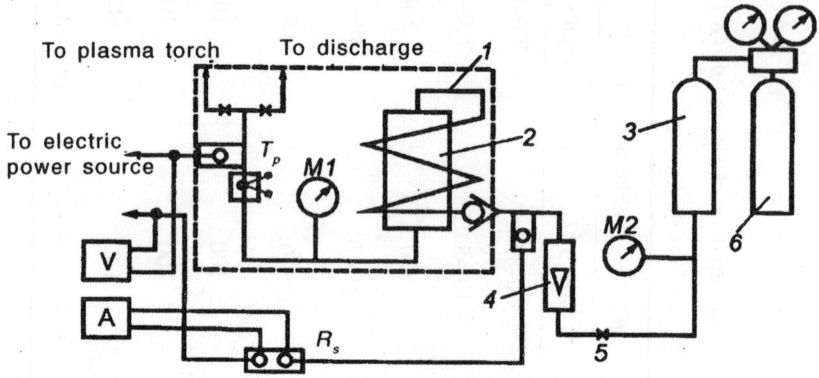

Fig. 7.11. Scheme of a steam generator. 1) tubular coil; 2) damper; 3) hydroaccumulator; 4) flow rate meter; 5) regulation valve; 6) cylinder with compressed air.

erator is continuously supplied. Electric power sources are represented by a regulated DC sources whose power and volt–ampere ratio corresponds to the characteristics of the given steam generator.

The main technical data of the direct-flow electrical steam generator (PGPE-3) are given below:

Initial medium	Distilled or chemically purified water
Resistance of tubular coil, ohm	0.7
Steam productivity, g/s	5...30
Steam temperature, °C	≤400
Steam pressure, 10^5 Pa	≤20
Power, kW	≤100
Dimensions, m	0.66×0.6×1.23
Weight, kg	90

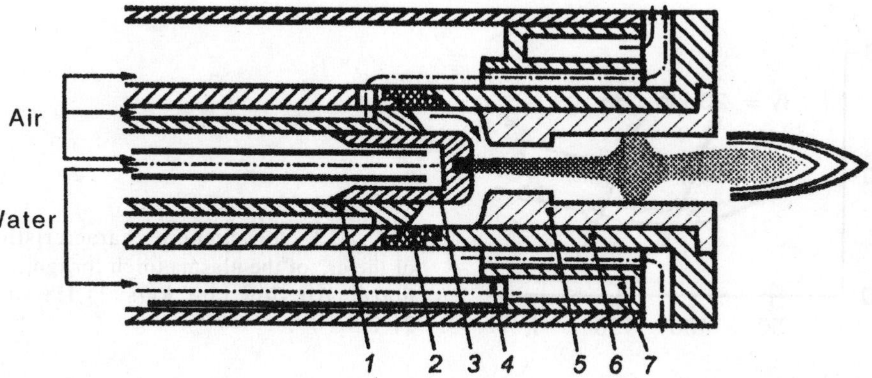

Fig. 7.12. Plasma torch for igniting a mazut jet (cooling with water or air). 1) cathode holder; 2) insulator; 3) cathode; 4) water supply pipe; 5) anode; 6) anode collar; 7) water cavity.

A plasma torch for igniting mazut

In some processes, for example, in plasma firing of the jet of mazut nozzles of an energy boiler, it is necessary to use several plasma torches simultaneously. Some of them are activated for a short period of time (operating time up to 1 min), and others for an operating time of several hours. Therefore, an advanced plasma torch has been developed. In the plasma torch, in long-term continuous operation, the thermally stressed sections are cooled by water, and in short-term operation with air [6].

The diagram of this plasma torch is shown in Fig. 7.12. Cooling of the anode *5* is indirect, i.e., only the collar *6*, containing the anode, is cooled with water or air. This makes it possible to avoid using sealing rings in the water jacket *7* of the anode, required for direct cooling. Cathode *3* with a thermochemical insert (standard component) is pressed into the copper cathode holder *1* also without sealing rings, thus increasing the service reliability of the section.

In water cooling, the water flows through the separate pipe *4* into the cavity *7* of the anode holder. Since the consumption of water for the cooling of the anode is low (up to 80 g/s), in igniting the mazut jet, the water is discharged into a firebox (although it is also possible to discharge the water in the same manner when supplying it through an additional pipe). The cathode is cooled with water by the conventional method of cooling the end electrode.

In cooling of the output electrode with air, the latter flows around the developed surface of the anode holder and then into the firebox. Similarly, the air flows around the cathode holder; part of the

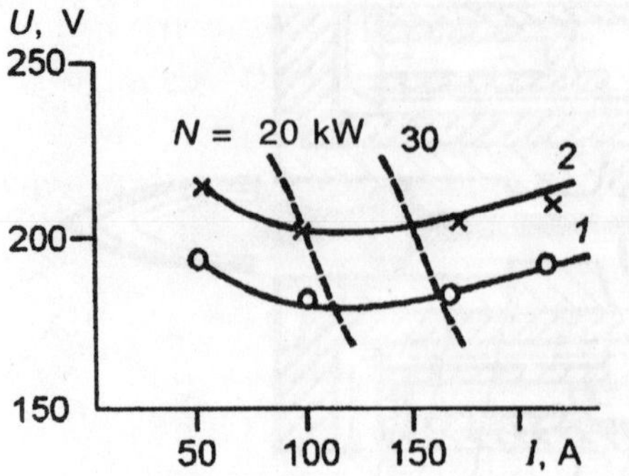

Fig. 7.13. Volt–ampere characteristics of the arc of the plasma torch for igniting mazut. Air flow rate, kg/s: 1) $1.8 \cdot 10^{-3}$; 2) $2.5 \cdot 10^{-3}$.

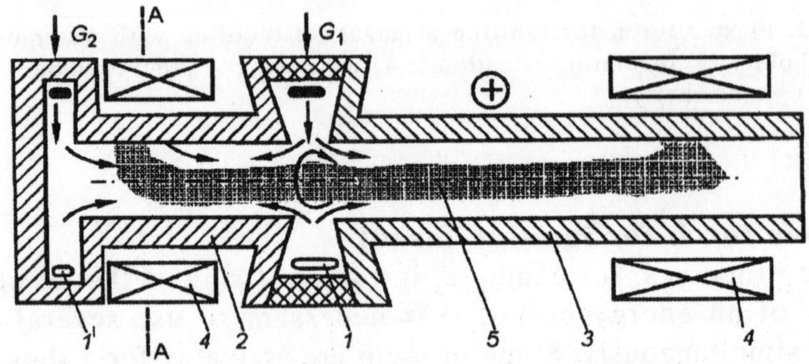

Fig. 7.14. Two-chamber plasma torch. 1,1') vortex chambers (G_1, G_2 are flow rates of the gas through the chambers); 2) cylindrical cathode; 3) anode; 4) solenoid; 5) electric arc.

flow travels into the discharge chamber, and the remaining part is discharged into the channel for supplying air for cooling the anode holder. Cooling air is also supplied inside the cathode section.

The anode is made of copper, is cylindrical, with a ledge, so that the VAC of the arc, shown in Fig. 7.13, has a rising section. The operating life of the electrodes with water cooling is up to 10 h. The thermal efficiency of the plasma torch is approximately 0.75.

7.2.2. The two-chamber plasma torch

In industry, in addition to the single-chamber plasma torches, the two-chamber linear plasma torch with tubular electrodes is also used (Fig. 7.14). To reduce the specific erosion of the electrodes, solenoids, intensifying the displacement of arc spots on the internal surface of the cylindrical electrodes, are installed on them. By varying the ratio of the flow rates of the gas G_1 and G_2, supplied into the vortex chambers, it is possible to displace the plane against the direction of the flows A–A and, consequently, the plane of rotation of the radial

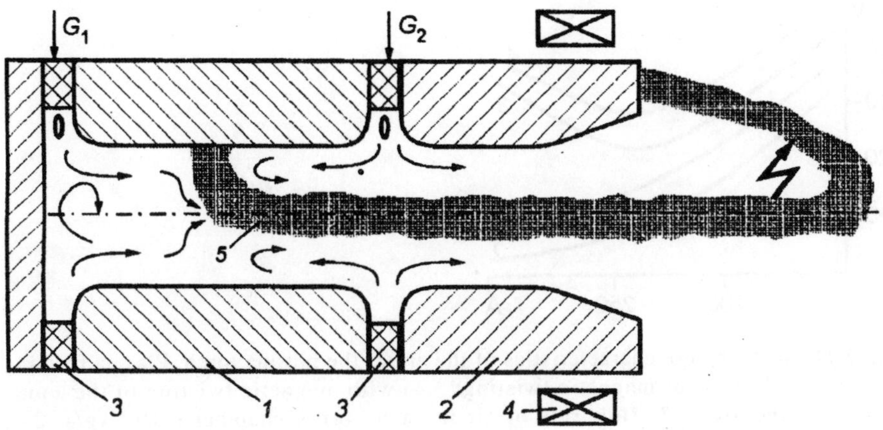

Fig. 7.15. EDP-212 two-chamber plasma torch. 1) cathode; 2) anode; 3) vortex chamber; 4) solenoid; 5) electrical arc.

section of the arc, thus ensuring the most uniform consumption of the electrode of on the length and greatly increasing its operating life.

The two-chamber plasma torch is less restricted in the selection of the type of working gas and arc current. The anode and cathode may be produced (depending on the type of gas) from copper, special cast iron, pressed material produced from powders of tungsten, copper, stainless steel and other compositions. The VAC of the arc is drooping.

7.2.3. The two-chamber plasma torch with an extended arc

The EDP-212 two-chamber plasma torch [7] is a further development of the EDP-199 plasma torch [8, 9]. It is characterised by the unusual configuration of the output electrode–anode (Fig. 7.15). It is a modified 'ledge' electrode (Fig. 2.9). In the electrode, the angle $\alpha < 90°$ but greater than 15°, indicating a possible collapse of the flow. This careful estimate is associated with the fact that the diffusor part of the channel is characterised by the supply of heat and, therefore, the angle of detachment-free gas flow may be greater than 15°. In addition to this, the absence of detachment of the flow is also indicated by erosion of a large part of the conical surface, determined by the arc spot. With the selected design, part of the arc in the form of a rotating loop is blown into the open space thus increasing the efficiency of interaction of the high-temperature electric arc zone with the surrounding medium.

The plasma torch has two vortex chambers (*3*) for the introduction of the plasma-forming gas, the internal tubular electrode (*1*), and

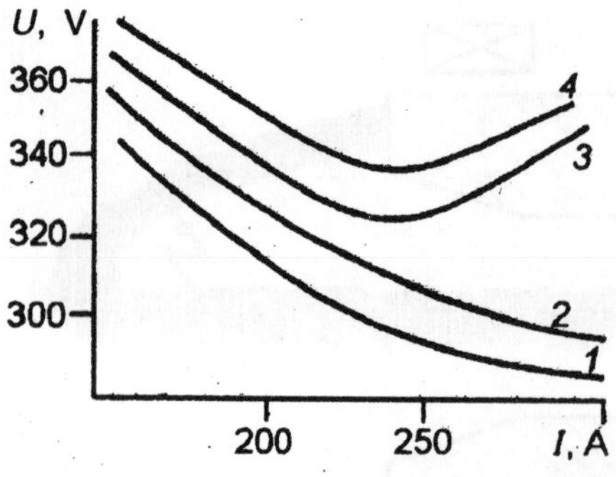

Fig. 7.16. Volt–ampere characteristics of the arc of the plasma torch. Plasma forming gas – air; 1,2) without magnetic twisting; 3,4) with magnetic twisting of the anode section of the arc. 1,3) flow rate of air in each vortex chamber $6 \cdot 10^{-3}$ kg/s; 2,4) $8 \cdot 10^{-3}$ kg/s.

modified 'ledge' electrode–anode (*2*). Both electrodes are produced from copper. Under the effect of the vortex flows of the plasma forming gas with the flow rate of G_1 and G_2, part of the arc is stabilised on the axis of the discharge chamber. In the zone of contact of these flows arriving from the vortex chambers, there is the radial (closing) section of the arc. The arc spot travels in the azimuthal direction on the surface of the cathode under the effect of gas-dynamic forces. In addition to gas-dynamic twisting of the anode spot, the arc may be magnetic as a result of the use of the electromagnetic coil (*4*) connected in series with the arc.

Without magnetic twisting, the VACs of the arc (plasma forming gas–air) are drooping (curves *1*, *2* in Fig. 7.16), and when the magnetic field is applied, they are U-shaped (curves *3*, *4*), and the falling part of the VAC of the arc is determined by the decrease of the length of the arc with increase of current as a result of shunting in the loop. The breakdown voltage in the loop decreases with increase of the current because the gas temperature increases. A slightly different situation is observed in in-series connection of the solenoid with the arc. The electrodynamic forces, acting on the arc, increase the speed of rotation of its near-electrode section and of the entire loop and, consequently, the elements of the arc 'impact' on the cold gas of the surrounding space, thus increasing the strength of the electrical field and the breakdown voltage in the arc loop and preventing the decrease of the length of the arc together with the appearance of the rising part of the VAC. In the rising part of the VAC, the arc burns in a stable manner without any ballast resistance. When the magnetic field is applied the specific erosion of the anode decreases. To increase the operating life of the cathode, it is convenient to use a number of a measures, described in detail in chapter 10.

The EDP-212 plasma torch is used at present in the plasma ignition of carbon dust fuel, in the processes of mazut-free firing of energy and water heating boilers, in the stabilisation of the combustion of the coal dust jet and stabilisation of the yield of the liquid slag in the boilers with liquid slag removal. Because of the complicated composition with the torch, the magnetic coil on the output electrode has not as yet been used. The practice of long-term service of the plasma torch in the conditions of thermal electric power stations shows that the operating life of the cathode is no less than 250 h, and that of the anode is twice as long. The thermal efficiency of the plasma torch is approximately 0.8.

7.3. PLASMA TORCH WITH THE MEAN ARC LENGTH FIXED WITH A LEDGE

Figure 7.17 shows the diagram of the simplest variant of such a plasma torch. The main special feature of this device is the step output electrode, consisting of two cylinders of different diameters, and $d_3 > d_2$, and the ratio $d_3/d_2 = 1.8$.

The sudden expansion of the channel creates aerodynamic conditions behind a ledge in which the preferential shunting of the arc takes place directly behind the zone of collapse of the flow (chapter 2). This results in a constant mean arc length in a relatively wide range of variation of the determining parameters, such as arc current, the gas flow rate and consumption (at fixed values of L_2 and d_2).

The VAC of the arc is U-shaped. Of greatest interest in practice

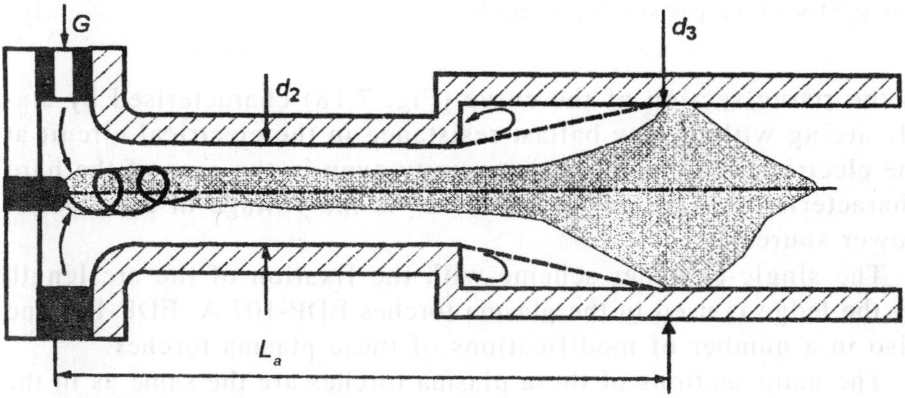

Fig. 7.17. A plasma torch with a 'ledge' and a fixed mean arc length.

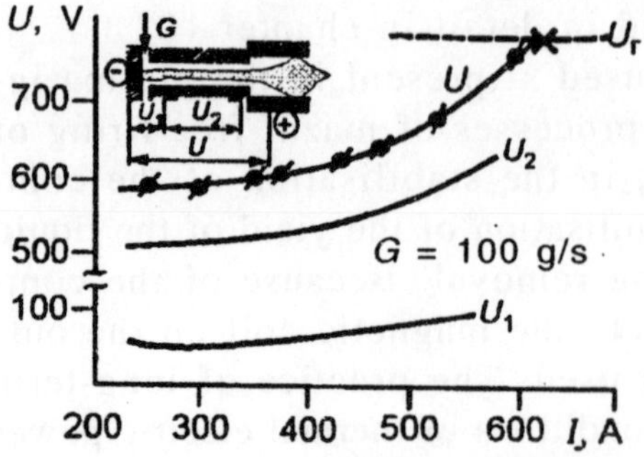

Fig. 7.18. Volt–ampere characteristic of the arc ($U = f(I)$) of a plasma torch with a ledge.

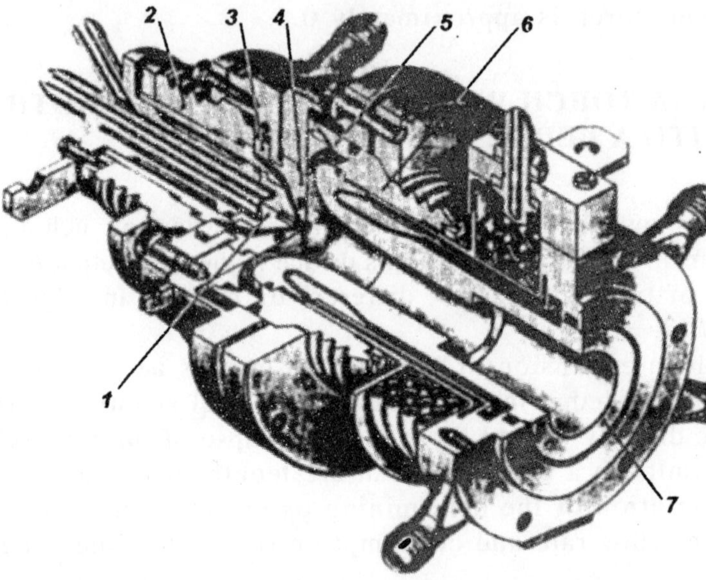

Fig. 7.19. Section through ED-210 plasma torch with a ledge, power up to 1000 kW. 1) cathode; 2,6) insulators; 3) shielding gas supply section; 4) inter-electrode insert; 5) working gas supply; 7) anode.

is the rising section of the curve (Fig. 7.18) characterised by stable arcing without any ballast resistance in the electrical circuit at the electrical efficiency close to unity even in the case of the hard characteristic of the power source U_s is the voltage of the electric power source).

The single-chamber scheme with the fixation of the arc length by the ledge is used in the plasma torches EDP-107 A, EDP-120 and also in a number of modifications of these plasma torches.

The main sections of these plasma torches are the same as in the plasma torches with the self-setting arc length. The output electrode is divided into two independent sections–spacers (in front of the ledge) and the anode. The sections may have the same electrical potentials, or maybe electrically insulated from each other.

The cathode is made of thoriated or lanthanised tungsten, and embedded in the copper holder flush with the edges, or in the form of a tungsten rod, clamped in a collett.

Plasma torches are used in pilot-plant and experimental systems for the heating of air, oxygen, nitrogen (EDP-107 A, EDP-135), hydrogen, a mixture of hydrogen with methane (EDP-120), and for plasma-chemical technological processes. Figure 7.19 shows the EDP-120 plasma torch (section) with a power of 1000 kW [2].

7.4. PLASMA TORCHES WITH THE MEAN ARC LENGTH FIXED BY THE INTER-ELECTRODE INSERT

These types of plasma torch [10] may be divided into three groups:

1. Plasma torches with the inter-electrode insert (IEI), consisting of a set of disks-sections, with part of the gas (flow rate g_i) blown into the gaps between the sections (Fig. 7.1);

2. Plasma torches with the inter-electrode insert produced from a porous material with part of the gas blown through the pores;

Fig. 7.20. GNP-1.5 plasma torch with an inter-electrode insert.

3. The plasma torches with a gas-dynamic inter-electrode insert.

Figure 7.20 shows the plasma torch with an inter-electrode insert GNP-1.5, with a power up to 1500 kW. It is designed for the heating of air, nitrogen, hydrogen and a mixture of hydrogen with methane. The electrodes and the sections of the insert are cooled with water. At a rate of up to 40 kg, the overall length of the plasma torch is 0.3–0.8 m.

In the first group of the plasma torches, the inter-electrode insert is assembled from sections (made of copper in most cases) electrically insulated from each other and from the electrode. Consequently, each section is under a specific potential in relation to the earth.

In the plasma torches of the second group, the inter-electrode insert (solid or sectional) is produced from a porous material obtained by sintering from the powders of ceramics, tungsten and other materials. Plasma torches with the inter-electrode insert of the first and second group permit control of the arc voltage by turbulisation of the gas flow in the initial section of the channel.

The special feature of the heaters of the third group is the presence of the vortex chamber with a large diameter. Consequently, there is no mass exchange between the axial gas flow, in which the arc burns, and the main gas flow. Therefore, the losses of energy through the

Fig. 7.21. Block-type plasma torch GNP-10 with a power of up to 10 MW.

walls of the chamber are determined mainly by radiation. The VAC of the arc may be both drooping and rising.

The block-type plasma torch

The experimental results were used in the development of the original design of a plasma torch with a block-type inter-electrode insert: a set of inter-electrode inserts and blocks may be used to produce plasma torches with the power of 1.5–10 MW (for plasma torches with the power smaller than 1 MW it is recommended to use the ledge scheme). Figure 7.21 shows the block plasma torch with the inter-electrode insert, type GNP-10, with the power of up to 10 MW, developed for metallurgical and chemical industries.

The equations derived in chapters 5 and 6 may be used for engineering calculations of plasma torches with the sectioned inter-electrode insert and the blowing of gas distributed along the insert, and also for calculating the plasma torches with a porous inter-electrode insert.

The calculations of the VAC and thermal characteristics of the arc for the plasma torches with the power of 0.75, 1.5, 5 and 10 MW, carried out using the equations in the chapters 5 and 6, were in good agreement with the experimentally measured values.

7.4.1. Plasma torches for heating hydrogen and water-containing media

The investigators at the ITPM Institute, in cooperation with the experts of the Scientific Research Institute of Chemical Engineering in Novosibirsk, have developed a series of plasma torches for heating hydrogen and hydrogen-containing gases. The range of the power of the investigated plasma torches is wide, from 100 kW to 10 MW. Two types of linear plasma torches were used: a plasma torch with the fixation of the mean arc length with a ledge, and a plasma torch with an inter-electrode insert. In the section, attention is given to several special features of these plasma torches.

Plasma torches with the mean arc length fixed with a ledge: EDP-109/200; EDP-114; EDP-120

The EDP-109/200 plasma generator is a single-chamber axial plasma torch with a step output electrode designed primarily for heating hydrogen and a mixture of hydrogen with methane [11]. This design has been developed to obtain the rising VAC of the arc for operation with the electric power source with a hard VAC.

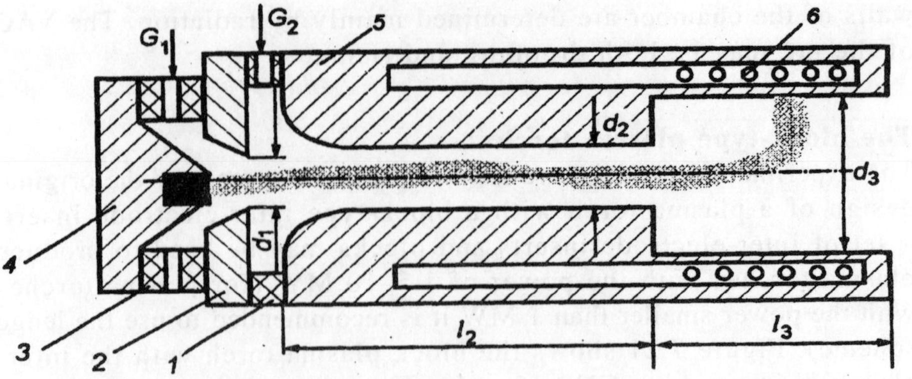

Fig. 7.22. EDP-109/200 plasma torch with the mean arc length fixed by a ledge.

The diagram of the plasma torch is shown in Fig. 7.22. The main elements of the plasma torch are: the cathode *4*, the ignition electrode *2*, the step anode *5*, the sections for the supply of the shielding *3* and main *1* gases. A thoriated or lanthanised tungsten rod is inserted flush into the copper holder of the cathode *4* and secured by brazing. The ignition electrode *2* is made of copper in the form of a disc with an internal diameter of $d_1 = 1.6$ cm. The dimensions of the copper anode: $d_2 = 0.8$ cm, $d_3 = 1.6$ cm, the length of the anode $l = l_2 + l_3 = 9–15$ cm. The solenoid *6* is installed to ensure uniform operation of the working surface of the output section of the anode.

Figure 7.23 shows the VAC of the hydrogen arc (curve *1*). For the values $l_2 \leq 3$ cm the *U–I* characteristic is rising and is positioned above the calculated characteristic for the arc with the self-setting length (curve *2*), and at a current intensity of more than 530 A it is situated higher. This is in qualitative agreement with the mechanism of formation of the VAC of the arc in a plasma torch with a step electrode, described in [12]. A further increase of current intensity

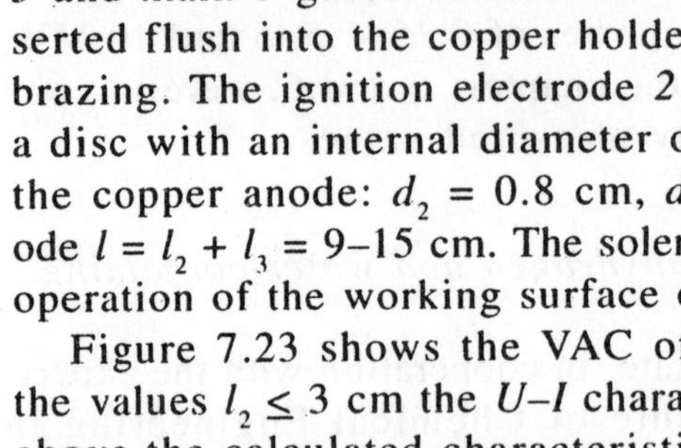

Fig. 7.23. Volt–ampere characteristics of the arc. Pressure at outlet of the plasma torch $1.4 \cdot 10^5$ Pa. Hydrogen flow rate 1 g/s; $d_2 = 0.8$ cm; $d_3 = 1.6$ cm. 1) $l_2 = 3$ cm; 2) calculated curve for a cylindrical anode with a diameter of 0.8 cm; 3) $l_2 = 3.9$ cm; 4) supply of mixture: 0.83 h/s hydrogen + 0.95 g/s methane at $l_2 = 5$ cm.

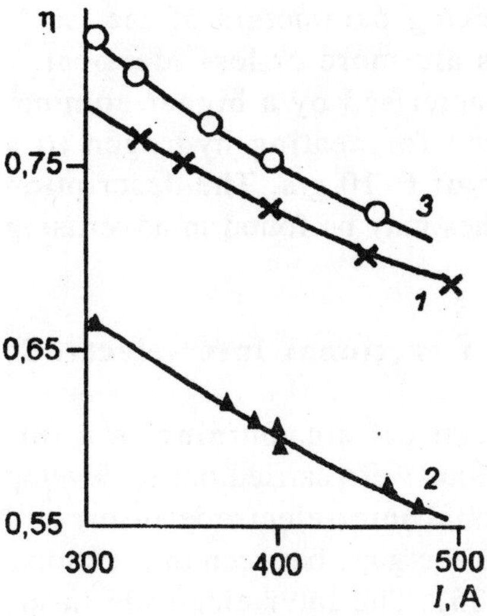

Fig. 7.24. Dependence of the thermal efficiency of the plasma torch on arc current intensity. $p = 1.4 \cdot 10^5$ Pa; d_2 = 0.8 cm; d_3 = 1.6 cm; l_2 = 5 cm, 2) G_{H2}= 1 g/s; l_2 = 9 and 11.6 cm, respectively; 3) G = 0.9 g/s H_2 + 0.8 g/s CH_4; l = 11.6 cm.

results in the merger of the curve *1* with curve 2. In the range 300–600 A at l_2 < 3 cm, the arcing is stable and shunting of the arc takes place only behind the ledge.

At l_2 = 3.9 cm, the VAC of the hydrogen arc (curve *3*) coincides almost completely with the calculated curve 2, because shunting of the arc already takes place in the section of the electrode with diameter d_2, i.e. in front of the ledge. In this case, the amplitude of the pulsations of current and voltage naturally increases.

If the working gases are represented by a mixture of hydrogen and methane, the VAC of the arc, burning in the plasma torch with l_2 = 5 cm is well above the characteristic of the hydrogen arc (compared curve *3* and *4*); arcing is stable.

Figure 7.24 shows the dependence of the thermal efficiency of the plasma torch on current intensity. At l = 9 cm and a current of I = 400 A, η = 0.72 (curve 1); the temperature of the hydrogen jet reaches 3400 K. However, if with other parameters being the same, the total length of the anode is l = 11.6 cm (curve *2*), the efficiency of the plasma torch decreases to η = 0.6. The addition of methane to the working gas (whilst maintaining a constant general volume flow rate) increases the value of η to 0.75 (curves *2* and *3*).

According to [2], the duration of continuous operation of the cathode is no less than 100 h, that of the anode no less than 300 h, at a current intensity of up to 500 A.

The design of the EDP-114 and EDP-120 plasma torches is the same. The special feature of the EDP-114 plasma torch is the absence of an intermediate (starting) electrode and the section for supplying the shielding gas. The working parameters of the EDP-109/200 and EDP-114 plasma torches are more or less identical.

The EDP-120 plasma torch is characterised by a higher nominal power (up to 1 MW) and is calculated for heating hydrogen to a temperature of 3200 K, at a flow rate of 6–10 g/s. The description and characteristics of these plasma torches may be found in advertising literature [2].

The EDP-119 plasma torch with a sectional inter-electrode insert

For physical investigations of the electrical arc, burning in a turbulent flow of different gases, investigations were carried out to develop a EDP-119 plasma torch with a sectional inter-electrode insert and blowing part of the working gas through the gaps between the sections, distributed along the channel (Fig. 7.25). The inter-electrode insert is a set of individually cooled disks–sections insulated from each other and from the electrodes. It makes it possible to vary the arc voltage in a wide range, determine the local electrical and thermal characteristics of the arc, carry out various optical investigations, etc. The design of the plasma torch has proved to be so reliable and successful in service of the of the torch has been used for many years as the main working plasma torch of pilot plant equipment for the pyrolysis of petrol and hydrocarbons, and also for the processing of organic chlorine production waste. It has been used for a number of investigations of the characteristics of the hydrogen arc.

The VAC of the arc at relatively low mean mass temperatures of

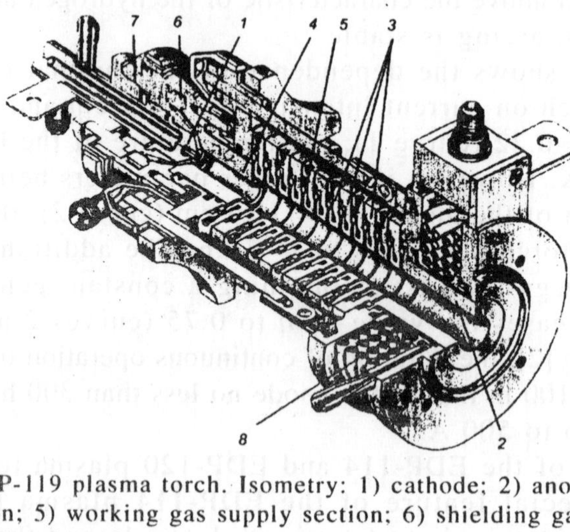

Fig. 7.25. EDP-119 plasma torch. Isometry: 1) cathode; 2) anode; 3) section; 4) starting section; 5) working gas supply section; 6) shielding gas supply section; 7) insulator; 8) solenoid.

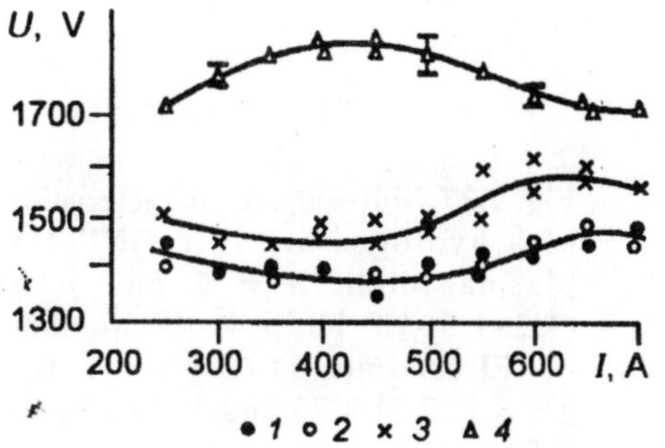

Fig. 7.26. Volt–ampere characteristics of the arc in EDP-19 plasma torch. 1) G, g/s = 7, $p \cdot 10^{-5}$, Pa = 1.1÷1.4; 2) 8 and 1.1÷1.4, respectively; 3) 5 and 1.2÷1.6; 4) 7 and 2.

hydrogen are hard or slightly drooping. The form of the characteristic for the mean mass temperature of hydrogen of 3500–3800 K is shown in Fig. 7.26. In this case at a current intensity of (600–700) A the plasma torch operates in the regime of anomalous values of the transfer coefficient. The addition of methane to hydrogen (up to 12 vol.%, curve *4*) increases the arc voltage by 10–15% and, correspondingly, the strength of the electrical field. The form of the *U–I* characteristics of the arc remains approximately unchanged.

The thermal efficiency of the plasma torch changes from 0.92–0.96 at a current intensity of 300 A to 0.8–0.9 at 700 A. The second digit in both cases was obtained after adding 8 vol.% of methane.

The duration of continuous operation of the electrode according to the test results at a current intensity of up to 700 A is: the end tungsten cathode 100–250 h; the cylindrical copper anode 200 h.

The isometry of the plasma torch and its technical characteristic may be found in [2].

GNP-1.5 plasma torch with a sectional inter-electrode insert

This torch was developed at the ITPM Institute in cooperation with the Scientific Research Institute of Chemical Engineering. The unified industrial electric gas heater with a sectional inter-electrode insert GNP-1.5 (Fig. 7.20) is designed for heating hydrogen and other gases, and also mixtures. The power is in the range 300–1500 kW. The principal circuit of the plasma torch does not differ from that of EDP-119. The plasma torch is an electric arc gas heater with vortex stabilisation of the arc and consists of a multiposition cathode [13], the anode section and the inter-electrode insert produced from individual sections electrically insulated from each other [14].

The typical VAC of the hydrogen arc in the GNP-1.5 plasma torch at a nominal power of 750 kW is shown in Fig. 7.27. It may be assumed that in the investigated range of the parameters, the *U–I* characteristic is hard. For comparison, the characteristic of the plasma torch in

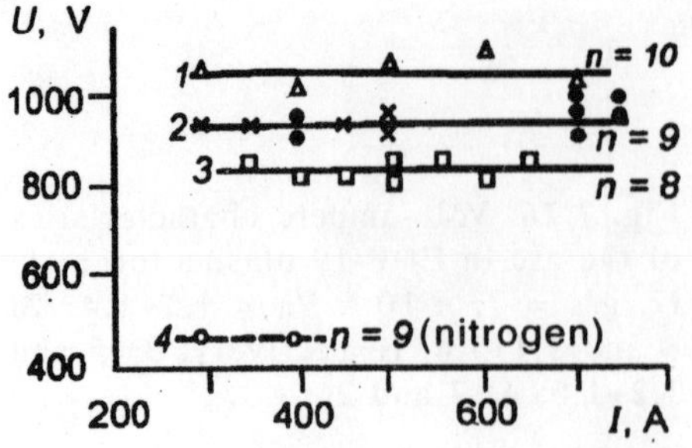

Fig. 7.27. Volt–ampere characteristics of a hydrogen arc in a GNP-1.5 plasma torch. d = 2 cm, p = (1.2÷1.5)·10^5 Pa; n_i is the number of IEI sections. 1) G, g/s = 4, G_0, g/s = 1; 2,3) 3.75 and 1; 4) 12 and 3.5.

operation with nitrogen is also shown (curve *4*). The voltage of the hydrogen arc is more than twice the voltage of the nitrogen arc.

The strength of the electrical field of the arc and the thermal characteristics of GNP-1.5 are shown in chapters 5 and 6.

The operating time of the multiposition cathode is longer than 1000 h. The operating life of the output electrode–anode of the GNP-1.5 plasma torch is approximately 200 h.

PR-05 plasma torch with the power up to 5000 kW

This plasma torch is designed for application in industrial plasma-chemical systems [2]. The design of GNP-1.5 and PR-05 takes into account the requirements of industrial service: there are no multiple hoses in water and gas supply systems, water and gas are supplied into the elements of the plasma torch through internal channels made in the body; measures are taken to ensure safety – electric power is connected through hermetic cable inputs. Figure 7.28 shows the VAC of the arc running in the PR-05 plasma torch. The working gases can be: air, nitrogen, hydrogen and natural gas.

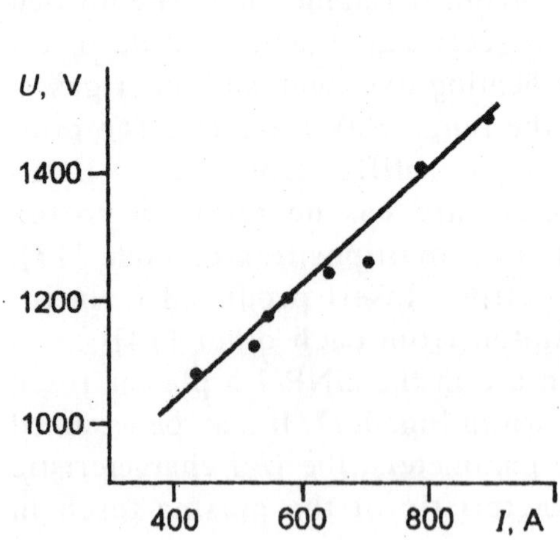

Fig. 7.28. Volt–ampere characteristic of the arc air in a PR-0.5 plasma torch. $G = 100\cdot10^{-3}$ kg/s; cathode shielded with argon, flow rate G_{Ar} = 7 nm^3/h; $d = 30\cdot10^{-3}$ m, $\bar{a}$ = 20.

High-power plasma torch GNP-10

Large-capacity processes in chemistry and metallurgy require plasma torches or blocks of plasma torches with the power of several tens of thousands of kilowatt. The construction and testing of these plasma torches requires solving a number of scientific and engineering problems [15]. These problems include: organisation of the ignition and stable burning in different gases of the electrical arc up to several metres long used for plasma-chemical processes, the continuous operating life of plasma torch, etc. The results of combined investigations of the plasma torches with a sectional inter-electrode insert, carried out at the ITPM Institute in cooperation with the Scientific Research Institute of Chemical Engineering, were used in the development of a prototype of GNP-10 plasma torch with a power of 10 MW. This plasma torch is an electric gas heater on the linear type with one-sided discharge of the plasma jet, with gas-vortex stabilisation of the arc, and with blowing of the working gas distributed along the length of the arc channel, and with magnetic rotation of the anode section of the arc. The plasma torch is designed for heating different gases and gas mixtures to temperatures of 2000–6000 K for further application in high-temperature technological processes.

From the designer viewpoint, the GNP-10 is a sectional inter-electrode insert, consisting of three blocks, connected together. The block of the sections of the first stage is the GNP-1.5 plasma torch. This block contains the cathode section, single or revolving. It is connected to the block of the sections of the second stage – GNP-5 plasma torch. The block of the sections of the third stage (GNP-10 plasma torch) contains the anode section with a solenoid for magnetic twisting of the arc. Each stage has individuals supplies of cooling water and the plasma forming gas (Fig. 7.21).

The power sources for the GNP-10 plasma torch are the standard high-voltage regulated rectifiers. Specially developed starting RC-circuits are connected to the electrical circuit of the plasma torch through the current supplies of certain sections of the inter-electrode insert, for reliable ignition of the plasma torch.

The typical VAC of the arc, running in the GNP-10 plasma torch and its stages in operation with hydrogen, is shown in Fig. 7.29. Since the plasma torch does not operate in the nominal conditions (reduced length of the inter-electrode insert, small gas flow rate, current intensity up to 500 A), the maximum power is 4 MW (curve *3*). The curve *4* in the same graph shows the thermal efficiency of the GNP-10 plasma torch, equal to 0.8–0.7.

In the process of testing the plasma torch, the erosion of the

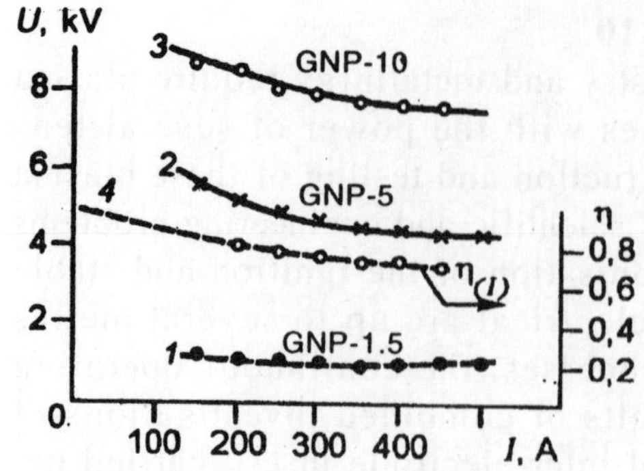

Fig. 7.29. Typical volt–ampere characteristics of the arc burning in hydrogen in GNP plasma torch. $p = 1 \cdot 10^5$ Pa; G, g/s: 1) 4; 2) 7.5; 3) 15; 4) thermal efficiency of GNP-10

electrodes was investigated. As already mentioned, the duration of operation of the revolving cathode is longer than 1000 h, the expected service life of the anode is up to 400 h.

EDB-185A plasma torch for heating hydrogen at higher pressures

In some plasma-chemical processes, it is convenient to use electric arc hydrogen heaters, operating and pressures of up to 10^6 Pa or higher. The preliminary tests carried out at higher pressures were conducted in EDP-119 plasma torches [16]. For this purpose, a nozzle is installed behind the plasma torch for regulation of the gas pressure in the channel of the plasma torch in the range $(2–6) \cdot 10^5$ Pa.

The integral characteristics of the arc will be discussed. The VAC of the arc is slightly drooping. In the plasma torch, containing 5 sections of the inter-electrode insert, and the starting section, the voltage decreases by 100–150 V with an increase of current intensity to 300–600 A. The increase of the gas pressure in the channel $(1–5)\cdot 10^5$ Pa almost doubles the arc voltage (800–1400 V). The dependence of the thermal efficiency of the plasma torch on current intensity and gas pressure at a hydrogen flow rate of $G = 3 \cdot 10^{-3}$ kg/s is presented in the following table:

Number of IEI sections	I, A	$p \cdot 10^{-5}$, Pa	η
5	300	2.56	0.87
	500	4.4	0.7
	600	5.3	0.54
10	300	2.9	0.7

Thus, with increase of the hydrogen pressure in the channel of the plasma torch, the thermal efficiency greatly decreases. At a pressure is higher than $5 \cdot 10^5$ Pa up to 50% of the heat generated by the arc is absorbed by the wall. Consequently, the density of the heat flow on the internal surface of the channel approaches the limiting value for the given cooling system. Therefore, the cooling of the elements of the plasma torch at higher pressures in the channel is of primary importance.

Taking into account the above results, the EDP-185A plasma torch was developed, designed for operation at higher pressures of hydrogen in the channel. The scheme of the plasma torch is identical with that of the EDP-119 plasma torch.

Technical characteristics of EDP-185A plasma torch

Nominal power, kW	750
Nominal current intensity, A	600
Hydrogen flow rate, g/s	4
Gas pressure in the channel of the plasma torch, Pa	up to $6 \cdot 10^5$
Thermal efficiency	higher than 0.5
Number of sections (n)	5

The VAC of the EDP-185A plasma torches shown in Fig. 7.30 for two values of pressure in the channel. It may be seen that at $I > 400$ A, the characteristic is almost completely hard or only slightly increases.

The dependence of the thermal efficiency of the plasma torch on the specific power supply to the gas is shown in Fig. 7.31. At a pressure in the channel of $3 \cdot 10^5$ Pa, the thermal efficiency at nominal parameters is not lower than 0.6.

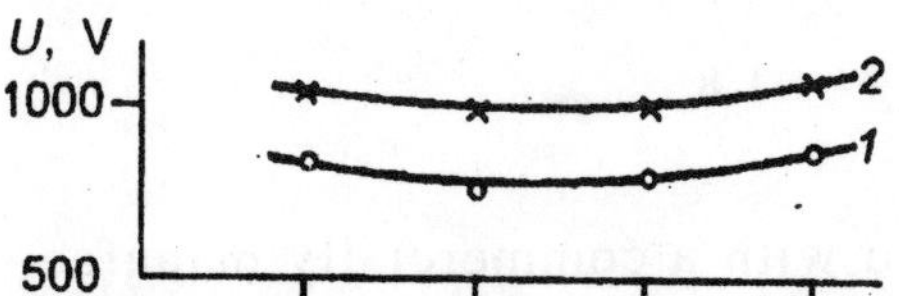

Fig. 7.30. Volt–ampere characteristics of the arc in EDP-185 A plasma torch at a hydrogen flow rate of $G = 4$ g/s and pressure $p = 2.2 \cdot 10^5$ Pa (curve 1) and $3 \cdot 10^5$ Pa (curve 2).

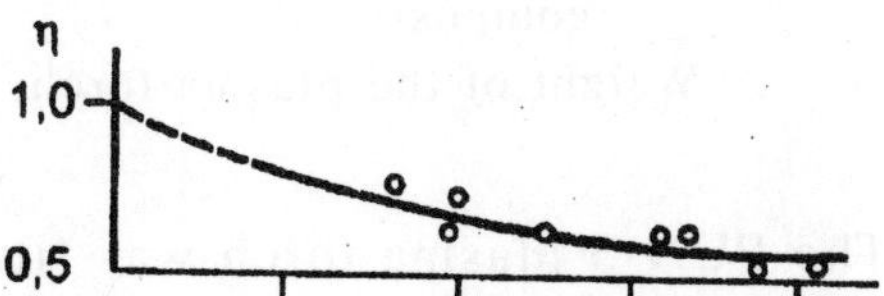

Fig. 7.31. Dependence of the thermal efficiency of EDP-185 A plasma torch on specific energy input to te gas.

The plasma torch operates in a stable manne in the given range of the parameters. However, the rod tungsten cathode does not withstand long-term operation at elevated pressures and, consequently, the duration of start-ups does not exceed 1–2 h. To develop industrial plasma torches, operating at higher pressures, it is necessary to solve the problem of stability of the cathode.

7.4.2. The unified plasma torch (PUN-3) for spraying

This plasma torch is designed for depositing coatings for various aplications on machine components and systems. The PUN-3 electric gas heater belongs to the group of linear plasma torches with inter-electrode inserts. The presence of the insert makes it possible to obtain the required power at low currents of the electric arc in comparison with the currently available spraying plasma torches produced in Russia and abroad, and greatly reduce the pulsations of velocity and temperature of the gas, discharged from the plasma torch. The PUN-3 plasma torch ensures high reproducibility of the results of spraying, may be used in automated systems and flow lines for the production of components with sprayed coatings [2].

Technical data

Nominal power, kW	30
Working current, A	170–200
Working gas	nitrogen, argon, helium and mixtures of these gases
Duration of continuous operation, h	40
Productivity using powders, kg/h	
metallic	13
ceramic	5
composite	7
Weight of the plasma torch, kg	1.8

The PUN-3 plasma torch was fitted with a commercially manufactured attachment for plasma spraying UMP-7.

7.5. PLASMA TORCHES WITH A SPLIT ARC

The increasing range of technical applications of the electric arc systems urgently requires solving a number of problems. One of the most important problems is the increase of the operating life of electrodes with a large increase of the current of the electric arc. Analysis of the literature sources shows several possible methods of solving the problem of increasing the working life of the electrodes associated with the splitting of the arc or of part of the arc into several current-conducting channels (elements of the arc), with the circuit of each channel containing or not containing a ballast resistance. In both cases, the constant condition is the stable burning of the split discharge.

The splitting of the arc with attachment of the support spots of its elements to the individual sectors of the electrodes enables:

–increase of the service life of the electrode as a whole;

–control of energy generation in a specific volume of the plasma;

–increase of the power of the plasma torch as a result of increasing the total arc current, etc.

Some of the principal circuits of these plasma torches will now be examined.

7.5.1. Plasma torch with longitudinal splitting of the arc in the output electrode

These systems are relatively simple from the viewpoint of technical realisation and were developed in the period of construction of plasma torches with inter-electrode inserts. The characteristic scheme of these torches is shown in Fig. 7.32 [17, 18]. Here the anode of the plasma torch and a number of sections of the inter-electrode insert are connected to the power source through ballast resistances. It is therefore possible to split the anode load along the arc, and by changing the value of the resistance and ratios between them to ensure stable functioning of the discharge.

The criterion of static stability for n parallel-connected arcs has the following form [19]:

$$K_n = (1/n)(\partial U_a/\partial I_i - \partial U_n/\partial I),$$

Here U_a is arc voltage, I_i is the current through the split part of the arc. Accepting that the voltage in each elementary arc is equal, it

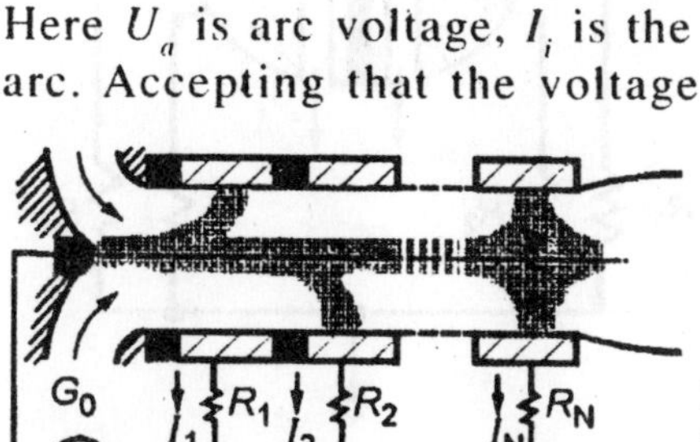

Fig. 7.32. Principal scheme of a linear plasma torch with longitudinal splitting of the arc in the output electrode.

is obtained that for stable operation of the split (distributed) arc it is necessary to ensure that the VACs of the individual elements of the arcs are increasing. If they are drooping, ballast resistances must be connected to their electrical circuit. However, these plasma torches are not used widely because of the complicated design of current supplies to the sections of the inter-electrode inserts; in addition to this, the ballast resistances greatly reduce the electrical efficiency of the plasma torch.

7.5.2. Plasma torch with a divided radial section of the arc

The systems with this division are used more widely in comparison with the previously described systems. The first device with radial splitting of the anode section was proposed in the middle of the 30s of the previous century [20]. The anode is produced in the form of longitudinal hollow bands uniformly distributed along the circumference and forming a discharge cavity. At entry on the axis of the chamber there is a rod cathode, and the plasma forming gas is supplied in the tangential direction from the side of the cathode. On the external side of the anode bands, there are magnetic coils forming a longitudinal magnetic field. The radial anode sections of the arc, interacting with the external magnetic field, rotate around the axis of the chamber. A relatively uniform plasma flow is produced at exit from the nozzle of the anode.

In the beginning of the 70s of the previous century, investigations were carried out into the system [21] identical with that described previously (Fig. 7.33). The output electrode was produced in the form of individual sectors closing the ends of the split arc. The sectors

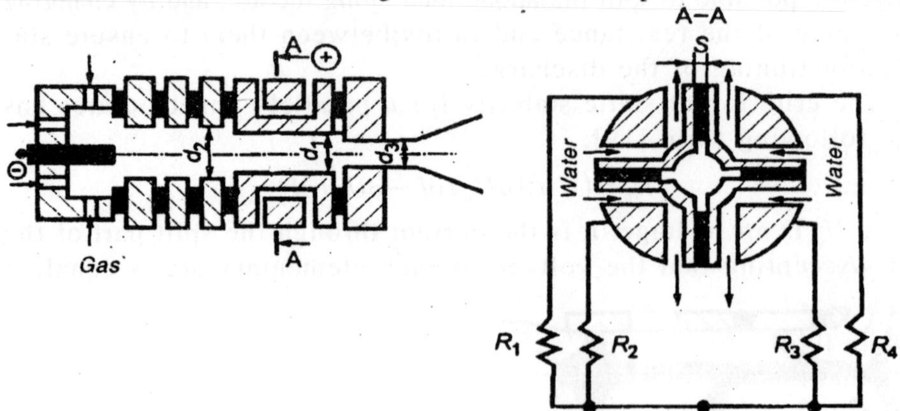

Fig. 7.33. Principal scheme of splitting of the anode end of the arc.

were insulated from each other, and the surface of the sectors facing the arc provided an additional channel to the cylinder. Each of the sectors was connected through a resistance to the power source. In contrast to the previously described scheme of the plasma system, the output of the anode sectors contains an electrically neutral nozzle, and the sections of the inter-electrode inserts were placed in the gap between them and the cathode section. The diameters of the orifices of the anode (d_1), sections of the inter-electrode insert (d_2) and the nozzle (d_3) were identical or connected together as $d_1 \geq d_2 \geq d_3$. The form of the anode sectors may greatly differ, for example in the form of pipes distributed symmetrically around the axis of the electric discharge chamber. The anode, the anode heads and the sections of the inter-electrode insert were produced mainly from copper and tungsten or from pseudo-alloys of tungsten with silver or copper.

To increase the stability of attachment of the near-electrode sections of the arc, it is sometime necessary to use the rod electrodes positioned radially in relation to the axis of the plasma torch, and the distributed electrodes are connected to the individual power source or in parallel to one of the terminals of the power source. The maximum current per electrode should not exceed 150 A. To ensure stable operation of the distributed arc, as in the previously examined schemes, the ballast resistance is connected in sequence with every rod cathode.

In the 70s, special attention was given to plasma generators with carbon distributed anodes, radially situated around the chamber at the angle of 120°. Torches of this type for technological applications were constructed by the companies Ionarc Smelters Ltd, Humphries Corporation, etc [22, 23]. The difference between them was mainly in the special features of introduction of the processed material into the electrical arc. The anodes were represented by carbon bars rotating around their axis, ensuring uniform wear of the anode rods and simultaneous stable feed of the reduction medium into the treatment zone of the material. These devices were used in construction of, for example, the technological process of extraction of the zirconium oxide from $ZrSiO_4$.

7.5.3. *Plasma torch with a split input cathode section of the arc*

These plasma torches are similar to a large extent to arc systems with the split anode end of the arc. The specific feature of the former is that the near-cathode part of the arc is blown with the cold plasma forming gas in contrast to the anode section which receives a relatively high-temperature gas flow.

Until recently, the cathode system with several electrodes was

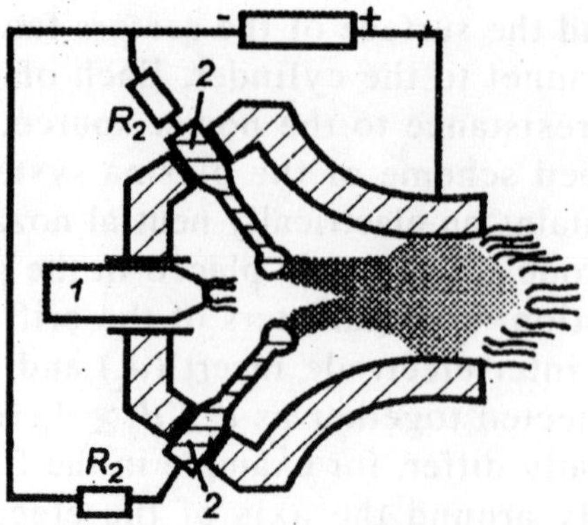

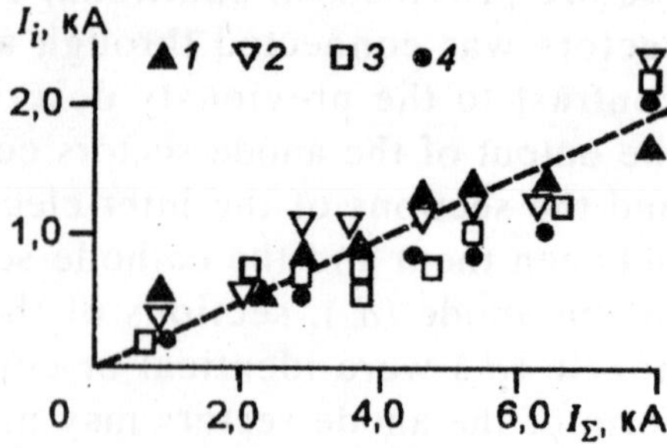

Fig. 7.34. Plasma torch with a split cathode section (2) and an auxiliary ignition plasma torch (1).
Fig. 7.35. (right) Current through individual rod electrodes/cathodes. 1–4) electrode numbers.

constructed using an additional gas ioniser, i.e., a plasma torch [24], ballast resistances with a single electric power source [25, 26] or by the formation of a discharge with a specific characteristic.

For example, laser pumping was carried out using a plasma torch with the rising section of the VAC of the arc in the area of anomalous glow discharge [27].

A variety of this cathode section is the multielectrode cathode. The plasma torch with a multielectrode cathode, described in [24] (Fig. 7.34) contains the auxiliary ignition plasma torch *1*, the electrodes *2* of the cathode system, produced in the form of variable-section bars from a refractory material with the smaller section between the working end and the base of the electrode. It should be mentioned that in specific conditions it is possible to ensure, according to the authors, diffusion attachment of the arc to the electrodes and the erosion rate can be reduced. Another important factor in this case is the relative position of the electrodes, namely the distance from the electrodes to the axis of the chamber of the plasma torch and the angle of inclination of the single electrode to the axis [28].

The experiments showed identical values of the current intensity for all four electrodes–cathodes (Fig. 7.35). It should be stressed that this is observed in the operation of the pilot arc which is used only for initiation facilitating ignition of the main high current arc, with smooth transition from zero current.

The highest stability of arcing is found if argon is used as the working gas. The presence of ballast resistances in the electrical circuit reduces electrical efficiency.

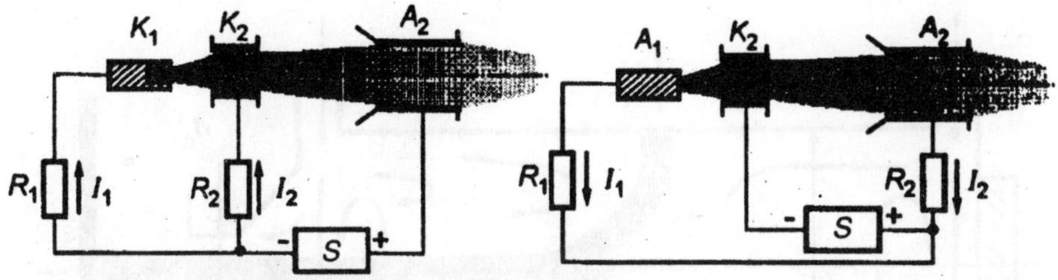

Fig. 7.35. Two schemes of plasma torches with diffusion attachment of the cathode section of the arc to the surface of the tubular electrode.

In this system, erosion of the cathode material takes place on several electrodes at a lower current intensity on each electrode in comparison with the current of the main arc. For a thermochemical cathode, for example, specific erosion increases exponentially with current, i.e. $\bar{G} \sim e^{kI}$ and, consequently, in division of the arc into several current-conducting elements, the total value $\bar{G} = \sum_i i\bar{G}_i = \sum_i e^{kI_i}$ is considerably lower than the specific erosion in the case of passage of the entire current through a single cathode.

7.5.4. A plasma torch with diffusion attachment of the cathode section of the arc to the surface of a tubular electrode

Figure 7.36 shows two schemes of the cathode section. The authors of [29, 30] noted the volume nature of the plasma flow in the region of the high-current cathode K_2, which determines the diffusion cathode attachment of the arc.

More complicated designs of multi-arc electrodes are also available, but they differ quite considerably.

7.5.5. Multi-arc cathode without ballast resistances in the electrical circuit

Many investigators have studied the process of non-stationary splitting of the arc discharge and, in particular, shunting, because the life-time of the simultaneously existing discharge channels was in the range 10^{-5}–10^{-3} s [25].

Evidently, the study [31] is one of the first describing the process of true division of the arc discharge. Examination of the shunting of the arc in the cavity of a cylindrical cathode of a two-chamber plasma torch in argon revealed the formation of two or more current-conducting radial channels which could merge into a single channel and then be divided again. Detailed investigations of the aerodynamics

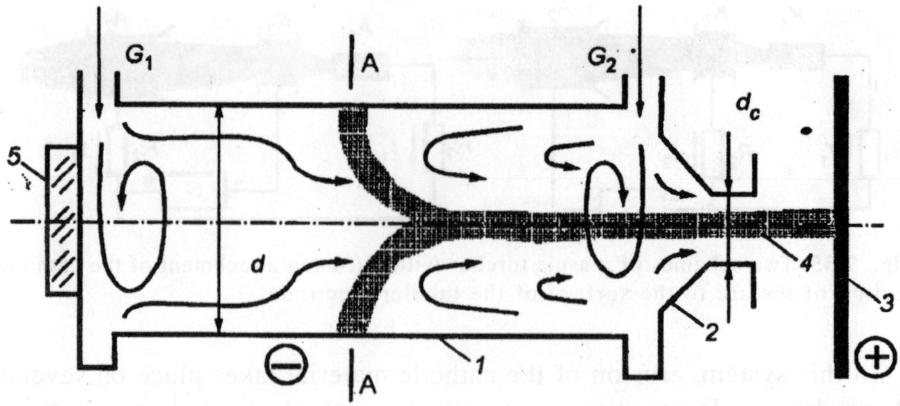

Fig. 7.37. Eperimental equipment. 1) cathode; 2) nozzle; 3) anode; 4) electrical arc; 5) inspection window.

of the twisted gas flow and of its effect on the characteristics of the electrical arc in these electrodes resulted in the conclusions on the possibilities of organising stationary, stable and controllable diffusion of discharge into several near-electrode sections in a hollow electrode under the same potential [32, 33].

The experiments carried out to examine the phenomenon of self-spontaneous division of the arc were conducted in a plasma torch whose circuit the shown in Fig. 7.37. The cathode is a hollow copper cylinder with the vortex chamber at the edges. On the one side, the cylinder is covered with a steel disc; in the experiments, the disc was replaced with glass for visualisation and filming of the near-electrode sections of the arc. The output section of the cylinder contained a nozzle through which the discharge was closed on the output electrode–anode. In some section A–A, determined by the ratios of the supplied gas flow rates G_1 and G_2, the two flows met. In the case of a relatively low discharge current, one radial section of the arc exists in a stable manner and carries out the relatively uniform rotational motion in the area of contact of the flows (Fig. 7.38a). With the increase of current, examination showed the process of non-stationary splitting of the arc into two (and more) radial sections (Fig. 7.38*b*, *c*). In some cases (*b*) the resultant sections of the discharge are almost completely stationary and exists for a long time, and in other cases (*c*) after the start of splitting, one section remains stationary and the other one continues to move catching up after a certain period of time with the first one and merging with it into a single current-conducting channel. This indicates directly the position of both radial sections

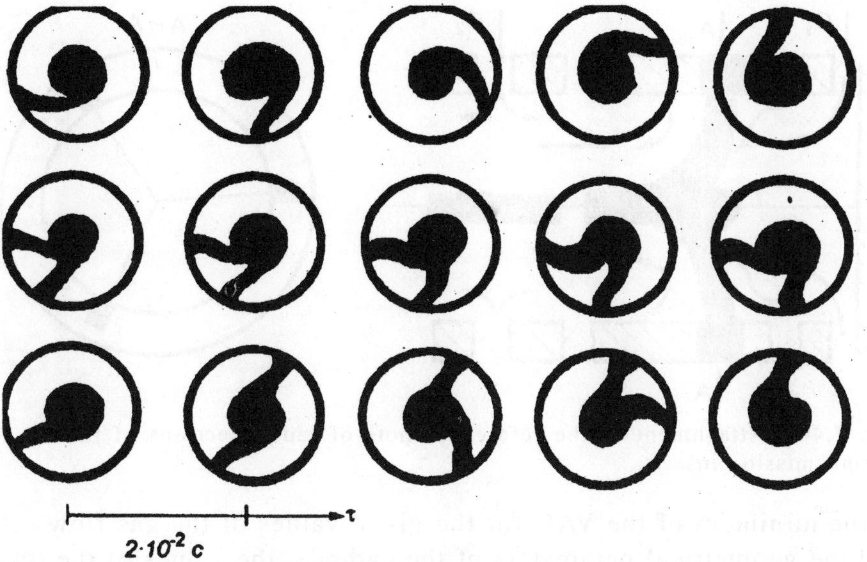

Fig. 7.38. Discharge conditions of a copper cylindrical cathode, d = 50 mm. $G_1 = G_2 = 3 \cdot 10^{-3}$ kg/s; gas – air. a) with a single radial section of the arc, I = 500 A; b,c) with two arc sections, I = 650–750 A.

of the arc in the same plane.

A special feature of the arcing regime with several unstable radial sections is the duration of existence, equal to approximately 0.1 s or even longer, which is considerably longer than the duration of a single act of shunting. Consequently, in the specific experimental conditions, the observed phenomenon of self-division of the discharge is characterised by high stability [34]. The VACs of the arcs, situated in the plane of the cathode (Fig. 7.39) have, according to the measurements, increasing and decreasing sections. With increase of the gas flow rate the minimum voltage is displaced in the direction of high current. Therefore, stable arcing of the parallel discharges, formed as a result of spontaneous division of the initial arc into, for example, two sections, having the same VAC, is possible only on the rising sections of the characteristics, when the total current $I > 2\,I_0$. Consequently, knowing the current intensity

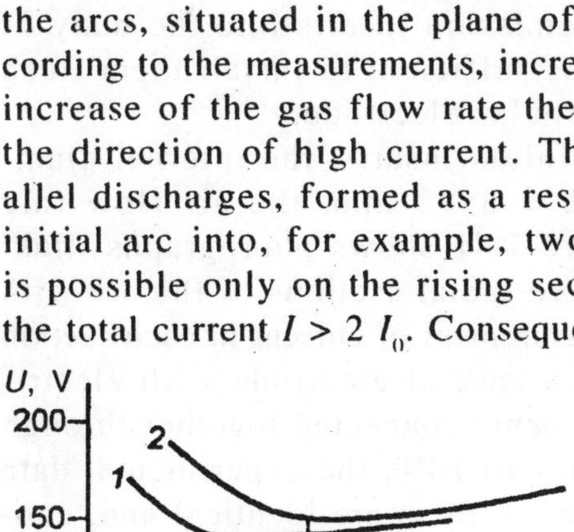

Fig. 7.39. VAC of the arc in the cathode cavity (in front of the nozzle). d = 250 mm; d_c = 20 mm. 1) $G_1 + G_2 = 6 \cdot 10^{-3}$ kg/s; 2) $7 \cdot 10^{-3}$; gas – nitrogen.

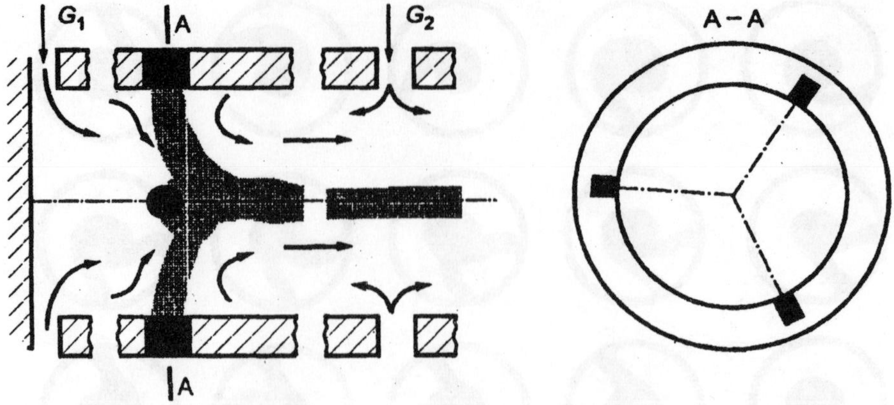

Fig. 7.40. Attachment of the reference spots of radial sections of the arc to thermoemission inserts.

in the minimum of the VAC for the given values of the gas flow rate and the geometrical parameters of the cathode, the values of the total current I can be used to determine the number of radial arcs in the stable arcing conditions: if $I \leq 2I_0$, only one radial section burns, if $2I < I < 3I_0$, two radial sections form, etc.

The VAC (Fig. 7.39), strictly speaking, are not real for each discharge because they also include the VAC of the general section of the arc which may influence the position of the minimum in respect of current. However, according to the experiments, these deviations do not exceed 15–20%.

To ensure stationary splitting of the discharge it is necessary to generate, on the internal surface of the electrode, local areas of preferential attachment of the arc in the zone of contact of the flows, for example, by installing thermoemission inserts into the body of a copper electrode (Fig. 7.40); it is sufficient to install them flush with the internal surface of the tubular electrode.

In the experiments with the hafnium inserts, the internal diameter of a copper electrode was d = 8–12 mm, the air flow rate $G_1+G_2 = (0.8\text{–}2) \cdot 10^{-3}$ kg/s. Figure 7.41 shows photographs illustrating the formation of 2, 3 or more radial sections of the arc with increasing current intensity. The measurement of current in each section of the discharge was taken using a special electrode with electrically insulated thermoemission elements connected together through measuring shunts. With the accuracy to 10%, the experimental data show that the currents flowing through them are identical and, consequently, the VAC are also identical. Investigations showed the presence of the rising section of the VAC in every section of the discharge in the stable regime (Fig. 7.39).

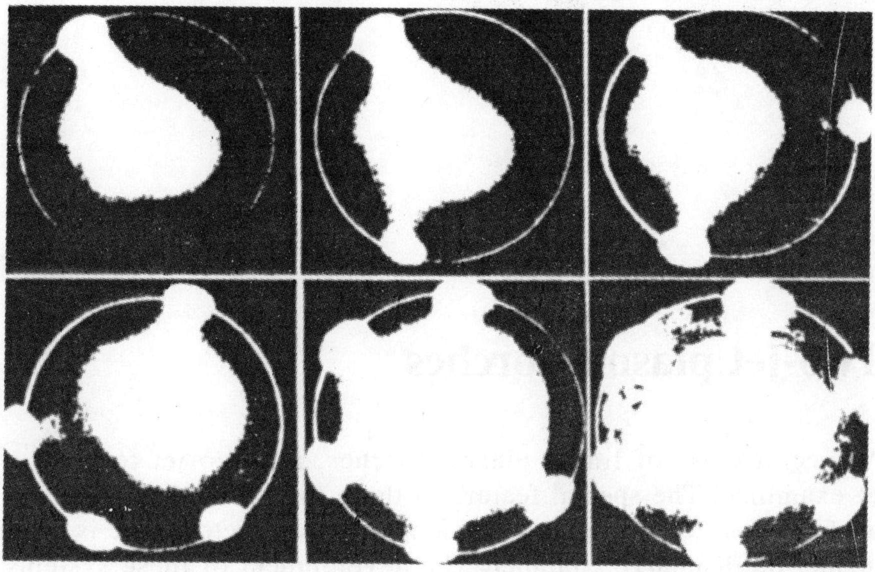

Fig. 7.41. Formation, with increasing current intensity, of two-, three- and more radial sections of the arc

In the development of plasma torches with a multi-arc cathode, based on spontaneous division of the arc, the need for supplying two gas flow rates and of controlling them results in a number of cases in the cumbersome structures of the system and complicated service conditions. However, the hydrodynamics of the flow of the twisted gas in the electrode, identical with that described previously, can also be generated in one-sided supply of the gas by appropriate selection of the geometrical dimensions of the internal cavity of the electrode. This was used as a basis for the development of high-current multi-arc cathode for plasma torches designed for cutting of metals and for general applications, for example. Active elements of the cathodes are hafnium inserts with a diameter of 2.5 mm. In electrodes, calculated for currents of up to 1000 A, there were six embedded hafnium inserts.

Chapter 8

Two-jet plasma torches

A special class of linear plasma torches, i.e. two-jet torches, will be examined. The special feature of these torches is that in all design modifications, a large part of the length of the electrical arc is situated inside the discharge chamber. The development of these systems of plasma torches was caused by the expansion of the range of technological processes in which it was convenient and more advantageous from the viewpoint of energy to exert the direct effect of the electrical arc on the process material or on the surface of the component, the powder, gas or dispersed liquid. The arcing conditions, including the electrical characteristics, are determined by a larger number of free parameters in comparison with the classic linear plasma torches: the distribution (distance) of the electrode sections in relation to each other, of the internal diameter of the output (from the electrode sections) diaphragms, the strength of the external magnetic field, acting on the immersed part of the arc, and a number of other factors. The plasma forming gases are represented by any gas required in the process.

The special feature of the two-jet plasma torch is not only the large part of the arc transferred into the open space, but also the independence of the gas supply to the the electrode sections and the working space.

The scheme of one of the first industrial variants of the two-jet plasma torch (developed in the 60s of the previous century) is shown in Fig. 8.1a [1, 2]. It is also important to mentioned earlier variants of the two-jet plasma torches [3, 4] with stabilisation of the arc by the magnetic field and with the mixing chamber of the plasma flows (Fig. 8.1b). Industrial verification was carried out at a power of the plasma torches of up to 7 MW and the atmospheric pressure. There are also systems working at high pressures.

The results of similar investigations of the thermophysical, gas-dynamic, electrical and erosion characteristics of the two-jet plasma

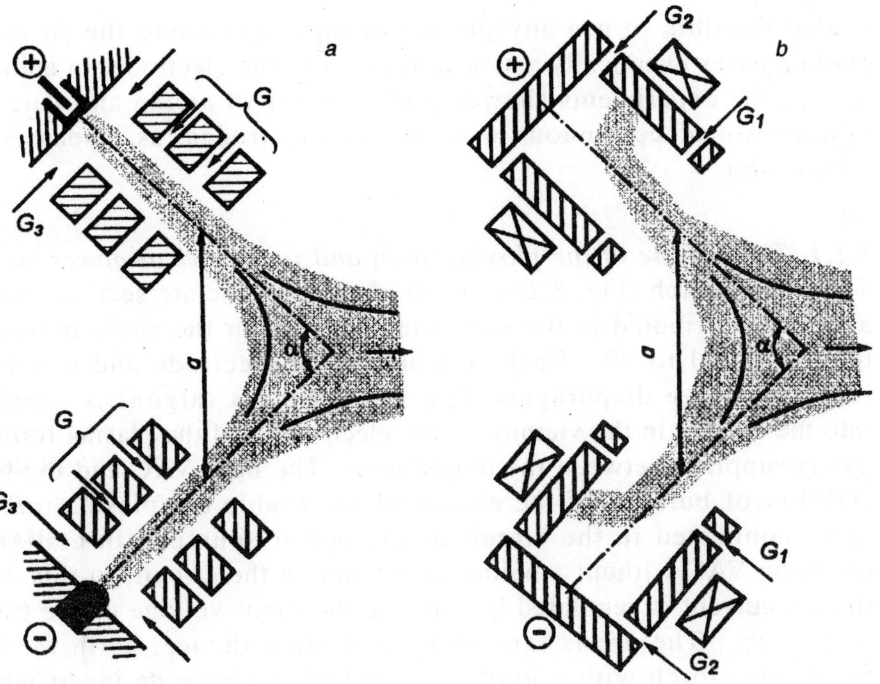

Fig. 8.1. The diagram of the two-jet plasma torch. a) with the stationary reference arc spot; b) with moving reference spots.

torches have made it possible to develop highly efficient systems operating with working gases, such as hydrogen, air, argon, oxygen, methane.

The results of investigations of the plasma torch, shown in Fig. 8.1a, are described in detail in [5]. Here, we describe briefly the principal scheme of this plasma torch and its main characteristics.

8.1. THE TWO-JET PLASMA TORCH WITH STATIONARY ARC SPOTS

In the investigated DGP-50 plasma torch, deveoped and investigated in 1972 [6], there are electrode sections with reference arc spots. The main characteristics of the torch are: the rated power 5–50 kW; arc current 50–250 A; the flow rate of the shielding gas (argon) 0.03–0.05 g/s, the flow rate of the plasma forming gas (air, argon, hydrogen, etc) 0.1–0.6 g/s. As in the linear plasma torch, the application of the additional shielding gas, introduced into the zone of interaction of the arc spot with the surfaces of the electrodes, makes it also possible to use any plasma forming gas along the flow, including gases chemically active in relation to the electrode. In addition to this, the experiments showed high stability of arcing and long operating life of continuous work of the electrodes, as discussed in detail later.

8.1.1. The scheme of the plasma torch and its electrical power supply

The plasma torch (Fig. 8.2) consists of two electric arc sections whose axes are distributed in the same plane and under the angle α smaller than or equal to 90°. Each section has an electrode and a nozzle, including three diaphragms. The shielding gas (argon) is supplied into the section in the vicinity of the electrode, and the plasma forming gas is supplied between the diaphragms. The hard VAC and the high stability of burning of the electrical arc enable the plasma torches to be connected in the circuit of the power source with a slightly drooping VAC without a ballast resistance in the circuit. In this case, the act current is regulated by varying the input voltage of the transformer Tp_1. The arc is ignited by a starting device, as in the linear plasma torch with a long sectioned inter-electrode insert where electrical capacitances are used for this purpose. The voltage from the rectifier B is supplied to the electrodes of the plasma torch. When pressing the 'start' button, a high-frequency discharge forms between the electrodes and the nearest diaphragms (in the circuit: capacitance C_1 is the first coil of the secondary winding of the transformer Tp_3– resistance R_1 – capacitance C_8). The discharge initiate electrical arcs in the gaps between the cathode (anode) and the first diaphragm. The current is closed through the capacitance C_1, the resistance R_1 and the first coil of the secondary winding of the transformer Tp_3 of the first circuit. The resultant plasma flows lead to the ignition of the electrical arcs between the electrodes and the second sections of the inter-electrode insert by means of a high-voltage pulsed voltage from the second coil of the secondary winding of the transformer Tp_3 using the capacitances C_2, C_8 and resistance R_2. Similarly, the arcs are closed from the electrodes to the service sections of the inter-electrode inserts, and, subsequently, from the plasma jets leaving the nozzles.

The intensity of the current of the auxiliary arcs is the restricted by the resistances R_1–R_3, the duration of the current by the capacitances C_1–C_3. The resistance is R_4–R_6 are used for taking the charge from the condensers C_1–C_3. When starting the plasma torch, the blocking condensers C_4–C_7 prevent the formation of a discharge and microarc's between the diaphragms of the nozzle. The LC_{10}-filter is

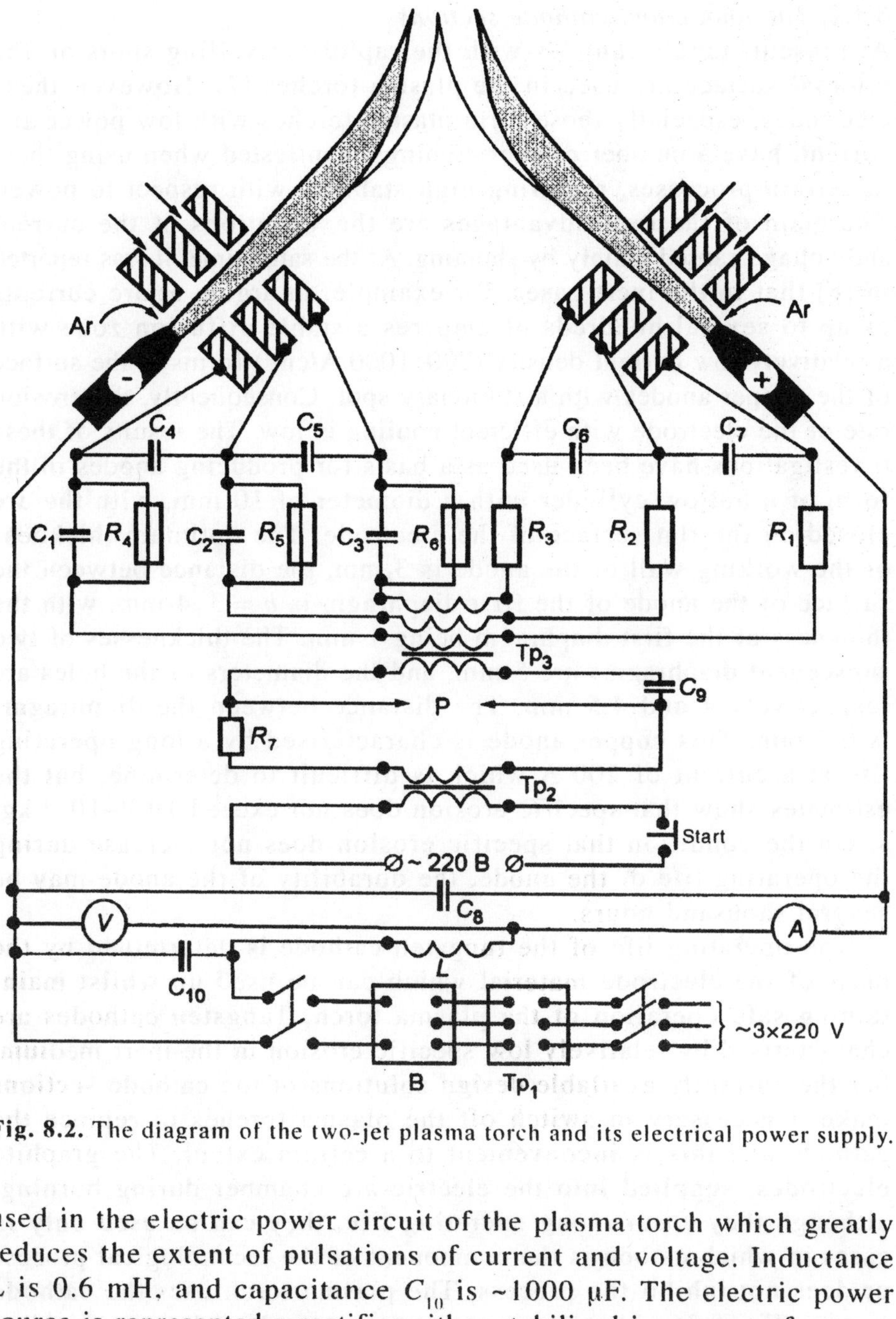

Fig. 8.2. The diagram of the two-jet plasma torch and its electrical power supply.

used in the electric power circuit of the plasma torch which greatly reduces the extent of pulsations of current and voltage. Inductance L is 0.6 mH, and capacitance C_{10} is ~1000 μF. The electric power source is represented a rectifier either stabilised in respect of current whose open circuit voltage should not be lower than 300 V, or by a non-stabilised rectifier, with a ballast resistance.

8.1.2. The anode and cathode sections

At present, tubular anodes with the rapidly travelling spots on the internal surface are used in the plasma torches [7]. However, these electrodes, especially those form plasma torches with low power and current, have a number of shortcomings manifested when using them in certain processes, requiring high stability with respect to power. The main of these disadvantages are the variations of the current and voltage caused mainly by shunting. At the same time, it was reported in [8] that in the inert gases, for example, in argon, at arc currents of up to several hundreds of amperes a stable diffusion zone with a relatively low current density (200–1000 A/cm^2) forms at the surface of the copper anodes with a stationary spot. Consequently, the erosion rate of the electrode with efficient cooling is low. The results of these investigations have been used as a basis for producing anodes in the form of a hollow cylinder with a diameter of 10 mm, with the arc closed on the flat surface of the electrode. The optimum thickness of the working wall of the anode is 3,mm, the distance between the surface of the anode of the first diaphragm is h = 3÷4 mm, with the thickness of the first diaphragm being 1 mm. The thicknesses of two subsequent diaphragms are 5 mm, and the diameters of the holes are respectively 4 and 4.5 mm. The distance between the diaphragms is 0.5 mm. This copper anode is characterised by a long operating life at a current of 200 A which is difficult to determine, but the estimates show that specific erosion does not exceed 10^{-13}÷10^{-14} kg/s. On the condition that specific erosion does not increase during the operating life of the anode, the durability of the anode may be several thousand hours.

The operating life of the tungsten cathode is determined by the mass of the electrode material which can be used up whilst maintaining safe operation of the plasma torch. Tungsten cathodes are characterised by relatively low specific erosion in the inert medium, but the currently available design solutions of the cathode sections make it necessary to switch off the plasma torches to replace the cathode and this is inconvenient to a certain extent. The graphite electrodes, supplied into the electric arc chamber during burning, although they ensure long operating life, they can be used only in cases in which carbon is the component of the technological process or does not inhibit the process. The plasma torch uses the cathode section (Fig. 8.3) with a moving tungsten electrode and argon shielding [19]. The torch is connected by a threaded joint with a copper holder so that the electrode can be moved in the presence of constant and reliable heat removal. The tungsten rod extends outside the plane

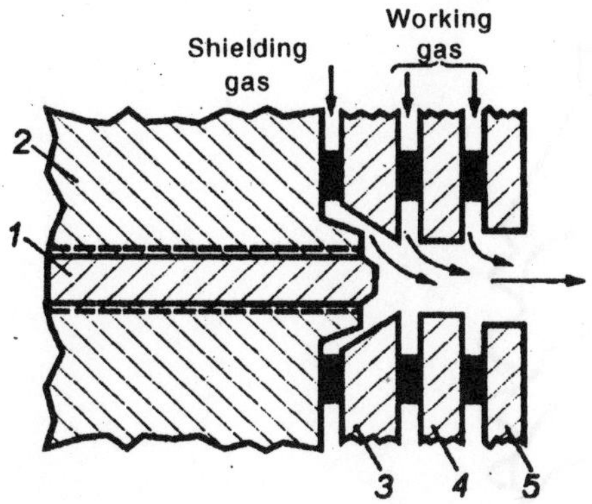

Fig. 8.3. The cathode section. 1) the tungsten cathode; 2) the copper collar of the cathode holder; 3–5) diaphragms.

of the copper cathode holder to a small distance. The electrode is displaced periodically from the holder without switching off the plasma torch thus ensuring a long operating life in continuous production. The argon medium and the low degree of specific erosion of tungsten guarantee a low degree of contamination of the plasma with tungsten vapours.

The cathode section is optimised. The diameter of the rod non-activated tungsten is 4 mm, the thread M4 × 0.7; extension from the copper holder l_k = 0.5–1.5 mm; the flow rate of the shielding gas 0.03–0.05 g/s, the flow rate of the plasma forming gas (air and others) 0.03 g/s; current intensity range 20–250 A. The components of the plasma torch are cooled by the flow of distilled water.

8.1.3. Service life characteristics of electrodes

The electrodes of the as matters were subjected to the tests to determine the service life. The alsatian was tested at optimum parameters and the our current of 200 A for 10 h. It was not possible to detect the wear of the copper electrode by weighing. There was no erosion when using, as the working plasma gases, hydrogen, air, nitrogen, oxygen, and others. Examination also showed no effect on the quality of the anode surface of sharp changes in the current both in arcing and in arc ignition. The flow rate of the shielding gas (argon) at the arc current of 50–250 A changed in the given range irrespective of the type and flow rate of plasma forming gas.

The cathode section was tested to determine the service life for 200 h at the arc current of 200 A, the flow rate of air of 0.3 g/s, and the flow rate of the shielding gas of 0.05 g/s. The initial profile of the electrode was in the form of a truncated: with the tip angle

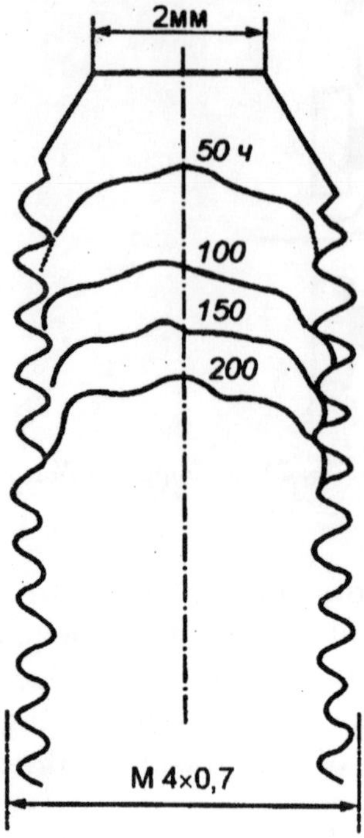

Fig. 8.4. The variation of the form of the electrode as a result of erosion in operation for 200 h.

of 60° in the area with the diameter of 2 mm (Fig. 8.4). After every 25 hours of continuous operation, the plasma torch was switched off to determine the degree of erosion and the form of the cathode contour. The specific erosion of the tungsten rod cathode changed only slightly with time and its average value was $5 \cdot 10^{-12}$ kg/C. Regardless of the fact that the cathode spot was situated in a stable manner in the centre of the electrode (the diameter of the molten zone did not exceed 2 mm), tungsten evaporated uniformly over the entire end surface resulting in the conservation of the almost constant form of the electrode with time. During operation for 200 h, the length of the electrode decreased by less than 4 mm. This means that the duration of continuous operation at the selected length of the tungsten rod may be several thousand hours.

8.1.4. Thermal and electrical characteristics

The heat flow into the electrode sections of the plasma torch was determined by calorimetric measurements. The results of the measurements in the anode section are presented in Fig. 8.5. The experimental points were obtained when the parameters were varied in the range: $h = 1 \div 2$ mm, $G_{Ar} = 0.03 \div 0.024$ g/s, $G_{air} = 0 \div 0.3$ g/s. The diameter of the first anode diaphragm was 3 mm.

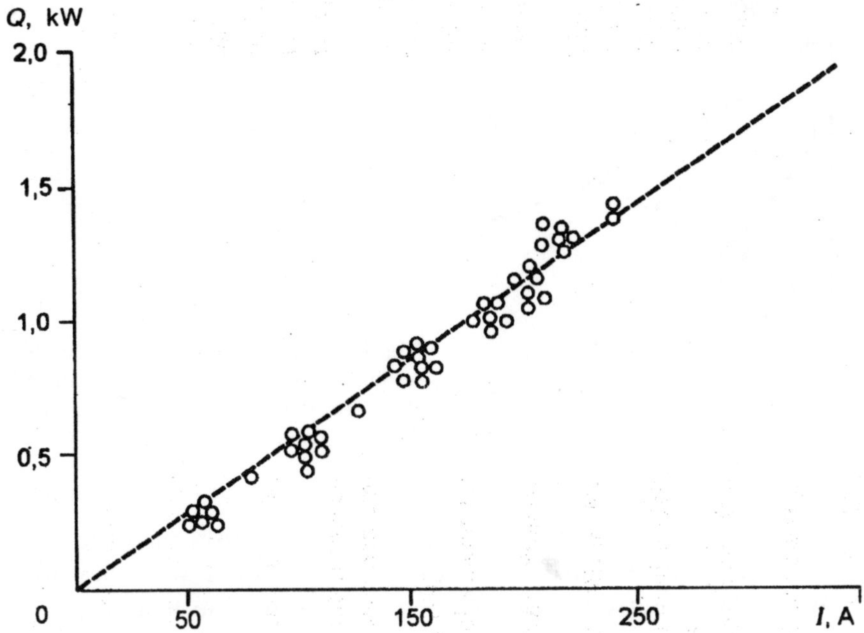

Fig. 8.5. The heat flow into the anode as a function of arc current.

Identical results were obtained for a tungsten rod anode with the diameter of 8 mm, brazed in flush into the copper holder [10]. These results are indicated by the broken line. In both cases, the dependence of the heat flow on the current in the range 40÷600 A is linear and the experimental data in good agreement, regardless of the fact that the anodes were produced from different electrodes. The heat flow into the anode is proportional to the arc current and is almost constant when the parameters vary in the given range. The volt equivalent of the heat flow is ~6 V. The comparatively low value of the heat flow may be explained by the absence of connected heat transfer because the gas flows along the surface of the electrode to the base of the arc and subsequently along the base, and also by a decrease of the anode voltage drop as a result of the increase of the plasma temperature at the surface of the anode during stop-down of the arc. Consequently, in the experiments, the heat flow was detected only through the arc spot.

The heat flow into the cathode holder and the diaphragm of the nozzle was determined at a flow rate of 0.05 g/s of argon and 0.3 g/s of air (Fig. 8.6). At a fixed extension of the electrode l_k, the heat flow increases with increasing current intensity in accordance with the linear law (curve 1), but in the current intensity range 50÷

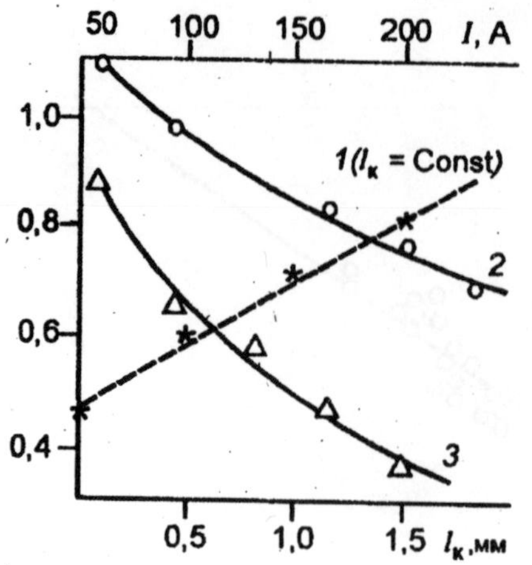

Fig. 8.6. The heat flow into the cathode holder $Q = f(I)$ at $l_k = 1.15$ (1) and heat flow into the cathode holder (2) and the first diagram $Q(l_k)$ at I = 200 A.

250 A, the heat flow is stronger than the heat flow into the tungsten cathode, pressed into the copper holder flush with the edges [10]. When the electrode extension is increased, the heat flow into the cathode holder decreases (curve 2), which is caused by the increase of heat removal from the size surface of the electrode, radiation and convection, and also by the loss of energy through the evaporation of material. The heat flow on to the diaphragm (curve 3) also decreases with increasing electrode extension. This is associated with the fact that the end of the electrode approaches the diaphragm and the electric arc, propagating from the cathode, fills the hole in the first diagram to a lesser extent.

The VAC of the arc and the efficiency of electrode sections will be examined. The results of measurements for two modifications of the two-jet plasma torch, which differ from each other by the dimensions of the diaphragms for both electrode sections, are presented below. Table 8.1 gives the values of the parameters of the two modifications of the plasma torch I and II, used in the experiments. Here Δl is the thickness of the diaphragm. The VAC were investigated for different operating conditions: the type of plasma gas, its flow rate and also the position of the plasma heads were varied.

Figure 8.7*a* shows the VAC of the arc in burning in a plasma torch of the first modification (I) with three gases. Curve 1 corresponds to the generation of purely argon plasma, curves 2 and 3 to the generation of air–argon and hydrogen–argon plasma at the maximum arc voltage obtained as a result of the separation of the plasma heads.

The VAC of operation of the plasma torch of the second modification (II) with three other gases is shown in Fig. 8.7*b*. These curves

Table 8.1. The values of the parameters of two modifications of the plasma torches in experiments

Modification	Diameter of hole and thickness of diaphragm, mm	Diaphragm 1	Diaphragm 2	Diaphragm 3	Plasma-forming gas G, g/s	Arc length l_a, cm
I	d	3.5	4.0	4.5	Argon 0.3	7.0
	Δl	1.5	5.0	5.0	Air 0.3	10.0
					Hydrogen 0.018	11.0
II	d	3.2	3.4	3.6	Nitrogen 0.24	15.0
	Δl	1.0	4.2	4.2	Helium 0.06	12.5
					Carbon dioxide 0.015	10.5

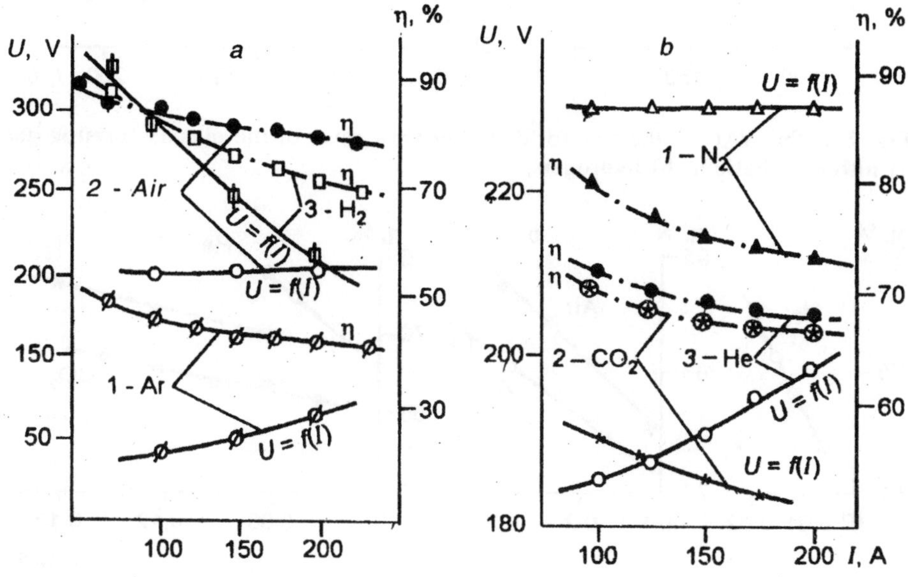

Fig. 8.7. The VAC of the arc and the efficiency of the plasma torch of I (a) and II (b) modification. Gas: a) argon, air, hydrogen; b) nitrogen, carbon dioxide, helium.

correspond to the generation of nitrogen–argon, CO_2–argon and helium–argon plasma.

The VACs, recorded for different flow rates of the plasma forming gas, are presented in Fig. 8.8. Arc voltage increases with increasing gas flow rate indicating the strong effect of the rate of the discharged gas jet on the conditions of the burning of the open electrical arc in the plasma torches of this type.

The dependence of the efficiency (η) of the two-jet plasma torch on the flow rate of the plasma forming gases is shown in Fig. 8.9. In the examined range of the parameters η is a linear function of

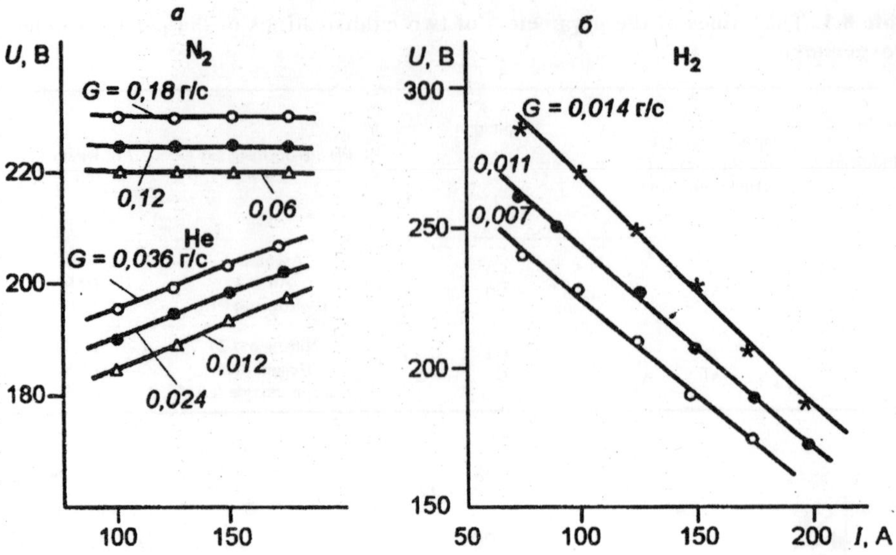

Fig. 8.8. The VAC of the Ark for different flow rates of the plasma-forming gas. a) nitrogen, helium; b) hydrogen.

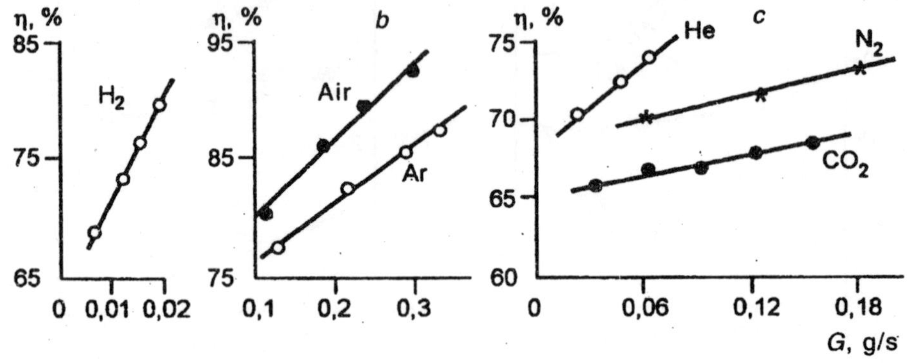

Fig. 8.9. Dependence of the efficiency of the two-jet plasma torch on the flow rate of the plasma forming gases. a) hydrogen; b) air, argon; c) helium, carbon dioxide.

the gas flow rate. The typical schlieren interferograms of the plasma flow of the two-jet plasma torch (I = 105 A) in the bands of the finite and infinite width at the flow rate of the plasma forming gas of 0.12 g/s are shown in Fig. 8.10.

8.1.5. The temperature field of the plasma flow

Measurements were taken of the radial profile of temperature for the anode and cathode jets in the sections whose distance from the outlet of the nozzle of the head of the plasma torch was 2.5; 10;

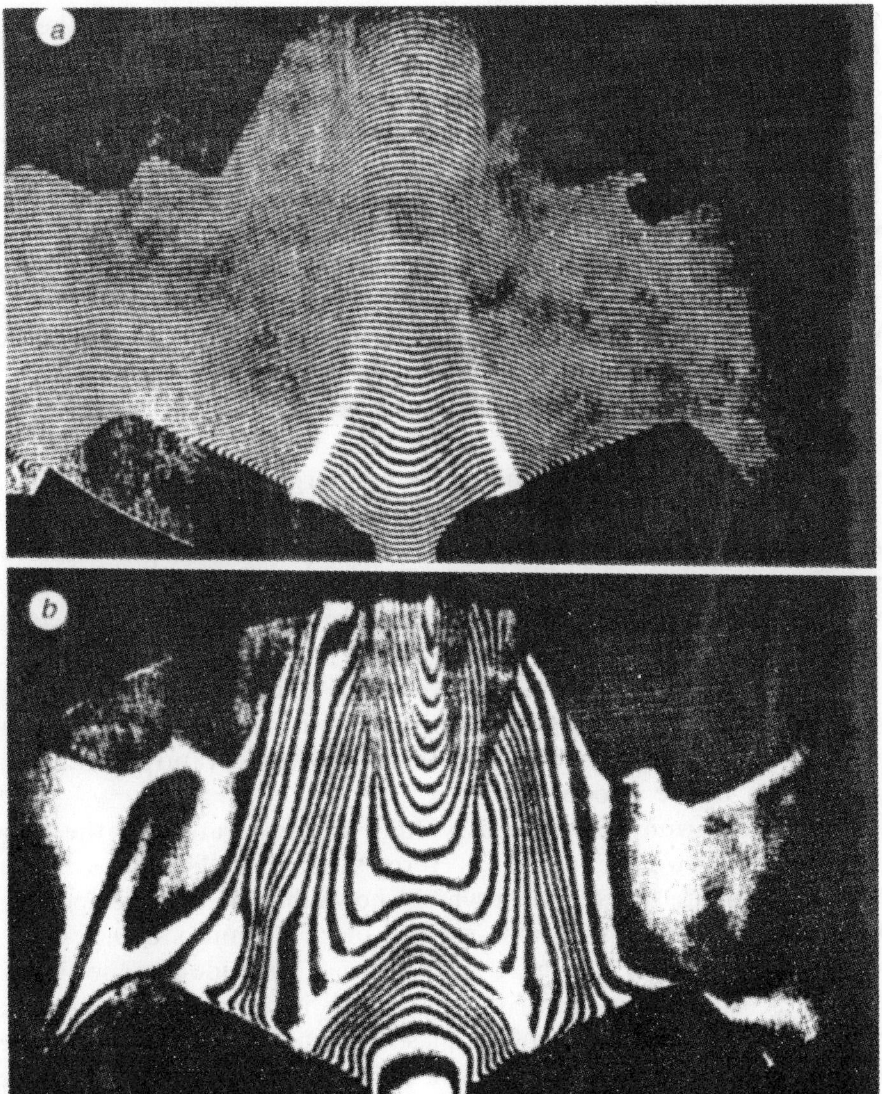

Fig. 8.10. Schlieren interferrograms of the plasma flow of the two-jet plasma torch (I = 105 A) in the fringes of the finite (a) and infinite (b) width at a flow rate of the plasma forming gas of 0.12 g/s (the photograph is published for the first time).

15; 20; 25 and 30 mm. The temperature was determined by the standard method of absolute intensity of the spectral line ArII 4806 Å and of the continuum in the range 4810 Å. The jets were regarded as axisymmetric, and the Abel equation was used for transition from the integral to local radiation intensities. In the determination of the temperature profile in the selected section, the plasma torch was placed in the position in which the axis of the jet was normal to the slit of the spectrograph so that it was possible to take into account the bending of the jets.

The temperature in the main flow was calculated on the basis of the ratio of the intensity of the pyrometric group from nine lines of tin, situated in the spectral range 2495–2863 Å. The values of the integral intensity of radiation were also used. This was determined by the complex form of the configuration of the main flow, formed by the merger of the jets, and by the absence of axial symmetry in the flow. In the temperature measurements, the local thermodynamic equilibrium in the gas discharge plasma was assumed [11–13]. Investigations were carried out at the current intensity of 105 A, arcing voltage of 145 V, the flow rate of the plasma forming argon of 0.12 g/s, and the angle of merger of the jets $\alpha = 60°$.

The results of measurements of the radial distribution of the temperature in the cathode (–) and anode (+) jets of the plasma torch in different cross sections of the torch along the axis z are presented in Fig. 8.11. The plasma jets to the area of merger are named electrode or cathode and anode jets.

The temperatures, calculated from the absolute intensity of the lines and continuum, coincide together within the limits of the measurement error. In the cathode jet, the radial temperature gradient is higher than in the anode jet.

The distribution of the axial values of temperature in the axial direction is shown in Fig. 8.12. Here, the graph also gives the error of temperature measurement which shows that the axial temperature in the cathode jet for the cross-section of z = 2 mm is 600 K higher than in the anode jet, and these values do not coincide together within the limits of the measurement error. With increase of the distance from the outlet of the nozzle of the plasma torch this difference decreases and in the section $\overline{z}$ = 30 mm, the axial temperatures of the electrodes are equalised. The steep gradient and the high axial temperature in the cathode jet are in good agreement with the literature data.

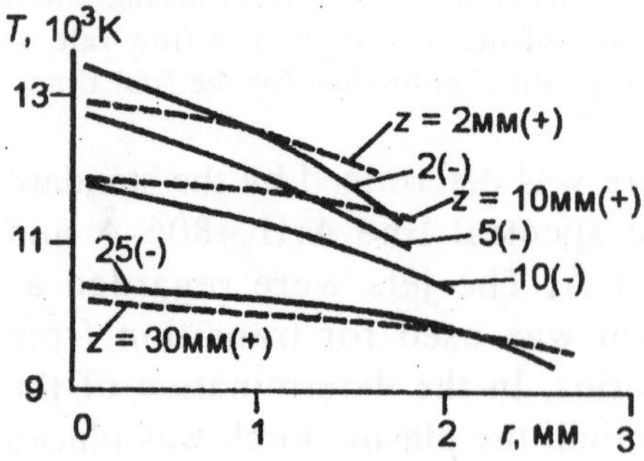

Fig. 8.11. Radial distribution of temperature in the cathode (–) and anode (+) jets in different sections along the axis z.

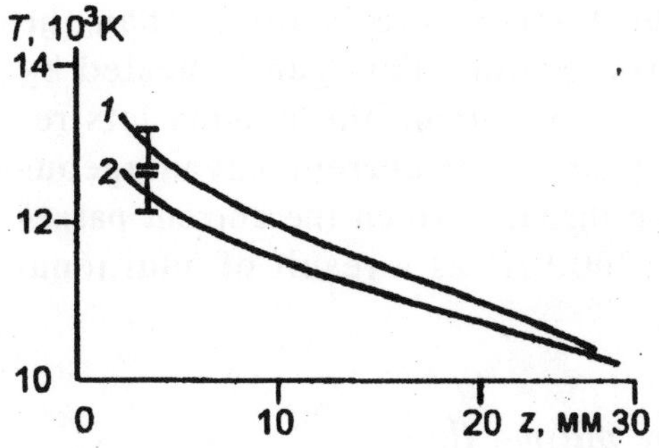

Fig. 8.12. The axial distribution of temperature in the cathode (1) and anode (2) plasma jets.

The plasma temperature at the anode is determined mainly by the transformation of electrical power, heat conductivity and by the radiation of plasma and, therefore, depend strongly on the constriction of the column at the anode. The arcs with the metallic anodes in the inert gases and in pure nitrogen show constriction at the anode only at low current (less than 30 A). At currents higher than 30 A, there is no constriction of the arc. A more or less flat diffusion region forms at the anode [8].

Figure 8.13 shows the total temperature field of the plasma flow of a two-jet plasma torch constructed on the basis of the measured true temperatures in the electrode jets and the effective temperatures in the main flow. The photograph of the plasma flow (Fig. 8.10) shows that on approach to the area of merger, the plasma jets are smoothly bent and gradually become parallel for the given operating regime of the plasma torch. In examination through an optically dense filter, the brightly glowing jets appeared to be completely independent (this is also indicated in Fig. 8.13). Measurements show that there is a

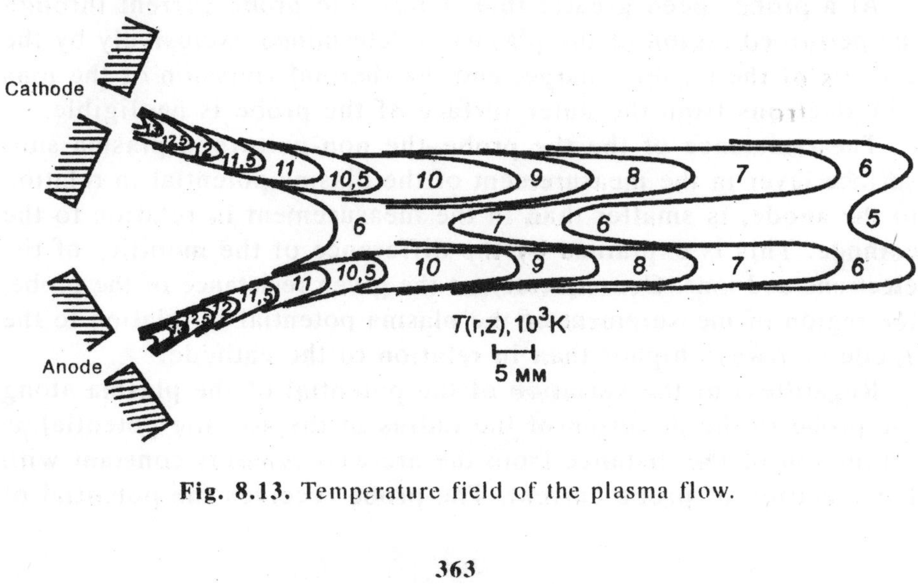

Fig. 8.13. Temperature field of the plasma flow.

gap between the jets in which the temperature is lower than the temperature of the jets in this cross-section. This gap is heated by the conductive and convective heat flows from the plasma jets resulting in the distribution of the passage of current advantageous from the energy viewpoint. The zone through which the current passes has a slightly higher temperature (7000 K) as a result of additional heating by Joule heat generation.

8.1.6. The electrical structure of the plasma flow

The electrical structure of the plasma flow of the two-jet plasma torch was investigated by the probe method. The results were used to determine the physical pattern of the process of passage of current between two current-conducting jets. The probe was in the form of a tungsten wire with a diameter of 0.2 mm. To prevent heating and evaporation of the probe material, the probe was moved through the plasma at a speed of 1–5 m/s. The dynamic perturbation, associated with the high rate, was not introduced into the gas flow.

The measurement of the potential is possible [14] when the electrical resistance of the probe circuit is considerably higher than the resistance of the plasma between the probe and the appropriate electrode. However, the latter depends on the position of the probe in the arc because the conductivity of plasma rapidly decreases from the axis to the periphery. Therefore, two regions should be separated on the oscillogram: the region of the arc in which the potential can be measured, and the region adjacent on the outside, in which the electrical resistance of the plasma between the probe and the appropriate electrode is higher than the resistance of the circuit of the measuring probe.

At a probe speed greater than 1 m/s, the probe current through the perturbed region of the plasma is determined exclusively by the carriers of the plasma charge, and the thermal emission of the ions and electrons from the outer surface of the probe is negligible.

The resistance of the the probe–the non-perturbed plasma subsurface layer in the measurement of the plasma potential in relation to the anode, is smaller than in the measurement in relation to the cathode. This is explained by the difference of the mobility of the electrons and ions. Consequently, at the given resistance of the probe, the region of measurement of the plasma potential in relation to the anode is always higher than in relation to the cathode.

Regardless of the variation of the potential of the plasma along the probe in the direction of the radius of the arc, the potential as a function of the distance from the arc axis remains constant with the variation of probe current. The probe receives the potential of the zone of the arc with the highest conductivity which is in contact with the probe.

The near-electrode areas are characterised by the presence of a radial electrical field with a high strength, and under the effect of the field the electrodes in the vicinity of the cathode tend to the discharge axis, disrupting the normal mechanism of ambipolar diffusion of the charge carriers. Thus, in the vicinity of the cathode, the discharge is constricted. On the other hand, at the anode, the radial field results in the expansion of the discharge.

By simultaneous oscillographic recording of the voltage and arc current it is possible to take into account the perturbing effect of the probe. It has been shown that the potential of the plasma point increases linearly with the resistance of the probe and at some values of the latter the potential ceases to depend on the resistance of the probe. The increase of the potential on the axis of the arc takes place until the resistance of the plasma between the probe and the cathode is higher than the resistance in the circuit of the probe; the potential ceases to increase when the resistance in the probe circuit becomes higher. Depending on the resistance of the probe, the width of the region of measurement of the potential is ~28 mm, although the actual diameter of the glowing part of the arc is not greater than 6 mm.

The results of probe investigations were used to construct the x-ray-potential lines of the arc and the flow, and the region of minimum resistance was determined by varying the probe resistance. For each current line, calculations were carried out to determine the distribution of the strength of the field and the lines of the same strength were constructed. Using the values of the strength of the field, the electrical conductivity as a function of mean temperature, the total current was calculated and it was established that diffusion current flow takes place between the jets.

The probe signal was supplied to the input of the oscilloscope through a mercury contact. To vary the probe current, a variable resistance with the maximum value of 1 ohm, equal to the input resistance of the oscilloscope, was connected parallel to the input terminals of the oscilloscope.

The linear speed of the probe was 5 m/s. The value of the speed was selected in the experiment taking into account the absence of thermal emission, evaporation of the probe material and the minimum dynamic perturbation of the plasma. The absence of thermal emission of the electrons and ions was controlled on the basis of the presence of a symmetric signal from the probe. The probe potential was measured in relation to the cathode. In connection with the anode, the amplitude of the signal was not changed. The results of investigations in the form of the field of the equipotentials are presented in Fig. 8.14. All the measurements were taken at I = 105 A, U = 145 V, the total flow rate of the plasma forming gas (argon) for both electrode sections was 0.12 g/s, the angle $\alpha = 60°$. Control of the arc voltage shows that in intersection of the arc by the probe, the voltage increased on average by 1–2 V. This variation may be ignored in comparison with the total value of voltage. In construction of the equipotentials, no account was made of the contact difference of the probe–plasma potentials because in the case of argon plasma, this difference is directly proportional to the temperature and does not exceed 6 V [15, 16]. To determine the region of passage of current, the minimum value of the probe resistance was selected by experiments in every section of the flow. At this value, the condition of constancy of the axial value of the potential is still fulfilled.

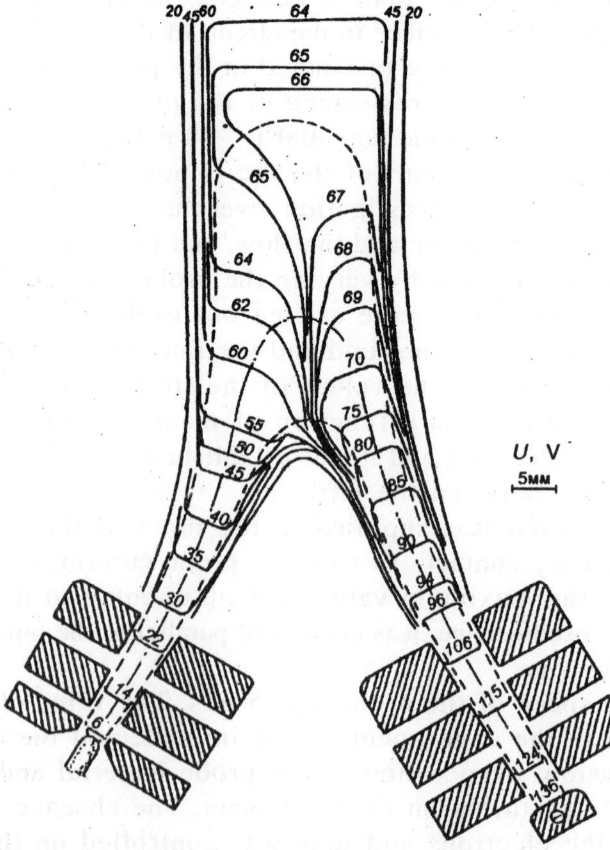

Fig. 8.14. Distribution of the electrical potential of the probe in the plasma flow (solid line). Broken line – the region of passage of current; the dot-and-dash line– the line of the lowest electric resistance of the plasma.

The region of passage of current was also determined using the current lines. The anode and cathode jets were arbitrarily divided into ring-shaped zones with a nonuniform step in the radius. Figure 8.15 shows the pattern of the section of the jets and the plasma flow in the axial direction. The boundaries of the ring-shaped zones in the form of (11) lines may be regarded conventionally as the jets of electric current. The current lines were constructed as normals to the equipotentials obtained from the probe measurements. The graph shows that the given lines are basically uniformly distributed in the direction of the height and in the gap between the jets, but rarefaction is detected along the flow. The distribution of the strength of the electrical field from the cathode to the anode was calculated along each line. Thin broken lines show the lines of equal strength of the field in the jets and in the plasma flow. As a result of the exami-

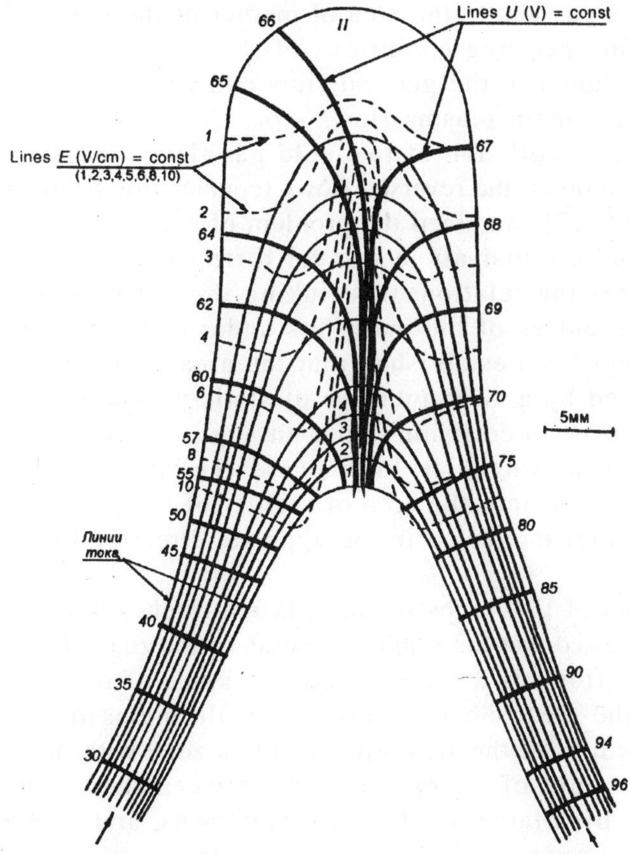

Fig. 8.15. The pattern of the flow lines (thin solid lines) and the lines of equal strength of the electrical field (broken lines) in the plasma flow, determined on the basis of probe measurements of the potential. The thick lines indicate the lines of equal potential.

nation of the temperature and electrical structures, the arc can be classified as a non-independent arc discharge.

8.1.7. Interaction between current-conducting plasma jets

The open arc, consisting of the cathode and anode jets, discharged from the electrode sections of the two-jet plasma torch distributed under an angle, is interesting as an object for examining the nature and character of the interaction of the colliding plasma conductors. From the practical viewpoint, the region of merger of the current-conducting jets should be examined to determine the conditions ensuring the high efficiency of introduction and heating of the material and also the operation of plasma electrodes used in currently available electric arc systems.

The experiments with the introduction of dust-like particles or fume-coloured gas jets to the area of merger of the plasma jets show that in certain operating conditions of the plasma torch, the latter are easily included in the general flow and penetrate into the central zone of the main plasma flow.

In the case of collision of two cold gas flows under an angle in relation to each other, the reversed flows (counter flows) always appear. The authors of [17] examined the problem of collision of two slightly non-isothermal circular air jets under different angles, determined by experiments the relationships linking the pulse of the reversed flow with the pulses of the main jets under different collision angles, and it has been established that the area of merger of the jets is characterised by a large increase of static pressure. The pulse of the reversed flow in collision under an angle of 30° may reach 4% of the pulse of the single jet. In this case, the introduction of a fine-dispersion material into the area of merger of the jets is ineffective because the latter is ejected in the opposite direction by the reverse the jet.

The authors of [7], investigating plasma torches with a displaced arc PVD-2, showed that the stable formation of the zone of displacement of the plasma flows is detected under the angle of $\alpha = 90°$ between the axes of the electrode sections. At small angles α, the cathode and anode sections of the arcs separate and a zone with the generation of a small amount of energy appears between them. The zone is characterised by a large number of breakdowns, and at $\alpha > 90°$ the size of the zone greatly decreases. It has also been established that at contact of the plasma flows, formed by every electrode section, the mixing zone is characterised by the formation of transverse plasma flows injecting the introduced material from the plasma [18]. It should be mentioned that this evidently takes place as a result of inefficient stabilisation and organisation of the flow of the plasma in the electrode sections of the PVD-2 plasma torch.

Investigations of the DGP-50 plasma torch show that as a result of the magnetic interaction in collision of two current-conducting plasma jets, the pattern of formation of the merged flow is slightly different. Figure 8.16*a* shows that at the gas flow rates higher than some value (for the selected angle α and the distance between the electrode sections) the jets are not distorted and collide under a large angle. In this case, it may be difficult to introduce the material into the central zone of the main flow because the appearance of a reversed flow is quite possible. When the gas flow rate is reduced (Fig. 8.16*b*), the current-conducting jets gradually bend and the area of merger is characterised by an almost zero collision angle. Further, the flow is formed by two parallel jets with a small space between them. With a further decrease of the flow rate of the plasma forming gas, the jets are greatly distorted and the flow is formed by two greatly diverging jets (Fig. 8.16*c*). In examination through an optically dense filter brightly glowing jets (Fig. 8.16 *a–c*) appeared to be completely independent.

The efficient stabilisation of the plasma electric arc jets and their magnetic interaction greatly change the pattern of the flow. In certain operating conditions of the plasma torch, the high pressure zone in the area of merger of the jet is characterised by the formation of a rarefaction zone supporting the formation of an injection effect which greatly facilitates the introduction of substances into the high-

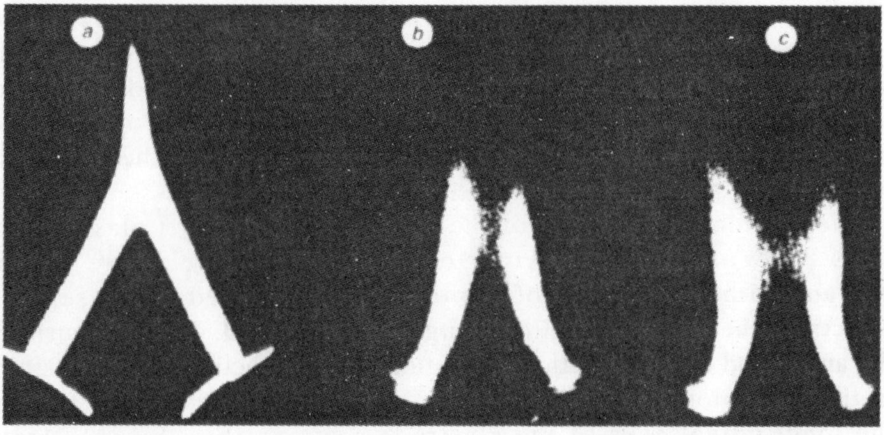

Fig. 8.16. Variation of the configuration of the plasma jet in relation to the flow rate of the plasma forming gas (argon).

temperature zone. Evidently, this phenomenon is typical not only of the given design of the two-jet plasma torch but should also be reflected in any multi-jet system with the identical direction of the flows and determine their technological efficiency.

8.2. THE TWO-JET PLASMA TORCH WITH A SCANNING ARC AND STATIONARY ARC SPOTS

In this type of DC plasma torch, the scanning (displacement) of the sections of the arc outside the electrode is carried out using the alternating magnetic field which enables the zone of merger of the plasma flows to be displaced from the electrode sections and expand the area of contact of the electric arc with the processed material or surface.

8.2.1. Electrical characteristics

In the sections, the results are presented of examination of the relationships governing the behaviour of the electric arc of the two-jet plasma torch subjected to the effect of a transverse alternating magnetic field, and the energy and amplitude–frequency characteristics of the arc in the process of spatial displacement are described [19–21]. Special attention is given to the effect of the frequency of the external magnetic field and the physical conditions of the surrounding medium on the amplitude values of the scanning arc with its energy parameters. The special features of the thermal interaction of the electric arc with the surface of the solid body in connection with the possibility of shunting part or entire current of the arc through the surface of the processed material, are discussed. The results of the investigations of the thermal and dynamic characteristics of the scanning arc of the two-jet plasma torch, obtained in a wide frequency range, may be used in the development or analysis of operation of other types of plasma torch with the displaced arc of the direct, alternating or pulsed current type.

We examine the scheme of the system of the two-jet plasma torch (Fig. 8.17*a*) and the behaviour of the electrical arc where the anode and cathode sections of the torch are subjected to the separate effects of the transverse alternating magnetic field. For this purpose, the anode and cathode sections are fitted with magnetic systems resulting in the deflection of both sections of the arcing plane passing through the longitudinal axes of the electrode sections (Fig. 8.17*b*). Argon was used as the plasma forming gas. The anode section of the plasma torch contained a flat copper end electrode, and the cathode section

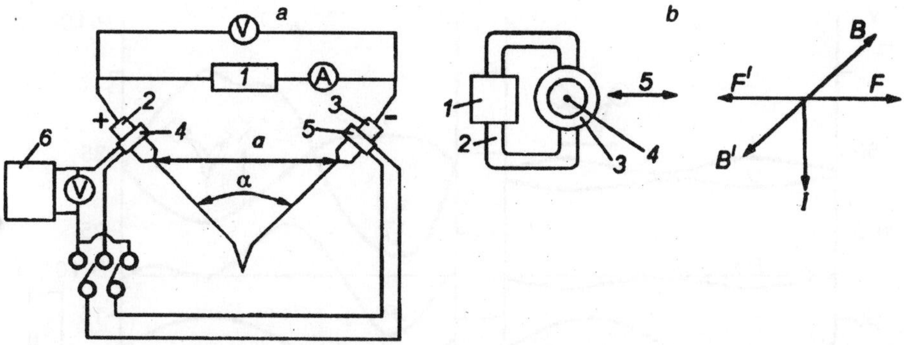

Fig. 8.17. Diagrams of experimental equipment (a) and electromagnetic system (b). a) 1 – specialised electric power source; 2, 3 – the anode and cathode sections; 4, 5 – magnetic system; 6) the power source of the induction coil; b) 1 – coil; 2– magnetic circuit; 3 – the body of the electrode section; 4 – the arc; 5 – the direction of movement of the plasma jet; B – induction of the magnetic field; I – arc current; F – the force acting on the arc.

a tungsten rod. The magnetic deflecting systems contained two coils installed on the magnetic circuits, which were secured to the casings of the sections of the plasma torch. The alternating sinusoidal voltage of industrial frequency was supplied to the coils of an autonomous power source.

In the deflection of the electrode sections of the arc in one direction, the total length of the arc changes only slightly and, consequently, the pulsations of current and voltage in the arc equal to 4–9% of the mean values (Fig. 8.18*a*). However, if the sections of the arc are deflected synchronously to opposite sides, the total length of the arc changes periodically to a considerably higher value and, correspondingly, the amplitude of the pulsations of the current and arc voltage greatly increase (20–45%), Fig. 8.18*b*.

Magnetic scanning makes it possible to regulate efficiently the power generated in the arc, and also control the movement and distributions of the sections of the arc in the area of merger. For example, the power of the electrical arc in the experiments was varied from 13 to 24 kW, and this change took place without using complicated electronic regulators usually connected to the circuit of the power source of the plasma torch. The amplitude of displacement of the sections of the arc of the investigated plasma torch and the areas of the merger reached 80–100 mm.

It should be mentioned that the oscillations of current may have a negative effect on the rate of erosion of the electrode, although the nature of these pulsations greatly differs from, for example, the pulsations caused by shunting. In the development of the circuit of

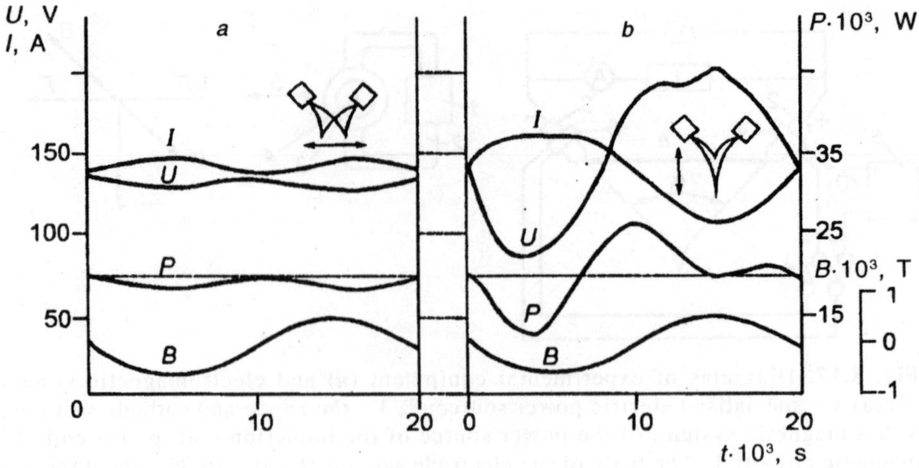

Fig. 8.18. Variation of current, voltage and of power, generated in the electrical arc, under the effect of the alternating magnetic field with the deviation of the electrode sections of the arc to one side (a) and to the opposite side (b).

a plasma reactor, this factor must be taken into account and special investigations should be carried out, if possible.

The effect of the frequency of the external magnetic field on the amplitude deviations of the scanning electrical arc will be examined [20]. It has been noted that the frequency characteristics of the electrical arc, burning in a channel, are masked by other effects and resonance phenomena and, consequently, separation in the pure form is difficult. Investigations were carried out at a current of the order of 100 A, a voltage of 140 V, the argon flow rate through every nozzle of 0.12 g/s, the distance between the output nozzles $d = 80$ mm, and the angle $\alpha = 90°$. The magnetic induction in the gap between the tips of the magnetic deflecting systems was maintained at 0.25 mT. Equipment (Fig. 8.17*a*) also contain. a GZ-33 sound generator for supplying alternating voltage of the given frequency in the range from 20 to $20 \cdot 10^3$ Hz to the coils of the magnetic deflecting systems.

Figure 8.19 shows the curves of variation of the voltage in the arc in relation to the frequency of the external magnetic field in synchronous deflection of the sections of the arc to different sides. The broken straight line shows the voltage in the arc without application of the magnetic field. The graph shows that the maximum of variation of the voltage in the arc is obtained in the frequency range 105–115 Hz. It is interesting to note that the frequency characteristics of the anode and cathode sections of the arc, obtained at separate oscillations of the sections, are similar to the frequency characteristic of the entire electrical arc.

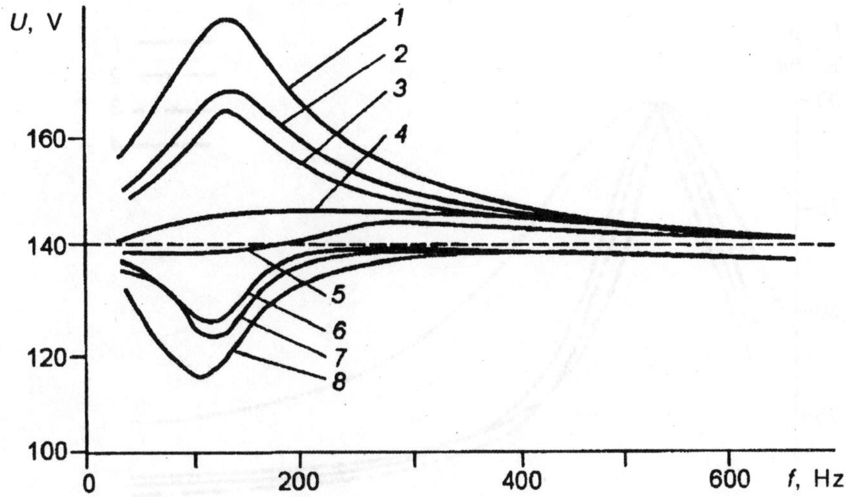

Fig. 8.19. Dependence of the voltage in the arc of the two-jet plasma torch on the frequency of the external magnetic field. 1, 8) maximum and minimum values of the voltage in the arc, respectively, with the deviation of the sections of the arc to opposite sides; 4, 5) the same, for the deviation to one side; 2, 7) the deviation of the anode section of the arc; 3, 6) the same, for the cathode section.

Figure 8.20 shows the curves characterising the relative variation of the amplitude and arc voltage during the total period of the variation of the external magnetic field in relation to frequency. At frequencies of 105–115 Hz there is a distinctive (similar to resonance [22]) maximum of the absolute and relative variation of the amplitude of oscillations and voltage in the electrical arc.

Additional experiments were carried out in which the interaction of the external magnetic field was varied from 0.12 to 0.50 mT, the flow rate of the plasma forming gas from 0.06 to 0.24 g/s, current intensity from 60 to 140 A, and the angle α from 60 to 120°. The amplitude of oscillations of the sections of the arc and the regions of the merger varied in direct proportion to the variation of the induction of the external magnetic field, inversely proportional to the flow rate of the plasma forming gas and current intensity in the arc. At the same time, examination showed the displacement of the position of the maxima of the absolute and relative variation of the amplitude of the oscillations of the sections of the arc and arc voltage on the appropriate frequency characteristics, i.e. within limits of the given ranges of variation of the parameters, determining the burning of the electrical arc of the two-jet plasma torch, and its frequency characteristics remained almost completely constant and independent of them.

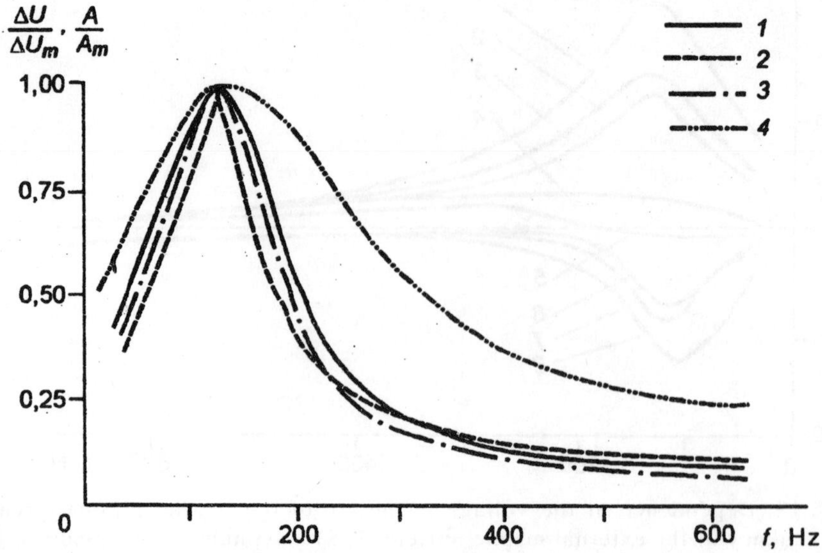

Fig. 8.20. Relative variation of the oscillations of voltage in the electric arc and the amplitude of deviation of the voltage in the alternating magnetic field in relation to frequency. 1) simultaneous oscillation of the anode and cathode sections of the arc to opposite sides; 2) the same, for the oscillation of only the anode section; 3) the same, for the oscillation of only the cathode section; 4) simultaneous deviation of the anode and cathode sections.

8.2.2. Interaction of the electrical arc with the surface of the solid

The surfaces of components are treated using plasma torches of different types, and the nature of effect of the plasma torches on the solid surface may greatly differ [21]. The indirect action plasma torches, in which the electrical arc burns in a gas-discharge chamber, are characterised by the relatively low density of the heat flow of the discharged jet of the order of $(2–6)\cdot10^6$ W/m^2 [7, 23, etc]. At the same time, the high gas-dynamic pressure, characteristic of the jet high-temperature flows, often complicates the application. The high-density of the heat flow to the substrate, up to $(1.5–2)\cdot10^7$ W/m^2, is characteristic of the direct action plasma torches in which the cathode or anode end of the arc rests on the treated surface. However, in the attachment area, the arc is usually constricted which, as mentioned previously, results in the failure of the surface layer of the treated material, and complicates the treatment of thin-wall structures and layers of protective coatings.

Investigations were carried out to examine the interaction, with the surface of the solid, of an arc plasma filament stabilised by the rotating cylinder in the direction parallel to the surface. In this case, the specific heat flow from the electric arc reaches $2 \cdot 10^7$ W/m^2 [24, 25].

The treatment of engineering non-conducting components of the brick and ceramic plates type is carried out using an extended constricted arc, and large engineering components, requiring arc powers of up to 70 kW are processed using plasma torches with electromagnetic compression of the arc to the treated surface [26–28].

When solving these problems, it is promising to use the two-jet plasma torch. In this case, the surface of the component can be treated by the plasma flow of the cathode and anode jets in the area of merger. The density of the heat flow in the zone is characteristic of the indirect action plasma torches. The direct contact of the arc with the treated surface may result in extremely high densities of the heat flow, characteristic of the direct action plasma torches. The treated surface may be both electrically conducting and non-conducting, flat or with a complicated relief, dense or porous.

The interaction of the arc of the two-jet plasma torch with the surface will be examined. The heat flows were determined using a copper water-cooled disc sensor with a diameter of 110 mm.

When treating the surface of the solid with the electric arc, several characteristic treatment conditions may be realised depending on the mutual position of the solid and the arc (Fig. 8.21). Of these, there are three main ones, presented below in the order of increasing intensity of treatment: 1) without direct contact between the treated surface and the electrical arc; 2) with contact between the electrical arc and the treated surface, but without current contact with the surface; 3) treatment directly (mainly) with the electric arc with partial or complete shunting of the arc current through the treated layer of the component. In the case of non-conducting components only the first two regimes can be used.

Prior to contact between the arc and the surface (1), the electrical characteristics of the arc (current and voltage) remain unchanged.

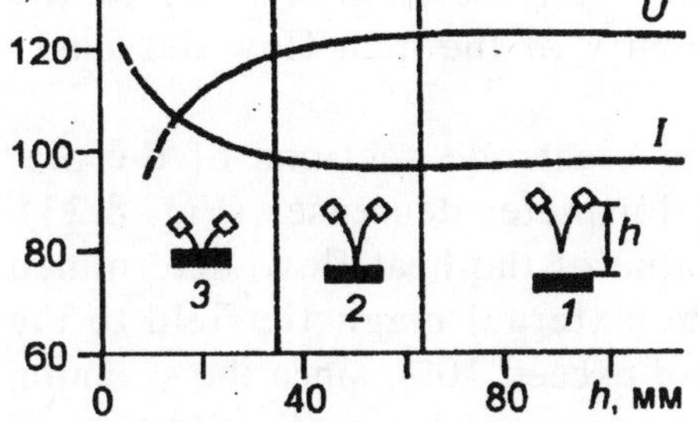

Fig. 8.21. Variation of the current intensity and arc voltage of the plasma torch in relation to the height of the torch above the surface of the solid.

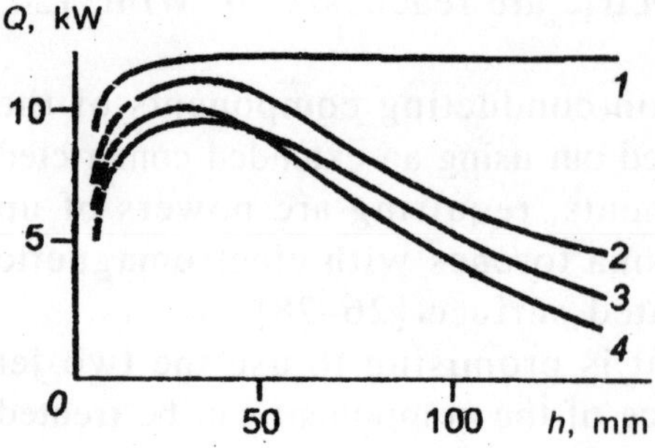

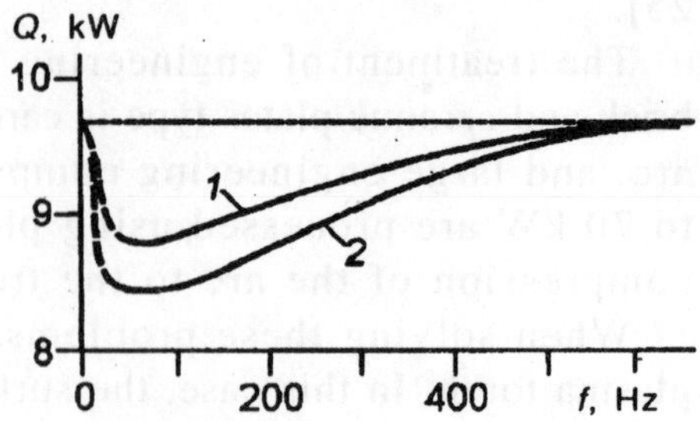

Fig. 8.22. Variation of the electrical power of the arc of the two-jet plasma torch (1) and the heat flow into a calorimetric sensor in relation to h without applying the external magnetic field to the arc (2) and with application of the field with synchronous oscillation of the anode and in cathode sections in the opposite sides (3) and to one side only (4). Argon, $G = 0.12$ g/s; $f = 100$ Hz.
Fig. 8.23. The dependence of the heat flow from the electrical arc into the calorimetric sensor on the frequency oscillations of the arc in the external magnetic field. 1, 2) for synchronous variation of the anode and cathode sections to opposite sides and to one side, respectively. $G = 0.12$ g/s; $h = 60$ mm, argon.

From the moment of contact (2), the region of merger of the cathode and anode jets is differentiated, resulting in an increase of current and a decrease of arc voltage. In particular, this is evident from the moment of shunting of the anode and cathode sections of the arc through the copper calorimeter (3).

Analysis of the results of measurement of the heat flows, presented in Fig. 8.22, shows that they depend strongly on many factors. Curve 1 corresponds to the power generated in the arc, curve 2 to the intensity of the heat flow into the calorimeter from the stationary electrical arc; curves 3, 4 are the power oscillating in the alternating magnetic field with the frequency $f = 100$ Hz. The form of the curves shows that on approach of the arc to the surface of the calorimeter, the intensity of the heat flows increases and reaches the maximum value for the region of transition from the regime of treatment of the surface without current contact of the electrical arc with the surface to the regime with shunting of the arc current through the treated layer of the components. Subsequently, the intensity of the heat flow decreases with a decrease of arc length.

In magnetic scanning of the near-electrode sections of the are, the heat flow from the arc to the calorimeter decreases (Fig. 8.23). The maximum deviation from the value of the heat flow, determined by the arc without application of the external magnetic field to the arc (broken line), is small and does not exceed 10%, since the scanning with the electrical arc results in the more uniform distribution of the heat flow on the surface.

The efficiency coefficient of the process of heat treatment of the surface, i.e. the fraction of arc energy, transformed into the heat flow into the solid, reaches: 90–95% in shunting of the arc current through the surface of the component, 70–80% in contact between the arc and the surface, and 40–50% without contact.

To continue discussing the subject of surface treatment of the materials, it is useful to mention briefly the special attention given to the development of other methods of surface hardening of steels using various concentrated energy sources, such as quenching with the electron beam, laser, high-frequency and pulsed quenching.

They are based on the general concept of rapid heating of the surface layer of the metal to high temperatures (almost up to melting) and subsequent rapid cooling (self-quenching) of this layer with the rates not lower than the rate of the reversed martensitic transformation.

In contrast to laser quenching with its low efficiency (~7–10%) and plasma quenching, in which the rate of heating of the layer is restricted of the top by the rate of heat transfer through the phase boundary, the methods of electron-beam quenching and high-frequency pulsed quenching (HPQ) are of the surface-volume type when the energy generated in the quenched layer and the heating rate are determined only by the design of the energy sources.

Special attention should be given to the method of high-frequency pulsed quenching. This method is the cheapest, may be based on standard high-frequency generators, is highly productive and easy to automate. Industrial systems for this method, using high-frequency generators with a power of up to 200 kW and frequencies of 66 and 440 kHz have already been developed. They make it possible to produce hardened layers on steel and cast iron components with a depth of up to several millimetres with the structure of fine-dispersion martensite [29].

At specific powers in the pulse of up to ~10^5 W/cm^2, it is possible to harden any structural, tool and other steels with the carbon content of [C] $\geq$ 0.3% [30), and the hardness of the quenched layer may reach more than 60 units of HRC$_e$. For low carbon steels, the method of hardening them using high-frequency pulsed quenching with simultaneous case hardening, has been developed [31].

There are a number of applications of high-frequency pulse quenching. The method is used for hardening the surfaces of shafts and bodies of revolution, internal surfaces of sleeves and cylinders, the surface of flat components (the guides of machines, cutting dies, etc). Deformation is extremely small (less than $8 \cdot 10^{-5}$ m/m) which is especially important when hardening long components (rods, shafts, etc).

Recently, high-frequency pulsed quenching has been used successfully in a number of plants in Russia.

8.3. TWO-JET PLASMA TORCH WITH TUBULAR ELECTRODES

The special feature of this plasma torch is that it can operate without a shielding inert gas supply to the zone of attachment of the arc spot to the electrodes. In addition to the fact that the argon is relatively expensive, to maintain a specific minimum flow rate of argon (this

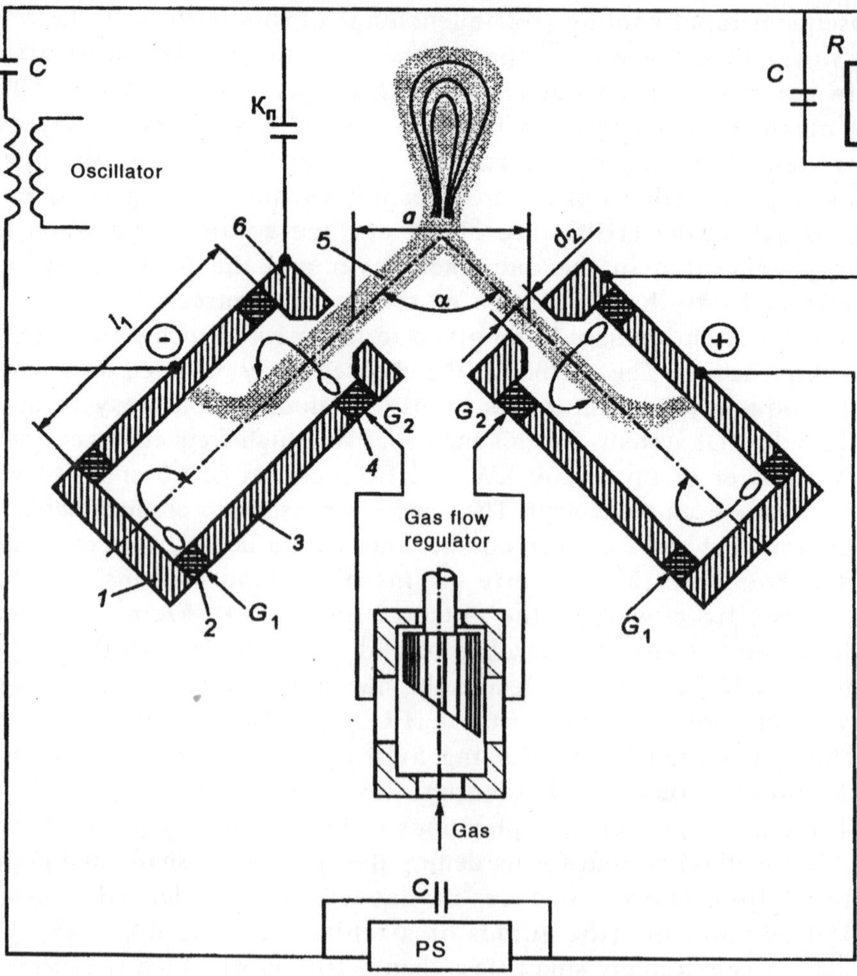

Fig. 8.24. The diagram of the two-jet plasma torch and its electric power supply.

has a controlling effect on the service life of the electrodes), special attention must be given to service of the plasma torch. The value of this factor increases in the conditions in which the operation of the plasma torch is not controlled continuously. The application of tubular copper electrodes greatly expands the working current range.

8.3.1. Design of the plasma torch and electrical circuit

The plasma torch (Fig. 8.24) consists of two identical electrode sections [32–36], each of which represents schematically the previously described two-chamber plasma torch. The electrode section contains: two vortex chambers 2 and 4 through which two flows of the plasma forming gas with the flow rate G_1 and G_2 twisted in the same direction are introduced; the electrode 3 with the cylindrical internal surface of which the arc is attached (5); the back cover 1; the nozzle 6 insulated electrically from the electrode and playing the role of the auxiliary electrode at the moment of starting up the plasma torch.

The electrode sections of the two-jet plasma torch are distributed in such a manner that their axes intersect under some angle α. The aerodynamics of the plasma forming gas in both electrode sections is almost identical to that of the gas in the internal electrode of the two-chamber plasma torch.

One of the possible circuits of supplying the electric power to the two-jet plasma torch is shown in the same figure 8.24. The plasma torch is started up by a single oscillator by means of simultaneous electrical breakdown of the gap between the electrode and the nozzle in every electrode section. The contactor K_s is closed and the starting current is restricted by the additional resistance R. When the stationary value of starting current is reached, the contactor K_s is switched off and the arc current is close to in respect of the plasma jets discharged from the nozzle. Other electrical circuits of connecting the plasma torch have been tested. If the power of the oscillator is not sufficient for simultaneous reliable initiation of two parallel arcs, it is necessary to use two oscillators. Both oscillators ignite the arc in the same electrode section. The arcs can be ignited in the electrode sections simultaneously or in the succession. In the case in which the voltage of the electric power source of the plasma torch is lower than the required voltage, two identical thyristor converters are activated. In this case, two oscillators are used and the nozzles of the electrode sections are connected to the common terminal of the power sources.

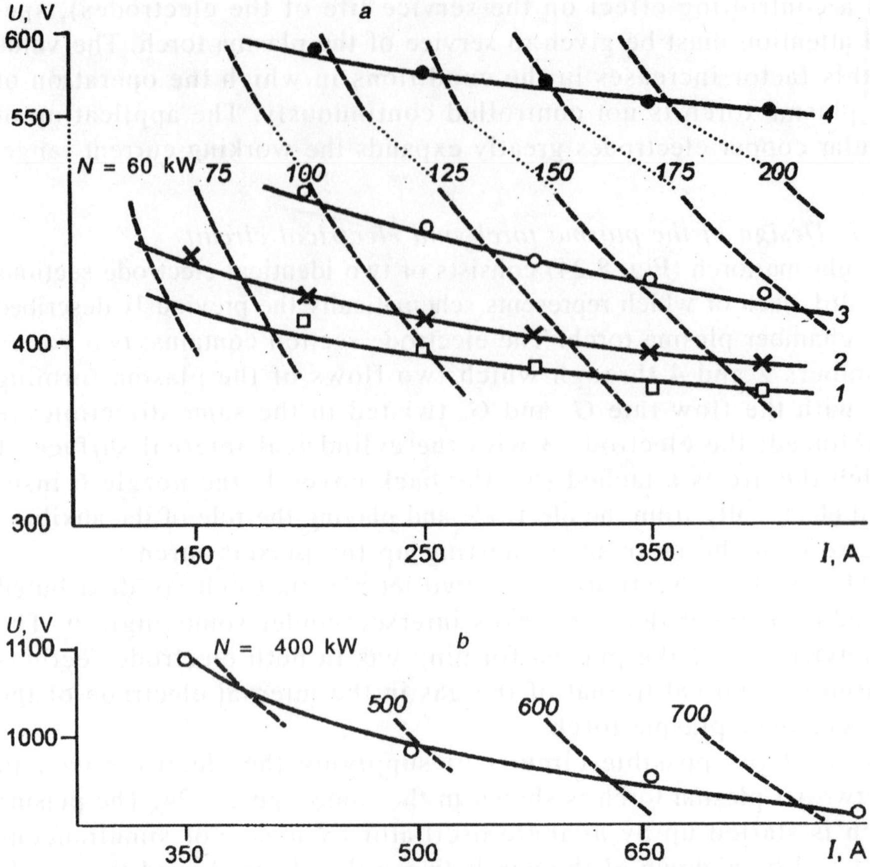

Fig. 8.25. The volt–ampere characteristics of the arc ($\alpha = 90°$). a) $d_2 = 18.5 \cdot 10^{-3}$ m: 1) $G_\Sigma = 6 \cdot 10^{-3}$ kg/s; 2) $8 \cdot 10^{-3}$; 3) $12 \cdot 10^{-3}$; 4) $d_2 = 8 \cdot 10^{-3}$ m; $G_\Sigma = 12 \cdot 10^{-3}$ kg/s; b) $d_2 = 25 \cdot 10^{-3}$ m; $G_\Sigma = 40 \cdot 10^{-3}$ kg/s. Broken lines – equal power lines.

8.3.2. The plasma torch characteristics

The VACs of the two-jet plasma torch, determined using air as the plasma forming gas, are slightly drooping (Fig. 8.25*a*). The curves 1–3 were constructed at different values of G_Σ of the total (for both electrode sections) air flow rates: here d_2 is the diameter of the orifice of the nozzle. Arc voltage is strongly affected by the angle between the axes of the electrodes α and the parameter d_2. The curves 2 and 4 differ in the diameter of the orifice of the nozzle d_2 by a factor of 2 at equal gas flow rates. Comparison shows that the decrease of d_2 increases the arc voltage by almost 200 V, i.e., approximately by a factor of 1.5 which is understandable, because $E \sim 1/d$. By analogy with the investigated plasma torch, a plasma torch with a higher power was also developed and tested: working current up to 800 A. Its

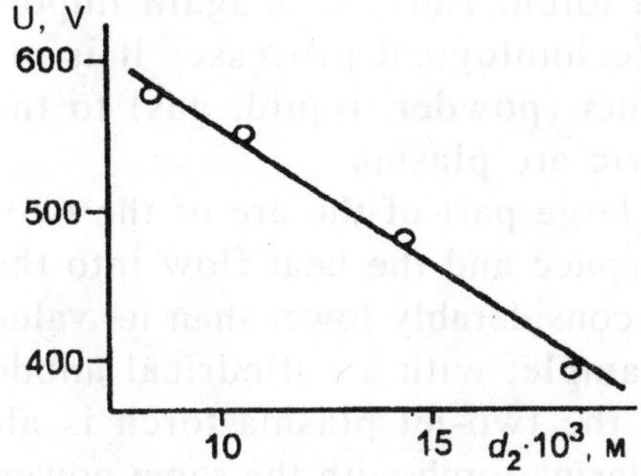

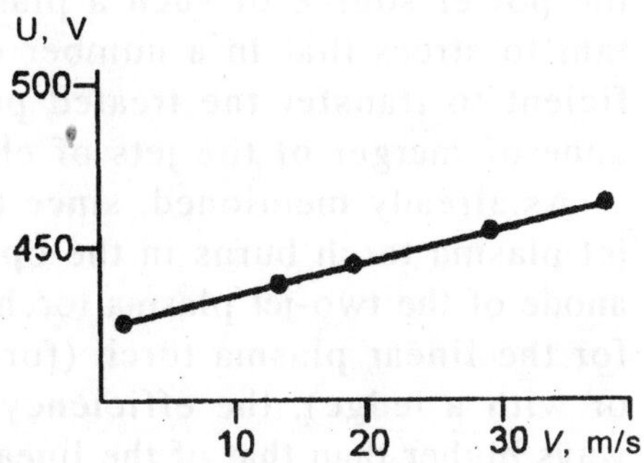

Fig. 8.26. The dependence of arc voltage of the diameter of the orifice of the nozzle d_2, $I = 300$ A, $G_\Sigma = 12 \cdot 10^{-3}$ kg/s.
Fig. 8.27. Dependence of the arc voltage on the rate of blowing the flow of air, discharged from a flat nozzle, on the zone of merger of the plasma jets.

characteristic parameters are: $d_2 = 25 \cdot 10^{-3}$ m, gas flow rate $G_\Sigma = 40 \cdot 10^{-3}$ kg/s. The VAC of the arc for this case is shown in Fig. 8.25*b*. The dependence of the arc voltage on the diameter of the orifice of the nozzle d_2 is shown in Fig. 8.26. The arc voltage for the plasma forming gas – air can be estimated using the equation:

$$U = 2 \cdot 10^3 \left(I^2 / (G_\Sigma d_2) \right)^{-0.2} (G_\Sigma / d_2)^{0.25} \times$$
$$\times (p \cdot d_2)^{0.35} \left[2 + a / (l_1 \sin(\alpha/2)) \right].$$

The geometrical parameters d_2, a, l_1 are presented in Fig. 8.24. The presented results were obtained for discharge from nozzles of the plasma jets into a stationary surrounding medium, and a large part of the arc (approximately 60%) was situated outside the channels of the electrode sections. In technological applications, this freely burning part of the arc may be placed in the flow of some medium, and the vector of the flow velocity may be characterised by different directions in relation to the plane in which the axis of the electrode sections are situated (and, consequently, the open part of the arc). The external flow has a specific effect on the arc voltage. As an example, Fig. 8.27 shows the variation of the arc voltage in blowing the air flow on the zone of merger of the plasma jets, with the airflow discharged from a flat narrow (width $4 \cdot 10^{-3}$ m) nozzle. The vector of flow velocity is situated in the plane of the plasma jets. In the specific case, the increase of arc voltage is not so large, but the conditions of additional blowing of the freely burning section of the arc should be taken into account when selecting the parameters of the power source of such a plasma torch. Here, it is again important to stress that in a number of technological processes it is efficient to transfer the treated product (powder, liquid, gas) to the zone of merger of the jets of electric arc plasma.

As already mentioned, since the large part of the arc of the two-jet plasma torch burns in the open space and the heat flow into the anode of the two-jet plasma torch is considerably lower than its value for the linear plasma torch (for example, with a cylindrical anode or with a ledge), the efficiency of the two-jet plasma torch is always higher than that of the linear plasma torch with the same power. In this variant, the efficiency of the plasma torch reaches 0.9.

Specific erosion and operating life of electrodes

In the electrode section of the two-jet plasma torch, the zone of attachment of the arc to the surface of the tubular electrode moves (as in the case of the two-chamber plasma torch) in the azimuthal direction under the effect of aerodynamic forces of the flows of the plasma forming gas, twisted in the vortex chambers. Comparison of the results of the relatively detailed measurements of the zone of attachment of the arc to the electrode and of the results of visualisation of 'cold' blowing shows that the attachment of the arc to the electrode takes place in the zone of contact of the twisted flows, discharged from two vortex chambers. For the constant ratio of the flow rates of the plasma forming gas in the twisting chambers, the electrode is characterised by the formation of a circular zone whose mean width is equal to $(6–8)\cdot 10^{-3}$ m. The specific erosion of the cathode is $\bar{G}_c = 2\cdot 10^{-9}$ kg/C, and that of the anode $\bar{G}_a = 10^{-9}$ kg/C [33, 35]. If the ratio of the flow rates of air supplied into the vortex chambers is changed, the zones of contact of the flows inside the electrode and, consequently, the zones of attachment of the arc to the electrode will be displaced in a specific direction. For example, if the gas flow rate G_1 increases and G_2 decreases (Fig. 8.24), the attachment of the arc to the electrode will be displaced in the direction of the nozzle, and vice versa.

However, it is relatively difficult to realise this type of distribution of the flow rate of the gas G_1 and G_2. A simpler solution of the scanning of the radial sections of the arc in the discharge chamber of the electrodes in the axial direction is shown in Fig. 8.24. In this variant, the flow rate G_1 into both electrodes remains constant, and the flow rate G_2 change periodically. The frequency of the process, optimum from the viewpoint of specific erosion, is 2–3 periods per minute.

In the regime in which the azimuthal displacement of the electrode spots of the arc is accompanied by scanning of the arcs by the previously mentioned method in the axial direction, measurements were taken of the specific erosion of both electrodes. Air was used as the plasma forming gas. The working current was selected equal to 300 A. The material of both electrodes was the same, copper. Initially, tests were carried out in a laboratory stand. The total operating time of the electrodes was 14 h. Subsequently, tests were carried out in the industrial conditions (TETs-2 Thermoelectric Power Station, Novosibirsk), where the electrodes operated for a total of 60 h. The results of both tests were almost identical.

After operation for 60 hours, the depth of the depression in the anode did not exceed 10^{-3} m and according to estimates (on the condition of maintaining specific erosion on the same level), the operating life of the anode may exceed 500 h. The operating life of the cathode was estimated at 200 h. Taking into account that the electrode sections of these two-jet plasma torch are identical, by changing the polarity of the electrodes (if the above condition is fulfilled), the continuous operating time of the plasma torch may equal 300–400 h.

The mean specific erosion of the cathode in the conditions of additional axial scanning of the near-cathode section of the arc was on the same level as without scanning, i.e. approximately $2 \cdot 10^{-9}$ kg/C. The specific erosion of the anode was more than an order of magnitude lower than the general value and equalled $(4-5) \cdot 10^{-11}$ kg/C. This difference of the values of the specific erosion of the electrodes is explained by different conditions of the surface layer of the metal in the zone below the reference part of the arc, as described in detail in chapter 10.

The two-jet plasma torch with the tubular electrodes has been used efficiently in tests of the plasma ignition of coal dust suspended mixtures in the process of mazut-free heating of a coal dust power boiler and may be used in a number of other technologies [36].

Chapter 9

Alternating current plasma torches using industrial frequency

In recent years, in addition to direct current plasma torches, alternating current (AC) plasma torches have been used more and more extensively in different types of plasma-chemical technology. In particular, this relates to technologies based on the application of high-power plasma generators. In this type of generator, the gas is heated by the energy of alternating current of industrial frequency. Instead of the ballast resistance, stabilising the direct current arc, the operating regime of the AC plasma torches is stabilised and regulated by inductance coils (reactors). This greatly simplifies the electrical power circuit and reduces its price. If necessary, the reactive power losses in the reactor may be compensated using a bank of condensers or a synchronous compensator (at high powers). The physical processes of the burning of the arc at direct and alternating currents are basically identical, but in many cases preference is given to AC plasma torches. This is explained by the fact that alternating current is more readily available and, using this current, it is easier to obtain the required parameters of the power sources and regulate them; the service of AC electrical engineering equipment is simpler, etc.

However, AC plasma torches are not yet used widely. This is caused by the fact that the application of alternating current is associated with additional difficulties caused by the variability with time of the electrical parameters of the power source and associated mainly with: 1) with ensuring continuous arcing at alternating current in the transition of current through zero in the linear plasma torches; 2) with non-correspondence (without taking special measures) of the form of the current and voltage curves, reducing the coefficient of utilisation of the power of the power source; 3) with the variability of the supply of energy in time in the single-phase system resulting in the appearance of time pulsations of the parameters of the heated gas and, consequently in the need to solve problems of both removal of these pulsations and the constancy of the gas parameters which is often an important requirement on plasma torches, used in technology and industry; 4) with the need to ensure uniform phase loading and using three-phase current for supplying the power to the plasma torch in order to avoid exerting a negative effect on the operation of other users of electric energy; 5) with the presence in the certain systems of three-phase plasma torches of special features when taking electrical measurements.

One of the advantages of AC plasma torches may be associated with the duration of operation. The point is that in the majority of DC plasma torches, the operating time of the electrodes greatly differs, sometimes by several orders of magnitude. Since the polarity of the electrodes in the AC plasma torch (cathode–anode) changes with the frequency of the electrical mains, the wear of the electrodes is more uniform and, with other conditions being equal, it should be expected that the operating life of the electrodes of the plasma torch is higher in comparison with the DC plasma torch. However, it should be mentioned that the change of the polarity of the electrodes with a high frequency may have a negative effect on the service life. This problem requires further detailed examination.

The results of a large number of investigations of DC and AC arcs show [1–3] that almost every DC plasma torch may be connected to the circuit of one of the phases of alternating current, if the problem of sustaining the arc in the break period of current is solved. This is a linear phase AC plasma torch. The connection of the plasma torches to the three-phase mains of industrial frequency may be carried out using different methods. Below, we examine the schemes of both single-phase and three-phase plasma torches.

9.1. SINGLE-PHASE AC PLASMA TORCH

9.1.1. Special features of powering the alternating current arc

Prior to examining the results of investigations of the AC plasma torches, it is necessary to examine the special features of ignition of the electrical arc in these plasma torches and the problems associated with supplying electrical power to AC arcs.

The initial ignition of the electric arc in the DC plasma torches is carried out by different methods:

– by means of short-circuiting the electrode with a wire, a moving bar (a starting electrode) or by some other similar means;

– by self-breakdown in the gap between the electrodes, if the voltage of the power source is sufficiently high (in a number of cases, especially at a large distance between the electrodes and at higher gas pressures, to reduce the breakdown voltage, the gap between the electrodes is irradiated);

–by shortening the gap between the electrodes by a gas-discharge channel formed as a result of supplying to the electrodes a high-voltage pulse with high-frequency from an oscillator.

In the case of AC heaters, the first method is obviously not suitable if no special measures are taken for ensuring further continuous burning of the electrical arc.

The most widely used method of igniting the AC electrical arc is the third method. The principal diagram of the oscillator is shown in Fig. 9.1. Its basic element is the high-voltage transformer with large scattering, with the oscillatory circuit, consisting of the inductance L_1, condenser C, and the spark discharger P_3, connected to the secondary winding of the transformer. The capacitance C is charged from the high-voltage transformer to the breakdown voltage of the discharger. After breakdown of the discharger, the L_1CR-discharge circuit is characterised by the formation of oscillations of voltage and current with the frequency, determined by the parameters of the circuit:

$$f_c = (1/2\pi)\sqrt{1/L_1C - R^2/4L_1^2}, \qquad (9.1)$$

where R is the ohmic resistance of the circuit.

The energy of the charged condenser is scattered in the form of heat in the discharger and in the plasma torch, and also in the form of electromagnetic radiation. High-frequency voltage is applied to the plasma torch through the connecting coil L_c.

There are two possible variants of connecting the oscillator to the plasma torch. Figure 9.1 shows the principal diagram of coil L_c with parallel connection to the arc. In this case, the following are added to the general electrical circuit: the condenser C_1, protecting the power source against shortening through the connecting coil, the choke coil L_z, calculated for the total arc current, and capacitance C_z, used as protection of the power circuit of the plasma torch from the high-voltage of the oscillator. However, this system has one considerable disadvantage.

In fact, according to equality (9.1), the natural frequency of the discharge L_1CR-circuit depends on the value of the ohmic resist-

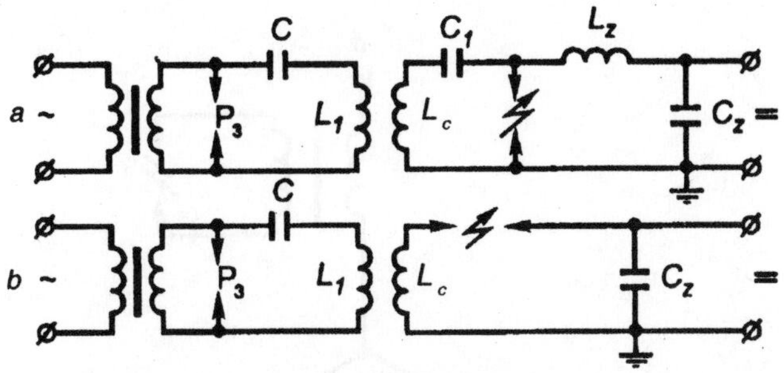

Fig.9.1. Connection of the oscillator to the arc of the plasma torch. a) parallel; b) in-series.

ance of the circuit R which consists of the ohmic losses of the condenser C, inductance L_1 and the ohmic resistance of the arc of the discharger, which changes during a discharge. Consequently, the oscillator generates a wide spectrum of frequencies in which the frequency, equal to the resonant frequency of the in-series circuit L_zC_z, may appear. In this case, this circuit shows the formation of a resonance of voltages and higher voltage penetrates into the power source which may cause disruption of the normal operation of the power source. This may be avoided by replacing the discharger of the oscillatory circuit with an element characterised by switching properties and a constant resistance. The undesirable effects may also be weakened by the in-series connection into the circuit of the discharger of an additional ohmic resistance, which is considerably higher than the resistance of the arc of the discharger. However, this is accompanied by additional losses, which require increasing the necessary oscillatory power.

In the second variant (see Fig. 9.1*b*) the connecting coil is connected in the circuit in sequence with the arc and is calculated for the passage of the total arc current from the power source. The small increase of the complexity of the circuit is compensated by excluding from the circuit the condenser C_1 and the choke coil L_z. The circuit protects reliably the power source against the effect of the oscillator.

The AC power circuit with a drooping static VAC will be examined. To ensure stability of arcing, the characteristics of the power source, as shown previously, should be steeply drooping. In laboratory practice, the steeply drooping characteristic is produced by successive connection into the circuit of the arc of active or reactive resistance, which ensures, for example, when using liquid or tubular rheostats, the smooth and deep regulation of the current flowing through the electrical arc. However, in this case, a large fraction of the power of the source is lost in the ballast resistance.

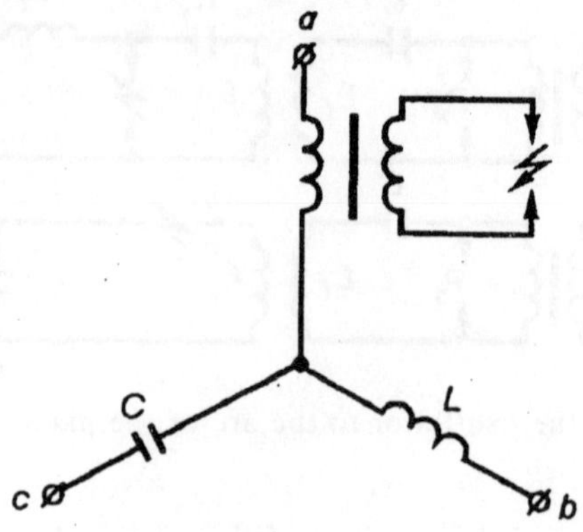

Fig.9.2. Electrical diagram of power for the arc for a three-phase AC arc.

In the case of high-power plasma torches, it is possible to use the power sources with hard stabilisation of the current regardless of the dependence on the variation of the low resistance, for example, the schemes with the saturation effect or transformers with magnetisation.

When using, as the power source, the alternating or direct current generators, the steeply drooping external characteristics may be generated by regulating the excitation of the generator.

It is interesting to examine the electrical circuit of the power source of the arc shown in Fig. 9.2. Here, the arc, burning in the plasma torch, is a load for one of the phases of the three-phase alternating current circuit, and the remaining phases are loaded one by one by the inductance and capacitance in such a manner that, on the whole, the circuit has a high power factor. The examined circuit has the property of stabilisation of the load current (the current of phase *a*), when the phase *b* contains inductance *L*, and the phase *c* the capacitance *C* with equal reactive resistances, i.e. $\omega L = (\omega C)^{-1}$. In this condition, the current of phase *a* is independent of the resistance of the load, and the load voltage is directly proportional to its resistance. The connection of the plasma torch in the phase through the transformer makes it possible to regulate the current, flowing through the arc, by changing the transformation factor.

When using the alternating current arc in the plasma torch, in addition to need of constructing a power source with a steeply drooping external characteristic, it is also necessary to connect into the circuit of the arc a reactive resistance which would maintain continuous burning of the electrical arc in the case when the current passes through zero. This impairs the cos φ of the circuit and makes it necessary to take special measures for improving this parameter. Consequently, it is necessary to carry out research in the direction of producing the rising static characteristic of the active charge and maintaining constant arcing when the current passes through zero. The connection, into the arc circuit, of an inductance for organising the continuous burning of the arc is not the best solution of the problem of the application of the alternating current power source for powering plasma torches. It is more efficient to use parallel-connected low-current high-frequency generators.

9.1.2. Combined burning of high current and high-frequency arcs

In the currently available linear generators of low-temperature plasma (plasma torches), using the high-current AC electrical arc as the source of thermal energy, the maintenance of the continuous (break-free) burning of the arc is ensured, as already mentioned, by connecting a corresponding inductance into the power circuit. The value of cos of the mains does not exceed 0.6÷0.7.

It is therefore necessary to solve the problem of creating suitable physical conditions in the electric discharge chamber of the AC linear plasma torch in which the high current arc not only burns continuously without any inductance in the circuit but also ensures the sinusoidal variation of voltage in specific conditions. These conditions can be created if, for example, an auxiliary high-voltage but low-current direct arc burns between the electrodes of the plasma torch, in addition to the main high current arc [4]. From the physical viewpoint, it is completely clear that in this case the working space of the plasma torch will contain, at any moment of time, an electrically conducting channel through which a high-current arc discharge may develop continuously with time. The complexity of this task is in the separation of the direct and alternating current mains.

Taking this into account, attention will be given to the possibility of using, as an auxiliary arc, a high-frequency high-voltage discharge whose purpose is to ensure, at any moment of time, the existence in the arcing chamber of an electrically conducting channel through which high-current discharge may develop. It may easily be shown that the protection of the high current power sources of both direct and alternating current from the high frequency voltage is possible in this case at any power of the high current discharge.

The frequency of the auxiliary charge is determined by the deionisation time of the gas after interrupting the supply of energy to the arc column. For the arcs, freely burning in air, the deionisation

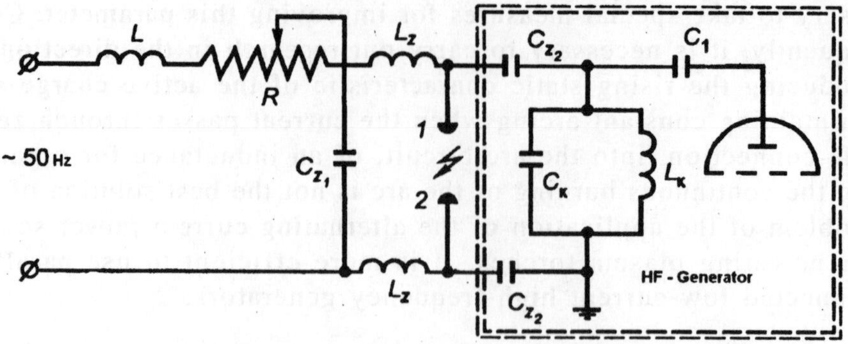

Fig.9.3. Principal diagram of connection to the plasma torch of two electric power sources: high-current with industrial frequency and low-current high frequency power source.

time varies in the range $10^{-3}\div10^{-4}$ s. If the arc burns in a channel with a gas flow, this time is even shorter. The experiments show that at frequencies of 10^6 Hz in the plasma torch with gas-vortex stabilisation and axial blowing of the high-frequency arc, the latter burns in a stable manner and continuously. Arc breaks (extinction) are detected at low frequencies. Thus, to maintain continuous burning of the high current AC arc with industrial frequency, it is sufficient in all likelihood to ensure that the generator, supplying the power to the auxiliary arc, generates voltage with a frequency of the order of 1 MHz.

Experience shows that the auxiliary arc with the power of several kilowatts ensure stable and break-free burning of the high current AC arc in plasma torches with gas-vortex stabilisation in a relatively wide range of the current and gas flow rate. When using the high-frequency source of high-voltage, it is desirable to use it as a means for the initial ignition of the arc. In the actual systems of the axial plasma torches, the minimum gap between the electrodes, determined by the flow rate of the gas and the permissible radial speeds, is 4÷7 mm. For the initial electrical breakdown of this gap, the ignition voltage should be 20–30 kV. As shown later, as a result of the small increase in the complexity of the electrical power circuit, it is possible to increase the open circuit voltage of the high-frequency generator to the required values.

Figure 9.3 shows the diagram of connection to the plasma torch of two electric power sources, the high current and high-frequency low current source. The voltage (for the arc) to the electrodes of the plasma torch 1, 2 from the powerful power source is supplied in the general case through the inductance, the additional ohmic resistance R and air inductance coils L_z; the high-frequency generator is connected through the condensers C_{z_2}, preventing short-circuiting of the power source through the air coil L_c of the oscillatory circuit of the generator. The capacitances C_{z_2} also play the role of the restricting value of the high-frequency current, passing through the plasma torch. The inductances L_z and the capacitance C_{z_1} form the divider of the high-frequency voltage, transmitting to the power source the voltage equal to the voltage drop in the condenser C_{z_1}. If necessary, for example, to ensure initial ignition, the voltage of the high-frequency generator can be increased several times selecting the condenser C_{z_2} in the inductance L_z in such a manner as to obtain a successive resonance (voltage resonance). When the arc discharge forms between the electrodes, the condition of successive resonance is violated and the arc is powered by the output voltage of the generator. When the discharge is extinguished, the condition of breakdown is automatically restored, if the arc breaking is not associated with disruption of operation of the high-frequency generator.

In the experiments examined below, in separate burning of the high-frequency and high current arcs, the length of the latter was smaller. Since the power of the high-frequency arc is low in comparison with the power arc and its energy contribution is small, it may be assumed that the physical conditions in the combustion chamber in the application of the high-frequency arc remained almost unchanged and determined by the power arc. The VAC of the AC arc is shown in Fig. 9.4.

The experiments show that the VAC of the high current AC arc (in the break-free regime, for example as a result of connecting and inductance) both with the high-frequency arc and without it differ only slightly from each other (Fig. 9.4a, curves 1 and 2). The VAC 2, obtained in the presence of the high-frequency arc, is slightly lower than the characteristic 1, burning without the high-frequency arc. At low powers ($I \approx 50$ A), the deviation in respect of power does not exceed 4%, and at higher powers ($I \approx 150$ A), it is smaller than 2%, which approximately corresponds to the increase of the power as a result of the high-frequency arc. With increase of the current, the high-frequency arc changes the VAC to an even lesser extent. It should be mentioned that with the variation of the current of the arc (variation of the load resistance from 0 to 10^3 ohm), the current, generated by the high-frequency generator, does not change by more than 20% because the latter operates in the current generator regime.

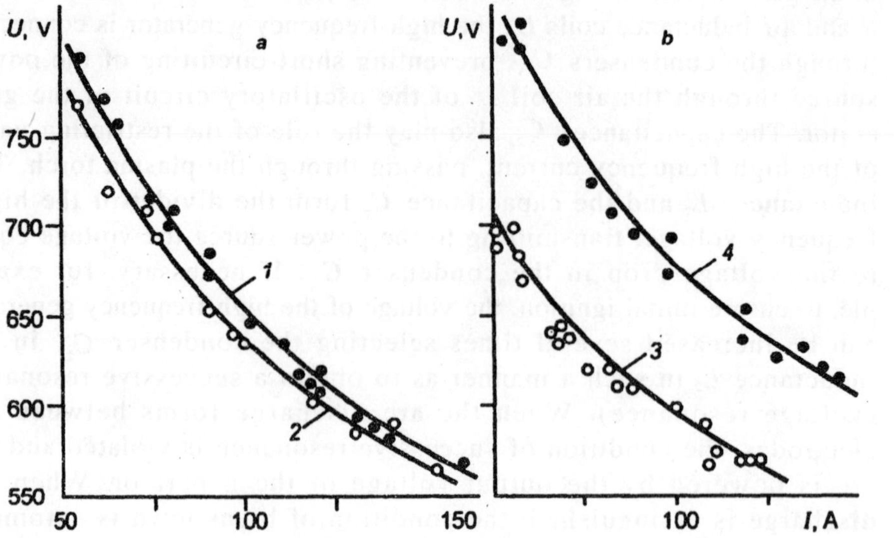

Fig.9.4. Volt–ampere charactertistics of the AC arc. $d = 2 \cdot 10^{-2}$ m, $p = 1 \cdot 10^{-5}$ Pa. a) $G = 7.45 \cdot 10^{-3}$ kg/s; 1) inductance is connected without the HF arc; 2) inductance and the HF arc are connected; b) $G = 9.15 \cdot 10^{-3}$ kg/s; 3) without inductance and HF arc; 4) without inductance but with the HF arc.

A completely different situation is observed in burning of the arc with current breaks (Fig. 9.4*b*). The curve 3 corresponds to the burning of the arc without the high-frequency arc and with current breaks, curve 4 corresponds to the case with the high-frequency arc. In both cases, no inductance was included in the circuit. The reason for such a large difference in the *U–I* characteristics can be easily understood examining the oscillograms of voltage and current (without inductance) for this case, shown in Fig. 9.5*a* without the high–frequency arc, and in Fig. 9.5*b* with the high-frequency arc. The connection of the high-frequency discharge eliminates current breaks and increases the effective voltage of the high current arc and this is also reflected in the characteristics. Thus, the experiments show that in the combined combustion of the high-frequency and high current arcs, it is possible to ensure break-free burning of the latter without inductance in the circuit at cos φ close to unity. In this case, the arc voltage is lower than amplitude value of the voltage of the power source.

As indicated by Fig. 9.4, the VAC of the AC arc is drooping. Therefore, for stable arcing, a ballast resistance was connected to the circuit of the arc. However, in certain conditions, the high-frequency arc maybe used to ensure not only continuous but also stable arc-

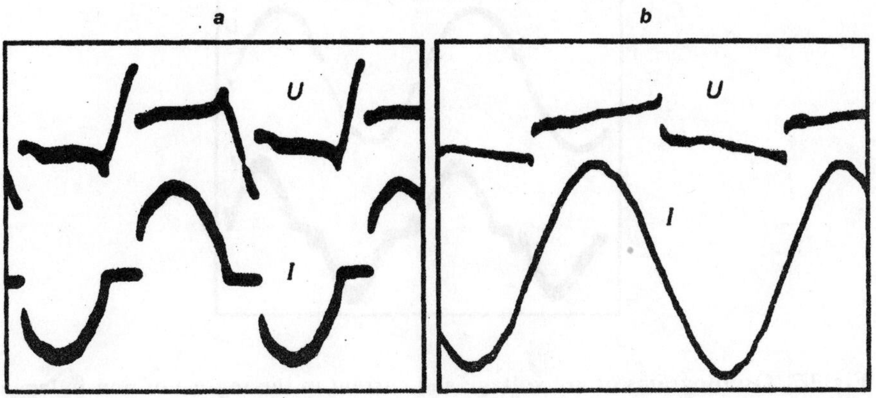

Fig.9.5 Oscillograms of voltage and arc current: a) without connected inductance and HF arc; b) without inductance but with HF arc.

ing without a ballast resistance. In this case, the dynamic characteristics of the current and arc voltage in the half cycle should be close to the sinusoids with efficient utilisation of the power of the power source.

If the static VAC of the electrical AC arc with the self-setting length (Fig. 9.6, curve 1) is such that the condition $U > U_m \sin\omega t$ is fulfilled, the arc in the plasma torch cannot burn independently because at any moment of time the voltage (curves 2, 3, Fig. 9.6) is lower than the required voltage. When the high-frequency arc is connected, the arc does exist and its voltage U should follow both in the value and form the available voltage of the source, i.e. it should be sinusoidal. This arc discharge is referred to as non-independent. In this case, the available power of the power source is completely

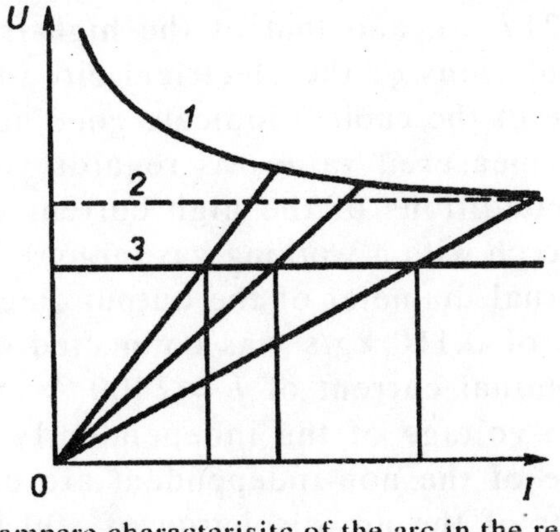

Fig.9.6. Static volt–ampere characterisitc of the arc in the regime of non-independent arcing. 1) static U–I-characteristic of the arc; 2) amplitude values of the voltage of the power source; 3) effective value of arc voltage.

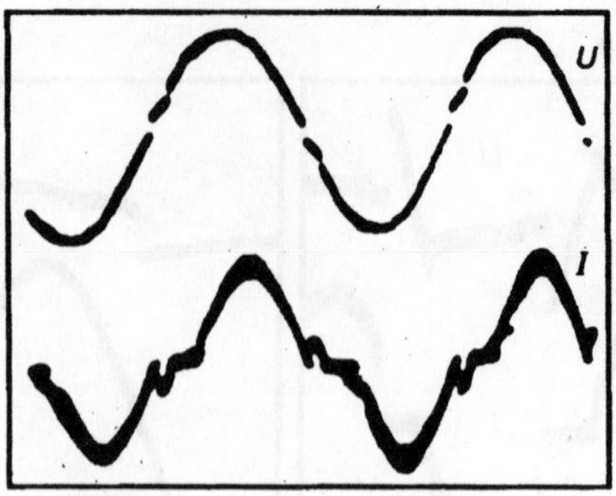

Fig.9.7. Oscillograms of arc voltage and current in the regime of non-independent arcing with HF arc.

utilised. The experimental verification of the existence of sinusoidal dynamic characteristics of the arc in respect of current and voltage was carried out on a single-phase plasma torch with the independent arcing regime at the following parameters: U = 400 V, at I = 240 A, the diameter of the output electrode $3.0 \cdot 10^{-2}$ m, the flow rate of argon $7 \cdot 10^{-3}$ kg/s, pressure $1 \cdot 10^{5}$ Pa. The plasma torch was connected to the AC mains with the voltage of 380 V (U_m = 537 V) of industrial frequency through an additional active resistance of 0.6 ohm, reducing the voltage, supplied by the mains, to the value considerably lower than the voltage required for the independent burning of the electrical arc.

The experiments showed: the main arc burns only with the high-frequency arc present; in the dynamics, the voltage and current change approximately in the sinusoidal manner (Fig. 9.7).

In the given experiments, the power of the high current arc discharge was 52 kW (U = 217 V), and that of the high-frequency arc was 0.75 kW. The calculations of the electrical circuit (taking into account the resistance of the cables) indicate good agreement between the calculated and measured values as regards voltage.

Similar arcing conditions of the high current arc were also realised in a plasma torch with a working gas (argon). The phase plasma torch with the internal diameter of the output electrode $5 \cdot 10^{-2}$ m at an argon flow rate of 0.110 kg/s was connected directly to 380 V mains with the nominal current of I = 2400 A. At an arc current of I = 1500 A, the voltage of the independently burning arc was 270 V. The voltage of the non-independent arc discharge was the same. At the power of the arc discharge of 300 kW, the required power of the high-frequency arc did not exceed 4 kW, i.e., was not greater than 1%.

In operation of a plasma torch with a non-independent arc it is possible to regulate the power, supplied to the arc, by changing the current passing through it from the high-frequency generator. Increasing the power of the auxiliary arc, it is possible to increase the power current to the nominal value. In this case, the efficiency of utilisation of the power of the power source is 100%.

Evidently, this regulation is possible in plasma torches using direct and alternating current. Experiments show that in the plasma torches with vortex argon stabilisation of the power, the variation of the power of the high-frequency arc from 2 to 4 kW increases the arc current from 240 to 2500 A (gas flow rate 0.110 kg/s).

Thus, in organisation of the combined burning of the high current and high-frequency arcs, the following processes are realised: 1) break-free burning of the arc even in the case in which the circuit does not contain any reactive resistance; 2) burning of the arc with dynamic characteristics in respect of current and voltage, similar to sinusoidal. In the latter case, the arc burns in a stable manner without a ballast resistance at any form of the external characteristic of the power source.

9.1.3. Volt–ampere characteristics of the AC arc, burning in a phase laminar vortex plasma torch

To calculate AC generators of low-temperature plasma, it is necessary to obtain data on their electrical and gas-vortex characteristics for the case in which the plasma torch with the gas-vortex turbulisation is characterised by the burning of the AC arc with industrial frequency, together with the high-frequency arc, reducing the duration of the current breaks of the independently burning arc.

Investigations [5] were carried out in a two-chamber plasma torch with the internal diameter of the output electrode of d = (10, 20, 30 and 50)·10^{-3}m. The parametric criterion $\bar{d} = d_p/d$ was maintained constant, and equal to 1.2. Here d_p is the internal diameter of the phase (end) electrode. The ratio $\bar{l}_p = l_p/d_p$ was approximately equal to 10, where l_p is the length of the phase electrode. The electrodes are made of copper and cooled with water. The working gas was air, supplied into the chamber of the arc through the main (flow rate G_1) and end (G_2) vortex chambers at the ratio $\bar{G} = G_1/G_2 = 3$, which was maintained constant in the experiments. The relative position of the end of the arc in the cavity of the internal electrode, determined on the basis of the trace left by the arc spot on the internal surface of the electrode, was constant and determined by the zone of contact of the two flows in it.

The total flow rate of the gas $G = G_1 + G_2$ through the plasma torch was varied in the range $(5 \div 120) \cdot 10^{-3}$ kg/s, the arc current intensity was 40–200 A. In the investigation of the plasma torches with the internal diameter of the output electrode of (30 and 50) $\cdot$ 10^{-3} m, the characteristics were determined at the pressure in the chamber of $(1 \div 10) \cdot 10^5$ Pa. In the experiments, the oscillograms of current and voltage were photographed. Some of them, corresponding to the operating regime of the plasma torch with the high-frequency arc, were presented previously in Fig. 9.5. The absence of the breaks in arcing was inspected visually by examining the oscillograms.

The typical VACs of the arc for $d = 30 \cdot 10^{-3}$ m and $p = 1 \cdot 10^5$ Pa, are presented in Fig. 9.8. In the investigated plasma torches, the processes in the discharge chamber are determined, as in the DC plasma torches, by the three main dimensional complexes:

$$K_1 = I^2 / Gd; \; K_2 = G/d; \; K_3 = pd.$$

The determined complex was $\bar{U} = Ud/I$. They are all dimensional parts of the appropriate dimensionless similarity criteria. In the experiments, the complexes, presented in the main units of the SI system, varied in the range:

$$5 \cdot 10^4 \leq K_1 \leq 9 \cdot 10^8, \; A^2 \cdot s/(kg \cdot m);$$
$$0.5 \leq K_2 \leq 12, \; kg/(m \cdot s);$$
$$10^3 \leq K_3 \leq 5 \cdot 10^4, \; Pa \cdot m.$$

Figure 9.9 shows, as an example, the dependence $\bar{U} = f(K_1)$ for $K_3 = 3000$ and several values of K_2. The graph shows that, as in the case of direct current, the relationship between $\bar{U}$ and K_1 is exponential, and the exponent may be regarded as constant in the examined range of variation of K_1. In addition to this, U depends not only on K_1, but also on K_2 and K_3.

By analogy with the DC plasma torches, the VAC of the electric arc, burning in the phase plasma torch, may be represented in

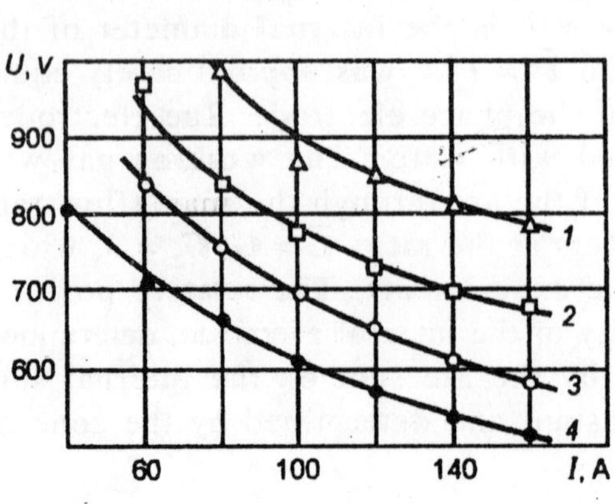

Fig.9.8. Volt–ampere characteristics of the AC arc with HF current. $d = 3.0 \cdot 10^{-2}$ m, $p = 10^5$ Pa; $G = 30 \cdot 10^{-3}$ kg/s (1); $20 \cdot 10^{-3}$ (2); $15 \cdot 10^{-3}$ (3); $10 \cdot 10^{-3}$ (4).

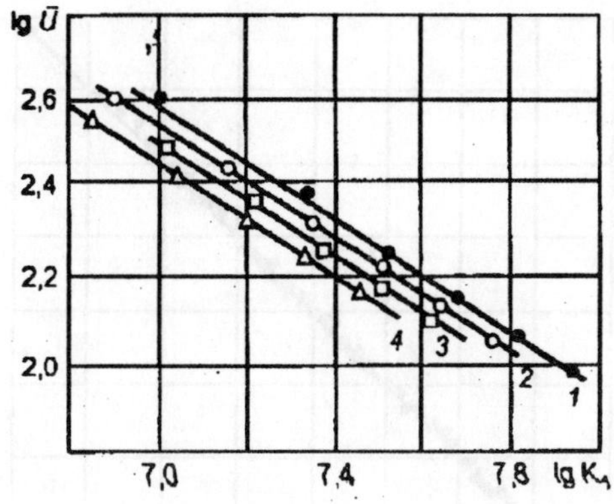

Fig.9.9. Dependence $\bar{U} = f(K_1)$. $d = 3 \cdot 10^{-2}$ m, $p = 1 \cdot 10^5$ Pa. 1) $G = 10 \cdot 10^{-3}$ kg/s; 2) 15; 3) 20; 4) 30.

the following form:

$$Ud/I = 2143\left(I^2/Gd\right)^{-0.655}\left(G/d\right)^{-0.345}\left(pd\right)^{0.20}. \qquad (9.2)$$

To show how the above equation describes the experiments, Fig. 9.10 shows the dependence of lg $\bar{U}_{exp}$ on lg $\bar{U}_{cal}$ where $\bar{U}_{cal}$ is the right-hand side of equation (9.2). Equation (9.2) in the investigated range of the parameters may be simplified, if the determining dimensional complex is represented by the arc voltage $U = \bar{U}\,I/d$, and the determining complex K_1 is represented by the complex $K_5 = G/I = (K_2/K_1)^{0.50}$. It should be mentioned that $K_4 = I/d$ is also the dimensional part of the dimensionless criterion.

After simple transformations, the following simplified equation may be recommended for calculating the voltage of the arc burning in the phase plasma torch accompanied by the auxiliary arc:

$$U = 2143\left(G/I\right)^{0.31}\left(pd^{0.20}\right). \qquad (9.3)$$

The equation was verified in the variation range $320 \leq K_5^{-1} \leq 40 \cdot 10^3$; $10^3 \leq K_3 \leq 5 \cdot 10^4$. The exponent γ at K_3 was determined on the basis of the effect of the variation of the electrode diameter on arc voltage (subsonic discharge of the gas from the nozzle of the plasma torch, $p = 1 \cdot 10^5$ Pa). Examination of the dimensional criteria shows that it can also be determined, maintaining the value of d constant, but changing p at the end of the discharge chamber of the arc. These investigations were carried out on plasma torches with d = (30 and 50) $\cdot\ 10^{-3}$ m. As an example, Fig. 9.11 shows the appropriate VAC of the arc. In particular, they show that with increase of pressure, voltage also increases. The dependence $U = f(K_3)$ at a constant ratio G/I, including the range of variation of K_3 both by means of

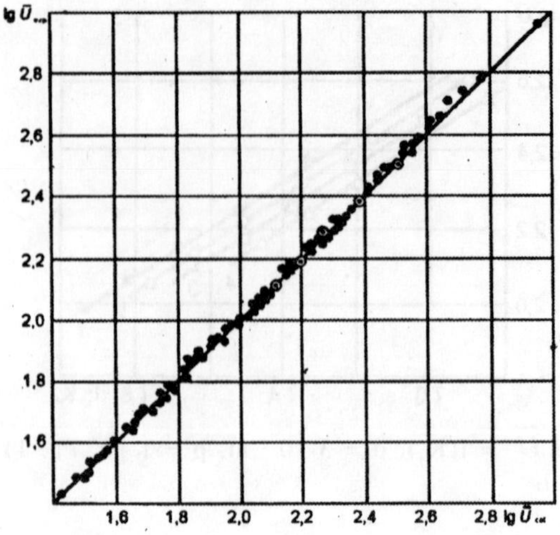

Fig.9.10. Correspondance of experiment with calculations using equation (9.2).

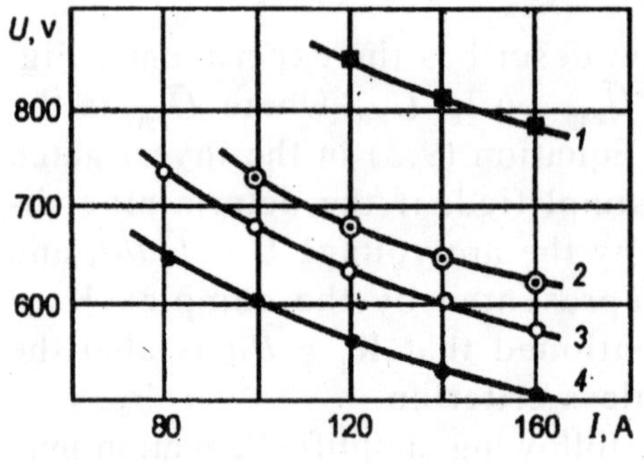

Fig.9.11. Volt–ampere characteristic of the arc. d = $30 \cdot 10^{-3}$ m. 1) G = $15 \cdot 10^{-3}$ kg/s, p = $(3.4–3.5) \cdot 10^5$ Pa; 2) G = $1o \cdot 10^{-3}$ kg/s, p = $(2.38–2.48) \cdot 10^5$ Pa; 3) G = $15 \cdot 10^{-3}$ kg/s, p = $1.5 \cdot 10^5$ Pa; 4) G = $10 \cdot 10^{-3}$ kg/s, p = $1 \cdot 10^5$ Pa.

d and *p*, is shown in Fig. 9.12.

In conclusion, it should be mentioned:

1. The investigations of the phase plasma torch with gas vortex stabilisation of the arc made it possible, in the investigated range of the variation of the determining complexes, to determine the equation for calculating the VAC of the arc in the criterial form, which may be used in calculating high-power plasma torches.

2. The high-current AC arc of industrial frequency accompanied by auxiliary high-frequency arcs burns in the phase plasma torch in a stable manner and without current breaks.

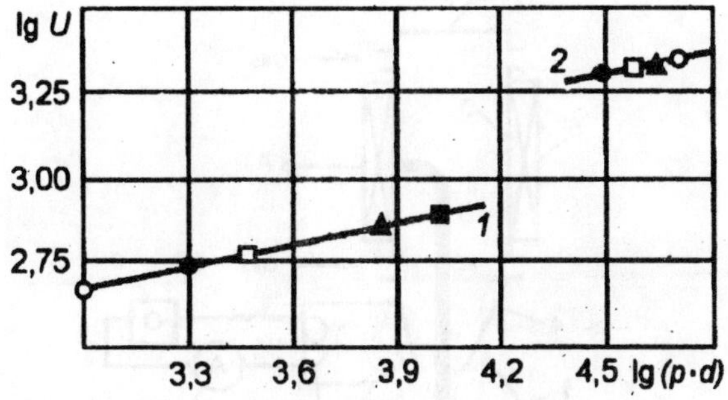

Fig.9.12. Dependence of lg U on lg (K_3) at K_5 = const. 1) O – $d = 10 \cdot 10^{-3}$ m, $p = 1 \cdot 10^5$ Pa; ● – respectively $20 \cdot 10^{-3}$, $1 \cdot 10^5$; □ – $30 \cdot 10^{-3}$; ▲ – $30 \cdot 10^{-3}$, $2.4 \cdot 10^5$; ■ – $30 \cdot 10^{-3}$, $3.5 \cdot 10^5$; 2) $d = 50 \cdot 10^{-3}$ m; ● – $p = 6 \cdot 10^5$ Pa; □ – $7.42 \cdot 10^5$; ▲ – $8.58 \cdot 10^5$; O – $9.83 \cdot 10^5$.

9.2. THREE-PHASE PLASMA TORCHES OF THE ZVEZDA TYPE

As already mentioned, almost any DC plasma torch may be used also for operation using AC. However, special features of supplying the power to the high-power plasma torches from the three-phase mains, in particular, the requirement to ensure symmetric loading of the phases, has made it necessary to develop a completely new scheme of the three-phase plasma torch referred to as Zvezda (star) [3].

9.2.1. The scheme of the plasma torch and operating principle

The type Zvezda plasma torch (Fig. 9.13) contains three identical arc chambers, situated under the angle of $2\pi/3$ in relation to each other, and the common mixing chamber 1. Each arc chamber has the end cover (backplate) 2, the chamber–electrode 3 and the confusor 4. The phases of the mains are connected to the electrodes. From the end cover and the confusor, the electrode is separated by electrical insulators. They are used for the tangential supply of the working gas ensuring gas vortex stabilisation of the arc on the axis of the discharge chamber.

The main amount of the gas is supplied between the electrode and the confusor, the additional amount (no more than 10% of the main amount) is supplied between the electrode and the backplate in order to prevent the short-circuiting of the end of the arc with the backplate. The heated gas exits the plasma torch through the nozzle of the mixing chamber, whose axis is perpendicular to the plane of the drawing.

Each electrode is fitted with solenoids 5, and the effect of the

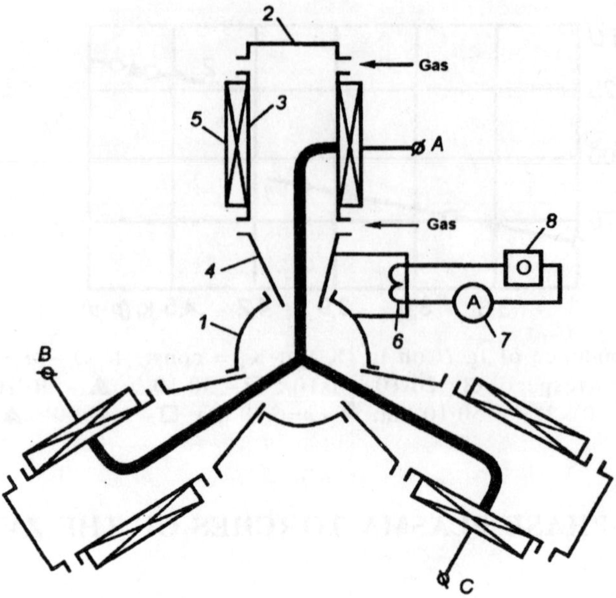

Fig.9.13. Diagram of the 'star' type plasma torch.

magnetic field of the solenoids results in the rotation of the near-electrode (radical) sections of the arc, causing an increase of the operating life of the electrode. The electrodes, confusors, the mixing chamber and the output nozzles are cooled with water.

The plasma torch is activated as follows. Initially, the systems for supplying the water and the working gas are activated. Subsequently, a voltage is supplied to the electrodes and, at the same time, the auxiliary high-frequency low-power discharge between the confusor and the needle-shaped tungsten electrode is ignited in each arc chamber using a special power source. The arc passes through the insulator situated between the confusor and the electrode, and protrudes by 5÷7 mm above the internal surface of the electrode. The high-frequency discharge closes the electrode–confusor gap; under the effect of applied voltage, a breakdown takes place in the gap with the formation of an arc. After igniting the main arc, the closing section of the arc, situated in the confusor, is moved downwards along the flow under the effect of the aerodynamic forces. Since the confusors of all three arc chambers are electrically connected with each other, the ends of the arc at the bottom of the flow are closed by the 'star' circuit with the zero point on the 'metal'. After the passage through the confusors, the ends of the arcs are extended into loops and are blown by the gas flow into the mixing chamber closing each other in the central region of the chamber, i.e. the arcing scheme is the star with the zero point in the centre of the mixing chamber (this is why these plasma torches are referred to by this name). The confusors and the mixing chamber become electrically neutral in this case.

Since the closure of the arc into the star with the zero point in the plasma is the principal feature of this plasma torch, the existence of this mechanism of interaction between the arcs was verified by experiments, using two methods. The first method is based on taking photographs of the expected region of closure of the arcs using a high-speed camera. The photographs confirmed the closure of the arc with each other (Fig. 9.14). It should be mentioned that the brightness of the arc discharges differs. This is the reflection of the general property of the symmetric three-phase system according to which the instantaneous values of current in each phase are not equal to each other at any moment.

However, the glowing regions, visible on the photographs, maybe simply jets of hot gas, discharged from the confusors, and not arc discharges. The second method was used to confirm the closing of the arc. The confusor was insulated electrically from the mixing chamber and connected, with the chamber, by the external current conductor (Fig. 9.13), passing through the current transformer 6. The secondary circuit of the transformer contains the ammeter 7 and the oscilloscope 8. If the arcs close with each other in the form of a star with the zero point in the plasma, the ammeter should show the absence of current, and the oscilloscope a straight line. In the case of arcing, with closure to the walls of the confusor, the ammeter records the total arc current and the oscilloscope records the sinusoidal curve. Intermediate cases are also possible in which the closing of the arcs with each other is periodically disrupted. In this case, the ammeter shows a fraction of the total arc current.

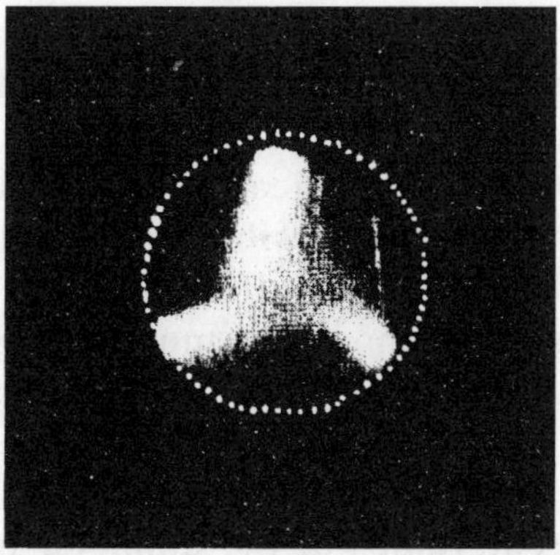

Fig.9.14. Photograph of arc discharge in a mixing chamber.

The experiments confirmed the closing of the arc with each other in the normal operating conditions of the plasma torch. This method is also highly useful for inspecting the operation of the plasma torch.

9.2.2. Volt–ampere and thermal characteristics of the arc

The temperature of the gas heated by the arc can be increased without increasing the current intensity by increasing the current density in the arc discharge. The enclosure of the arc in a sectional channel of a small diameter is a widely used method of reaching this target. In this case, because of the considerable design difficulties and more complicated conditions of arc discharges, the method cannot be used. Therefore, experiments were carried out with the effect of increasing current density in the arc by constricting the arc in the confusor channel with optimum parameters.

Firstly, the profile of the confusor should be capable of constricting the arc in such a manner as to ensure the maximum energy contribution to the arc. Secondly, the output diameter of the confusor d_c should not be very small, otherwise the output cross section may show the shunting of the arc on the wall and, consequently, the principle of interaction of the arcs by the 'star' scheme may be disrupted. A large number of experimental investigations on simulation single-phase equipment was carried out to solve these problems.

We examine the characteristics of a Zvezda-6 plasma torch, with the rated power of 6 MVA. The plasma torch is powered by the three-phase mains with a voltage of U_c = 6 kW, the nominal current intensity I = 600 A. To ensure stable arcing and regulation of current intensity, an inductance coil (reactor) with the maximum inductive resistance of 6 ohm is connected in series with the arc to each phase. The discrete variation of the resistance of the reactor was ensured by unsoldering.

The geometry of the flow part of the plasma torches may be described as follows: the internal diameter of the electrode d_e = 50 mm, the diameter of the output section of the confusor d_c = 20 mm, the diameter of the cylindrical mixing chamber was relatively small, 90 mm, so that it was possible to reduce the heat losses and obtain a relatively high thermal efficiency factor.

Figures 9.15 and 9.16 show the most important characteristics of the plasma torch – the dependence of the mean mass temperature T of the gas in the mixing chamber, pressure p, the specific energy contribution to the arc P_{sp} and the thermal efficiency η on the gas flow rate G at different values of the diameter of the critical section of the output nozzle d_{cr} (working gas–air). Temperature T was determined

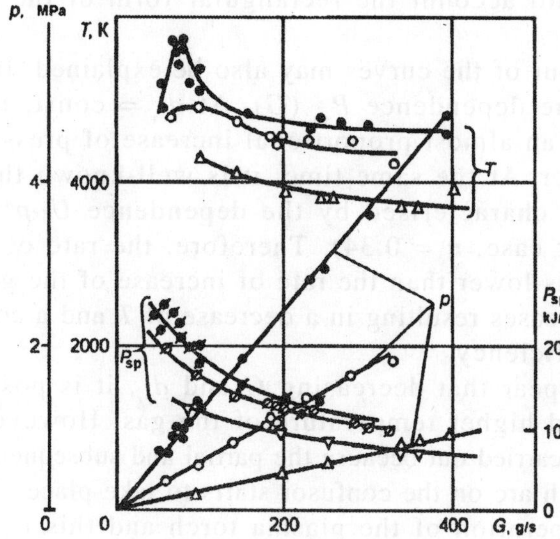

Fig. 9.15. Thermal efficiency of Zvezda-6 plasma torch. •, ⌀ d_{cr} = 14 mm; o, ⌀ 20 mm; △.▽30 mm.

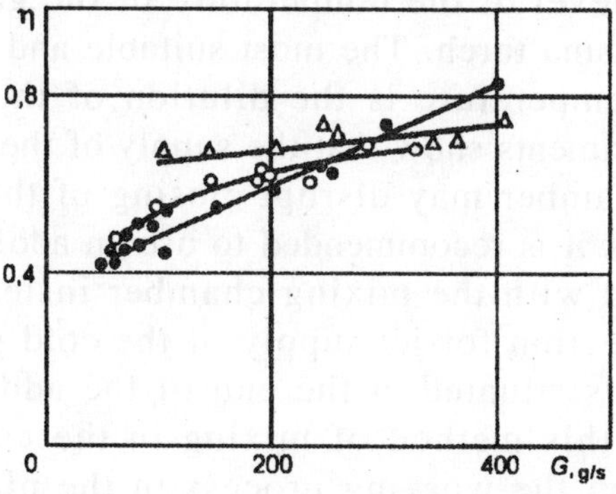

Fig.9.16. Characteristic of Zvezda-6 plasma torch. ● – d_{cr} = 14 mm; o – 20 mm; △ – 30 mm.

by the energy balance method with the error not exceeding 5%. The application of the method in the supersonic gas discharge regime from the nozzle of the mixing chamber is justified by the experimentally verified homogeneity of the flow field in the output section of the nozzle and the almost complete absence of oscillations of the pressure in the mixing chamber. The specific energy contribution was calculated from the equation $P_{sp} = N/G$, where $N = 3\ KUI$ is the total power of all discharges, and $K = 2\sqrt{2}/\pi$ is the coefficient which takes into account the rectangular form of the arc voltage curve U.

The behaviour of the curves may also be explained. In particular, we examine the dependence P_{sp} (G). At d_{cr} = const, the increase of G indicates an almost proportional increase of pressure p in the mixing chamber. At the same time, it is well-known that the electrical arcs are characterised by the dependence $U \sim p^n$, and $n < 1$ (in the present case, $n = 0.34$). Therefore, the rate of increase of the voltage U is lower than the rate of increase of the gas flow rate G, and P_{sp} decreases resulting in a decrease of T and a corresponding increase of efficiency.

It would appear that decreasing G and d_{cr}, it is possible to obtain higher and higher temperature of the gas. However, this cannot be actually carried out because the partial and subsequently complete shunting of each arc on the confusor starts to take place. The shunting disrupts the operation of the plasma torch and this is reflected in a large scatter of the experimental data, a decrease of U and, consequently, of P_{sp} and T (the solid symbols on the curve T in Fig. 9.15).

To solve a number of scientific and technical problems, it is necessary to ensure a lower level of the temperature of the gas than the level produced in the plasma torch. The most suitable and efficient method of reducing this temperature is the dilution of the hot gas with a cold one. The experiments show that the supply of the cold gas directly into the mixing chamber may disrupt closing of the arcs with each other and, therefore, it is recommended to use an additional cylindrical chamber connected with the mixing chamber instead of the nozzle and fitted with a section for jet supply of the cold gas. In this case, the output nozzle is situated at the end of the additional chamber. The advantage of this method of mixing-in the cold gas is that it has no influence on the working process in the plasma torch.

After the development and testing of the Zvezda-6 plasma torch, the same approach was used for the construction of a more powerful plasma torch Zvezda-20 (rated power 20 MVA). However, the further increase of power within the framework of the scheme is difficult because of the need to increase the arc current intensity and this is restricted by the stability of the electrodes. In the development of the Zvezda-50 (50 MVA) plasma torch, the principal circuit was modified: instead of three, the plasma torch consists of six arc chambers situated in the same plane (6-ray star), i.e. two 3-ray plasma torches operate with the common mixing chamber. The sequence of alternation of the phases is of no importance (for example

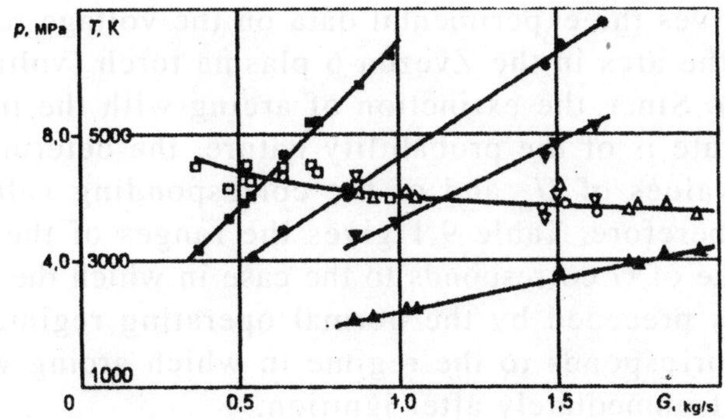

Fig.9.17. Characterisitcs of Zvezda-50 plasma torch at I = 700÷900 A. Open symbols – temperature T, solid symbols – pressure p. □, ■ – d_{cr} = 14 mm; ○, ● – 17 mm; ▲△ – 30 mm.

ABCABC or AABBCC). The thermal efficiency factor resulting from the application of the compact mixing chamber is not lower than that of the Zvezda-6 plasma torch.

The Zvezda-50 plasma torch is powered with a voltage of 10 kV (nominal current I = 1400 A) from the three-phase mains and operates in a stable manner at a pressure in the mixing chamber of up to 20 MPa.

As an example, Fig. 9.17 shows the dependence of pressure p and the mean mass temperature T on the gas flow rate G at I = 700÷900 A, recorded in the Zvezda-50 plasma torch at different values of d_{cr}. Attention should be given to the relatively small decrease of $T = f(G)$ with an increase of d_{cr}.

The Zvezda-50 plasma torch was used as the starting point for the development of the most powerful plasma torch, Zvezda-100 (100 MVA). It consists of two Zvezda-50 plasma torches connected together in such a manner that they form a common mixing chamber with the output nozzle situated on the axis of the chamber. The Zvezda-100 plasma torch is characterised by the same value of pressure and gas temperature as the Zvezda-50 plasma torch, but the flow rate of the gas is twice as high.

In the previous section, it was shown that at low gas flow rates and high current intensity the arc can be shunted on the wall of the confusor. The effect on the behaviour of the arc of the increase of the flow rate and pressure of the gas and of a decrease of current intensity will be explained. The results show that the increase of G and p and a decrease of I are restricted by the increase of arc voltage U to the values at which the arcing in the plasma torch becomes unstable.

Table 9.1 gives the experimental data on the voltage resulting in extinction of the arcs in the Zvezda-6 plasma torch (voltage of arc extinction U_m). Since the extinction of arcing with the increase of the gas flow rate is of the probability nature, the determination of the accurate values of U_m and of the corresponding values of G_m is difficult. Therefore, Table 9.1 gives the ranges of these values. The lower value of G corresponds to the case in which the extinction of the arc was preceded by the normal operating regime, and the upper value corresponds to the regime in which arcing was extinguished almost immediately after ignition.

Table 9.1 shows that the maximum extinction voltage $U_m \approx 2.8$ kV. This result may be interpreted as follows.

The theory of the AC arc shows that continuous (without breaks) burning of the arc in the single-phase circuit with any inductance connected in series may take place only if the effective value of the voltage in the arc U is no more than 0.70 on of the EMF of the power source $\mathcal{E}$. In a symmetric three-phase circuit this voltage increases to ~0.84ε. In the case examined in Table 9.1 ε = 3.5 kV and the condition of break-free arcing has the form: $U \leq$ 2.9 kV. This value is close to the maximum value for the extinction voltage of arcing. Consequently, the phenomenon of arc extinction at $U > U_m$ may be explained by the fact that during the breaks in current the discharge gap is rapidly deionised and its electrical strength increases. Consequently, after passage of current (and voltage) to zero, repeated arc ignition is not possible.

It should also be mentioned that the condition $U_m/\mathcal{E} \leq 0.84$ is valid only for arcs which do not require high peaks of ignition voltage, i.e., for high current arcs. With increase of the peak of the ignition voltage, the extinction voltage of arcing U_m decreases.

Thus, knowing the mains voltage, it is possible to determine approximately the maximum attainable voltage in the arc and estimate the limiting operating conditions of the Zvezda plasma torch.

We examine briefly the problem of the maximum power factor of the plasma torch K_{max} which is the ratio of the power N, generated in arc discharges, to the power of the power source N_s. If it is assumed that the voltage in the arc is rectangular, and current

Table 9.1 Extinction voltage of arc at different G and d_{cr}

d_{cr}, mm	20	20	14	14
G, g/s	326÷354	325÷370	290÷300	250÷260
U_m, kV	2.5÷2.6	2.1÷2.3	2.8	2.7

sinusoidal, the mean power of the arcing half cycle can be calculated from the equation $N = (2\sqrt{2}/\pi)\ UI$, where I is the effective current. At $U = 0.84\mathscr{E}$, $N_{max} = 0.76 \cdot \mathscr{E}I = 0.76\ N_s$, i.e. the maximum power factor of the plasma torch $K_{max} = 0.76$. In the experiments, the maximum value was ~0.7.

9.2.3. Generalised working characteristics of plasma torches

For analysis of the experimental data in the plasma torches with vortex stabilisation of the arc, the authors of [3] used the generalised criterial dependence:

$$UI/Gh_0=f(I^2/Gd\sigma_0 h_0,\ pd^2/Gh_0^{1/2}), \tag{9.4}$$

where d is the characteristic size; σ_0, h_0 are the characteristic values of electrical conductivity and enthalpy of the gas.

Multiplying the determining criteria, we obtain the new criterion $\Pi = I^2pd/G^2\sigma_0 h_0^{3/2}$, which includes all the parameters influencing the dependent parameter, i.e. arc voltage. As shown previously, the values of σ_0 and h_0 may be accepted as constant, at least for one type of gas, and transferred to the coefficient. For transition to the dimensional complexes we denote through $K = UI/G$, $K_0 = I^2pd/G^2$. The expression for the generalised volt–ampere characteristic has the following form $K = f(K_0)$. Complex K_0 is the product of three previously mentioned complexes, reflecting the energy criterion and the Reynolds and Knudsen numbers:

$$K_0=I^2\ pd/G^2 = (G^2 = (I^2/Gd)(d/G)(pd). \tag{9.5}$$

In turn, the complex

$$K = UI/G = U\sqrt{(I^2/Gd)(d/G)} \tag{9.6}$$

also includes the energy criterion and the Reynolds number. The characteristic dimensions d in the estimates is represented by the mean value between the diameters of the electrode and of the output cross section of the confusor $d_m = (d_e+d_k)/2$. The flow rate of the gas is considered through one phase plasma torch $G_1 = G/3$.

In the generalisation of the experimental material, we use the following dimensions of the quantities: $|U|$ = V, $|I|$ = A, $|G_1|$ = g/s, $|p|$ = MPa, $|d_m|$ = cm.

Figure 9.18 shows the results of processing experimental data obtained on plasma torches of different schemes: the Zvezda (star) type, on the simulation single-phase plasma torches and on two plasma torches of the vortex scheme using direct current with a confusor channel and approximately constant arc length. It may be seen that all points fit quite accurately a single curve. The deviations of the large majority of the points does not exceed +15%. The range of the variation of the determining complex K_0 and of the quantities included in the complex is as follows:

K_0=16÷3.2·10^5; I = 0.27÷3.0 kA;
G_1=1÷330 g/s; p=0.01÷12 MPa; d_{cr} = 3÷7.5 cm.

Thus, the complex K_0 reflects adequately the effect of different factors on the arc voltage.

It should be mentioned that the single dependence also expressed the points corresponding to the operation of the Zvezda-Thai plasma torches using both air and nitrogen, helium and mixtures of nitrogen, helium and carbon dioxide, and also the points obtained on the plasma torches using direct and alternating current. This indicates, in particular, that the non-stationarity, determined by alternating current of industrial frequency, has no significant effect on the working complex in the plasma torches with vortex arc stabilisation.

The dependence, shown in Fig. 9.18, is expressed by the equation:

$$P_{sp}=1.84\cdot 10^3\ (I/G_1)^{0.68}\ (pd_m)^{0.34}. \tag{9.7}$$

This shows that:

$$U=1.84\cdot 10^3\ (G_1/I)^{0.32}\ (pd_m)^{0.34}. \tag{9.8}$$

Equation (9.8) shows that the VAC of the arc U ~$I^{-0.32}$ is slightly drooping, and the dependence of voltage and pressure has the form U ~ $p^{0.34}$. Special experimental examination, carried out in [3], shows that for a long AC arc, stabilised on the axis of the channel, this dependence U(p) is retained at least up to p = 100 MPa. If we return to the main criterial complexes, used in chapter 5, the dependence (9.8) is transformed to the form:

$$U=1.84\cdot 10^3\ (I^2/Gd)^{-0.16}\ (G/d)^{0.16}\ (pd)^{0.34}. \tag{9.9}$$

In addition to this, if we transfer to the main units of the SI a system, the equation (9.9) is presented in the following form:

$$U=732.5\ (I^2/Gd)^{-0.16}\ (G/d)^{0.16}\ (pd)^{0.34}. \tag{9.10}$$

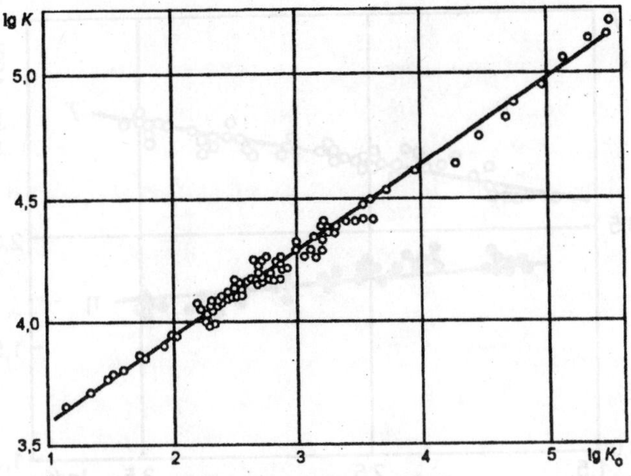

Fig.9.18. Generalised volt–ampere characteristic.

it may easily be seen that the exponents at the dimensional complexes are close to those presented in section 5.1 in the generalisation of the VAC of the AC and DC arcs. This again confirms the conclusions according to which the physical processes in the DC and AC arcs are the same.

To calculate the output parameters of the plasma torch (in particular, gas temperature) it is insufficient to use only one VAC and it is important to know also thermal efficiency η, i.e. the fraction of the heat losses in the walls from the power supply to the arc discharges. The results of processing the appropriate experimental data (a) show that the efficiency of the Zvezda-type plasma torches may also be represented in the form of a dependence on complex K_0 (Fig. 9.19). Here, the scatter of the points is greater than in the case of the generalised VAC, and reaches 20–25%. The equation for the efficiency has the following form:

$$\eta = (I^2 p d_m / G^2_{12})^{-0.09}. \tag{9.11}$$

and shows that the efficiency may be increased only by increasing G_1. With other conditions being equal, the latter results in a decrease of gas temperature. Naturally, the equation (9.11) presented in this form is valid only in the examined range of variation of the dimensional complexes and of the values included in this equation because in formal examination η exceeds 1 at $G \to \infty$. It should also be mentioned that the determining complex does not include the parametric criterion l/d_m, where l is the length of the phase plasma torch. This

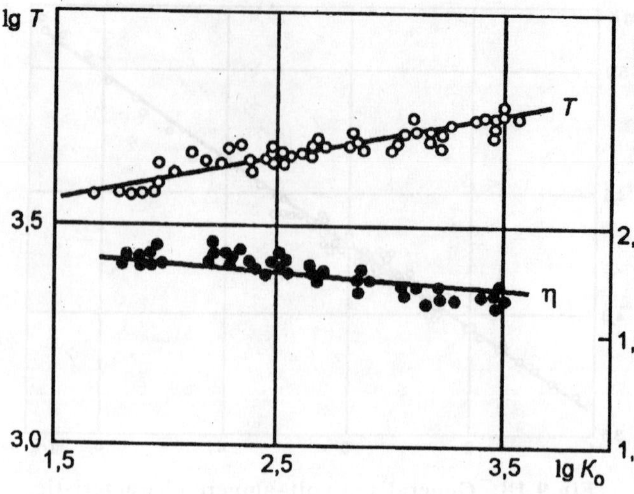

Fig.9.19. Generalised experimental data for temperature and thermal efficiency

criterion determines the total value of the heat losses, and its absence indicates that the examined plasma torches of different powers and schemes were geometrically similar because the observed scatter of the experimental data is determined by this factor.

According to definition for the three-phase plasma torch $\eta = 3G_1 c_p T / N_\Sigma$, where N_Σ is the electrical power supply to the arc discharges, whose value depends on K_0. Thus, it may be assumed that not only the efficiency but also the gas temperature are determined mainly by the complex K_0. The results of processing the experimental data are also presented in Fig. 9.19. The scatter of the points does not exceed ±10%. The appropriate equation in the examined range of the parameters has the form:

$$T = 2.6 \cdot 10^3 \, (I^2 d_m p / G^2)^{0.095}. \qquad (9.12)$$

This equation shows that the large increase of T may be achieved only by a large increase of the complex K_0. For example, when the value of K_0 is increased 10 times, temperature increases by only 25%.

9.3. THREE-PHASE PLASMA TORCHES WITH THE TRIANGLE-TYPE CONNECTION

The presence of several simultaneously burning AC arcs in a single chamber makes it possible to produce simple and reliable plasma torches converting the energy of electrical current into thermal energy with a high efficiency (0.8÷0.9). To obtain these results, a large number of scientific research investigations and design studies was carried out at the Institute of Problems of Electrophysics of the Russian Academy of Sciences [6–12].

All the developed and investigated structures have one common feature – three electrodes in a single chamber. The only difference is that inert gases, nitrogen and hydrogen are heated using rod-shaped electrodes produced from tungsten or tungsten-containing materials, and oxidation media are heated by water cooled copper tubular electrodes.

9.3.1. Plasma torches with rod electrodes

The plasma torches with the rod electrodes are divided into two series, depending on power: plasma torches of the PPT type with the power of up to 140 kW, and EDP-type plasma torches consisting of three basic models with a nominal power of 2, 10 and 80 MW. From the design viewpoint, the two types of plasma torch identical and have three main sections: the casing, the arc chamber (nozzle) and the electrode block [8]. In the EDP-type torches (Fig. 9.20), the casing and the arc chamber are produced as single sections. The electrodes are produced from tungsten with additions. The arc chamber and

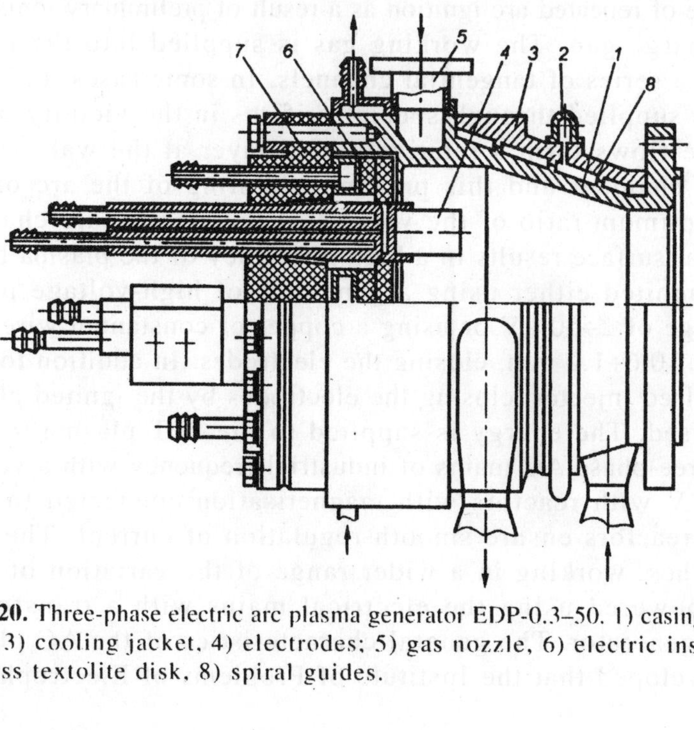

Fig.9.20. Three-phase electric arc plasma generator EDP-0.3–50. 1) casing; 2) corrugated joint, 3) cooling jacket, 4) electrodes; 5) gas nozzle, 6) electric insulation insert, 7) glass textolite disk, 8) spiral guides.

Table 9.2

Parameter	PPT-10.30	PPT-3/100	PPT-3/100M
Power, kW	to 50	to 100	to 140
Current (actual), A	50÷350	50÷450	50÷150
Voltage (actual) V	25÷100	35÷140	40÷150
Pressure in chamber, MPa	to 0.6	to 0.6	to 0.1
Working gas	H_2, He, Ar, N_2	H_2, He, Ar, N_2	H_2, He, Ar, N_2
Gas flow rate, kg/s	to 0.02 (Ar)	to 0.04 (Ar)	to 0.05 (Ar)
Gas enthalpy, MJ/kg	2.4÷5.4	1.2÷4	1.2÷4.5
Efficiency, 5	to 80	to 80	to 80
Dimensions, 10^{-2} m	28×20×15	30×25×20	32×27×27
Nozzle diameter, 10^{-2} m	2	5	5
Length of arc chamber, 10^{-2} m	15	12	12
Volume of the chamber, 10^{5} m^3	330	690	710
Surface area of the chamber, 10^{-2} m^2	1.46	2.85	3.10
Type of electrode	Rod	Rod	Coupled rods
Cross section of the electrode 10^{-4} m^2	0.78	0.78	1.56
Power source	Three-phase mains	Three-phase mains	Three-phase mains

the electrodes are cooled with water. In some latest designs, the electrodes are also cooled with a gas. The three-phase arcing regime in the chamber of the plasma torch makes it possible to use a low voltage of repeated arc ignition as a result of preliminary ionisation of the discharge gap. The working gas is supplied into the chamber through a series of tangential channels. In some cases, the axial flow is also supplied through special orifices in the vicinity of the walls. These flows form a relatively cold layer at the walls of the electric arc chamber and this prevents shunting of the arc on the wall. The optimum ratio of the volume of the electric arc chamber to its internal surface results in a high efficiency of the plasma torch. The arc is ignited either using a generator of high-voltage pulses with a voltage of 2÷50 kV or using a copper or constantan wire with a diameter of 0.6÷1.2 mm, closing the electrodes. In addition to this, a special pulsed injector, closing the electrodes by the ignited plasma blob, was used. The energy is supplied to the PPT plasma torches using the three-phase AC mains of industrial frequency with a voltage of 220/380 V with reactors with magnetisation connected to each phase. The reactors ensure smooth regulation of current. The EDP plasma torches, working in a wider range of the variation of voltage, were powered using the electrical mains with a transformer and a turbogenerator. The general characteristics of the AC plasma torches, developed that the Institute of Problems of Electrophysics

Table 9.2 (continued)

PPT-6	EDP-0.2	EDP-2	EDP-5	EDP-5M	EDP-80
to 100	to 200	to 2000	to 6000	to 10000	to 80000
30÷050	80÷570	2000÷6400	2000÷9000	2000÷9000	10000÷26000
25÷200	20÷240	250÷450	400÷1200	400÷1200	500÷2000
to 0.6	to 0.1	to 2.5	to 3.0	to 5.0	to 2.5
He, Ar, N_2	H_2, He, Ar, N_2	H_2, He, Ar, N_2	H_2, He, Ar, N_2	H_2, He, Ar, N_2	H_2, He, Ar, N_2
to 0.02 (Ar)	to 0.04 (Ar)	to 1.0 (N_2)	to 3 (N_2)	to 3 (N_2)	to 20 (Ar) 10 (N_2) 0.5 (He)
2.0÷5.8	2.0÷6.0	1.2÷3.3	1.2÷54	1.2÷5	1÷11
to 85	to 80	to 70	to 90	to 90	to 84
22×20×18	25×27×30	38×35×32	67.5×35×32	67.5×47×65	61×85×65
2	2.5	14	18	18	20
10	13	18	28	28	25
240	210	2600	5300	5300	12000
1.25	1.50	7.4	12.5	12.5	18.8
tubular	tubular	compound rod	compound rod	compound rod	compound rod
0.78	2.2	19.8	19.8	19.8	37.2
Three phase mains from separate choke coils	Three phase mains	Three phase mains, transformer. Generators	Three phase mains, transformer. Generators	Three phase mains, transformer. Generators	Turbo generator

of the Russian Academy of Sciences are presented in Table 9.2.

The PPT-10/30 plasma torch [6] is designed for the rated power in the range 3–50 kW and for operation with nitrogen, hydrogen and inert gases. The plasma torch consists of three sections: the casing, the nozzle and the electrode block. The casing of the plasma torch is cylindrical, has a cooled jacket and a flange for connecting the nozzle. The casing contains a chamber for the tangential feed of the working gas which travels along three spiral channels directly into the arcing zone. In addition to this, the casing contains the electrode block consisting of three electrodes, electrode holders and a heat-resisting electrically insulating insert. The electrode of the plasma torch are produced from tungsten with a diameter of 8÷10 mm in the form of rods, with the eccentrically positioned leg, so that it is possible to regulate the distance between them. The electrodes are inserted into the electrode holders positioned under the angle of 120° in relation to each other.

The results of long-term tests were used in the development of the design of a heat-resisting electrically insulating insert which fully satisfies technical requirements.

The nozzle is connected through a bolted joint to the flange of the casing of the plasma torch. The flange has a groove into which a sealing rubber insert is placed ensuring a leaktight joint between the casing and the nozzle part of the plasma torch. The nozzle has a cooling jacket with a spiral channel organising the flow of the cooling liquid and 'thermal decoupling', compensating thermal expansion of the nozzle.

PPT-6 six-electrode plasma torch

The PPT-6 six-electrode plasma torch was developed for more uniform heating of the gas and increasing electrical power. The plasma torch consists of two three-phase plasma torches, situated in the same casing and having the common arc chamber, and also the same system for supplying the working gas. The PPT-6 plasma torch consists of the same main sections as the PPT 10/30 plasma torch. The electrodes of the six-electrode plasma torch are produced from a tungsten attachment, brazed using a silver brazing alloy into a copper pipe with a diameter of 10 mm.

PPT-3/100 and PPT-3/100 M plasma torches [7, 8]

The PPT-3/100 plasma torch has a power of up to 100 kV · A, and the PPT-3/100 M plasma torch a power of up to 140 kV · A. The two plasma torches consist of the same main sections as the PPT-10/30 plasma torch, and differ from it by the geometrical dimensions of the arc chamber and the casing, the cross-section of the gas channels of the chamber for the tangential supply of the working gas and the design of the electrode section. Because of the increase of the working current in the circuit of the plasma torch, the cross-section of the electrodes was also increase. In addition to this, an electrode section with cooled electrode holders was developed. This section is used for installation in each unit of the paired electrodes made of two tungsten rods.

EDP-type plasma torches (EDP-0.2; EDP-2; EDP-5; EDP-80)

The series of the plasma torches with a power of 200 kVA, 2 MVA, 5 MVA and 80 MVA is designed for station heating of the inert gases, and also nitrogen and hydrogen (Fig. 9.20).

The specific feature in comparison with the PPT series is that the cooling jacket of the casing is section into two parts with subsequent connection by a corrugated ring. The jacket has two nozzles through which the cooling water is supplied closer to the thermally most heavily stressed section of the arc chamber – the outlet of the nozzle. To protect the internal surface of the wall chamber, it is necessary to use an additional gas flow directed along the axis and concentrated in the vicinity of the walls. In a number of electrode designs, sets of rods or tungsten wires are used. All the electrodes with the internal end part are cooled with water and some of them also with a gas flow directed in the axial direction. The gas is also blown for the heat-insulating electrode disc. For better protection against the radiant heat flow and thermal shocks, a metallic electrode unit was developed with water cooling of the end surface and a gas screen, preventing the short-circuiting of the arc with the surface. Thus, in addition to the main gas flow, this type of plasma torch also uses additional independent gas supplies: through the electrode, blowing through the slit along the electrode holder, and also through the bottom of the electrode unit. The gas flow in the chamber of the plasma torch is a complicated process. The blowing of gas through the electrode increases the arcing voltage and makes it possible to obtain the maximum possible power at the given parameters of the electric power supply system. However, in the case of long-term operating regime, wear of the electrode is accompanied by the formation of an unstable regime determined evidently by a decrease of the electrode length (approach of the arc to the bottom of the electrode unit). In long-term operation of the plasma torch, it is necessary to have a reserve in the mains voltage to ensure stability when the geometrical and physical parameters of the electrodes change.

The EDP-80 plasma torch with a power of up to 80 MW is designed for operation during several seconds (up to 5 s). Because of the short-term operating time and the design of the plasma torches, the arc chamber, the casing and the electrodes are not cooled. The EDP-80 plasma torch consists of the arc chamber produced from Cr18Ni10Ti stainless-steel with the wall thickness of 18 mm, and a casing. The casing and the arc chamber are connected together by means of 12 pins, passing through the flanges of the casing and the arc chamber. A ring with a conducting gas nozzle is placed between the flanges using a copper gasket. The ring and the casing of the plasma torch form a circular chamber for the supply of the working gas. The casing contains two rows of the tangential distributed orifices, through which the working gas is supplied to the arc chamber of the plasma torch. The heat-resisting electrically insulating insert consists of a disc produced from Al_2O_3, insulating pipes, which are also installed on a Al_2O_3 disc. The discs in the insulating pipes are connected together by high-aluminium segment directly in the casing of the plasma torch. Copper electric holders (cylindrical) are inserted into the isolating pipes and have a flat contact area for securing the current-conducting busbar and a threaded socket for installing the electrode. The electrode, of the compound structure, consists of tungsten rods with the diameter of 10 mm brazed into a copper cup with a threaded tail. The insulating disc, receiving pressure, is secured to the flange of the casing of the electrode.

9.3.2. AC plasma torches with rail tubular electrodes

This type of plasma torch, in contrast to those described previously, is designed for operation with oxidation media and, in particular, for heating of air. For this purpose, a plasma torch with a power of 0.1÷1 MW with tubular copper electrodes was developed (Fig. 9.21). The plasma torch consists of the four basic parts: the casing 1, the output flange with the nozzle 2, the electrode system 3 and the injector 4. The casing is produced from stainless steel and has the form of a cylinder transferring into a truncated cone. The casing is cooled with water. Along the length of the casing there are three rings with tangential orifices through which the working gas is supplied into the discharge chamber. The supply of the gas to each ring is independent. The output flange is also made of the stainless-steel (12Cr18Ni10Ti) and is cooled with water, supplied from the casing through brass sleeves, sealed with rubber rings. The electrodes are U-shaped and are produced either from the copper pipes with a diameter of 10 mm or from a copper bar with a diameter of 20 mm with the internal orifice 8 mm in diameter. Brass nozzles are brazed into the electrodes and used for securing the electrode in the casing and for the supply of cooling water. The electrode is introduced into the casing through a ceramic insulator and a fluoroplastic sleeve. Current-conducting bars are connected to the nozzles, closer to the injector. The rear part of the casing of the plasma torch (Fig. 9.21) contains the injector connected from the side of the truncated cone. The output nozzle of the injector is directed into the gap between the electrodes (3÷5 mm).

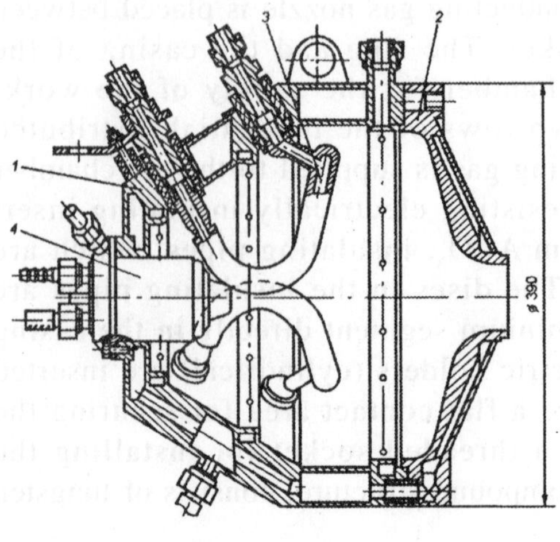

Fig. 9.21. Three-phase plasma torch of the PPT series. 1) casing; 2) nozzle; 3) electrode; 4) injector.

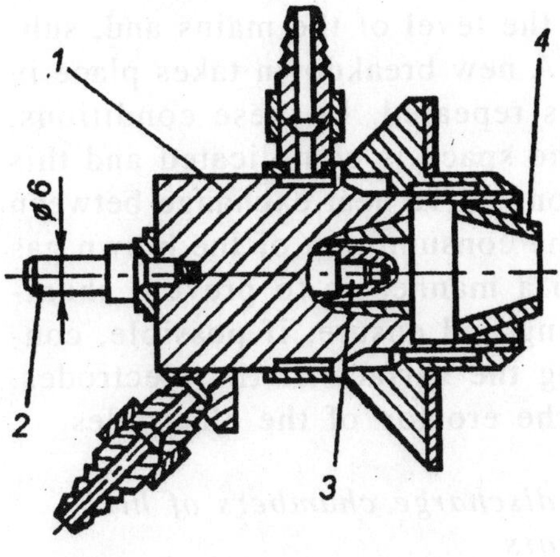

Fig.9.22. High voltage single-phase plasma-injector. 1) casing, 2) insulator, 3) replaceable electrode for nozzle.

The injector is in the form of a high-voltage single-phase AC plasma torch with a power of up to 3 kW (Fig. 9.22). The injector consists of the casing, a ceramic nozzle and two electrodes. The casing is produced from stainless steel and is watercooled. The casing contains two cylindrical channels merging in the discharge chamber under the angle of 15°. Tangential blowing of the gas, supplied from the common chamber, is provided through every channel. The discharge chamber ends with the ceramic nozzle secured by a nut. Each electrode has the form of a brass cylinder through which a brass rod passes. A ceramic sleeve is placed between the rod and the internal sleeve of the cylinder. The brass rod has a thread into which the conical copper tip with the cylindrical insert pressed into it is wound. The insert is produced from different materials (copper, cermet, etc).

The operating principle of the injector may be described as follows: alternating voltage with the amplitude of 6 kV is supplied to the electrodes, the effect of the voltage results in the electrical breakdown between the wall of each electrode channel and copper conical tips. The resultant two short arcs are blown by the gas flow to the ends of the electrodes between which the arc burns in the chamber. If the arc is extinguished, the process is repeated.

The sequence of operation of the plasma torch with the rail-type tubular electrodes is as follows. A plasma, generated by the injector, is situated constantly in the narrowest part between the U-shaped electrodes. When a voltage is applied, and arc forms between the electrodes, and the arc moves in the direction of expansion of the distance between the electrodes under the effect of the electromagnetic force, generated by the intrinsic magnetic field and the hydrodynamic force, generated by the gas flow. When the arc length is increased, the voltage in the arc increases to the level of the mains and, subsequently, the arc is extinguished. A new breakdown takes place in the narrow part and the process is repeated. In these conditions, the form of the arc discharge in the space is complicated and this greatly increases the intensity of convective heat exchange between the arc and the surrounding gas. The consumption of the blown gas is selected in experiments in such a manner as to prevent shunting of the arc on the metallic casing and ensure, if possible, continuous movement of the arc along the surface of the electrodes. The latter has a strong effect on the erosion of the electrodes.

9.3.3. Main physical processes in discharge chambers of high-power three-phase plasma generators

In all plasma generators, including AC generators, the electrical energy is converted to thermal energy. From the viewpoint of the thermodynamics, this process may take place with the efficiency equal to unity. However, since the heating of the gas in the chamber of the plasma torch is carried out by the thermal interaction of the gas with the arc discharge, losses of various types are unavoidable. To reduce the losses, it is necessary to control the main energy flows in each specific design of the plasma torch. This is possible only when the relationship is established between the nature of the main physical processes, taking place in the chamber of the plasma torch, and the parameter such as consumption, gas pressure, current intensity, and arcing voltage. A number of measuring systems was developed for solving this problem in three-phase plasma torches [9].

A system was developed for recording arc radiation to analyse the nature and dynamics of radiation inside the chamber and determine the arc geometry. The system includes a special optical scheme consisting of a lens, whose central part contains a circular non-transparent screen. This optical system makes it possible to eliminate the effect of radiation of the jet in analysis of radiation inside the chamber of the plasma torch and determine the discharge geometry. In the radiation from the arcs inside the plasma torch, passing through the nozzle, is projected onto the screen using this optical system. The discharge, projected onto the screen is photographed with a high-speed camera. Using optical filters, the same camera is used for taking photographs of the electrodes and of the arcs burning between them (photographs were taken through the nozzle orifice).

The system for recording the electron concentration and plasma temperature

The electron concentration was determined from the intensity of the continuous spectrum, temperature from the ratio of the intensities of two lines. The spatial localisation of the measured region was carried out using the previously mentioned optical system. In addition to this, the plasma area was not closer than 4÷5 cm from the surface of the electrode because at a smaller distance the image can already be eliminated by the radiation of the electrode. At a larger distance, the oscillations of the plasma filament may be so high that at some moments of time the plasma does not penetrate into the analyzed volume. Consequently, the error of determination of electron concentration increases.

All the measurements of the concentration of the electrons and temperature were taken with time sweep sufficient for resolving in time the oscillations of intensity as a result of turbulence. The radiation from the plasma torch was projected, using the optical system, to the input of special devices which generated both the overall (in respect of time) spectrum of plasma radiation inside the plasma torch and also the time dependence of the radiation intensity at different frequencies. An SI-8-200 lamp was used as the image source. The concentration of the electrons and plasma temperature were determined inside the plasma torch on the basis of the relative intensity of the lines of tungsten and of the continuous spectrum. The error of the measurement of concentration was 50%, temperature 15%.

As already mentioned, in all designs of the plasma torches, the gas, moving into the working chamber, forms a relatively cold gas screen at the walls where the conductivity is almost zero. Consequently, the positive arc column does not touch the walls. Under the effect of the flow of the working gas and the electrodynamic forces, the arcs are elongated in the direction of exit from the electric arc chamber.

The experiments showed the existence of two arcing regimes: diffusion and constricted (Fig. 9.23).

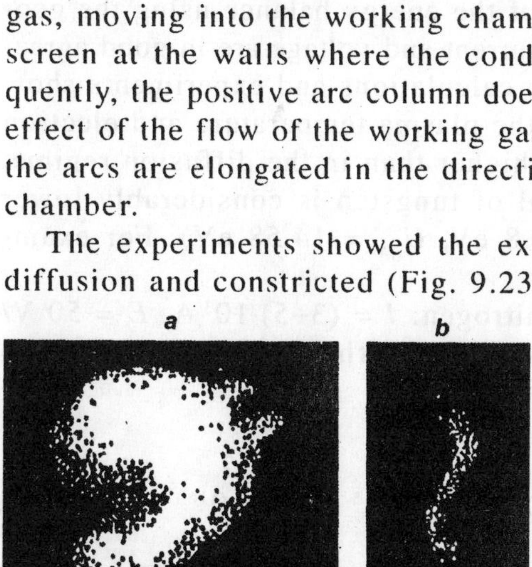

Fig.9.23. Arcing regimes in three-phase plasma torch. a) diffusion; b) constricted

In the diffusion regime, the arcs occupy a large part of the volume of the chamber. The discharge is clearly turbulent. Pulsating plasma blobs together with the oscillations of the plasma pressure and voltage oscillations are detected. This regime is maintained in the pressure range in the discharge chamber of 0.1÷0.35 MPa at a gas flow rate of 1÷10 kg/s (nitrogen) (depending on the type of plasma torch). The actual value of current varies in the range 1÷20 kA (in the EDP-type systems). In small systems (PPT-type), the current is in the range 0.1÷0.5 kA, consumption 1÷6 g/s. To increase the pressure, the discharge is transferred to the constricted regime. For example, for nitrogen, this pressure is 3.5 MPa. In this case, the diameter of the filament rapidly decreases to the size close to the diameter of the emitting electrode surface. The density of arc in the current increases and arc voltage decreases. The arc temperature is considerably higher than in the diffusion regime.

The main reason for the existence of the two discharge regimes [1] is associated with the effect of tungsten vapours on the electrical properties of the arc. In the diffusion regime (similar to the plasma with additions of alkali metals), the conducting properties of the plasma are determined by the ionisation of tungsten vapours which, at a lower gas pressure, diffuse into the large volume of the chamber. With increasing pressure, the contribution of the vapours to conductivity rapidly decreases. In addition to this, the absorption of radiation becomes more intensive. Conditions typical of superheating instability are developed, and this also results in constriction of the discharge. The evaluation of the main parameters of the plasma: the temperature concentration of the electrons – on the basis of the energy balance using the geometrical size of the arc and the current and voltage are in good agreement with the experiments. The calculations and experiments show that in the constricted regime, the plasma temperature and electron concentration are considerably higher than in the diffusion regime, because the ionisation potential of tungsten is considerably lower than that of nitrogen ($U_{i,W}$ = 7.8 eV, $U_{i,N}$ = 14.58 eV). For example:

The diffusion discharge of nitrogen: $I = (3 \div 5) \cdot 10^3$ A, E = 50 V/cm, the area of the current section of the discharge zone S = 50 cm^2, p = (0.2÷0.3) MPa

$T \cdot 10^3$, K	4	5	6	7
n_e, cm^{-3}	$1.3 \cdot 10^{15}$	10^{15}	10^{15}	10^{16}

the constricted discharge: $I = (4 \div 5) \cdot 10^3$ A, $E \approx 70$ V/cm; S =

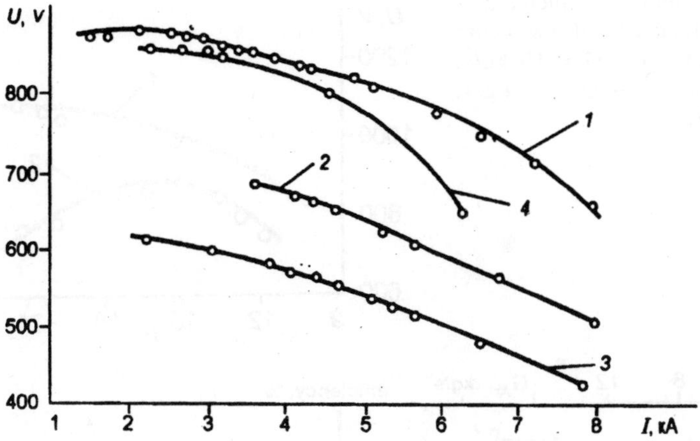

Fig.9.24. Volt–ampere characteristics of EDP-5 plasma torch 1–3) working gas – nitrogen; 1 –G = 2.5 kg/s, p=1.7 MPa; 2 –G = 2.1 kg/s, p=0.5 MPa; 3 – G = 1.9 kg/s, p= 0.22 MPa; 4) Working gas – helium; G = 0.15 kg/s, p=1.1 MPa.

1 cm^2, p = 0.6 MPa:

$T \cdot 10^3$, K	8	9	12
n_e, cm^{-3}	$5.2 \cdot 10^{16}$	$4.6 \cdot 10^{16}$	$4.3 \cdot 10^{16}$

Comparing the results, it may be concluded that in the examined conditions to ensure the required value of n_e, the temperature in the constricted discharge should be $T > 10^4$ K, and in the diffusion discharge $T < 7 \cdot 10^3$K.

As shown by estimates [9], the nature of heat exchange between the arc and the surrounding gas in the diffusion regime is determined by turbulent heat conductivity and convection, and in the constricted regime by convection and radiation.

The VAC of the arc at low current and gas flow rate are drooping, and with increasing current (I > 10 kA) rising (Fig. 9.24, 9.25). In all cases, the reason for the increase of the voltage is the increase of the flow rate, and the increase of efficiency is caused by the increase of the gas flow rate and power (Fig. 9.26 *a*, *b*). The rising nature of the VAC at high currents is caused by the fact that the plasma temperature is such that the electrical conductivity depends only slightly on temperature (Coulomb scattering).

The increase of the gas flow rate increases the extent of removal of heat and, consequently, temperature and electrical conductivity remain constant, and to increase current intensity, it is necessary to increase voltage.

Fig.9.25. Volt–ampere characteristics of EDP-80 plasma torch (working gas – nitrogen). 1 – G = 10 kg/s, p = 0.5 MPa; 2 – G = 5 kg/s, p = 0.2 MPa;

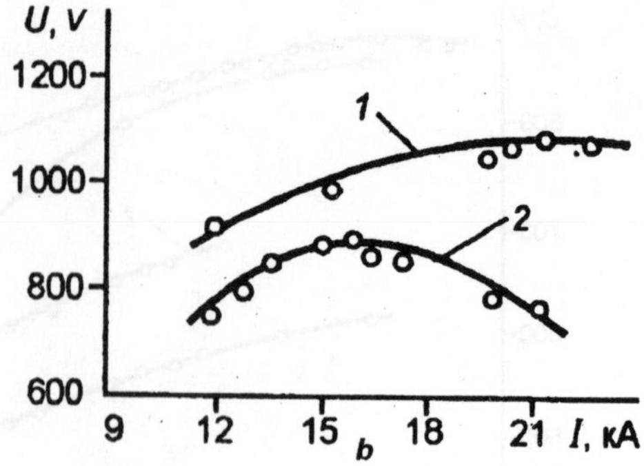

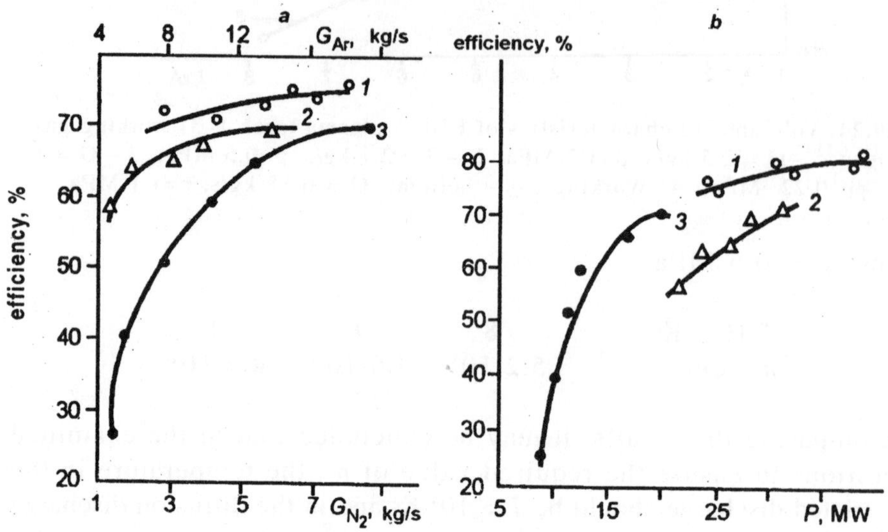

Fig.9.26. Dependence of efficiency on gas flow rate (a) and power (b) of EDP-80 plasma torch. 1) I = 30 kA, working gas – nitrogen; 2) I = 16 kA, nitrogen; 3) I = 17 kA, argon.

9.3.4. Near-electrode processes

One of the most important characteristics, which determine the industrial application of plasma torches, is the operating life of the plasma torch. As in the DC plasma torches, the critical area of the AC plasma torches, which determine the duration of operation, are the electrodes. In particular, the electrodes are subjected to high thermal loading, mainly in the areas of arc attachment. Depending on the electrode material, current intensity, and cooling conditions, attachment of the arc maybe of the diffusion or constriction type. Naturally, for the same value of current intensity, the constricted regime (arc spot) is most heavily thermally stressed. For the cathode, the nature of current attachment is determined to a large extent by the work function of the cathode material.

The erosion of the electrodes is also strongly affected by the chemical properties of the working gas. The presence of oxygen in cases in which strong oxide coatings form, for example, in hafnium and zirconium, results in a decrease of the degree of erosion because of a decrease of the work function of the electrons. However, at high temperatures (high currents), the presence of oxygen results in the processes identical to combustion and this greatly increases erosion for the same materials (hafnium, zirconium). In many cases, the erosion process is caused not only by the evaporation of the material but also by the removal of molten metal by the gas flow. For correct theoretical description of the erosion process, there are no actual mathematical models of the present time. The main method of examining erosion is the experimental method. In the experiments, the dependence of specific erosion (g/C) on the current intensity, electrode material and design is determined. The nature of arc attachment and the surface temperature of the electrode are also investigated. To examine the erosion of rod electrodes of AC plasma torches, the authors of [9] developed a method of measuring the surface temperature of the electrodes using an optical circuit in which the enlarged image of the electrodes is projected either on the screen or the slit of the monochromator. The image on the screen may be used to fix the nature of arc attachment and, when using a pyrometer, determine the brightness temperature of the image of the electrode. To determine the true temperature of the electrode, a reference light source with the known temperature is placed in the area of the electrode. The electrodes are photographed on the screen using a high-speed camera placed in the area of the pyrometer.

To separate the linear spectrum of the glow of plasma from the continuous spectrum of radiation of the electrode, it is necessary to use a system of filters and a lens with a non-transparent screen. In addition to this, the surface temperature of the electrode is measured also at the moment of rapid disconnection of current (in accordance with the decrease of the intensity of plasma radiation).

In experiments with media not containing oxygen, trials were carried out using thoriated, lanthanised and yttrium-doped tungsten. The experimental results show that for the alternating current with a frequency of 50 Hz at $I \geq 200$ A, a spot forms on the surface of the electrode. Increasing current results in a transition to the diffusion regime (emission takes place from the entire surface). This transition depends only on the surface temperature and occurs suddenly. In the presence of a spot, the temperature outside the spot is equal to 2000 K and lower and is almost constant over the entire surface,

Table 9.3. Erosion characteristics of electrodes of PPT′ and EDP series plasma torches

Plasma torch	Working gas	Pressure in chamber, MPa	Electrode material	Electrode diameter, cm	Area of the electrode, cm^2
PPT	Nitrogen, Argon	0.15÷0.25	Thoriated tungsten VT-10, 1% Th_2O	1	$2\cdot10^{-2}$
	As above	0.15÷0.25	As above	1	0.8
PPT	x				
EDP-5	Nitrogen	0.12 0.2	Lanthanised tungsten VL, 1 % La_2O_3	5	0.8
	x				
	x				
	As above	0.3÷.7	As above	5	0.8
	As above	0.65÷1.2	"	5	0.8
	As above	0.4÷0.6	"	5	0.8
EDP-80	Hydrogen	0.2÷0.7	"	7	3
EDP-5	Hydrogen	0.12÷0.32	"	5	0.8
EDP-2	Nitrogen	0.1÷0.8	Yttrium-doped tungsten SVI, 2.01 % Y_2O_3	5	0.8

i.e. the contribution of the area of the electrode outside the spot to the emission current is negligible. The radius of the spot is 0.6÷0.8 mm. The effective mean temperature of the spot varies in the range 3200÷3400 K. In the centre of the spot, there is a region of the melt with the radius of r = 0.1÷0.2 mm, where the temperature reaches the values of $\geq$3800 K.

The most probable admission mechanism in this case is thermal-auto electron emission (*T–F*-emission) capable of ensuring the observed current density in the spot.

The increase of current as a result of heating by Ionic current the electrode surface at a temperature of 2800÷3000 K, transition takes place from the spot to the diffusion regime. In the transition region, the formation of two spots on the same electrode was detected in a number of cases. The transition to the regime without the spot takes place in all types of plasma torches when the appropriate temperature is released. If the electrodes are made of the material containing the additions of substances with increase the emission capacity (thorium, yttrium, lanthanum), the temperature of transition to the regime without the spot maybe reduced. This results in a large decrease of the extent of erosion. The characteristic special feature of operation of these electrodes is that an increase of the surface temperature results in the failure of the emission surface (depletion with additions). However, as a result of diffusion from the players of the material of the additions, the emitting layer is restored. There is a temperature range in which the rates of these processes are equal and this greatly reduces the extent of erosion of the electrode (Table 9.3). The table shows that at working currents

Table 9.3 (continued). **Erosion characteristics of electrodes of PPT and EDP series** plasma torches

Emitting surface, emission	Actual current, A	Current Density, A/cm^2	Temperature of surface ???, K	Specific Erosion, g/C
Spot, T–F - emission	100÷200	(5÷10)·10^3	3200÷3800	(1÷3)·10^{-7}
Th$_2$O, T - emission	200÷500	(2.5÷6)·10^2	2800÷3000	(0.6÷2)·10^{-7}
La, T - emission	(3÷4)·10^3	(3.5÷5)·10^3	3400÷3600	10^{-4}
W, T - emission	(3.5÷5)·10^3	(4.5÷6)·10^3	3500÷3800	(3÷5)·10^{-4}
W, T - emission	(2.5÷3.3)·10^3	(3÷4)·10^3	3600÷4000	(5÷7)·10^{-4}
W, T - emission	(4÷8)·10^3	(5÷10)·10^3	3900÷4500	(8÷9)·10^{-4}
La, T - emission	(10÷15)·10^3	(3.3÷5)·10^3	3300÷3400	10^{-4}
W, T - emission	(3÷5)·10^3	(3.8÷6.2)·10^3	3400÷3700	(3÷4)·10^{-4}
Y, T - emission	(2÷3.5)·10^3	(2.5÷4.4)·10^3	3200÷3400	(1÷5)·10^{-5}

not exceeding 200 A, in the absence of the arc spot, the specific erosion of the rod electrodes made of tungsten with the additions is small, $10^{-7}÷10^{-6}$ g/C (the working gas does not contain oxygen).

If air is used as the working gas, the degree of erosion of the electrodes made of tungsten and tungsten-containing alloys is doubled. In this case, it is efficient to use water-cooled copper electrodes working in the arc spot regime moving along the surface. This movement maybe ensured by the rail gun effect or by the effect of hydrodynamic forces [10]. The continuous movement of the arc spot along the surface of the electrode restricts the time during which the electrode is in the given zone and, consequently, reduces erosion. As shown by the calculations of non-stationary heat exchange of the arc with the surface of the electrode (taking into account the evaporation and melting processes), minimum erosion is ensured when the attachment time of the arc does not exceed the duration of heating to the melting condition in the area of attachment of the arc [11]. For example, for currents of $I = 10^2$A and the diameter of the arc spot d_s = 0.5 mm, this time is equal to 10^{-4} s. The experiments show that in these conditions, the specific erosion of the copper tubular water cooled electrodes (I = 500 A, flow rate of air 30 g/s) is 10^{-6} g/C [12]. At this value of erosion in the laboratory experiments it was possible to obtain a long operating life of the electrode prior to replacement at a power of the plasma torch of N = 300÷500 kW. Since all other elements of the plasma torch (chamber, injector) were completely capable of efficient operation, the results were used in the development of AC plasma torches designed for long-term operation at a power of up to 1 MW.

9.4. High-voltage multi-electrode plasma torch

The main characteristics of the high-voltage multielectrode AC plasma torch are the following:

–the high voltage of the power source (~10 kV), which makes it possible, at relatively low arc currents (~10÷20 A) to realise the arcing conditions of the single-phase arc with the power up to ~20÷30 kW. At these currents, the arc is characterised by a low pulse flow as a result of intrinsic electromagnetic forces and, consequently, it is extended in the flow to the length of ~1 m with the voltage in the arc decreasing to ~1÷2 kV at a low degree of erosion of the electrode;

–the realisation of burning of the multi-electrode (three- and six)-arcs increases the total power and also the volume of the plasma and improves the stability of arcing as a result of the mutual thermal effect;

–the application of the alternating current of industrial frequency enables a simple electric power source to be used.

The principal diagram of a three-electrode plasma torch is shown in Fig. 9.27 [13–16]. The electrode holders 3 are installed, through the fluoroplastic insulators 2, on the casing 1 which contains the copper electrodes 4 with the working arc gap δ_2 and the breakdown gap δ_1 in the narrowest area between the electrodes and the additional conical electrode 6 which may travel along the axis of the plasma torch, changing the gap between them and the electrodes. The channel 5 is used for supplying the fuel aeromixture; 7– is one of the branches of the electrical arc; 8 – the flow of the aeromixture; 9 – the airflow.

The multi-electrode high-voltage AC plasma generator is powered by a step-up three-phase transformer TM 40010/04 with the power of 400 kW with the open circuit voltage at the secondary windings of 10 kV. The power is supplied to the six-electrode plasma torch by the circuit with the dismantled 'star', and for the three-electrode system, the secondary windings are connected by the 'star' scheme with the isolated neutral.

Arc current was restricted and arcing was stabilised using inductance ballast resistances which makes it possible to reduce the losses of the active power because of the low ohmic resistance and, consequently, increase the efficiency of the system as a whole, simplify the cooling system, and also efficiently stabilises arc current [17].

The geometrical characteristics of the electrodes have a significant effect on the dynamics of discharge and on all plasma parameters. In the examined multi-electrode plasma torch they include: the cross-section and the length of the electrodes, the initial (breakdown) and final (working) discharge gaps. The experiments showed that of the diameter of the electrodes may be close to the diameter of the arc (d_a ~1 cm). In this case, the arc is stable in the entire examined current range, $I \approx 5 \div 25$ A. The length of the working part of the electrodes influences the ratio of the duration of movement of the arc along the electrodes and the arcing time at the ends of the electrodes, and also the operating life of the electrodes.

The nature of movement of the arc between the electrodes depends on the geometry of the discharge gap which is determined by the form and position of the rod electrodes in space. It also includes the narrowest gap δ_1 (Fig. 9.27) between the conical and the main electrodes in which the initial breakdown takes place, the central part and the final working gap δ_2, which determines the distance between the ends of the electrodes. As shown previously, the initiation of the arc starts by the electrical breakdown in the gap δ_1. Subsequently, the arc is ignited between the remaining electrodes and blown to the separated ends.

As shown by the investigations, carried out at the airflow rate of ~10 m/s and the arc current of ~10÷20 A, depending on the value of δ_1 there may be different regimes of burning of the high-voltage arc. In the case of very small gaps δ_1 ~ 1÷2 mm, the electrical discharge after a breakdown remains in the narrow gap. With increase of δ_1 the arc travels to the ends of the electrodes, but if the gap is smaller than ~4 mm, then the gap is broken through with increasing arc voltage. This is accompanied by shunting of the arc, elongated at the ends of the electrodes, by the resultant short discharge, and by the pulsed arcing regime in which evolution of the arc is repeated after every breakdown.

Electrically more suitable is the regime with long-term burning at the ends of the electrodes of the elongated arc with the maximum possible length [18]. The highest possible voltage, close to the extinction voltage, is reached in the arc and, consequently, maximum energy generation in the arc is found here. For this purpose, the value of δ_1 should be such that the breakdown takes place at the voltage close to the maximum instantaneous voltage of the power source (~10 kV). The experimentally selected gap in which a reliable breakdown took place and there was no arc shunting was

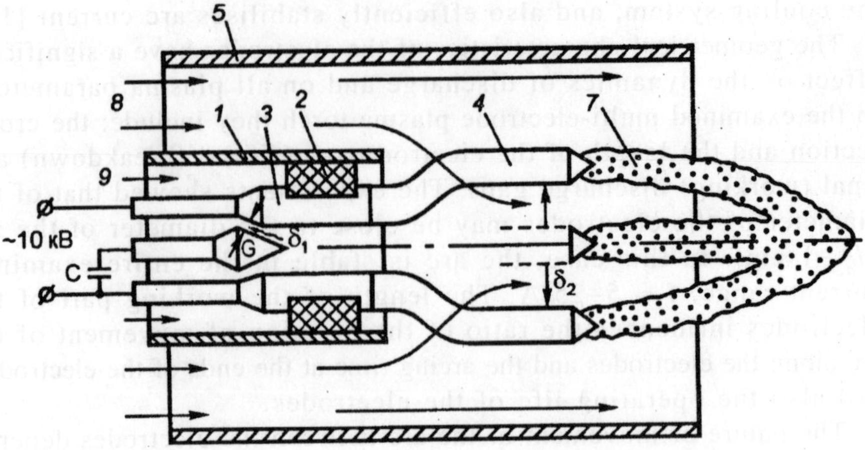

Fig.9.27. Diagram of a high-voltage plasma torch.

$\delta_1 \approx 4 \div 5$ mm.

Figure 9.28 shows one of the possible methods of moving the area of attachment of the arc to the electrodes. One of the ends of the arc (lower end of the figure) together with the adjacent section of plasma is moved further on the surface of the electrode and, consequently, the form of the arc changed as a whole. Consequently, the sections of the arc close to the other end moved closer to the electrode surface and, subsequently, the arc was short-circuited and its end moved in a jump to the new area of attachment, situated higher along the flow. After moving to the ends of the electrodes, the arc is elongated along the flow with a further increase of the length to $l_a \sim 1$ m.

Figure 9.29 shows the film frames of high-speed filming of the dynamics of a three-electrode 6 A arc from the moment of breakdown to reaching the maximum length. The electrodes were in the horizontal position in the free space and blown with the horizontal flow of cold the, with the velocity of the flow at the outlet of the electrodes being ~5 m/s. Under the effect of the flow, the arc is elongated to the length of the order of 1 m with the mean transverse size of the glowing channel of $d_a \approx 0.8 \div 1$ cm. The duration of continuous arcing in the quasi-stationary regime was $2 \div 4$ s (Fig. 9.30). Extinction of the arc was followed by a new breakdown and the process was repeated.

Examination of the electrical characteristics was carried out for the quasi-stationary arcing regime [13, 17]. One of the main characteristics of the plasma system is the static volt–ampere characteristic (VAC) of the arc, which determines the dependence of the

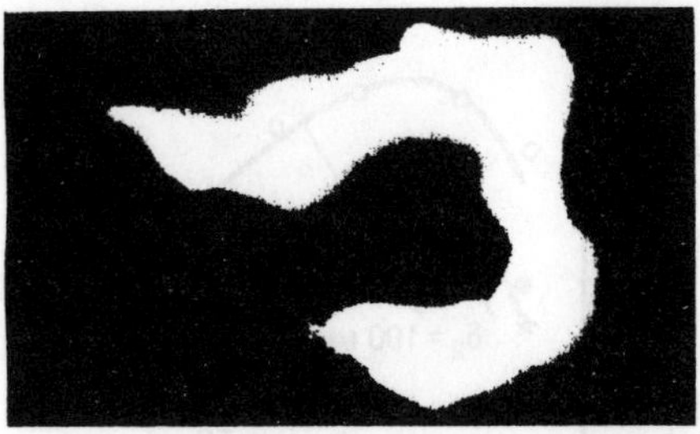

Fig. 9.28. High-voltage 6 Arc in a air flow

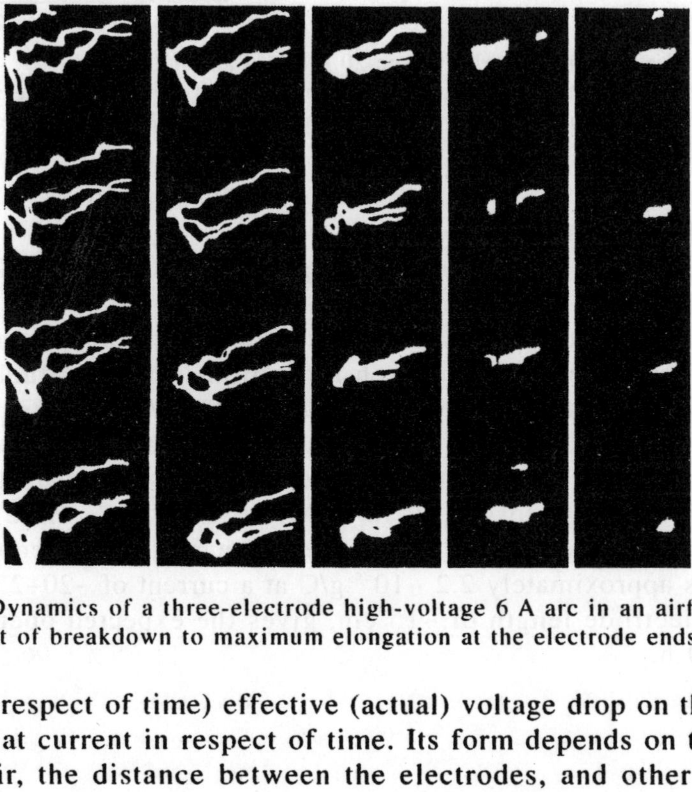

Fig.9.29. Dynamics of a three-electrode high-voltage 6 A arc in an airflow from the moment of breakdown to maximum elongation at the electrode ends.

mean (in respect of time) effective (actual) voltage drop on the mean effective at current in respect of time. Its form depends on the flow rate of air, the distance between the electrodes, and other parameters.

The dependences of the active power of the three-electrode arcs on the flow rate of air u_a at the amplitude of the phase current of 24 A and different distances between the electrodes δ_2, presented in Fig. 9.31. It may be seen that with a decrease of δ_2, the maximum of the curves of the arc power $N(u_a)$ is slightly displaced in the direction of higher velocities because of the increased stability of arcing, but the absolute value of N decreases because of a decrease of arc length. The maximum power of the plasma, obtained in the investigated range of the parameters, is 50 kW for the three-electrode arc and 110 kW for the six-electrode arc at $u_a \approx 8 \div 10$

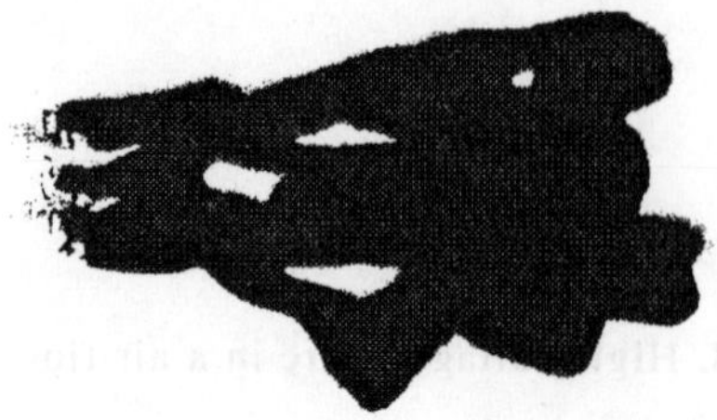

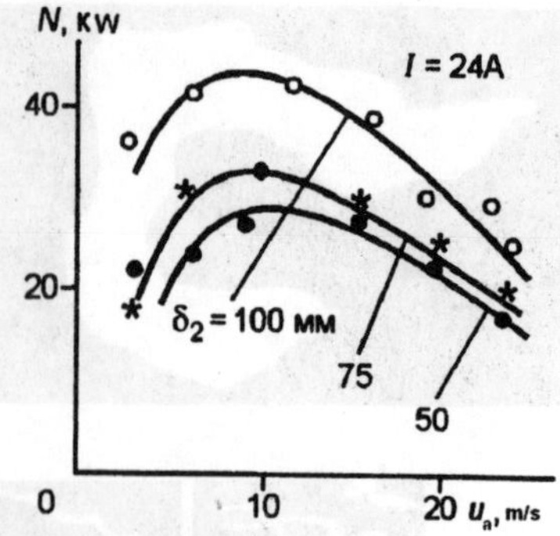

Fig.9.30. Six-electrode high-voltage 20 A arc in an airflow.
Fig.9.31 (right). Dependence of the active power of three-electrode arcs on the flow rate of air u_a at the amplitude of phase current of 24 A and different values of δ_2.

m/s and $\delta_2 \approx 10$ cm.

The theoretically estimated temperature of the arc column was 7000–8000 K.

This plasma torch was used in the development of a high-voltage multi-electrode plasma torch for igniting coal dust fuel [19–21]. The tests show that at a power of ~35÷40 kW, the ignition of the entire fuel flow, passing through the torch, takes place at a flow rate of the coal through the channel with the arc of up to ~1 t/h, at relative energy losses of $\overline{N} = N/N_t \approx 0.3 \div 0.5\%$. Here N_t is the power of the torch. The specific erosion of the non-cooled electrodes is approximately $2.2 \cdot 10^{-5}$ g/C at a current of ~20÷25A which, at the electrode length of ~15 cm, gives the expected operating life of ~500 h.

Chapter 10

Near-electrode processes and methods of reducing electrode erosion

As shown in the previous chapters and in a large number of publications, at present plasma technology is occupying more and more important positions in greatly differing branches of industry. This is explained not only by the extensive possibilities and advantages of plasma technology but also by the development of highly efficient plasma equipment satisfying the conditions of industrial service, linear plasma torches of any power and design, and also electric power sources and sources for the ignition of the electric arc in plasma torches, and by the possibilities of preliminary calculations of the electrical and thermal characteristics on the basis of criterial equations.

The increase of the scale of technical applications of electric arc plasma torches leads to more stringent requirements on the reliability of these systems associated with the need for further increases of the operating life of the most heavily thermally stressed elements – electrodes, especially at high currents.

The service life of the electrodes is determined by the electrophysical, aerodynamic and thermal processes in electrode regions of the arc discharge, and on the surface of the electrode and inside the crystal lattice of the metal from which the electrode is produced. Methods of solving the problem of the service life of the electrodes differ greatly and also depend on the service conditions, applications for which plasma torches are used, i.e., on the technological process and the working gas, heated in the plasma torch, and on the current intensity and gas pressure in the discharge chamber.

Starting in the eighties, as a result of theoretical and experimental investigations it was possible to improve greatly the quantitative parameters of specific erosion of tungsten cathodes, reducing erosion of tungsten rods pressed flush into the copper water cooled collar, and the stationary arc spot in the argon medium to extremely small values of 10^{-13} kg/C at a current intensity not exceeding 1 kA. Significant successes have also been achieved in reducing specific erosion of the copper tubular output electrode–anode in operation with different working gases and with a moving arc spot. The value of $\overline{G}$ was reduced from $1 \cdot 10^{-9}$ kg/C to $4 \cdot 10^{-11}$.

A fundamental phenomenon – recirculation of the ions of the electrode material in the zone of the stationary cathode arc spot was discovered in 1970. This phenomenon was described briefly for the first time in the monographs in [1, 2] and was subsequently studied further in a number of investigations. This will be discussed in the appropriate sections of this chapter. It was found that part of the evaporating cathode material in the zone of the arc spot or on the side surface of the cylindrical cathode, penetrating into the column of the electric arc, is ionised and then returned by the electrical field to the end surface of the cathode in the zone of arc attachment. In the case examined in [2], the cathode was partially restored.

The authors of the above studies examined only partially the unknown phenomenon associated with the formation of the cathode section with the constant restoration of the thermal emission insert, i.e. with an infinite operating life. Starting in 1973, publications already appeared describing the conditions of operations of the cathode of the high-current arc in the regime of constant renewal, i.e. the first experimental results appeared confirming the possibility of formation of a cathode section with an infinite operating life.

The problem of explaining the mechanism of self-restoration and its theory are far from solved. Extensive investigations have been carried out into this phenomenon to explain the effect of various parameters, such as the pressure of the gas medium, the geometry of the cathode section and the cooling rate of the section, the composition of the gas mixture, the surface temperature of the cathode and the number of other parameters.

Successes have also been achieved in reducing the erosion rate of copper cooled cylindrical anodes with the moving arc spot. Usually, the mean value of the specific erosion of anodes of this type is 10^{-9} kg/C at a current intensity in the range 0.1÷4 kA, a pressure of 10^5 Pa and for a wide range of gases (air, nitrogen, oxygen, hydrogen). A large decrease in the extent of specific erosion and, consequently, the increase of the service life of the copper tubular output anode have been achieved in the two-jet plasma torch with the axial gas-dynamic scanning of the radial section of the arc with the length of $6 \cdot 10^{-2}$ m along the axis of the tubular copper anode with the frequency of 5–6 pulsations per minute and in the presence of twisting of the gas flow (air). The extent of specific erosion decreases from the mean value of $1 \cdot 10^{-9}$ kg/C to $4 \cdot 10^{-11}$ kg/C, i.e., by almost 2 orders of magnitude at a current of 200 A.

It is important to mention another important circumstance, i.e., large-scale shunting did not take place in these conditions.

One should also mention the so far incomplete investigations of the 'diffusion' attachment of the moving anode end of the arc to the surface of the output electrode. This could be realised on the internal surface of the copper tubular output anode of a linear plasma torch with an inter-electrode insert. To realise the 'diffusion' attachment, the anode was protected against the working gas (commercial nitrogen, air) by a small amount of argon oR natural gas (propane–butane). The resultant value of specific erosion was $6 \cdot 10^{-12}$ kg/C. However, the mechanism of 'diffusion' attachment is not clear because the uniform erosion of the surface of the anode may also be explained by another phenomenon, i.e. the existence of a large number of microarcs, formed in the process of burning of the electric arc in the near-anode space and changing their position in the space with high frequency.

The diffusion attachment of the anode stationary arc spot in argon may also be related to the unique phenomenon, observed in 1985 [3]. With special profiling of the surface of a copper anode (for example, a depression on the flat surface in the form of a hemisphere) on which the end of the arc, stabilisers by the gas flow (with the circumferential component of velocity) rests, specific erosion in the range of variation of current of 200–1000 A (according to indirect estimates, because the instrumental method cannot be used to determine this value), did not exceed 10^{-17} kg/C.

At the end of the 70s, attention was given to the possibility of increasing the service life of electrodes by longitudinal or radial splitting of the electrical discharge. The devices used for this purpose are relatively simple to produce and appeared simultaneously with the development of plasma torches with inter-electrode inserts. However, only the cathode sections with the radial splitting of the arc (without connecting ballast resistances into the electrical circuit) have been used in practice.

At the end of the last century, special attention was given to investigations of the changes in the structure of the material of copper electrodes subjected to the effect of the electrical arc. The assumption on the controlling role of the crystal structure of the material and the grain size has been confirmed. An important moment of the process of failure is the formation of a network of cracks in the depth of the electrode material. Undoubtedly, in this case, one of the main mechanisms of failure of the electrode are thermal elastic stresses formed in the material as a result of the nonuniform thermal expansion of the material. The formation of cracks greatly reduces the removal of heat to the liquid, cooling the electrode, and, therefore, the temperature of the working surface of the electrode increases and the erosion rate is also higher. In computer investigations in the complete formulation of the problem, calculations were carried out of the characteristics of pulsed temperature fields in the anode and also, in modelling formulation, the adjoint characteristics of the stress state of the subsurface and the players. The aim of these investigations was to explain in detail the observed patterns of the microstructure and the possibility of predicting the structure of the material which would minimise the effect of pulsed thermomechanical stresses [4].

The increase of the dispersion and homogeneity of both the inclusions and of the material of the base of the matrix (electrode) should reduce the rate of the process of formation and development of dislocations. Experimental verification of operation of the electrode, produced from a copper single crystal, shows that in the limiting case, the body of the cathode does not contain cracks.

It is also important to find methods of refining the structure of cast copper, reducing the degree of chemical and physical heterogeneity by modification of the material with ultrafine powders of refractory compounds, with the size of $0.1 \div 0.5$ μm.

Taking this into account, the investigators should pay attention to the collected scientific material, reflecting the effect of specific factors on the rate of erosion.

It is important to stress that the physical nature of the near-electrode processes is very complicated, and the characteristics of these processes greatly depend on the material, form and method of cooling of the electrodes, pressure, temperature, the nature of flow and type of gas, and also on the method of organising at discharge in the gas-discharge system. In addition to this, at high gas pressures (of the order of atmospheric pressure or higher), the thickness of the near-electrode layers is very small which in the case of high current densities greatly complicates the experimental examination of near-electrode processes. In this connection, in addition to the experiments, investigators have also used theoretical methods of examining the near-electrode processes which greatly expands the possibilities of explaining the nature of these processes and determining special features of the entire process.

Thus, the break in the solution of the problem of increasing the service life of electrodes and in the development of high-efficiency low-temperature plasma arc generators, i.e. plasma torches with a wide spectrum of power and for various technological gases, maybe based on the complex search of different methods of solving the problem at the interface of various sciences.

10.1. Heat flows into the electrodes through arc spots

Electrode erosion is of mostly thermal nature. Prior to examining the problem of the intensity of heat flows into the electrodes, it is important to mention the main processes taking place in the areas of contact of the arc with the electrodes.

The presence of current (transfer of electricity) between the electrodes of the active charge requires from the cathode either introduction into plasma of electrons (in the amount I/e per second) or, which is equivalent, receiving from the arc column positive ions in the amount $I/(N \cdot e)$, where N is the degree of ionisation of the atoms. The anode should therefore either receive I/e electrons from the plasma or supply $I/(N \cdot e)$ ions to the arc column.

The efficiently cooled metallic (usually copper) anodes, used in the plasma torches, do not supply the ions to the arc column. This is indicated by the value of the specific erosion of the anode in the plasma torches which is 10^{-9} kg/(A · s) and less. Even in the cases in which the atoms leaving the anode are completely once ionised, the ion current is less than a small fraction of the value required for sustaining arcing. Consequently, charge transfer in the near-anode region takes place only by electrons. In contrast to the anode, the cathode should supply charged particles. When using materials with high thermal emission capacity, the thermal electrons are capable (because of the tunnelling effect) of ensuring the current densities sufficient for sustaining the discharge at temperatures lower than the melting point of the electrode. A significant role in reducing the potential barrier, preventing the exit of the electrons from the cathode, is played by different additions or surface films, i.e., the products of chemical reactions of the cathode material with the surrounding gas medium.

The heat balance in the cathode and the anode will be examined. Three theories are available for cathode phenomena: thermal electron emission, auto-electron emission, and ion currents. The first two explain the charge transfer between the cathode and the plasma of the arc column mainly by electrons, the third one mainly by ions. The heat balance should be examined considering the possibilities of the two process mechanisms.

The ions transfer the kinetic and potential energy to the cathode. Since it is assumed that the length of the zones of the electrode potential drop is approximately equal to the length of the free path of the plasma particles, it may be assumed that the ions not collide in the zone and reach the surface of the cathode with the kinetic energy required for acceleration in the region of the cathode potential drop. Denoting the decrease of the potential in the cathode zone by U_c, and the ion current by I_i, we obtain the equation for the kinetic energy transferred by the ions to the cathode per unit time: $U_c I_i a_i$. Here a_i is the coefficient of accommodation of the ions, characterising the extent of transfer of energy to the cathode and equal to the ratio of the difference of the energies prior to and after collision with the surface of the cathode to the energy prior to collision. In complete reflection of the ions from the surface $a_i = 0$, and in complete absorption $a_i = 1$. Thus, the ion, supplied to the cathode surface, may be reflected and scatter its energy on other particles in the gas increasing the temperature in the cathode region or may be absorbed by the surface of the cathode, scattering the kinetic energy in the crystal lattice of the cathode and increasing its temperature.

In both cases, the ion may also generate, on the cathode, the neutralisation energy, i.e. the potential energy stored during ionisation. For neutralisation of the ion, the surface of the cathode should be reached by the electron which takes away from the cathode the yield energy which depends on the properties of the cathode but is usually smaller than the ionisation energy of the plasma atom. Therefore, neutralisation is characterised by the generation of energy $(U_i - \varphi) I_i a_{in}$, where a_{in} is the coefficient of accommodation of the ion, which transforms into a neutral atom.

In addition to the previously mentioned processes which determine the supply of energy to the cathode, an important role is played by the transfer of heat to neutral particles by conventional heat conductivity. In this case, the particles may generate on the cathode the energy of dissociation, joining into molecules (for many atomic molecules) and excitation energy.

The cathode is also heated as a result of the absorption of plasma radiation but the fraction of radiation in the total balance is usually very small.

The energy loss by the surface of the cathode takes place as a result of removal of heat into the body of the cathode by heat conductivity Q_c, thermal emission of the electrons, radiation from the area of contact with the plasma; part of the energy is used for the evaporation of the material of the surface. The fraction of the heat losses through radiation and evaporation is usually smaller and is not taken into account in the balance, especially at high current density, typical of the cathodes of the plasma torches.

Thus, the equation of energy balance has the following form:

$$Q_c = U_c \cdot I_i a_i + (U_i - \varphi) I_i a_{in} - I_e \cdot \varphi + Q', \qquad (10.1)$$

where φ is the work function; I_e is the thermoelectronic current; Q' is the heat flow of the processes not taken into account (primarily, the heat conductivity of the gas).

Introducing the ion component of total current $S = I_i/I$, assuming that $a_i = a_{in} = 1$, and also taking into account that $I_e + I_i = I$, we obtain the expression for the heat flow, supplied into the body of the cathode:

$$Q_c = I\left[S(U_c + U_i) - \varphi\right] + Q' \qquad (10.2)$$

This simple approximate equation is suitable for analysis but, in addition to the work function φ and easily measured values of I and Q_c, the other terms are not determined. Cathode voltage drop U_c may also be measured, although this is associated with certain difficulties.

In addition to the ions of the heated gas, the ions of the cathode metal, formed from the atoms, evaporated from the cathode surface, may take part in the process of current transfer. It is possible to estimate the amount of electricity which can be transferred by the evaporated and ionised atoms of the electrode material in the near-cathode region, from the experimental values of erosion. This estimate can be used for concluding that, in the present case, the number of metals ions is far from sufficient for transfer. However, it should be taken into account that the ionised atoms of the electrode material may return to the surface of the cathode under the effect of the cathode electrical field [1]. This phenomenon is referred to as the recirculation of atoms and has played a significant role in the development of tungsten cathodes with a very low value of specific erosion. The latter is caused by the low ionisation potential of the metal vapours in comparison with the ionisation potential of the surrounding working gas. For example, for copper U_i = 7.72 eV, for tungsten 7.8, iron 7.90, oxygen 13.6, argon 15.8 eV. Correspondingly, the probability of ionisation of the metal vapours is considerably higher than the probability of ionisation of the gas atoms.

Falling on the cathode, the metal ions are neutralised and then evaporate, circulating in the cathode region of the discharge. Consequently, the number of atoms, evaporated from the surface of the cathode, may be considerably higher in comparison with the estimate obtained from the experimental value of erosion. The amount of electricity, transferred by the metal ions in the cathode region, may represent a large fraction of the ion component of current. This effect does not make it possible to determine accurately the value of the ionisation potential (the atom of the metal or surrounding gas) when calculating neutralisation energy.

In the case of thermal cathodes, it is more efficient to represent U_i in the balance equation by the value of the ionisation potential of the gas in which the arc burns. For the 'cold cathode', on the other hand, it is more accurate to use the value of U_i, relating to the vapours of the cathode material.

Although the fraction S of ion current on the cathode may differ in different cases ($S = 0.03 \div 0.3$ and more), it is always considerably higher than the fraction of ion current in the arc column ($S < 0.01$). This results in the formation of a spatial charge in front of the cathode which determines the value of U_c (in contrast to the arc column characterised by quasi-neutrality, and the concentration of the ions in the near-cathode region is usually considerably higher than the electron concentration).

For the thermal cathode, the role of total current is in heating of the cathode to the temperature at which the conductivity electrons have the thermal energy higher than the level of the potential barrier at the boundary of the cathode. For the 'cold' cathode in auto-electron emission, the role of the ions is the development of the electrical field 'pulling out' the electrons from the cathode.

In this case, the term $I_e \cdot \varphi$ is removed from the balance equation which has the following form:

$$Q_c = IS\left(U_i - \varphi + U_c\right) + Q' \tag{10.3}$$

Equations (10.2) and (10.3) show that the fraction of ion current S is very important for determining the amount of energy, transferred by the charged particles.

The mechanism of auto-emission applied to the arc is doubted by many investigators. It is assumed that the more probable mechanism for the 'cold' cathode is the mechanism of thermal auto-emission in which the vapours of the cathode material play a significant role. They are at the temperature equal to the boiling point of the material of the cathode and at the pressure higher than the pressure

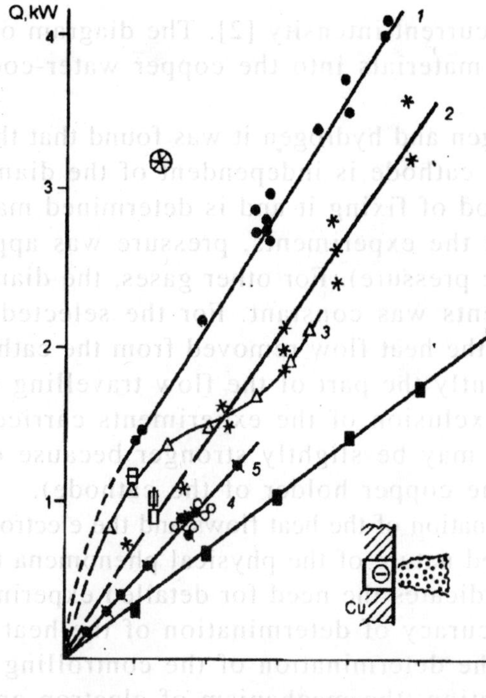

Fig.10.1 The heat flow into the body in the cathode in relation to current itensity. The base of he cathode made of tungsten: 1) nitrogen, 2) argon, 3) helium, 4) hydrogen, 5) air. The base of the cathode made of graphite: 6) mixture of CH_4 +CO_2 (regeneration regime, calculations [9]): □ – mixture of CH_4 + Ar, $G_\Sigma = 6 \cdot 10^{-4}$ nm^3/s, $I = 250$ A; $Q_c = 0.9$ kW [34]. The base of the cathode made of copper: ⊟ – mixture of CH_4 + Ar, $G_\Sigma = 2.2$ g/s, $I = 200$ A; $Q_c = 1.17$ kW [37]. The base of the cathode made of graphite: ⊟ – mixture of CH_4 + Ar, $G_\Sigma = 2.2$ g/s, $I = 250$ A; $Q_c = 1.0$ kW [37].

of the surrounding medium.

The experimental data obtained in the measurement of Q_c for the rapidly moving cathode spot of the air arc give the value of the volt equivalent of 13.5 V of the atmospheric pressure. According to equation (10.3), at $Q' = 0$, the value of the volt equivalent is only 4.1 V for $S = 0.35$. Since the value of Q' is usually small in comparison with the ion contribution to the thermal balance, to eliminate the balance deficit, it is necessary to propose either a considerably higher (close to unity) fraction of the ion current or a large contribution of the effect of emission heating (Nottingham effect). The experiments carried out in [5] show that the term Q' plays a significant role only in the case of low currents.

Figure 10.1 shows the experimental dependence of the heat flow into the cathode working in nitrogen, argon, helium, hydrogen, air and other gases, on current intensity [2]. The diagram of fixing the rod-shaped cathode materials into the copper water-cooled holder is also shown there.

When using nitrogen and hydrogen it was found that the heat flow into the body of the cathode is independent of the diameter of the rod and of the method of fixing it and is determined mainly by the current intensity (in the experiments, pressure was approximately equal to atmospheric pressure). For other gases, the diameter of the rod in the experiments was constant. For the selected scheme of the cathode section, the heat flow removed from the cathode section reflects most efficiently the part of the flow travelling through the arc spot (with the exclusion of the experiments carried out in argon where the flow may be slightly stronger because of the radiant heat flows on the copper holder of the cathode).

Even a brief examination of the heat flows and the electrode processes shows the complicated nature of the physical phenomena taking place in these areas and indicates the need for detailed experiments aimed at improving the accuracy of determination of the heat balance on the electrodes and the determination of the controlling parameters (the ion current fraction, the mechanism of electron emission, accommodation coefficients of the ions, temperature profile, etc).

The electrophysical processes, associated with current transfer in the near-anode region differ from those described previously.

The equation of the energy balance for the anode, if the radiant heat flows between the plasma and the arc spot are ignored, together with the losses of energy through evaporation, may be represented in the following form:

$$Q_a = I\left[U_a + \varphi + 5kT_e / 2e\right] + q_\tau \cdot f_a . \tag{10.4}$$

The first term IU_a is the kinetic energy, transferred to the anode by the electrons per unit time. The second term $I \cdot \varphi$ is the potential energy, i.e. the energy of neutralisation of the electron arriving on the surface of the anode and neutralising the positive ion of the metal in implantation in the lattice. The enthalpy of the electrons, falling on the surface of the anode, corresponds to the temperature of the electrons in the plasma at the boundary of the region of the anode voltage drop plus the energy acquired by the electrons in movement inside this region. The last type of energy $(5/2)\cdot(IkT_e/e)$, reflecting the contribution of the thermal energy of the electrons [2], slightly differs from the thermal energy of the electrons $(3/2)\cdot(kT_e)$, because the electrons cannot retain the Maxwell distribution in the zone of the near-anode drop.

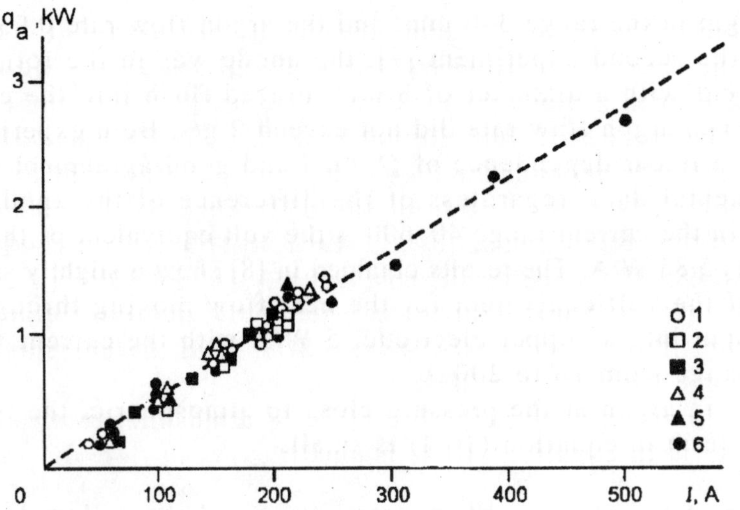

Fig. 10.2. Dependence of the heat flow into the electrode–cathode on the intesity of current in argon. *1 – 5*) data from [6]; *6*) from [7].

The last term includes the inflow of energy to the anode from the plasma of the arc column as a result of electrical conductivity, recombination, and may be expressed for high-temperature gradients in the near-wall layer (and these gradients form in the majority of cases) through the difference of enthalpies [1]:

$$q_t = \left(\lambda_w / c_{pw}\right)\left[\left(h^* - h_w\right)/Z\right], \tag{10.5}$$

where h^* is the specific enthalpy of the plasma at which the conductivity converts to zero in the approximation of the true dependence $\sigma^*(h)$; h_w is the enthalpy of the gas at the wall temperature; Z is the distance from the surface of the anode along the normal on which the enthalpy changes from h^* to h_w.

If the contribution of the last term in equation (10.3) is small, there is a linear relationship between the heat flow into the anode and the electrical arc current. If necessary, equation (10.5) may include radiant heat flows which travel to the anode from the plasma and leave the anode as a result of radiation of the surface into the surrounding medium. The results of several experiments will now be examined.

The data on the intensity of the heat flow into the body of the electrodes of the anode spot in argon at the atmospheric pressure may be obtained from the experimental results presented in Fig. 10.2. In one of the experiments [6], the anode was in the form of a copper sheet, and the distance of the diaphragm from the surface of the anode was varied in the range 1–2 mm, the internal diameter of the diaphragm in the range 3–6 mm, and the argon flow rate 0.03–0.12 g/s. In the second experiment [7], the anode was in the form of a copper bar with a diameter of 8 mm, brazed flush into the copper holder. The argon flow rate did not exceed 2 g/s. Both experiments showed a linear dependence of Q_a on I and good agreement in the experimental data, regardless of the difference of the anode materials. In the current range 40–600 A the volt equivalent of the heat flow was 5.85 W/A. The results obtained in [8] show a slightly smaller value of the volt equivalent for the heat flow moving through the anode spot into a copper electrode: 5 W/A with the current varied in the range from 10 to 200 A.

Thus, in argon at the pressure close to atmospheric, the role of the last term in equation (10.4) is small.

10.2. The form of the eroded surface of a rod thermal cathode with a stationary arc spot

The most important final result of the effect of the arc spot on the working surface of the electrode from the technical viewpoint is the value of the specific erosion of the surface as a consequence of complicated thermal, electrical, chemical and other processes in the near-electrode region, on the surface and inside the body.

For better understanding of some important processes, we start with examining the form of the eroded surface of a tungsten rod cathode after a relatively long service life.

Figure 10.3 shows the scheme of experimental equipment used for investigating special features of erosion of a relatively long cylindrical tungsten thermal cathode in argon. The diameter of the rod *3* was d = 3 mm. The rod was pressed into a copper water cooled collar *1*. The flow rate of argon was 5 g/s, current intensity 100 A. The arc spot *4* was stabilised on the end surface of the cylinder

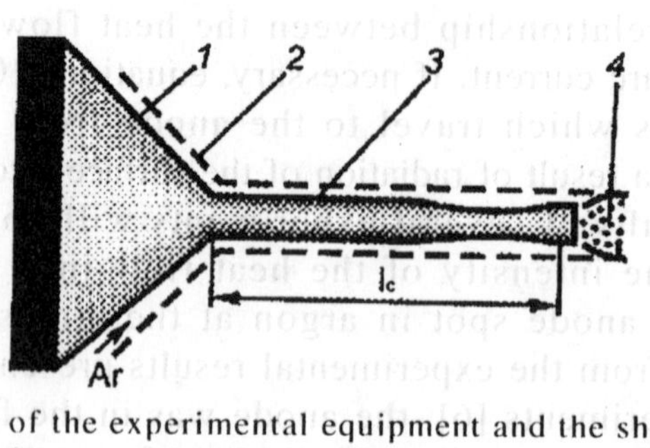

Fig.10.3. Diagram of the experimental equipment and the shape of the cylindrical tungsten cathode after operation for 1 h. Current intesntiy 100 A; working gas – argon.

by a vortex flow. The operating time was 1 h. The broken line *2* shows the compression diaphragm.

Two characteristic the results will be noted [2]: 1. The highest degree of erosion is not detected at the end of the electrode, with the arc resting on the surface, but in the more remote part, which in the figure is clearly visible as a 'neck'; 2) the initial length of the cathode l_c during operation for one hour did not decrease and even slightly increased as a result of the formation of a 'growth'.

Analysis of the structure of the end part of the surface of the rod shows that it is characterised by the formation of a growth similar to part of the hemisphere, with the maximum thickness of up to $0.6 \cdot 10^{-3}$ m and with an uneven surface. The material of the hemisphere could form only as a result of a small part of the evaporated tungsten atoms from the side surface of the cylinder. These atoms, penetrating into the electric arc, are ionised and directed by the electrical field to the end surface of the cathode. The experiments showed for the first time one of the possible methods of restoring the length of the cathode, i.e. achieving a low value of specific erosion. The same experiments indicate that the optimum length of the cylindrical cathode outside the copper water cooled holder should be equal to zero ($l_c = 0$). In fact, at $l_c > 0$ the attachment of the end of the arc to the cathode may be of the diffusion type, and the surface temperature is lower than the melting point, but the area of evaporation of tungsten is several orders of magnitude larger in comparison with the area of the constricted spot, i.e. specific erosion should be higher. However, if the spot is constricted, the temperature of tungsten in the zone of the arc spot is already closer to the boiling point of the metal and not to the melting point. However, since in the case of tungsten the rate of evaporation from the unit area increases by an order of magnitude with temperature increasing in 100°C steps, it is clear that the regime with $l_c > 0$ is not efficient. It is also important to note the possibility of ejection of metal droplets during boiling which may even increase further the value of the specific erosion of tungsten.

At $l_c = 0$ the situation is different. The ark spot is constricted and stationary in the space and with time, the surface temperature of tungsten is high, possibly of the order of the melting point, but not the boiling point, because the cooling of the rod is relatively efficient. In this case, part of the evaporated metal is ionised and returned to the surface. As shown later, at $l_c = 0$ the value of specific erosion was the lowest, equal to $1 \cdot 10^{-13}$ kg/C. To obtain this value, it is necessary to ensure efficient thermal contact between the tungsten rod and the copper holder. If this condition is not satisfied, the ex-

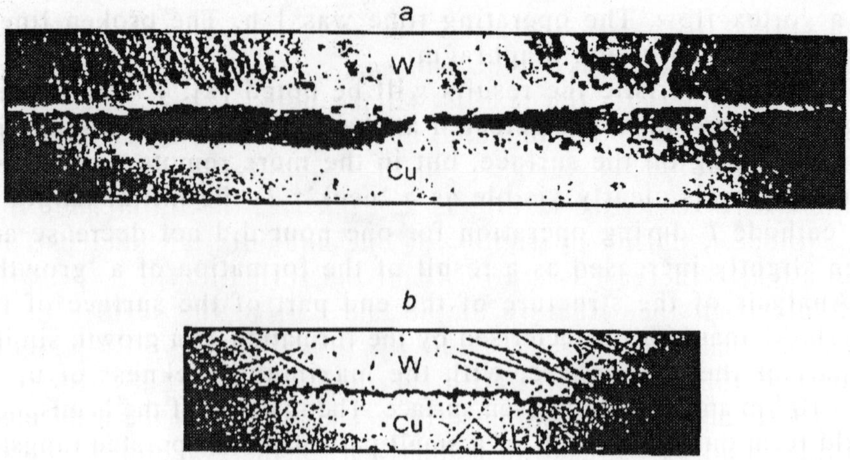

Fig.10.4. Sections of the surface of the zone of contact of tungsten with copper. The quality of thermal contact is ensured: *a*) only by brazing with a brazing alloy; *b*) by pressing in a special oxygen–free atmosphere.

perimental points are scattered in respect of the value of specific erosion by up to 2–3 orders of magnitude.

Figure 10.4 (magnification 300) shows the section of the surface of the contact zone of copper with tungsten when the quality of contact is determined only by brazing with a brazing alloy. There are clearly visible large cavities not filled with the brazing alloy and this reduces the extent of heat removal from the tungsten rod to the copper holder and increases the specific erosion of tungsten to a certain degree, depending on the quality of contact. Efficient thermal contact is obtained by pressing in a special oxygen-free atmosphere; the gap produced in this case does not exceed $3 \cdot 10^{-6}$ m (Fig. 10.4*b*).

The form of erosion of the surface of a rod cathode ($l_c > 0$) at high currents will be examined [9, 10]. Figure 10.5 shows photographs of two rod cathodes produced from alloy tungsten after 10 (*a*) and 3 (*b*) hours of operation at a discharge current of the order of 800÷1000 A. In the first case (*a*) the discharge burns in the atmosphere of commercial nitrogen: the arc spot is constricted, the growth on the end is distributed locally, forming 1–2 quite large projections.

It should be added that the rate of erosion is determined not only by high temperature in the zone of the constricted arc spot with the accompanying physical processes, but also by intensive oxidation of tungsten with oxygen with the formation of volatile oxides on the larger area of the side surface of the cylinder (commercial nitro-

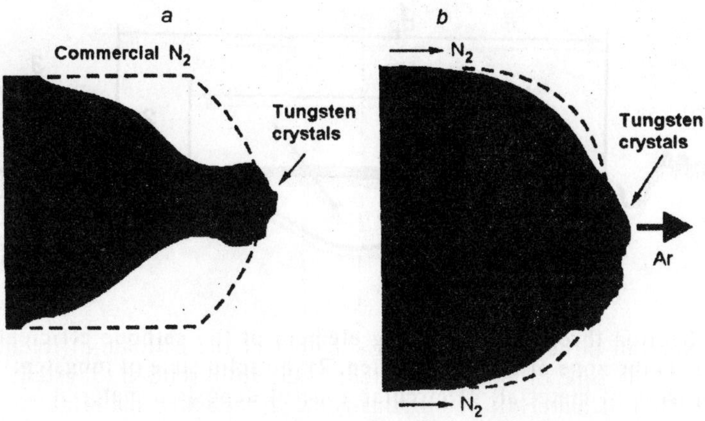

Fig.10.5. Erosion of the surface of a tungsten rod cathode. *a*) after operation for ten hours; *b*) after operation for three hours, current ~800÷1000 A.

gen contains up to 1% volume units of oxygen).

In the case *b* argon was supplied ($0.5 \cdot 10^{-3}$ kg/s) to the working surface through an orifice in the body of the cathode, and the side surface was blown with commercial nitrogen with a relatively low oxygen content (no more than 0.2% volume units). The attachment of the arc took place on the edge of the axial orifice, drilled in the body of the cathode (d_o = 1.5 mm); this region shows a tungsten growth of irregular shape. According to the results of spectral analysis, in both cases the projections consist of pure tungsten. The profile diagrams of the working surface of both cathodes show that the growth of tungsten in the area of attachment of the arc to the cathode exceeds the previous level of the working surface (i.e., the cathodes are 'longer') by more than $(1 \div 2) \cdot 10^{-3}$m. The growth of the projections is especially clearly evident on the cathodes produced from activated tungsten in helium. In all likelihood, the latter is associated with the fact that the strength of the electrical field in the helium medium is considerably higher than in argon. For example, on a cathode made of VL–10 tungsten (d_c = 3 mm, l_c = 25 mm) 'whiskers' with the size of $(0.5 \div 1.5)$ 10^{-3} m grow on the end surface of the cathode at I = 150 A [11].

Thus, if $l_c > 0$, the evaporated atoms from the side surface of the tungsten rod and the atoms which penetrated into the electrical arc are partially ionised and under the effect of the electrical field are directed to the end surface of the cathode, increasing the cathode length.

At $l_c = 0$ the optimum value of the diameter of the tungsten rod d_c and efficient thermal contact of the rod with the copper holder

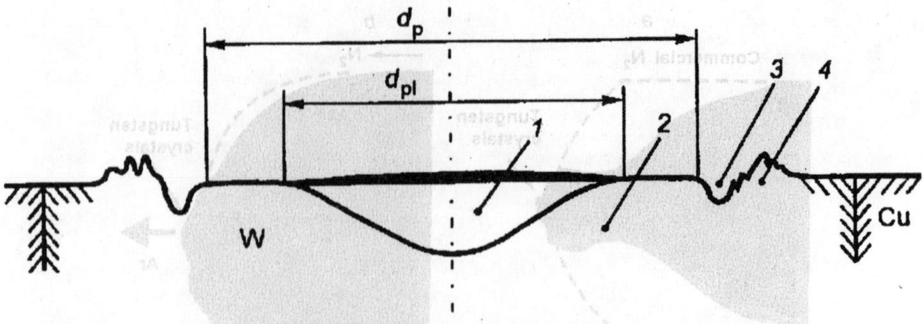

Fig.10.6. Section through the working element of the cathode efficiently cooled with water. 1) the zone of molten tungsten; 2) the solid state of tungsten; 3) circular zone of erosion of material; 4) circular zone of deposited material

resulted, as shown previously, in the minimum level of specific erosion of the cathode, equal to 10^{-13} kg/C.

The surface area of tungsten heated to high temperature and situated outside the zone of attachment of the arc (outside the recycling zone or the zone of recirculation of metal atoms) is considerably smaller than in the case of the cathode rod and, consequently, electrode erosion is also considerably smaller. Because of the design special features of the cathode with $l_c = 0$, this cathode is always characterised by the realisation of the constricted attachment of the arc, and the zone on the surface of the cathode is characterised by the formation of temperatures close to even higher than the melting point of tungsten. It would appear that this circumstance should result in an increase of the erosion of the cathode in both the steam and liquid phases. However, this does not takes place. Thus, the experimental data, presented in [2], contradict the current views on the effect of the type of attachment of the arc on the working efficiency of the thermal emission cathode.

What is the actual pattern of erosion, observed at $l_c = 0$? As shown in [10], it may be described as follows. The main amount of the material is removed from the region outside the arc spot and the spot itself shows even an increase of the amount of the cathode material. Figure 10.6 shows the section of the working element of the efficiently cooled cathode (the schematic was drawn on the basis of the photograph of a section published in [10]), which worked at a current of 400 A in argon for 10 h. There are several characteristic zones.

1. In the centre of the cathode 2 characterised by the attachment of the arc, there is the melt zone *1* with a diameter d_m, indicating a high level of temperatures in this zone. The depth of the melt reaches the value $d_m/2$. However, the main process of removal

of the material the cathode, which determines cathode erosion, does not take place from the central region of the surface of the cathode characterised by the maximum temperature, about from the more remote zone *3* from the centre of the cathode with a considerably lower temperature. In the process of operation of the cathode, a circular zone of erosion forms at some distance $d_a/2$ from the centre of attachment of the arc. The duration of formation of the this zone and the rate of deepening and, consequently, the specific erosion of the cathode depend strongly on the presence of active (in relation to tungsten) components in the gas, flowing around the cathode. There is also the zone of deposited tungsten *4* in the form of a circular bead.

Regardless of the recirculation andregeneration of the atoms of the cathode material in the spot, in the long-term operation of the cathode, erosion also takes place in the area of the arc spot. This takes place in connection with the increase of the depth of the circumferential zone of erosion around the spot, leading to less efficient heat removal from the zone of arc attachment; the uniformity of heat removal from the entire region of the spot is also disrupted and this results in de-stabilisation of attachment of the arc and, consequently, the increase of the rate of erosion in the very area of the spot.

The specific erosion of the cathode with $l_c = 0$ is strongly affected by the presence of oxygen in argon. Figure 10.7 shows the experimental dependences of the value of specific erosion of a rapidly cooled cathode on the oxygen concentration of argon, flowing around the cathode surface.

The working element of the cathode is a rod made of lanthanised tungsten VL-10, with the diameter $d_c = 5$ mm. In both cases, the length of the working elements was 10 mm, and the diameter of the water cooled holder 30 mm [10].

The graph clearly shows not only the large increase of the value of specific erosion with increasing oxygen content of argon but also a large increase of the specific erosion at a constant oxygen content of argon with the increase of the diameter of the tungsten rod and, consequently, the oxidation area.

These results will be briefly analysed. On the basis of physical considerations one can expect differences in the processes of mass transfer in passage through the boundaries of the arc spot because the zone of the arc spot in the near-cathode region is characterised by the presence of a strong electrical field whose effect results in the situation in which the metal atoms, which left the sur-

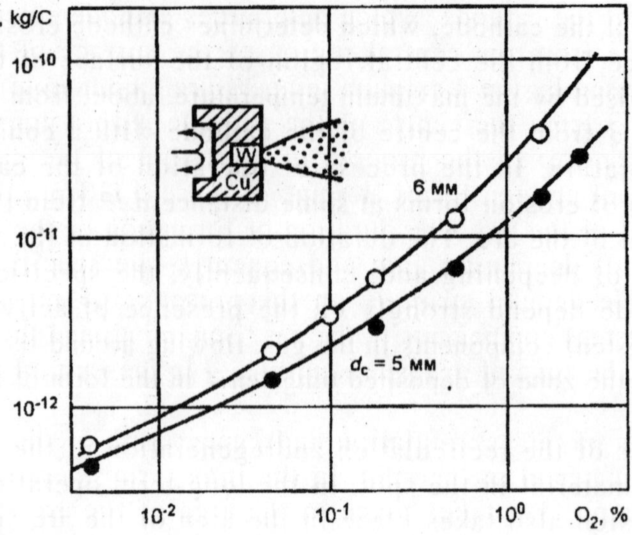

Fig.10.7. Dependence of the specific erosion of the cathode on the oxygen concentration in argon (I=200 A).

face of the cathode and were ionised, return back to the surface. In the evaporation of metal atoms, their chemical interaction with active components of the working gas and the formation in both cases of the ions, the absence of a strong electrical field (outside the arc spot) enables the atoms to leave the cathode in all cases.

It may also be asserted that the erosion of the thermal emission cathode is determined by the removal of the material from the surface area in the immediate vicinity of the arc spot, where temperature is still sufficiently high, and there are already valid reasons for the return of the atoms on the surface. In all likelihood, the spot is characterised by the complete circulation of the atoms of the cathode material. However, part of the atoms, leaving the surface of the cathode outside the spot because of thermal motion, penetrate, as shown by the experience, into the region of the arc discharge where, after ionisation, they return under the effect of the forces of the electrical field to the surface of the cathode in the zone of the arc spot and, consequently, this determines the increase of the mass of cathode material in the region of the spot observed in, for example [9].

Thus, analysis of the mechanism of failure of thermal emission cathodes makes it possible to conclude the presence of the recirculation of the atoms of the cathode material in the cathode spot, and also the regeneration of the part of the atoms leaving the surface of the cathode outside the spot. The occurrence of the process of recirculation of the atoms of the cathode material in the near-cathode region is indicated by experiments in [12]. In [9], the authors calculated the amount of material leaving the cathode surface during arcing. Calculations show that the dependence of Q_c on current intensity is linear, i.e. it corresponds to the results of experiments, but is considerably lower than the experimental data. This requires further investigations.

10.3. Specific erosion of tungsten thermal cathodes

Figure 10.8 shows the data on the erosion of intensively cooled tungsten rod thermal cathodes in different conditions of work in the inert medium. The experimental points of specific erosion $\bar{G}$ in helium (vertical rectangles) and in argon (solid circles), obtained at I = const, but at different values of l_c, indicate the strong effect of the extension length l_c of the tungsten rod on erosion [13, 14]. The specific erosion of the lanthanised cathode in special purity nitrogen (the oxygen content not higher than 0.001%), for the case in which $l_c \neq 0$ and the cathode is sharpened into a cone, and the current intensity of approximately 270 A is equal to $(1.3 \div 2.0) \cdot 10^{-12}$ kg/C (see the symbol Δ) [15]. In comparison with commercial nitrogen (oxygen content up to 0.5%), specific erosion decreased 2–3 times.

Attention will be given to the decrease of erosion $\bar{G}$ with increase of current, according to curve *1* in Fig. 10.8. This is associated with the fact that in the structure of the cathode section of the investigated plasma torch, the authors utilised the concept of the distribution of current by splitting of the arc into several current-conducting arc channels with attachment to the end surfaces of the tungsten rods (d_c = 2 mm, l_c = 6 mm). The uniform division of current is achieved by auxiliary heating of the cathodes from an additional electric arc plasma torch [16]. As already mentioned in chapter 7, splitting of the arc takes place only if the split section of the arc is characterised by the rising section of the VAC. As the current increases, the arc can be split into two or more current-conducting channels. In this case, with increase of current and of the number of current-conducting channels, each channel is characterised by a lower value of current and this is also the reason for the formation of the decreasing section of the $\bar{G}$–I-characteristic. The general level of the specific erosion of tungsten remains very high, regardless of diffusion attachment of the end of the arc to the thermal cathode.

At $l_c > 0$ we obtain the optimum value of current intensity at which erosion is minimum (curve *2* in Fig. 10.8 [17]). In both cases (curves *1* and *2*), the attachment of the arc is of the diffusion type, but the larger surface of evaporation of tungsten does not make it possi-

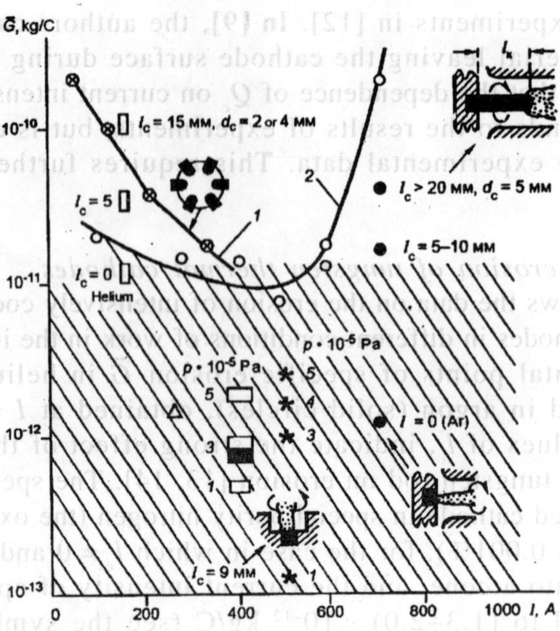

Fig.10.8 Dependence of the specific erosion of the tungsten cathode of current at different length l_c (argon).

ble to obtain lower values of specific erosion, detected at $l_c = 0$.

The experimental values of $\overline{G}$ in argon, nitrogen and hydrogen for the zero electrode extension length ($l_c = 0$) are characterised by a large scatter and, consequently, are described by the cross-hatched area (Fig. 10.8). The value of $\overline{G}$ is influenced mainly by the quality of thermal contact between the tungsten rod and the copper compression water cooled section, the content of oxygen in the plasma forming gas, and the recirculation of the tungsten vapours in the cathode region.

As shown in [18], the effect of the presence of oxygen in nitrogen at the atmospheric pressure on the erosion of the tungsten cathode (d_c = 4 mm, l_c = 0) may be described as follows: at of the oxygen concentration up to 0.5% and the current intensity $I = 250 \div 300$ A, the value of is in the range $(2 \div 5) \cdot 10^{-12}$ kg/C. Starting at the oxygen concentration of 0.7%, the value of $\overline{G}$ rapidly increases and reaches $(2 \div 4) \cdot 10^{-8}$ kg/C when the oxygen content is increased to 1.5%. The results of metal physics and x–ray diffraction analysis of the longitudinal sections of the tungsten cathode show that in the tested specimens and the oxygen concentration in nitrogen of 1.0÷1.5% of the boundaries of the structural formations at a depth of 0.5÷0.8

mm characterised by the presence of films of WO_3 oxides. These films result in stresses in the electrode material and consequently, the large block structure is disrupted as a whole together with the formation of a random grain structure. The effect of gas pressure on the specific erosion of the electrode $\bar{G}$ is clearly evident in Fig. 10.8 (stars). With increasing pressure, specific erosion rapidly increases. The graph also gives the data for the specific erosion of the cathode in argon at $I = 400$ A and in the same range of variation of pressure (horizontal rectangles), but in this case the geometrical dimensions of the vortex chamber remained constant [19]. The experimental point, corresponding to the solid horizontal rectangle, relays to the case in which the pressure in the vortex chamber is $p = 5 \cdot 10^5$ Pa, and the diameters of the orifices in the twisting ring are reduced in comparison with $p = 1 \cdot 10^5$ Pa. This improves the stabilisation of the act spot is on the surface of the cathode and greatly reduced specific erosion.

Erosion of the rapidly cooled thermal cathode depends greatly on the diameter of the tungsten insert [17]. That is the optimum value of the diameter d_c at which the value is minimum (Fig. 10.9).

Thus, for the rapidly cooled thermal cathodes ($l_c = 0$) in the optimum working regime and the current intensity of up to 1000 A, it is possible to obtain the value of specific erosion $\bar{G} = 1 \cdot 10^{-13}$ kg/C as a result of recirculation [17, 20].

10.4. Specific erosion of thermal chemical cathodes

In the previous section, we discussed the tungsten cathode, pressed flush with the surface of the copper cathode holder and working in inert media. Oxygen-containing gases are also used in practice. In this case, the zirconium cathodes, referred to as the thermochemical cathodes, have been used in plasma torches for different applications for more than 25 years. Because of the low heat conductivity of these materials, the rods produced from these materials are of small diameter (1÷3 mm) and are pressed into the copper holder flush with the surface, i.e. with zero extension. The maximum permissible current intensity I in standard zirconium or hafnium cathode sections does not exceed 300 A, because if the value is higher, the erosion rate increases. The experimental data for $\bar{G}$ of the standard (basic) cathodes produced from zirconium (hafnium), manufactures by industry for air-arc cutting torches [8, 21], are shown in Fig. 10.10 (curve *1*).

In the cathode section, using more efficient water cooling [22, 23] (the arc burns in steam), it is possible to reduce specific ero-

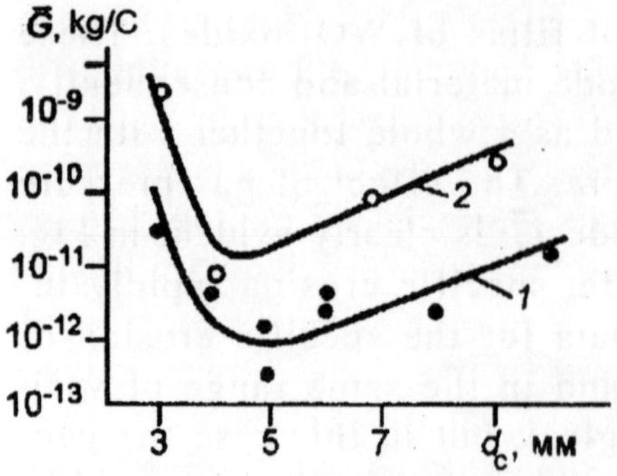

Fig.10.9 Dependence of $\overline{G}$ on the diamter ofd the tungsten rod d_c. 1) I = 370÷ 400 A, H_2; 2) I = 1000 A, N_2

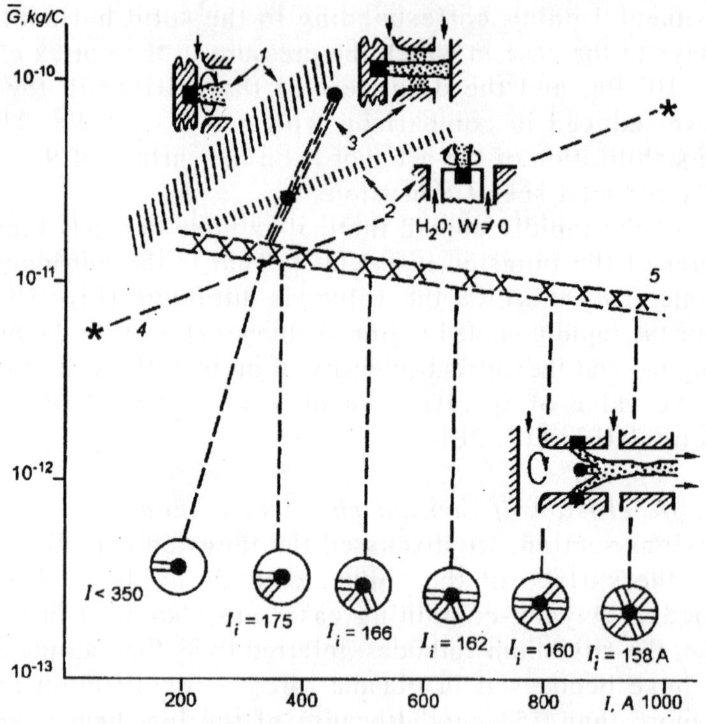

Fig.10.10. Dependence of specific erosion of different types of thermochemical cathodes on arc current intensity.

sion. This is especially important in the case of high current (curve 2); with increasing current intensity, the value $\overline{G}$ increases, but the rate of increase is considerably smaller in comparison with the basic cathodes.

The data presented in Fig. 10.10 relate to the long-term regimes of continuous operation of the cathodes. However, as shown by the experiments, the erosion of the cathodes depends greatly on the cyclic nature of operation. Figure 10.11 shows the dependence of the specific erosion of the cathode on the number n of activations of the arcs

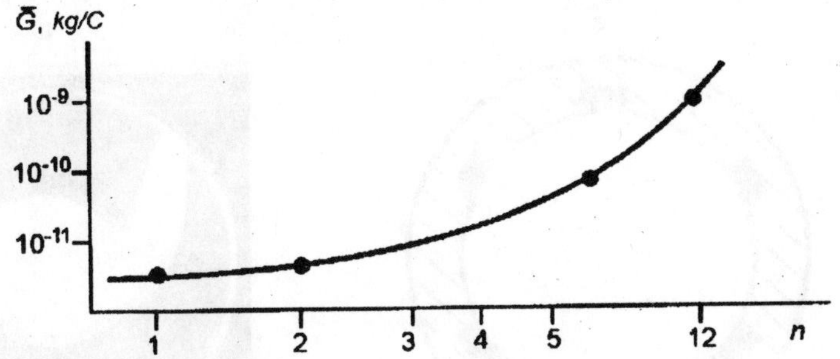

Fig.10.11 Dependence of $\bar{G}$ of the thermal chemical cathode on the number of arc ignitions n.

for the same total operating time, 1 h. Analysis of the graph shows that electrode erosion takes place mainly at the moment of arc ignition.

Curve *3* (Fig. 10.10) shows the experimental points, corresponding to the cathode section *3*, in which, in addition to the constricted attachment, there is diffusion current transfer to the cylindrical surface, produced from zirconium [24]. In this case, the total specific erosion also decreases in comparison with the basic curves *1*, corresponding to the cathode section *1*.

Considerable successes have been achieved in the area of manufacture of the thermal emission insert from special powder materials, based on the oxides of the titanium subgroup with different positions [25]. The powder mixture was pressed into a blind hole in a copper watercooled holder. The sintering of the mixture and formation of the 'powder alloy' already took place after the first act of ignition of the arc. The high melting point, the high heat resistance of the film of the alloy, high emission properties, low evaporation rate and a sufficiently high electrical conductivity made it possible to widen the range of the values of current intensity (10÷1000 A) and increase the operating life of the cathode (Fig. 10.10, broken line *4*). There is a well-formed arc column and stable arcing in different gas media.

Regardless of certain successes, the specific erosion of the investigated cathodes remained high at I = 500÷1000 A, and the operating life of the cathode is insufficient because of the small mass of the cathode.

The problem of increasing the operating life at high values of current intensity has been solved by splitting the cathode section of the arc in the hollow cylindrical electrode into several arcs with attachment of the arc spots to the thermal emission cathode inserts, placed around the circumference. The principal possibility of stable splitting without

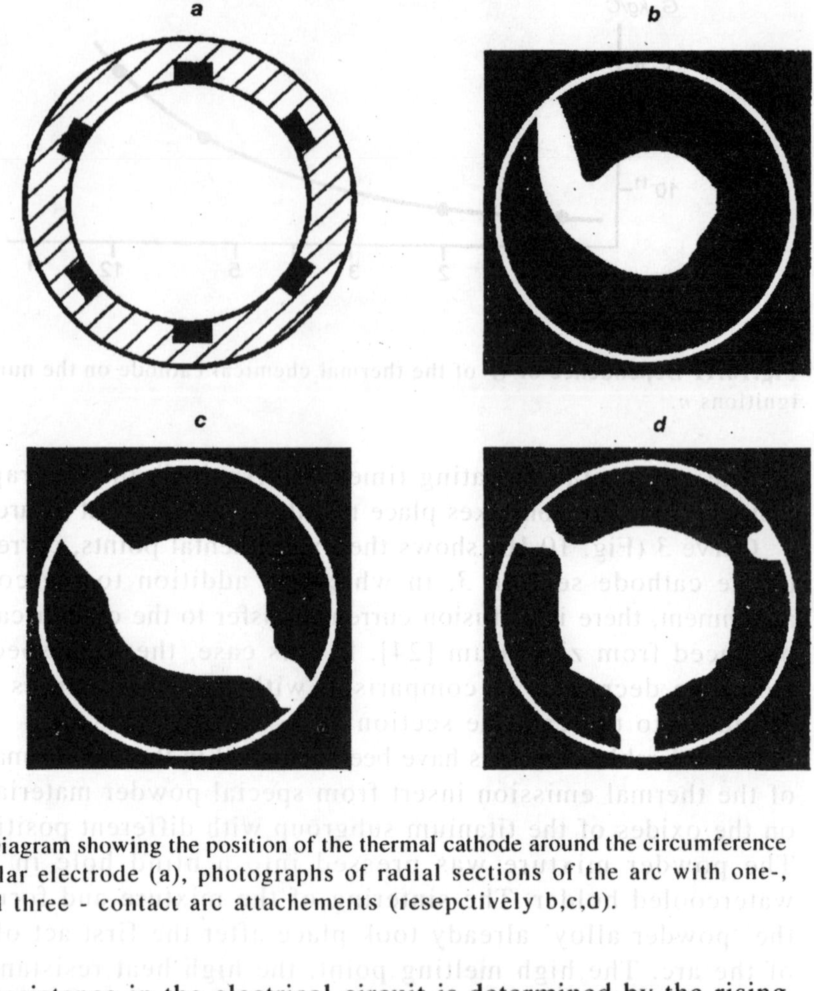

Fig.10.12. Diagram showing the position of the thermal cathode around the circumference of the tubular electrode (a), photographs of radial sections of the arc with one-, two – , and three - contact arc attachements (resepctively b,c,d).

a ballast resistance in the electrical circuit is determined by the rising section of the VAC in the radial section of the arc [26, 27]. It should be mentioned that previously, in chapter 7, we discussed the control splitting of the arc. For this purpose it is necessary to fulfil two conditions: 1. Existence of the attachment points of the support spots of the current-conducting elements of the arcs; this role is played by thermal emission inserts, pressed into the copper cathode section; 2. The stable position in the space of the 'plane' of rotation of the radial section of the arc in the zone of distribution of the thermal emission inserts, ensured by the appropriate organisation of the gas flow in the cavity of the cylinder.

Figure 10.12 shows the scheme of the distribution of the thermal cathodes around the circumference of the tubular electrode and the photographs of the radial sections of the arc with single-, two-

and three-contact attachment of the arc, burning in the air, to the hafnium inserts. The number of attachments increases spontaneously with increasing current intensity. Figure 10.10 (curve *5*) shows the dependence of the specific erosion of the cathode on the total arc current. There is not only a relatively low specific erosion (10^{-11} kg/C) in a wide range of variation of the current but, which is most interesting, there is a tendency for a decrease of specific erosion with increasing total current. This is associated with the fact that increasing current intensity increases the number of current-conducting channels and the current intensity for the single thermal emission insert decreases.

The structure of the material of the zirconium cathode inside will be described briefly. The authors of [26, 28] investigated the structure of the material below the cathode spot in stationary attachment of the arc on the surface of a zirconium cathode. In accordance with the temperature field, the processes of gas diffusion from the cathode region into the melt and the solid in chemical reactions and phase transformations, a specific structure of the material forms below the melt of the cathode. The authors of these studies noted (I = 200 A, and gas flow rate 10 g/s) three characteristic zones with sharp boundaries the multilayered structure below the attachment spot of the arc. The first zone, as viewed from the surface of the cathode, with a thickness of 20÷30 μm consists of crystalline zirconia with the monoclinic structure; it is followed by the second zone, consisting of a glassy phase, thickness 150÷300 μm. In the cathodes with the short operating time (5 to 300 s) there was also the third the zone, and the size of the zone can be used in estimate the true size of the attachment spot of the arc. The zone *3* widens with time, occupying a specific layer, representing initially the liquid melt of the oxides and sub oxides of zirconia and, subsequently, changing its chemical and phase composition with the formation of nitrides and oxides.

10.5. Structure of the internal surface of the cylindrical hollow tungsten cathode

The cathode is a cylindrical pipe of polycrystalline tungsten with the wall thickness of 1.5 mm, internal diameter 10 mm, and 90 mm long (Fig. 10.13*a*). The discharge bronze between the rod-shaped cathode, produced from tungsten, and the internal surface of the pipe–cathode, the plasma forming gas is argon, current intensity 250 A, gas pressure in the discharge chamber 0.1 MPa. The distribution of temperature along the external surface of the pipe is identical to the tem-

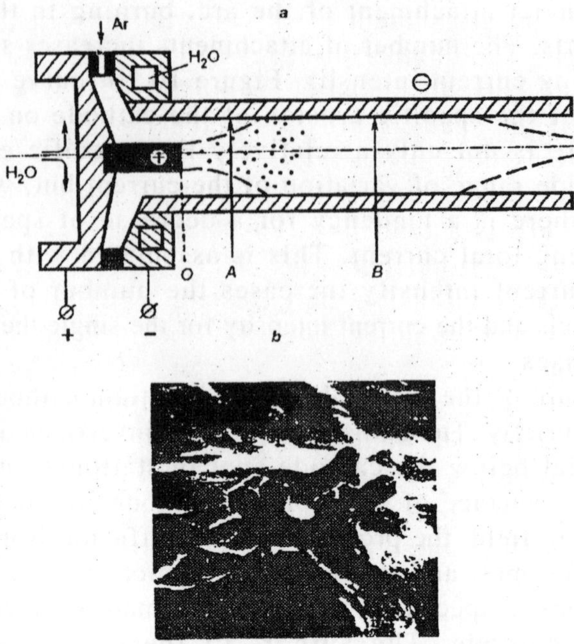

Fig.10.13. Cylindrical hollow tungsten cathode. a) diagram of euipment; b) photograph of a group of tungsten crystals, formed on the surface of the cathode at the end of operation.

perature distribution on the surface of the classic hollow cathodes working at a reduced pressure [29], i.e. the temperature of the wall is maximum not at the end of the pipe but at a distance of 2–3 diameters from the end of the pipe. It is interesting to examine the structure of the internal surface of the tungsten tubular cathode in different sections up to the zone of the effect of the arc (*OA*) on the side of the water-cooled end of the cathode and in the zone of the effect of the attachment spot of the arc (*AB*).

The temperature of the section (*OA*) of the surface of the cathode, as a result of cooling with water, is, according to the measurements, not higher than 1000 K. The section is not visited by the reference spot of the arc, as confirmed by photographs of the surface made at a high magnification (×170): there are traces of machining and sharp boundaries of the polycrystals of tungsten. This corresponds to the initial structure of the cathode material.

A completely different picture is obtained in the section of the surface (*AB*) with which the arc is in contact. The zone contains the maximum of surface temperature, which is equal to approximately 2700 K.

As shown by the photographs of the surface of the cathode in the sections *OA* and *AB*, produced using a Cambridge scanning microscope, the surface underwent large changes: crystalline structures are clearly visible, the cracks were 'healed', the surface became wavy. There are groups of tungsten crystals, formed, in all likelihood, as a result of the deposition of tungsten from the gas phase. Figure 10.13 shows a fragment of the group of crystals; the photograph was made at a magnification of ×1130.

10.6. Special features of the structure of the working surface of rod tungsten under the effect of the reference spot of the arc.

To conclude discussion of date and information for the thermal cathodes, we shall describe several special features of the changes in the structure of the working surface of a rod-shaped tungsten cathode ($l_c > 0$) under the effect of the spot of the electrical arc in different gases.

The reader who would like to obtain more information on the material presented below, should turn to the studies [11, 14, 30, 31].

The cathode section is a tungsten rod, brazed into a watercooled copper holder. The characteristic parameters of the rod are: extension length l_c, diameter d_c. The characteristics of the surface produced in different gases and mixtures will be examined. The plasma forming gas is helium, d_c = 3 mm, l_c = 3 mm, gas pressure in the discharge chamber 0.1 MPa, I = 120 A. The surface temperature of the working cathode 4000 K. What is the general appearance of the surface of the cathode after operation for 15 minutes? In the zone of the effect of the arc spot, the cathode surface is hemispherical, with the tip melted. With increase of the distance from the tip, the surface temperature decreases. The boundary of the alloys characterised by the presence of spherical formations, whose form is identical to the Fermi surface for tungsten (it is possible that they represent the nuclei of tungsten crystals). In the vicinity of the base of the cathode, where the surface temperature is low, there are structures of random form characteristic of tungsten oxides.

The addition of 25% nitrogen into helium reduces the size of the molten part of the surface of the end of the cathode, increases its temperature and current density. A further increase of the nitrogen content of helium (up to 75%) reduces even more the size of the area of the molten part of the and surface of the cathode and increases its temperature. The use of 100% nitrogen results in the formation of a constricted cathode spot with a high current density and high-temperature. This is accompanied by the formation, growth and subsequent breakdown of a 'bubble' of molten metal.

We shall examine the effect of oxygen, present in helium, on the changes in the structure of the near-cathode region of the arc discharge. All the parameters of equipment remain the same. At an oxygen content of 0.1%, the end of the cathode is molten and the surface of the melt shows the formation of a bubble whose size increases with time. Subsequently, the bubble bursts and a new bubble form in its position. When the oxygen content in helium is increased to 0.5%, particles of molten metal start to 'fly' from the surface of the cathode. The bubble does not manage to grow. At the oxygen content of 5%, the sub-surface processes change qualitatively in comparison with previous processes. The discharge is constricted in the radial direction and the melt occupies only a small part of the cathode surface. The centre of the melt is characterised by the formation of a bubble expanding upwards. With time, the bubble bursts. The gases, penètrating through the tip of the 'projection', carry with them of the main material and this greatly increases the extent of cathode erosion. The molten mass, trapped by the gas, travels along the helical trajectory along the axis of the discharge. A further increase of the oxygen content (up to 100%) results in even greater constriction of the reference spot of the arc and this reduces the size of the molten zone. The liquid metal separates in the form of droplets from the cathode.

What are the processes taking place on the surface of the cathode if hydrogen is used as the plasma forming gas? In this experiment, $d_c = 6$ mm, $l_c = 0$, current intensity 400 A, operating time approximately 7 min, gas pressure 1.1 MPa. Almost the entire working surface is in the molten condition. The arc spot is constricted and moves.

Investigations were also carried out into the effect of the plasma forming gaps-commercial nitrogen with 1% oxygen, on the variation of the form of the end surface of the cathode. The experiments were conducting using VL-10 lanthanised tungsten with the lanthanum content of 1% At current intensity 150 A, $d_c = 3$ mm, $l_c = 4$ mm. The investigated part of the cathode surface was in the molten condition and was characterised by the formation of a porous structure in solidification. A spherical growth appears in the centre of the cathode on the conical projection. Afteı the end of arcing, the lanthanum content of the material of the gross increase of 50% in comparison with the initial content. Possibly, this takes place as a result of the evaporation of lanthanum from the heated surface of the cathode (outside the discharge), inclusion of the part of lanthanum in the column of the electrical arc, ionisation and deposition under the effect oí the electrical field on the surface of the cathode in the zone of the arc spot.

As shown in the previous sections, in the conditions of high temperatures, realised in the electrodes of the plasma torches, a significant role is played by the processes of evaporation and oxidation of the material because the inert working gas is usually not spectrally pure. The rate of evaporation of the material in vacuum is determined by the Langmuir law:

$$P = m\sqrt{2\pi RT / M},$$

where P is the pressure of saturated vapours, m is the rate of evaporation, R, M is the gas constant and the molecular weight of the vapours.

The evaporation rate of tungsten in the vacuum in the solid and liquid phases [32] is presented in Fig. 10.14. The curves *1* and *2* show of that at the atmospheric pressure of the neutral gas, for example, N_2, the rate of evaporation of tungsten at T = 3200 K changes in by approximately three orders of magnitude in comparison with vacuum as a result of the decrease of the diffusion rate of the tungsten vapours through nitrogen. If these data are compared with the value of the specific erosion of tungsten in nitrogen in the presence of the electrical arc (Fig. 10.8), the significance of the recirculation of the atoms of the electrode material in decreasing the erosion rate becomes evident.

It is important to mention another fact, which has a strong effect on the increase of the erosion rate. It is the presence of the oxides.

In oxidation of tungsten, the following oxides are formed: WO_2, WO_3, and others. The melting point of WO_2 is 1540 K, that of WO_3 is 1750 K and, in addition to this, WO_3 is characterised by a very high evaporation rate, as indicated by the curves *3, 4, 5* in Fig. 10.14. In transition from pure tungsten to the oxide WO_3 the volume of the oxide increases by more than three orders of magnitude and this results in the formation of stresses in the oxidised film which fracture the film and ensure further access of oxygen to the metal. This is accompanied by the increase of the rate of oxidation of tungsten.

The resultant values of the rate of removal of the mass of the cathode from VL-10 lanthanised tungsten in atmospheric helium of high purity (the oxygen impurity ~10^{-3} vol%) are considerably higher than the rate of evaporation in vacuum; it may be concluded that at T < 3000 K, the controlling factor are the oxidation processes (Fig. 10.14, curves *8*, *9* and *2*).

The form of the temperature dependence of the rate of removal

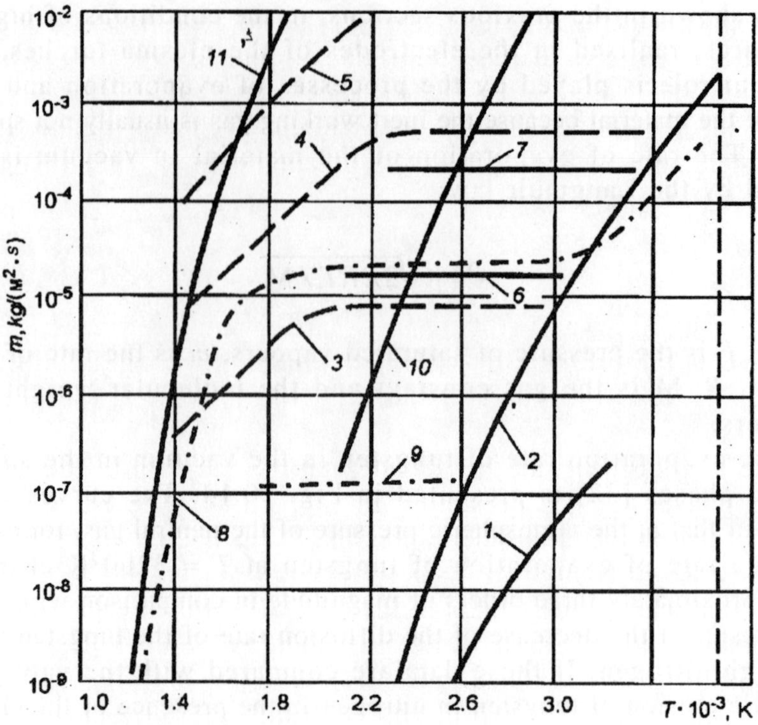

Fig.10.14. Dependence of the rate of removal of the mass of the electrode material on its temperature.

Curve No	Material	Working Gas
1	W	N_2(p =10^5 Pa)
2	W	Vacuum
3	W	O_2(p =0.1 Pa)
4	W	O_2(p =10 Pa)
5	W	O_2(p =10^3 Pa)
6	W	Ar (p =10^5 Pa) + O_2 (p =10 Pa)
7	W	Ar (p =10^5 Pa) + O_2 (p =10^3 Pa)
8	VL–10 Tungsten	He (p =10^5 Pa) + O_2 (p =10^3 Pa)
9	VL–10 Tungsten	He (p =10^5 Pa) + O_2 (p =1 Pa)
10	ThO_2	Vacuum
11	WO_3	" "

of the mass (curves *9*, *6*, *7*, Fig. 10.14) is determined by the fact that at T > 2000 K, the tungsten oxides form a diffusion barrier on the surface of the metal and, consequently, inhibit the axis of oxygen to the electrode surface. In this range, the rate of the removal of the mass m is independent of temperature.

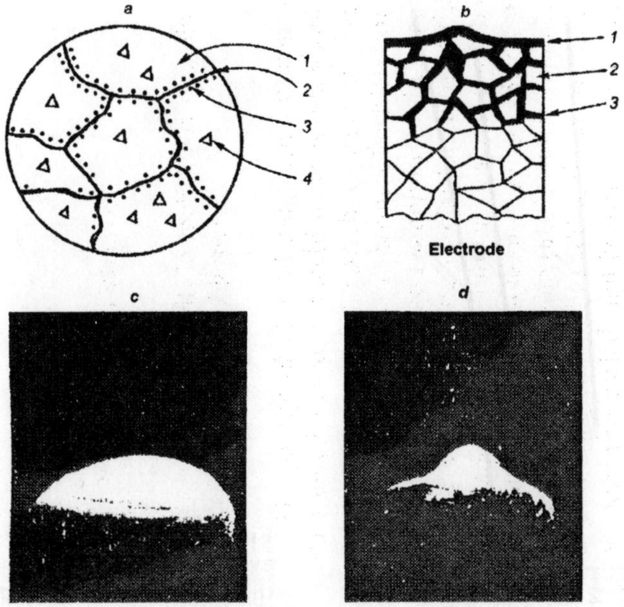

Fig.10.15 Changes in the structure of a tungsten cathode. a) structure of crystalline tungsten; 1) tungsten crystallite (grain); 2) grain boundary; 3) inclusions at the grain boundaries; 4) inclusions in the volume of the grain; b) structure of the tungsten electrode: 1) liquid metal film with a bubble; 2) tungsten grain; 3) weakened grain boundary; c) photograph of the film of the melt on the surface of the electrode, d) formation of the bubble.

In air, intensive oxidation of tungsten takes place at a temperature of 1100–1400 K. In heating in the medium of hydrocarbons to 1400–1500 K, examination showed the formation of semi-carbides and carbides with the melting point in the range 2900–3150 K. Carburisation of tungsten is very intensive at $T > 1900$ K.

At high temperatures, tungsten does not form stable nitrides with nitrogen.

The operating efficiency of the tungsten electrode is greatly affected by the structure of the material and the presence of nature of the impurities in the electrode.

This detailed examination of the behaviour of tungsten is the result of the fact that this element is used widely for the cathode of the plasma torches because of its unique properties–maximum values of the melting and boiling points, minimum rate of evaporation, high values of the latent heat of melting, the strength properties, and many others.

The tungsten, used for the production of electrodes, is a polycrystalline material consisting of crystals–grains of metals of different dimensions

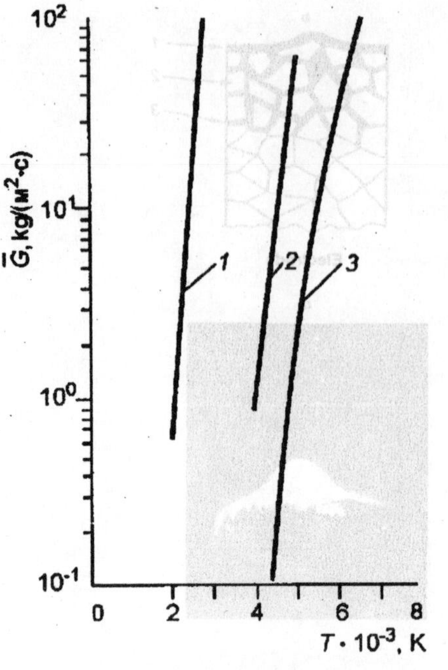

Fig.10.16. Dependence of the rate of evaporation of the electrode material in the liquid state. 1) copper; 2) carbon; 3) tungsten.

and shapes. The grain boundaries are characterised by the concentration of various inclusions: oxides, nitrides, carbides, other refractory compounds and also intermetallic compounds and other impurities. Some of the impurities remained inside the grains; in certain conditions, they may play a negative role. This pattern may be described by the scheme, shown in Fig. 10.15 a, b.

In the process of heating to a specific temperature, the impurities (oxides of calcium, potassium, aluminium, iron, silicon, etc) start to evaporate. The melting point of WO_2 is 1570 °C, boiling point 1850 °C, and the values for WO_3 are approximately the same.

The formation of the gases is equivalent to the increase of the volume by approximately a factor of 10. If this takes place inside the solid, extremely high pressures form. The liquid film formed on the surface of the electrodes is not fractured by the gas in the initial moment of the formation of the metal melt (Fig. 10.15*c*) because there are significant surface tension forces, but the film starts to bend, forming a bubble (Fig. 10.50*d*). With time, the size of the bubble increases and the bubble breaks. A new bubble forms in its area. The accumulated gases penetrate, at a very high velocity, through the bubble and carry with themselves particles of the molten metal and individual grains whose boundaries were extremely weakened.

Oxygen (nitrogen) penetrates into the intercrystalline lattice of tungsten along the cracks in the grains, forming oxides (nitrides) of tungsten. The resultant oxides (nitrides) evaporate and the processes in repeated.

Reducing the grains size of tungsten and other electrode materials, it is possible to reduce the rate of failure of the electrode material.

We describe briefly graphite and copper, used widely in the plasma torches as the cathode and anode materials (Fig. 10.16). In air, graphite starts to oxidise at $T = 720 \div 770$ K forming at $T < 870$ K CO_2, and CO at $T > 870$ K. In nitrogen, carbon is stable up to 3300 K. Erosion of the carbon electrodes in air takes place mainly as a result of the formation of oxides and cyanide compounds.

In the heating of copper (melting point 1356 K, boiling point 2873 K) in air surface oxidation takes place: the copper oxide CuO (black) forms in the range $460 \div 650$ K, and this oxide dissociates at $T > 1070$ K. A two-layer scale forms in the temperature range $650 \div 1370$ K characterised by incomplete oxidation. The surface layer of the scale contains CuO, in the internal layer the Cu_2O oxide (red-brown colour). The melting point of the latter is 1500 K. The copper oxide Cu_2O is characterised by high electrical and thermal resistance. Even at high temperatures copper does not react with hydrogen, nitrogen and carbon, and gases such as CO, CH_4 and O_2, reduce Cu_2O to Cu.

10.7. Review of studies of self-restoring cathodes

Examination of the special features of the erosion of the surface of tungsten cathodes in argon and nitrogen made it possible to confirm the existence of recirculation of tungsten atoms in the zone of attachment of the arc and deposition of part of the atoms on the cathode surface in this range.

Historically, this phenomenon was observed for the first time in 1965 in the Chemical Company at Borzesti (Romania) in the examination of failed electrodes used in the process of electrocracking of natural gas to acetylene. The phenomenon of random formation (with subsequent development) of a carbon growth (pyrocarbon) on the surface of a tubular cast iron internal electrode–cathode of a linear plasma torch in the zone of movement of the arc spot was found. With time, the size of the growth increased in the radial direction with subsequent transition to the axis of the electrode. The end of the cathode section of the arc bordered with the tip of the carbon growth. At the moment of formation of the growth, the movement of the end of the radial section of the arc was interrupted. A similar phenomenon was also observed at the Saratov Chemical Company in electrocracking of natural gas to acetylene.

In both cases, the formation of the growth resulted in undesirable consequences because this would reduce the arc voltage and, consequently, the power of the plasma torch. In addition to this, the growth reduced the size of the effective cross-section, changed the aerodynamics in the discharge chamber and increased the hydraulic resistance of the system. In the final analysis, the electrode failed in operation. Later, it became clear that the formation of pyrocarbon on the internal surface of the cast iron cathode was associated with the deposition of carbon ions from the gas phase on the internal surface.

Taking into account the very high applied significance of this phenomenon and the results obtained in the development of self-restoring cathodes in appropriate gas media, it would be useful to present a review of studies into the subject.

In the study published in 1973 [33] it was reported that in burning of a constricted DC arc in saturated hydrocarbons a constantly renewed cathode is produced from the material of the gas phase, i.e., in carbon in the given case. The initially installed cathode plays only the function of the substrate on which the carbon, forming the cathode in the later stages, precipitates. In the initial stage of the process, the thickness of the cathode increases. Subsequently, the increase of the thickness of the cathode is interrupted and the heat flow into the cathode is stabilising. This occurs at the moment when the rate of evaporation of the atoms and of the deposition of carbon atoms become identical. Figure 10.17 shows the curve of variation of the heat flow into the cathode on time [34].

X-ray diffraction was used for the investigation of the carbon content of both the central and peripheral sections of the cathode. A large difference between them was already detected in visual examination. If the carbon in the central section with a diameter of 3–4 mm is granular, in the peripheral section it has the form of large flakes. Examination by x-ray diffraction showed that the carbon in both sections has the distinctive structure of graphite. The carbon, precipitated in the pyrolysis process outside the cathode (in the nozzle, on the reactor walls) has an almost amorphous structure, characteristic of soot.

In [35, 36], further investigations were carried out into the concept proposed in [2] of the formation of a tungsten 'growth' on the end surface of a cylindrical welding electrode as a result of evaporation, in all likelihood, of the tungsten oxides from the side surface. The formation of oxide is associated with the presence of oxygen in the working gas which is a mixture of argon and oxygen. As shown previously, the rate of evaporation of the tungsten oxides is sev-

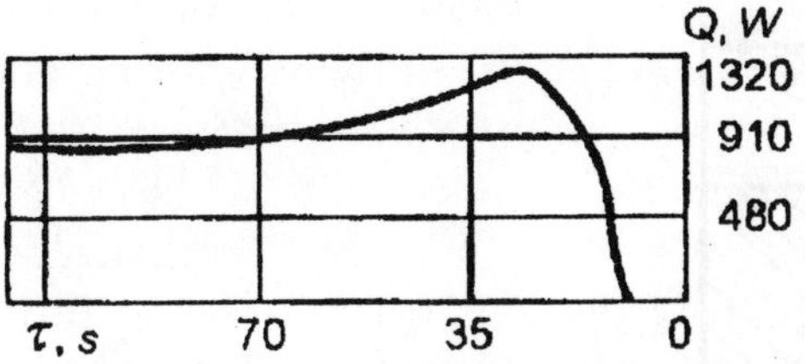

Fig. 10.17. Variation of the heat flow into a graphite cathode with time. I = 250 A: plasma forming gas – the mixture of methane with argon ($G_\Sigma = 6 \cdot 10^{-4}$ nm³/s).

eral orders of magnitude higher than the rate of evaporation of tungsten at the same temperature. The authors accepted the mean temperature of the electrode within the limits of the evaporation zone which is 4–5 length gages of the electrode length, equal to 1500 K. However, oxidation of tungsten in oxygen the takes place at a temperature of approximately 600 °C. The formation of a 'growth' from tungsten may also take place in pure argon [2]. X-ray diffraction microanalysis showed [36] that the 'growths' on the surface of lanthanised tungsten rods consist of pure tungsten and inclusions of a complicated composition (tungsten with lanthanum). Along the height, the 'growth' includes the same chemical elements as tungsten rod.

In the experiments with a non-consumable electrode (tungsten, molybdenum, titanium, etc) and in burning of an arc in gas mixture consisting of saturated hydrocarbons and argon [36], examination shows the formation on the active surface of the cathode (regardless of the cathode material) of an object in the form of a circular cup, consisting of fine dispersion tungsten which does not include the elements present in the composition of the investigated cathode. Similar investigations were continued in [37], in particular, attention was given to a system consisting of copper (cathode substrate) and the hydrocarbon mixture with argon. A special feature of work with the cathode is a high heat flow into the cathode (3.6 kW at a current of 250 A) in the first stage of arcing (during 30 s). After approximately 90 s the heat flow becomes lower in transition of the cathode to the regime of constant renewal to 1.15 kW (for example, Fig. 10.17).

In all investigated cases, the recirculation of the carbon atoms and the formation of unique 'growth' forms are clearly evident to various degrees. This indicates the self-restoration of the cathode in the carbon-containing gases and, consequently, the unlimited increase of its operating life.

The authors of [38] investigated the scheme of a constantly restored erosion-free cathode. As previously, the investigations were carried out using a compound cathode (Fig. 10.18) with the active insert *1* with zero extension. In this case, the active insert remains un-

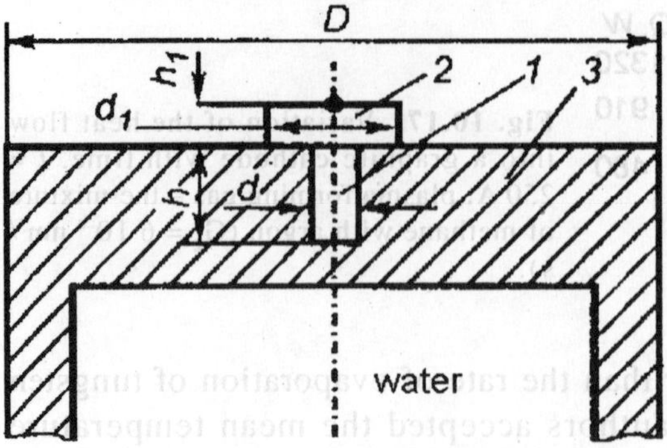

Fig.10.18. A compound cathode and the formation of a 'growth' on a passive cathode.

changed only in the first seconds after igniting the arc up to the formation of the real cathode 2 with a diameter d_l produced from graphite, precipitated from a carbon-containing gas atmosphere. Subsequently, the insert *1* becomes passive and plays,, like the copper water cooled casing *3*, only the function of the heat-transferring member from the working surface of the anode to cooling water.

As shown by the direct measurements, in the investigated range of the parameters (current intensity 500÷800 A, the flow rate of the mixture of the natural and carbon gases 3÷5 g/s at a mass ratio from 1:4 to 1:6.5), the thickness of the growth h_l and the diameter of the real cathode, working in the condition of constant renewal, remain unchanged and equal to approximately ~0.2 and 3.0 mm.

Further investigations of the renewable cathode in the carbon-containing atmosphere [39] at currents up to 750 A show that cathode renewal is possible only if a number of conditions are fulfilled, including the maintenance of the specific concentration of free carbon in the gas atmosphere of the arc. In this case, it is necessary to ensure the transport of carbon to the near-cathode region of the arc and its localisation on the working surface of the cathode. Some requirements are also imposed on the composition and flow rate of the plasma-forming mixture, and the electric power source. The compulsory condition is to produce the active insert from the material ensuring the optimum working regime of the compound cathode as a whole. This general formulation is not made more accurate and is not interpreted in [39, 40, etc] but the need for ensuring a short time of formation of the real constantly renewable cathode, the strong bond of the substrate with the carbon of the real cathode, etc are discussed; the material of the substrate should have high thermophysical and thermomechanical characteristics. According to the view of the author of [39], the need for fulfilling these requirements is not very strong, and the range of the investigated materials is restricted to the carbide-forming elements of the groups IV and VI of the periodic system of elements and graphite. These materials are used in the form of roads with a diameter of 1.5÷2.0 mm and the length of 3÷4 mm. The rods are pressed into copper holders, forming compound cathode. The thermoemission insert of the compound cathode in operation maybe characterised by the formation of high thermal stresses and strains [41, 42] capable of causing thermomechanical failure. The results of analysis of the inserts produced from carbide-forming metals show that in the initial ignition of the arc prior to the formation of the real carbon cathode, a large volume of the metal of the insert is melted and saturated with carbon up to the formation of carbides of various composition. Consequently, the initially homogeneous insert becomes laminated along its length with different chemical composition and properties and this results in the formation of thermal stresses and strains up to the failure of the insert and disruption of its thermal contact with the copper casing.

When using copper cathodes (closure of the arc with the copper casing), the formation of the real carbon cathode is delayed, as shown previously, by tens of of seconds and is accompanied by the random displacement of the arc spot. This leads to the formation of craters in the area of melted-out copper and low bonding strength of the real cathode with copper. In turn, this causes the separation of the real cathode from copper, especially in the transition conditions (activation and disconnection of the arc, changes of current, etc) and instantaneous (in the case of high currents) thermal failure of the cathode.

The cathode with the graphite insert also consists of layers but the chemical composition of these layers is identical. The homogeneity of the composition and, consequently, of various characteristics of the cathode, such as strength, the coefficient of linear and volume expansion, heat conductivity, etc, throughout the entire arcing period prevents the formation of thermomechanical stresses and strains.

The combination of graphite of all positive (from the viewpoint of the formation and operation of the constantly renewable cathode) characteristics enabled the author of [39] to recommend this material as the active insert of the compound cathode in burning of the arc in a carbon-containing plasma-forming medium. This has been confirmed by experiments. At currents of 500÷700 A, the graphite insert, cooled with water with the flow rate of 0.3÷0.35 kg/s, operated for tens of hours in the erosion-free conditions at the current density through the real cathode of $(8\div10)\cdot10^3$ A/cm^2.

Thus, the stable renewal of the graphite or other emitter is possible only if the gas atmosphere of the arc contains carbon (or metal) in the amount ensuring the equilibrium of the processes of precipitation of the material in the form of positive ions and neutral atoms, on the one side, evaporation and the removal, in the form of chemical compounds, if they form, on the other hand.

When using multicomponent chemically active (in relation to the compound cathode) gas mixtures, the selection of the material for producing the elements of the compound cathode and of the cooling conditions should be carried out taking into account the chemical reactions between the working gas and the cathode surface [43].

The possibility of working in the regime of constant renewal of thermal cathodes produced from refractory compounds was investigated in [44–46]. The working media, designed for self-renewal of the cathode, include titanium tetrachloride ($TiCl_4$), which is liquid at room temperature and gaseous at temperatures above 150°C, and other chlorides of refractory metals.

In [45], the authors presented the calculated data on the current density in a renewable cathode reduced from tungsten in the current intensity range 300–10 A. It has been established that the temperature of the cathode increases with a decrease of current from (3555÷3525) to (3815÷3775). There is also a large increase of the current density from (7.6÷4.1) · 10^7 A/m^2 to (52.8÷19.2)·10^7 A/m^2 and the pressure of saturated vapours p_{mc} (T_c) above the cathode. The resultant values of the parameters for tungsten are also similar for cathodes made of Ta, Hf, and Zr.

As in the case of the graphite cathode, a decrease of current increases T_c and the pressure of saturated metal vapours above the cathode. However, the pressure is four orders of magnitude lower than the pressure of the saturated carbon vapours above the real graphite cathode. Correspondingly, the content of the refractory metal in the plasma, corresponding to the regime of the constant renewal of the cathode, in the range 10÷1000 A is 10^{-9}÷10^{-7} kg/s, and for carbon it is 10^{-6}÷10^{-4} kg/s. This means that the constant renewal of the cathode made from refractory metals is possible at the extremely low pressure of compounds of these metals in the gas atmosphere of the arc and its existence, in contrast to the real graphite cathode, is practically not limited in the range of low currents.

In [47] experiments confirmed the possibility of renewal of the tungsten cathode from the gas phase.

Up to the end of the 80s, a very large amount of experimental and theoretical material was collected explaining the recycling process, i.e. the return of a large part of the atoms of the material of the emitter on the surface which they left as a result of cathode sputtering or sublimation for a cylindrical thermal cathodes with $l_c > 0$.

In [48], investigations were carried out into the recycling of ions in a hollow cathode, working in the arc discharge regime. For rod cathodes, used in the conditions of atmospheric discharge, the authors propose to one-dimensional model of the return of evaporated particles in the form of atoms [49] or ions [50]. The one-dimensional application is basically not suitable for the hollow cathode and, consequently, the authors of [48] used a two-dimensional mathematical model of recycling. The proposed mathematical model of recycling makes it possible to calculate the number of the atoms of the emitter returning to the wall of the channel of the electrode after their departure from the surface under the effect of cathode sputtering or sublimation and, consequently, to determine the resultant erosion of the emitter at every point of the latter.

Special features of the renewal of the graphite hollow cylindrical cathode were investigated in [51]. In the experiments, the internal diameter of the cylinder was 20 mm, the wall thickness 5 mm, current 300 A; the working mixture was $CH_4 + 0.5O_2$, the cathode section was cooled with water. According to the authors, the nature of reduction of carbon on the substrate is a relatively complicated process. In high-temperature pyrolysis of hydrocarbons, not only individual ions and atoms but also crystals and macroscopic polymer structures may take part in this process. The authors noted a high level of specific erosion of the cathode, equal to $2 \cdot 10^{-8}$ kg/C, i.e., no self-renewal can be considered.

In [52], a small amount of information is provided on the experimental examination of a graphite cathode in the regeneration regime. The plasma-forming gas was represented by a mixture of hydrocarbons (methane, propane, butane) and the oxidation agent (carbon dioxide, oxygen, air).

The experiments with the determination of the conditions corresponding to full regeneration, were carried out in equipment containing a cathode-nozzle section with an optical window for examination of the cathode region of the arc discharge. The range of the current was 300÷1200 A. The dependence of the diameter of the cathode spot of the arc on current $d_s = 0.25 \cdot 10^{-3} \cdot I^{0.34}$ m and on the specific heat flow through the arc spot $q_s = 0.36 \cdot 10^{8} \cdot I^{0.32}$ W/m² was obtained.

The metallographic and x-ray diffraction analysis of the emitting surface shows that the entire surface below the arc spot is covered with the layer of anisotropic pyrocarbon, with a thickness of 0.1÷0.2 mm, or 5÷10% of the thickness of the graphite insert.

It was also found that to ensure long-term efficiency of the cathode, it is necessary to cool the cathode efficiently and, consequently, ensure high-quality contact between graphite and the water cooled copper holder. Efficient contact was obtained using a lead-titanium brazing alloy. This alloy efficiently wets both copper and graphite. The melting point of the alloy is lower than the melting point of copper. If cracks appear in graphite, the alloy fills the cracks, ensuring stable thermal and electrical contact at the copper–graphite interface.

10.8. The rate of increase of the mass of the cathode in a carbon containing medium

On the basis of the data on the erosion of tungsten rod thermal cathodes with efficient cooling in different working conditions (Fig. 10.8) and considering the data presented in section 10.7, it may be concluded that the given cathode has an infinite operating life. In addition to the presented data, the results will be discussed of the experimental investigations of operation of graphite rod cathodes with the diameter of d_c = 5 mm, pressed flush into a copper watercooled holder [53]. The working medium was fluoromethane CF_4 supplied into the gas-discharge chamber of the plasma torch with twisting ensuring the stable position of the arc spot on the axis of the cathode rod. In the first stages of arcing (Fig. 10.19) in the current intensity range 300÷900 A the surface of the graphite cathode was characterised by deposition of carbon leading to the formation of a relatively thick (approximately 1 mm) film of pyrographite of the cylindrical shape with a high mechanical strength and efficient adhesion with the substrate (see the diagram of the cathode section in Fig. 10.19). In the process of further operation of the plasma torch with multiple starting up, the thickness of the resultant 'growth' did not change indicating the relatively rapid establishment of the regime of dynamic equilibrium between the mass of the carbon ions (deposited under the effect of the electrical field of the arc) arriving in the arc and also as a result of the diffusion of the atoms, and the mass of the carbon atoms, evaporated from the cathode surface. The self-renewal regime was also observed in short-term activation (2–3 min) at a current intensity of 200 A on CF_4, the mixture CF_4+ C_2F_6 in the ratio of 1/1 volumes, and pure C_2F_6.

10.9. Erosion of copper cold tubular electrodes

In this section, attention will be given to the reasons determining

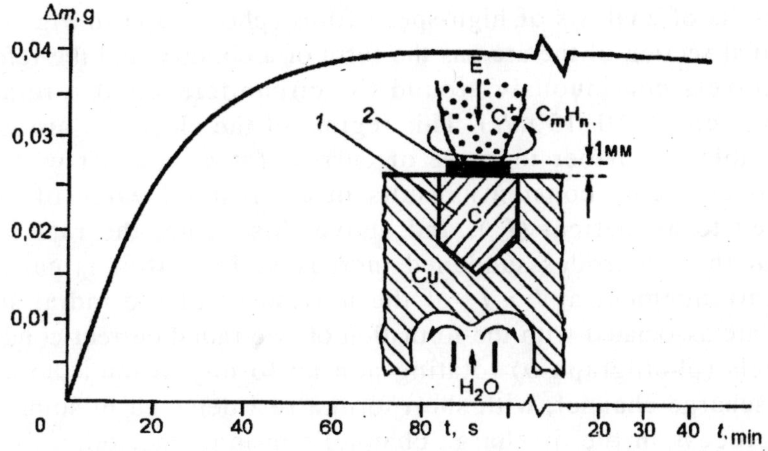

Fig.10.19. Dependence of the mass of carbon deposited from the CG_4 gas phase on the graphite cathode, on the operating time of the plasma torch. *1*) a carbon insert; *2*) A real newly formed cathode produced from pyro-graphite – carbon.

the rate of erosion of the electrodes in the presence of a moving arc spot.

The anode and cathode spots of the electrical arc on cold tubular electrodes, produced from materials with a low melting point, are characterised by a very high level of the heat flow ($10^6 \div 10^7$ W/m^2). To ensure the acceptable level of electrode erosion, the near-electrode sections of the arc are artificially moved along the internal cylindrical surface of the tubular electrode, acting on the surface by the aerodynamic and electromagnetic forces [54–56].

Practical experience with the application of these electrodes shows their high reliability and promising nature, especially in heating oxygen-containing media.

10.9.1. Dependence of specific electrode erosion on current

The simplest case – the plane of rotation of the closing radial section of the arc of a linear plasma torch is normal to the axis of the cylinder and fixed in space. Consequently, the arc spot travels around the circumference, and the with of the erosion area of metal (in the case of copper) is not greater than several millimetres. The dependence of the specific erosion of the cathode (anode) $\bar{G} = G_m/(I \cdot t)$ on current in this case is shown in Fig. 10.20.

As shown by the investigations of the tubular copper electrodes, there are two distinctive arcing regimes. In the first regime, the value of $\bar{G}$ is practically independent of current intensity. According to the results of analysis of high-speed films (photograph in Fig. 10.20), the radial section of the arc has the form of a comma, and the reference spot travels continuously around the circumference at a relatively high velocity (~10÷15 m/s). This regime of the electrical arc is most favourable. A further increase of current (at d_1 = const with other parameters being constant) results in a threshold value of current referred to as critical (I_{cr}), and above this value, the rate of erosion of the electrodes saddened increases. The latter is caused by the rearrangement at $I > I_{cr}$ of the movement of the radial section of the arc associated with the formation of two radial current-conducting channels (photograph *b*) rotating non-uniformly around the axis of the discharge channel, with short-term arrest detected in some cases. This process in the discharge channel continues periodically causing extensive failure of the electrode surface. This phenomenon has not as yet been unambiguously explained. At the same time, it is well-known that the value of I_{cr} depends on electrode diameter d_1. This is clearly indicated by the data presented in Fig. 10.20. It is also important to note the effect of the physical–mechanical properties of the electrode material, and the nature and flow rate of the gas [29, 31, 57, 58] on the value of I_{cr}.

In more detailed examination of the photograph of the surface area of the arc (Fig. 10.20*a*) in the case of the subcritical arcing conditions ($I < I_{cr}$), we can see the splitting of the arc spot indicating its jump-like displacement in the direction of rotation of the gas, caused by 'the wall element of the radial section of the arc–electrode surface' small-scale shuntin. With increase of current intensity ($I > I_{cr}$) examination shows the formation of a powerful cathode (anode) jet creating favourable conditions for the 'electrode surface–axial section of the arc' large-scale shunting, and the pattern of movement of now already two radial sections of the arc greatly changes. Photograph *b* in Fig. 10.20 shows clearly the resultant cathode jet.

In the subcritical range of current intensity, the mean value of specific erosion is in the range $(1÷3)\cdot 10^{-9}$ kg/C and slowly decreases in the direction of increasing current. As indicated by [59], the additional application of the axial magnetic field, and also the increase of $d_1 = d_2$ to 100 mm and of the flow rate result in the displacement of I_{cr} to (3÷4) kA.

The authors of [60] published data on the specific erosion of a copper tubular cylindrical cathode of a coaxial plasma torch with the diameter $5 \cdot 10^{-2}$ m in the presence of an axial magnetic field with B = 0.025 T. The data show unambiguously that, starting at some critical current, cathode erosion rapidly increases. At $I < I_{cr}$

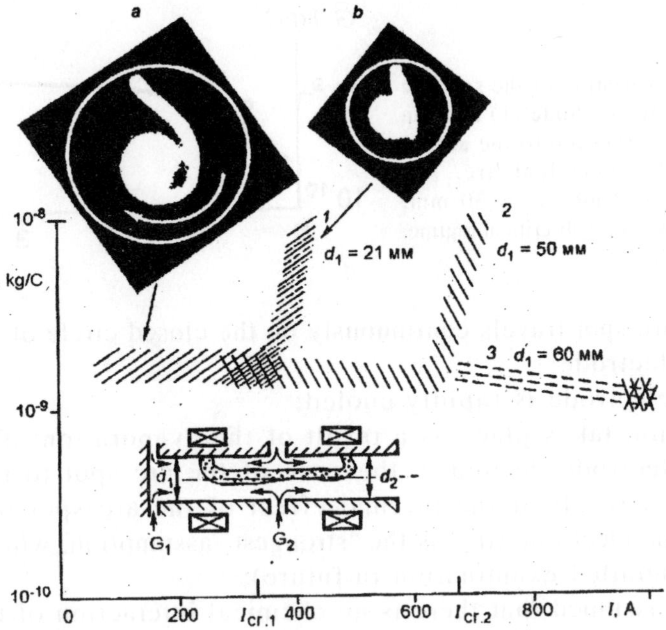

Fig.10.20. Conditions (1, 2 – d_1 = 21 and 50 nm respectively) of erosion of copper tubular cathodes with moving arc spot in relation to the arc current intensity and the data for d_1 = 60 mm (3).

the value of $\bar{G}$ is independent of current intensity and equals (3÷4) 10^{-9} kg/C, i.e., is approximately in the previously mentioned range.

The results of classic experiments published in [61] confirmed the constancy of the maximum level of specific erosion of the electrodes in long–term operation ($I < I_{cr}$) and rotation of the radial section of the arc in one plane (the absence of axial displacement of the arc both as a result of large–scale shunting and electro-aerodynamic forces of the pulsation nature). The cathode and the anode were identical. The internal diameter $d = 3 \cdot 10^{-2}$ m; $G_1 + G_2 = 6 \cdot 10^{-3}$ kg/s, the air flow rate was also constant; the speed of movement of the arc spot $W_s \leq 9$ m/s. Figure 10.21 shows the dependence of $\bar{G}$ on the operating time t. Attention should be given to the fact that the specific erosion increases only in the first 1÷1.5 h of operation of the plasma torch; after this time, the process of formation of the subsurface structure of the metal (mainly of the formation and development of dislocations in the metal), development and formation of the oxide film is completed, and the period of stable erosion stars.

In [62] the authors investigated for the first time in detail the theory of electrode erosion in non-stationary spots of the electrical arc using a number of assumptions. Some of these assumptions are:

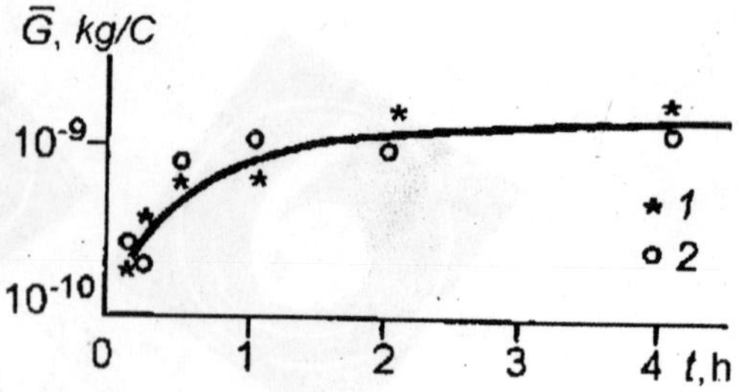

Fig.10.21. Variation of the specific erosion of the cathode (1) and the andoe (2) in relation to the arcing time of the electrical arc. I = 200 A, $W_s \leq 9$ m/s, d_1 = 30 mm, working gas – air; sub-critical regime.

–the arc spot travels continuously on the closed circle of the surface of the electrode;

–the electrode is rapidly cooled;

–erosion takes place as a result of the evaporation of the metal of the electrode, heated in the zone of the arc spot to the boiling point as a result of the thermal effect of the arc spot on the surface of the electrode (this is the 'strongest' assumption, which requires further detailed examination in future);

–it is assumed that there is no chemical interaction of the plasma with the electrode leading to the formation of volatile compounds; this requires discussion, because copper electrodes operate in most cases in oxidising media.

Nevertheless, this work was the first attempt to cast light on the extent of specific erosion under these assumptions.

10.9.2. Effect of the speed of travel of the radial section of the arc and of its axial scanning on specific erosion

The first case will be initially examined. The plane of rotation of the arc is constant. Figure 10.22*a* shows the dependence $\bar{G} = f(W_s)$ for a copper anode. At low values of W_s (< 9 m/s), the level of specific erosion is ~10^{-9} kg/C which is in completely satisfactory agreement with the really is data. With increase of the speed W_s, the value of specific erosion decreases, and at W_s> 30 m/s specific erosion is equal to ~$2 \cdot 10^{-11}$ kg/C.

In the case of the cathode, the situation is different. The value of $\bar{G}$ is almost completely independent of the speed of travel of the arc spot and its mean value is 10^{-9} kg/C (the broken line). This behaviour of erosion may be described by simple physical interpretation. The cathode should supply electrons to the arc in order to maintain continuous arcing and this is possible only (in the case of the copper anode) at temperatures close to the melting point of copper or even exceeding this temperature.

What is the nature of erosion of the electrodes in superposition,

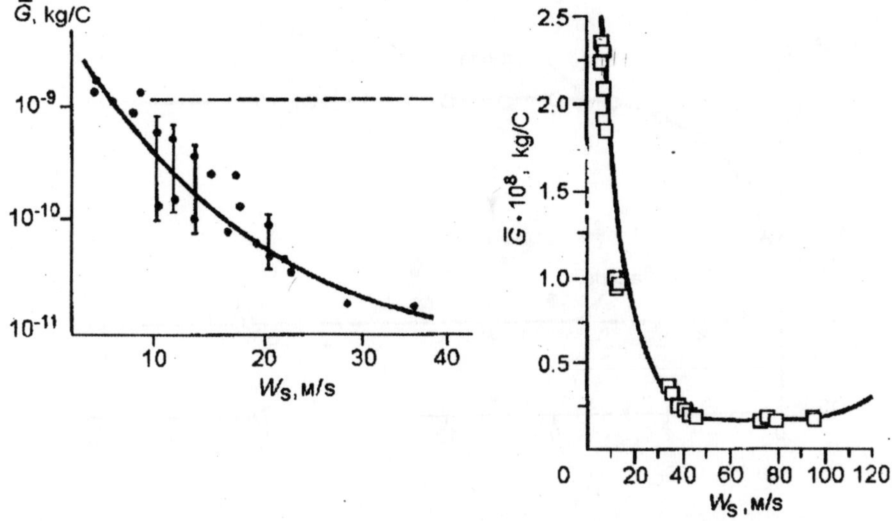

Fig.10.22. Dependences: a) specific erosion of a copper cylindrical electrode on the speed of movement of the arc spot W_s around the circumference (I~250 A) (solid curve – anode, broken curve – cathode); b) specific erosion of a copper cylindrical cathode [64] on W_s.

on the rotational movement, of the radial section of the arc of translational movement along the axis with the frequency of 4÷6 Hz in relation to some plane normal to the axis of the electrode? In this case, the length of the zone of displacement of the arc spot along the electrode (scanning) is 2–3 diameters of the channel; in displacement of the arc spot around a circle with a speed of 15 m/s the displacement of the spot along the axis in a single rotation is 1÷2 mm, i.e. the arc spot during its movement along a helix is displaced, in every rotation, by 1–2 mm to the cold party of the electrode surface.

As indicated by Fig. 10.23, in the first hours of operation of the plasma torch there is an increase (in the case of the cathode) of the value of $\bar{G}$ from the values of ~10^{-10} kg/C to ~10^{-9} kg/C; reaching the value of ~10^{-9} kg/C, the value of specific erosion remains constant over more than 60 hours of operation of the plasma torch. The results are identical with those presented in Fig. 10.21.

The level of specific erosion of the copper cathode in the first minutes of operation ($\bar{G}$~10^{-10} kg/C) is evidently basic and cannot be reduced by any increase of the speed of travel of the near-electrode section of the arc W_s because it is determined by the mechanism of functioning of the cathode spot.

The situation in the case of the anode is different. For the same mean speed of movement of the near-anode section of the arc

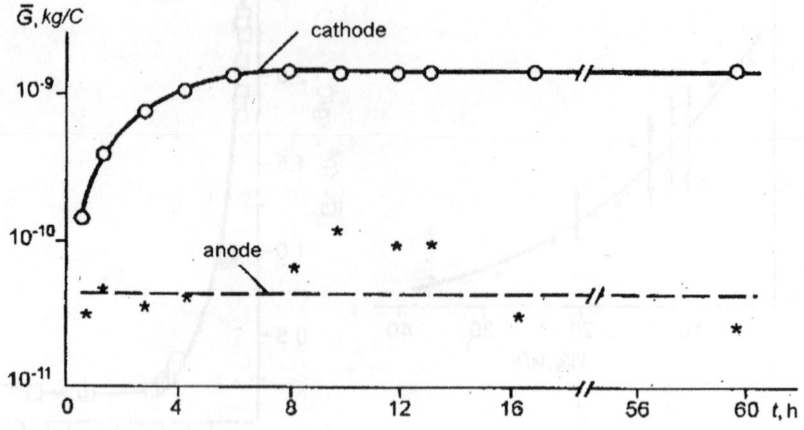

Fig.10.23. Dependence of $\bar{G}$ on the operating time of the electrodes of a two-jet plasma torch in translational movement of the radial section of the arc under the effect of aerodynamic forces (I = 250 A, $W_s \approx$ 15 m/s).

W_s = 15 m/s, the specific erosion decreased by more than an order of magnitude and equalled $\bar{G} \sim 6 \cdot 10^{-11}$ kg/C. It should be mentioned that this value remains constant from the very first minutes of operation of the plasma torch.

After operation for 60 hours of a two-jet plasma torch, microsections were produced from the longitudinal section of the tubular cathode and anode (Fig. 10.24). Examination shows that the extent of erosion in the anode is considerably smaller than in the cathode which was characterised by the barrel-shaped form.

It is also important to mention the following: the eroded surface of the anode is characterised by a lower degree of oxidation in comparison with the cathode, indicating the lower surface temperature of the anode.

In a shorter study [63], the authors presented the satisfactory generalisation of the experimental material [64] for the erosion of cold electrodes under the effect of non-stationary arc spot on the bases of the application of simple one-dimensional or quasi-dimensional models of heat conductivity [65, 66]. The authors of [63] examine the application of these models for the cases of the continuous or jump-like movement of the arc spot.

Figure 10.22*b* shows the dependence of the rate of erosion of a copper electrode on W_s for the case of jump-free movement of the anode arc spot [64]. Unfortunately, in [63] it was not mentioned

a b

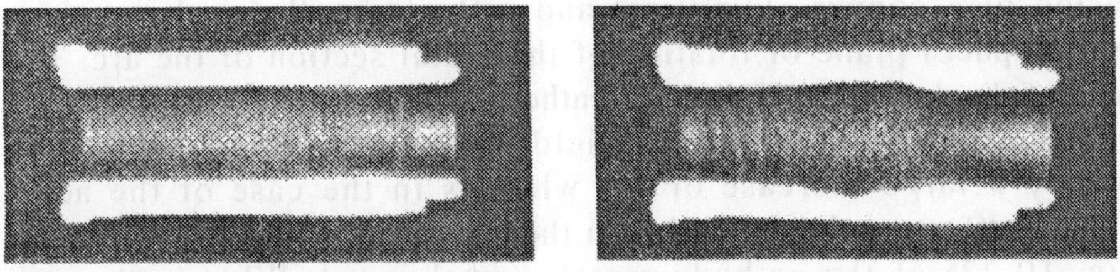

Fig.10.24. Contours of the meridional section through working electrodes. *a*) anodes; *b*) cathodes; d = 30 mm, l = 120 mm.

whether the arc spot moves along a closed circle or in a straight line on the surface of the electrode with constant temperature. Similarly, no reports have been published on the operating time of the electrode in the determination of the erosion rate. As shown previously, there is a large difference in the values of the specific erosion of the cathode and the anode.

10.9.3. Effect of axial magnetic induction on the erosion rate

Figure 10.25 (curve *1*) shows the dependence $\overline{G} = f(B_z)$ for the output copper cylindrical anode (with a ledge) obtained in the following conditions: working gas – steam, arc current intensity 400 A, large-scale shunting [67]. Regardless of the pulsations of the radial section of the arc in the axial direction and of the effect of the axial magnetic field on the closing radial section of the arc, specific erosion decreased only by a factor of 5 on reaching the value B_z = 0.1 *T*. In addition to this, there was a tendency for an increase of $\overline{G}$ with a further increase of B_z. This may be explained by the increase of the surface temperature of the electrode as a result of the effect, on the surface of the electrode, of strong radiant heat fluxes from the radial (closing) section of the arc, because an increase of Bz results in stretching of the sections along the surface of the electrode. The presence of large-scale shunting does not ensure that the value $\overline{G} \sim 2 \cdot 10^{-11}$ kg/C is reached, as was the case in the two-jet plasma torch without shunting.

In operation of a plasma torch using dry steam, the surface of the copper cylinder in the zone of the effect of the arc spot retains its bright colour which is explained, in all likelihood, by the absence of oxidation of the metal because of a shortage of free oxygen at the electrode wall. However, if air is used as the working gas, the surface is always black because of the formation of the copper oxide Cu_2O.

Curve 2 in Fig. 10.25 characterises the dependence of the specific erosion of a copper cylindrical and cathode on B_z for the stationary (in space) plane of rotation of the radial section of the arc. The speed of displacement of the near-cathode section of the arc is increased by the effect of the magnetic field on the arc. This is accompanied by a large increase of $\overline{G}$, whereas in the case of the aerodynamic effect on the arc and with the increase of W_s, specific erosion (Fig. 10.22) of the cathode remains unchanged. What is the reason for this? In the application of the magnetic field, the arc, as in the case of the aerodynamic effect, has the form of a comma. However, under the effect of the magnetic field, the arc is stretched more extensively along the surface of the electrode, as clearly indicated by Fig. 10.26 [61]. The effect of the external magnetic axial field on the variation of the form of the 'radial' section of the arc was described in [67]. The greatly bent form of the arc increases the

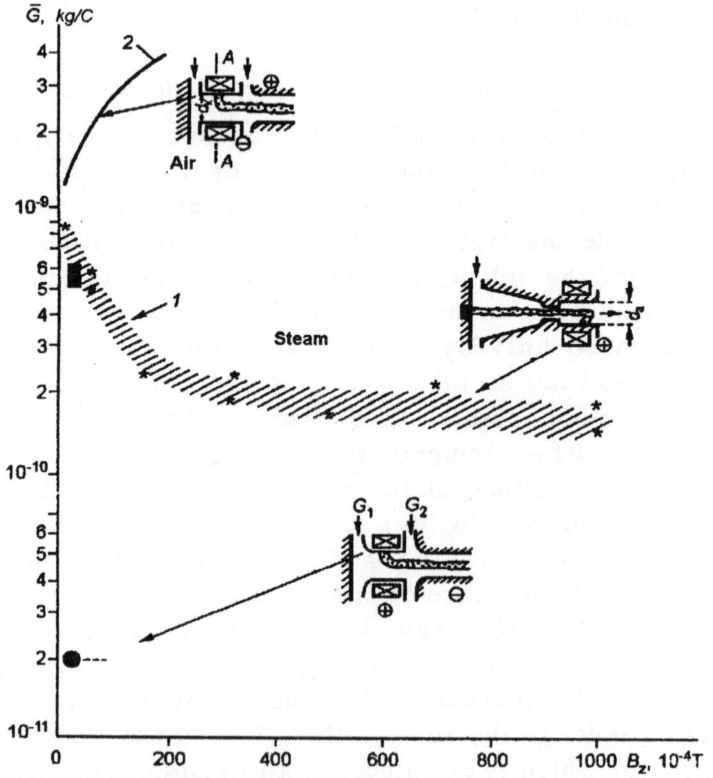

Fig.10.25. Dependence of specific erosion of electrodes on B_z. Curve 1 – dependence $\overline{G}= f(B_z)$ for the output anode. The plasma forming gas – steam. Steam flow rate 5 g/s; d_a = 18 mm; I = 400 A. Curve 2 – dependence $\overline{G} = f(B_z)$ for the end copper tubular cathode. Working gas – air. G = 6 g/s; d_c = 50 mm; I =600 A.

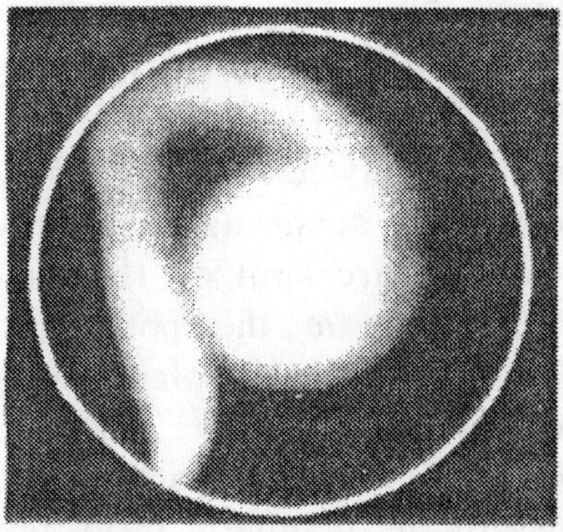

Fig.10.26. Radial section of the arc in an end cylindrical cathode. The plane of rotation of the arc is fixed. B_z = 0.02 T. I = 600 A.

intensity of the heat flow into the ring-shaped band on which the arc spot moves. In this case, the surface temperature of the copper electrode in the zone of action of the arc spot increases and the rate of evaporation of the material is also higher. Figure 10.60 shows the magnitude of the increase of the rate of erosion with increase of the temperature of molten copper.

These considerations show that to reduce the rate of erosion of the copper tubular anode, it is necessary to prevent bending of the radial section of the arc; this can be realised most efficiently by producing, in the tubular electrode in the plane of rotation of the arc, an axial magnetic feed of the appropriate topology which ensures uniform rotation of the investigated section in the form of a radial 'wheel spoke'.

10.9.4. Aeromagnetic axial scanning of the radial section of the arc in the internal tubular electrode-cathode

Comparison of the two results – increase of specific erosion with increasing electrode temperature [68] and the increase of the surface temperature of the ring-shaped groove in periodic 'visits' of the arc spot [69, 70] leads to the conclusion on the efficiency of translational displacement (scanning) of the arc spot on the surface of the electrode to ensure that the spot always moves on the pre-cooled surface. This makes it possible, firstly, to reduce the extent of specific erosion and, secondly, increase the surface area of the electrode, affected by the arc spot. Both factors increase the service life of the electrode.

The scanning of the arc spot may be carried out most efficiently and simply by the application of the external alternating axial effect on the rotational movement of the radial section of the arc, determined by the gas vortex. Depending on the nature of the forces, causing translational movement in the longitudinal direction, the methods of scanning are subdivided into three types: gas [1], magnetic [55, 56] and aeromagnetic [54]. Changing the ratio between the frequency of rotation of the attachment of the arc and the scanning frequency *z*, we obtain three types of trajectory of the arc spot on the surface of the cylindrical anode (Fig. 10.27*c*). At $\omega_\varphi > \omega_z$, the spot moves along the helical trajectory. When $\omega_\varphi < \omega_z$, scanning takes place along a snaking-like path. If the frequencies are equal $\omega_\varphi - \omega_z$ the scanning process is interrupted and the spot moves along a narrow path which, however, is not circular as in the absence of the magnetic field, but elliptical. Naturally, this possibility should not be allowed. Of special interest is the variant of similar frequency $\omega_\varphi - \omega_z \pm \Delta\omega$, where $\Delta\omega \ll \omega_z \sim \omega_\varphi$. In this case, the ellipse is open and itself rotates around the axis *z* with the frequency $\Delta\omega$.

In the variant of aeromagnetic scanning (Fig. 10.27*a*), the radial part of the arc, in addition to the rotational movement in the field of the gas vortex, also carries out translational movement along the axis. If the arc moves to one side under the effect of ponderomotive force F_m, then it moves in the reverse direction under the effect of the axial aeromagnetic force F_a, determined by the circulation

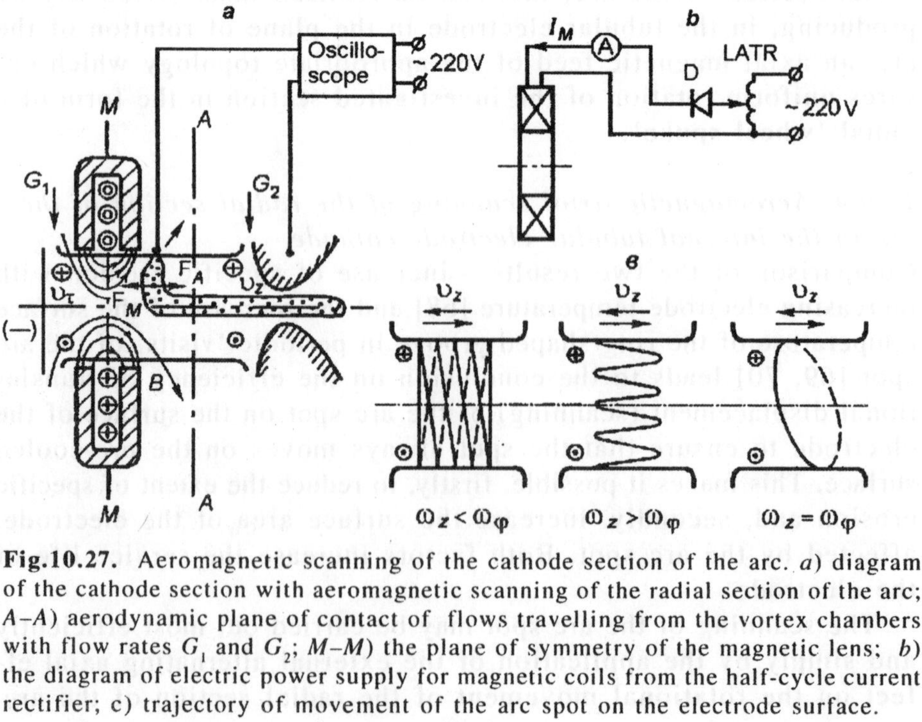

Fig.10.27. Aeromagnetic scanning of the cathode section of the arc. *a*) diagram of the cathode section with aeromagnetic scanning of the radial section of the arc; *A–A*) aerodynamic plane of contact of flows travelling from the vortex chambers with flow rates G_1 and G_2; *M–M*) the plane of symmetry of the magnetic lens; *b*) the diagram of electric power supply for magnetic coils from the half-cycle current rectifier; c) trajectory of movement of the arc spot on the electrode surface.

flow of the gas with the axial component with velocity v_z, always directed in the direction of the plane *A–A*. The section in which scanning takes place, is situated between the aerodynamic plane *A–A* (the area of contact of two circulation flows) and the plane *M–M*, generated by the field of the magnetic lens [54].

Aeromagnetic scanning is effective only if the connection of the magnetic lens is matched with gas twisting. Analysis of the situation resulted in the determination of a simple rule which has the following form for a tubular cathode: the direction of current in the winding of the magnetic lens may be regarded as inverse in relation to the direction of the circumferential velocity of the gas vortex. In Fig. 10.27, the directions of the gas and the current are denoted by the standard method: using the circles with the plus sign ⊕ or with the dot ⊙. The dot indicates movement towards the reader, and the plus away from the reader behind the plane of the page. If the rule of the signs is not fulfilled, the magnetic lens will eject radial attachment instead of pulling it in.

In the case of a tubular anode, in order to match the effect of the magnetic field with the vortex flow, the directions of the current in the lens and of gas twisting should be identical.

The magnetic lens operates in the pulsed regime, is powered from the single-phase mains through a single half-cycle rectifier (Fig. 10.27*b*). The scanning frequency of the section of the arc in the experiments was 50 Hz. As a result of scanning, the reference spot does not move around the circumference and, consequently the heated trace, but it moves along a spiral, i.e. the cold surface of the tubular electrode. This makes it possible, firstly, to reduce the extent of specific erosion and, secondly, increase by approximately an order of magnitude the area of the working surface of the electrode 'swept' by arc spot. The duration of continuous operation of the end tubular electrode–cathode in aeromagnetic scanning greatly increases. In the experiments for a copper cathode with the internal diameter of $3 \cdot 10^{-2}$ m, in air, at a current of I = 300 A, the resultant value of $\bar{G} = (5 \div 7) \cdot 10^{-10}$ kg/C. The mean value of magnetic induction did not exceed 50 gauss. In Fig. 10.25, this value of $\bar{G}$ is indicated by the solid rectangle.

For the internal tubular electrode–anode with the same geometrical parameters as for the cathode, the value of $\bar{G}$ decreased almost 80 times in comparison with the mean level of $\bar{G}$ for the output copper anode with the self-setting arc length, i.e., reached the value $2 \cdot 10^{-11}$ kg/C (the solid circle in Fig. 10.25). This is comparable with the value of $\bar{G}$ in aerodynamic scanning (Fig. 10.23). The application of aeromagnetic scanning in the EDP-212 two-chamber plasma torch, used in coal dust torches for plasma ignition of the boilers of thermal electric power stations, makes it possible to increase greatly their service life.

The oscillograms of the voltage (Fig. 10.28) of the section of the arc in the internal tubular electrode show that it changes with the scanning frequency.

During the service life of the plasma torch in aeromagnetic scanning of the cathode section of the arc in the internal tubular electrode, oscillographic recording was carried out of the pulsations of the difference of the potential of the arc between the tubular cathode and the inter-electrode insert (Fig. 10.27*a*). Assuming that the insert has the potential equal to the potential of the section of the hour, situated in the vicinity of the arc or close by, oscillograms were recorded of the pulsations of the difference of the potentials, and of special interest was the nature of pulsations and not the potential difference. Figure 10.28 shows the oscillograms indicating that the range of the variations of the voltage reaches 70÷80 V. This takes place as a result of changes in the length of the section of the arc in the cylindrical electrode in the section with the length Δl_z ~32 mm during the period $\Delta t \sim 0.01$ s.

We shall evaluate some other aspects of this case. The mean speed of displacement of the arc along the electrode $\upsilon_z \sim l_z/\Delta t$ ~3.2 m/s. In this case, the attachment of the arc moves along the helical line with the step $\Delta z_0 = \upsilon_z \cdot \pi d/\upsilon_\varphi$ ~15 mm (at υ_φ = 20 m/s). For

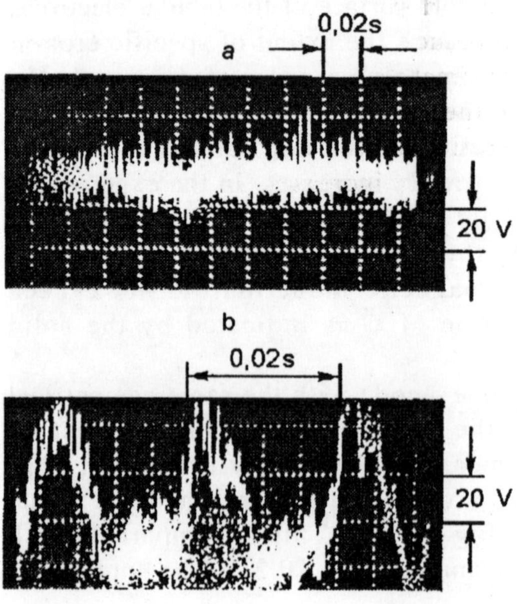

Fig.10.28. Oscillograms of arc voltage in the internal electrode. *a*) without scanning; *b*) with aeromagnetic scanning.

efficient scanning, it is sufficient to ensure that the scanning step is not smaller than the diameter of the arc spot, i.e. $\Delta z_{0min} = d_s$. The minimum frequency of the translational movement in this case is $f=\upsilon_{z\ min}/2\Delta l_z=0.4/(2\cdot32\cdot10^{-3})\sim6$ Hz.

Here $\upsilon_{z\,\min} = \upsilon_\varphi \dfrac{d_s}{\pi d} = 20\cdot\dfrac{2\cdot10^{-3}}{\pi\cdot30\cdot10^{-3}} \sim 0.4\text{m/s}$, $d_s = 2\cdot10^{-3}$ m, $d = 30\cdot10^{-3}$ m. It may be seen that the estimates of the scanning frequency are in satisfactory agreement with the experimental values.

Since the increase of the pressure of the working gas in the channel of the plasma torch reduces the speed of the gas flow whilst maintaining the gas flow rate, it is difficult to influence the attachment of the arc by zero dynamic methods. In this case, it is recommended to use the methods of magnetic scanning [55, 56] in which the axial alternating force effect on arc attachment in the tubular electrode takes place by means of the magnetic fields. They form by means of magnetic lenses, installed along the axis of the electrode and connected, in a specific sequence, to a power source.

Magnetic scanning may be continuous or discrete. In discrete scanning, each of the magnetic lenses is connected alternately. Single-phase, three-phase, etc, connection is possible. The principal special feature of continuous scanning [55] is the presence of the shift angle of the phases equal to $\pi/2$ between the two lenses, and the formation as a result of this of a magnetic wave, carrying out continuous translational movement along the axis and moving away the arc attachment. This results in the most uniform working of the electrode.

Regardless of certain advances in reducing the degree of the specific erosion of the copper tubular cathode, the problem of reducing the this parameter to an even greater extent (by orders of magnitude) has not been solved.

10.9.5. Effect of surface temperature of the copper electrode on specific erosion

Investigations in [68] were carried out on a three-chamber plasma torch with a ledge without magnetic lenses and solenoids (Fig. 10.29*a*). The temperature of the working surface of the wall T_w was regulated by changing the thermal resistance of the three-layer electrode (Fig. 10.29*b*), consisting of the casing *1* with the internal insert *2* and the air interlayer *3* between them. The thermocouple *4* was pressed into the central zone of arc attachment. In one case (internal tubular cathode) the arc spot travels around the circumference through a narrow ring-shaped band and the working surface of this electrode is small, and in the other case (output step electrode) the surface

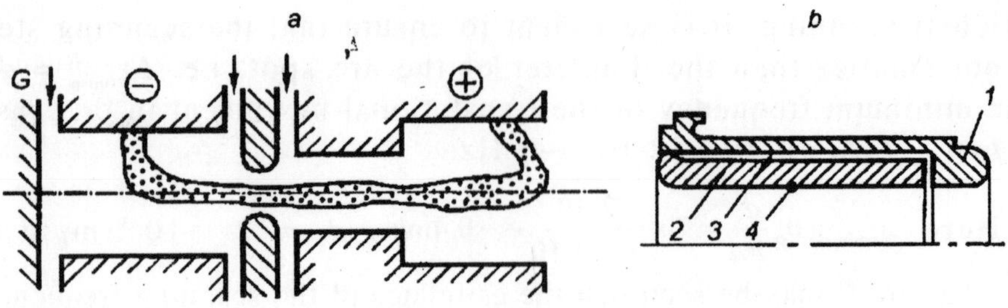

Fig.10.29. Diagram of the three-chamber plasma torch (*a*) and the three-layer electrode (*b*).

of the electrode, 'swept' by the spot, is considerably larger because of large-scale longitudinal shunting of the arc. In the first case, the thermocouple averages-out the wall temperature of the cathode along a narrow band along which the arc spot travels, and in the second case the situation is more complicated. The thermocouple is installed in the centre of the shunting zone whose length, as shown previously, is equal to 2–3 diameters of the channel. Since the longest holding time of the arc spot is found in the centre of the shunting zone, the thermocouple should show the maximum temperature in the selected regime. This is confirmed indirectly by the photograph of the eroded surface of the electrode–anode (Fig. 2.31, chapter 2).

Figure 10.30 shows the experimental results of measurements of the specific erosion of the anode (*1*) and the cathode (2) at a gas pressure of 10^5 Pa and different values of the electrode temperature. When temperature is increased from 100 to 600 °C, the value of $\bar{G}$ is almost doubled. The data for the cathode were recorded at a current of 120 A. Each experimental point was determined in the time equal to 30 min of operation of the plasma torch.

Thus, the search for the methods of intensification of the cooling of the anode walls or methods of preventing the increase of the temperature of the wall, for example, as a result of reducing the rate of formation of microcracks or shortening the time during which the arc spot is in the stationary state, is still important.

10.9.6. Magnetic control of the behaviour of the radial section of the arc in the plasma torch

The rate of erosion of the electrode surface in the arc plasma torch is more or less controlled by the dynamic behaviour of the section of the arc in the vicinity of the area of attachment to the electrode (arc spot) and by the jump-like nature of displacement of the spot

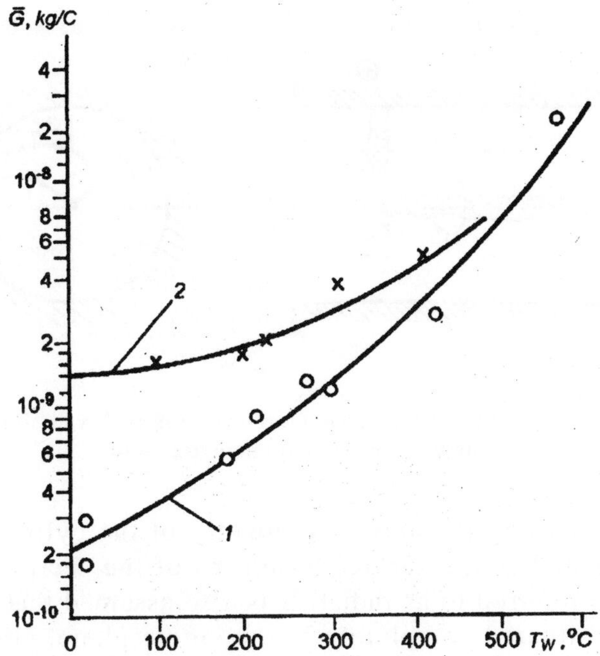

Fig.10.30. Dependence of dependence $\bar{G}$ on the surface temperature of the copper electrode T_w (working gas – air). 1) Step anode, shunting in the output electrode; d_a = 26 mm; I = 300 A; 2) cylindrical cathode; plane of rotation of the arc is stationary. d_c = 26 mm; I = 120 A.

on the electrode surface. It is assumed that the specific force effect on the appropriate section of the arc [55, 56] may remove the shortcomings associated with the natural non-stationary behaviour of the arc spot and, consequently, may result in a large decrease of erosion and increase of the service life of the plasma torch.

Below, we describe an analytical problem of the control of the dynamic behaviour of the radial section of the arc in a cylindrical discharge chamber of a plasma torch with vortex stabilisation of the arc using the external magnetic field. Estimates are obtained for the scale of the field (several tens of gauss for selected characteristic parameters of the plasma torch) and its spatial distribution, ensuring the uniform rotation of the investigated section of the arc in the form of a radial 'wheel spoke'. The mechanism of ensuring continuous movement of the arc spot on the surface of the electrode is discussed.

The relevant elements of the plasma torch are shown schematically in Fig. 10.31a. It has been assumed that the stabilised arc is straight

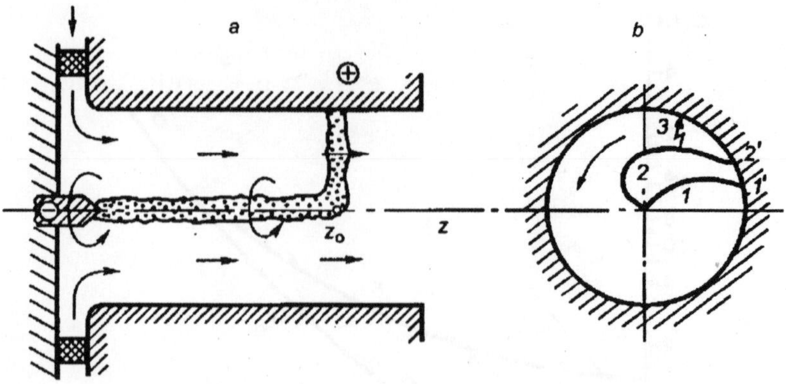

Fig.10.31. Schematic image of the plasma torch (*a*) and the dynamics of the radius section of the arc (*b*) in the absence of the magnetic field.

and is situated along the axis of symmetry of the cylindrical anode and its connected to the surface by means of the section of the arc which will be referred to as radial. It is also assumed that the section of the arc is situated completely in the single plane, normal to the axis of symmetry (in Fig. 10.31*a* it is denoted as the plane $z = z_0$). Of course, because of the rotation of the gas the 'radial' section cannot be purely radial, i.e. rotation of the arc prevents the appropriate section of the arc from bending and stretching, and the arc occupies in two consecutive moments of time the positions indicated schematically in Fig. 10.30*b* the numbers *1* and *2*. In principle, in transfer of the arc from position *1* into position *2*, the spot of the arc also slightly moves (points *1*', *2*') because of the operation of the diffusion and locally collective mechanisms of displacement of the spot. However, the relative elongation of the arc is important. If the arc length is sufficiently large, there is a difference of the potential between the electrode and the appropriate point of the arc, capable of breaking the relatively heated gas gap (in the figure, the breakdown is indicated by number *3*). The arc spot moves in a jump into a new position, and the evolution of the arc starts again.

This schematically described 'jumping' and bending (in the plane $z = z_0$) section of the arc is associated with the well-known shortcomings, and overcoming the shortcomings is also the subject of magnetic control examined in the study. Firstly, because of the existence of the cold boundary layer on the electrode, the arc spot may travel in a continuous manner on the surface of the electrode only at a low speed. This speed of the diffusion and locally convective origin is not associated in any way with the speed of rotation of the core of the flow in the channel and leaves unavoidably to the mechanism of displacement of the spot by means of successive shunting. This circumstance, associated with the delays of the arc spot in a single area, results in the melting of the electrode and erosion of the electrode surface. Secondly, the presence of the radial section of the electrical arc, which is almost parallel to the electrode surface at a small distance from the surface, results in the formation of an additional heat flow to the electrode and this is also undesirable.

To overcome these shortcomings, it is proposed to use the appropriately distributed external magnetic field to ensure that under the effect of the field, the radial section of the arc transforms into a uniformly rotating purely radial 'wheel spoke' whose end slides continuously on the surface of the electrode.

For the approximate solution of the problem of determination of the magnetic field, essential for maintaining the spoke-like form of the arc, it is proposed to:

1. The longitudinal component of the speed of the flow in the chamber of the plasma torch in the plane $z = z_1$, in which the investigated radial section of the arc is established, is equal to zero, i.e. $v_z|_{z=z_0}=0$;

2. The rotational movement of the gas in this plane is given by the known distribution of the circumferential component of the speed $v_\varphi(r)$ (here and in the rest of the book r, φ, z are the cylindrical coordinates, v, v_z, B_z are the appropriate components of the vector fields v and B);

3. The force acting on the element of the arc from the side of the gas flow may be assumed to be equal to the force of resistance of the transverse flow around the solid cylinder of the appropriate diameter in the investigated flow.

Thus, it is assumed that under the effect of the forces from the side of the flow and electromagnetic volume forces, the arc channel has the form of a wheel spoke and rotates with the angular velocity. For this purpose, the external magnetic field in the plane $z = z_0$ should have the z-component, which depends on r, i.e. $B|_{z=z_0}=B(r)e_z$, e_z is the unit vector on the axis z. To determine the required distribution $B(r) = B_z(r)$ the transfer to a non-inertial system, associated with the rotating arc. In the system, the radial arc channel is stationary. The inertia forces are radial and equalised by the pressure gradient. Consequently, to ensure that every element of the arc with the length dr is in the rest condition, the total φ-component of the force, acting on this element, should be equal to zero:

$$dF_\varphi + df_\varphi = 0 . \tag{10.1}$$

Here $dF_\varphi = C_d(\mathrm{Re})\, d(1/2)\rho\, (\upsilon_\varphi - \Omega r)^2\, \mathrm{Sign}\, (\upsilon_\varphi - \Omega r)\, dr$ and the forces of hydrodynamic resistance, acting on the element dr; $df_\varphi = I_r B_z(r)\, dr$ is the force from the side of the magnetic field. In these equations, C_d (Re) is the drag coefficient which depends on the Reynolds number; d is the diameter of the plasma channel; $I_r = -I$, I is the arc current intensity, ρ is the density of the external flow; $(\upsilon_\varphi - \Omega r)$ is the speed of the gas in relation to the plasma channel. It should be mentioned that in the case of a relatively strong applied field B, the speed of displacement of the arc is itself determined by the strength of the field and, consequently, the drag coefficient C_d should not be regarded as the known value independent of B. Consequently, it is assumed (see study [71]) that $C_d = C_d(B)$, and we can determine empirically the given dependence and, on this basis, determine the speed of the arc in the field B. In this study, we examine the inverse case, when the field B is used to ensure movement with the given speed and, consequently, the coefficient C_d is regarded as given.

From the condition (10.1) we obtain the required distribution of the field B_z ensuring the rotation of the arc with the angular velocity in the form of a 'wheel spoke':

$$B_z(r) = \frac{d}{I} C_d(\mathrm{Re}) \frac{\rho}{2} \left[\upsilon_f(r) - \Omega r\right]^2 \mathrm{Sign}\left(\upsilon_f - \Omega r\right). \qquad (10.2)$$

Initially, we are interested in the order of the value of the required field. To estimate the strength, we accept the following values for the characteristic parameters, which determine the required quantity:

$$\upsilon_{\max} = 20\,\mathrm{m/s},\ I = 300\,\mathrm{A},\ \mathrm{d} = 4 \cdot 10^{-3}\,\mathrm{m},$$

$$\rho = 1\ \mathrm{kg/m^3},\ \nu = 1.5 \cdot 10^{-5}\ \mathrm{m^2/s}$$

Here $\upsilon_{\max}$ is the maximum speed of rotation of the gas; ν is the kinematic viscosity of the flow.

The drag coefficient to the flow around the cylinder depends on the Reynolds number and, as indicated by [72], it differs only slightly from unity in a wide range of Re from 10^2 to 10^5 C_d. For the examined parameters:

$$\mathrm{Re} = \frac{\upsilon_{\max} \cdot d}{\nu} = 5.3 \cdot 10^3$$

and, therefore, it maybe assumed that $C_d = 1$. From equation (10.2) we obtain an estimate for the strength of the required field:

$$B_0 = \frac{d}{I} C_d(\mathrm{Re}) \frac{\rho}{2} \upsilon_0^2 \qquad (10.3)$$

Representing υ_0 by the value υ_{max} = 20 m/s, we have B_0 = 27 G. It may be seen that the very weak magnetic field is capable of competing with dynamic forces; consequently, the search in the direction of magnetic control is highly promising.

We now turn to function $B_z(r)$. The density of the gas ρ, included in equation (10.2), is assumed in the first approximation to be constant in the radius in the vicinity of the electrode surface where the required magnetic field is also concentrated because here the gradients of the gas temperature are small, and the pressure gradient, determined by rotation, is also small because the Mach number is low. Consequently, the distribution $B_z(r)$ is determined by the profile of the speed of rotational movement of the gas $\upsilon_\varphi(r)$ and by the angular velocity Ω of rotation of the 'wheel spoke'. The rotational movement may be represented in the form of quasi-solid rotation with the angular velocity in the core of the flow and the boundary layer at the electrode surface, and Ω is a free parameter; it may be determined on the basis of design considerations. Depending on the ratio Ω/ω, the distribution of the required magnetic field along the radius will have a specific type. In a partial case $\Omega = \omega$, i.e. when the 'wheel spoke' rotates together with the core of the flow, the magnetic field should be concentrated in the zone of the boundary layer in the vicinity of the electrode surface (Fig. 10.32). Here the graph *a* shows the functions $\upsilon_\varphi(r)$ and the quantity $|\upsilon_\varphi(r) - \Omega r|$, included in (10.2) (vertical hatching), and the graph *b* shows schematically the required distribution of the *z*-component of the magnetic field along the radius. The maximum value of the field B_* (the field on the internal surface of the anode) is determined from equation (10.3) in which υ_0^2 is represented by the quantity $(\omega R)^2 = (\Omega R)^2$, which is slightly higher than υ^2_{max} (here R is the internal radius of the anode). It is clear that B^* has the order of magnitude of B_0 determined previously. For all other cases $\Omega/\omega <$ (or $>$) 1, and also $\Omega/\omega < 0$ (the arc rotates in the direction opposite to the rotation of the gas), the required field is far from being homogeneous in the radius, and in all cases B_z on the axis of symmetry is equal to zero. (This is the principal moment. In the presence of a strong magnetic field on the axis of symmetry, the regime of the rotating 'wheel spoke' cannot be realised. Instead of this, the radial section of the arc will carry out non-regular oscillations with the amplitude increasing with increase of the strength of the field). Consequently, the magnetic system, required for the generation of the field, differs from the conventional solenoid. Without paying special attention to this problem, we note that the fields of several tens of gauss with the given distri-

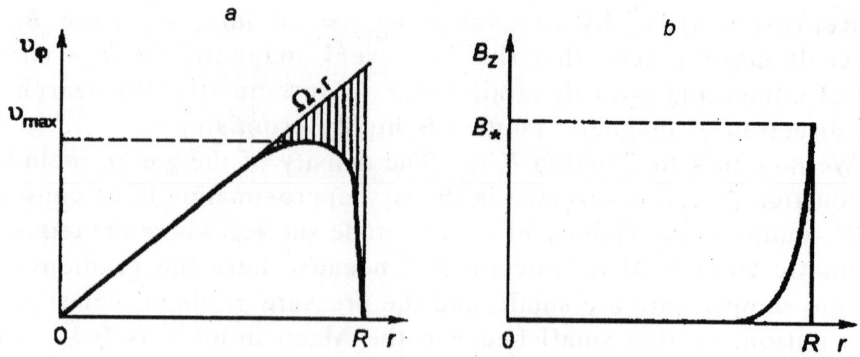

Fig.10.32. Qualitative behaviour of the function $\upsilon\varphi(r)$, $|\upsilon\varphi\ (r) - \Omega r|$ (*a*) and the distribution of the magnetic field B_z (*r*) for the partial case $\Omega = \omega$ (*b*).

bution $B_z(r)$ in the fixed plane z = const may be generated by means of the simplest iron-free magnets in the form of a short solenoid with the specific distribution of the current density along the length of the solenoid. In particular, the distribution $B_z(r)$, shown schematically in Fig. 10.32, can be realised by means of only three circular turns, with the total current equal to 0.

The above considerations show that the problem of maintaining the radial section of the arc in the form of a straight 'wheel spoke' may be solved by relatively simple means. However, it must be ensured that the described magnetic field also shows the mechanism of continuous displacement of the arc spot with the required velocity. This problem arises in the connection with the following circumstance. Until now, we discuss the uniform rotation of the radial section of the arc in the form of a straight spoke, bearing in mind that the forces, acting on every elementary section of the arc, are mutually equalised (see equation (10.1)). However, now we must pay attention to the element of the arc in the immediate vicinity of the surface of the anode. In addition to the forces, included in (10.1), the end of the element is subjected to the effect of a non-equalized surface viscosity force from the side of the solid surface of the electrode. This force tends to cause a delay of the investigated element in relation to the uniformly rotating spoke and, consequently, bend the spoke at the electrode. The problem is the further evolution of the bent end of the 'spoke'. In the absence of the magnetic field, the bent end of the arc which grows in length, results in the previously mentioned mechanism of shunting and in a 'jump' of the art spot. However, in the presence of the near-electromagnetic field with the intensity, determined by equation (10.2), the section of the arc, bent at the surface of the electrode is pressed to the electrode restoring the straight form of the arc and pulling the arc spot to the required position.

Thus, the magnetic field, which moves the arc in the form of the spoke, also ensures itself both the stability of the straight form of the arc in relation to the distortions, associated with the delay of the near-electrode sections of the arc, and the intensity of movement of the arc spot on the electrode with the required speed.

The results obtained here may be regarded as the initial material for the formulation of appropriate experiments with the magnetic control of the arc.

10.9.7. Role of oxygen in reducing the operating life of the electrode

The degree of specific erosion of a copper cylindrical electrode is determined to a large extent by the presence of oxygen in the working gas. This is clearly indicated by the curve of the dependence of specific erosion on the operating time of the anode (Fig. 10.33). Commercial nitrogen, supplied at the start of the experiments, is displaced with time by special purity argon. With a decrease of the oxygen content of the gas, supplied into the discharge chamber, the degree of specific erosion of the anode decreases and, in the final analysis, changes by more than an order of magnitude. In pure nitrogen, the extremely low value of specific erosion was recorded for the copper cylindrical anode [31].

In the case of operation of a copper step output anode in a pure oxygen medium [73], the surface of the copper electrode in the zone of arc shunting is characterised by the rapid formation of the film of Cu_2O and CuO which are known as efficient thermal and electric insulators. The erosion of the electrode in oxygen rapidly increases with increasing current in comparison with air and at a current intensity

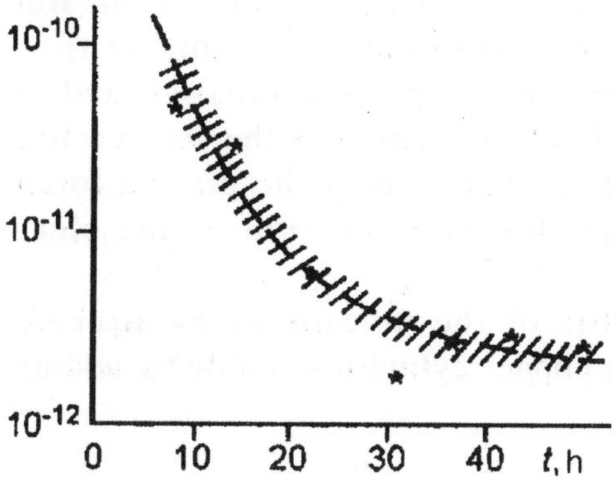

Fig.10.33. Dependence of the specific erosion of a copper output electrode – anode on the operating time of the plasma torch *t*, during which the content of oxygen (in percent) in the working gas – nitrogen decreased. *I* = 180 A.

of I = 700 A reaches the value $\bar{G} = 10^{-9} \div 10^{-8}$ kg/C. It is important to note a certain special feature in the first 40–60 hours of operation of the plasma torch: with the growth of the dense layer of the scale on the surface of the electrode, the time during which the attachment spot of the arc remained stationary increases; consequently, the probability of formation of cavities in the metal also increases and the formation of cavities is accompanied by a rapid increase of the rate of erosion of the anode material and by the transfer of the copper oxides by the gas flow from the arc shunting zone; the appearance of oxides and cavities is responsible for the decrease of the stability of arcing, i.e. the amplitude of oscillations of the mean length of the arc in the output electrode and, consequently, current and voltage increase; the presence of the cavities disrupts the symmetry of the fields of temperature and speed. Regardless of these shortcomings, the linear plasma torch with a step output electrode showed high efficiency in the oxygen medium.

Superpure gases (argon, nitrogen, helium) are relatively expensive. Are there no other methods of obtaining positive results in reducing the specific erosion of the copper anode? We present several experimental data which would make it possible to use a different approach in a number of cases to solving the problem of increasing the service life.

Interesting results were obtained in the operation of a plasma torch with an output electrode–anode with a ledge in which the working gas was a gas–oxygen mixture [74]. In operation with pure oxygen, the anode in the zone of attachment of the arc was characterised, already after operation for 1 h, by the formation of a film of CuO increasing the arc length and also arc voltage up to arc extinction. In order to prevent high rate oxidation of the anode, natural gas was added to oxygen to produce, at the surface of the anode, a medium close to neutral. Natural gas was supplied in the corner behind the ledge with twisting. Already in the case of small additions of natural gas ($K = G_{CH4}/G_{O2} = 0.1 \div 0.15$, where K is the ratio of the volume flow rates), the rate of oxidation of the anode rapidly decreased, and at $K = 0.35 \div 0.4$ oxidation was completely interrupted; the service life of the anode greatly increased; after operation of the plasma torch for 48 hours, there were no visual traces of erosion of the electrode.

We shall examine several results of the investigations directed at reducing the specific erosion of a copper cylindrical anode by adding carbon dioxide to air.

In [77] the results were presented of investigations of the operation of a spraying plasma torch with an inter-electrode insert. It was reported that the addition of hot gas to air (unfortunately, the mass of volume ratio of the gases was not given) rapidly increases the resistance of the anode. In the first 6–10 h of operation at currents up to 300 A, erosion was almost negligible as a result, evidently, of the formation of a graphite (pyrographite) film on the surface of the anode. Photographs of the arc at the end of the anode of the plasma torch indicate the transition from the constricted attachment of the anode spot (in air) to mainly diffusion attachment in the case of the gas mixture. Possibly, there is no diffusion attachment, and there are only numerous simultaneous acts of microshunting of the closing turbulent section of the arc on the high-temperature layer of carbon. In this case, the voltage of the electrical breakdown between the arc and the wall rapidly decreases.

The positive role of carbon, deposited in operation, in reducing the rate of erosion of the copper anode, was reported in [40]. The anode was cylindrical with a ledge; the ratio d_3/d_2 ~1.6. The deposited carbon in the channel of the anode is localised behind a ledge in the form of a thin continuous layer of constant thickness at a distance of (0.5÷0.7) d_3 from the ledge. This is equal to approximately seven times the height of the ledge and this corresponds to the length of the zone of breaking up of the flow behind the ledge; soot may build up in this zone. Further behind the ledge, according to the data presented by the author, the continuous soot layer changes to helical bands (lines) of graphite.

It is well-known that the length of the arc shunting zone behind the breaking up zone is smaller than in smooth cylindrical electrodes and, in addition to this, in this case there is a graphite layer with the temperature higher than in the case of the copper substrate; both factors result in a large decrease of the pulsations of voltage from 100÷120 to 1÷1.5 V with a frequency of 1.5÷2.0 kHz.

The large decrease of specific erosion was recorded in protection with argon of the initial section of the surface of the output copper anode in the zone of holding of the arc spot. The principal diagram of the plasma torch with such an anode section is shown in Fig. 10.34. The graph also gives the dependence $\bar{G} = f(\tilde{G})$, where $\tilde{G} = G_{N_2} \cdot [G_{N_2}+G_{Ar}]^{-1} \cdot 100\%$. Curve *1* was recorded when the mixture of commercial nitrogen and argon ($G_1=G_{N_2}+G_{Ar}$) is supplied only through the main vortex chamber at the end of the cathode, and the curve *2* – in the case of separate supply of the gases: nitrogen was blown through the main chamber, argon through the vortex chamber in front

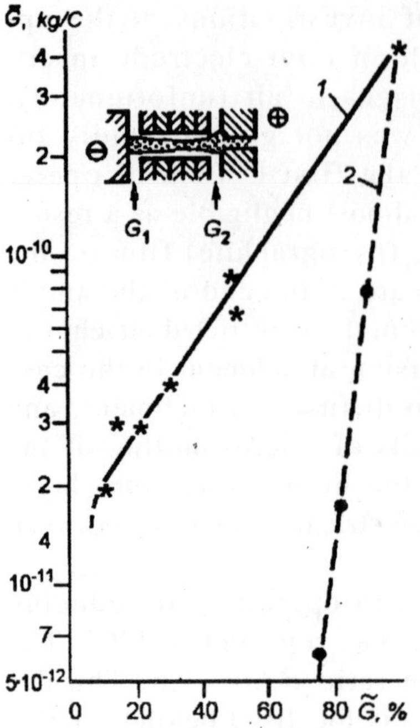

Fig.10.34. Dependence $\bar{G}$ on the parameter $\tilde{G}$, taking into account the content of argon (in percent) supplied for anode protection.

of the anode with the flow rate G_2. In the case of separate supply of the gases, there is sometimes the possibility of the diffusion attachment of the anode end of the arc to the surface of the electrode. However, the mechanism of diffusion attachment has not been proved because the uniform erosion of the surface of the anode may be explained also by other phenomena – simultaneous existence of a large number of microarcs, formed in the process of burning of the turbulent arc in the near-anode space and changing the position in the space with a high-frequency (tens of kilohertz).

Satisfactory results were obtained by replacing the shielding gas argon by propane-butane, with air used as the working gas. Figure 10.35 shows the dependence of the specific erosion of the copper anode $\bar{G}$ (circles) on the flow rate of propane G supplied at the anode. For comparison, the graph also shows the dependence of erosion on the flow rate of argon at the anode (stars) when commercial nitrogen was used as the working gas.

Figure 10.36 shows that the dependence of the specific erosion of the anode on the coordinate of blowing the shielding gas which was propane. The experiment time was up to 1.5 h.

Thus, the protection of the surface of the electrode–anode by pure argon, nitrogen, helium or natural gas reduces anode erosion. In all likelihood, this is based on the prevention of the oxidation of the surface

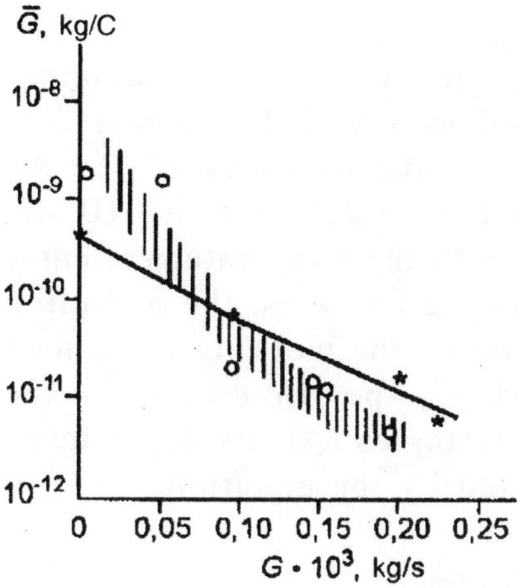

Fig. 10.35. Dependence of $\overline{G}$ on the flow rate of propane G, supplied at the anode (circle); I = 180 A. Stars – shielding gas argon, I = 200 A, working gas – air.

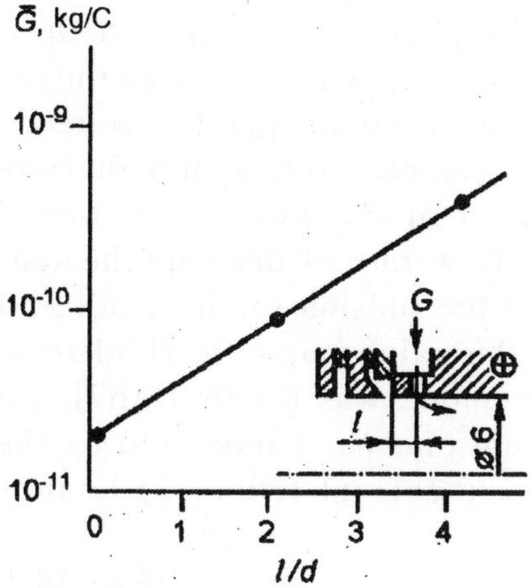

Fig. 10.36. Dependence of $\overline{G}$ of a copper output electrode – anode on the coordinate of blowing the shielding gas – propane. $d_a = 6\cdot10^{-3}$ m; I = 200 A, the flow rate of propane $G = 0.1\cdot10^{-3}$ kg/s, working gas – air.

of copper, i.e. the formation of oxide films or restoration of these films as a result of formation, as in the case of application of natural gas. The latter process may be even more complicated, namely: carbon may be deposited on the surface of the electrode and this may result in most significant consequences – preventing erosion completely.

The results indicate that the effect of oxygen on the extent of specific erosion is very strong. This has also been confirmed by a large number of data obtained by other authors.

10.9.8. Integral characteristic of specific erosion of the output copper tubular anode

General information on the dependence of the specific erosion of a copper anode of the current intensity in different gases (air, hydrogen, nitrogen, steam) is presented in Fig. 10.37 (cross-hatched region *1*- the data for air). The relatively large scatter of the experimental points is associated with a number of factors: the difference in the physical-chemical properties of the working gases, the difference in their properties and structure of the materials of the electrodes and the cooling conditions, the presence of current pulsation, etc. These experiments were carried out on the linear plasma torches with vortex stabilisation of the arc. In this case, the displacement of the radial section of the arc in the space was determined by the circumferential component of the aerodynamic force and by the process of large-scale and small-scale shunting.

The experiments to examine the specific corrosion of a copper anode in steam plasma were conducted in a EDP-193 steam vortex plasma torch with a thermochemical hafnium cathode. The diameter of the working section d_3 behind the ledge was $18 \cdot 10^{-3}$ m, the flow rate of dry superheated steam with the temperature at entry into the plasma torch of 300±50°C was $2.2 \cdot 10^{-3}$ kg/s, the induction of the axial magnetic field in the region of the working section of the anode was 0.026 T [67]. In Fig. 10.37, specific erosion in the steam plasma is indicated by the open triangles (Δ). Its dependence on arc current (curve 2) is approximated by the equation:

$$\bar{G} = 1.78 \cdot 10^{-10 + \frac{I}{670}}, \text{ kg/C}$$

Region 3, delineated by the broken lines, characterises the erosion of the copper cylindrical anode in hydrogen in the range I = 300÷1000 A at the atmospheric pressure. The internal diameter of the anode was varied from $2 \cdot 10^{-2}$ to $4 \cdot 10^{-2}$ m. The magnetic field of the solenoid, placed on the anode, was varied in the range 0.06÷0.1 T. In these working conditions of the plasma torch, the specific erosion of the copper anode was in the range (10^{-10}–10^{-11}) kg/C.

In experiments with the Ar–He mixture, the value of $\bar{G}$ for a step output copper electrode with the diameter of $2.8 \cdot 10^{-2}$ m at I = 1.9÷2.3 kA in the presence of an axial magnetic field was (1.9÷2.5) 10^{-11} kg/C (the cross-hatched rectangle 4 in Fig. 10.37).

In the UMP-6 standard plasma torch with indirect cooling of the copper anode (channel diameter 8 mm) in the nominal working regime (I = 270 A) in commercial nitrogen (oxygen content up to 0.5%), specific erosion of the electrode was $4.5 \cdot 10^{-11}$ kg/C (this value is indicated by the circle 5 in Fig. 10.37). In special purity nitrogen (oxygen content not higher than 0.001%) the value $\bar{G}$ = (2.4÷2.6) $\cdot 10^{-11}$ kg/C, i.e. there was no large decrease of erosion (point 6). It important to note one important fact: in this plasma torch, designed for spraying powders, there is no flow twisting.

In coaxial-axial type plasma torches with a partially displaced arc, resting by its end on the end of the copper anode, when the axial magnetic field is applied to the closing section of the hour, specific erosion at I = 300–600 A (long crosshatched rectangle 7) is almost constant and its average value is (1.5÷2.05 $\cdot 10^{-9}$ kg/C) [75].

The same graph shows several experimental points, indicating the possibility of a further decrease of the specific erosion of the copper anode. For example, in a two-jet plasma torch with axial scanning of the radial section of the arc along the axis of the tubular anode

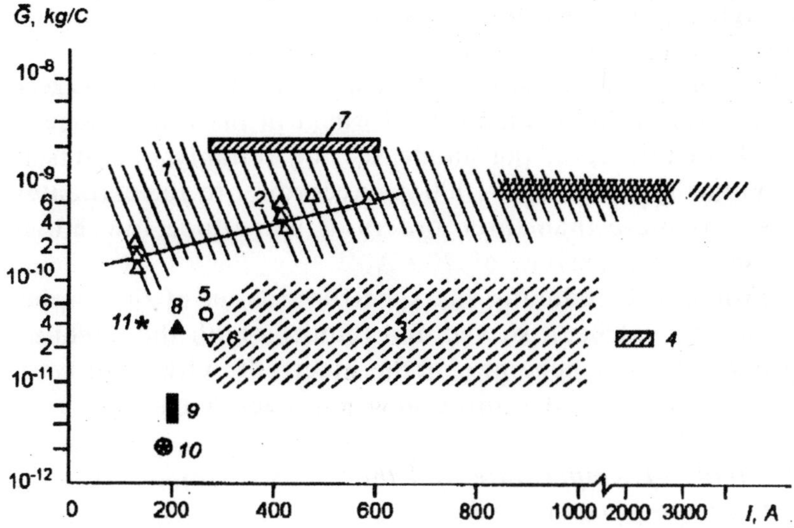

Fig.10.37. Dependence of the specific erosion of the copper anode on current intensity in different gases. 1) air; 2) steam; 3) hydrogen; 4) mixture of Ar and He; 5) commercial nitrogen (O_2 – 0.5%); 6) specialpurity nitrogen (O_2 < 0.001 %); 7) air (coaxial plasma torch); 8) air (internal tubular anode with aeroscanning of arc attachement); 9) air with a gas screen of argon; 10) air with a gas screen of propane – butane; 11) air (anode made of stainless steel).

in the section with the length of $6 \cdot 10^{-2}$ m with a frequency of 5–6 pulsations per second and in the presence of twisting of the airflow, the value of $\bar{G}$ did not exceed $4 \cdot 10^{-11}$ kg/C (solid triangle *8*). It is again important to mention an important factor: there was no large-scale shunting in these conditions.

In shielding the surface of the copper output anode with argon in the zone of holding of the arc spot (as mentioned previously), the value of $\bar{G}$ decreased to $5 \cdot 10^{-12}$ kg/K (solid rectangle *9*).

Satisfactory results were also obtained in replacing the shielding gas argon by propane–butane (the circle with the star inside *10*), using air as the working gas.

Attention should also be given to two experimental results, associated with specific erosion of the copper anode with the stationary anode spot, stabilised by the vortex flow of argon. In [76], the results are presented for the erosion of a copper thin flat wall of the anode with a stationary spot resting on it. At a current of 200 A, continuous operation for 10 h and the optimum wall thickness (approximately 3 mm), the value of $\bar{G}$, according to estimates, did not exceed 10^{-14} kg/C. A decrease of the wall thickness results in the burning-through of the wall at the moment of starting up, and the increase (to more than 5 mm) in the melting.

Even lower specific erosion of the copper anode in argon is recorded if the arc rests on the hemisphere, formed in the flat surface of the anode [3]. In this case, the anode section reliably operates in the current intensity range 200–1000 A. According to the estimates, the value of $\bar{G}$ is more than three orders of magnitude lower than for the flat sheet at a current of 200 A.

The problem of reducing the specific erosion of the copper anode has not as yet been solved and requires both the development of qualitatively new schemes of the electrode with improved erosion characteristics and finding new approaches to service.

10.9.9. Fields of temperature and thermal stresses in the electrode of the plasma torch

The tubular copper anode of the plasma torch is usually referred to as 'cold' because it is rapidly cooled with water. However, the temperature of the internal surface of the anode in the zone of action of the arc spot may reach the melting point ($T_m \approx 1083$°C) and higher, approaching the boiling point (T_{boil} ~2600 °C). Solid material is found below the molten layer. At temperatures close to the melting point, the structure of the metal may be characterised by the occurrence of complicated physical–chemical processes resulting in the formation of pores and cracks, reducing heat conductivity and mechanical strength. The processes, taking place inside the metal and determined by the presence of the alloying elements, modifiers and impurities, are identical with those described in section 10.6. Less extensive damage is found in high-purity metallic materials in the single crystal condition, but the macrostructure of these materials changes at high temperature of the material because of the presence of permanent micro-impurities (in the case of copper, the effect of oxygen and hydrogen is most significant [70, 78]).

In addition to this, an important factor influencing the efficiency of the 'cold' electrode is the alternating stress state of the material resulting from the temperature gradients both along the length and between the internal heated (r_1) and external cooled (r_2) surfaces, and also from the variations of the temperature field in the electrode as a result of the displacement of the arc spot (Fig. 10.38) [69, 78–80]. The displacement of the spot (forced or random) results in most cases in the formation of cracks and the cracking zone–discontinuities in the thickness of the electrode leading mainly to a decrease of the mean heat conductivity and heat-accumulated properties of the internal layers of the material. The result is the increase of the thickness of the liquid film in the zone of action of the arc spot, more extensive removal of the material as a result of evaporation oxidation and a decrease of the efficiency of the material [62].

The characteristics of the thermal stress state of the 'cold' electrode and of the processes of formation of cracks were investigated in [70, 79] by computer modelling. The investigations were based on the well-known physical–mathematical models: Fourier heat conductivity equations, Stefan melting and solidification equations, Hooke equation for the stress state of the material.

On the whole, the task of calculations–temperature T (r, z, φ) [70, 78] and stress $\sigma(r, z, \varphi)$ states of the electrodes is complicated, mathematically adjoint and multiconnected. In addition to this, it is necessary to examine the actual material whose thermophysical and mechanical properties depend on temperature. Without this, it is not possible to obtain reliable data on the characteristics of the processes, taking place in the material, which continuously operates in the conditions of the uniquely wide temperature range and very high temperature gradients.

The special feature of the investigated processes is also the fact that the heat of the electrical arc acts through the arc spot (flow q_s (τ)) on the material of the electrode in the form of pulses and locally: during a short period of holding time τ_0 the arc spot in the 'stationary' state. In addition to this, the mean size of the spot is many times smaller than the size of the internal surface of the electrode affected by the spot (Fig. 10.38). Correspondingly, the fields of temperature T (r, z, φ) and thermal stresses σ (r, z, φ) in the thickness of the electrode are non-stationary and three-dimensional. This circumstance, together with the need to take into account the temperature dependence of the properties of the material, makes it necessary to use the numerical methods of solving systems of differential equations with variable coefficients and other methods of computer modelling.

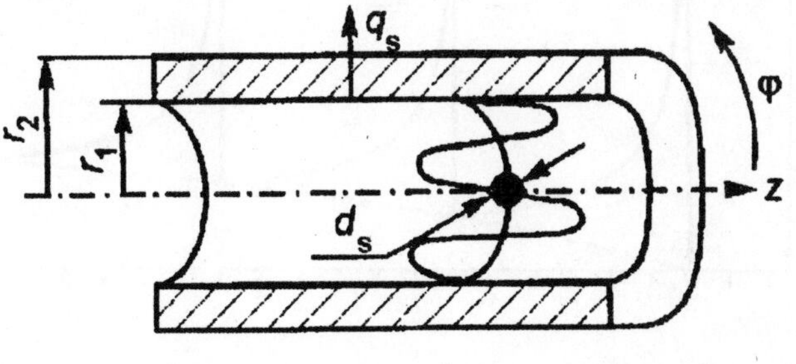

Fig. 10.38. Diagram of the tubular anode of a plasma torch used in calculating the temperature field and the field of thermal stresses.

Investigations of the heat processes in the electrodes of the plasma torches in the pulsed formulation started only recently [65, 69, 78, 80]. In earlier studies attempts were made for the analytical solution of mainly the heat conductivity problem and evaporation of the material [62, etc]. Although the study, described in the section, was started relatively recently [70, 78], and the general complicated nature of the problem requires solution in stages, the results may be used for characterisation of the special features of the fields $T(\tau, r, z, \varphi)$ and $\sigma(\tau, r, z, \varphi)$ in the dynamics. This is of special interest in solving the problem of the operating life of the electrodes.

We examine special features of the temperature field in the conditions of pulsed heating with the arc moving along the closed ring. The graphs shown here illustrate some of the results of the numerical modelling of the non-stationary effect of the act spot on the electrode material. The spatially one-dimensional problem of determination of the temperature field $T(r,\tau)$ was solved. For this problem, $q_s(\tau)$ is the pulsed-periodic heat source (Fig. 10.38). The temperature fields were determined for two heating conditions: *1*–without melting, *2*–with melting. In both cases, the rectangular form of $q_s(\tau)$ was selected. The period of repetition of the pulses of the flow q_s and the duration of its continuous effect τ_0 were determined by the speed of movement of the arc spot w_s and the arc length $2\pi \cdot r_1$. However, for the first regime, the pulse time q_s was restricted by the additional condition in which the values of the temperature of maximum heating T_{1h} of the internal surface did not exceed T_m (Fig. 10.39).

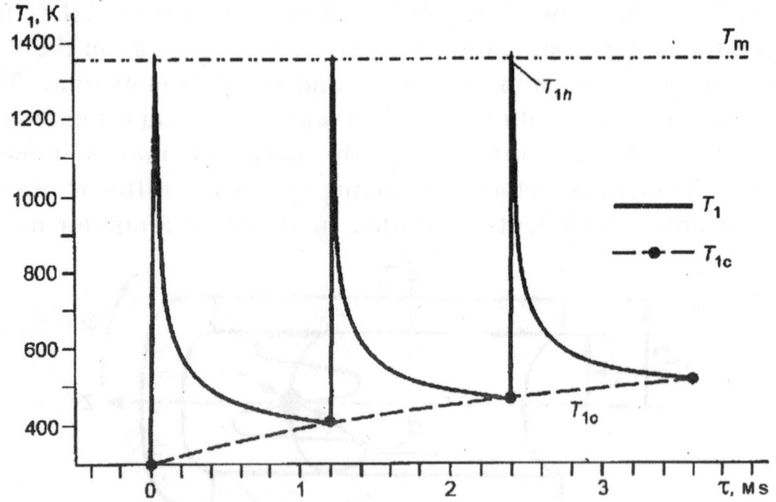

Fig.10.39. Time dependence of the first three modelling 'non-melting'. pulses $T_1(r) \le T_m$; $r_0 \le r_0^{max}$; $q_s = 5 \cdot 10^9$ W/m^2.

During the first rotation of the arc, the starting temperature $T_{1.0}$ was equal to 25°C for all points in the trace. The cooling temperature at each point of the trace $T_{1s} > T_{1.0}$ and increased in a subsequent rotation with general heating of the anode (Fig. 10.39). The upper point of the temperature pulse T_{1h} (in accordance with the given regime *1*) remained constant at equal to T_m and, consequently, the boundary of the solid material ($r = r_s$) was always found at T_m. At the speed of rotation of the arc w_s = 40 m/s, the pulsed heat flow $q_s = 5 \cdot 10^9$ W/m² [79] in the first heating cycle does not lead to melting of copper (the duration of action of the heat pulse $q_s(\tau)$ at the spot d_s = 2 mm does not exceed the critical value $\tau_{0\ max}$ = 50 μs) [62, 79].

In the wall of the anode the temperature pulse $T(\tau)$ transforms to a temperature wave (Fig. 10.40) and rapidly attenuates if the anode is produced from pure defect-free copper with high heat conductivity.

As indicated by Fig. 10.39, T_{1c} rapidly increases and already in the third rotation of the arc, the third actual pulse $q_s = 5 \cdot 10^9$ W/m², τ_0 = 50 μs, not restricted by the condition of maximum heating $T_{1c} < T_m$, starts to melt the wall layer of the copper electrode. However, at all $r > r_s$, where the material remained solid, the form of the wave $T(r, \tau)$ corresponds almost completely to Fig. 10.40. In this case, the liquid/solid interface forms at the radius r_s inside the thickness of the anode. At this interface, the temperature does not exceed T_m.

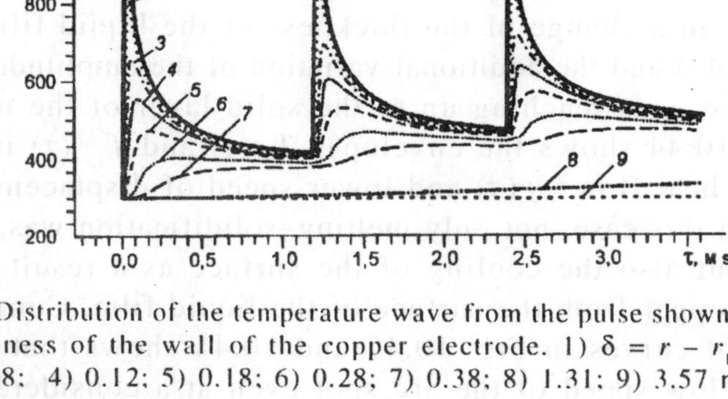

Fig.10.40. Distribution of the temperature wave from the pulse shown in Fig.10.39. in the thickness of the wall of the copper electrode. 1) $\delta = r - r_1$ = 0 mm; 2) 0.04; 3) 0.08; 4) 0.12; 5) 0.18; 6) 0.28; 7) 0.38; 8) 1.31; 9) 3.57 mm.

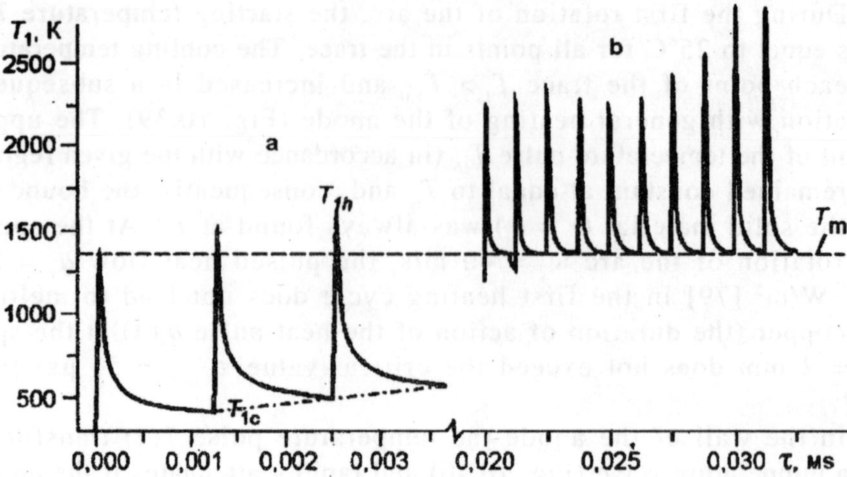

Fig.10.41. Time dependence of simulation melting pulses with the parameters of intensity and duration corresponding to Fig.10.39.

The process of heating and cooling of the internal surface of the anode in the conditions of regime *2* with the temperature higher than T_m is shown in Fig. 10.41. Figure 10.41*a* shows the result of action of three initial rotations of the arc, when T_{1h} only starts to exceed T_m. After the 16th revolution, temperature T_1 of the internal surface of the copper anode becomes lower than T_m only for a very short period of time (a single peak at $T_1 < T_m$), and starting from the 17th revolution, $T_1 > T_m$ (Fig. 10.41*b*). The envelope of the maximum temperatures $T_{1h}(\tau)$ and $T_{1c}(\tau)$ (Fig. 10.42) shows that already after several passages of the arc spot through the examined point on the surface of the electrode, the temperatures $T_{1h} > T_m$ and $T_{1c} > T_m$ appear, consequently, a molten zone forms (liquid film). Thus, in the investigated regime at $q_s = 5 \cdot 10^9$ W/m^2 and $w_s = 40$ m/s the surface of the anode in the circular trace from the effect of the spot remains always liquid. If there was no evaporation of the melt in this case, this would result only in a change of the thickness of the liquid film $\Delta l = r_s - r_1$ (Fig. 10.43) and the additional variation of the amplitude and form of the pulse $q_s(\tau)$ reaching up to the solid layer of the material.

Figure 10.44 shows the envelopes $T_{1h}(\tau)$ and $T_{1c}(\tau)$ in the case of smaller heat flows $q_s(\tau)$ and lower speed of displacement of the act spot. In this case, not only melting–solidification was taken into account, but also the cooling of the surface as a result of evaporation of copper from the surface of the liquid film. Comparison of the pairs of curves in Fig. 10.42 and 10.44 shows that at a comparatively low speed of the arc spot even at a considerably lower

Fig.10.42. Development of a simulation pulse process. Special features of the envelope lines $T_{1h}(r)$ and $T_{1c}(r)$ are related to the formation of the liquid film.

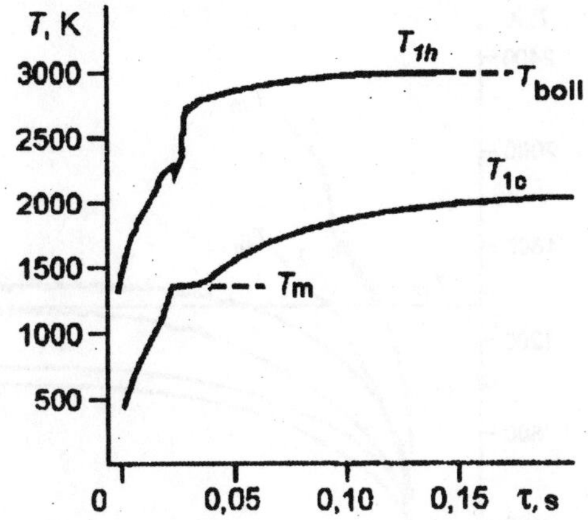

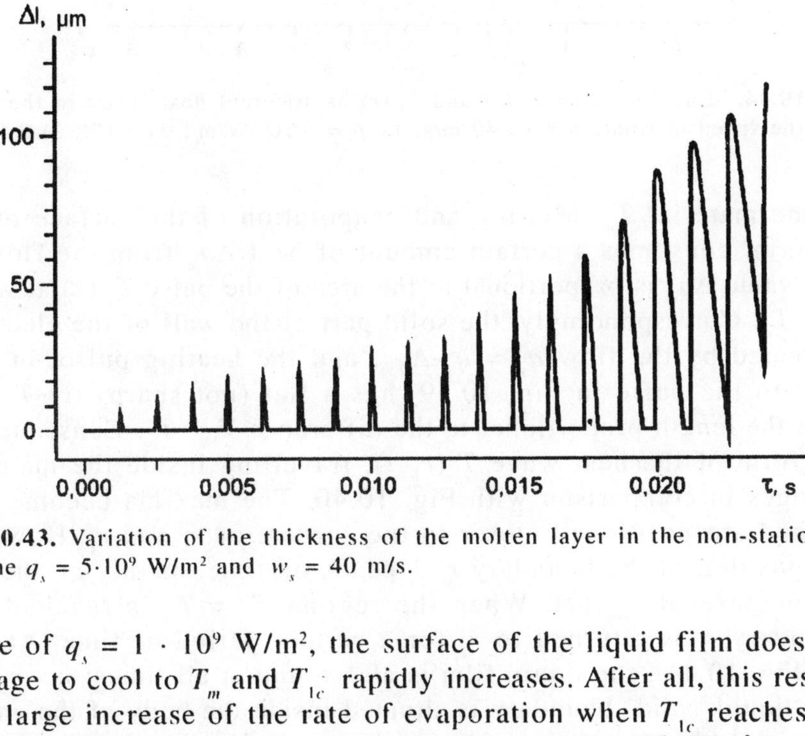

Fig.10.43. Variation of the thickness of the molten layer in the non-stationary regime $q_s = 5 \cdot 10^9$ W/m² and $w_s = 40$ m/s.

value of $q_s = 1 \cdot 10^9$ W/m², the surface of the liquid film does not manage to cool to T_m and T_{1c} rapidly increases. After all, this results in a large increase of the rate of evaporation when T_{1c} reaches the boiling point of copper T_{boil}. For the flow $q_s = 2 \cdot 10^8$ W/m² (curves *3*, Fig. 10.44) melting obviously does not take place but these heat flows are more characteristic of diffusion attachment of the arc in the anode (investigations of this process requires a different formulation of the boundary conditions in the physical–mathematical model of the problem).

As indicated by Fig. 10.41, the 'melting' temperature pulses $T_1(\tau)$ differ from 'non-melting' pulses mainly by the fact that the temperature of the end of heating T_{1h} is higher than the melting point

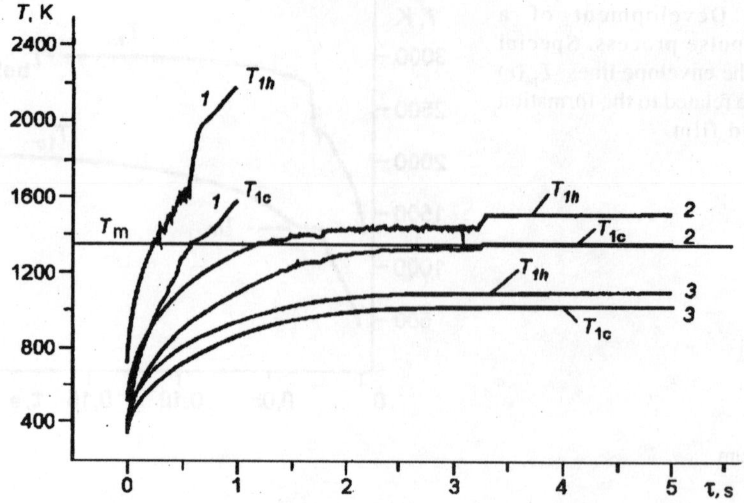

Fig.10.44. Envelope line $T_{1h}(r)$ and $T_{1c}(r)$ at different heat flows in the pulse and the speed of rotation w_s = 40 m/s. 1) $q_s = 1 \cdot 10^9$ W/m²; 2) $5 \cdot 10^8$; 3) $2 \cdot 10^8$.

of the material T_m. Melting and evaporation of the surface of the material consumes a certain amount of heat Δq_m from the flow q_s. The value Δq_m is proportional to the area of the pulse T_1 (τ) at which $T_1 > T_m$. Correspondingly, the solid part of the wall of the electrode is heated by the flow $q_w = q_s - \Delta q_m$, and the heating pulse, in contrast to the pulse in Fig. 10.39, has a flat (not sharp) tip $T = T_m$ with the length proportional to the difference $T_{1h} - T_m$. Consequently, the form of the heat wave T (r, τ), travelling inside the material, changes in comparison with Fig. 10.40. The maxima become flatter and are positioned closer to the surface of r_1 and r_s. However, the position of the boundary r_s depends on time, mainly on the rate of increase of T_{1c} (τ). When the regime $T_{1c} = T_m$ is reached, the boundary r_s is returned to r_1 for a certain period of time. At $T_{1c} > T_m$ (Fig. 10.44), the liquid film is also found at all times: $r_s > r_1$, i.e. the 'liquid–solid' boundary is situated inside the body of the anode. However, this has only a slight effect on the distribution $T(r)$ at $r > r_s$, and the form of the waves $T(r)$ remains almost the same as in Fig. 10.40, only they become wider. Curves *2* in Fig. 10.40 show that the quasi-stationary process may also take place, in which the boundary r_s slightly oscillates and moves continuously inside the wall. In this case, the distribution of temperature $T(r)$ approaches a stationary distribution, whose characteristic feature is the relatively small variations $\pm\Delta T_{1c}$ in a narrow solid zone in the vicinity of the surface r_s, with the gradient up to $dT/dr \approx 1000$ K/mm (Fig.

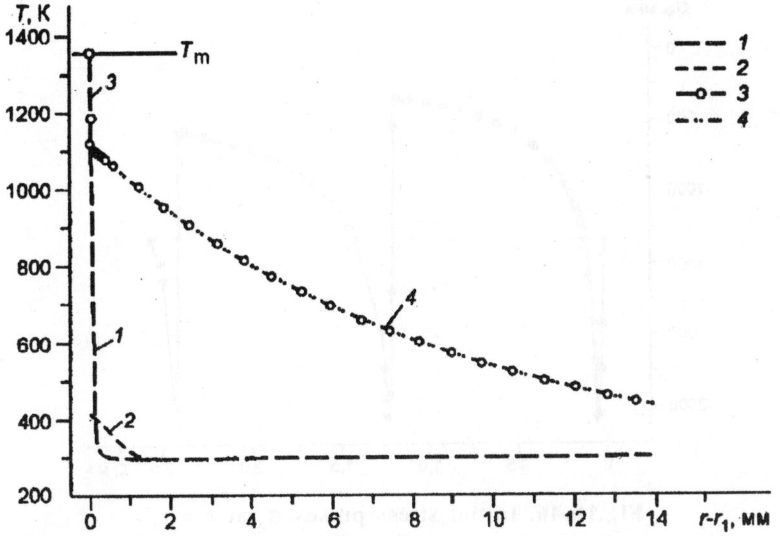

Fig.10.45. Distribution of temperature in the body of the anode. 1) at the end of the front of the first pulse; 2) at the end of the cooling stage after the first heat pulse; 3) at the end of the cooling stage after the pulse in the quasi-stationary regime (for the 'non-melting' pulses); 4) at the end of the front of any pulse in the quasi-stationary regime.

10.45). The spatial areas with such temperature gradient are the areas of the highest temperature stresses.

Figure 10.45 shows the distribution of temperature in the body of the anode. The distribution was used to calculate thermal stresses in the electrode of the plasma torch during movement of the arc spot. The problem of calculating the stresses in the material was solved (in the first stage) in the 'thermoelastic' formulation for a cylindrical thick-wall pipe (the geometrical model of the anode, see Fig. 10.38). It is assumed that irreversible strains do not form in the material and no pores, cracks or shear phenomena appear. (For the volume of the material in the immediate vicinity of the arc spot and subjected to high stresses, the thermoelastic model of the mechanical behaviour of the material may be only the first approximation and a transition to the elastoplastic model is essential). However, already the first results of the calculations of the pulsed process make it possible to draw important conclusions.

For analysis of the main special features of the stress field, investigations were carried out using the one-dimensional solutions $T(\tau)$ and $T(r)$ (Fig. 10.39–10.45). Correspondingly, the thin layer of the material in the vicinity of the surface r_1 is characterised by the occurrence of the pulsed thermoelastic process (Fig. 10.46) changing

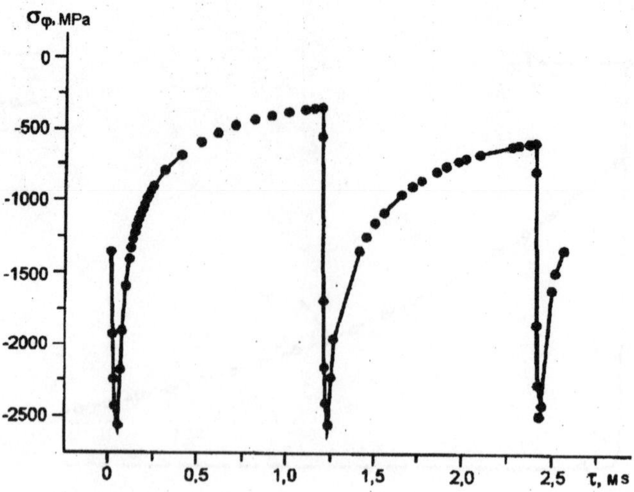

Fig.10.46. Initial stress pulses σ_φ at $r = r_1$.

to the wave process at some depth (as in the case of Fig. 10.40). As indicated by Figs. 10.41–10.44, the amplitude of the variations of temperature in the solid part of the material after the start of the process rapidly decreases because of the increase of T_{1c}. The amplitude of the oscillations of thermomechanical stresses decreases by the same margin.

After the establishment of the quasi-stationary temperature field (Fig. 10.45, curves *3*, *4*) the distribution of the thermal stresses (r) in the anode wall corresponds to $T(r)$. As indicated by Fig. 10.47, in the conditions of the regime 1, i.e., the action of 'non-melting' pulses $T_1(\tau)$, as in Fig. 10.39, 10.40, the normal radial stresses r $\sigma_r(r)$ are always compressive ($\sigma < 0$) but relatively low. The main role in the stress state of the anode is played by the 'circular' stresses $\sigma_\varphi(r)$ and 'axial' stresses $\sigma_z(r)$. The sign of the stresses changes in the thickness of the anode: from compressive stresses $\sigma < 0$ on the hot internal side, to the tensile stresses $\sigma > 0$ in the cold side of the anode. On the hot side, the azimuthal stresses σ_φ, σ_z are approximately three times higher than in the cold side.

The assumptions of mechanical failure of the material the electrode follow from the comparison of the acting σ_r, σ_φ, σ_z and fracturing stresses σ_B, σ_c (Fig. 10.47, 10.48). In particular, it is important to ensure that the level of the azimuthal stresses in the vicinity of the internal surface of the copper anode is close to the handbook value of the ultimate compression strength of copper (σ_c). Unfortunately, the reference literature for the mechanical properties of copper does

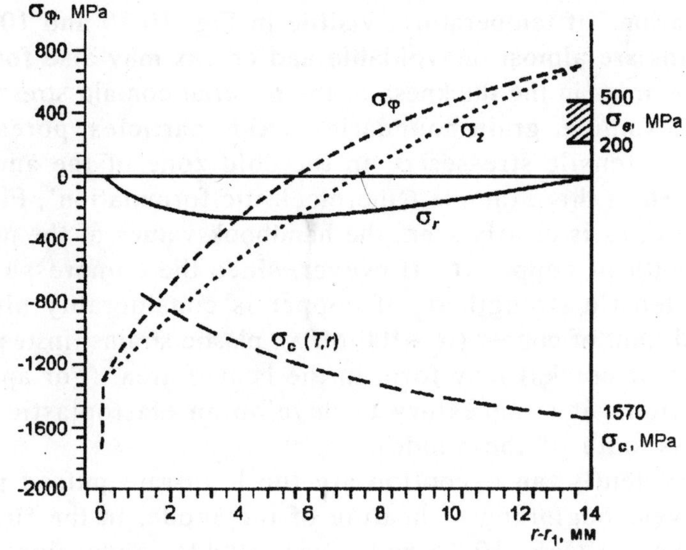

Fig.10.47. Distribution of stresses in the quasi-stationary regime (see Fig.10.45, curve *4*).

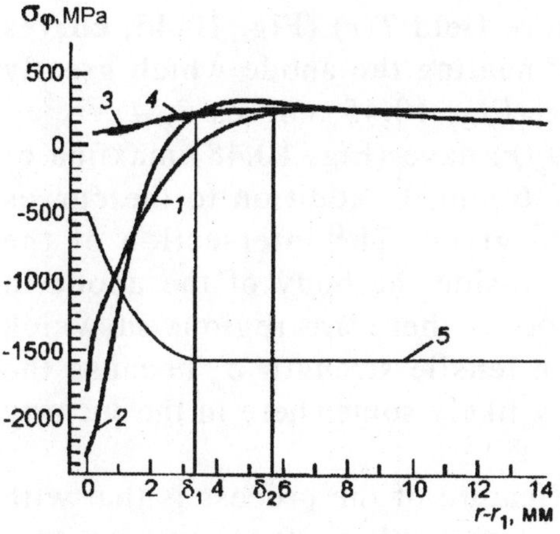

Fig.10.48. Initial (at the moment of time, when $\sigma_\varphi(r) \geq \sigma_B$) distribution of stresses σ_φ in comparison with the distributions of strength σ_B, σ_c. 1) for the regime without melting, Fig.10.39; 2) with melting of copper; 3,4) distribution of σ_B for the first and second regime; 5) distribution of σ_c.

not contain data on the temperature dependence of σ_c; it is only possible to assume that, for example, σ_c decreases in heating in approximately the same manner as the Brinell number. Using this analogy, Fig. 10.47 shows the curve $\sigma_c(T, r)$ describing the quasi-stationary distribution $T(r)$ in Fig. 10.45 (curve *4*). Analysis of the curves $\sigma_\varphi(r)$, $\sigma_z(r)$ and $\sigma_c(r)$ in Fig. 10.47 makes it possible to assume that formation of shear strains is possible somewhere below the spot, at a depth of approximately δ~1 mm from the internal surface of the anode, even in the stationary regime of operation of the anode. In the zone of variations of temperature, visible in Fig. 10.39 and 10.45, the shear strains are almost unavoidable and cracks may also form, especially if the areas in the thickness of the material contain stress concentrators (for example, grain boundaries, oxide particles, pores).

The tensile stresses σ_φ in the cold zone of the anode wall, calculated in this example ('thermoelastic formulation', Fig. 10.48) may exceed, as is clearly seen, the handbook values of the ultimate tensile strength of copper σ_φ. However, since the compression strength σ_c and tensile strength σ_B of copper is considerably higher than the yield limit of copper (δ_T ~100 MPa), plastic strains (instead of formation of shear cracks) may form in the heated area. (For appropriate calculations, it is necessary to develop an elastoplastic model of the stress state of the anode).

Evidently, an exception are the high-rate pulsed phenomena at the very beginning of heating of the anode, in the first revolutions of the arc (Fig. 10.39 and 10.41–10.43). According to the results of the calculations (Fig. 10.48), the distributions $\sigma_\varphi(r)$ for the first several tens of revolutions (losses) are of different nature in comparison with the quasi-stationary process (Fig. 10.47). This is associated with a different nature of the temperature field $T(r)$ (Fig. 10.45, curves *1*, *2*) at the start of the process of heating the anode which greatly differs from the stationary regime (Fig. 10.45, curves *3*, *4*).

The appropriate distributions $\sigma_\varphi(r)$ have (Fig. 10.48) maxima of the tensile stresses in the range 3–6 mm. In addition to the curves $\sigma_\varphi(r)$, the curves $\sigma_B(T, r)$ are also given. The intersection of the curves $\sigma_\varphi(r)$ and $\sigma_B(r)$ shows that inside the body of the anode in the initial stage of the pulsed process there are regions in which the tensile stresses are higher than tensile strength σ_B because the formation of tensile cracks is highly likely somewhere in the vicinity of the internal surface (region δ_2–δ_1).

However, the important special feature of the process is that with heating of the anode, i.e. the increase of T_{1c} (Fig. 10.41) and straightening of $T(r)$, the tensile maxima on the curves $\sigma_\varphi(r)$ decrease and disappear, and the intersection of the curves $\sigma_\varphi(r)$ and $\sigma_B(r)$ is displaced to the cold part of the anode wall in the direction of the cooled surface. After all, the situation becomes approximately the same as the one shown in Fig. 10.47 with the quasi-stationary temperature distribution in the body of the anode (Fig. 10.45, curves *3*, *4*).

Both distributions (Fig. 10.47 and 10.45) still require clarification. But the main features of the processes are represented quite accurately owing to the fact that the temperature dependence of the properties was initially specified in the problem. Therefore, it is necessary to draw conclusions regarding the course of the quasi-stationary process, when the internal surface of the anode melts and evaporates (Fig. 10.41–10.44).

When the solid part of the anode ($r > r_s$) is subjected to the effect of the heat pulse $q_w = q_s - \Delta q_m$, which passed through the liquid film, its initial form greatly differs from rectangular: it has the form of a wave similar to the form of the temperature waves inside the material (Fig. 10.40) at $r > r_1$. (Similar but greater changes take place as a result of the formation of the oxide film on the hot surface of the copper electrode). Differences are represented by the smaller curvature of the front and the decrease of the wave in comparison with the 'sharp-tip' pulse (Fig. 10.39), and also by the lower value of the initial amplitude $T_m - T_{1c}$. Therefore, on the distributions $\sigma_\varphi(r)$ and $\sigma_z(r)$, the maximum of tensile stresses in the first pulses with melting initially increases (Fig. 10.48, curve 2) but with heating to the stationary state the maximum disappears and the intersection of the curves $\sigma_{\varphi,z}$ (r) and $\sigma_B(T, r)$ is displaced to the cold zone and new cracks cannot form. However, the cracks formed in the initial pulses (as in Fig. 10.40) can no longer disappear. In this case, generally speaking, the material of the anode is not continuous (sound) and the distributions $T(r)$ have steps [70]. The distributions $\sigma_r(r)$, $\sigma_\varphi(r)$ and $\sigma_z(r)$ of the radial, axial and circumferential stresses, change correspondingly. In the stage of investigations of the thermally stressed state of the anode, these phenomena have not as yet been examined. The main conclusion from the already completed investigations is the one following from the results presented in Fig. 10.47 and 10.48: the cracks in the subsurface layers of the copper anode form in all likelihood in the period when heating of the anode only started as a result of the direct effect of the electric arc and in subsequent stages they do not appear; possibly, they result in the formation of new cracks as a result of significant changes in the structure of the material and in temperature distributions.

10.9.10. Structure of the material of the subsurface layer of a tubular electrode

We examine special features of the structure of the subsurface layer of a copper tubular electrode subjected to the cyclic effect of high specific heat flows from the side of the arc spot moving along the surface.

Investigations were carried out in a two-jet plasma torch so that it was possible to examine erosion of both cathode and anode sections in the same conditions. The radial section of the arc 'rotated' around the axis of the tubular electrode and also carried out translational movement along the axis (in relation to some plane) with a frequency of 4–5 pulsations per second under the effect of the corresponding aerodynamic forces.

In the displacement of the arc spot on the surface of the electrode, the latter is subjected to cyclic thermal shocks and is characterised by the formation of structural defects (cracks). In long-term service, this results in mechanical failure and a decrease of heat and electrical conductivity.

Microsections of the meridional section of polycrystalline copper cathodes and anodes, working under the effect of the arc spot for several tens of hours, indicate the formation of a high-density network of cracks over a depth of approximately 2 mm and mechanical failure of the electrode material in the thin layer of the working surface of the electrode.

Figure 10.49 shows another fragment which has not separated from the electrode, i.e. 'detachment'. The cross-section of the fragment contains more than 10 grains. Structural changes are especially clearly evident on the cathode. The longest cracks form at a depth of the order of 1.6÷1.8 mm from the working surface, and not on the working surface. Figure 10.50 shows isolated cracks at a depth of 300 μm from the electrode surface. The final stage of failure is delayed by easy stress relaxation in the high-temperature subsurface layer. With time, the process of build up of dislocations in the radial direction is completed and this is followed by stationary erosion. The rate of this process is many times higher than that of the initial process because

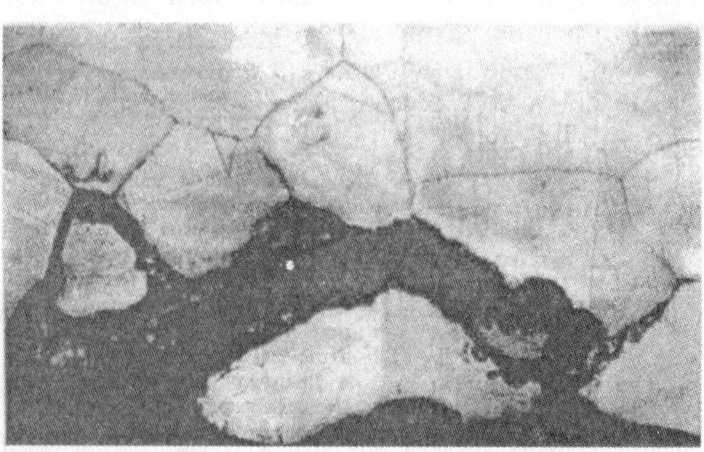

Fig.10.49. Cathode made of polycrystalline copper; 'detachment of fragment' at the surface (×200).

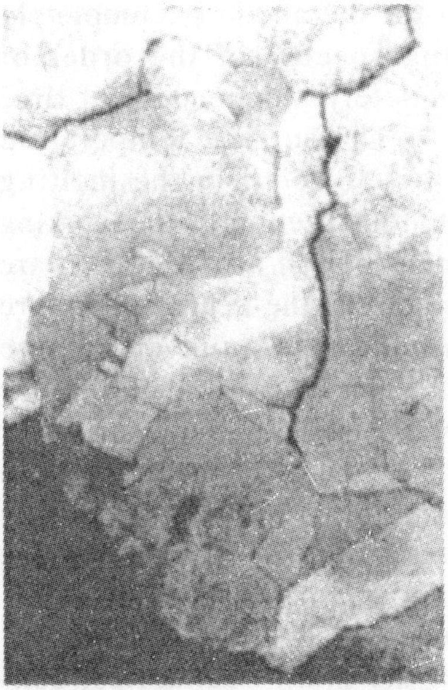

Fig.10.50. Distinctive isolated crack at the depth of 300 μm from the working surface of the electrode

the cracked subsurface layer is characterised by a considerably lower heat conductivity and this results in an increase of both the surface temperature of the electrode and the rate of erosion.

We examine the situation in greater detail. We return to the case in which the arc spot moves along a closed band. As shown by the experiments and the calculations, presented in the previous sections, in this case, the temperature of copper within the limits of the band reaches the melting point for both the cathode and the anode already after several rotations of the arc spot. It is natural to expect the same values of the specific erosion of both electrodes and this was also confirmed by the experiments (Fig. 10.21). In the first minutes of burning of the electrical arc, the value of $\overline{G}$ is relatively low because the structure of the material is not yet damaged. With time, the crack formation process is completed and the 'stationary' operating regime of the electrode starts to operate, and the specific erosion reaches the limiting value. The copper material in the zone of the band is in the molten state and, in all likelihood, its properties are identical for both cases.

What can be expected as a result of the introduction of additional axial scanning of the radial section of the arc? For the cathode, the dependence of $\overline{G}$ on the operating time of the arc remains constant (Fig. 10.23) because to ensure the required electronic emission from the surface, the temperature should not be lower than the melting point.

The nature of the dependence $\bar{G} = f(t)$ for the anode is completely different. Already at low axial scanning speeds (of the order of 10^{-2} m/s), specific erosion is almost completely independent of time and decreases by almost a factor of 1.5. The surface temperature in the zone of the effect of the arc spot is lower than the melting point of copper. The contours of the longitudinal section of the working electrodes, which operated for approximately 60 h at I = 200 A, are shown in Fig. 10.24. In the case of the anode (*a*) the degree of failure is small, difficult to see, and in the cathode it is quite distinctive (*b*). It should be mentioned that in the zone of action of the arc spot, the low-melting impurities in copper (lead, bismuth) transfer to the molten condition. This greatly reduces the mechanical strength of bonding of the individual grains because in the process of melting the impurities change the volume (expand), and the position of the individual grains also changes. In displacement of the spot of attachment of the arc, shrinkage takes place in the cooling and solidifying inclusions and cracks appear at the grain boundaries; consequently, the heat conductivity of metal decreases, and in subsequent arrival of the attachment spot of the arc in the same area the copper grains may separate and, consequently, the erosion rate may increase.

The appearance of the melting zone under the attachment spot of the arc is also the reason for the increase of the duration of holding of the attachment in the stationary condition. It is promising to use composite materials characterised by high stability under thermal loading, including fibre-reinforced composites, and also materials containing inclusions of ultrafine powders with the grain size smaller than 1 μm [81].

The increase of the dispersion and homogeneity of both the inclusions and the base material of the matrix results in improvement of the uniformity of distribution of current of the surface of the electrode and, consequently, in a decrease of the mean current density, a decrease of the extent of erosion failure of the surface and in more uniform wear of the surface.

Experimental verification of the work of the electrodes, produced from copper single crystals, shows that in this case, the body of the cathode does not contain any cracks. There are only spot-type heterogeneities at a distance of 0.5–0.7 mm from the surface. In all likelihood, these are gas cavities.

Thus, the main reason for the formation and development of the of the mechanism of failure of the electrodes are the thermal stresses formed as a result of the steep radial and axial temperature gradients in the electrode and also oxidation of the electrode at the grain boundaries. The processes of formation of dislocations are most intensive in the body of the cathode.

10.9.11. Methods of reducing the erosion rate of copper tubular electrodes

With increase of the dispersion and homogeneity of the structural components of the metal, the physical and thermal mechanical properties of the metal improve. The limiting case is the single crystal which does not contain any structural of chemical heterogeneities. This has a strong effect on the increase of the service characteristics of the electrode produced from such a material.

Taking into account the importance of the problem of increasing the service life of the electrodes, it is convenient to examine in greater detail a new method of improving the structure of polycrystalline metal. As shown by a large number of investigations aimed to find the methods of increasing the dispersion and homogeneity of the material, the required results may be obtained by adding to the metal melt ultrafine powders (UFP) with the grain size smaller than 0.1 μm in the amount of 0.01÷0.05 wt%. This will be examined on the example of using ultrafine powders as modifiers in steels, cast iron and aluminium alloys, because investigations of this type with copper specimens are still in the initial stage, despite the fact that good results are expected. It should be mentioned that for a number of plasma–chemical processes, based on the application of, for example, carbon dioxide, the tubular cathode is sometimes produced from cast iron.

The introduction of the ultrafine powders is aimed at ensuring the resistance of the material to the long-term effect of high temperatures, mechanical loading and chemically active media. It is well-known that the quality of metal in the cast condition is controlled by its primary structure [82, 83]; the control of the structure only by the thermal physical effect on the solidification processes does not ensure the required properties of the material. However, the application of expensive alloying elements for improving quality is not always economically justified.

Recently, special methods have been developed for the preparation introduction into the melt of modifying additions based on ultrafine particles a refractory compounds. One of the promising methods of producing at her dispersed powders is plasma synthesis. This process is carried out in the flow of nitrogen, argon or helium plasma at 5000÷8000 K and with a steep axial temperature gradient. Under the effect of high-temperature the initial condensed substance is at transferred to the vapour state. The reactions of formation of the target progress take place in the gas phase with high rates, and the specific conditions of condensation enabled them to be produced in the form of ultrafine chemically pure powders [84].

We describe some of the results of laboratory and pilot plant investigations characterising the effect of the ultrafine powders on the properties of cast metal [83, 84]. Modification particles were prepared using ultrafine powders of refractory compounds–Ti(CN), Nb(CN), SiC, etc, synthesised by the plasma–chemical method, with the dispersion (estimated on the bases of the specific surface) from 10 to 100 m^2/g. Analysis by electron microscopy showed that the mean size of the particles of Ti(CN) is approximately 0 .05 μm, and the dispersion limits are in the range 0.01÷0.10 μm. The ultrafine particles were subjected to vacuum degassing and solid-phase activation and were subsequently pressed into briquettes using a protecting substance.

Taking into account the fact that the superheating of the melts is accompanied by jump-like changes of their structure-sensitive properties (electrical conductivity, viscosity, etc) in the ranges 1600÷1650 and 1780÷1830 °C, referred to as the first and second critical points, investigations were carried out into the effect of the temperature of addition of the modifiers on the efficiency of modification of the alloys. Efficiency was estimated on the bases of the degree of refining of the macrograins and also on the basis of the morphology and topography of the carbide phase.

The results obtained for the experimental melts with sampling of reference samples at different temperatures shows that in the case of a ZhS-6K nickel alloy at 1400÷1600 °C, the conditions of nucleation support the formation of the cast structure with equiaxed grains. In the specimens of this alloy, the grain size after the addition of a ultrafine powder modifier Ti(CN) at any temperature in the range 1400÷1600 °C was approximately 4 times smaller in comparison with the non-modified alloy. Superheating this alloy, modified at 1400°C, increases the grain size by the rate of increase of the grain size is half the rate in the case of the non-modified alloy. It is also important to mentioned that the stability of the dimensions of the grains in the alloys, modified at different temperatures, indicates the high stability of the given modifier in the melt and the possibility of casting alloys with high superheating without any risk of increasing the grain size of the structure.

Examination of the morphology of the carbides for different holding times of the modified alloy shows that in this case the carbides become equiaxed and are distributed more uniformly in the volume of the grain, in contrast to the non-modified alloys, in which they are straightened into chains and have the form of 'Chinese hieroglyphics'. The duration of holding has only a slight effect on the morphology of the MeC carbides, and is also indicates the stability of the modification effect.

The process of modification of cast iron is based mainly on the variation of the degree of dispersion and structure of the phase components as a result of the introduction into the melt of small amounts of dispersed and additions which changed the nature of solidification. At present, there are a large number of methods and means of modification of cast iron about most of them have certain shortcomings. Therefore, experimental investigations were carried out into the effect of ultrafine powders of refractory compounds on the car structure and mechanical characteristics of grey cast iron in simulation and industrial castings [84, 85].

Analysis of the micrographs, produced from the specimens of the customs, modified with the ultrafine powders, indicates the refining of the graphite inclusions and changes in their morphology. In addition to the plate-shaped form, typical of grey cast irons, the morphology becomes flaky or globular. This is accompanied by a corresponding increase of the mechanical characteristics of castings: the tensile strength *B* increases by 30–50%, relative elongation by 20–40%.

Thermal cycling tests (50 thermal cycles) were carried out on specimens produced from standard unmodified castings. Examination of the micrographs of sections of the specimens shows that modified cast iron is characterised by a lower rate of growth of graphite inclusions. Consequently, treatment of the cast iron with the ultrafine powder also increases the resistance of its structural and phase components to the high-temperature effects and, consequently, it may be expected of the stability of the mechanical and physical–mechanical properties of the castings would increase.

The efficiency of the effect of the ultrafine powder on the structure of aluminium alloys was investigated in casting semi-continuous ingots with a diameter of 420 mm, produced from AMg6 alloy. The amount of the introduced ultrafine powder did not exceed 0.05 wt%. Investigations of the structure on the template of the cross-section of the ingots show that if in casting by standard technology the mean size of the grain cross-section is 0.322 mm^2, then as a result of the addition of the ultrafine powder SiC, BN and TiN, this parameter decreases

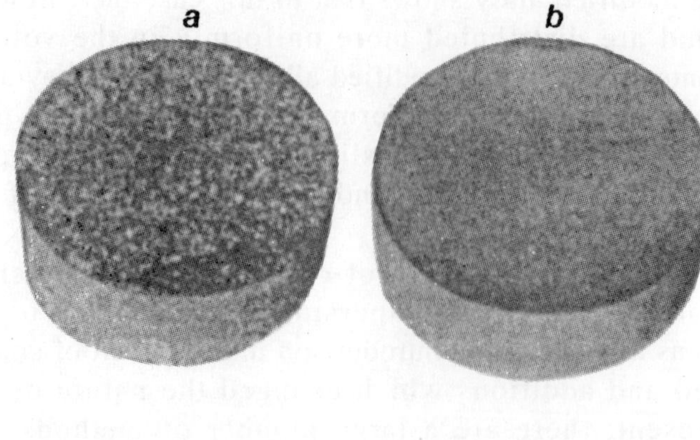

Fig.10.51. Cross sections of ingots of an aluminium alloy. a) the ingot modified by standard technology; b) modified by ultrafine powder.

to respectively 0.123, 0.146 and 0.0 78 mm^2 (i.e., approximately 2.6, 2.2 and 4.1 times). Figure 10.51 shows the photographs of two metallographic sections of the cross-section of the ingots of the aluminium alloy: in the first photograph (*a*) modification of the aluminium alloy was carried out by standard technology, in the second photograph (*b*) by the addition of the appropriate ultrafine powder. Even visual comparison shows that the grain size in the second case is 7–10 times smaller. It is well-known [86] that modification of copper with the ultrafine powders of SiO_2 and Al_2O_3 increases the stability of the properties in relation to high temperatures (in particular, high hardness is retained in a wide temperature range).

Chapter 11

Plasma reactors

In the last couple of decades, the industry of many countries of the world underwent a crises in further improvement of the traditional metallurgical, chemical and other production. It has become necessary to apply completely new processes which would reduce the metal and energy requirement, increase the extent of processing the initial material into the final product and not increase the number of production stages. In addition to this, it was necessary to improve greatly the economic circumstances, i.e., as a result of new processes reduce harmful admissions into the atmosphere, greatly reduce the area for storage of production waste, and ensure complete automation of the entire technological cycle. One of the methods of solving these problems is the application in new processes of low temperature plasma, generated in electric arc plasma torches or plasma technological reactors.

We shall describe several schemes of electric reactors, slightly differing from each other. The specific features of the applications of these systems is based on producing the maximum economical, ecological and social effects in processing chemical – metallurgical and energy materials.

11.1. MULTIJET REACTORS

11.1.1. Kinematic scheme

The extensive application of plasma torches in chemical and metallurgical industries, in testing heat shielding coatings of aerospace systems, in aerodynamic investigations, and for a number of other purposes have revealed a number of problems, such as the construction of powerful electric arc reactors with a long operating life, the temperature and velocity field uniformly distributed in the cross-section, with a high thermal and electrical efficiency [1].

Usually, at the exit from the axial plasma torch, especially with the self-setting arc length, the temperature field is non-uniform and high-frequency pulsations of power, pressure and temperature reach several tens of percent. Therefore, the plasma chemical system should include a mixing chamber which improves the kinematic and dynamic characteristics of the flow in the system. In plasma chemical reactors, to ensure that process takes place in the maximally favourable conditions, it is also necessary to ensure efficient mixing of the working body (chemical starting materials) with the plasma jet. In this case it is not possible to continue without using an efficient mixing chamber.

With the expansion of the area of application of electric arc heaters, there is a tendency for increase of the power of technological systems. The power of a single plasma torch has already exceeded tens of mega watts. However, these high powers can at present be reached only by using high currents and this reduces this thermal resistance of electrodes in the zone of the effect of the attachment spot of the arc. The role of the erosion rate, which determines the operating life of the electrodes, is especially important in stationary systems design for continuous operation for hundreds and in some cases thousands of hours. The transition to using reactors with a mixing chamber to which several plasma torches are connected, makes it possible to solve the problem of plasma technological systems of almost any power with an efficient temperature field and a long operating life. Consequently, in the last twenty years, special attention has been given to the development of multi-jet preheating systems in which at a high total power, the unit power of the plasma torch may be reduced in proportion to the number of the plasma torches connected to the mixing chamber which is a natural element of the system in this case.

The problem is solved by designing a multi-jet preheating system with a general mixing chamber which one part of the required gas flow rate is supplied through plasma torches and the other one (gas, mixture of gases, mixture of gas with powder) is supplied through the end of the chamber directly into the reactor.

This kinematic scheme of gas supply makes it possible to ensure, using relatively simple means, the required correspondence between the available voltage of the standard power source and the required arcing voltage. The additional possibility of supplying the gas, by passing the plasma torches, simplifies the regulation of the reactor parameters. The plasma torches of the multi-jet pre-heater are connected in parallel to the electric power circuit and consequently, it is possible to regulate in a simple manner the power supply to the gas, and maintain a constant total gas flow rate with a symmetric distribution of the plasma torches around the perimeter of the chamber.

We examined the simplest scheme of the reactor – a cylindrical mixing chamber of a multi-jet pre-heater whose kinematic flow diagram is shown in Fig.11.1. Part of the cold gas is supplied directly into the chamber, by passing the plasma torches, the other part is heated in the torches. When supplying the high-temperature gas in the radial direction, to ensure efficient mixing of the gas it is necessary to achieve deep penetration of the jets into the cold flow (to more than half the channel diameter). The depth of penetration of the whole jet discharge from the nozzle of the plasma torch in the radial direc-

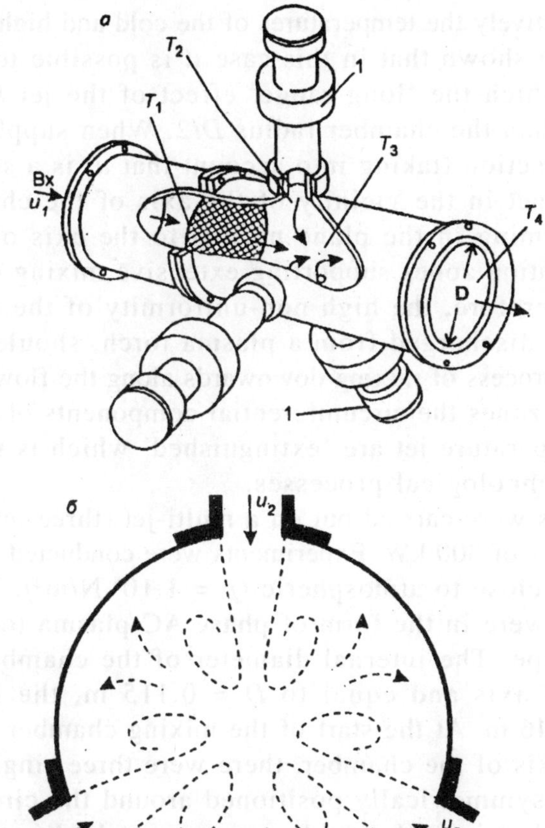

Fig.11.1. Diagram of the mixing chamber of a multi-jet electric arc pre-heater (a) and the mixing diagram of the jet (b). 1) phase plasma torches; 2) mixing chamber; Bx – input of the cold gas.

tion with the velocity u_2 into the cold axial flow moving at velocity u_1, is determined by the equation

$$h = K d_p \sqrt{\rho_2 u_2^2 / (\rho_1 u_1^2)}$$

Here h is the range of the radial (transverse) jet; d_p is the diameter of the discharge jet assumed to be equal to the internal diameter of the nozzle of the output electrode of the plasma torch; K is the coefficient which depends on the angle of contact of the flows (K= 2.0 at α =90°, which was observed in the experiments); r_1, r_2 are the appropriate densities of the cold and hot gases. Assuming that this equation also holds for the higher temperature jet, penetrating into the flow restricted by the walls, at $\bar{T}>3$, where $\bar{T} = T_2/T_1$, T_1 and T_2 are respectively the temperatures of the cold and high temperature gases, it may be shown that in this case it is possible to satisfy the conditions in which the 'long range' effect of the jet h is considerably greater than the chamber radius $D/2$. When supplying the jet in the radial direction (taking into account that u_1 is a small value), they make contact in the vicinity of the axis of the chamber (colliding jets), forming in the plane normal to the axis of the chamber large circulation zones supporting extensive mixing of the flows (Fig.11.1*b*). Therefore, the high non-uniformity of the temperature field of the jet, discharged from a plasma torch, should be rapidly removed in the process of mixing downwards along the flow. In addition to this, in these zones the circumferential components of the velocity of the high temperature jet are 'extinguished' which is important in a number of technological processes.

Investigations were carried out on a multi-jet (three-jet) pre-heater with a total power of 300 kW. Experiments were conducted at a pressure in the chamber close to atmospheric ($p = 1 \cdot 10^5$ N/m²). The electric arc pre-heaters were in the form of phase AC plasma torches of the two-chamber type. The internal diameter of the chamber was constant along the axis and equal to D = 0.115 m, the length was: L = 0.23 and 0.46 m. At the start of the mixing chamber in the plane normal to the axis of the chamber, there were three single-phase AC plasma torches symmetrically positioned around the circumference. In front of the hot jets, the cold gas was supplied into the chamber through a pipe of the same diameter as that of the mixing chamber (Fig.11.1). The wall of the chamber was efficiently cooled with water.

As already mentioned, one of the requirements imposed on the mixing chamber of the multi-jet pre-heater is the high efficiency of the mixing, the high temperature jet entering the chamber at the shortest distance downwards along the flow from the zone of contact of the jets. No chemical reaction takes place in the examined chamber and, therefore, the measure of efficiency of mixing is represented by the uniformity of the temperature field in the core of the flow at exit from the chamber (0.85 D), evaluated by the lower rms deviation σ of the temperature of the gas flow.

As an example, we examine several characteristic temperature fields of the gas T_4 at exit from the mixing chamber into two mutually perpendicular cross sections for different temperature gradients $\bar{T}$ and the relative length of the chamber $\bar{L} = L/D = 2$ (Fig. 11.2). The flow rate of the cold gas G_1 changes from $3 \cdot 10^{-3}$ to $60 \cdot 10^{-3}$ kg/s and the total flow rate of the high temperature gas G_2 remains constant and equal to $30 \cdot 10^{-3}$ kg/s. As indicated by the temperature fields, shown in the graph, and also according to the total cycle of investigations, the mixing chamber of the given kinematic scheme is highly efficient. The intensity of the process of energy and mass exchange between the jets is so high in the entire volume that already at a distance $\bar{L} = 2$ the temperature field has a high degree of uniformity. A further increase of the relative length of the mixing chamber ($\bar{L} = 4$) increases the strength of the effect of the wall boundary layer and the 'dip' of the temperature profile in the direction of the chamber wall. The relative length of the chamber $\bar{L} = 2$ is in all likelihood close to the optimum value because in a wide range of variation $\bar{T} = 5–11$ and the ratio of the flow rates of the cold gas to the high temperature gas (G_1/G_2) rms deviation σ in the selected section of the flow core does not exceed 2%, which is an efficient indicator of the intensity of mixing of the gas (Fig.11.3). The increase of the gas temperature does not cause any significant changes in the quantitative characteristic of the mixing process, as reported in a number of studies. This is in agreement with the conclusions made previously. An identical conclusion is obtained by examining the experimental results of mixing the cold jets of different densities (air + methane).

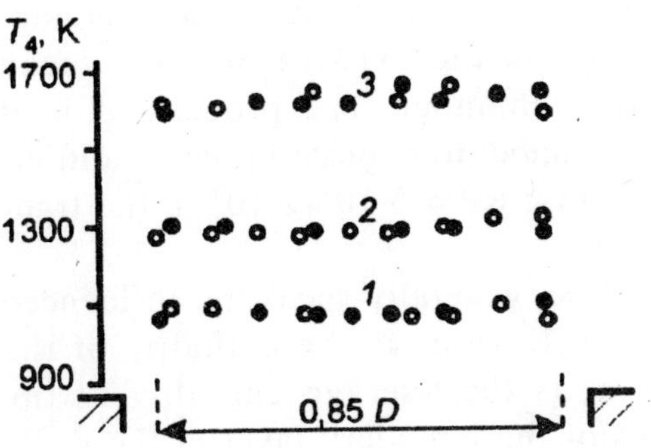

Fig.11.2. Temperature field of the gas at exit from the mixing chamber. $\bar{T}$ =2.0; G_2 = $30 \cdot 10^{-3}$ kg/s; 1 – $\bar{T}$ = 9.3; G_1 = $60 \cdot 10^{-3}$ kg/s; 2 – $\bar{T}$ = 11.3; G_1 = $60 \cdot 10^{-3}$ kg/s; 3 – $\bar{T}$ = 10.5; G_1 = $3 \cdot 10^{-3}$ kg/s.

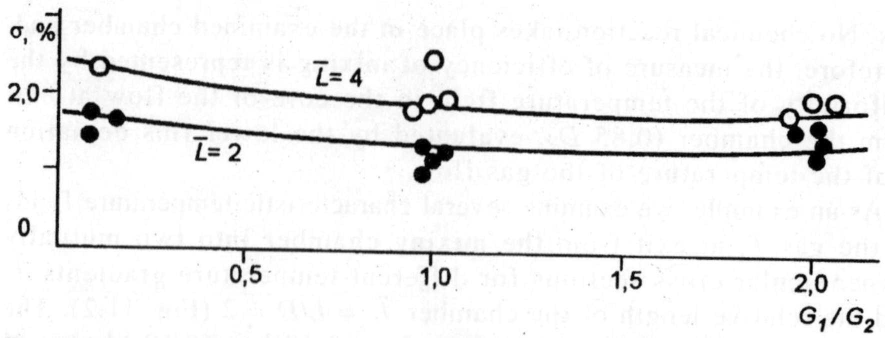

Fig.11.3. Dependence of the rms deviation of the temperature of the gas flow on the ratio of the flow rates for two values of $\overline{L}$.

Thus, if the high temperature jets operate in the collision regime when the depth of penetration $h > D/2$, efficient mixing is already obtained at low relative lengths of the chamber.

11.1.2. Thermal efficiency

We examine the second important characteristic of the mixing chamber, i.e. the thermal efficiency, determined by the ratio of the heat losses through the wall to the content of the flow at entry. The theoretical solution of this problem is very difficult because heat exchange between the hot gas and the wall takes place in the presence of strong and very complicated initial turbulence of the flow, and the high non-uniformity of the temperature and velocity fields of the gases discharged from the plasma torch into the chamber. Using the method of criterial generalisation of the experimental data which makes it possible to determine, with a certain degree of approximation, the relationship between the efficiency and the determining criteria. The analysis shows that the efficiency of the mixing chamber is a function of two criteria: the Reynolds number Re and the dimensionless length $\overline{L}$.

The value of Re is calculated from the average value of temperature T_3, static pressure P_3, equal to 1 atm in the experiments, the mean velocity u_3 at entry into the mixing chamber. The process of heat exchange in the investigated mixing chamber took place in the conditions corresponding to the Reynolds numbers: Re = $5\cdot10^3$–$2\cdot10^4$ in the transition flow region.

The heat exchange process should be, generally speaking, influenced by the enthalpy factor $\overline{h}_3 = (h_3/h_w - 1)$. Here h_w is the enthalpy of the gas at the wall temperature, and h_3 is the average enthalpy of the gas. In the absence of perturbations of the boundary layer at the walls

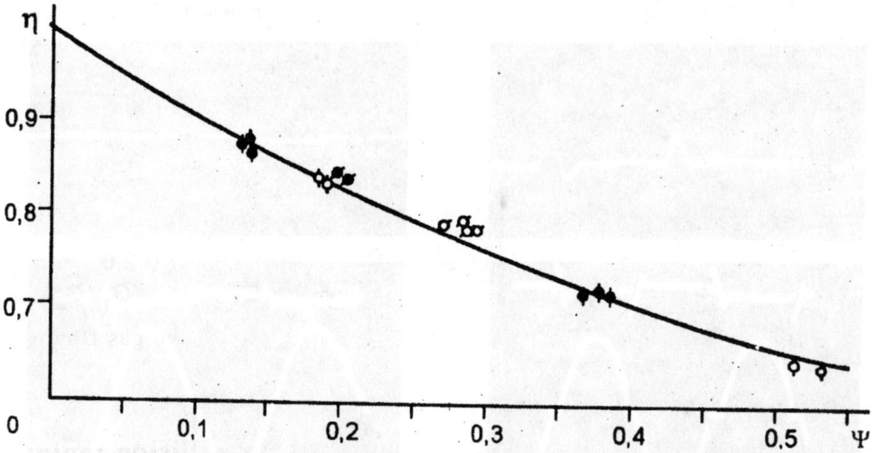

Fig.11.4. Thermal efficiency of the mixing chamber.

of the channel and at $h_w \ll h_з$ the local value of the Stanton number, and consequently, the thermal efficiency depend only slightly on the enthalpy factor in the region of laminar and turbulent boundary layers [2]. The experimental results show that in a more complicated case of the gas flow in a pipe, the effect of this factor (varies from 3 to 17) is small. We ignore the variation of the Prandtl criterion and also the radiation of the gas, because of its small contribution to the heat losses in the given experimental conditions.

The required dependence for η was determined in the form

$$\tilde{\eta} = (1-\eta)/\eta = a\bar{L}^{\alpha}\,\mathrm{Re}^{\beta}.$$

Processing the experimental data gives the approximate equation

$$(1-\eta)/\eta = 145\bar{L}^{0.50}\,\mathrm{Re}^{-0.75} = \Psi.$$

The correspondence between the calculations of Ψ using the equation and the experimental values η is shown in Fig.11.4.

Thus, the cylindrical mixing chamber of the multi-arc pre-heater with the investigated kinematics scheme is characterised by a high efficiency of mixing of the jets at different temperatures and the uniform temperature field at exit from the chamber at $\bar{L}$ = 2.0. Under specific conditions, these chambers also have a high value of thermal efficiency.

11.1.3. Pulsations of total pressure

Taking into account the fact that the investigated phase AC plasma torches with a self-setting arc length were studied, it is interesting to examine the results of investigations of the processes of 'smoothing'

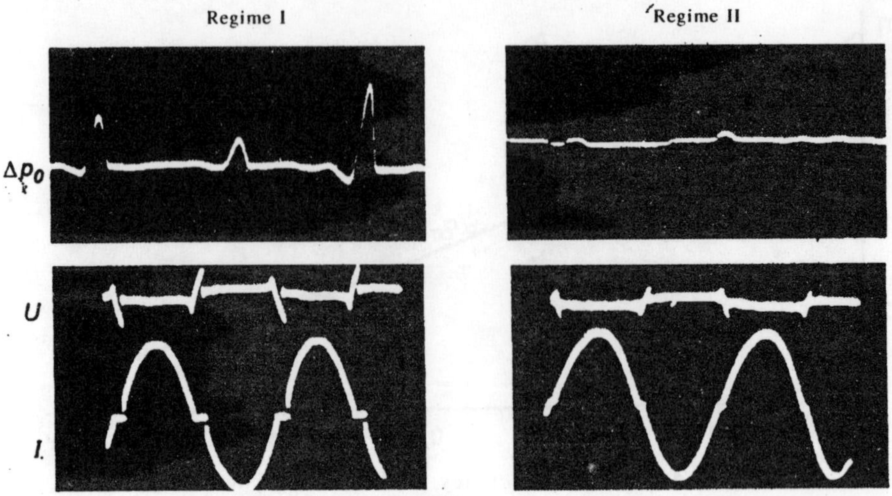

Fig. 11.5. Oscillograms of pulsations of total pressure Δp_0 at exit from the mixing chamber and also voltage U and current intensity I for two arcing regimes. I – with long current breaks (no HF discharge); II – with short current breaks (with HF discharge).

of pulsations of total pressure in the mixing chamber caused by the time dependence of the power of the phase plasma torch, the breaks in arc current in passage of current through zero, shunting of the arc, fluctuations of the gas flow rate and other reasons.

For this purpose, a total pressure pipe with a capacitance sensor installed in it was placed in the output section of the system. The pipe–loop pulsed system was calibrated in advance in a special acoustic stand, so that it was possible to obtain the required amplitude–frequency characteristic for decoding the oscillograms.

Figure 11.5 shows the oscillograms of pulsations of total pressure at the end of the mixing chamber Δp_0 and pulsations of voltage U and arc current I for two different arcing conditions in the phase plasma torch. Regime I is characterised by long current breaks, and, consequently, breaks in power. This leads to large pulsations of the total pressure reaching (by the maximum value) 30–40 %. The frequency was 150 Hz. If a high frequency current is superimposed on the power AC arc, the current breaks can be shortened or completely removed. Regime II corresponds to the case of a large decrease of the breaks, as indicated by the current and voltage oscillograms. In this regime, small pulsations of the total pressure, reaching 5–7% at the same frequency of 150 Hz, remain. In the normal arcing conditions with no breaks in current and only the time dependence of the power of the phase plasma torches, and also high-frequency voltage and current pulsations, caused by shunting of the arc (with the frequency $f \approx 10^4$ Hz), no total pressure pulsations were detected.

Since the pulsed sensor, used in the experiments, records the pressure pulsation exceeding 1%, the form of the oscillograms shows that in the absence of breaks in arc current the pressure pulsations at the end of the mixing chambers are smaller than 1% [3].

The results are in agreement with the conclusions reached in studies by other authors made on the basis of powering an electric arc preheater with five coaxial DC plasma torches assembled in the form of a star-like engine and operating with a very short mixing chamber. The high frequency oscillations, excited by the arcs, are almost completely extinguished in the mixing chamber; the rotational movement of the gas, determined by one of the plasma torches, is compensated by the rotation movement of the jets discharged from other plasma torches; the maximum variations of the pressure are approximately an order of magnitude smaller than those detected in single plasma torches.

The mixing chamber is also characterised by the extinction of pulsations of temperature and gas velocity.

Efficient mixing of the jets with different temperatures in the cylindrical chamber and the sufficiently satisfactory uniformity of the temperature field at exit from the chamber can also be obtained when introducing radial cold jets into the axial plasma flow [4]. If the jets meet in the vicinity of the axis, i.e. they are in the collision regime, then sufficiently efficient mixing takes place at relatively small lengths of the chamber. However, it should be mentioned that the best mixing of the gases takes place in the case of radial introduction of the hot gas into the cold axial flow.

11.2. HYDRODYNAMIC AND THERMAL ENGINEERING CHARACTERISTICS OF A THREE-JET REACTOR

The multi-jet plasma chemical reactors have the form of combined mixing chambers and are used widely in the chemical technology of processing production waste, in producing ultrafine power materials, and other purposes.

However, the need for efficient cooling of the reactor walls results in the deviation of the characteristics of these reactors on the characteristics of single-type ideal reactors and this complicates the solution of purely engineering problems of apparatus formulation of the processes. It is therefore necessary to carry out experiments to examine reactors of this type.

The investigations were conducted on a three-jet plasma chemical reactor with the following main geometrical, hydrodynamic and thermal engineering parameters:

Internal diameter, m	0.036; 0.04; 0.046
Length in m; in length gages	1; 15
Volume, m^3	0.001–0.002
Volume flow rate of gas-heat carrier (Nitrogen), nm^3/h	9–23
The range of temperature of the gas along the length of the reactor, K	5700–1600
Wall temperature range, K	440–300
Temperature factor and Reynold's number of the flow, Re_D	750–1400

The plasma technological processes are realized in the flow of argon, hydrogen and nitrogen (their mixtures) [5], natural gas, water steam, and other gases.

The gases, used as heat carriers, greatly differ in their energy characteristics. Therefore, when selecting the plasma forming gas, it is necessary to take into account the following parameters: possibility of obtaining high enthalpy values; use as a chemical reagent; inertness in relation to target products.

Enthalpy is one of the main parameters of the plasma flow and greatly determines the technological capacities of the reactor. Gases such as nitrogen, hydrogen at relatively low temperature (4000–8000 K) are characterised, as the result of molecule dissociation, by high enthalpy values which are almost an order of magnitude higher than the enthalpy of monatomic gases (argon, helium).

The use of high-enthalpy molecular gas as a plasma forming gas is convenient because it results in high values (60–80%) of the efficiency of the plasma torches, whereas for the plasma torches working with argon or helium this parameter is considerably lower [6].

In relation to nitrogen, hydrogen is characterised by higher enthalpy and heat conductivity ensuring the utilisation of the thermal energy generated in the recombination of hydrogen plasma, and in a number of processes hydrogen maybe used also as a heat carrier and a reduction agent [7]. Nitrogen, like hydrogen, may also be used as a heat carrier and a chemical reagent, for example, in the processes of nitriding in production of nitrides [8], or in processes of carbidization in producing carbide and carbonitrides [9, 10].

11.2.1. Some apparatus schemes of high-temperature synthesis reactors

The reactors for producing ultrafine powders of refractory compounds will be discussed. Of considerable importance, together with the selection of the optimum technological parameters, are the engineering problems of the apparatus schemes.

The designed apparatus should ensure:

– introduction of the mass of the starting material into the zone of the jet (flow) heated to the highest temperature and its maximum possible processing;

– stability of parameters in the reaction zone, and also prevention (it necessary of oxidation of high dispersion products when the trapping system is opened.

However, in the conditions of service of actual systems, the fulfilment of this requirement is complicated by factors such as the wide grain size composition of the starting product, the use of chemical reactors with water cooled walls in which thermophoresis and direct contact of the immersed jet with the walls results in the formation of a dense layer of a deposit on the walls reducing the cross-section of the reactor, and also in the often uncontrollable interaction of high dispersion synthesized materials with the atmosphere, etc.

The set of these problems and the natural tendencies to retain the compact form of plasma equipment have resulted in the development of a number of designs of plasma reactors which have been reviewed in, for example, [11].

We shall mention only some the investigated designs of the reactors for different technological variants of the jet processes:

1. A reactor with parallel jets (Fig. 11.6*a*) in which several plasma torches are situated in the same plane ensuring efficient processing of the polydispersed starting material;

2. A reactor with a fluidised layer (Fig.11.6*b*) in which the starting material is supplied into the high temperature zone of the jet from the surrounding fluidised layer;

3. The reactor with a plasma torch connected coaxially with it including with the counter supply of this starting material (Fig.11.6 c).

4. Multi-arc reactor in which several plasma jets are supplied into the flow of the initial material. This approach makes it possible to construct high power systems.

Thus, the reactors for jet-plasma processing may operate in the technological chain of systems and they have already become traditional and chemical technologies. However, one of the main problems

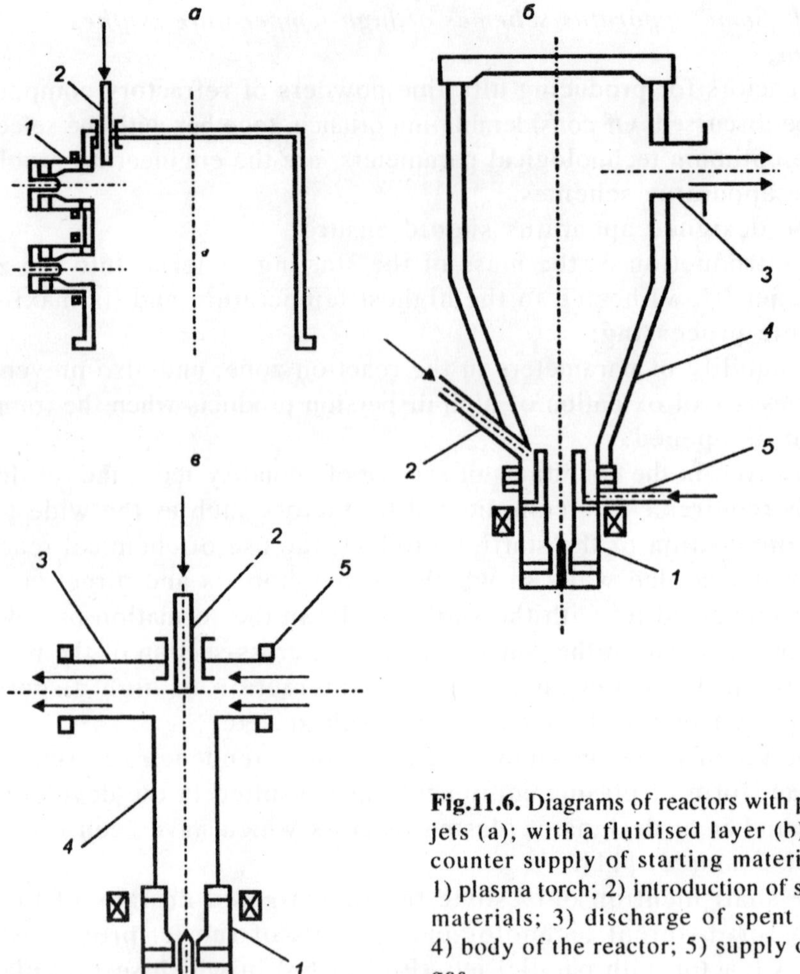

Fig.11.6. Diagrams of reactors with parallel jets (a); with a fluidised layer (b); with counter supply of starting material (c). 1) plasma torch; 2) introduction of starting materials; 3) discharge of spent gases; 4) body of the reactor; 5) supply of cold gas.

of applied plasma chemistry has not as yet been solved, i.e. the development of scientific fundamentals of plasma chemical reactor construction and, consequently, the selection of the reactor or design and development of new processes are carried out mainly by empirical methods. An exception are represented by reactors homogeneous plasma chemical processes [12].

The problems of designing reactors for heterogeneous processes were examined to some extent in [13–20] and generalized in [21, 22] but further work is required for a final solution. Nevertheless, even in the absence of a united approach to selecting reactors for heterogeneous processes, it may be concluded that multi-arc [multi-jet] reactors have considerable technological possibilities. The resultant level of the electrical power (150–300 kW) and productivity (from 0.3–0.5 t per annum of powder of refractory compounds to 80 t/annum of high dispersion oxide catalysts) indicates that these reactors have found a special place in the technology of precision inorganic synthesis. It would be important to examine their design special features and technological variants.

11.2.2. Reactors based on a multi-jet mixing chamber

In addition to simple design, which is the main advantage of the investigated system, it is important to mention a significant shortcoming – the presence of a steep radial temperature gradient in the wall layer. This reduces the technological possibilities of these reactors. Although these shortcomings have been eliminated in counter flow reactors with combined two-stage heating, the literature contains no data on the use of the later in laboratory or experimental systems. There are also no data on the possibilities of improving thermal engineering characteristics of direct flow reactors or for example using thermal shielding.

According to [22] the multi-jet direct-flow reactors are produced with different angles of inclination of the plasma torches to the axis of the reactor – from 15 to 90°, and this greatly changes the hydrodynamic and thermal engineering characteristics of the reactors. Consequently, analysis of the relationship of the angle of inclination of plasma jets with reactor characteristics is evidently essential when preparing recommendations for designing reactors of this type. Here we present the main data on the hydrodynamic and thermal engineering characteristics of multijet reactors.

The hydrodynamics of plasma flows in three-jet reactors was investigated in [23–28]. In [23], the hydrodynamics of the flows was investigated on the cold model of the reactor in the form of a cylindrical chamber with an internal diameter of 0.08 m into which three air jets situated under the angle of 120° in relation to each other were introduced under the angle of 30 and 45° in the axis of the chamber through a lid. The Reynold's number of the flow corresponds to the Reynold's number of the transition flow from laminar to turbulent. It was also established that already at a distance of 0.33 gages from the lid of the reactor, the axial velocity is described by the laws of propagation of the three jets. In this case, in contrast to analysis and the data presented in [27], the maximum velocity is displaced into the near-axial zone as a result, according to the authors, of the asymmetry of collision of the jets and special features of development of turbulence in the zone of collision of the jets. In the near-axial zone above the point of collision of the jet under the angle of inclination of the jets in relation to the reactor axis, there is a region of rising flows [29], which disappears when the angle of inclination of the jets is reduced to 30°. There are also differences in the near-wall region of the circulation zone, the increase of the intensity of turbulence in the direction from the axis to the chamber wall at a weak dependence on the flow rate of the gas – heat carrier, and the angle of its introduction, and a twisted flow may also fall. The results of investigations carried out on the cold model are on the whole confirmed by the measurements of the velocity and temperature of the non-dusted plasma flow formed by three jets.

The authors of [25] examined the total and static pressures, and the temperature in different zones of a three – jet reactor with the angle of introduction of plasma jets of 45°. The results of investigations of the velocity field in the different section of the reactors, presented in Figs.11.7 and 11.8 using the data from [25], and also the variation of dynamic pressure, static pressure and temperature both along the cross section and the length of the reaction channel show that in the initial sections of the reactor there are large zones of reversed flows and re-circulation region determined by the non-isobaric and non-isothermal nature of the high temperature flow. The presence of steep temperature and velocity gradients in both plains with the high temperature flow result in high intensity heat and mass transfer and, consequently, the transverse dimension of the flow rapidly increases in the direction of movement, and the profile of the hydrodynamic gravatus is 'smoother'. Already at a difference of 1.5 gages from the origin of the reactor, the distribution of temperature and total pressure in the cross section has the form typical of the flow moving in cylindrical channels.

In [26] investigations were carried out into the temperature field of a reactor with the angle of inclination of plasma torches of 90° at a distance of two gages from the origin of the reactor in the Reynold's number range of the flow of 450–600. The authors noted the absence of a radial temperature gradient in the central part of the reactor and the presence of a large wall low temperature zone.

The special features of heat exchange in the channels of the direct flow three-jet plasma reactors will be examined. The heat exchange in such a reactor may be regarded as the interaction of the plasma-dispersed materials – reactor walls system, usually sub-divided by the heat exchange of the plasma flows with the walls of the reactor and the heat exchange of the particles of the dispersed material with the plasma flows [21].

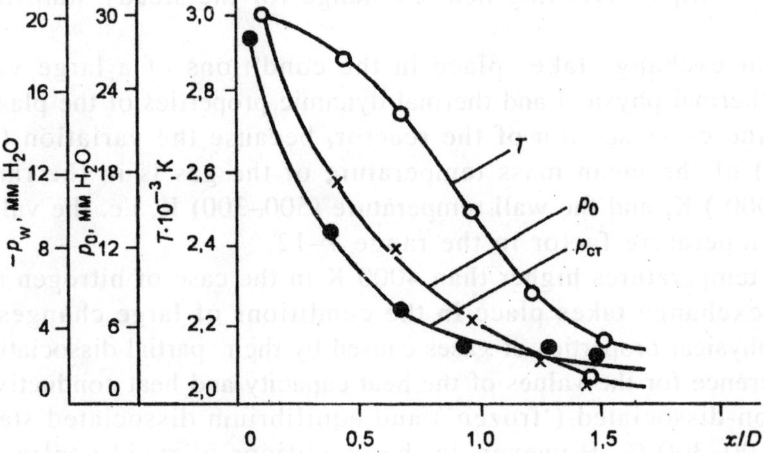

Fig.11.7. Variation of temperature, static and total pressure along the reactor axis.

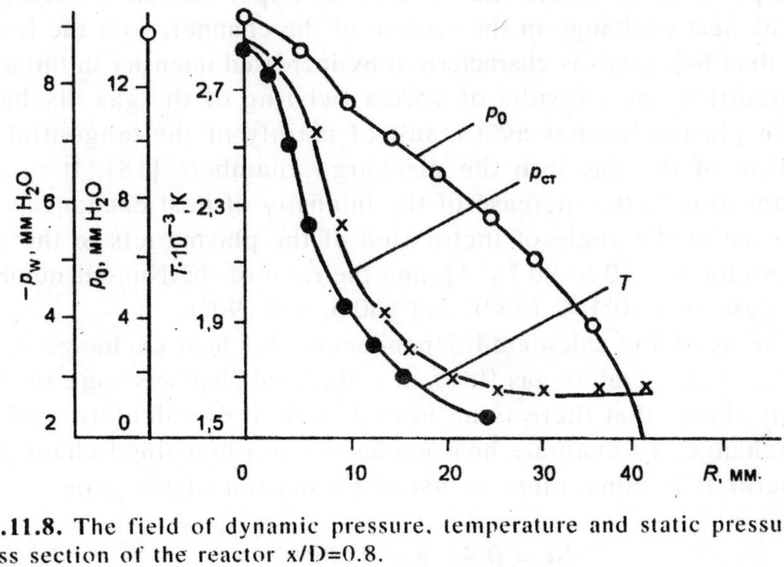

Fig.11.8. The field of dynamic pressure, temperature and static pressure in the cross section of the reactor x/D=0.8.

The heat exchange of the high-temperature gas flows with the cooled walls of the cylindrical channels in laminar flow conditions was investigated in [15, 16, 18, 29, 30]. It has been established that the heat exchange in the channels of the plasma reactors is characterised by the following special features which are not found or have no significance in the flow in pipes of slightly heated gas flow.

1. Heat exchange takes place at simultaneous formation of the thermal and hydrodynamic boundary layers and, consequently, the relationship of heat exchange in the initial section greatly differ from

the relationships governing heat exchange for the steady stabilized flow.

2. Heat exchange takes place in the conditions of a large variation of thermal physical and thermal dynamic properties of the plasma flow in the cross-section of the reactor, because the variation (for nitrogen) of the mean mass temperature of the gas is in the range (6000–2000) K, and the wall temperature (500–300) K, i.e. the values of the temperature factor in the range 7–12.

3. At temperatures higher than 4000 K in the case of nitrogen and air heat exchange takes place in the conditions of large changes in thermal physical properties of gases caused by their partial dissociation: The difference for the values of the heat capacity and heat conductivity of the non-dissociated ('frozen') and equilibrium dissociated states exceed 100–300 %. However, in the conditions of rapid cooling of the flow, the degree of deviation from the equilibrium state is almost impossible to determine because of experimental difficulties.

4. The heat exchange in the section of the channel, with the length smaller than 6–8 gages is characterised by increased intensity in turbulent flow conditions as a results of vortex twisting of the gas discharge from the plasma torches as a result of mainly of the tangential introduction of the gas into the discharge chambers [18]. It is also important to note the increase of the intensity of heat exchange with an increase of the angle of inclination of the plasma jets to the axis of the reactor from 0 to 90 ° [31], and the ratio of the Nusselt numbers in this case is 1 (0°), 2 (45°), 3.3 (60°), 4.4 (90°).

5. The available calculated dependencies for heat exchange in the flow of high temperature gas flows in cylindrical channels were studied. Analysis shows that there is no united method of calculation of the heat exchange. To evaluate heat exchange in a non-lined channel of the reactor it is convenient to use the equation of the type

$$St = 0.44\ Re^{-0.53}\ Pr^{-0.67}.$$

proposed in [15].

6. The intensity of heat exchange may be reduced either by lining the channel of the reactor, described in [21] (no calculation relationships have been published for this case), or by adding into the flow a dispersed material whose mass concentration in the process varies from 0 to 4.4 % [18]. In this case, the intensity of the heat flow into the wall decreases as a result of the transfer of heat from plasma to the dispersed material and this reduced the temperature of the plasma flow and the temperature pressure between the plasma

and the reactor walls. The decrease of the intensity of the heat flow is taken into account by adding, to the calculation dependence for the Stanton number, the correction coefficient ???$_\mu$ determined in the range of the flow rate mass concentrations of the dispersed material 0.15–0.2% as follows

$$\varepsilon_\mu = 0.7\left[G_p/(G_t + G_g)\right]^{-0.128}$$

where G_p, G_t, and G_g are respectively the mass flow rate of the dispersed material, the transport gas and the plasma forming gas, kg/s.

The heat exchange between the plasma and the particles of the dispersed material has been examined in sufficient detail in [32], and the heat exchange in the channels of a three-jet plasma chemical reactor was studied in [21], were it is reported that the efficiency of inter-component heat exchange and the nature of the dependence of calculations are determined by the concentration of the particles in the flow, the shape of the particles, the surface roughness, relative velocity, the presence of temperature gradients in the boundary layer of the particle, and the flow regime. In [21], the inter-component heat exchange of the particles, moving in the plasma flows, was calculated using the following criteria of dependence

$$\mathrm{Nu} = 2\lambda_\varepsilon/\lambda_g + 0.78\ \mathrm{Re}^{0.5}\ \mathrm{Pr}^{0.33}\ (\rho_g\ \mu_g/(\rho_s \cdot \mu_s))^{0.2}\ \varepsilon_\beta.$$

The correction coefficient is determined from:

$$\varepsilon_\beta = 7.82 \cdot 10^{-8}\ \beta^{-2.1}.$$

The mean volume concentration of the dispersed material in the calculations section of the reactor Δx_i can be determined from the relationship:

$$\beta = \left(\sum_0^{x_i} \beta_i \Delta x_i\right) / \sum_0^{x_i} \Delta x_i,$$

and the local volume of concentration as

$$\beta_i = (G_p/G_g)\ (\rho_{gi}/\rho_{pi})(\upsilon_{gi}/\upsilon_{pi})\ (D^2/D_{ci}^{\ 2}).$$

Analysis of the heat exchange of the single- and two-component of the plasma flow with the reactor walls and inter-component heat exchange indicate not only the complicated nature of the process of the problem but also indicates a number of interesting problems in respect of scientific and applied aspects. The detailed solution of these problems requires a large amount of experimental and theoretical information. These problems include, for example, the derived dependencies for calculating heat exchange of one- and two- component flows with the reactor walls in the case of thermal shielding of the walls by different linings. It is also necessary to examine the problem of the effect on the heat exchange with the walls of the channel of the two components gas flow, the composition and dispersion of the solid phase. With regard procedures, it is still necessary to solve the problem of the processing of experimental data obtained for heat exchange in the temperature range above 4000 K using nitrogen or air as a gas and heat carrier.

In addition to this analysis of the literature data on high temperature heat exchange in plasma-technological systems enables the optimum (from the technological viewpoint) design of the reactor to be selected: it is the structure with the angle of inclination of 30 or 45°.

11.2.3. Thermal engineering characteristics of a three-jet direct flow reactor

The abundance of the data on the apparatus formulation of the plasma processes and selection of gas indicates the large variety of design of the reactors determined by differences in the problems of chemical technology, solved using plasma heating. Gas selection is also determined by the nature of the solved problems, but in all cases preference is given to molecular gases - nitrogen and hydrogen, characterised by the required by the thermal physical properties, or to their mixtures with inert gases. Thermal dynamic calculations have confirmed the inertness of nitrogen in respect of many refractory carbides and borides at temperatures above 2300–2600 K and, therefore, nitrogen and similar gases are used widely in the processes of processing dispersed initial materials characterised usually by high endothermic properties. The type of reactor is determined not only by the physical-chemical characteristics of the starting material (dispersion, specific mass, electrical and magnetic properties, melting point, evaporation point, chemical transformations) but also by the required power level. As regards the required power and, consequently, productivity in processing the dispersed initial material, one of the leading positions is occupied by multijet direct-flow reactors which have been developed to pilot plant and industrial levels.

Analysis of the data on the operation of reactors of this types, special features of service, and hydrodynamic and thermal engineering characteristics enables the following conclusions to be drawn on the optimum combination of the design elements, thermal engineering and technological possibilities, and determine the tasks requiring special attention in the direction of improvement of the reactor and development of fundamentals of calculation and design of these reactors.

1. The optimum design is the three-jet reactor with uniform distribution of plasma torches around the circumference with the angle of inclination of the plasma jets to the axis of the reactor or 30–45° and thermal shielding of the rapidly cooled walls; this reactor is characterised by the maximum operating life of the mixing chamber, the high uniformity of the radial distribution temperature and velocity at the minimum loses of the starting material and thermal energy.

2. To improve the characteristics of the reactors of this type it is essential to examine the special features of heat exchange in the reactor and the methods of improving thermal shielding of the walls.

The following results will be described:

– investigations of the thermal engineering characteristics of a three-jet direct-flow reactor with the angle of introduction of the plasma jets of 30° with different variants of thermal insulation of the walls;

– investigations of heat exchange in the thermally insulated channel of the reactor an determination of the relationship for calculating and design.

The experimental object is shown in Fig.11.9. The plasma flow was generated using three electric arc pre-heaters (plasma torches) EDP-104A with the power of up to 50 kW each [33] installed in the mixing chamber under the angle of 30° to the axis of the reactor. The mixing chamber is connected to a sectional water cooled cylindrical channel with an internal diameter of 0.046 m.

The electrical power and ignition circuits of the three plasma torches, operating with a mixing chamber from a single power unit, is shown in Fig.11.10. The power source consists of the separator P, the power unit (PU), including a rectifier constructed on the basis of the Larinov circuit, with the mean value of rectified voltage of 540 V from which electric power is supplied to each of the plasma torches Pl_1, Pl_2, Pl_3, through the system of water cooled ballast. Resistance Br_1 and Br_3 using breakers P_1, P_2, P_3 and magnetic starters, 1S–6S. Each plasma torch is activated by a system of electrical ignition consisting of high voltage transformers HO220/6 kW Tp and BT, capacitances C_1, C_2 and C_3, the discharger Pa, fuses F_1 and F_2 and starting buttons K_1 and K_3. The electric power source and control and measuring devices are protected against high voltage of the initial system using the voltage choke coil CK. The total power of plasma torches is regulated in steps by balanced resistances in the range 30–140 kW .

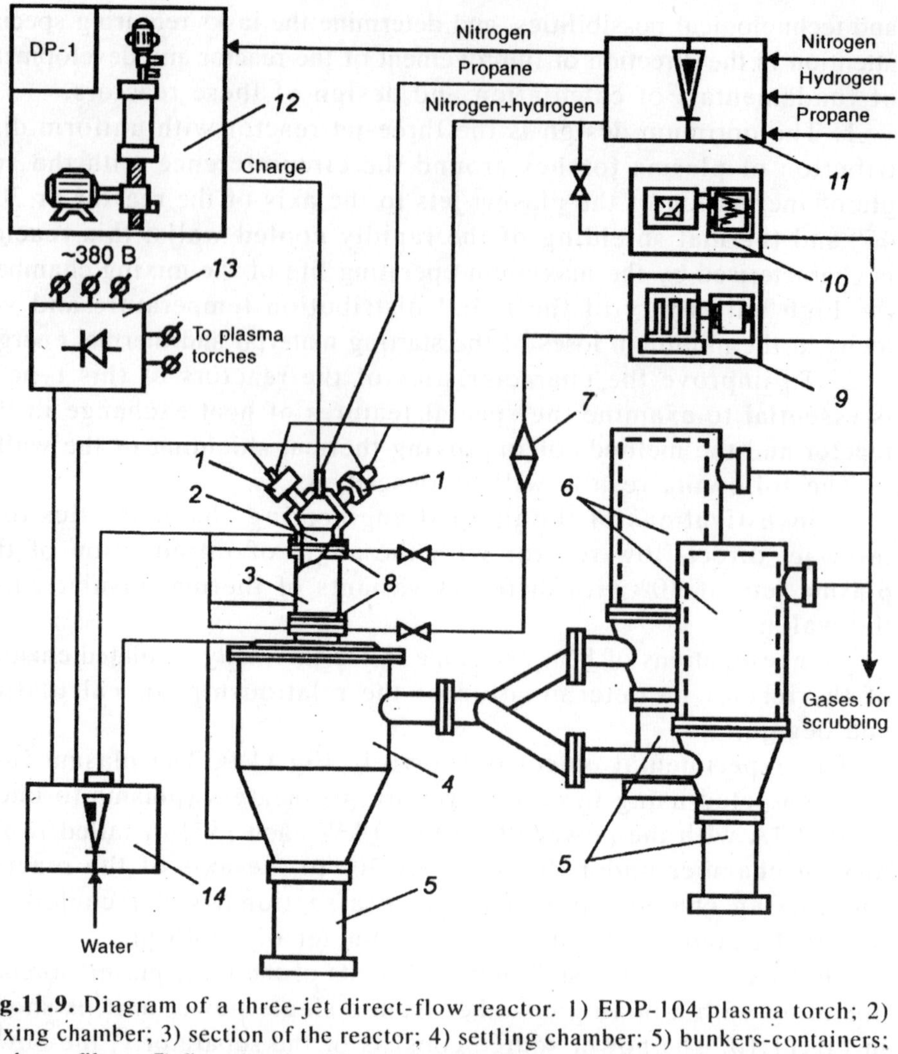

Fig.11.9. Diagram of a three-jet direct-flow reactor. 1) EDP-104 plasma torch; 2) mixing chamber; 3) section of the reactor; 4) settling chamber; 5) bunkers-containers; 6) sleeve filters; 7) fine cleaning filters; 8) probe for taking samples; 9)chromotograph LKhM-VMD (Tvet-1); 10) Tsirkon-M gas analyser; 11) gas supply system; 12) power dosing device; 13) power source for the plasma torch; 14) water supply system.

The stand is fitted with all devices required for analysis of the outgoing gases at temperatures of 1400–3500 K and filters for removing the dust. The dispersed initial material is supplied into the reactor using a device for dosing the powder–gas mixture DP-1 ensuring stable supply of the material in the range ±2% [34]. The dust-gas mixture, formed in the dosing device, is introduced along the axis of the reactor in the zone of collision of the high temperature gas jets discharged from the plasma torches, using a water cooled lance

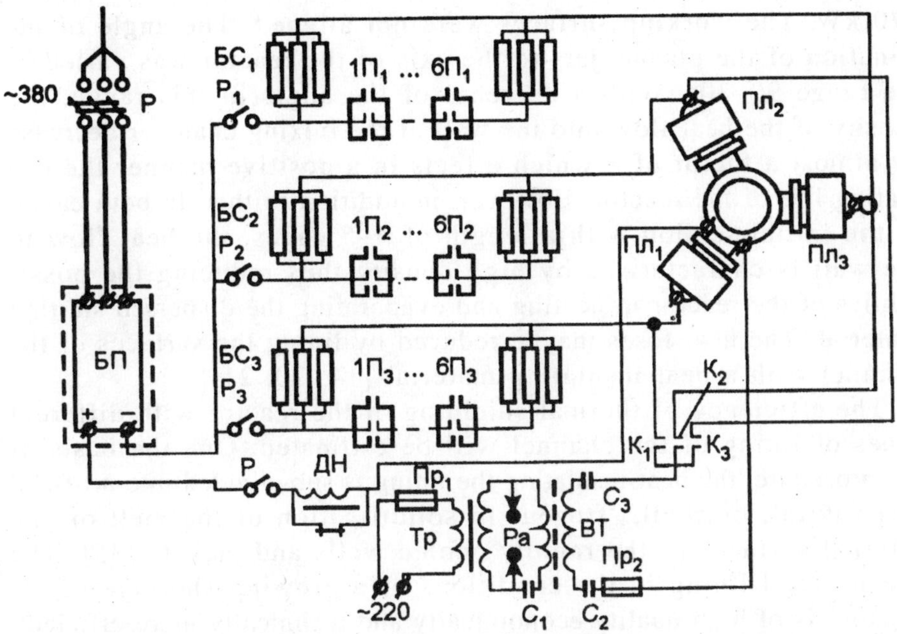

Fig.11.10. The circuit of the electric power source and arc emission in three-jet reactor plasma torches.

installed in the mixing chamber. The rate of supply of the powder into the reactor at a constant flow rate of the transport gas is regulated by changing the internal diameter of the lance using attachments. The lance is also used for supplying a gaseous reduction agent into the reactor. The materials, treated in the plasma, are quenched using a quenching ring based at the exit from the reactor and producing the form of a hollow metallic disk with a thickness of 8 mm and the internal diameter 46 mm with four orifices with a diameter of 1 mm distributed uniformly around the circumference was supplying the cold gas (nitrogen) into the reactor. The condensed treatment products, taken away from the reactor by the out going gases, are trapped in the settling chamber and in two sleeve filters. The filtration cloth is represented by stainless steel mesh with twill weaving. The surface area of settling in the chamber is 1 m^2, in the sleeve filter 3 m^2 resulting in the rate of filtration in the range 0.001–0.002 m^3/(m^2·s). If necessary, the dust–gas flow is cooled to the working temperature of the filters (800–900 K) in the heat exchanger TK with the surface area of 1 m^2.

11.2.4. Energy balance of the reactor

Investigations were carried out in a reactor with a total power of

~70 kW. The working surfaces were not aligned. The angle of inclination of the plasma jets to the axis of the reactor was varied in the range 90–30°. With a decrease of the angle in this range, the density of the heat flow into the wall of the mixing chamber decrease by almost a factor of 3 which effects in a positive manner the operating life of the reactor. However, in addition to this, in both cases, in the initial section with a length of 4–5 gages, the heat flow to the wall is characterised by high density thus reducing the possibilities of the reactor in heating and evaporating the dispersed starting material. The heat loses maybe reduced by lining the surfaces of the channel with a heat insulating material [14, 20, 21].

The efficiency of thermal shielding of the reactor with different types of lining of the channel will be estimated. One the basis of the procedure, the heat insulating the lining is sub–divided into artificial and natural, or skull, 'frozen' in solidification of the melt on the internal surfaces of the rapidly cooled walls and jackets [35]. The use of skull lining in the conditions of a growing shortage of refractory is of high quality economically and technically more efficient.

The application of lining with heat insulating layers, reduces the density of heat flow and increases the temperature of the internal surface of the wall thus, making sure that the characteristics of the reactor are as close as possible to the characteristics of the appropriate ideal reactor. Taking into account the high heat conductivity of graphite screens, it may be assumed that the application, as lining, of the materials with lower heat conductivity in comparison with graphite increases the temperature of the internal surface of the line channel [36, 37]. Infact, the substitution of graphite by corundum at an initial enthalpy of the flow of $5.55 \cdot 10^3$ kJ/kg reduces the heat flow into the wall and increases the temperature of the wall on average of 10–20%. However, artificial lining of plasma chemical reactors are difficult to produce, especially when using materials based on fused oxides, and according to service experience they are also insufficiently stable. In fact, the artificial lining fails quite rapidly in the initial section of the channel of the reactor along the length of several gages, i.e. in the zone with the heat flows in the wall of the reactor are the highest. The technological efficiency of skull lining based on oxides of chromium, vanadium, titanium, zirconium and silicon used as the starting material in the processes of reduction synthesis of the appropriate boride and carbides, are shown in [19] in plasma reduction synthesis of boron carbide.

A special feature of plasma reduction processes in comparison with the processes in which the apparatus has skull lining, is the possibility of controlling in a wide range the phase composition of the skull and changing this composition in relation to the technological parameters of synthesis from the initial oxide to the target product (usually carbide or boride). However, comparison of the thermal physical properties of substances included in the composition of the skull, shows that it is more efficient to produce the skull from oxide. The reactor with the angle of inclination of the plasma jets of 30° and the skull lining of the walls is shown in Fig.11.11*a* and *b* shows the reactor with the angle of inclination of the plasma torches of 90°. The formation of the crown confirms the previously mentioned relative special features of hydrodynamics of the reactor of similar design predetermining that they should not be used in a more developed state. The longitudinal distribution of the rate of growth of the skull, the temperature of the wall of the non-lined reactor and the temperature of the internal surface of the skull layers for the two operating regimes of the plasma torches is shown in Fig.11.12. The origin of the coordinate *x* is at the point of intersection of the axis *x* of the cylindrical chamber with its diameter at entry to the chamber. The completion of the processes of skull formation depends on the minimum rate corresponding to the length of the reactor of 7–8 gages. It maybe seen that in all cases the skull lining increases

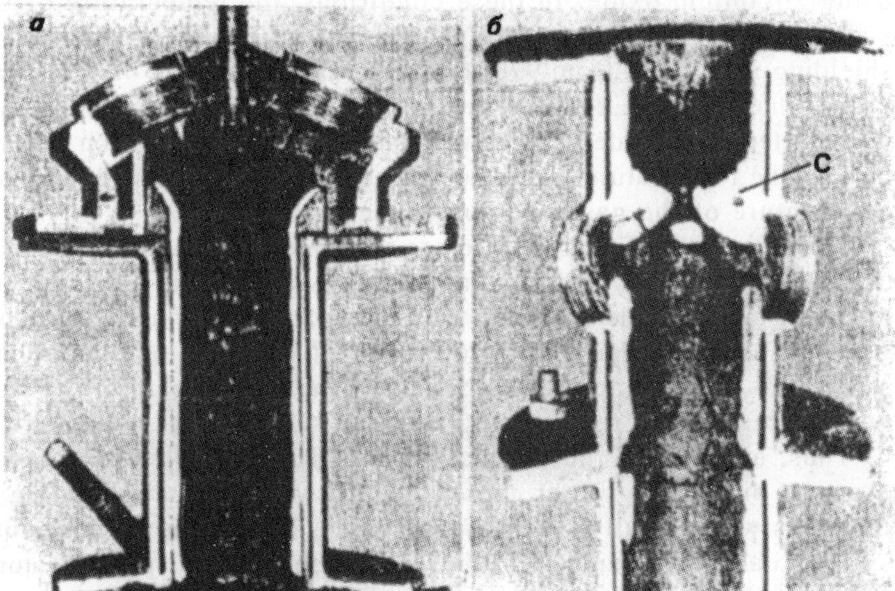

Fig.11.11. Three-jet direct flow reactor with a skull lining. a) angle of inclination of plasma jets $\alpha = 30°$; *b* – 90°.

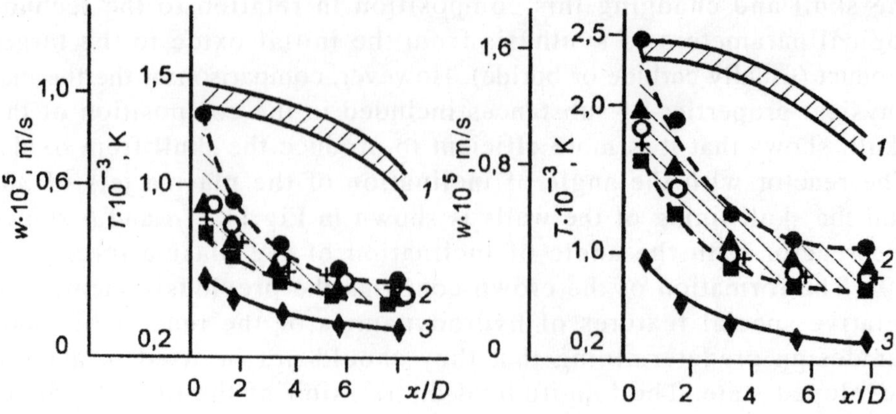

Fig.11.12. Longitudinal distribution of the rate of growth of skull *w* (1); temperature *T* of the internal surface of skull lining (2) and the temperature of the non-lined wall of the reactor (3) at a power of the arc discharge of 30 kW (a) and 80 kW(b).

the temperature of the wall, and the maximum increase, by a factor of 2–2.25 (the power of the arc discharge 80.0 kW) obtained when using zirconium oxide. In these conditions, the temperature of the wall changes from 2500 to 1000 K, whereas in the absence of thermal shielding it changes from 970 to 400 K. The formation of 'hot' wall causes that the zone of the reactor with the highest thermal stresses is characterised by a decrease of the density of the heat flow on average by 20°, the temperature factor decreases by 100% and the mean mass temperature of the gas-heat carrier increases by 15%.

The temperature fields in the cross-section of the reactor (Fig.11.13) are characterised by equidistant curves whose form indicates the relatively rapid completion of the process of formation of a homogeneous gas flow along the length of the reactor of less than one gauge and a decrease of the temperature gradient in the central path of the flow with increase from the starting sector of the reactor .

Thus, analysis of the thermal engineering characteristics of the three-jet direct flow water cooled reactor with thermal shielding of the walls with a skull made of fused oxides of vanadium, chromium, silicon, titanium and zirconium shows the technological special features of using skull lining in the process of reduction synthesis.

When examining the three-jet direct flow reactor the data published in the literature on the rational selection of plasma forming

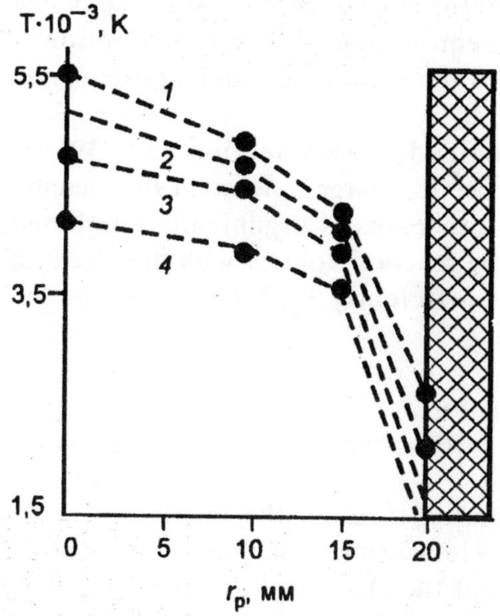

Fig.11.13. Temperature fields of the gas-heat carrier in the cross section of the reactor at x/D = 0.8 (1); 1.80 (2); 3.5 (3); 5.8 (4) and power of arc discharge of 80.8 kW. (ZrO_2 skull).

media, apparatus systems of jet plasma processes, hydrodynamic and thermal engineering characteristics of the plasma reactors and special features of heat exchange in them were analysed in detail; the efficiency of thermal shielding of the channel of the reactor by lining of different types was investigated. It was shown to be efficient to use skull lining; the heat exchange of the plasma flow with the walls of the channel at Reynold's numbers of 750 – 1400 was examined. Criterial relationships were determined for calculating the heat transfer coefficient.

11.3. COMBINED DC REACTOR WITH ELECTROMAGNETIC CONTROL

The decrease of the quality of natural resources and increasing volume of production require investigation and development of not only different designs of plasma technological reactors but also of different electrical-gas dynamic effects on the process material. Consequently, it is possible to select the optimum reactor for a specific technological process, ensuring high efficiency of extraction of useful components from the processed starting material. The concept of the optimum reactor includes: long operating life of the electrodes, high thermal efficiency of the reactor, efficiency of processing the starting material, simple automation and changes in the operating regime of the reactor, small metal requirement at high productivity. It should also be mentioned that the plasma technological reactor for processing solid (powder) materials are subject to another requirement – low consumption of the working gas phase in relation to the solid phase and, consequently, low rate of the two-phase flow.

In the combined reactor, examined in this section [38, 39] the possibility of simultaneous occurrence in a large volume of the chamber of chemical and electro physical processes is organically combined, together with efficient 'filling' of the reaction volume with the electrical arc, moving in the space with a sufficiently high rate, low gas and flow rates.

11.3.1. Principal circuit of the reactor

Figure 11.14. shows the schematic of the reactor. The following sections should be mentioned:

Two graphite electrode 1, inserted through the insulators in the lid; the cylindrical (ellipsoidal) electric arc chamber 2, assembled from longitudinal sections 10 ; lid of the chamber 9; with four orifices for introducing the powder; the electromagnetic system consisting of a direct current solenoid 4 and alternating current coils 3, 5; main and side poles 7, 8; magnetic circuits 6.

The cylindrical reactor 2 with a height of 250 mm and a diameter of 160 mm is assembled from the section 10 produced from stainless steel with a thickness of 1.5 mm. These sections are electrically insulated from each other. The use of the sectional chamber makes it possible to prevent (or weaken) the process of shunting of the electrical arc to the steel wall of the chamber and prevent its failure.

The electromagnetic circuit of the reactor has four poles: two main ones 7 and two side ones 8. On the external side, the poles are enclosed in the right angled magnetic circuits 6. The main poles carry four coils. Two of them 4 are connected together in series and generate a transverse magnetic field B_1 parallel to the axis 0z (Fig.11.15) with a maximum with the axis of the chamber. The coils 3 and 5 are connected to the three phase electrical circuit generating a magnetic field B_2 with variable strength.

In the majority of investigations of the electro-physical and thermal characteristics in the reactor of this type, the working gas was air. The same reactor was used later for developing technological processes of melting powder materials.

11.3.2. Electromagnetic method of forming a rising volt–ampere characteristic of the arc

As shown previously, magnetic field B_1 in the reactor, generated by

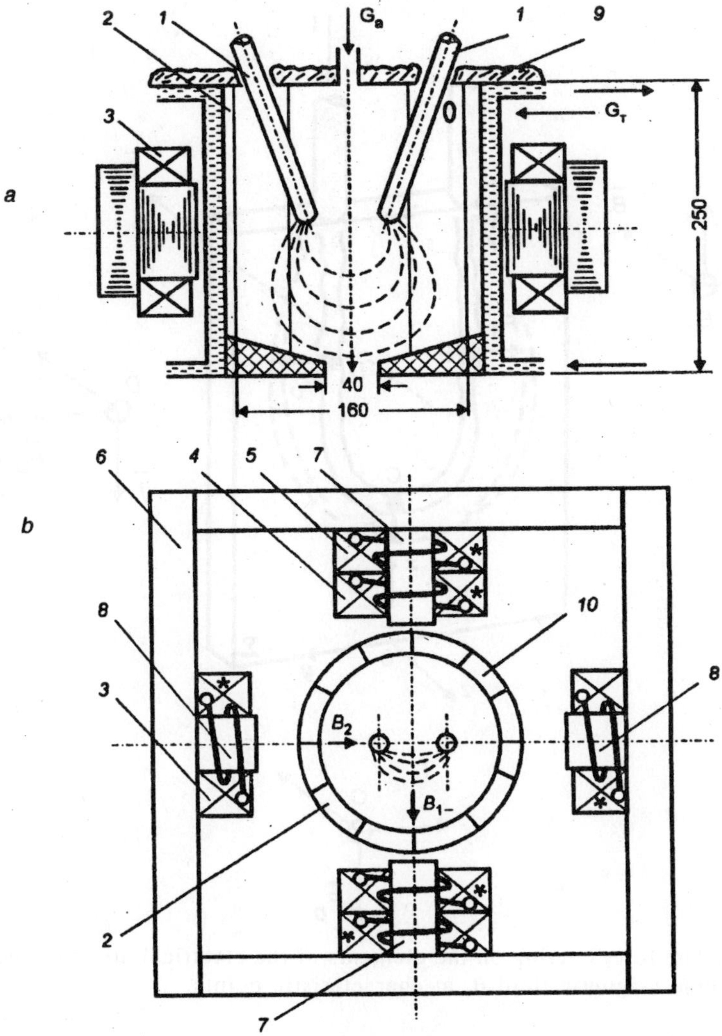

Fig.11.14. Diagram of a plasma technological DC reactor. a) side view. b) top view. 1) electrode; 2) sectional electric arc chamber; 3,5) AC coils; 4) DC coil; 6) magnetic circuits; 7,8) the poles of the magnets; 9) lid of the reactor; 10) section.

the solenoid, is normal to the plane containing the electrodes and, consequently, normal to the electrical arc situated in the same plane [40].

Figure 11.15. shows the diagram of interaction of the magnetic field B_1 with the elements of the electrical arc. At the initial moment of ignition of the arc when the arc runs between the ends of the electrodes along the line A, the magnetic field 'pushes out' the arc from the gap between the electrodes. The effect of the forces F in three characteristic points of the arc B, C, D is directed to

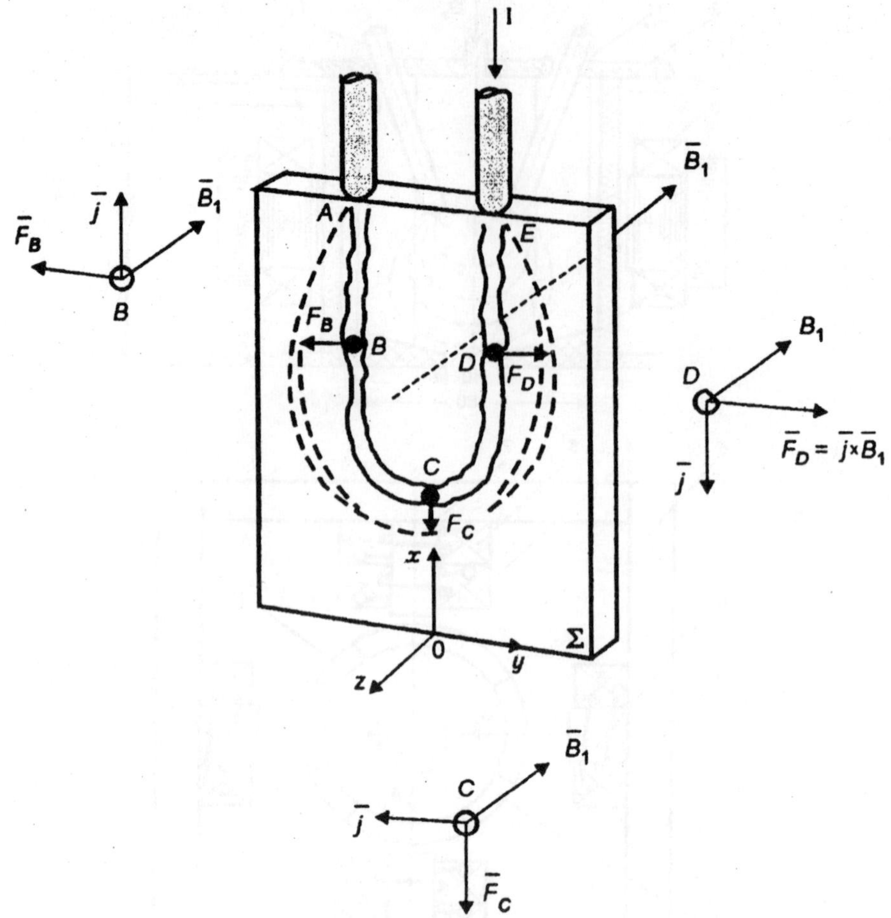

Fig.11.15. The forces acting on the elements of the electrical arc from the side of the transverse magnetic field B_1 at characteristic points.

expanding the contour of the arc (indicated by the broken line). If the induction of the magnetic field is the same at any point of the plane Σ, and the diameter of the chamber is not sufficiently large, then it is natural that the arc may at any moment of time reach the wall of the reactor and, consequently, short circuit the casing of the reactor resulting in rapid failure of the latter. The short circuit of the arc with the wall is also possible because of the fact (Fig.11.16) that the wall of the chamber under a floating potential equal to the potential of one of the points of the arc. To prevent this, the alternating magnetic field B_2 is superposed on the arc discharge (Fig.11.14). As shown by the experiments, an increase of B_2 results in a large chain in the form of the arc discharge (individual examination of the working

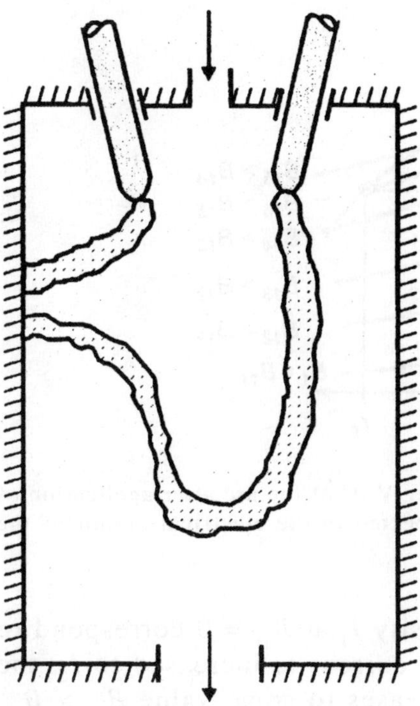

Fig.11.16. Diagram of shunting of the arc on the wall and formation of a cascade arc.

reactor); there is a transition from the distinctive constricted discharge to a deconcentred volume discharge resulting in more efficient heat exchange between the arc and the surrounding medium and increasing the arc voltage and power and also resulting in the uniform distribution of the gas temperature in the volume of the chamber.

We now examine the mechanism of formation of a rising VAC of the arc with the superposed transverse magnetic field (in the absence of the alternating magnetic field), functionally linked with the arc current.

If the induction of the magnetic field $B_1 = B_{11}$ is equal to zero, then the VAC of the arc is drooping (Fig.11.17), curve 1). This is explained by the arc-arc shunting at the end of the arc loop. With increasing current, the shunting in the arc loop which increases intensity reduces the length of the loop, i.e. voltage decreases. At specific values $B_1 = B_{11}$, B_{12}, B_{13} ... the situation remains unchanged only the VAC curves of the arc are lifted higher because an increase of B_1 whilst maintaining the specific induction value constant, is accompanied only by an increase of the initial length of the arc and, consequently, of its voltage. However, if the transverse magnetic field is matched with arc current, the VAC of the arc will increase. This process will be examined. It is assumed that at the given moment of arcing the

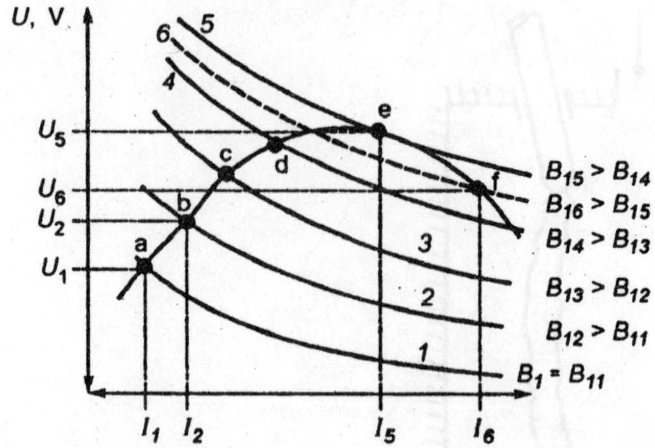

Fig.11.17. Diagram of formation of the rising VAC of the arc after application of the magnetic field B_1. The solenoid is connected in the electrical circuit of the arc.

voltage U_1 and the arc current intensity I_1 at $B_{11} = 0$ correspond to the point *a* (curve 1). When the arc current is increased to I_2, the induction of the magnetic field increases to some value $B_{12} > B_{11}$, which increases the forces increasing the arc length, i.e. the voltage increases to U_2. Point *b* is situated on the drooping VAC of the arc, corresponding to B_{12} = const. A further increase of current is described by the points *c*, *d*, *e* on the curves 3, 4, 5. Consequently, the required VAC of the arc passes through the points *a,b,c,d,e*, located in the appropriate VAC characteristics of the arc at constant values of the induction of the magnetic field B_{11}, B_{12}, B_{13}, B_{14}, B_{15}.

With a further increase of current intensity ($I_6 > I_5$) and magnetic induction ($B_{16} > B_{15}$) we obtain the regime in which arc elongation is interrupted and, in addition to this, the arc is constricted as a result of dominance of the shunting process, which reduced the length of the arc, in comparison with the process increasing the arc length. The VAC of the arc (curve 6) drops below the curve 5 and the point *f*, corresponding to the current I_6, is situated below the point *e*, i.e. $U_6 < U_5$.

This results in the formation of a rising and a drooping section of the VAC of the arc in the combined reactor of the examined type.

To fill a large working space of the reactor with the arc it is necessary to apply the alternating magnetic field B_2 (Fig.11.14*b*) whose vector can be parallel to both the axis 0*y* and 0*z*. The variable forces, acting on the elements of the arc in this case, move the arc with a high rate throughout the entire volume of the chamber.

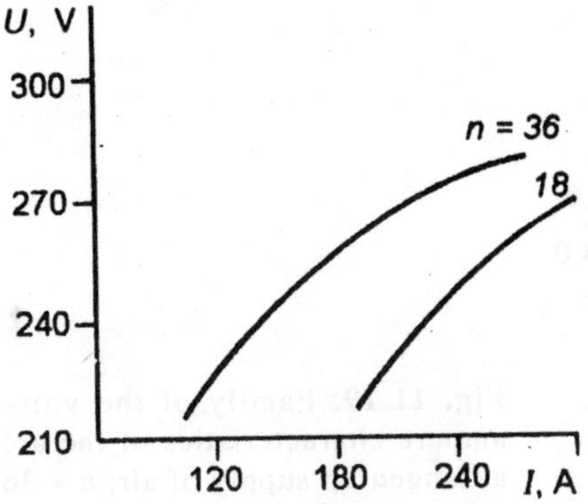

Fig.11.18. Variation of the VAC of the arc in relation to the number of turns of a series coil n forming the magnetic field, B_1. Gas supply – tangential with flow rate $G_t = 2.6$ g/s.

Figure 11.8 shows the VAC of the arc for two initial values of $B_1 = B_{11}$, determined by the number of turns n per 1 cm of the length of the coil. The strength of the magnetic field inside the coil (in the reactor between the poles of the magnetic circuits (Fig.11.14, position 7) is determined from the equation $B = 0.4\pi \cdot n \cdot I$, Oe, where n is the number of turns.

In the experiment, the mean strength of the alternating magnetic field was $B_2 =$ const. The graph shows that in a wide range of variation of current intensity (I = 90–300 A) the VAC of the arc rapidly rises and the voltage increases with an increase of the number of turns n. It should be mentioned that in this experiment the supply of gas with the flow rate $G_t = 2.6 \cdot 10^{-3}$ g/s was carried out along the tangent to the circumference.

The investigations also showed a high sensitivity of the VAC of the arc to the change in position of the ends of the series coils in relation to the external surface of the electric arc chamber. The gas was introduced into the reactor in the tangential direction. The broken lines in Fig. 11.19 show the VAC of the arc for the case of tight contact of the ends of the coils with the surface of the reactor. When the ends of the coils were 20 mm from the surface of the chamber, the scattering of the magnetic flux weakened the effect of the flux on arc discharge. Consequently, the arc voltage at I = const and identical flow rates decrease. There was a tendency for the appearance of a drooping section of the VAC of the arc (solid curves) at lower current intensities.

11.3.3. Effect of the gas flow rate and the method of introduction of the gas into the reactor of the volt–ampere characteristic of the arc

In axial supply of the gas through the orifices in the lid of the reactor it is natural to expect an increase of voltage with an increase of the gas flow rate whilst maintaining constant current intensity,

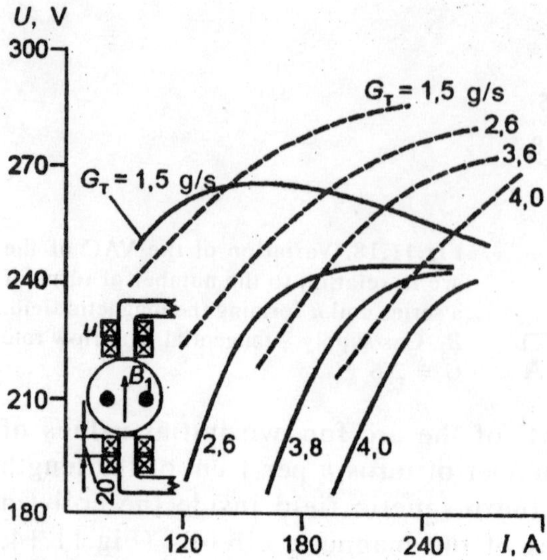

Fig. 11.19. Family of the volt–ampere characteristics of the arc at tangential supply of air; $n \approx 36$ turns. The broken lines relate to the regime in which the ends the coils are tightly pressed to the cylinder of the reactor, and the solid line – the distance between the surfaces in 20 mm.

because the arc-breakdown voltage in the loop increases resulting in a general elongation of the loop (Fig. 11.20, the two solid curves). With increase of the gas flow rate from $1.2 \cdot 10^{-3}$ to $1.8 \cdot 10^{-3}$ kg/s (by 50%) the voltage increases, although only slightly. This is associated with a low flow rate of the gas because of the large cross section of the reactor. At the mean mass temperature of the gas $T \sim 3000$ K the axial velocity of the gas is well below 1 m/s. However, for a number of processes these low velocities are essential because they ensure deeper heating of the solid fractions and, consequently, more efficient treatment of the material, and its melting point is easily reached.

The VAC of the arc is strongly affected by the change of the axial G_a blowing of the gas to the discharge chamber to tangential blowing G_t. Figure 11.20 shows, for comparison, the VAC of the arc (broken curve) corresponding to tangential blowing and a high gas flow rate ($2.6 \cdot 10^{-3}$ kg/s). Nevertheless, this characteristic is also lower than the two previous VAC of the arc. This effects is explained quite simply. An increase of the gas flow rate increases its tangential velocity and. Consequently, the archimedes force contracting the high-temperature gas and elements of the arc to the axis which reduces the arc-arc breakdown voltage, i.e. reduces the arc-length and the voltage in the arc.

It should also be mentioned that the tangential velocity of the gas may also change the dimensions of the area of the cross section of the orifices to which the gas is introduced. This is achieved easily by, for example, changing the number of orifices that operate

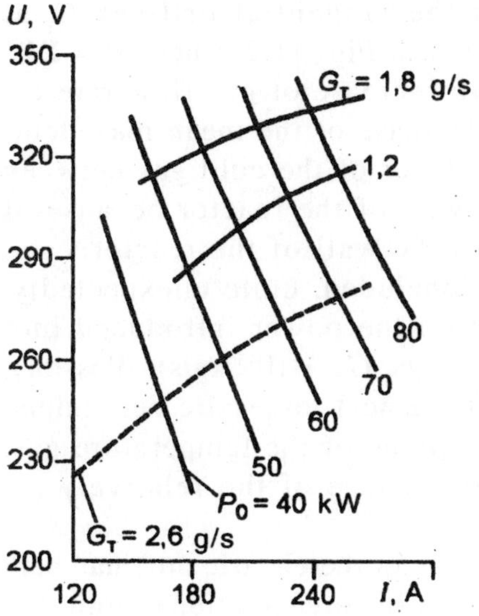

Fig. 11.20. Volt–ampere characteristics of the arc for axial air flow (solid lines) and tangential air flow (broken lines), number of turns 36.

at a constant gas flow rate.

Thus, the examined scheme of the combined plasma chemical reactor is characterised by extensive possibilities in varying the power of the electrical arc and the gas temperature in it so that the reactor can be used for greatly differing the plasma technological processes.

11.3.4. Thermal characteristics of the reactor

We examined the thermal loses T_l in the cylindrical sectioned wall of the reactor, determining them by the generally accepted procedure. In the reactor, the two adjacent sections are connected in series in water and form a single calorimeter with the individual supply and discharge of the cooling water. The error of measurement of the heat loses did not exceed 5%. The experiments were carried out with the simultaneous supply of nitrogen through the tangential axial orifices of the reactor. Some of the experiments were conducted with a diaphragm in the dished end of the reactor, with the orifice having the diameter $d = 40 \cdot 10^{-3}$ m.

The distribution of the relative heat losses $P_i/\Sigma P_i$ in the doubled sections of the arc chamber for different powers of the arc P_0 is shown in Fig.11.21. Here $\Sigma P_i = P_l$, P_i are the losses into the doubled section. With the accuracy of ±6% they can be regarded as identical on average and equal to 0.14. This indicates indirectly the large uniformity of the temperature field in different sections of the reactor.

The dependence of the relative heat losses into the chamber wall

P_l/P_0 on the gas flow rate through the tangential orifices G_t at $G_a = 1.\cdot10^{-3}$ kg/s and $P_0 = 72$ kW shown in Fig. 11.22; here $P_l = \Sigma P_i$. The heat losses slightly decrease with increase of gas flow rate G_t. This is caused by two reasons: 1) a decrease of the mean mass temperature of the gas; 2) growth of the layer of the cold gas between the high temperature region and the wall of the reactor because of applying the gas along the tangent to the wall of the reactor.

Examining Fig. 11.23, it may be concluded, quite unexpectedly, that the ratio, P_l/P_0 does not depend on the power introduced into the reactor through the arc ($G_t = 1.8\cdot10^{-3}$ kg/s, $G_a = 1\cdot10^{-3}$ kg/s). Possibly, this may be explained by the fact that the heat losses are determined mainly by the radiation of the large volume of the temperature gas, and the convective losses are small because of the relatively efficient gas insulation of the wall.

What is the temperature field of the gas in the electric arc chamber? Figure 11.24 shows temperature profiles in three sections along the height of the chamber x = 10; 70 and 110 mm. The origin of the coordinates is shown in Fig. 11.15: counting was carried out from the lower outlet of the chamber in the direction of electrodes. Experiments were carried out in the absence of a diaphragm. In all three investigated sections the temperature field in the diameter was relatively uniform. This is especially important for efficient realization of technological processes. The non-uniformity of the temperature did not exceed ±12%.

The distribution of the gas temperature along the axis of the chamber x for the installed and removed diaphragm in the reactor is shown in Fig. 11.25.

It is interesting to explain the dependence of temperature at a selected point on the axis of the chamber (x = 70 mm) on power P_0 supplied

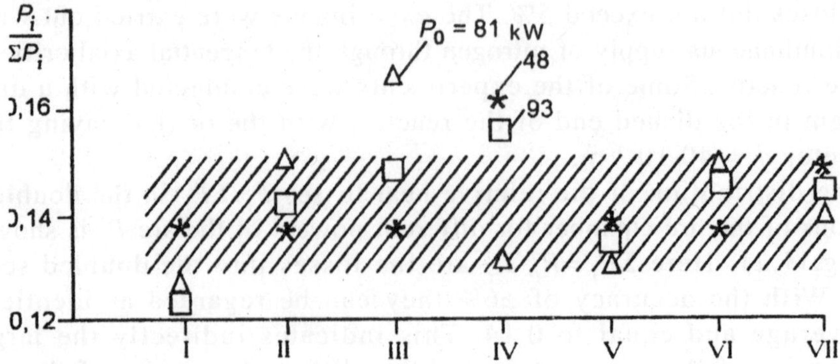

Fig. 11.21. Distribution of heat losses in sections of the reactor for three values of arc power and gas flow rate $G_t = 1.8\cdot10^{-3}$ kg/s, $G_a = 1.0\cdot10^{-3}$ kg/s.

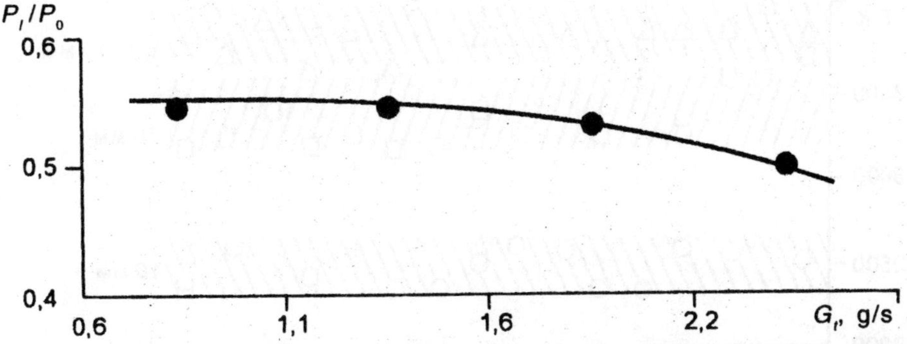

Fig. 11.22. Distribution of relative heat losses into the wall of the chamber P_l/P_0 of the flow rate of gas through tangential orifices G_t. $G_a = 1 \cdot 10^{-3}$ kg/s, $P_0 = 72$ kW.

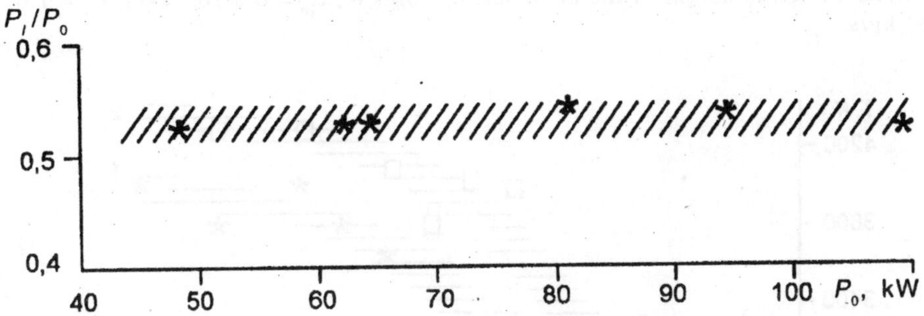

Fig.11.23. Dependence of P_l/P_0 on power P_0 introduced into the arc at $G_t = 1.8 \cdot 10^{-3}$ kg/s, $G_a = 1 \cdot 10^{-3}$ kg/s.

into the arc (Fig.11.26). The dependence is linear in the investigated power range.

We examine thermal losses in sections of the reactor, for example, the wall of the reactor, the lid and electrode feed mechanism. Measurements were taken with the supply of the ZrO_2 powder into the electric arc chamber with a flow rate of 80–100 kg/h. The variation of the heat losses during 90 min of operation of the reactor is shown in Fig.11.27.

It is important to note the characteristic special features of the curves corresponding to one of the zirconia melts. Curve 1 reflects the variations of the supplied power P_0 over a long period of time in the range 230–240 kW, curve 2 the heat losses into the chamber wall. At the initial moment of melting in the absence of a skull on the chamber walls, the losses were equal to approximately 100 kW (~42%) With increase of the thickness of the skull the losses decrease and already after holding for 30 min they are less than

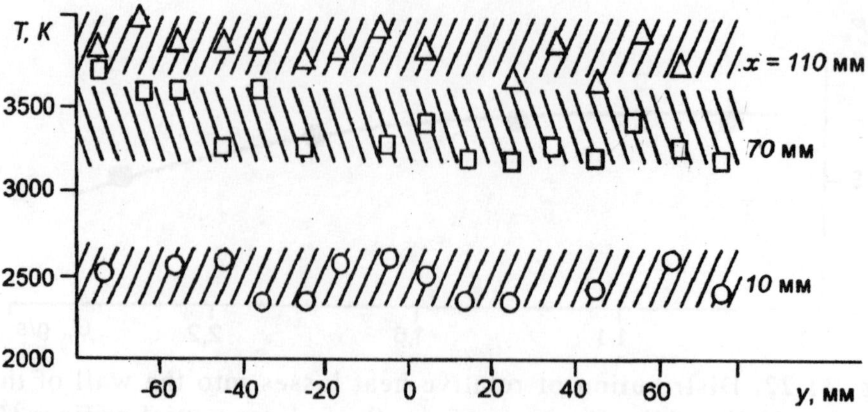

Fig.11.24. Profiles of the gas (nitrogen) temperature along the axis y in three cross sections along the height of the chamber. P_0= 80 kW; G_a = 0.8·10^{-3} kg/s; G_t = 1.6·10^{-3} kg/s.

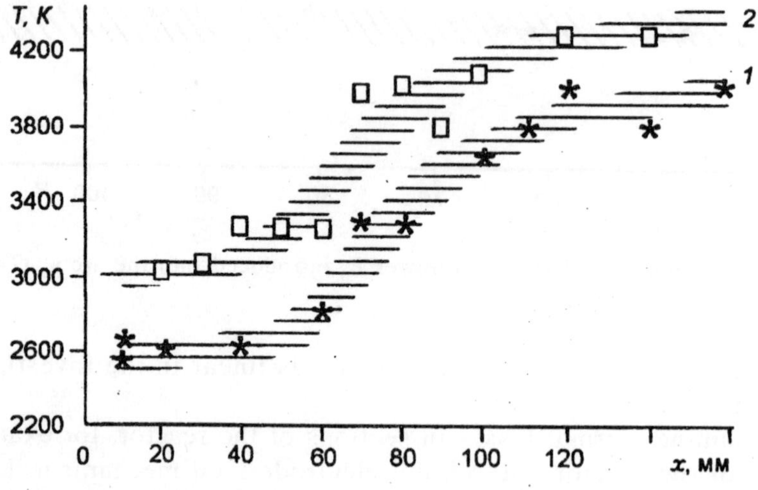

Fig.11.25. The distribution of gas (nitrogen) temperature along the axis of the chamber. G_a = 0.8·10^{-3} kg/s; G_t = 1.6·10^{-3} kg/s; P_0= 80 kW. 1) no diaphragm, 2) installed diaphragm with d = 40 mm.

50 kW, i.e. they are more than halved. At this moment of time in operation of the reactor, the thickness of the skull reached 30–50 mm and the process of growth of the skull was interrupted and the system reached the working regime. The heat losses in the lid (curve 3) increased with time from 25 to 40 kW which, evidently, maybe explained by the increase of the mean mass temperature in the volume of the reactor as a result of a decrease of the heat losses

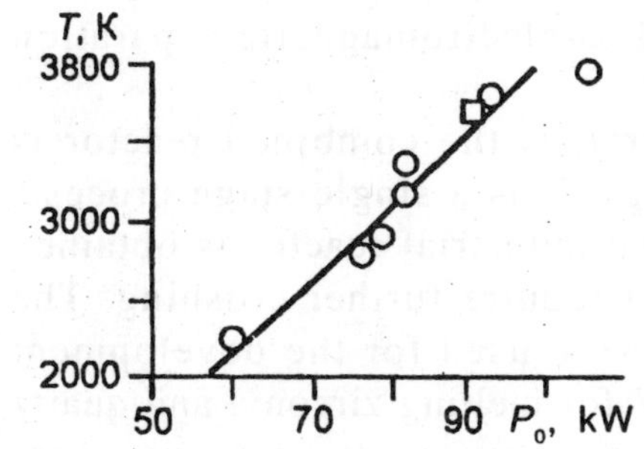

Fig.11.26. Dependence of the gas temperature at the axis of the chamber at a point x = 70 mm on arc power.

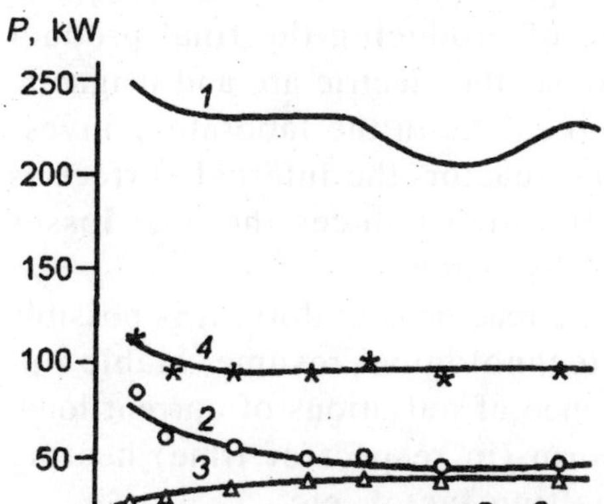

Fig.11.27. Variation of the heat losses in the section of the reactor with time. 1) arc power; 2–6) heat loss: 2) into the cylindrical wall of the reactors; 3) into the lid; 4) total loss; 5,6) into the feed mechanism of the anode and cathode.

through the reactor wall. The heat losses in the feed mechanisms of the anode *5* and cathode *6* were small (5–6 kW). The total heat losses are described by curve 4.

11.3.5. 400 kW industrial reactor for producing melted zirconium

At present zirconia is melted in furnaces with a power of the order of 400 kW [41,42]. The paths of the furnace is lined with graphite plates. In the process of melting the zirconia block, the furnace is warmed up at a voltage in the electrical arc of 80–90 V and a current intensity of 3000–3600 A. The total load of the material into the furnace is approximately 2.5–3.0 t. The melting of the block continues until the entire volume of the jacket of the bath is filled. 30 –40 min prior to the end of melting, loading of the material into the furnaces interacted and melting of the charge starts to take place. After cooling the block, the unmelted skin is knocked away manually, from the surface of the block. The total losses of the material in dressing are equal to 40–50 %. Subsequently, melted zirconia is initially refined in a screw crusher to the 40 mm fraction and smaller, and then crushing continues in a hammer crusher to the fraction smaller than 8 mm.

The resultant granules are subjected to electromagnetic separation and to rinsing to remove iron.

The plasma method of melting ZrO_2 is the combined reactor is free from many of these shortcomings. It is a single stage process, and the target product at exit from the industrial reactor is obtained in the form of granules which do not require further crushing. The results of laboratory investigations were used for the development of equipment with a power of 400 kW for melting zirconia and quartz materials. Transition from one technological process to another requires only development of a new scheme of producing the final product outside the outlet of the reactor, whereas the electric arc and magnetic circuit of the system remain unchanged. As in the laboratory investigations, in industrial service of the reactor, the internal surface is covered with a dense layer of skull which reduces the heat losses into the wall of the reactor, cooled by water.

If the supply of the material into the reactor is uniform it is possible to ensure: smooth changes in the technological resume; stable operation of the electrical arc, i.e. absence of pulsations of current load, which, in turn, results in the uniform (in respect of time) heating of the supplied material, a high melting factor, etc.

Figure 11.28. shows schematically the design of equipment with a power of 400 kW. The system consists of: 1) a bunker for supplying materials; 2) graphite electrode with feed mechanisms 5, 3) the lid of the reactor; 4) the electric arc chamber; 6) electromagnetic system; 8) the rolling track for collecting the melted product 9. The skull 7, formed in the process of preparation of equipment for melting, is also shown.

After igniting the arc between the electrodes, a powder is supplied through a drum feeder with the chamber of the reactor. In the reactor chamber, the power melts. When the melt falls into the product collector, filled with water, it is granulated. In the first 30 mins of operation of the reactor the initial material is supplied at a reduced rate to the formation on the walls of the arc chamber of a skull with a thickness of 30–50 mm. When this value is reached, the process of growth of the skull is interrupted. The thermal regime is stabilised and the equipment reaches the working regime corresponding to the productivity in respect of the starting material of 150 kg/h. Subsequently, the melt starts to flow from the reactor. The design of the lid of the reactor enables visual control of the position of the ends of the electrodes and also measurements of the melt temperature.

The electrical and technological parameters of the process are as follows: arc current intensity 1200 K, voltage 300 V, power 360 kW,

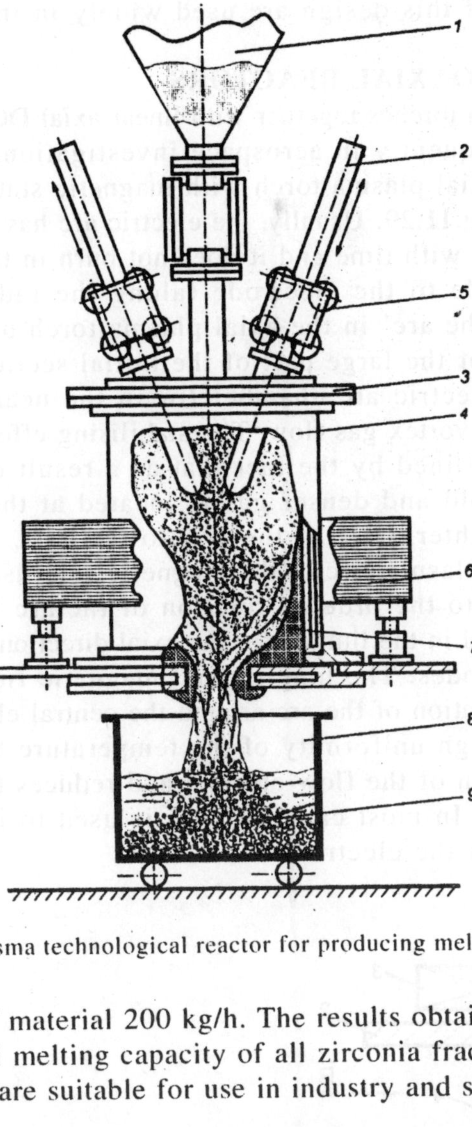

Fig.11.28. The plasma technological reactor for producing melted stabilised zirconia.

consumption of material 200 kg/h. The results obtained with the melts confirm the full melting capacity of all zirconia fractions. The melting target products are suitable for use in industry and satisfy the technical requirements.

The specific energy losses determined from the results of a large number of melts for zirconia were equal to on average to 1.8 kW h/kg, 2.5 times less than in melting in furnaces. This thermal efficiency of the reactor was equal to 0.6.

The tests show that the burning rate of graphite anodes is 2 times faster than that of the cathode and, consequently, it is supplied into the reactor at a rate twice as high as the cathode, The specific erosion of the anode at a current of 1100 A was equal to $2 \cdot 10^{-7}$ kg/C that of the cathode $1 \cdot 10^{-7}$ kg/C, which corresponds to the data published by other authors.

The systems of this design are used widely in industry.

11.4. PLASMA COAXIAL REACTORS

The coaxial plasma torches together with linear axial DC plasma-torches are used very frequently in aerospace investigations. The simplest diagram of a coaxial plasma torch with magnetic stabilisation of the arc is shown in Fig.11.29. Usually, the electric arc has the complicated form which varies with time and it does not burn in the shortest path from the electrode to the electrode (along the radius). The term 'stabilisation of the arc' in the axial plasma torch usually refers to the stabilisation of the large part of the initial section of the arc at the axis of the electric arc chamber (or in the near-axial region), carried out by the vortex gas flow. The stabilising effect of the vortex on the arc is explained by the fact that as a result of the centrifugal forces, the cold and denser gas is located at the wall displacing the heated lighter gas, i.e. the arc, to the axis.

In the coaxial plasma torch, 'the magnetic stabilisation of the arc discharge' refers to the ordered rotation of the arc by the longitudinal magnetic field in the limited (in the axial direction) space between the coaxial electrodes. The longitudinal magnetic field is produced by a solenoid. Rotation of the arc around the central electrode ensures the sufficiently high uniformity of the temperature field of the gas in the cross section of the flow channel and reduces the erosion rate of the electrodes. In most cases, a wire is used to ignite the electrical are between the electrodes.

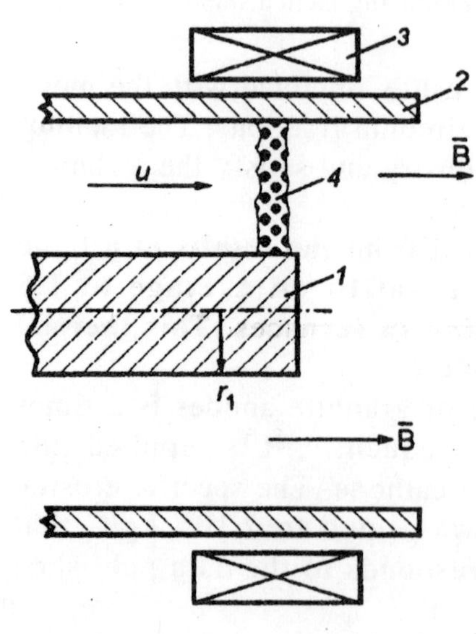

Fig.11.29. Diagram of a coaxial plasma torch with a solenoid. 1,2) coaxial electrode; 3) solenoid; 4) arc.

11.4.1. Coaxial electric arc DC plasma torch

In many cases of industrial application of the low temperature plasma it is necessary to generate high temperature flows with the maximally uniform distribution of temperature in the cross section of the channel and a low flow rate of the gas and the initial material in the axial directions. [43]. However, the plasma flows, discharged from the linear electric arc preheater with gas vortex stabilisation of the arc, contain a distinctive high temperature core, so that there is a problem with equalization of the temperature field. This requires installation of damping or volumes and, consequently, additional losses of energy or care.

One of the devices with a sufficiently uniformed field of the temperature of the discharged plasma flow is a coaxial plasma torch with magnetic stabilisation of the arc.

We examine the simpler scheme of such a plasma torch-reactor (Fig.11.29): two coaxial reactors, 1,2 inserted into the solenoid, 3, generating the external magnetic field in the zone of arcing. Under the effect of the magnetic field, the arc starts to move and disappears at rotation. To explain the kinematics of movement of the arc, we examine the scheme of the plasma torch without a gas flow, when $u = 0$ (Fig.11.30a). Fig.11.30*b* shows a section of the circular strip with a thickness dr situated at a distance of r from the axis of the plasma torch. The element of the arc a, situated in this strip at the moment of time $t_1=0$, subjected to the effect of the electromagnetic force ($I \times B$), direct in the direction normal to the element, occupies a moment $t_2 = t_1+dt$ position b and its position in the strip is occupied by the arc element c which is in position d at time t_1. If the actual velocity of movement of the element of the arc is w, then the element passes the path $w{\cdot}dt$ within the limits of the strip, and in the tangential direction (in the direction of apparent rotation) the path $\omega{\cdot}r{\cdot}dt$. In geometrical relationships it is easy to find the link between the angular velocity of 'rotation' of the arc ω in the form of the arc channel (assuming that movement is steady and I = const, B = const, i.e. w = const). It has the form:

$$r d\varphi = \sqrt{(\omega{\cdot}r / w)^2 - 1}. \tag{11.1}$$

The integration of equation enables us to represent the form of the arc by the following equation:

$$\begin{aligned}\varphi = &\sqrt{(\omega{\cdot}r / w)^2 - 1} - \arccos\left(w / (\omega{\cdot}r)\right) - \\ &-\sqrt{(\omega{\cdot}r_1 / w)^2 - 1} - \arccos\left(w / (\omega{\cdot}r_1)\right).\end{aligned} \tag{11.2}$$

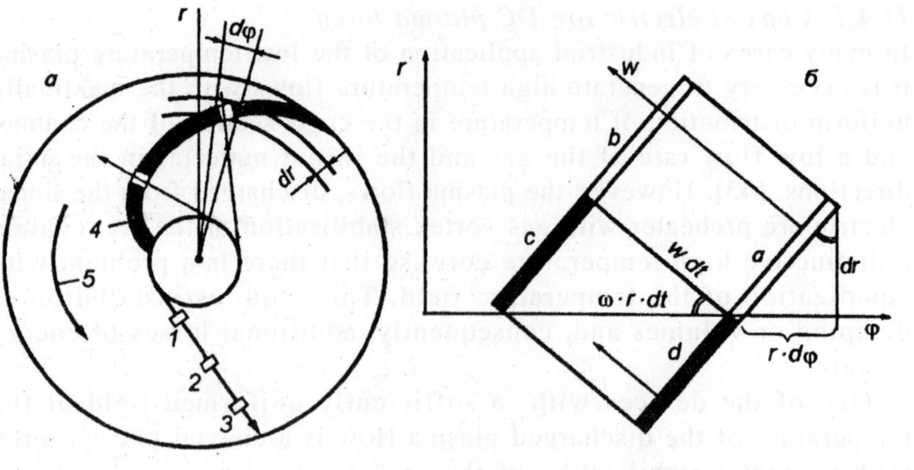

Fig.11.30. Scheme (*a*) and the diagram of movement of the arc (*b*) in a coaxial plasma torch in the presence of an axial magnetic field. 1,2,3) successive positions of the arc elements; 4) instantaneous position of the arc; 5) direction of movement of the arc.

Counting is carried out from $r = r_1$, where the value of the angular coordinate $\varphi = 0$ is assumed. The resultant equation (11.2) describes the instantaneous form of the arc. The displacement of the fixed element of the arc with time maybe determined from another geometrical relationship:

$$\sqrt{(\omega r)^2 - w^2} \cdot dt = \sqrt{1+(r \cdot d\varphi / dr)^2} \cdot dr,$$

which is integrated using the previous equation (11.2) for the case of the steady rotation of the arc, i.e ω =const. As a result of integration we obtain

$$\omega \cdot t = \sqrt{(\omega \cdot r / w)^2 - 1} - \sqrt{(\omega \cdot r_1 / w)^2 - 1}. \qquad (11.3)$$

The time is counted from the moment when the element is at the radius r_1.

It may be shown that the movement of the arc element in the plasma torch with a homogeneous magnetic field takes place from the internal to external electrode. It is assumed that at the initial moment of time, the arc is situated strictly along the radius. When the magnetic field is applied, the entire arc column assumes the same linear velocity, i.e. the angular velocity of the arc in the vicinity of the internal electrode must be higher. The arc appears to twist around the internal electrode so that the convexity of the line, describing the shape of the arc, will be directed from the axis of the electrodes. This form is established because of the displacement of the elements of the arc from the internal to the external electrode (see the diagram in Fig.11.30). Assuming that the arc should be normal to the surface of the internal electrode, and setting as r_1 the radius of the internal electrode, from (11.1) we obtain

$$\omega \cdot r_1/w = 1 \tag{11.4}$$

This is the condition of determination of the angular velocity of the 'rotation' of the arc.

Taking into account equation (11.4), the equation (11.2) and (11.3) have the form:

$$\varphi = \sqrt{(r/r_1)^2 - 1} - \arccos(r_1/r); \tag{11.5}$$

$$r/r_1 = \sqrt{1 + (\omega t)^2}. \tag{11.6}$$

The equivalent form, described by equation (11.5) is confirmed quite satisfactorily by arc photographs.

One of the most important special features of arcing in a coaxial plasma torch is the displacement of the individual elements along the column of the arc to the outer electrode. In the presence of the gas flow in the channel of the plasma torch, the kinematic of movement of the arc becomes more complicated but the qualitative pattern remains unchanged [44,45]. The arc-wall shunting phenomenon, especially on the outer surface of the electrode, corrects the shape of the arc and influences the speed of movement of the near-electrode sections of the arc along the electrode surface.

General considerations show that the strength of the electrical field of the arc subjected to the effect of the external magnetic field, should be higher than in linear plasma torches with gas-water stabilisation with other conditions being equal (current intensity, gas flow rate, pressure), especially in the initial section of the electric discharge chamber. This is associated with different mechanism of heat exchange of the arc with the surrounding medium. If in the arc with longitudinal blowing heat transfer to the gas flow takes place mainly by means of heat conductivity, in the coaxial plasma torch the convective heat transfer is more important.

Detailed descriptions of the investigations of coaxial DC plasma torches with a single central electrode-cathode was published in [1,43–45].

11.4.2. Coaxial plasma torch–reactor

In a number of processes in processing of material it is necessary to use high temperature conditions with the controlled composition and flow rate of the gas which do not affect the characteristics of the heater characterised by a higher concentration of power and a high utilisation factor. This regime is ensured by the coaxial DC plasma torch – reactors [46] with a non-cooled graphite electrodes (Fig 11.31). The outer electrode – cathode 1, with a diaphragm below it, forms the working zone for processing the material. The supply of the material for processing and transfer of the material into the zone of the heat working gas takes place through the internal cavity of the anode 2. The gap between the electrodes is 0.2 m. A DC arc is under the effect of axial component of the induced magnetic field B_0 of the solenoid 3. The cylindrical tubular-cathode is thermally insulated by soot 4, and fire clay lid 5, filling the heart resisting non-magnetic casing 6. The arc travels in the gap between the electrodes under the effect of the magnetic field and heats the electrodes to 2000 – 2600 K. In addition to the heat insulation of the arcing zone, this ensures the formation, as reported by the authors, of a volume charge whose appearance was recorded using a signal from the probes 7 and the current conductor 8. In constriction of the discharge the signal varied in the amplitude (because shunting is possible. The value of B_0 was varied in the experiments [46] from 0.01 to 0.02 T. The Larmour radius was in the range from several units to tens of per cent of the free path length of the electrons, resulting in the formation of the tangential component of the velocity of the electrons in the volume discharge conditions.

The existence of the volume discharge is attributed by the authors only to the fact that the value of the signal from the probes is constant with time. However, a different interpretation is also possible, namely: the absence of large scale shunting at a high temperature

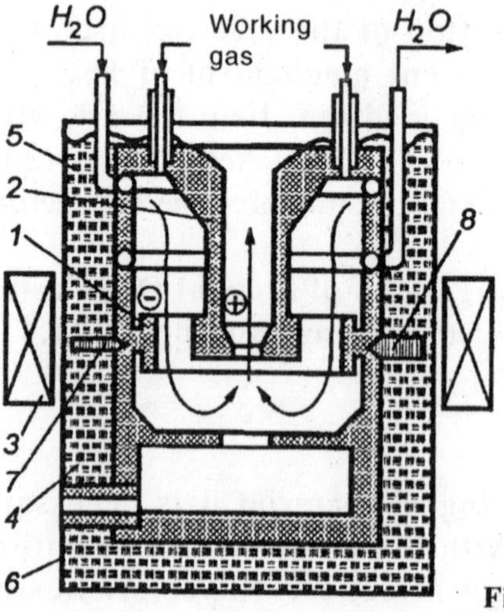

Fig.11.31. Diagram of the coaxial plasma reactor.

of the cathode wall and, consequently, the absence of pulsations of current and voltage. In chapter 2 dealing with the breakdown voltage between the arc and the wall it is concluded that the voltage rapidly decreases with increasing temperature of the cathodes surface, and in the examined case, the cathode was made of carbon, heated to a high temperature. In all likelihood, the electrical arc, it is still constricted, rotates, in the space between the electrodes under the effect of the magnetic field B_0. The reference spots of the arc travel continuously on the surface of the electrode, without jumps, so that the signals from the probes are constant in respect of time. The volt–ampere characteristics for this case are shown in Fig.11.32. They are rising resulting in stable arcing without any ballast resistance in the electrical circuit and with the electrical efficiency coefficient close to 1.

The power of the coaxial reactor in the experiments were varied from 100 to 400 kW, the discharge current intensity from 200 to 800 A; the flow rate of the working gas (nitrogen, mixture if air with methane) varied from 0 to 10 mm^3/h and did not effect the VAX of the arc. The efficiency of equipment increased with increasing power and reached 0.9. The rate of erosion of the electrodes was determined only by the rate of the process of evaporation of the material. The operating conditions of the plasma torch were stable, the temperature of the gas in the working zone reached 3000 K.

11.5. COAXIAL DC REACTOR WITH ELECTROMAGNETIC CONTROL

The reactor working with three phase DC will be examined [47–50]. The authors of these studies investigated two different circuits of the three phase coaxial reactor: with 2 rod electrodes positioned parallel and symmetrically in relation to the axis of the cylindrical reactor at some distance from each other and with three rod electrodes positioned in the tips of the equilateral triangle whose centre is located on the axis of the reactor. In both variants, one of the electrodes was always represented either by a narrow cylindrical strip made of graphite or, in the majority of experiments, the entire internal wall of the reactor made of the same material. The region of burning of the electrical arc, formed between the electrode, was subjected to the effect of the axial magnetic field generated by DC solenoid embracing the outer surface of the reactor. The electrical arc was ignited with a wire.

The majority of the experiments conducted on a three-phase reactor

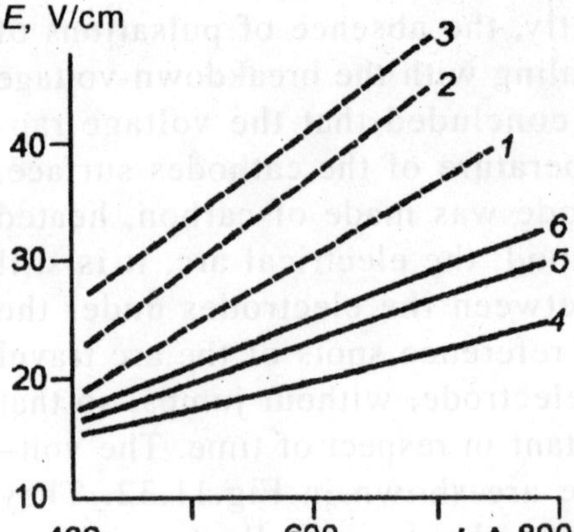

Fig.11.32. The volt–ampere characteristic of the arc. T = 2300 K; 1) B = 0.03 T; 2) 0.06; 3) 0.08; T = 2600 K: 4) B = 0.03 T; 5) 0.06; 6) 0.08.

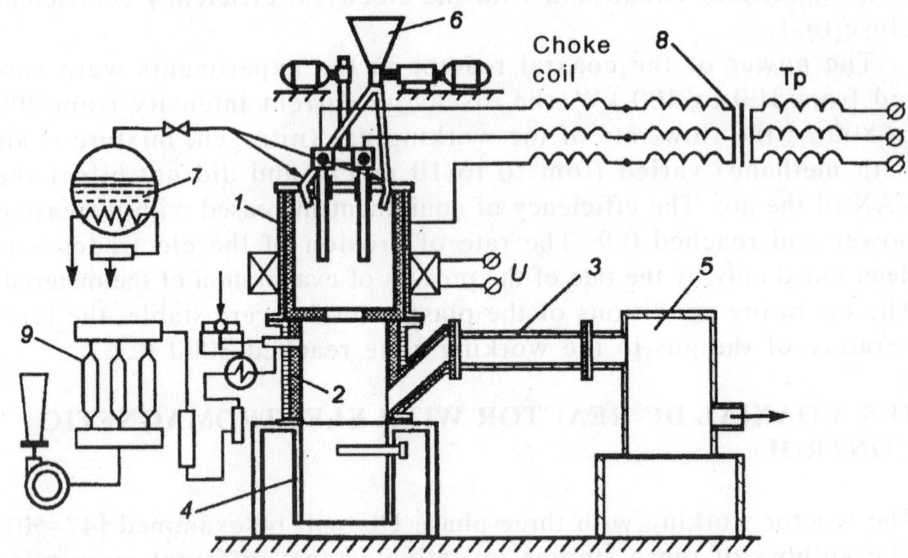

Fig.11.33. Three-phase reactor: 1) reactor; 2) container for receiving the gas and slag; 3) pipe for discharging gas; 4) slag collector; 5) gas cooler; 6) dust feeder; 7) gas supply; 8) electric power supply; 9) gas analyser.

with two rod electrodes. The diameter of the cylindrical chamber was d = 100, 150 and 200 mm, the diameter of the rod electrodes was respectively 20, 25 and 30 mm, the height of the chamber H = 200 mm (Fig.11.33). The electrical arc ran in a specific sequence between all electrodes.

In this reactor circuit it is important to discuss the restricted rotation of the arc in the plane normal to the axis of the chamber, or on the conical surface under the effect of the magnetic field and aero dynamic forces. It was not possible to describe the movement of this type, as carried out for the classic coaxial DC reactor–plasma torch when the arc is effected only by the magnetic field. The general diagram of the examined experimental system is shown in Fig.11.33.

In test of the single-phase reactor, where one of the rod electrodes is positioned along the cylinder, the supply of the powder on the upper lid was accompanied by the formation of a vortex gas flow. The resultant rotation of the arc both in the plane and normal to the axis in the meridional plane caused the circular motion of the powder particles (sol of Kuzbass coal, particle size d = 100 μm). This was supported indirectly by the electrical arc. The viscosity of the case of the gas in the arc is considerably higher than the viscosity of the surrounding medium.

A large part of the powder was not only heated in the high temperature gas flow to the melt condition but also displaced by the centrifugal forces to the reactor wall where it formed a solid skull at contact of the melt with the cold water cooled wall. Subsequently, the molten powder flowed downwards along the skull into the slag collector. Small changes in the thickness of the skull, measured in the direction of the circumference of the reactor at different heights of the later indicate the sufficiently high uniformity of the field of the circumferential velocity of the gas in the selected plane of the reactor situated below the electrodes. The presence of the skull increased the operating life of the graphite cylindrical electrode and increased the thermal efficiency of the reactor.

On the basis of the results of the experiments carried out by the authors of [50], the VAC of the arc is calculated by the criteria of dependence of the arc voltage drop in relation to the current intensity I, magnetic induction B, the flow rate of the reagent G_p and the diameter of the chamber D:

$$U=1.79\cdot10^{-3}\ (I/D)\cdot(10I\cdot B/G_p)^{0.113} \qquad (11.7)$$

The determining dimension criterial complexes were I/D and $I\cdot B/G_p$. The later complex characterises the interaction of the magnetic field with the arc: $G_p = \rho \upsilon F$ is the mass flow rate of the process materials; υ is its velocity; F is the cross-sectional area of the reactor; ρ is the density of the material.

Comparison of the experiments with the calculations using the generalized equation for three diameters of the reactor is shown in Fig.11.34. It is interesting to note the relatively large scatter of the experimental values (points) whose absolute value is 50%. We believe that the reasons for the scatter are not random. The dimensional criterial

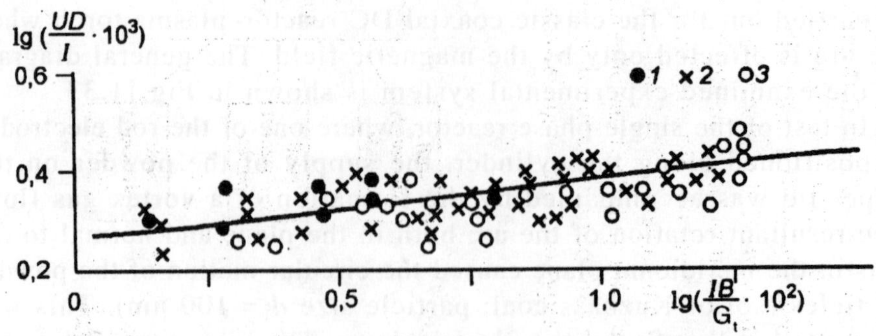

Fig. 11.34. Generalised dependence $UD/I = f(I{\cdot}B/G_t)$ for three values of the chamber diameter. D: 1) 100 mm; 2) 150; 3) 200.

complex I/D can be written in the form:

$$I/D = (I^2/G_pD)^{0.5}(G_p/D)^{0.5},$$

i.e. it is associated with the energy complex and the Reynold's number. In addition to this, the experiments were carried out at three values of the diameter of the reactor D. Therefore, regardless of the constant gas pressure in the discharge chamber, it is also necessary to take into account the complex (pD) proportional to the Knudsen criterion.

If generalisation is carried out taking into account separately all complexes, the scatter of the experimental points greatly decreases and the equation (11.7) has a different form.

We now present the dependence of the thermal efficiency of the reactor $\eta_r = P_t/P$, determined as the ratio of the thermal energy P_t, transferred to the material, to the arc power P. The flow rate of the transport gas is low in comparison with the mass of the processed material in unit time. It is difficult to determine P_t and, therefore, the author when calculating η_r used in all likelihood the experimental data into on the heat loses in the water cooling the reactor.

Processing of the experimental material gives the following criterial equation for geometrically similar cylindrical reactors:

$$\eta_r = 1.4\ (10^2 I{\cdot}B/G_p)^{-0.266} \qquad (11.8)$$

The equation holds, as also noted by the author of [50], only in the current range I = 100–500 A, the flow rate of the reagent of 3–60 kg/h and at a constant value of the ratio I/D (Fig.11.35). Restricted extent of application of this equation can clearly be seen, for example, at $B \to 0$ or $G_p \to \infty$.

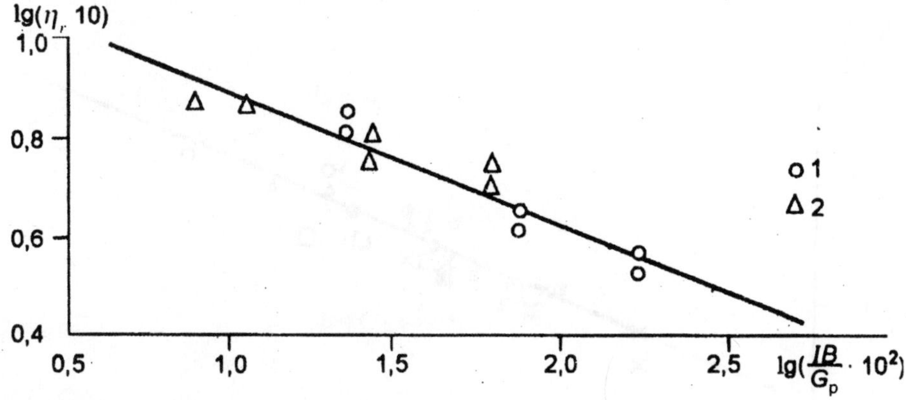

Fig.11.35. Generalized dependence of the thermal efficiency of the reactor $\overline{G}_p$ of the complex IB/G_p. G_p, kg/h: 1) 3.3; 2) 8.7.

The authors of [47–49] evaluated the effect of productivity of a reactor, determined by the dimensionless parameter $\overline{G}_{ef} = G_w/G_p$, where G_w is the amount of molten powder material settled on the wall and falling into the slag collector, because the powder material is the final product in this technology. Generalization of the experimental data and also analysis show that the technological efficiency of the investigated reactor is described by the equation:

$$\overline{G}_{ef} = G_w/G_p = 12.12\left(10^2 \cdot IB/G_p\right)^{0.433} \quad (11.9)$$

Figure 11.36 shows the results of experimental examination of the technological efficiency for single-phase and three-phase reactors. For the single phase arc G_p=9.4–19.2 kg/h, I = 320–370 A, and for the three phase arc the consumption of the powder was 19.2 kg/h, arc current 340 A. The powder was supplied into the reactor chamber through a jet under the effect of the gravitational force or with a dispersed jet using a transport gas. The satisfactory agreement of the empirical dependence (11.9) with the experiment shows that the approach produced is efficient in generalisation of the experimental data for the examined combined-type reactors [49].

11.6. A reactor based on a linear plasma torch for pyrolysis and processing chemical production waste

The current production of petrochemical products is realized mainly using olefin initial materials – natural gas and oil fractions. The technology of industrial production of olefins (ethylene, propylene butadiene) is based on the thermal destruction of hydrocarbons, included in the composition of the processed starting material, using the processes

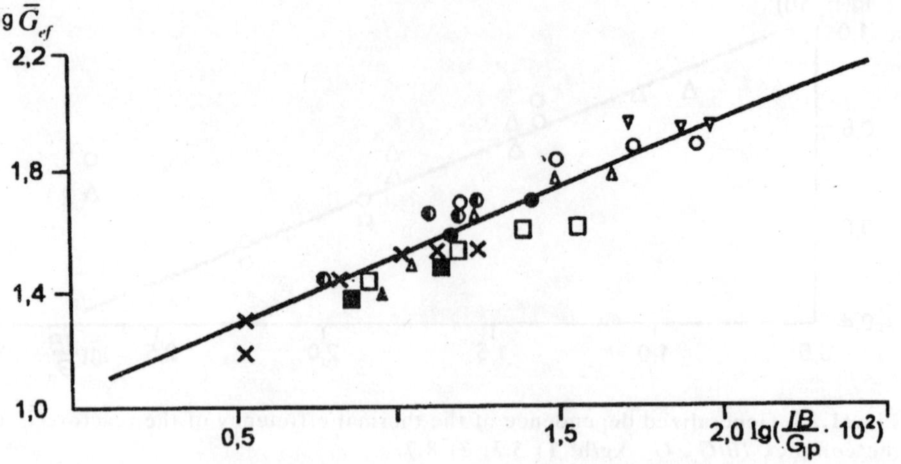

Fig.11.36. The generalized dependence of the technological efficiency $\overline{G}_{ef}$ on the complex IB/G_p for the single phase and three phase reactors.

such as thermal cracking and pyrolysis. However, the pyrolysis of all fractions in tubular furnaces is limited by the maximum temperature which can be reached (1173 K).

Plasma chemical technology has considerable possibilities because the temperature of the heat carrier may reach 3000–5000 K, and chemical transformations of hydrocarbons are accelerated at high temperatures. In this process, organic substances are destructed by the energy of the low temperature of plasma of the reaction gas (hydrogen, mixture of hydrogen with methane).

11.6.1. Production of acetone and ethylene from oil products

The technology of pyrolysis of low-octane benzene in hydrogen plasma was refined in pilot plant equipment at the Kaustik company, Sterlitamaksk. Acetylene and ethylene were produced from benzene. In the optimum conditions, the degree of transformation of benzene to acetylene was 75 wt.%. The consumption of electric energy in production of acetylene was 7.8 kWh/kg, and the total consumption for acetylene and ethylene 5.7 kW h/kg. The quality of acetylene and ethylene was good enough to use then for the synthesis of vinyl chloride, trichlorethylene and other products.

The results of experimental studies were used for technical and economical substantiation of the production of acetylene and ethylene from benzene by the plasma chemical method. The main arguments and justifications will now be examined.

At present, acetylene is produced by oxidation pyrolysis of natural gas, homogeneous pyrolysis of benzine and from calcium carbide.

The currently available pyrolysis methods have a number of shortcomings: low yield of acetylene, increases consumption of starting materials, low specific productivity of the reactor. The production of acetylene by the carbide method is time consuming and 30% of the initial carbon is lost in the form of carbon oxide. The technology requires a large amount of electrical energy (10–11 kW h/kg of acetylene and contaminates the environment (emissions of carbon oxides and dust in the atmosphere, formation of mineralized effluents). The low productivity of all existing methods of production of acetylene also increases the cost of chemical products based on acetylene and impairs the ecology of the environment. In plasma chemical technology, the consumption of initial materials per unit production decreases 2.2 to 2.8 times, and the yield of secondary harmful products decreases. According to the estimates, the general consumption of energy per unit mass of olefins in the plasma chemical method of processing is on the level of the energy consumption of the process of thermal pyrolysis [51, 52]. It is well known that benzene, kerosene and diesel fractions of oil are scarce fuels and, consequently, high-boiling fractions of oil (gas oil, mazut) are preferred for thermal pyrolysis. However, the processing of these fractions into olefins is difficult because of thermal dynamic and kinetic restrictions.

Therefore, experiments were carried out to examine the pyrolysis of vacuum gas oil and mazut in hydrogen plasma. The results show that the degree of transformation of gas oil to the sum of acetylene, ethylene and propylene reaches 75 wt.%, and the degree of transformation of mazut to the sum of acetylene, ethylene and propylene is 50 wt.% which is higher than in high temperature thermal pyrolysis in the flow of a homogeneous heat carrier [53].

11.6. Processing organic and chlorine organic chemical production waste

The plasma chemical technology uses, as the initial material, various organic waste because they contain a large amount of accumulated hydrocarbons. For example, at the currently available technology of production of chlorine-organic products, because of the low selectivity of the processes of processing the starting materials, the fraction of the chlorine organic waste is 0.5–60% of the produced final product and, therefore, the utilization of waste and return back to processing is an urgent task. The application of high temperature plasma chemical technology makes it possible to decompose the organic waste of any phase state (gaseous, liquid and solid) and also the waste in the form of inflammable and non-inflammable formations (highly chlorinated compounds of the type of hexachlorobenzoil, hexachloroethane and others). In pyrolysis of chlorine-organic and organic waste in hydrogen plasma, gas and soot are obtained. The composition of the gas contains acetylene, methane, hydrogen and also hydrogen chloride. Acetylene and hydrogen chlorine are the starting materials for production of vinyl chloride, and ethylene chloride may be produced from ethylene and hydrogen chloride. Plasma chemical soot is not inferior to thermal soot as regards quality.

The utilisation of the liquid base of chlorine organic products includes the stages of pyrolysis, cleaning of gases to remove homologues or acetylene and hydrocarbons C_3, C_4 and the process of synthesis of chlorine-organic products (Fig.11.37). The waste is pyrolysed in the main apparatus–the plasma system consisting of the plasma torch 2, the plasma chemical reactor 3, and the quenching device 9. The power to the plasma torch is supplied (direct current) for the electric power source 1.

The plasma system operates as follows. In the plasma torch 2, the plasma forming gas is heated by the electric arc to the mean mass temperature of 3000–5000 K. The gas in the condition of low temperature plasma travels into the plasma chemical reactor 3 where it is mixed with the initial material, and this is followed by heating and evaporation of the starting material with subsequent pyrolysis to produce acetylene, hydrogen chloride, methane and hydrogen. Pyrolysis gas is gas subjected to high speed quenching in the quenching device 9, and the produced pyrogas is then additionally cooled in the heat exchanger 8. The cooled pyrogas is compressed with a compressor 7, and supplied into the cleaning (scrubbing) reactor 4, where the pyrogas is cleaned to remove the homologues of acetylene and hydrocarbons C_3, C_4 by selective chlorination.

The process is realized in the bubbling system in a solvent in the presence of a catalyst. Undesirable impurities, chemically bonded with chlorine, are returned by the system to the pyrolysis stage. The scrubbed pyrogas, whose quality satisfies the technical requirements of olefin hydrocarbons, and is used for synthesis of organic products, is directed into the synthesis reactor 5, from where the reaction mass travels to separation 6, where the completed product is separated. The residue from the column 6 is used as the starting material in pyrolysis.

The technical effect from the process of processing waste by the plasma chemical method is based on the following:

– as a result of more efficient utilisation of the elements, present in the starting material, the yield of target product increases;

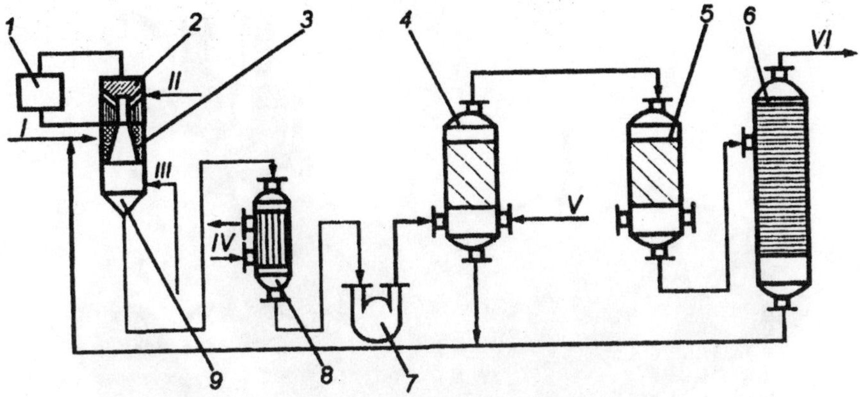

Fig.11.37. Technological diagram of equipment: 1) power source; 2) plasma torch; 3) reactor; 4) selective cleaning reactor; 5) synthesis reactor; 6) separation; 7) compressor; 8) heat exchanger; 9) quenching device. I – waste; II – plasma forming gas; III – quenching agent; IV – cooling agent; V – chlorine; VI) organic product.

– the process is closed and waste free;

– the use, as the heat carrier, of hydrogen plasma with a short contact time and high rates of the decomposition process ensures high specific productivity of the reaction volume and miniaturisation of equipment in pyrolysis.

Technology of plasma chemical processing of waste in hydrogen plasma was verified extensively on pilot plant equipment. The degree of transformation of the hydrogen into acetylene was estimated at 70 wt.%. The consumption of energy for the decomposition of waste was upto 2 kW h/kg. Twenty eight types of toxic waste were processed [54]. The test results were used for the development of a plasma chemical module with a productivity in respect of the initial material (waste) of 375 and 750 kg/h (PKh M-375 and PKh M-750), including:

– GNP-0.75 or GNP-1.5 plasma torch with a power of 750 and 1500 kW, respectively;

– Plasma chemical reactor;

– Quenching device;

– Disk filter for removing soot from pyrogas.

GNP-0.75 and GNP-1.5 relates to the class of linear plasma torches with gas vortex stabilisation of the arc and gas supply distributed along the length of the discharge chamber. The general view of the GNP-1.5 plasma torch is shown in Fig.11.38.

The multiple position tungsten cathode makes it possible to change the working section without switching the arc off and this has a positive effect on the duration of failure free operation of the plasma torch. A sectioned inter-electrode insert is placed between the cathode and

Fig.11.38. Industrial plasma torch GNP-1.5 with IEI (1) with a power of upto 1500 kW. For comparison, the photographs of the manual spraying plasma torch (2) and high current (up to 1 kA) plasma torch for cutting thick metal are shown (3).

the copper tubular anode and is used for changing the arc length in a wide range. Technical solutions were used to develop the unified design of plasma torches of different power [55, 56]/ The standard convertor PVT 2-800/8 is used as a power source for the plasma torch.

Conclusions

The monograph, presented to the reader, contains the results of 40 years of research, carried out mainly in the Department of Plasma Dynamics of the Insitute of Theoretical and Applied Mechanics (ITPM) of the Siberian Division of the Russian Academy of Sciences. A number of chapters have been written by scientists of some other institutes and they are published with their agreement.

In the monograph, special attention is given to a number of problems which are important in the development of highly efficient plasma torches. This concept includes: the high thermal and electrical efficiency of the plasma torch; the possibility of selecting the optimum system of the plasma torch with special reference to specific technology and in accordance with the availability of electric power sources; long service life of the most heavily thermally stressed sections of the plasma torch, i.e. the electrodes. On the basis of the phenomenon of recirculation of atoms (ions) of the electrode material in the near-cathode region of the stationary arc spot, it was possible to develop self-restoring cathodes at currents of up to 1 kA and at the atmospheric pressure. Consequently, it has been possible to develop cathodes with the infinite service life.

Of special importance are the problems of increasing the operating life of copper tubular electrodes and also explanation of the mechanisms having a negative effect on this parameter. There are several such mechanisms: the first one is associated with the formation of oxide films, the second one with the formation of dislocations in the near-surface working layer of the material as a result of high thermal stresses, caused by high heat flows through the arc spot and the cyclic nature of passage of the spot on the surface of the electorate. Preliminary theoretical and experimental investigations indicate two possible methods of solving the second problem: 1. The axial scanning of the radial section of the arc with a specific frequency in the given section of the tubular end electrode in the absence of large-scale shunting; 2. Improvement of the structure and physical-mechanical characteristics of the metal of the electrode by the introduction into the metal (in casting) of ultrafine powders with a specific structure and composition. According to the results of calculations and experiments, this reduces the extent of cracking and specific erosion (in comparison with the initial value) by a factor of 1.5–2.0. For the anode, it is fully realistic to reduce further the erosion rate if both factors taken into account.

Attention to the electrical–aerophysical processes in the electric arc chamber of the plasma torch is associated not only with the generalisation of experience but also with the need to provide material for discussion in the formulation of new tasks associated with the increase of the efficiency of operation of linear plasma torches. The group of the currently important tasks include the search for the methods of reducing heat losses in the 'ledge' output electrode-anode, with the main fraction of the losses occurring in the recirculation zone. The zone is also characterised by maximum erosion of the material of the electrode. The first evaluation investigations show already that in a number of cases it has been possible only to reduce the erosion rate of the anode. Special attention to the system of the plasma torch with the 'ledge' output electrode is explained by the stability of arcing; 100% electrical efficiency, because the VAC has the rising section; simple design and a wide range of power.

Attention has also been given to the classification of linear electric arc plasma torches and various design solutions have been studied extensively with special reference to heating of different gases. Steam plasma torches have also been discussed. These systems are used on an increasing scale because of the efficient ecological parameters and low cost of the working medium.

Taking into account the fact that several monographs have been published on the AC plasma torches in the last couple of years, the authors described only briefly the main characteristics of these plasma torches.

The monograph ends with the chapter on plasma-processing reactors. This is an important area because the reactors of different circuits are used widely in the industry of many countries of the world.

We shall formulate several priority problems requiring urgent solution.

1. In the development, mainly by means of experiments, of the self-restoring cathodes, successes have been achieved, especially in carbon-containing gas media. However, this is only the first step on the road to the development of long-life cathodes because the problem is characterised by the effect of a large number of parameters and it is necessary to take into account chemical reactions, taking place in the near-cathode region. The theoretical solution should deter-

mine the optimum values of the concentration of the working (shielding) gas in the near-cathode region resulting in self-restoration. It is also important to explain the effect of gas pressure, current intensity, the cooling rate of the cathode section and other parameters on the instability of the self-restoration process.

2. In the monograph, the authors present theoretical and experimental material for the substantiation of the controlling role of thermal stresses, formed in the subsurface layer of tubular and rod-shaped electrodes, on the formation and propagation of dislocations in relation to the density of the heat flow, travelling through the arc spot into the body of the electrode, the speed and nature of displacement of the spot on the surface. It has been shown, in particular for the tubular copper nodes, that it is possible to reduce the specific erosion in comparison with the mean value by more than an order of magnitude by selecting the optimum speed of displacement of the arc spot and the trajectory of movement on the working surface of the electrode.

Aeromagnetic scanning of the radial section of the arc and its special form, formed in the organisation of the appropriate topology of the magnetic field, should result in a further decrease of the specific erosion at occurrence of up to 1–2 kA. This is the second problem in the group of problems of increasing the efficiency of the electrodes.

3. The third problem is closely linked with the second problem but already relates to the development of new electrode materials with less extensive cracking and propagation of the cracks under the effect of non-stationary high-intensity heat flows. Theoretical and experimental advances have been made in the area of ferrous metals (steel, cast iron) indicating the nature of improving the physical-mechanical characteristics of the metal as a result of the addition of the ultrafine powders with the particle size smaller than 0.1 μm. The next task is to apply the theory to copper and carry out experimental verification.

4. It is also necessary to solve the problem of the electrophysical and aerodynamic mechanism of ensuring the extremely low specific erosion of copper cooled rod-shaped electrodes with a stationary arc spot at the end of the electrode in argon. The problem must be solved theoretically together with the possibility of extending the given effect to the gases, and also metals – cast iron, steel and other metals, which do not form non-conducting films (of the type of oxide films on the surface of copper in operation in, for example, air).

5. Investigations of the conditions of stable splitting of the radial section of the arc in the tubular electrode–cathode into several

radial current-conducting channels with the distinctive and stable attachment of the arc spot to the thermal emission inserts (Zr, Hf, W) has resulted in the development of unique cathode sections for various applications, operating for long periods of time in air, oxygen, nitrogen, and steam gas media. The experiments show the effect of electrical non-independent discharge on the copper holder in the gap between the thermal emission inserts. In further investigations, it is desirable to explain the nature of the discharge. This may lead to unexpected results which would make it possible to expand the direction of search for the methods of increasing the operating life of the cathode section.

6. In this monograph and in a number of other studies, data have been presented which have been obtained in the investigations of the protection of the walls of the discharge chamber of the plasma torch, especially with the inter-electrode insert, from the effect of high-intensity convective heat flows. At the moment, the practical results are highly positive. However, the protection of the walls from radiant heat flows, associated with the arc, especially in the case of high currents, still requires solution, and the losses may greatly exceed of the convective losses.

One of the solutions of the problems is associated, in all likelihood, with the application of porous materials with high thermal conductivity not subjected to corrosion and formation of oxide films, for the manufacture of the walls of the channel of the plasma torch. The solution of the problem should be available because the unit power of the plasma torches in a number of technological processes, used in the industry, has exceeded 1 MW. It is possible that because of advances in the manufacture of thermally and electrically conducting ceramics, the solution is in this area.

The problems examined in the monograph are found mainly and the interface of sciences and they can be solved only by the application of a complex approach and by cooperation of various directions of science and practice.

References

Chapter 1

1 Zhukov M.F., et al, Applied dynamics of thermal plasma, Novosibirsk, Nauka, 1975.

2 Zhukov M.F., et al, Electric arc gas heaters (plasmatrons), Moscow, Nauka, 1973.

3 Koroteev A.S., et al, Plasmatrons, design, characteristics, calculations, Moscow, Mashinostroenie, 1993.

4 Zhukov M.F. (ed), Electric arc plasma torches, Novosibirsk, 1980.

Chapter 2

1 Benkston K.A., Teploperedacha, 1970, **92**, No.4, 569-579.

2 Abramovich G.N., et al, Industrial aerodynamics, mechanics in the USSR in 50 years, Vol.2, Moscow, Nauka, 1970.

3 Khalatov A.A., et al, IFZh, 1977, **33**, No.2, 224-232.

4 Zhukov M.F., et al, Electric arc generators with inter-electrode inserts, Novosibirsk, Nauka, 1981.

5 Sukomel A.S., et al, IFZh, 1977, **33**, No.5, 816-821.

6 Zas'shkin I.M. and Popok N.I., Izv. Sib. Otd-niya AN SSSR, No.8, Ser. Tekhn. Nauk., 1979, No.2, 50-56.

7 Zhukov M.F., et al, Applied dynamics of thermal plasma, Novosibirsk, Nauka, 1975.

8 Vedernikov G.A. and Uryukov B.A., In: Problems of physics of low-temperature plasma, Nauka i Tekhnika, Minsk, 1970.

9 Dubrovni V.M., et al, PMTF, 1971, No.5, 17-23.

10 Grigaitis Yu.P., Arc as the sensor of turbulence of the flow, Proc. of 10th Nat. Conf. on Low-Temperature Plasma Generators, Vol.1, Minsk, 1986.

11 Zhukov M.F. and Koroteev A.S. (eds), Theory of thermal electric and plasma, Vol.2, Non-stationary processes and radiation heat exchange in thermal plasma, Novosibirsk, Nauka, 1987.

12 Olsen H.N., et al, J. Quant. Spectr. Rad. Transfer., 1968, **8**, 1419-1430.

13 Benenson D.M. and Cenkner A.A., Trans. ASME. Ser. C.J. Heat Transfer, 1970, **92**, No.2, 276-284.

14 Benenson D.M. and Baker A.J., AIAA Journal, 1971, **9**, No.8, 1441-1446.

15 Mel'nikova T.S. and Pikalov V.V., In: Meteorological physics for measuring high temperatures and plasma parameters, Scientific Research Institute of Meteorology, Khar'kov, 1979.

16 In: Proceedings of 8th Nat. Conf. on Low-Temperature Plasma Generators, Institute of Thermal Physics, Vol. 2, Novosibirsk, 1980, pp.213-216.

17 Sebald N., Appl. Phys., 1980, **21**, No.3, 221-236.

18 Plessl A., Appl. Phys., 1981, **21**, No.4, 377-328.
19 Tiller W., Appl. Phys., 1981, **25**, No.3, 317-328.
20 Filonin O.V., et al, In: 10th Siberian Conf. on Spectroscopy, Tomsk State University, Tomsk, 1981.
21 Mel'nikova T.S. and Pickalov V.V., Randon inversion in emission tomography of non-stationary plasma, Novosibirsk, Nauka, 1982.
22 Mel'nikova T.S. and Pickalov V.V., Examination of the parameters of electrical arc using a plasma tomograph, Novosibirsk, Nauka, 1983.
23 Preobrazhenskii N.G. and Pickalov V.V., Unstable problems of plasma diagnostics, Novosibirsk, Nauka, 1982.
24 Melnikova T.S. and Pickalov V.V., Beitr. Plasmaphys., 1982, **22**, No.2, 171-180.
25 Biberman L.M. and Horman G.E., J. QSRT, 1963, **3**, 221-245.
26 Schultz-Guide E., Z. Phys., 1970, **230**, No.5, 449-459.
27 Devoto B.S., ARL 71-0075, 1971.
28 Pakhomov E.P., Electrical arc, stabilised by the wall: shape, regions of existence, characteristics, dissertation, Moscow, 1980.
29 Shlikhting G., Theory of boundary layer, Moscow, Nauka, 1974.
30 Pakhomov E.P. and Yartsev I.M., Teplofizika Vysokikh Plazmy, 1977, **15**, No.5, 949-957.
31 Pickalov V.V. and Mel'nikova T.S., Plasma tomography, Vol.13, Theories, low-temperature plasma, Novosibirsk, Nauka, 1985.
32 Lebedev A.D., Examination of the effect of the gas low conditions, the characteristics of the electrical arc, Dissertation, Institute of Thermal Physics, Novosibirsk, 1972, pp.160.
33 Kolonina L.I. and Uryukov B.A., Siberian Division, AN SSSR, 1968, No.13, (3), 23-25.
34 Kolonina L.I. and Uryukov B.A., Siberian Division, AN SSSR, 1968, No.13, (3), 28-32.
35 Smolyakov V.Ya., PMTF, 1967, No.1, 151-157.
36 Dautov G.Yu., et al, PMTF, 1969, No.2, 67-72.
37 Mick G. and Krigs G., Electrical breakdown in gases, Moscow, IL, 1960.
38 Braun S., Elementary processes in gas discharged plasma, Gosatomizdat, 1961.
39 Dandaron G.-N.B., et al, PMTF, 1970, No.1, 138-141.
40 Dautov G.Yu., et al, ZhPS, 1975, **22**, No.4.
41 Aleshin N.F., et al, Heat and mass transfer: investigations and development, Minsk, 1981.
42 Lukashov V.P., et al, In: 9th Nat. Conf. on Low-Temperature Plasma Generators, Frunze, Ilim, 1983.
43 Zhukov M.F., et al, Siberian Division, AN SSSR, 1987, No.4, (No.1), 50-54.
44 Gol'dshtik M.A., Vortex flow, Novosibirsk, Nauka, 1981.
45 Eaton J.K. and Johnston J.P., AIAA J., 1981, **19**, No.9, 1093-1100.
46 Etheridge D.W. and Kemp P.H., J. Fluid Mech., 1978, **86**, No.3, 545-566.
47 Makiola B. and Ruck B., In: Proc. Int. Symp. Eng. Turbulence and Measurements, Dubrovnik, 1990, pp.487-496.
48 Ota T., et al, Turbulence, Heat and Transfer, Vol.1, New York, Begell House, 1995.
49 Driver D.H., et al, AIAA J., 1987, **25**, 914-919.
50 Chzhen P., Detachment flows, Moscow, Mir, 1973.

51 Richardson P.D., Estimation on the heat from the rear of the immersed body to the region of separated flow, ARL 62423, Brown University, 1960.
52 Richardson P.D., Chem. Eng. Sci., 1963, **18**, 149-155.
53 Ota T. and Nishiyama H., Int. J. Mass Trans., 1987, **30**, No.6, 1193-1200.
54 Yang J.T. and Tsai C.-H., Int. J. Mass Trans., 1996, **39**, No.11, 2293-2301.
55 Mori Y., et al, In: Proc. 8th Heat Transfer Conf., San Fransisco, Vol.3, 1986, 1083-1088.
56 Nishiyama H., et al, Warme-Stoffubertrag., 1988, **23**, 275-281.
57 Kiya M. and Sasaki K., J. Fluid Mech., 1983, **137**, 83-113.
58 Corin A.V. and Sikovsky D.Ph., Russian J. Eng. Thermophysics, 1995, **5**, 145-164.
59 Gorin A.V. and Sikovskii D.F., PMTF, 1996, **37**, No.3, 83-96.
60 Bek, et al, Teploperedacha, 1972, No.1, 124-133.
61 Gadd G.E., et al, Arg R&M 3148, Oct. 1958.

Chapter 3

1 Engel'sht V.S., et al, Low-temperature plasma, Vol.1, Theory of the column of the electrical arc, Novosibirsk, Nauka, 1990.
2 Zhukov M.F., et al, Theory of thermal electric arc plasma, Novosibirsk, Nauka, 1987.
3 Engel'sht V.S., et al, Mathematical modelling of the electrical arc, Frunze, Ilim, 1983.
4 Zhukov M.F. (ed), Theory of the electrical arc in the conditions of forced heat exchange, Novosibirsk, Nauka, 1977.
5 Granovskii V.L., Electrical current in the gas, steady current, Moscow, Nauka, 1971.
6 Kopansky J., In: Proc. 10th Intern. Conf. Phenom. in Ionized Gases, Oxford, 1971, pp.182-184.
7 Nazarenko I.P. and Panevin I.G., Modelling and method of calculating the physical-chemical processes in low-temperature plasma, Moscow, Nauka, 1974.
8 Zhukov M.F., et al, Applied dynamics of thermal plasma, Novosibirsk, Nauka, 1975.
9 Raizer Yu.P., Teplofizika Vysokikh Temperatur., 1972, **10**, No.6, 1152-1157.
10 Peters Th., In: Proc. 5th Intern. Conf. Ionised Phenom. in Gases, Amsterdam, 1962, pp.885-896.
11 Zarudi M.E., Izv. CO AN SSSR, 1967, No.3(1), 8-14.
12 Maecker H. and Stablein H., In: Proc. 10th Intern. Conf. Phenom. Ionised Gases, Oxford, 1971, p.178.
13 Whitman A.N. and Cohen I.M., J. Appl. Phys., 1973, **44**, No.4, 1552-1556.
14 Finkel'nburg I. and Mekker G., Electrical arcs and thermal plasma, Moscow, IL, 1961.
15 Gurovich V.Ts., et al, Teplofizika Vysokikh Temperatur., 1980, **18**, No.2, 256-265.
16 Zarudi M.E., Zhurn. Tekh. Fiziki., 1971, **41**, No.4, 734-743.
17 Mekker G., In: Moving plasma, Moscow, IL, 1961, pp.438-477.
18 Belousova L.E., Teplofizika Vysokikh Temperatur., 1971, **9**, No.6, 1131-1138.
19 Lowke J.J., J. Appl. Phys., 1970, **41**, No.6, 2588-2599.
20 Desyatkov G.A., et al, In: Proc. 13th Intern. Conf. Phen. Ionised Gases, Berlin,

1977, pp.513-514.
21 Moizhes B.Ya., et al, Zhrn. Tekhn. Fiziki., 1976, **46**, No.7, 1427-1431.
22 Artemov V.I., et al, Izv. SO AN SSSR, 1980, No.13, No.3, 18-20.
23 Levitan Yu.S., et al, In: Stability of the electrical arc, Novosibirsk, Nauka, 1973.
24 Kim D.Ch., Proc. of 8th Nat. Conf. on Low-Temperature Plasma Generators, Novosibirsk, 1980, pp.69-72.
25 Gurovich V.Ts., et al, Proc. of 8th Nat. Conf. on Low-Temperature Plasma Generators, Novosibirsk, 1980, pp.16-23.
26 Novikov O.Ya., Stability of the electrical arc, Leningrad, Energiya, 1978.
27 Maecker H.H. and Stablein H.G., IEEE Trans. Plasma Sci., 1986, **PS-14**, 291-299.
28 Desyatkov G.A., et al, In: Proc. 10th Intern. Symp. on Plasma Chemistry, Bochum, 1991, pp.1-6.
29 Desyatkov G.A., et al, J. High Temp. Chem. Proc., 1992, **1**, No.3, 291-298.
30 Maecker N., Z. Physik., 1955, **141**, 198-216.
31 Lelevkin V.I. and Otorbaev D.K., Experimental and theoretical models in the physics of non-equilibrium plasma, Frunze, Ilim, 1988.
32 Shoek P.A., In: Current problems of heat exchange, Moscow, Energiya, 1966, pp.110-139.
33 Pfender E. and Ekker E.R.G., Izv. SO AN SSSR, 1973, No.13 (3), 3-26.
34 Olsen H.N., Phys. Fluids, 1959, **2**, 614-620.
35 Busz-Peurkert C. and Finkelnburg W., Z. Physik., 1955, **140**, 540-546.
36 Sanders N., et al, J. Appl. Phys., 1982, **53**, 4136-4145.
37 Kolesnikov V.N., In: Physical optics, Moscow, Nauka, 1964.
38 Asinovskii E.I. and Pakhomov E.P., Teplofizika Vysokhikh Temperatur., 1968, **13**, No.6, 333-336.
39 Incroperra F.P., IEEE Trans. Plasma Sci., 1973, **PS-1**, 3-9.
40 Chen D.M. and Pfender E., IEEE Trans. Plasma Sci., 1981, **9**, 265-274.
41 Uryukov B.A., SO AN SSSR, 1981, No.3(1) 87-88.
42 Levitan Yu.S., SO AN SSSR, 1984, **10**, 117-137.
43 In: Turbulence. Principles and applications, Moscow, Mir, 1986.
44 Numerical modelling of turbulent phenomena in electric arc plasma, Bishkek, Ilim, 1993.
45 Kurochkin Yu.V., et al, Teplofizika Vysokhikh Temperatur., 1975, **13**, No.6, 1220-1224.

Chapter 4

1 Kirpichev M.V. and Mikheev M.A., Modelling of plasma devices, Publishing House of the Academy of Sciences of the USSR, Moscow and Leningrad, 1936.
2 Minaev A.N., Theory of similarity and its applications in thermal engineering, Moscow and Leningrad, Gosenergonizdat, 1959.
3 Minaev A.N., Theory of dimensionality of quantities and similarity and their application in thermal engineering, N.K. Krupskaya Institute, Moscow, 1968.
4 Gukhman A.A., Introduction into similarity theory, Moscow, Vysshaya Shkola, 1983.
5 Sedov L.I., Similarity methods and dimensions in mechanics, Moscow, Nauka, 1977.
6 Dautov G.Yu. and Zhukov M.F., PMTF, 1965, No.2, 97-105.

7 Dautov G.Yu. and Zhukov M.F., PMTF, 1965, No.6, 111-114.
8 Kutateladze S.S. and Yas'ko O.I., IFZh, 1964, **7**, No.4, 25-27.
9 Zhukov M.F., et al, Applied dynamics of thermal plasma, Novosibirsk, Nauka, 1975.
10 Koroteev A.S., et al, Plasma torches: design, characteristics, calculations, Moscow, Mashinostroenie, 1993.
11 Kutateladze S.S. and Yarygin V.N., In: Selected studies of S.S. Kutateladze, Novosibirsk, Nauka, 1989.
12 Kurosh A.G., Lectures in high algebra, Moscow, Nauka, 1971.
13 Kochin N.E., Vector calculus and origins of tensor calculus, Moscow and Leningrad, ONTI, 1937.
14 Mikhailov B.I., Proc. of 4th Nat. Scientific and Technical Conf.: Application of low-temperature plasma in industry, Czestochowa-Kokotek, Poland, No.5-8, 1979, 80-84.

Chapter 5

1 Dautov G.Yu. and Zhukov M.F., PMTF, 1965, No.2, 97-105.
2 Dautov G.Yu. and Zhukov M.F., PMTF, 1965, No.6, 111-114.
3 Kutateladze S.S. and Yas'ko O.I., IFZh, 1964, **7**, No.4, 25-27.
4 Zhukov M.F., et al, Applied dynamics of thermal plasma, Novosibirsk, Nauka, 1975.
5 Zhukov M.F. and Sukhinin Yu.I., Izv.Sib. otd-niya AN SSSR, Ser Tekhn.Nauk, AN SSSR, 1969, No.3(1), 55-60.
6 Zhukov M.F., et al, Izv.Sib. otd-niya AN SSSR, Ser Tekhn.Nauk, AN SSSR, 1970, No.3(1), 30-34.
7 Zhukov M.F. (ed), Fundamentals of calculations of linear plasma torches, Novosibirsk Institute of Thermal Engineering, AN SSSR, 1979.
8 Kutateladze S.S. and Yarygin V.N., In: Selected studies of S.S. Kutateladze, Novosibirsk, Nauka, 1989.
9 Zhidovich A.I., et al, In: Low-temperature plasma generators, Moscow, Energiya, 1969, pp.219-232.
10 Brilhac J.-F., et al, Plasma Chemistry and Plasma Processing, 1995, **15**, No.2, 231-255.
11 Brilhac J.-F., et al, Plasma Chemistry and Plasma Processing, 1995, **15**, No.2, 257-277.
12 Zhukov M.F. and Panin V.E. (eds), New materials and technologies. Extreme technological processes, Chap.7, Plasma classification of coal, Novosibirsk, Nauka, 1992.
13 Mikhailov B.I., IFZh., 1984, **16**, No.2, 325-326.
14 Zhukov M.F. (ed), Electric arc plasma torches advertising literature, Institute of Thermal Engineering, Novosibirsk, 1980.
15 Hadlestone R., Plasma diagnostics, Moscow, Mir, 1967.
16 Kolonina L.I. and Smolyakov V.Ya., In: Low-temperature plasma generators, Moscow, Energiya, 1969, pp.209-218.
17 Maecker H., Zeit. fur Physik., 1960, **158**, No.4, 392-404.
18 Edel's Kh., and Kimblin S.V., In: Low-temperature plasma, Moscow, Mir, 1967.
19 Dautov G.Yu. and Sazonov M.I., PMTF, 1967, No.4, 127-131.
20 Zhukov M.F., et al, Electric arc generators with inter-electrode inserts, Novosibirsk, Nauka, 1982.

21 Engel'sht V.S., et al, Diagnostics of low-temperature plasma, Novosibirsk, Nauka, 1995.

22 Zhukov M.F., et al, Izv.Sib. otd-niya AN SSSR, Ser Tekhn.Nauk, 1987, No.11(3), 25-51.

23 Artemov V.I., et al, Instabilities and turbulence in low-temperature plasma, Moscow Energy Institute, Moscow 1993.

24 An'shakov A.S., et al, Izv.Sib. otd-niya AN SSSR, Ser Tekhn.Nauk, 1970, No.8 (2) 3-11.

25 Zhukov M.F. and Timoshevskii A.N., Izv.Sib. otd-niya AN SSSR, Ser Tekhn.Nauk, 1973, No.13 (3), 66-70.

26 Dautov G.Yu. and Sazonov M.I., PMTF, 1967, No.4, 127-131.

27 Mustafin G.M., PMTF, 1968, No.4, 124-129.

28 Asinovskii E.I. and Zeigarnik V.A., TVT, 1974, **12**, No.6, 1278-1291.

29 Zhukov M.F., et al, In: Proc. 13th Intern. Conf. on Phenomena in Ion Gases, Berlin, 1977, Part 2, Leipzig, 539-540.

30 Zhukov M.F., et al, PMTF, 1979, No.6, 11-16.

31 Lukashov V.P. and Poednyakov B.A., SO AN SSSR, 1976, No.13(3???), 104-107.

32 Kutateladze S.S. and Leont'ev A.I., Heat, mass exchange and friction in a turbulent flow, Moscow, Energiya, 1972.

33 Kolonina L.I. and Uryukov B.A., Izv.Sib. otd-niya AN SSSR, Ser Tekhn.Nauk, 1968, No.13 (3), 23-27.

34 An'shakov A.S., et al, In: Proc. 6th Nat. Conf. on Low-Temperature Plasma Generators, Frunze, Ilim, 1974, 86-89.

35 Lukashov V.P. and Pozdnyakov B.A., In: Problems of hydrodynamics and heat exchange, Novosibirsk, 1976, pp.231-237.

36 Zhukov M.F. (ed), Theory of thermal electric arc plasma, Novosibirsk, Nauka, 1987.

37 Engel'sht V.S., et al, Theory of electrical arc, Novosibirsk, Nauka, 1990.

38 Uryukov B.A., Izv.Sib. otd-niya AN SSSR, Ser Tekhn.Nauk, 1975, No.3 (1), 3-10.

39 Shkarofksy I.P., ARL 73-0133, 1973, p.123.

40 Frind G. and Damsky B.L., ARL 70-0001, 1970, p.76.

41 Runstadler P.W., Harward University, Dept. Eng. and Appl. Phys., Techn. Rep., No.22, 1965.

42 Polaka L.S. (ed), Lectures on physics and the chemistry of low-temperature, plasma, Moscow, Nauka, 1971.

43 Uryukov B.A., Izv.Sib. otd-niya AN SSSR, Ser Tekhn.Nauk, 1973, No.13 (3), 48-59.

44 Kurochkin Yu.V. and Pustogarov A.V., Experimental investigations of plasma torches, Novosibirsk, Nauka, 1977.

45 Karabut A.B., et al, TVT, 1979, **17**, No.3, 618-625.

46 Arzamastsev A.N., et al, In: Proc. of 8th Nat. Conf. on Low-Temperature Plasma Generators, Part. 3, Novosibirsk, 1980, pp.4-7.

47 Bobrovskaya R.S., et al, Ibid, pp.8-12.

48 In: Proc. of 9th Nat. Conf. on Low-Temperature Plasma Generators, Frunze, Ilim, 1983, pp.80-81.

49 Karabut A.B., et al, TVT, 1979, **17**, No.3, 618-625.

50 Pustogarov A.V., et al, TVT, 1986, **24**, No.4, 639-643.

51 Pustogarov A.V., et al, Proc. of 11th Nat. Conf. on Low-Temperature Plasma Generators, Vol.1, Novosibirsk, 1989, 171-172.

52 Ibid, pp.169-170.

53 Zhukov M.F., et al, TVT, 1988, **26**, No.1, 1-9.

54 Zasypkin I.M., et al, Proc. of 9th Nat. Conf. on Low-Temperature Plasma Generators, Frunze, Ilim, 1983, pp.76-77.

55 Uryukov B.A., IIzv.Sib. otd-niya AN SSSR, Ser Tekhn.Nauk, 1981, No.3 (1), 87-98.

56 Kutaeladze S.S., Analysis of similarity in thermophysics, Novosibirsk, Nauka, 1982.

57 Steinberger S., Zeitschrift fur Physik, 1969, **223**, No.1, 1-18.

58 Nizovskii V.L. and Shabashov V.I., In: Proc. of 6th Nat. Conf. on Low-Temperature Plasma Generators, Frunze, Ilim, 1974, pp.102.

59 Knyazev Yu.R., et al, ZhTF, 1967, **37**, No.3, 523-532.

60 Dement'ev V.V., et al, In: Physics, technology and application of low-temperature plasma, Alma-Ata, 1970, pp.334-338.

61 Kushnarenko I.P, et al, In: Proc. of 6th Nat. Conf. on Low-Temperature Plasma Generators, Frunze, Ilim, 1974, pp.82-85.

62 Zakharkin R.Ya., et l, In: Proc. of 7th Nat. Conf. on Low-Temperature Plasma Generators, Vol.1, Alma-Ata, 1977, pp.94-97.

63 Yankovskii A.I., In: Examination of complicated heat exchange, Novosibirsk, AN SSSR, 1978, pp.138-146.

64 Zasypkin I.M., Electric arc hydrogen heaters. Thermal plasma and new materials technology, Vol.1: Investigations and design of thermal plasma generators, Cambridge Interscience Publishing, 1994, pp.424-468.

65 Barkov A.P., In: Proc. of 8th Nat. Conf. on Low-Temperature Plasma Generators, Vol.3, Novosibirsk, 1980, pp.21-24.

66 Grigor'ev M.A. and Rutberg F.G., In: Powerful generators and low-temperature plasma methods of examining their parameters, Scientific Research Institute of Electrical Engineering, Leningrad, VNIIelektromash, 1984, pp.43-51.

67 Grigor'ev M.A., et al, In: Plasma generators and methods of diagnostics, Scientific Research Institute of Electrical Engineering, Leningrad, VNIIelektromash, 1984, pp.129-133.

68 Peinter J.Kh. and Kroutil J.K., RTiK, 1980, **18**, No.3, 143-145.

69 Kurochkin Yu.V., et al, TVT, 1978, **16**, No.1, 195-197.

70 Zasypkin I.M., et al, In: Proc. of 9th Nat. Conf. on Low-Temperature Plasma Generators, Frunze, Ilim, 1983, pp.117-119.

71 Kravchenko S.K., et al, In: Proc. of 10th Nat. Conf. on Low-Temperature Plasma Generators, Vol.1, Minsk, 1986, 49-50.

72 Mikhailov B.I. and Shatokhin V.G., In: Proc. of 8th Nat. Conf. on Low-Temperature Plasma Generators, Vol.3, Novosibirsk, 1980, 64-67.

73 Mikhailov B.N., Ibid, pp.68-71.

74 Zasypkin I.M. and Mikhailov B.I., In: Physics of low-temperature plasma, Vol.2, Proc. of 8th Nat. Conf., Minsk, 1991, 67-68.

Chapter 6

1 Zhukov M.F., et al, Applied dynamics of thermal plasma, Novosibirsk, Nauka, 1975.

2 Dautov G.Yu. and Zhukov M.F., PMTF, 1965, No.2, 97-105.

3 Zhukov M.F. and Sukhinin Yu.I., Izv. AN SSSR, 1969, No.3 (1), 55-60.

4 German V.O. and Morozov M.G., Teplofizika Vysokikh Temperatur., 1965,

3, No.5, 765-770.
5 Zhukov M.F. (ed), Plasma torches. Investigations. Problems, Novosibirsk, Izd-vo SO RAN, 1995.
6 Smith R.T. and Folck I.L., AFFDL-TR-69-6, 1969, p.67.
7 Mortseva G.I., et al, In: Proc. of 4th Nat. Conf. on Low-Temperature Plasma Generators, Alma-Ata, 1970, pp.413-416.
8 Belyanin N.M. and Zyrichev N.A., IFZh, 1969, **16**, No.2, 212-217.
9 Yurevich F.B., et al, IFZh, 1967, **12**, No.6, 711-717.
10 Ganz S.N., et al, In: Production of bonding nitrogen in plasma, Kiev, Naukova dumka, 1967.
11 Zasypkin I.M., Electric arc hydrogen heaters/Thermal plasma and new materials technology. Vol. 1: Investigations and design of thermal plasma generators, Cambridge Interscience Publishing, 1994, pp.424-468.
12 Zhukov M.F., et al, Izv.Sib. otd-niya AN SSSR, Ser Tekhn.Nauk, 1970 No.3 (1), 30-34.
13 Zhukov M.F. (ed), Fundamentals of calculating linear plasma torches, Novosibirsk, 1979.
14 Brilhac J.-F., et al, Plasma Chemistry and Plasma Processing, 1995, No.2, 231-255.
15 Brilhac J.-F., et al, Plasma Chemistry and Plasma Processing, 1995, No.2, 257-277.
16 Nutsch G., In: 3rd European Congress on Thermal Plasma Processes, 1994, Aachen, Dusseldorf, VDI-Verlag GmbH, 1995, pp.201-209.
17 Zhukov M.F., et al, Electrical arc generators with inter-electrode inserts, Novosibirsk, Nauka, 1981.
18 Zhukov M.F., et al, Izv.Sib. otd-niya AN SSSR, Ser Tekhn.Nauk, 1980, No.13 (3), 77-85.
19 Dautov G.Yu., et al, PMTF, 1967, No.1, 172-176.
20 Zhukov M.F., et al, Izv.Sib. otd-niya AN SSSR, Ser Tekhn.Nauk, 1973, No.3 (1), 18-24.
21 Zasypkin I.M. and Urbakh E.K., In: Proc. of 9th Nat. Conf. on Low-Temperature Plasma Generators, Frunze, Ilim, 1983, pp.74-75.
22 Michael G. and Rothhardt L., Contributions to Plasma Physics, 1982, **22**, No.6, 477-484.
23 Ivlyutin A.I., et al, In: Proc. of 5th Nat. Conf. on Low-Temperature Plasma Generators, Vol.1, Novosibirsk, 1972, pp.205-207.
24 Zhukov M.F., et al, Electric gas heaters (plasma torches), Moscow, Nauka, 1973.
25 Zhukov M.F., et al, Theory of thermal electric arc plasma, Vol.2, Novosibirsk, Nauka, 1987, p.287.
26 Engel'shta V.S. and Uryukova B.A. (eds), Low-temperature plasma. Vol.1, Theory of the column of the electrical arc, Novosibirsk, Nauka, 1990, p.376.
27 Ambrazyavichyus A., Heat exchange in quenching of gases, Vil'nyus, Mokslas, 1983.
28 Shashkov A.G., et al, Heat exchange in electric arc gas heaters, Moscow, Energiya, 1974.
29 Ambrazevicius A. and Zukauskas A., In: 6th Intern. Heat Transfer Conf., Toronto, Canada, 1978, Vol.2, pp.559-563.
30 Marlotte G.I., et al, ARL 64-69, 1964.
31 Zhukov M.F., et al, Izv.Sib. otd-niya AN SSSR, Ser Tekhn.Nauk, 1979, No.8 (2), 61-66.

32 Ambrazyavichus A. and Zhukauskas A.S., In: Generation of electric arc plasma flow, Novosibirsk, 1987.
33 Ambrazyavichyus A.B., et al, In: Proc. of 6th Nat. Conf. on Low-Temperature Plasma Generators, Frunze, Ilim, 1974, pp.94-97.
34 Zarudi M.E. and Edel'baum I.S., In: Transfer phenomena in low-temperature plasma, Minsk, Nauka i Tekhnika, 1969, pp.82-87.
35 Zhukov M.F., et al, Izv.Sib. otd-niya AN SSSR, Ser Tekhn.Nauk, 1975, No.8 (2), 15-20.
36 Lukashov V.P. and Pozdnyakov B.A., Izv.Sib. otd-niya AN SSSR, Ser Tekhn.Nauk, 1976, No.3 (1), 8-11.
37 Painter J.H., AIAA Paper, 1973, No.74-731, 11.
38 Kutateladze S.S., Fundamentals of heat exchange theory, Novosibirsk, Nauka, 1970.
39 Kutateladze S.S. and Leont'ev A.I., Heat, mass exchange and friction in turbulent boundary layers, Moscow, Energiya, 1972.
40 Leont'ev A.I. and Volchkov E.P., In: Experimental investigations of plasma torches, Novosibirsk, Nauka, 1977.
41 Dyban E.P., et al, IFZh, 1971, **20**, No.2, 294-298.
42 Ambrazyavichyus A.B., et al, In: Heat exchange in high temperature gas flows, Vil'nyus, Mintis, 1972.
43 Vargaftik N.V., A handbook of the thermophysical properties of gases and liquids, Moscow, Nauka, 1972.
44 Zhukov M.F., et al, Fluid mech., Sov. Res., 1982, **11**, No.3, 75-84.
45 Painter J.H., et al, AIAA Journ., 1971, No.12, 2307-2308.
46 Lebsak V.A., et al, Obzor, TsAGI, No. 566, 1979.
47 Zhukov M.F., et al, Plasma chemical processing of coal, Moscow, Nauka, 1985.
48 Glebov I.A. and Rutberg F.G., Powerful plasma generators, Moscow, Energoatomizdat, 1985.
49 Fogel'son I.B., Transitor thermal sensors, Moscow, Sov. Radio., 1972.
50 Anyshakov A.S., et al, PMTF, 1967, No.1, 161-166.
51 Zheenbaev J., et al, In: Proc. of 5th Nat. Conf. on Low-Temperature Plasma Generators, Vol.2, Novosibirsk, 1972, pp.60-62.
52 Zhukov M.F., et al, In: Authors certificate No: 430801 (USSR), 1974, Bulletin No.19, 1976.
53 Zhukov M.F., et al, In: 11th Intern. Conf. on Phenomenoa in Ionised Gases, Prague, 1973, p.226.
54 Kharder and Kahn, RTiK, 1970, **8**, No.12, 132-140.
55 Kenon and Keis, Teploperedacha, 1969, **91**, No.2, 127-132.
56 Zhukov M.F., et al, In: Proc. of 6th Nat. Conf. on Low-Temperature Plasma Generators, Frunze, Ilim, 1974, 108-111.
57 Dobrinskii E.K., et al, Izv.Sib. otd-niya AN SSSR, Ser Tekhn.Nauk, 1979, No.8 (2), 42-49.
58 Yankovskii A.I., Examination of complicated heat exchange, Novosibirsk, 1978, 138-145.
59 Kosinov V.A. and Yankovskii A.I., In: Proc. of 7th Nat. Conf. on Low-Temperature Plasma Generators, Vol.1, Alma-Ata, 1977, 203-206.
60 Cann G.L., ARL 73-0043, 1973, p.131.
61 Kurochkin Yu.V. and Pustogarov A.V., In: Experimental investigations of plasma torches, Novosibirsk, Nauka, 1977, pp.82-104.
62 Karabut A.B., et al, Izv.Sib. otd-niya AN SSSR, Ser Tekhn.Nauk, 1976, No.8

(2), 10-13.

63 Pustogarov A.V.,,et al, Izv.Sib. otd-niya AN SSSR, Ser Tekhn.NaukR, 1976, No.8 (2), 7-9.

64 Kutateladze S.S. and Yarygin V.N., Experimental methods in dynamics of rarified gases, Novosibirsk, 1974.

65 Mironov B.P., Experimental investigations of plasma torches, Novosibirsk, Nauka, 1977.

66 Leont'ev A.I., et al, Thermal shielding of plasma torched walls. In: Low-temperature plasma, Vol.15, Novosibirsk, 1995.

67 Heberlein J., et al, ARL 70-007, 1970, p.111.

68 Kurochkin Yu.V., Izv.Sib. otd-niya AN SSSR, Ser Tekhn.Nauk, 1977, No.8 (2), 97-102.

69 Kutateladze S.S., et al, Experimental examination of wall turbulent flows, Novosibirsk, Nauka, 1975.

70 Kutateladze S.S. and Mironov B.P., PMTF, 1970, No.3, 162-166.

71 Kutateladze S.S., et al, PMTF, 1966, No.5, 123-125.

72 Onufriev A.T. and Sevast'yanenko V.G., Zhurn. Prikl. Mekhaniki i Tekhn. Fiziki., 1969, No.2, 17-22.

73 Vetlutskii V.N. and Sevast;yanenko V.G., Zhurn. Prikl. Mekhaniki i Tekhn. Fiziki., 1969, No.1, 136-138.

74 Soloukhin R.I., et al. Optical characteristics of hydrogen plasma, Novosibirsk, Nauka, 1973.

75 Zhukov M.F. (ed), Properties of low-temperature plasma and methods of diagnostics, Novosibirsk, Nauka, 1977.

76 Shidlauskas V.A., Radiation and complicated heat exchange in the hydrogen plasma arc, Dissertation, Kaunas, Moscow, 1980.

77 Kon'kov A.A., Tepolofizika Vysokikh Temperatur., 1979, **17**, No.4, 678-684.

78 Barkov A.P., et al, In: Proc. of 7th Nat. Conf. on Low-Temperature Plasma Generators, Vol.2, Alma-Ata, 1977, pp.204-207.

79 Mikhailov B.I. and Voichak V.P., In: New materials and technologies, Vol.1, Extreme technological processes, Novosibirsk, Nauka, 1992.

80 Mikhailov B.I., In: Proc. of 9th Nat. Conf. on Low-Temperature Plasma Generators, Frunze, Ilim, 1983, pp.336-337.

81 Mikhailov B.I., In: Proc. of 9th Nat. Conf. on Low-Temperature Plasma Generators, Frunze, Ilim, 1983.

Chapter 7

1 Zhukov M.F., et al, Applied dynamics of thermal plasma, Novosibirsk, Nauka, 1975.

2 Zhukov M.F. (ed), Electric arc plasma torches (advertising leaflet), Institute of Thermal Physics, Novosibirsk, 1980.

3 Romanovskii G.F., et al, Nikolaev, 1981, No.181, 3-6.

4 Mikhailov B.I., and Voichak V.P., In: New materials and technologies. Extreme technological processes, Novosibirsk, Nauka, 1992.

5 Mikhailov B.I., In: 4th Nat. Conf.: Application of Low-Temperature Plasma in Industry, Czestochowa-Kokotek, Poland, 1979, pp.79-82.

6 Peregudov V.S., et al, Energetik, 1997, No.2, 13-14.

7 Zhukov M.F., et al, Plasma mazut-free heating of boilers and stabilisation of burning of coal dust jets, Novosibirsk, Nauka, 1995.

8 Zhukov M.F., et al, In: High temperature dust-laden jets in plasma tech-

nology. Utrecht. Tokyo. VSP. 1990. pp.531-541.

9 Peregudov V.S., et al. Proc. of Int'l Conf.: High-Temperature Dusted Jets in the Processes of Treatment of Powder Materials, Novosibirsk. 1988. pp.167-173.

10 Zhukov M.F., et al. Electric arc generators with inter-electrode inserts. Novosibirsk. Nauka. 1981.

11 Volodin I.L., et al. Khim. Prom-st'. 1975. No.3. 211-215.

12 An'shakov A.S., et al. Izv.Sib. otd-niya AN SSSR, Ser Tekhn.Nauk, 1970, No.8 (2). 3-11.

13 Izinger Yu.V., et al. In: Proc. of 9th Nat. Conf. on Low-Temperature Plasma Generators. Frunze. Ilim. 1983. pp.174-175.

14 GNP-1.5/3M plasma torch. Service instructions. Novosibirsk. 1982.

15 Examination of the stability and reliability of operation of the section of 5-10 MW plasma torches. recommendations for construction of the industrial plasma torch. Novosibirsk. 1984.

16 Zasypkin I.M., et al. In: Proc. of 9th Nat. Conf. on Low-Temperature Plasma Generators. Frunze. Ilim. 1983. pp.118-119.

17 Patent 31337 (East Germany).

18 Dautov G.Yu., et al. In: Generators of low-temperature plasma. Proceedings. Vol.2. Minsk. 1986. pp.13-14.

19 Donskoi A.V. and Klubnikin V.S., Electric plasma processes and equipment in engineering. Mashinostroenie. Leningrad. 1979.

20 Patent 2052796 (USA).

21 Patent 107369 (East Germany)

22 Patent 3708409 (USA). 1248595 (Great Britain).

23 Patent 3661764 (USA). 1317918. 1317919 (Great Britain).

24 Author's certificate No. 527843 (USSR).

25 Zhukov M.F., et al. Thermochemical cathodes. Novosibirsk. 1985.

26 Egorov V.M. and Novikov O.Ya.. In: Theory of the electrical arc in the conditions of forced heat exchange. Novosibirsk. 1977. pp.163-173.

27 Patent 3666982 (USA).

28 Gavryushenko B.S. and Pustogarov A.V., In: Near-electrode processes and erosion of electrodes with plasma torches. Novosibirsk. 1977.

29 Bolotov A.V., et al. In: Proc. of 8th Nat. Conf. on Low-Temperature Plasma Generators. Vol. 2. Novosibirsk. 1980. pp.28-31.

30 An'shakov A.S., et al. Ibid., pp.24-27.

31 Dautov G.Yu., et al. TVT. 1967. **5**. No.3. 500-504.

32 Timoshevskii A.N. and Sudarev A.I., In: Thermophysical investigations. Novosibirsk. 1977.

33 Sudarev A.I. and Timoshevskii A.N., In: Proc. of 8th Nat. Conf. on Low-Temperature Plasma Generators. Vol. 2. Novosibirsk. 1980. pp.120-123.

34 Novikov O.Ya., et al. Multi-arc systems. Novosibirsk. Nauka. 1988.

Chapter 8

1 Product Engineering. 1967. **38**. No.16. 62-68.

2 Patent 3472995 (USA).

3 Patent 20872 (USA).

4 Patent 20873 (USA).

5 Zheenbaev Zh. and Engel'sht V.S., Two-jet plasma torch, Frunze. Institute of Physics and Mathematics, 1983.

6 Konabko R.I., et al, In: Proc. of 6th Nat. Conf. on Low-Temperature Plasma Generators, Frunze, Ilim, 1974, pp.155-158.

7 Donskoi A.V. and Klubnikin V.S., Electric plasma processes and equipment in engineering, Leningrad, Mashinostroenie, 1979.

8 Finkel'nburg V. and Mekker G., Electrical arcs and thermal plasma, Moscow, IL, 1961.

9 Zheenbaev Zh., et al, Izv.Sib. otd-niya AN SSSR, Ser Tekhn.Nauk, 1976, No.3 (1), 17-20.

10 Zhukov M.F., et al, Applied dynamics of thermal plasma, Novosibirsk, Nauka, 1975.

11 Gol'dfarb V.M., TVT, 1973, **11**, No.1, 180-191.

12 Ovsyannikov A.A., In: Physics and chemistry of low-temperature plasma, Moscow, Nauka, 1971.

13 Gol'dfarb V.M. and Dresvin S.V., TVT, 1965, **3**, 333.

14 Abdrazakov A. and Zheenbaev Zh., In: Investigations of the electrical arc in argon, Frunze, Ilim, 1966.

15 Zheenbaev Zh. and Engel'sht V.S., Laminar plasma torch, Frunze, Ilim, 1975.

16 Rother W., Possibilities of probe measurements of oxygen in plasma jets at atmospheric pressure, Dissertation, Meiningen, 1963.

17 Nugmanov A.M. and Temirbaev D.J., Tekhnicheskie Nauki, Alma-Ata, 1975, No.19, 44-51.

18 Author's certificate No. 683868. Class. V23K9/04, 1979.

19 Polyakov S.P., et al, IFZh, 1984, **46**, No.3, 476-480.

20 Polyakov S.P., et al, IFZh, 1986, **50**, No.2, 245-249.

21 Livitan N.V. and Polyakov S.P., IFZh, 1986, **50**, No.4, 635-640.

22 Physical encyclopaedic dictionary, Moscow, Sov. Entsiklopediya, 1965.

23 Zhukov M.F., et al, Plasma torches. Investigations.Problems. Novosibirsk, 1995.

24 Zolotovskii A.I., et al, IFZh, 1982, **46**, No.4, 604-607.

25 Demidovich A.B., et al, IFZh, 1982, **46**, No.3, 461-466.

26 Volokitin G.G., et al, In: Generators of low-temperature plasma, Proc. of 10th Nat. Conf., Kaunas, 1986.

27 Volokitin G.G., Izv.Sib. otd-niya AN SSSR, Ser Tekhn.Nauk, 1989, No.13 (3), 90-94.

28 New materials and technologies. Extreme technological processes, Vol.9: Plasma treatment of surface of engineering components, Novosibirsk, 1992, pp.1525-166.

29 Zhukov M.F., et al, Fizika i Khimiya Obrab. Materialov, 1994, No.6, 98-108.

30 Kostylev V.P. and Marusin V.V., Russian Federation Patent No. 1708872, 07.06.1993, A method of hardening steel components.

31 Zhukov M.F., et al, Russian Federation Patent No. 2057200, 08.04.1994, A method of treatment of components of low-carbon steels.

32 Anyshakov A.S., et al, In: Proc. of 11th Nat. Conf. on Low-Temperature Plasma Generators, Vol. 1, Novosibirsk, 1989, pp.145-146.

33 Urbakh E.K., et al, In: Proc. of 11th Nat. Conf. on Low-Temperature Plasma Generators, Vol. 2, Novosibirsk, 1989, pp.46-47.

34 Zhukov M.F., et al, In: High temperature dust-laden jets in plasma technology, Utrecht, Tokyo, VSP, 1990, pp.531-541.

35 Anyshakov A.S., et al, In: Automated electric technological systems, Novosibirsk, 1991, pp.39-44.

36 Zhukov M.F., et al, Plasma mazut-free firing of furnaces and stabilisation of burning of the coal dust jet, Novosibirsk, Nauka, 1995.

Chapter 9

1 Zhukov M.F., et al, Electric arc gas heaters (plasma torches), Moscow, Nauka, 1973.
2 Zhukov M.F., et al, Applied dynamics of thermal plasma, Novosibirsk, Nauka, 1975.
3 Koroteev A.S., et al, Plasma torches. Design, characteristics, calculations, Moscow, Mashinostroenie, 1993.
4 Bogatyreva V.A., et al, PMTF, 1968, No.3, 86-89.
5 Botagyreva V.A., et al, Izv.Sib. otd-niya AN SSSR, Ser Tekhn.Nauk, 1967, No.13 (3), 159-162.
6 Rutberg F.G., In: Some problems of investigation of gas discharge plasma and the formation of some magnetic fields, Leningrad, Nauka, 1970.
7 Kiselev A.A. and Rutberg F.G., In: Generators of plasma jets and high-current arcs, Leningrad, Nauka, 1973.
8 Kiselev A.A. and Rutberg F.G., TVT, 1974, **12**, No.4, 827-834.
9 Glebov I.A. and Rutberg F.G., Powerful plasma generators, Moscow, Energoatomizdat, 1985.
10 Rutberg F.G., et al, In: Physics of low-temperature plasma, Proc. of Conf., Petrozavodsk, 1995, Vol.3, 422-424.
11 Goryachev V.L. and Khodakovskii A.M., In: In: Physics of low-temperature plasma, Proc. of Conf., Petrozavodsk, 1995, Vol.3, 186-188.
12 Rutberg F.G. and Safronov A.A., In: Fourth European Conf. on Thermal Plasma Processes, Athens, 1996.
13 Engel'sht V.S., et al, In: Plasma activation of combustion of coal, Proc., Alma-Ata, KazNIIE, 1989.
14 Desyatkov G.A., et al, In: Plasma jets in Developm. New Material Techn. Proc. Intern. Workshop, Frunze, 1990, Utrecht, Netherlands, VSP, 1990, pp.499-509.
15 Desyatkov G.A., et al, In: 20th Intern. Conf. Phenom. Ionised Gases. Contrib. Pap., II, Cioccco, Barga, Italy, 1991.
16 Engel'sht V.S., et al, In: Physics of low-temperature plasma, Proc., Minsk, 1991.
17 Trapitsin N.F., High-voltage AC arc as a source of light for emission spectrum analysis, Frunze, Ilim, 1986.
18 Engel'sht V.S., et al, Plasma soldering iron, Author's certificate 1651041, USSR, No. 4700478/06, B.I. 1991, No.19.
19 Engel'sht V.S., et al, In: Problems of saving energy, Proc., Vol.1, Kiev, 1991, 57-58.
20 Engel'sht V.S., et al, Ignitor, Author's certificate No. 1483186 USSR, No. 4333663/28-06, B.I. 1989, No.20.
21 Desyatkov G.A., et al, Ignitor, Author's certificate No. 1636647 USSR, No. 4655409/06, B.I. 1991, No.11.

Chapter 10

1 Zhukov M.F., et al, Electric arc gas heaters (plasma torches), Moscow, Nauka, 1973.

2 Zhukov M.F., et al, Applied dynamics of thermal plasma, Novosibirsk, Nauka, 1975.
3 Dandaron G.-N.V., et al, Author's certificate No. 1748616, Priority invention 24.08.1989.
4 Basin A.S. and Ipat'eva O.S., In: Scientific Proc. of the Conf.: Electric Technology, Today and Tomorrow, Cheboksary, 1997, pp.68-69.
5 Mortseva G.I., et al, In: Physics, technology and application of low-temperature plasma, Alma-Ata, KazPTI, 1970, pp.68-69.
6 Zheenbaev Zh.Zh., et al, Izv.Sib. otd-niya AN SSSR, Ser Tekhn.Nauk, 1973, No.3 (1), 3-6.
7 Dandaron G.-N.B. and Urbakh E.K., In: Proc. of 5th Nat. Conf. on Low-Temperature Plasma Generators, Vol. 1, Novosibirsk, 1972, pp.40-43.
8 Bykhovskii D.G., Plasma cutting. Cutting arc and energy equipment, Leningrad, Mashinostroenie, 1972.
9 Stenin V.V., In: Proc. of 9th Nat. Conf. on Low-Temperature Plasma Generators, No.6, Frunze, Ilim, 1981, pp.170-171.
10 Stenin V.V., TVT, 1985, **23**, No.5, 858-862.
11 Gordeev V.F., et al, Avt. Svarka, 1981, No.6, 48-50.
12 Zhukov M.F., et al, Dokl. AN SSSR, 1981, **260**, No.6, 1354-1356.
13 Zhukov M.F., et al, Izv.Sib. otd-niya AN SSSR, Ser Tekhn.Nauk, 1978, No.3 (13), 3-20.
14 Gavryushchenko B.S. and PUstogarov A.V., In: Near-electrode processes and erosion of electrode plasma torches, Novosibirsk, 1977, pp.85-122.
15 Gonopol'skii A.M. and Korablev V.A., Izv.Sib. otd-niya AN SSSR, Ser Tekhn.Nauk, 1983, No.3 (1), 69-71.
16 Zakharkin R.Ya., et al, In: Proc. of 7th Nat. Conf. on Low-Temperature Plasma Generators, Vol. 2, Alma-Ata, 1977, pp.208-211.
17 Zhukov M.F., et al, In: Near-electrode processes and erosion of electrode plasma torches, Novosibirsk, 1977, pp.123-148.
18 Vasil'ev R.A., et al, In: Proc. of 10th Nat. Conf. on Low-Temperature Plasma Generators, Vol. 1, Minsk, 1986.
19 Stenin V.V., et al, In: Proc. of 11th Nat. Conf. on Low-Temperature Plasma Generators, Vol. 2, Novosibirsk, 1989, pp.4-5.
20 Dzyuba V.L. and Mazuraitis I.S., In: Proc. of 11th Nat. Conf. on Low-Temperature Plasma Generators, Vol. 2, Novosibirsk, 1989, pp.18-19.
21 Rossomakho Ya.V., Development of a thermochemical cathode based on hafnium for plasma cutting in oxygen, Dissertation, Leningrad, 1984.
22 Zakshever O.P., et al, In: Proc. of 9th Nat. Conf. on Low-Temperature Plasma Generators, Frunze, Ilim, 1983, pp.338-339.
23 Zhienbekov G.R., In: Proc. of 9th Nat. Conf. on Low-Temperature Plasma Generators, Frunze, Ilim, 1983, pp.156-157.
24 Bolotov A.V., et al, In: Proc. of 8th Nat. Conf. on Low-Temperature Plasma Generators, Vol. 2, Novosibirsk, 1980, pp.28-30.
25 Paton B.E., et al, Avt. Svarka, 1988, No.2, 43-46.
26 Zhukov M.F., et al, Thermochemical cathodes, Novosibirsk, Nauka, 1985.
27 Novikov O.Ya., et al, Multi-access systems, Novosibirsk, Nauka, 1991.
28 Zhukov M.F., et al, Sib. Fiz.-Tekhn. Zhurn., 1991, No.4, 113-117.
29 Vashchenko S.P., et al, Sib. Fiz.-Tekhn. Zhurn., 1992, No.1, 98-106.
30 Zhukov M.F., et al, Near-electrode processes in arc discharge, Novosibirsk, Nauka, 1982.
31 Zhukov M.F., et al, Plasma torches. Investigations. Problems, Novosibirsk,

1995.
32 Fesenko V.V. and Bolgar A.S., Evaporation of refractory compounds, Moscow, Metallurgiya, 1966.
33 Fridlyand M.G., TVT, 1973, **11**, No.2, 414.
34 Fridlyand M.G., TVT, 1976, **14**, No.1, 21-25.
35 Vykhovskii D.G. and Dryakov V.V., Fizika i Khimiya Obrab. Materialov, 1984, No.6, 48-56.
36 Bykhovskii D.G. and Kryakov V.V., Avt. Svarka, 1982, No.8, 42-43.
37 Fridlyand M.G., Avt. Svarka, 1977, No.1, 16-18.
38 Fridlyand M.G., et al, TVT, 1981, **19**, No.5, 1095-1097.
39 Fridlyand M.G., Izv.Sib. otd-niya AN SSSR, Ser Tekhn.Nauk, 1981, No.1 (3), 121-125.
40 Fridlyand M.G., Avt. Svarka, 1980, No.11, 60-62.
41 Zhukov M.F. and Nikiforovskii V.S., In: Experimental investigations of plasma torches, Novosibirsk, Nauka, 1977.
42 Fridyland M.G., Plasma torches with a constantly renewed cathode, Leningrad, 1986.
43 Fridyland M.G. and Portov A.B., Elektrotekhnicheskaya Promyshlennost'. Ser. Elektrosvarka, 1979, No.5, 11-13.
44 Moizhes B.Ya. and Nemchinskii V.A., ZhTF, 1972, **42**, No.5, 1001-1009.
45 Fridyland M.G., In: Proc. of 11th Nat. Conf. on Low-Temperature Plasma Generators, Vol. 2, Novosibirsk, 1989, pp.86-87.
46 Fridyland M.G. and Nemchinskii V.A., Izv.Sib. otd-niya AN SSSR, Ser Tekhn.Nauk, 1987, No.13 (3), 52-58.
47 Kryakov V.V., Fizika i Khimiya Obrab. Materialov, 1988, No.1, 72-78.
48 Kovalev V.N., et al, In: Proc. of 11th Nat. Conf. on Low-Temperature Plasma Generators, Vol. 2, Novosibirsk, 1989, pp.143-144.
49 Dokukin M.Yu., et al, In: Proc. of 9th Nat. Conf. on Low-Temperature Plasma Generators, Frunze, Ilim, 1983, pp.194-195.
50 Moizhes B.Ya. and Nemchinskii V.A., ZhTF, 1984, **54**, No.8, 1076-1084.
51 Taran E.N., et al, In: Proc. of 10th Nat. Conf. on Low-Temperature Plasma Generators, Vol. 1, Minsk, 1986, pp.97-98.
52 Chumakov A.N., et al, In: Proc. of 10th Nat. Conf. on Low-Temperature Plasma Generators, Vol. 1, Minsk, 1986, pp.99-100.
53 Zhukov M.F., et al, Teplofizika i Aeromekhanika, 1997, **4**, No.2, 159-170.
54 Mikhailov V.I., et al, Russian Federation Patent No. 1641179: A method of controlling the displacement of the arc spot on the internal surface of the cylindrical electrode of the electric arc plasma torch.
55 Zhukov M.F., et al, Russian Federation Patent No. 1641179: Electric arc gas pre-heater.
56 Mikhailov V.I., et al, Russian Federation Patent No. 1503673: Equipment for electric arc heating of gas.
57 An'shakov A.S., et al, Izv. SO AN SSSR, 1988, No.7(2???), 65-68.
58 Timoshevskii A.N., et al, In: Proc. of 11th Nat. Conf. on Low-Temperature Plasma Generators, Vol. 2, Novosibirsk, 1989, pp.50-51.
59 Koroteev A.S., TVT, 1986, **24**, No.5, 980-983.
60 Nevskii A.P., et al, Interaction of the arc with plasma torch electrodes, Minsk, Nauka i Tekhnika, 1982.
61 Urbakh E.K., et al, In: Proc. of 11th Nat. Conf. on Low-Temperature Plasma Generators, Vol. 2, Novosibirsk, 1989, pp.46-47.
62 Uryukov B.A., Experimental investigations of plasma torches, Novosibirsk,

Nauka, 1977.
63 Marotta A., et al, Proc. of Conf.: Physics of Plasma and Plasma Technologies, Minsk, 1997, pp.86-88.
64 Szente R.N., et al, J. Phys. D: Appl. Phys., 1987, **20**, 754.
65 Marotta A. and Sharakhovsky L.I., J. Phys. D: Appl. Phys., 1996, **29**, 2395-2403.
66 Krylovich V.I. and Shabolts A.S., Izv. AN BSSR, 1973, No.1, 93.
67 Iokhimovich Ya.B. and Mikhailov B.I., In: Proc. of 10th Nat. Conf. on Low-Temperature Plasma Generators, Vol. 1, Minsk, 1986, pp.111-112.
68 Mikhailov B.I., et al, Izv.Sib. otd-niya AN SSSR, Ser Tekhn.Nauk, 1985, No.10 (2?), 69-73.
69 Mikhailov B.I., TVT, 1985, **23**, No.5, 1000-1003.
70 Basin A.S., et al, Teplofizika i Aerodinamika, 1998, **5**, No.4, 583-592.
71 Esipchuk A.M., et al, In: Physics of plasma and plasma technologies, Minsk, Vol.1, 1997, 89-91.
72 Zhukauskas A.A. and Zhyugshda I.I., Heat transfer of the cylinder in the transverse flow of a liquid, Vil'nyus, Moksla, 1979.
73 Anyshakov A.S., et al, Izv.Sib. otd-niya AN SSSR, Ser Tekhn.Nauk, 1976, No.3 (1), 12-15.
74 Andreev A.A., et al, In: Proc. of 10th Nat. Conf. on Low-Temperature Plasma Generators, Vol. 2, Minsk, 1986, pp.47-48.
75 Ibraev Sh.Sh. and Sakipov S.B., Electric arc combined type reactors and methods of calculating them, Alma-Ata, Gylym, 1991.
76 Zheenbaev Zh.Zh. and Engel'sht V.S., Two-jet plasma torch, Grunze, Ilim, 1983.
77 Saakov A.G., et al, In: Proc. of 11th Nat. Conf. on Low-Temperature Plasma Generators, Vol. 2, Novosibirsk, 1989, pp.16-17.
78 Ipat;eva O.S. and Vasin A.S., Current problems of thermophysics and physical hydrodynamics, Novosibirsk, 1998.
79 Zhukov M.F., et al, Proc. of 2nd Int'l School-Seminar: Evolution of defective structure in metals and alloys, Barnaul, 1994, pp.109-110.
80 Gellert B. and Egli W., J. Phys. D.: Appl. Phys., 1987, **21**, 1721-1725.
81 Dorodnov A.M., et al, TVT, 1970, **8**, No.3, 492-499.
82 Babaskin Yu.E., Structure and properties of cast steel, Kiev, Naukova dumka, 1980.
83 Gol'shtein Ya.E. and Mizi V.G., Modification of cast iron and steel, Moscow, Metallurgiya, 1986.
84 Saburov V.P., et al, Plasmochemical synthesis of ultrafine powders and their application for modification of metals and alloys, Novosibirsk, 1995.
85 Sherepanov A.N., et al, In: 3rd M.V. Mokhosoev memorial international seminar of new materials. ISMM-96, Irkutsk, Russia, 1996, p.9.
86 Kana R.U. and Khaazena P.T. (eds), Physical metallurgy (translated from English), Edited by R.W. Kahn and P.T. Haasen, Vol.2, Phase transformations in metals and alloys and in alloys with special physical properties, Moscow, Metallurgiya, 1987.

Chapter 11

1 Zhukov M.F., et al, Electric arc gas heaters (plasma torches), Moscow, Nauka, 1973.
2 Kutateladze S.S., Fundamentals of heat exchange theory, Moscow, Atomizdat,

1979.
3 Bogatyreva V.A., et al, PMTF, 1968, No.3, 86-89.
4 Mortseva G.I., et al, In: Physics, technology and application of low-temperature plasma, Alma-Ata, 1970.
5 Polak L.S. (ed), Theoretical and applied plasma chemistry, Moscow, Nauka, 1975.
6 Dembovskii V., Plasma metallurgy (translated from Czech), Moscow, Metallurgiya, 1981.
7 Tsvetkov Yu.V. and Panfilov S.A., Low-temperature plasma in pre-conditioning processes, Moscow, Nauka, 1989.
8 Gavrilko V.P., et al, In: Physics-chemistry and technology of discrete powders, Kiev, IPM AN USSR, 1984.
9 Pershin V.D., et al, Izv.Vuzov SSSR. Ser. Khimiya i Khim. Tekhnologiya, 1981, **24**, 11, 15-19.
10 Galevskii G.V. and Krutskii Yu.L., In: High intensity processes in chemical technology, Proceedings, Lensoveta, 1988, pp.102-106.
11 Krapivina S.A., Plasmochemical technological processes, Leningrad, Khimya, 1981.
12 Onis'kova O.V., et al, Modelling reactors of plasma chemical homogeneous processes, Kiev, Naukova dumka, 1982.
13 Zhukov M.F. and Sukhinin Yu.I., Izv.Sib. otd-niya AN SSSR, Ser Tekhn.Nauk, 1970, No.8 (2), 12-19.
14 Galevskii G.V., In: Investigations of plasma processes and systems, Minsk, 1978, pp.117-125.
15 Litvinov V.K., et al, In: Investigations of plasma processes and systems, Minsk, 1978, pp.126-138.
16 Galevskii G.V., et al, In: Investigations of plasma processes and systems, Minsk, 1978, pp.36-41.
17 Burov I.S., In: Investigations of plasma processes and systems, Minsk, 1978, pp.42-48.
18 Burov I.S., In: Investigations of plasma processes and systems, Minsk, 1978, pp.49-61.
19 Krutskii Yu.L. and Nozdrin I.V., Proc.: Production of ferroalloys, Novosibirsk, 1986.
20 Galevskii G.V., et al, Proc.: High-temperature and plasmochemical processes, Leningrad, 1984.
21 Mosse A.L. and Burov I.S., Treatment of dispersed materials in plasma reactors, Minsk, Nauka i Tekhnika, 1980.
22 Ambrazyavichus A.B. and Litvinov V.K., High temperature heat exchange in plasma technological systems, a textbook, Sverdlovsk, UPI, 1986.
23 Bolotov A.V., et al, In: Proc.: Electrophysics, electromechanics and applied electrical engineering, Alma-Ata, 1982, pp.17-25.
24 Asanaliev M.K., et al, Fizika i Khimiya Obrab. Materialov, 1977, No.5, 111-116.
25 Soroka P.I. and Vereshchak V.G., In: Investigations of plasma processes and systems, Minsk, 1978, pp.62-67.
26 Massier P.F., et al, Trans. ASME, Heat transfer, 1969, No.1, 77-85.
27 Berbasov V.V., et al, In: Plasmochemical processes, Moscow, 1979.
28 Mosse A.L., et al, Inzh.-Fiz. Zhurn., 1987, **52**, No.3, 439-443.
29 Mosse A.L. and Burov I.S., In: Physics, technology and application of low-temperature plasma, Alma-Ata, KazPI, 1970, pp.28-37.

30 Burov I.S., et al, Izv.Sib. otd-niya AN SSSR, Ser Tékhn.Nauk, 1977, No.13 (3), 21-32.
31 Ambrazyavichyus A.B., et al, Mokslas i Tekhnika, 1977, No.2, 45-49.
32 Panfilov S.A. and Tsvetkov Yu.V., In: Physico-chemical investigations in metallurgy and metal science using computers, Moscow, Nauka, 1974.
33 Zhukov M.F. (ed), Electric arc plasma torches, advertising literature, Novosibirsk, 1980.
34 Rechkin V.N. and Gusev A.A., Author's certificate No. 1204518 USSR. Equipment for dosing powder-gas mixtures, Otkrytiya. Izobret., 1986, No.2, 83.
35 Ershova V.A. (ed), Electrothermal processes in chemical technology, a textbook, Leningrad, Khimiya, 1984.
36 Kosolapovoi T.Ya. (ed), Properties, production and application of refractory compounds, Moscow, Metallurgiya, 1986.
37 Krzhizhanovskii R.E. and Shtern Z.Yu., In: Thermophysical properties of non-metallic materials, Leningrad, Energiya, 1973.
38 Shevtsov V.P. and Voichak V.P., In: Proc. of 9th Nat. Conf. on Low-Temperature Plasma Generators, Frunze, Ilim, 1983, pp.362-363.
39 Shevtsov V.P., et al, In: Plasma metallurgy, Proceedings, Institute of Metallurgy, Moscow, 1985.
40 Shevtsov V.P., et al, In: Physics and chemistry of plasma metallurgical processes, Moscow, 1985.
41 Shevtsov V.P. and Zhumatov A.A., In: Proc. of Nat. Conf. on Low-Temperature Generators, Minsk, 1994.
42 Shevtsov V.P., Thermal plasma and new material technology. Vol.2, Cambridge Interscience Publishing, 1995, pp.473-492.
43 Uryukov B.A., PMTF, 1969, No.1.
44 Koroteev A.S., Electric arc plasma torches, Moscow, Mashinostroenie, 1980.
45 Koroteev A.S., et al, Plasma torches: design, characteristics, calculations, Moscow, Mashinostroenie, 1993.
46 Yukhimchuk S.A. and Vovchenko S.I., In: Proc. of 9th Nat. Conf. on Low-Temperature Plasma Generators, Frunze, Ilim, 1983, pp.370-371.
47 Sergeev P.V., et al, In: Problems of thermal power engineering and applied thermophysics, Alma-Ata, No.10, Nauka KazSSR, 1975, pp.157-163.
48 Ibraev Sh. Sh., et al, In: Problems of thermal power engineering and applied thermophysics, Alma-Ata, No.11, Nauka KazSSR, 1979, pp.135-142.
49 Ibraev Sh.Sh. and Sakipov Z.B., Electric arc combined type reactors and methods of calculating them, Alma-Ata, Gylym, 1991.
50 Ibraev Sh.Sh., Plasma technological processes and equipment of the combined type, Dissertation, Almaty, 1996.
51 Zhukov M.F. and Panin V.E. (eds), New materials and technologies. Extreme technological processes, Novosibirsk, Nauka, 1992.
52 Tukhvatullin A.M., et al, In: Proc. of 2nd Nat. Conf. on Plasmochemical Technology and Equipment Construction, Moscow, 1977, pp.244-246.
53 Tukhvatullin A.M. and Izinger Yu.V., In: Proc. of 9th Nat. Conf. "Khimreaktor-9", Vol.2, Grodno, 1986, pp.51-55.
54 Tukhvatullin A.M., et al, Khim. Prom-st'., 1986, No.9, 61.
55 Zhukov M.F., et al, Electric generators with inter-electrode inserts, Novosibirsk, Nauka, 1981.
56 Zhukov M.F. (ed), Fundamentals of calculating linear plasma torches, Novosibirsk, 1979.

Index

10